Modern Organic Synthesis:
An Introduction

MODERN ORGANIC SYNTHESIS:
An Introduction

George S. Zweifel

Michael H. Nantz

University of California, Davis

W. H. FREEMAN AND COMPANY

New York

The marine alkaloid norzoanthamine, whose energy-minimized structure is depicted on the front cover, exhibits interesting pharmacological properties, particularly as a promising candidate for an antiosteoporotic drug. It was isolated from the genus *Zoanthus*, commonly known as sea mat anemone. The alkaloid possesses a complex molecular structure; its total synthesis was accomplished in 41 steps by Miyashita and coworkers (*Science* **2004**, *305*, 495), a brilliant intellectual achievement. [Cover image by Michael Nantz and Dean Tantillo]

Publisher	CRAIG BLEYER
Senior Acquisitions Editor	JESSICA FIORILLO
Marketing Manager	ANTHONY PALMIOTTO
Media Editor	VICTORIA ANDERSON
Associate Editor	AMY THORNE
Design Manager	BLAKE LOGAN
Cover and Text Designer	CAMBRAIA FERNANDES
Senior Project Editor	GEORGIA LEE HADLER
Copy Editor	KAREN TASCHEK
Production Coordinator	PAUL W. ROHLOFF
Composition	MACMILLAN INDIA LTD
Printing and Binding	RR DONNELLEY

Library of Congress Cataloging-in Publication Data

Zweifel, George S.
   Modern organic synthesis: an introduction/George S. Zweifel, Michael H. Nantz.
      p. cm
   Includes index.
   ISBN 0-7167-7266-3
1.   Organic compounds - - Synthesis.   I. Nantz, Michael H.   II. Title.
   QD262.Z94 2006
   547′. 2 - - dc22

                                                                              2005044706

ISBN-13: 978-0-7167-7266-8
ISBN-10: 0-7167-7266-3

Second Printing

W. H. Freeman and Company
41 Madison Avenue
New York, NY 10010

Houndmills, Basingstoke RG 21 6XS, England

www.whfreeman.com

We dedicate this book to our former mentors at Purdue University,

Professor Herbert C. Brown
Professor Philip L. Fuchs

who have inspired our passion for organic chemistry

About the Authors

George S. Zweifel was born in Switzerland. He received his Dr. Sc. Techn. degree in 1955 from the Swiss Federal Institute of Technology (E.T.H. Zürich, Professor Hans Deuel) working in the area of carbohydrate chemistry. The award of a Swiss-British Exchange Fellowship in 1956 (University of Edinburgh, Scotland, Professor Edmund L. Hirst) and a Research Fellowship in 1957 (University of Birmingham, England, Professor Maurice Stacey) made it possible for him to study conformational problems in the carbohydrate field. In 1958, he became professor Herbert C. Brown's personal research assistant at Purdue University, undertaking research in the new area of hydroboration chemistry. He joined the faculty at the University of California, Davis, in 1963, where his research interest has been the exploration of organometallics as intermediates in organic synthesis, with emphasis on unsaturated organoboron, organoaluminum and organosilicon compounds.

Michael H. Nantz was born in 1958 in Frankfurt, Germany. In 1970, he moved with his family to the Appalachian Mountains of Kentucky. He spent his college years in Bowling Green, Kentucky, and earned a Bachelor of Science degree from Western Kentucky University in 1981. His interest in natural product synthesis led him to work with Professor Philip L. Fuchs at Purdue University, where he received his Ph.D. in 1987. Over the next two years, he explored asymmetric syntheses using boron reagents (Massachusetts Institute of Technology, Professor Satoru Masamune). In 1989, he joined the faculty at the University of California, Davis, and established a research program in organic synthesis with emphasis on the development of gene delivery vectors. His novel DNA transfer agents have been commercialized and have engendered a start-up biotechnology company devoted to nonviral gene therapy. In 2006, he joined the Chemistry Department at the University of Louisville as Distinguished University Scholar.

CONTENTS

Preface   ix

CHAPTER 1   **SYNTHETIC DESIGN**   **1**
            1.1   Retrosynthetic Analysis   1
            1.2   Reversal of the Carbonyl Group Polarity (*Umpolung*)   8
            1.3   Steps in Planning a Synthesis   14
            1.4   Choice of Synthetic Method   23
            1.5   Domino Reactions   26
            1.6   Computer-Assisted Retrosynthetic Analysis   26

CHAPTER 2   **STEREOCHEMICAL CONSIDERATIONS IN**
            PLANNING SYNTHESES   **31**
            2.1   Conformational Analysis   31
            2.2   Evaluation of Nonbonded Interactions   35
            2.3   Six-Member Heterocyclic Systems   40
            2.4   Polycyclic Ring Systems   41
            2.5   Cyclohexyl Systems with sp^2-Hybridized Atoms   44
            2.6   Significant Energy Difference   46
            2.7   Computer-Assisted Molecular Modeling   46
            2.8   Reactivity and Product Determination as a
                   Function of Conformation   47

CHAPTER 3   **THE CONCEPT OF PROTECTING FUNCTIONAL**
            GROUPS   **58**
            3.1   Protection of NH Groups   58
            3.2   Protection of OH Groups of Alcohols   60
            3.3   Protection of Diols as Acetals   69
            3.4   Protection of Carbonyl Groups in Aldehydes and Ketones   71
            3.5   Protection of the Carboxyl Group   78
            3.6   Protection of Double Bonds   82
            3.7   Protection of Triple Bonds   82

CHAPTER 4   **FUNCTIONAL GROUP TRANSFORMATIONS:**
            OXIDATION AND REDUCTION   **88**
            4.1   Oxidation of Alcohols to Aldehydes and Ketones   88
            4.2   Reagents and Procedures for Alcohol Oxidation   89
            4.3   Chemoselective Agents for Oxidizing Alcohols   93
            4.4   Oxidation of Acyloins   96
            4.5   Oxidation of Tertiary Allylic Alcohols   97
            4.6   Oxidative Procedures to Carboxylic Acids   98
            4.7   Allylic Oxidation of Alkenes   99
            4.8   Terminology for Reduction of Carbonyl Compounds   101
            4.9   Nucleophilic Reducing Agents   103
            4.10   Electrophilic Reducing Agents   109
            4.11   Regio- and Chemoselective Reductions   112

	4.12	Diastereoselective Reductions of Cyclic Ketones	115
	4.13	Inversion of Secondary Alcohol Stereochemistry	117
	4.14	Diastereofacial Selectivity in Acyclic Systems	118
	4.15	Enantioselective Reductions	124

CHAPTER 5   **FUNCTIONAL GROUP TRANSFORMATIONS: THE CHEMISTRY OF CARBON-CARBON π-BONDS AND RELATED REACTIONS**   **139**

	5.1	Reactions of Carbon-Carbon Double Bonds	139
	5.2	Reactions of Carbon-Carbon Triple Bonds	193

CHAPTER 6   **FORMATION OF CARBON-CARBON SINGLE BONDS VIA ENOLATE ANIONS**   **213**

	6.1	1,3-Dicarbonyl and Related Compounds	213
	6.2	Direct Alkylation of Simple Enolates	223
	6.3	Cyclization Reactions—Baldwin's Rules for Ring Closure	231
	6.4	Stereochemistry of Cyclic Ketone Alkylation	234
	6.5	Imine and Hydrazone Anions	236
	6.6	Enamines	238
	6.7	The Aldol Reaction	240
	6.8	Condensation Reactions of Enols and Enolates	256
	6.9	Robinson Annulation	260

CHAPTER 7   **FORMATION OF CARBON-CARBON BONDS VIA ORGANOMETALLIC REAGENTS**   **273**

	7.1	Organolithium Reagents	273
	7.2	Organomagnesium Reagents	283
	7.3	Organotitanium Reagents	286
	7.4	Organocerium Reagents	287
	7.5	Organocopper Reagents	288
	7.6	Organochromium Reagents	298
	7.7	Organozinc Reagents	300
	7.8	Organoboron Reagents	305
	7.9	Organosilicon Reagents	312
	7.10	Palladium-Catalyzed Coupling Reactions	322

CHAPTER 8   **FORMATION OF CARBON-CARBON π-BONDS**   **359**

	8.1	Formation of Carbon-Carbon Double Bonds	359
	8.2	Formation of Carbon-Carbon Triple Bonds	396

CHAPTER 9   **SYNTHESES OF CARBOCYCLIC SYSTEMS**   **412**

	9.1	Intramolecular Free Radical Cyclizations	412
	9.2	Cation-π Cyclizations	417
	9.3	Pericyclic Reactions	421
	9.4	Ring-Closing Olefin Metathesis (RCM)	433

EPILOGUE   **THE ART OF SYNTHESIS**   **443**

Abbreviations		445
Answers to Select End-of-Chapter Problems		446
Index		469

Modern Organic Synthesis: An Introduction is based on the lecture notes of a special topics course in synthesis designed for senior undergraduate and beginning graduate students who are well acquainted with the basic concepts of organic chemistry. Although a number of excellent textbooks covering advanced organic synthesis have been published, we saw a need for a book that would bridge the gap between these and the organic chemistry presented at the sophomore level. The goal is to provide the student with the necessary background to begin research in an academic or industrial environment. Our precept in selecting the topics for the book was to present in a concise manner the modern techniques and methods likely to be encountered in a synthetic project. Mechanisms of reactions are discussed only if they might be unfamiliar to the student. To acknowledge the scientists whose research formed the basis for this book and to provide the student access to the original work, we have included after each chapter the relevant literature references.

The book is organized into the following nine chapters and an epilogue:

- Retrosynthetic analysis: strategies for designing a synthetic project, including construction of the carbon skeleton and control of stereochemistry and enantioselectivity

- Conformational analysis and its effects on reactivity and product formation

- Problems for dealing with multiple functionality in synthesis, and their solutions

- Functional group transformations: classical and chemoselective methods for oxidation and reduction of organic substrates, and the availability and utilization of regio-, chemo-, and stereoselective agents for reducing carbonyl compounds

- Reactions of carbon-carbon π bonds: dissolving metal reductions, conversions to alcohols and enantiomerically pure alcohols, chemo- and enantioselective epoxidations, procedures for cleavage of carbon-carbon double bonds, and reactions of carbon-carbon triple bonds

- Formation of carbon-carbon single bonds via enolate anions: improvements in classical methods and modern approaches to stereoselective aldol reactions

- Methods for the construction of complex carbon-carbon frameworks via organometallics: procedures involving main group organometallics, and palladium-catalyzed coupling reactions for the synthesis of stereodefined alkenes and enynes

- Formation of carbon-carbon π-bonds: elaboration of alkynes to stereodefined alkenes via reduction, current olefination reactions, and transposition of double bonds

- Synthesis of carbocyclic systems: intramolecular free-radical cyclization, the Diels-Alder reaction, and ring-closing metathesis

- An epilogue featuring selected natural product targets for synthesis

Print Supplement

Modern Organic Synthesis: Problems and Solutions, 0-7167-7494-1
This manual contains all problems from the text, along with complete solutions.

We wish to express our gratitude to the present and former Chemistry 131 students at the University of California at Davis and to the teaching assistants of the course, especially Hasan Palandoken, for their suggestions and contributions to the development of the lecture notes. We would also like to thank our colleague Professor Dean Tantillo for his helpful advice. Professors Edwin C. Friedrich (University of California at Davis) and Craig A. Merlic (University of California at Los Angeles) read the entire manuscript; their pertinent comments and constructive critiques greatly improved the quality of the book. We also are indebted to the following reviewers of the manuscript:

Amit Basu, Brown University
Stephen Bergmeier, Ohio University
Michael Bucholtz, Gannon University
Arthur Cammers, University of Kentucky
Paul Carlier, Virginia Polytechnic Institute and State University
Robert Coleman, Ohio State University
Shawn Hitchcock, Illinois State University
James Howell, Brooklyn College
John Huffman, Clemson University
Dell Jensen, Jr., Augustana College
Eric Kantorowski, California Polytechnic State University
Mohammad Karim, Tennessee State University
Andrew Lowe, University of Southern Mississippi
Philip Lukeman, New York University
Robert Maleczka, Jr., Michigan State University
Helena Malinakova, University of Kansas
Layne Morsch, DePaul University
Nasri Nesnas, Florida Institute of Technology
Peter Norris, Youngstown State University
Cyril Párkányi, Florida Atlantic University
Robin Polt, University of Arizona
Jon Rainier, University of Utah
O. LeRoy Salerni, Butler University
Kenneth Savin, Butler University
Grigoriy Sereda, University of South Dakota
Suzanne Shuker, Georgia Institute of Technology
L. Strekowski, Georgia State University
Kenneth Williams, Francis Marion University
Bruce Young, Indiana-Purdue University at Indianopolis

We wish to thank Jessica Fiorillo, Georgia Lee Hadler, and Karen Taschek for their professional guidance during the final stages of writing the book.
Finally, without the support and encouragement of our wives, Hanni and Jody, *Modern Organic Synthesis: An Introduction* would not have been written.

CHAPTER 1

Synthetic Design

In character, in manners, in style; in all things, the supreme excellence is simplicity.

Henry Wadsworth Longfellow

C hemistry touches everyone's daily life, whether as a source of important drugs, polymers, detergents, or insecticides. Since the field of organic chemistry is intimately involved with the synthesis of these compounds, there is a strong incentive to invest large resources in synthesis. Our ability to predict the usefulness of new organic compounds before they are prepared is still rudimentary. Hence, both in academia and at many chemical companies, research directed toward the discovery of new types of organic compounds continues at an unabated pace. Also, natural products, with their enormous diversity in molecular structure and their possible medicinal use, have been and still are the object of intensive investigations by synthetic organic chemists.

Faced with the challenge to synthesize a new compound, how does the chemist approach the problem? Obviously, one has to know the tools of the trade: their potential and limitations. A synthetic project of any magnitude requires not only a thorough knowledge of available synthetic methods, but also of reaction mechanisms, commercial starting materials, analytical tools (IR, UV, NMR, MS), and isolation techniques. The ever-changing development of new tools and refinement of old ones makes it important to keep abreast of the current chemical literature.

What is an ideal or viable synthesis, and how does one approach a synthetic project? The overriding concern in a synthesis is the yield, including the inherent concepts of simplicity (fewest steps) and selectivity (chemoselectivity, regioselectivity, diastereoselectivity, and enantioselectivity). Furthermore, the experimental ease of the transformations and whether they are environmentally acceptable must be considered.

Synthesis of a molecule such as pumiliotoxin C involves careful planning and strategy. How would a chemist approach the synthesis of pumiliotoxin C?[1] This chapter outlines strategies for the synthesis of such target molecules based on *retrosynthetic analysis*.

E. J. Corey, who won the Nobel Prize in Chemistry in 1990, introduced and promoted the concept of retrosynthetic analysis, whereby a molecule is disconnected, leading to logical precursors.[2] Today, retrosynthetic analysis plays an integral and indispensable role in research.

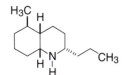

Pumiliotoxin C, a *cis*-decahydroquinoline from poison-dart frogs, *Dendrobates pumilio.*

1.1   RETROSYNTHETIC ANALYSIS[3]

The following discussion on retrosynthetic analysis covers topics similar to those in Warren's *Organic Synthesis: The Disconnection Approach*[3a] and Willis and Will's *Organic Synthesis*.[3g] For an advanced treatment of the subject matter, see Corey and Cheng's *The Logic of Chemical Synthesis*.[3b]

Basic Concepts

The construction of a synthetic tree by working backward from the *target molecule* (TM) is called *retrosynthetic analysis* or *antithesis*. The symbol $\Longrightarrow$ signifies a reverse synthetic step and is called a *transform*. The main transforms are *disconnections*, or cleavage of C–C bonds, and *functional group interconversions* (FGI).

Retrosynthetic analysis involves the disassembly of a TM into available starting materials by sequential disconnections and functional group interconversions. Structural changes in the retrosynthetic direction should lead to substrates that are more readily available than the TM. *Synthons* are fragments resulting from disconnection of carbon-carbon bonds of the TM. The actual substrates used for the forward synthesis are the *synthetic equivalents* (SE). Also, reagents derived from *inverting the polarity* (IP) of synthons may serve as SEs.

Synthetic design involves two distinct steps[3a]: (1) retrosynthetic analysis and (2) subsequent translation of the analysis into a "forward direction" synthesis. In the analysis, the chemist recognizes the functional groups in a molecule and disconnects them proximally by methods corresponding to known and reliable reconnection reactions.

Chemical bonds can be cleaved *heterolytically, homolytically,* or through *concerted transform* (into two neutral, closed-shell fragments). The following discussion will focus on heterolytic and cyclic disconnections.

heterolytic cleavage			
homolytic cleavage			
concerted transform			

Donor and Acceptor Synthons[3c,g]

Heterolytic retrosynthetic disconnection of a carbon-carbon bond in a molecule breaks the TM into an acceptor synthon, a carbocation, and a donor synthon, a carbanion. In a formal sense, the reverse reaction — the formation of a C–C bond — then involves the union of an *electrophilic* acceptor synthon and a *nucleophilic* donor synthon. Tables 1.1 and 1.2 show some important acceptor and donor synthons and their synthetic equivalents.[3c]

| Table 1.1 | Common Acceptor Synthons |

Synthon	Synthetic equivalent
R^+ (alkyl cation = carbenium ion)	RCl, RBr, RI, ROTs
Ar^+ (aryl cation)	$Ar\overset{+}{N}_2\ X^-$
$H\overset{+}{C}{=}O$ (acylium ion)	$H\overset{O}{\overset{\|}{C}}{-}X$ (X = NR$_2$, OR)
$R\overset{+}{C}{=}O$ (acylium ion)	$R\overset{O}{\overset{\|}{C}}{-}X$ (X = Cl, NR$_2'$, OR')
$HO{-}\overset{+}{C}{=}O$ (acylium ion)	CO_2
$\overset{+}{C}H_2CH_2\overset{O}{\overset{\|}{C}}{-}R$	$CH_2{=}CH\overset{O}{\overset{\|}{C}}{-}R$ (R = alkyl, OR')
$\overset{+}{C}H_2{-}CH_2C{\equiv}N$	$CH_2{=}CHC{\equiv}N$
$\overset{+}{C}H_2OH$ (oxocarbenium ion)	HCHO
$R\overset{+}{C}H{-}OH$ (oxocarbenium ion)	RCHO
$R_2\overset{+}{C}{-}OH$ (oxocarbenium ion)	$R_2C{=}O$

^{a} Note that α-halo ketones also may serve as synthetic equivalents of enolate ions (e.g., the Reformatsky reaction, Section 7.7).

| Table 1.2 | Common Donor Synthons |

Synthon	Derived reagent	Synthetic equivalent
R^- (alkyl, aryl anion)	RMgX, RLi, R$_2$CuLi	R$-$X
$^-$CN (cyanide)	NaC$\equiv$N	HCN
RC$\equiv$C$^-$ (acetylide)	RC$\equiv$CMgX, RC$\equiv$CLi	RC$\equiv$CH
(enolate)	(M = Li, BR$_2$)	
$Ph_3\overset{+}{P}{-}\overset{/}{\underset{\backslash}{C}}$ (ylide)	$\left[Ph_3\overset{+}{P}{-}\overset{/}{\underset{\backslash}{C}}{-}H\right] X^-$	$H{-}\overset{/}{\underset{\backslash}{C}}{-}X$
$R{\overset{-}{\diagdown}}NO_2$ (α-nitro anion)		$R{\diagdown}NO_2$

Often, more than one disconnection is feasible, as depicted in retrosynthetic analyses A and B below. In the synthesis, a plan for the sequence of reactions is drafted according to the analysis by adding reagents and conditions.

Retrosynthetic analysis A

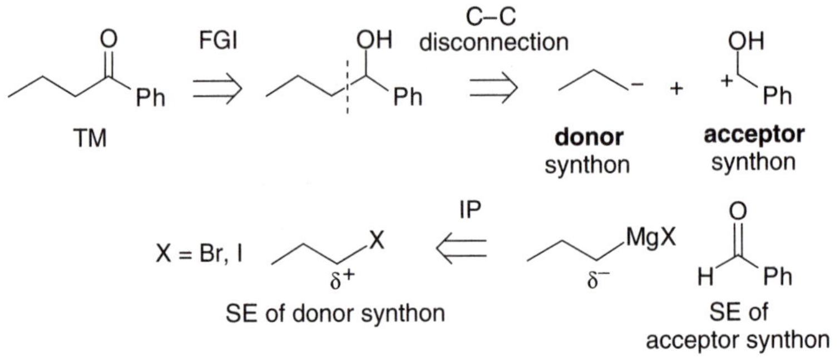

Synthesis A

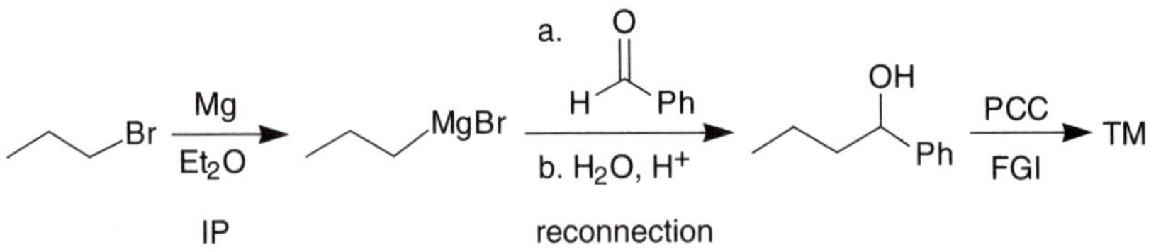

Retrosynthetic analysis B

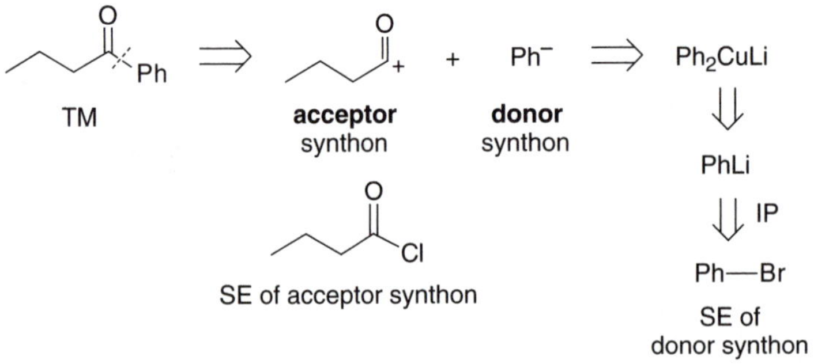

Synthesis B

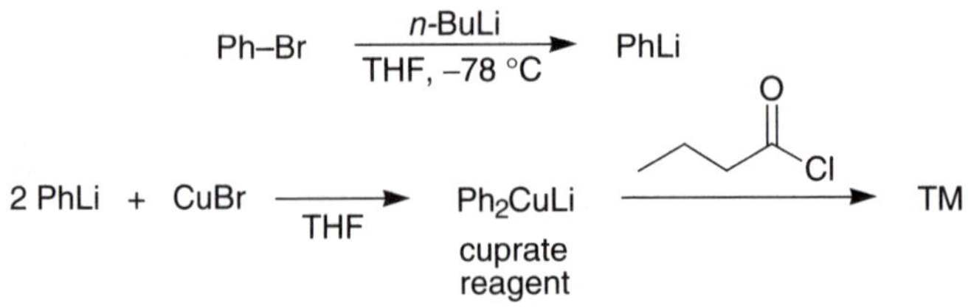

Alternating Polarity Disconnections[3g,4]

The question of how one chooses appropriate carbon-carbon bond disconnections is related to functional group manipulations since the distribution of formal charges in the carbon skeleton is determined by the functional group(s) present. The presence of a heteroatom in a molecule imparts a pattern of electrophilicity and nucleophilicity to the atoms of the molecule. The concept of *alternating polarities* or latent polarities (imaginary charges) often enables one to identify the best positions to make a disconnection within a complex molecule.

Functional groups may be classified as follows:[4a]

E class: Groups conferring electrophilic character to the attached carbon (+):
$-NH_2$, $-OH$, $-OR$, $=O$, $=NR$, $-X$ (halogens)

G class: Groups conferring nucleophilic character to the attached carbon (−):
$-Li$, $-MgX$, $-AlR_2$, $-SiR_3$

A class: Functional groups that exhibit ambivalent character (+ or −):
$-BR_2$, $C=CR_2$, $C\equiv CR$, $-NO_2$, $\equiv N$, $-SR$, $-S(O)R$, $-SO_2R$

The positive charge (+) is placed at the carbon *attached* to an E class functional group (e.g., $=O$, $-OH$, $-Br$) and the TM is then analyzed for *consonant* and *dissonant* patterns by assigning alternating polarities to the remaining carbons. In a consonant pattern, carbon atoms with the same class of functional groups have matching polarities, whereas in a dissonant pattern, their polarities are unlike. If a consonant pattern is present in a molecule, a simple synthesis may often be achieved.

Consonant patterns: Positive charges are placed at carbon atoms bonded to the E class groups.

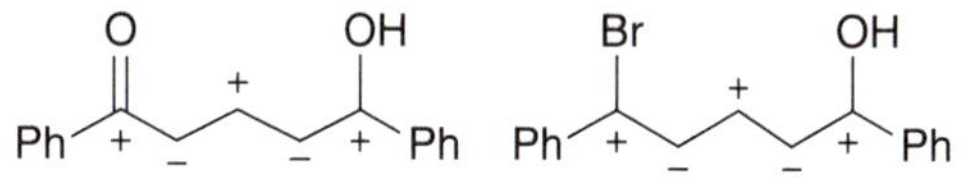

Dissonant patterns: One E class group is bonded to a carbon with a positive charge, whereas the other E class group resides on a carbon with a negative charge.

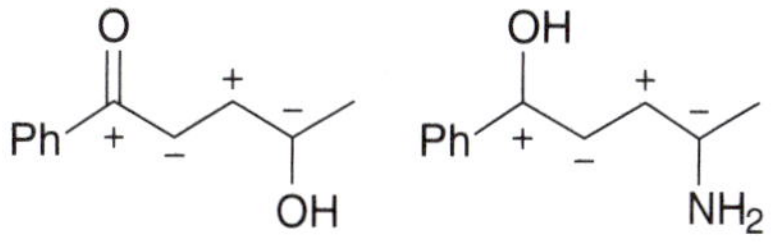

Examples of choosing reasonable disconnections of functionally substituted molecules based on the concept of alternating polarity are shown below.

One Functional Group

Analysis

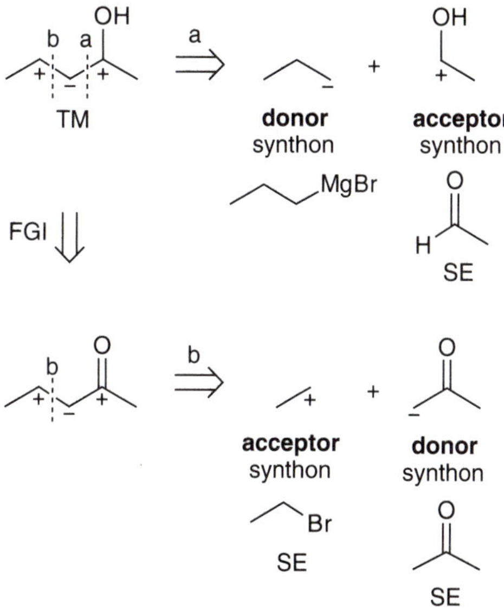

Synthesis (path a)

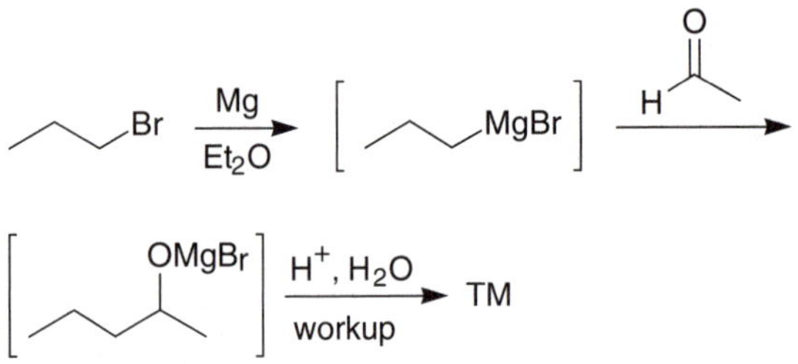

In the example shown above, there are two possible ways to disconnect the TM, 2-pentanol. Disconnection close to the functional group (path a) leads to substrates (SE) that are readily available. Moreover, reconnecting these reagents leads directly to the desired TM in high yield using well-known methodologies. Disconnection via path b also leads to readily accessible substrates. However, their reconnection to furnish the TM requires more steps and involves two critical reaction attributes: quantitative formation of the enolate ion and control of its monoalkylation by ethyl bromide.

Two Functional Groups in a 1,3-Relationship

Analysis

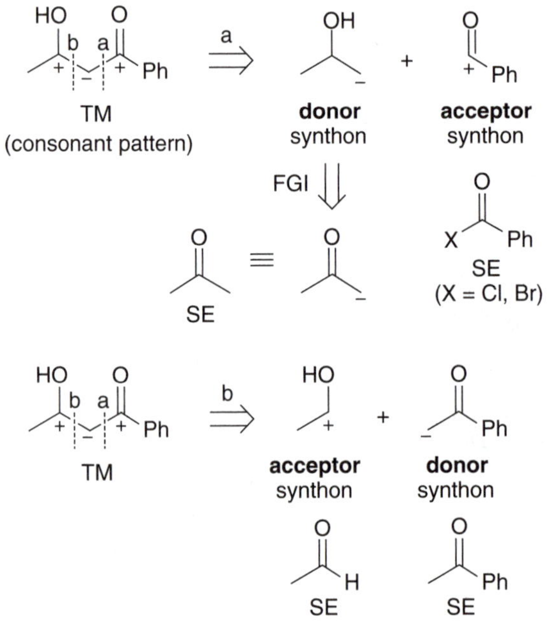

Synthesis (path a)

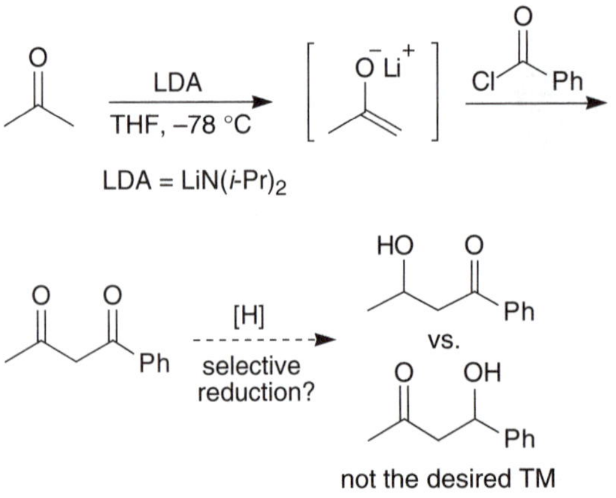

Synthesis (path b)

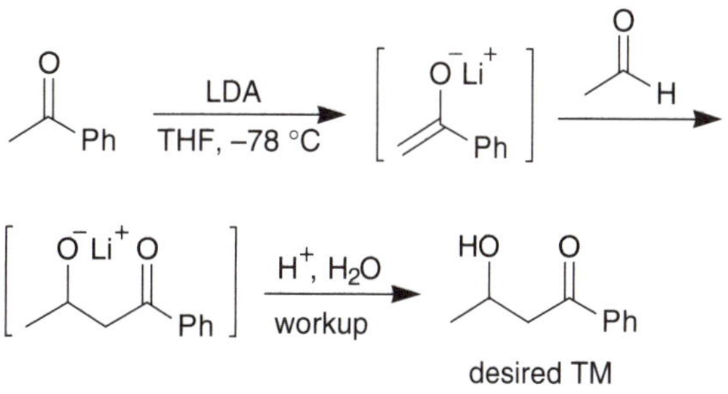

The consonant charge pattern and the presence of a β-hydroxy ketone moiety in the TM suggest a retroaldol transform. Either the hydroxy-bearing carbon or the carbonyl carbon of the TM may serve as an electrophilic site and the corresponding α-carbons as the nucleophilic sites. However, path b is preferable since it does not require a selective functional group interconversion (reduction).

Two Functional Groups in a 1,4-Relationship

Analysis

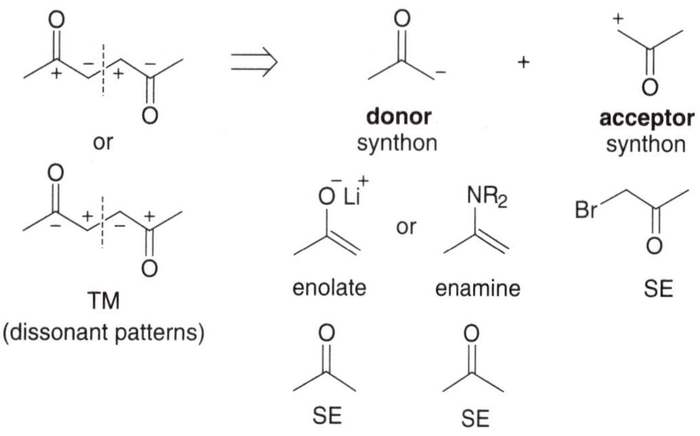

Synthesis

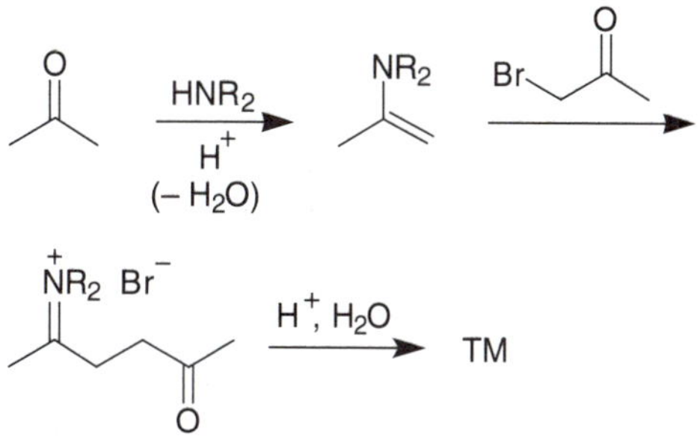

The *dissonant* charge pattern for 2,5-hexanedione exhibits a positive (+) polarity at one of the α-carbons, as indicated in the acceptor synthon above. Thus, the α-carbon in this synthon requires an *inversion of polarity* (*Umpolung* in German) from the negative (−) polarity normally associated with a ketone α-carbon. An appropriate substrate (SE) for the acceptor synthon is the electrophilic α-bromo ketone. It should be noted that an enolate ion might act as a base, resulting in deprotonation of an α-halo ketone, a reaction that could lead to the formation of an epoxy ketone (*Darzens condensation*). To circumvent this problem, a weakly basic enamine is used instead of the enolate.

In the case of 5-hydroxy-2-hexanone shown below, *Umpolung* of the polarity in the acceptor synthon is accomplished by using the electrophilic epoxide as the corresponding SE.

Analysis

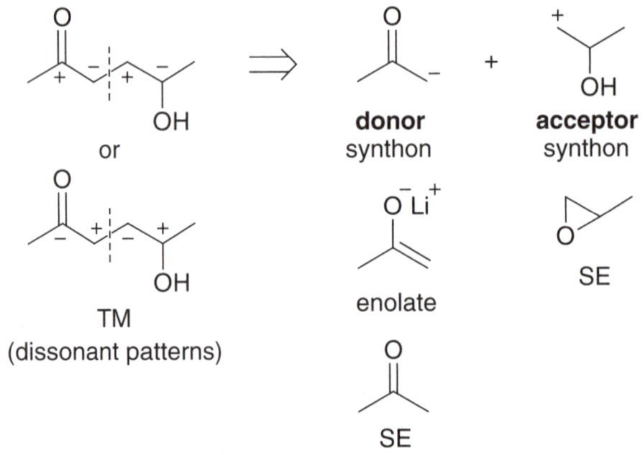

Synthesis

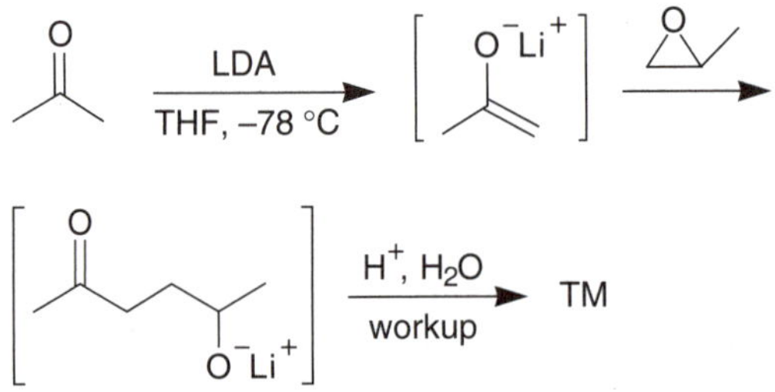

The presence of a C–C–OH moiety adjacent to a potential nucleophilic site in a TM, as exemplified below, points to a reaction of an epoxide with a nucleophilic reagent in the forward synthesis. The facile, regioselective opening of epoxides by nucleophilic reagents provides for efficient two-carbon homologation reactions.

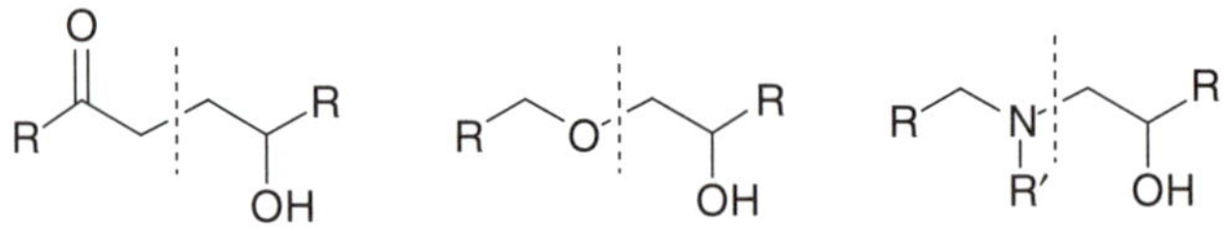

1.2 REVERSAL OF THE CARBONYL GROUP POLARITY (*UMPOLUNG*)[5]

In organic synthesis, the carbonyl group is intimately involved in many reactions that create new carbon-carbon bonds. The carbonyl group is electrophilic at the carbon atom and hence is susceptible to attack by nucleophilic reagents. Thus, the carbonyl group reacts as a *formyl cation* or as an *acyl cation*. A reversal of the positive polarity of the carbonyl group so it acts as a *formyl* or *acyl anion* would be synthetically very attractive. To achieve this, the carbonyl group is converted to a derivative whose carbon atom has the negative polarity. After its reaction with an electrophilic reagent, the carbonyl is regenerated. Reversal of polarity of a carbonyl group has been explored and systematized by Seebach.[5b,c]

Umpolung in a synthesis usually requires extra steps. Thus, one should strive to take maximum advantage of the functionality already present in a molecule.

"traditional" approach

Umpolung approach　(E$^+$ = electrophile)

formyl anion when R = H
acyl anion when R = alkyl

The following example illustrates the normal disconnection pattern of a carboxylic acid with a Grignard reagent and carbon dioxide as SEs (path a) and a disconnection leading to a carboxyl synthon with an "unnatural" negative charge (path b). Cyanide ion can act as an SE of a negatively charged carboxyl synthon. Its reaction with R–Br furnishes the corresponding nitrile, which on hydrolysis produces the desired TM.

Since formyl and acyl anions are not accessible, one has to use synthetic equivalents of these anions. Several reagents are synthetically equivalent to formyl or acyl anions, permitting the *Umpolung* of carbonyl reactivity.

Formyl and Acyl Anions Derived from 1,3-Dithianes[5b,c,f]

The most utilized *Umpolung* strategy is based on formyl and acyl anion equivalents derived from 2-lithio-1,3-dithiane species. These are readily generated from 1,3-dithianes (thioacetals) because the hydrogens at C(2) are relatively acidic (pK_a ~31).[6] In this connection it should be noted that thiols (EtSH, pK_a 11) are stronger acids compared to alcohols (EtOH, pK_a 16). Also, the lower ionization potential and the greater polarizability of the valence electrons of sulfur compared to oxygen make the divalent sulfur compounds more nucleophilic in S_N2 reactions. The polarizability factor may also be responsible for the stabilization of carbanions α to sulfur.[6]

1,3-dioxane (an acetal)
pK_a ~ 40

1,3-dithiane (a thioacetal)
pK_a = 31

The anions derived from dithianes react with alkyl halides to give the corresponding alkylated dithianes. Their treatment with $HgCl_2$–HgO regenerates aldehydes or ketones, respectively, as depicted below.

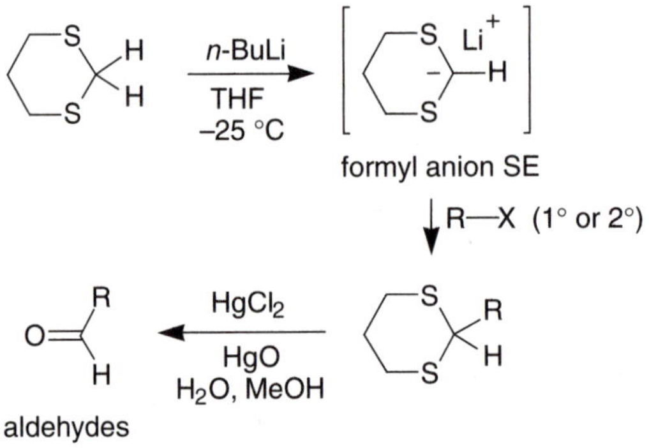

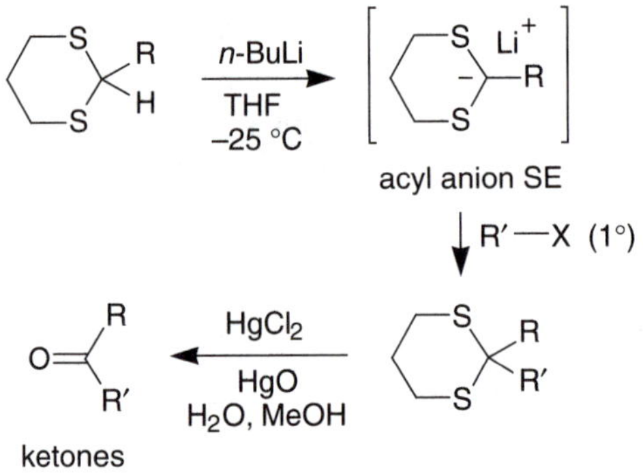

Dithiane-derived carbanions can be hydroxyalkylated or acylated to produce, after removal of the propylenedithiol appendage, a variety of difunctional compounds, as shown below. In the presence of HMPA (hexamethylphosphoramide, $[(Me_2N)_3P{=}O]$), dithiane-derived carbanions may serve as Michael donors.[7] However, in the absence of HMPA, 1,2-addition to the carbonyl group prevails.

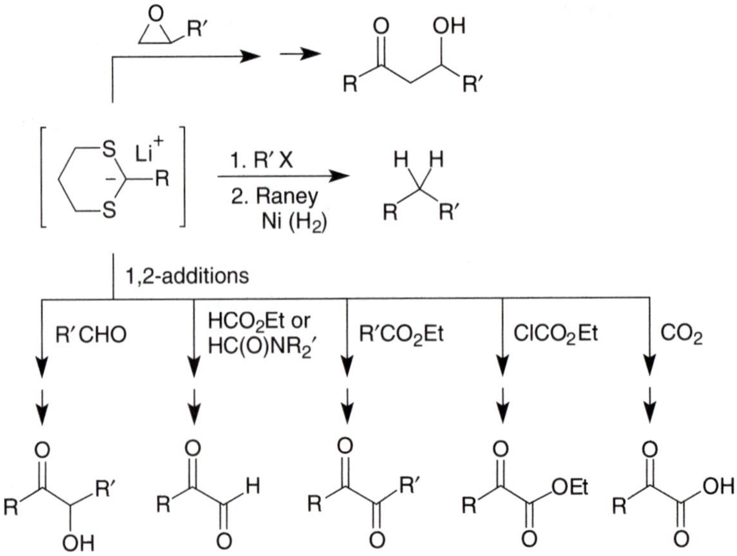

An instructive example of using a dithiane *Umpolung* approach to synthesize a complex natural product is the one-pot preparation of the multifunctional intermediate shown below, which ultimately was elaborated to the antibiotic vermiculin.[8]

TMEDA = *N,N,N',N'*-tetramethylethylenediamine (Me₂NCH₂CH₂NMe₂); used to sequester Li⁺ and disrupt *n*-BuLi aggregates.

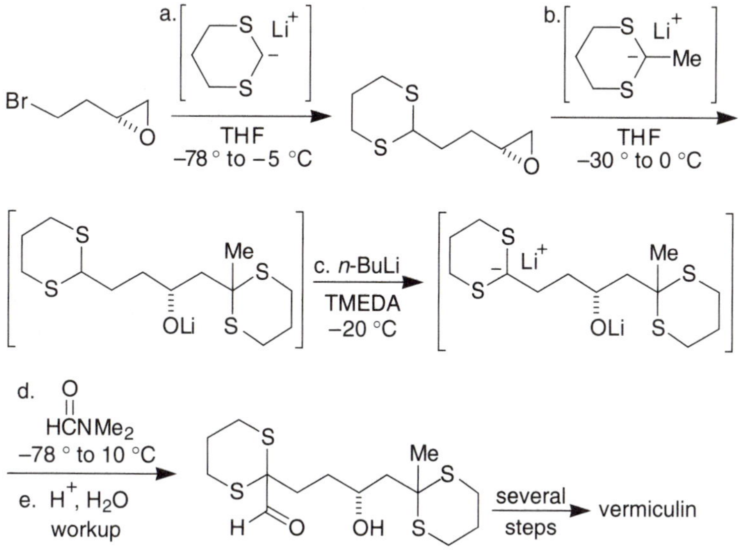

Acyl Anions Derived from Nitroalkanes[9]

The α-hydrogens of nitroalkanes are appreciably acidic due to resonance stabilization of the anion [CH_3NO_2, pK_a 10.2; $CH_3CH_2NO_2$, pK_a 8.5]. The anions derived from nitroalkanes give typical nucleophilic addition reactions with aldehydes (the Henry-Nef tandem reaction). Note that the nitro group can be changed directly to a carbonyl group via the Nef reaction (acidic conditions). Under basic conditions, salts of secondary nitro compounds are converted into ketones by the pyridine-HMPA complex of molybdenum (VI) peroxide.[9b] Nitronates from primary nitro compounds yield carboxylic acids since the initially formed aldehyde is rapidly oxidized under the reaction conditions.

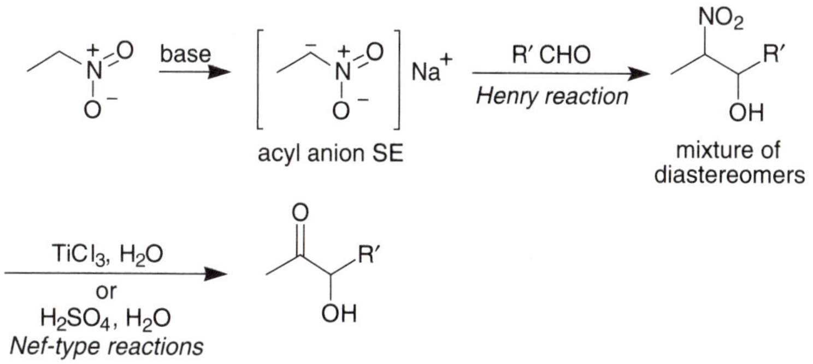

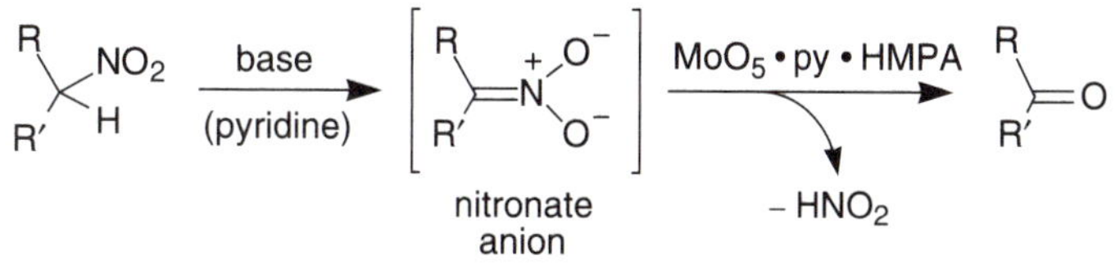

An example of an α-nitro anion *Umpolung* in the synthesis of jasmone (TM) is depicted next.[9a]

Analysis

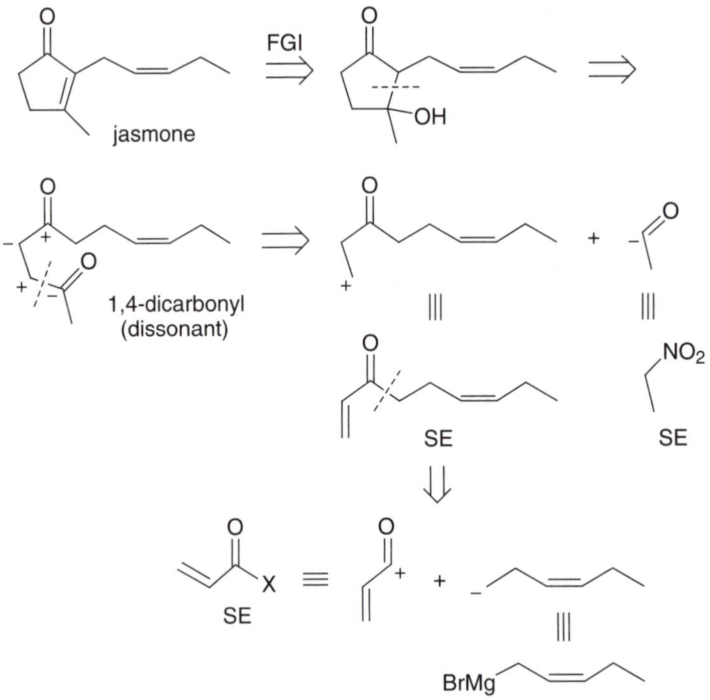
FGI
OH
jasmone
1,4-dicarbonyl
(dissonant)
NO2
SE
SE
X
SE
BrMg

Synthesis

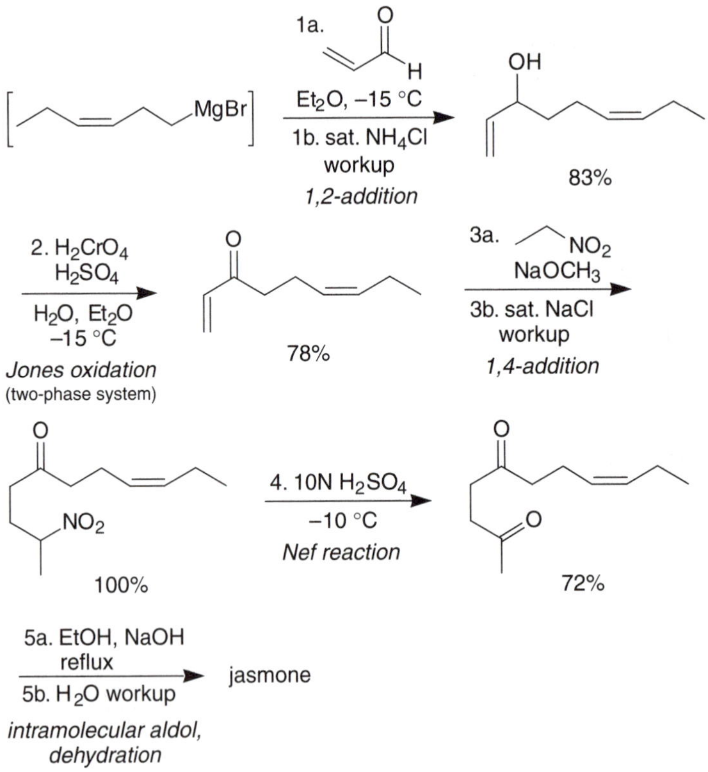
MgBr
1a.
H
Et2O, −15 °C
1b. sat. NH4Cl
workup
1,2-addition
OH
83%
2. H2CrO4
H2SO4
H2O, Et2O
−15 °C
Jones oxidation
(two-phase system)
78%
3a.
NO2
NaOCH3
3b. sat. NaCl
workup
1,4-addition
NO2
100%
4. 10N H2SO4
−10 °C
Nef reaction
72%
5a. EtOH, NaOH
reflux
5b. H2O workup
intramolecular aldol,
dehydration
jasmone

Acyl Anions Derived from Cyanohydrins[10]

O-Protected cyanohydrins contain a masked carbonyl group with inverted polarity. The α-carbon of an *O*-protected cyanohydrin is sufficiently activated by the nitrile moiety [CH_3CH_2CN, pK_a 30.9][11] so that addition of a strong base such as LDA generates the corresponding anion. Its alkylation, followed by hydrolysis of the result-ant alkylated cyanohydrin, furnishes the ketone. The overall reaction represents alky-lation of an acyl anion equivalent as exemplified for the synthesis of methyl cyclopentyl ketone.[10a]

An attractive alternative to the above protocol involves the nucleophilic acylation of alkylating agents with aromatic and heteroaromatic aldehydes via trimethylsilyl-protected cyanohydrins.[10b]

Acyl anion synthons derived from cyanohydrins may be generated catalytically by cyanide ion via the *Stetter reaction*.[10c,d] However, further reaction with elec-trophiles is confined to carbonyl compounds and Michael acceptors.

$$\left[\begin{array}{c} OH \\ | \\ R-C^- \\ | \\ CN \end{array}\right] + R'-CHO \longrightarrow \left[\begin{array}{cc} HO & O^- \\ | & | \\ R-C-C-R' \\ | & | \\ CN & H \end{array}\right] \rightleftharpoons \left[\begin{array}{cc} O & OH \\ \| & | \\ R-C-C-R' \\ | & | \\ CN & H \end{array}\right]$$

$$\longrightarrow \begin{array}{cc} O & OH \\ \| & | \\ R-C-C-R' \\ | \\ H \end{array} + CN^- \quad \text{(catalyst)}$$

cat. NaCN / DMF

1,4-addition

93%

Acyl Anions Derived from Enol Ethers

The α-hydrogens of enol ethers may be deprotonated with *tert*-BuLi.[12] Alkylation of the resultant vinyl anions followed by acidic hydrolysis provides an efficient route for the preparation of methyl ketones.

enol ether

a. *t*-BuLi, THF, –65 °C

acyl anion equivalent

b. R—X
c. H+, H2O

Acyl Anions Derived from Lithium Acetylide

Treatment of lithium acetylide with a primary alkyl halide (bromide or iodide) or with aldehydes or ketones produces the corresponding monosubstituted acetylenes or propargylic alcohols. Mercuric ion-catalyzed hydration of these furnishes methyl ketones and methyl α-hydroxy ketones, respectively.

$$R-Br + [Li-C\equiv C-H] \longrightarrow R-C\equiv C-H \xrightarrow[H_2SO_4, H_2O]{\text{cat. HgSO}_4} R-C(=O)-CH_3$$

(1° only)

$$RCHO + [Li-C\equiv C-H] \xrightarrow[-78\,°C]{THF} R-CH(OH)-C\equiv C-H$$

(or ketone)

$$\xrightarrow[H_2SO_4, H_2O]{\text{cat. HgSO}_4} R-CH(OH)-C(=O)-CH_3$$

1.3   **STEPS IN PLANNING A SYNTHESIS**[2,3]

In planning an organic synthesis, the following key interrelated factors may be involved:

- Construction of the carbon skeleton
- Functional group interconversions
- Control of relative stereochemistry
- Control of enantioselectivity

Construction of the Carbon Skeleton

Reactions that result in formation of new carbon-carbon bonds are of paramount importance in organic chemistry because they allow the construction of complex structures from smaller starting materials. Important carbon-carbon-bond-forming reactions encountered in organic syntheses are summarized in Table 1.3 and include

- Reactions of organolithium and Grignard reagents, such as RLi, RC≡CLi, RMgX, and RC≡CMgX, with aldehydes, ketones, esters, epoxides, acid halides, and nitriles
- Reactions of 1° alkyl halides with ⁻C≡N to extend the carbon chain by one carbon
- Alkylations of enolate ions to introduce alkyl groups to carbons adjacent to a carbonyl group (e.g., acetoacetic ester synthesis, malonic ester synthesis)
- Condensations such as aldol (intermolecular, intramolecular), Claisen, and Dieckmann
- Michael additions, organocuprate additions (1,4-additions)
- Friedel-Crafts alkylation and acylation reactions of aromatic substrates
- Wittig reactions, and Horner-Wadsworth-Emmons olefination
- Diels-Alder reactions giving access to cyclohexenes and 1,4-cyclohexadienes
- Ring-closing olefin metathesis

Table 1.3 Summary of Important Disconnections[3c]

TM	Synthons	SE (substrates)	Reaction type
			1,2-addition
			1,3-addition
			S_N2
			aldol condensation
			1,2-addition (Claisen condensation)
			1,4-addition (Michael addition)
			Wittig reaction
			Diels-Alder cycloaddition

Below are summarized some important guidelines for choosing disconnections of bonds. Thus, the initial stage of the retrosynthetic analysis *key fragments* are recognized, which then can be recombined in the forward synthetic step in an efficient way.[3]

- Disconnections of bonds should be carried out only if the resultant fragments can be reconnected by known and reliable reactions.

TM   straightforward disconnection

bad disconnection

Disconnection via path a leads to synthons whose SEs can be reconnected by a nucleophilic attack of phenoxide on the propyl bromide to furnish the desired TM. On the other hand, disconnection via path b would require either attack of *n*-PrO⁻ on bromobenzene to reconstruct the TM, a reaction that is not feasible, or displacement of a benzenediazonium salt by *n*-PrO⁻ M⁺.

- Aim for the *fewest* number of disconnections. Adding large fragments in a single reaction is more productive than adding several smaller fragments sequentially (see Section 1.4, convergent vs. linear synthesis).

- Choose disconnections in which functional groups are close to the C–C bonds to be formed since the presence of functional groups often facilitates bond making by a substitution reaction.

- It is often advantageous to disconnect at a branching point since this may lead to linear fragments that are generally more readily accessible, either by synthesis or from a commercial source.

- A preferred disconnection of cyclic esters (lactones) or amides (lactams) produces hydroxy-carboxylic acids or amino-carboxylic acids as targets. Many macrocyclic natural products contain these functional groups, and their syntheses often include a macrocyclization reaction.

X = O, NR

- Functional groups in the TM may be obtained by functional group interconversion.

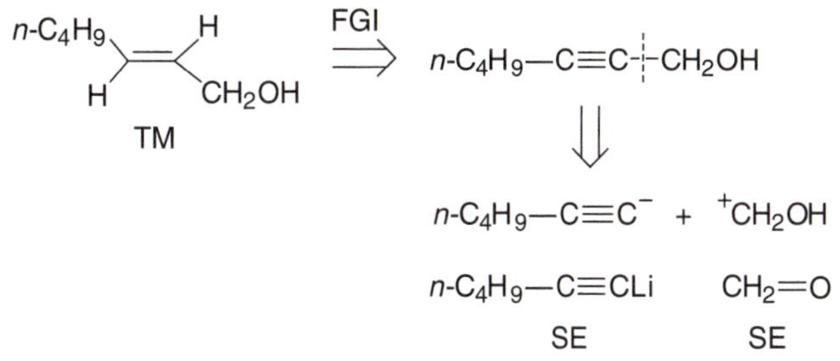

- Symmetry in the TM simplifies the overall synthesis by decreasing the number of steps required for obtaining the TM.

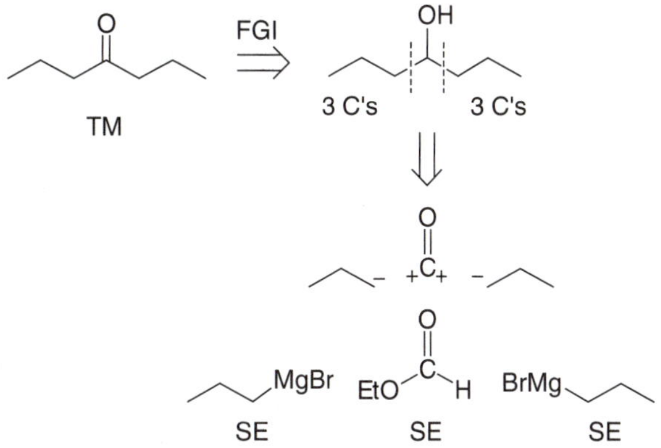

- Introduction of an *activating* (auxiliary) functional group may facilitate carbon-carbon bond formation. This strategy works well for the synthesis of compounds exhibiting a dissonant charge pattern. After accomplishing its role, the activating group is removed.

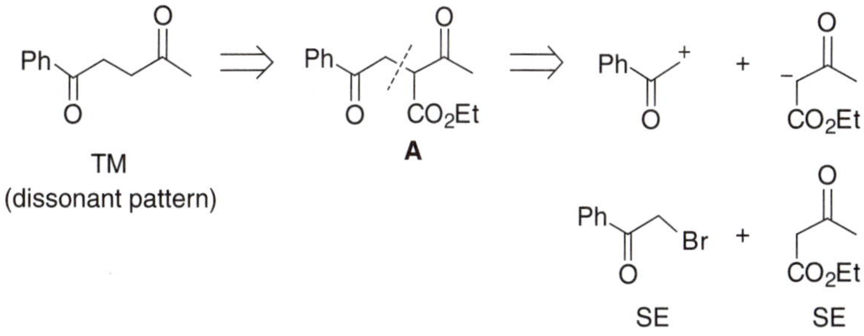

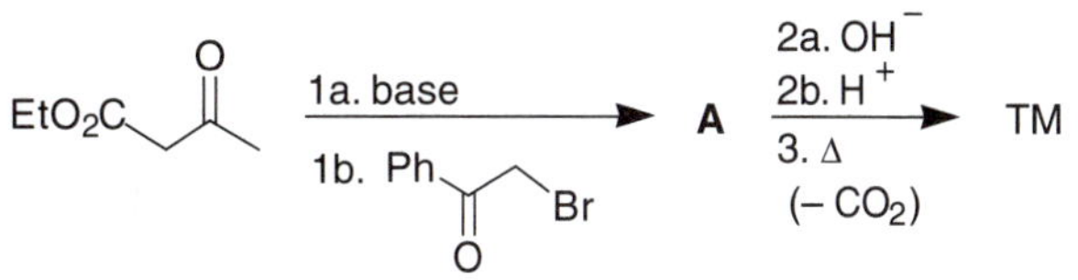

- There is no simple way to disconnect the TM shown below (dissonant charge pattern). However, the presence of a 1,6-dioxygenated compound suggests opening of a six-member ring. A variety of cyclohexene precursors are readily available via condensation and Diels-Alder reactions or via Birch reductions of aromatic compounds.

- Disconnection of an internal (*E*)- or (*Z*)-double bond or a side chain of an alkene suggests a Wittig-type reaction or an alkylation of a vinylcuprate, respectively.

- The presence of a six-member ring, especially a cyclohexene derivative, suggests a Diels-Alder reaction.

- The structural feature of an α, β-unsaturated ketone or a β-hydroxy ketone in a six-member ring suggests a *double disconnection* coupled with functional group interconversions [Michael addition followed by intramolecular aldol condensation (*Robinson annulation*)].

Functional Group Interconversions (FGI)[13]

Functional groups are the keys to organic synthesis. They can be converted into other functional groups by a wide variety of transformations such as by substitution, displacement, oxidation, and reduction reactions. Also, they may be used to join smaller molecular fragments to form larger molecules or to produce two smaller molecules from a large one. A number of selected functional group interconversions often encountered in organic synthesis are shown in Table 1.4a–k.

Table 1.4 Selected Functional Group Interconversions

a. Alkyl Chlorides

$$R-Cl \implies ROH \begin{cases} SOCl_2 + ZnCl_2 \\ PCl_3 \text{ or } PCl_5 \\ [Ph_3P + Cl_2] \longrightarrow [Ph_3\overset{+}{P}-Cl] \ Cl^- \\ [Ph_3P + CCl_4] \longrightarrow [Ph_3\overset{+}{P}-CCl_3] \ Cl^{-a} \end{cases}$$

[a] No HCl is formed; high yields of 1° and 2° alkyl chlorides; *Angew. Chem., Int. Ed.* **1975**, *14*, 801.

b. Alkyl Bromides

$$R-Br \implies \begin{cases} ROH \begin{cases} PBr_3 \\ [Ph_3P + Br_2] \longrightarrow [Ph_3\overset{+}{P}-Br] \ Br^- \\ [Ph_3P + CBr_4] \longrightarrow [Ph_3\overset{+}{P}-CBr_3] \ Br^- \end{cases} \\ RCOOH \text{ ------- } HgO + Br_2{}^b \end{cases}$$

$$RCH_2CH_2-Br \implies \begin{cases} RCH{=}CH_2 \text{ ------- } HBr + \text{free-radical initiator} \\ RCH{=}CH_2 \text{ ------- } a.\ BH_3{\cdot}THF;\ b.\ Br_2 + NaOCH_3 \end{cases}$$

[b] *J. Org. Chem.* **1961**, *26*, 280; see also Hunsdiecker reaction, *Org. React.* **1957**, *9*, 332.

c. Allylic and Propargylic Bromides

Allylic bromide $\implies$ alkene ------- NBS + free-radical initiator

$$RCH{=}CHCH_2Br \implies \begin{cases} RCH{=}CHCH_2OH \text{ ------- } NaBr, BF_3{\cdot}OEt_2{}^c \\ RCH{=}CHCH_2OH \text{ ------- } NBS, Me_2S{}^d \end{cases}$$

$$RC{\equiv}CCH_2Br \implies RC{\equiv}CCH_2OH \text{ ------- } CBr_4, Ph_3P$$

[c] *Tetrahedron Lett.* **1985**, 26, 3863. [d] *Tetrahedron Lett.* **1972**, *13*, 4339.

d. Alkyl Iodides

$$R-I \implies \begin{cases} ROH \text{ ------- } [(PhO)_3P + CH_3I] \longrightarrow [(PhO)_3\overset{+}{P}-CH_3]^{-e} \\ RBr \text{ ------- } NaI \text{ in acetone, heat}^f \\ ROTs \text{ ------- } NaI \end{cases}$$

[e] *J. Chem. Soc.* **1953**, *3*, 2224. [f] Finkelstein reaction: *Chem. Ber.* **1910**, *43*, 1528; see also *J. Org. Chem.* **1977**, *42*, 875.

(*continued*)

| Table 1.4 | Selected Functional Group Interconversions (*continued*) |

e. Nitriles

$RCH_2-CN \Longrightarrow$
- RCH_2-X ------ $NaC\equiv N$ in DMSO
- RCH_2CONH_2 ------ oxalyl chloride + DMF + $CH_3C\equiv N$ (Vilsmeier reagent)

$R-CN \Longrightarrow$
- $RCONH_2$ ------ $(CF_3SO_2)_2O$ [g] or $(CHO)_n$ + HCOOH [h]
- $RCH=NOH$ ------ TBSCl, imidazole [i]

$R-CN \Longrightarrow R'_2C=O$ ------ $TsCH_2NC$, t-BuOK [j]

---

[g] *Synthesis* **1999**, *64*. [h] *J. Org. Chem.* **1996**, *61*, 6486. [i] *Synth. Commun.* **1998**, *28*, 2807.
[j] *Org. React.* **2001**, *57*, 417.

---

f. 1° and 2° Alcohols

$RCH_2CH_2OH \Longrightarrow$
- RCH_2COOH ------ a. $LiAlH_4$; b. H_3O^+ or BH_3, THF
- RCH_2COOR' ------ a. $LiAlH_4$; b. H_3O^+
- RCH_2CHO ------ $NaBH_4$, EtOH
- $RCH=CH_2$ ------ a. BH_3, THF; b. NaOH, H_2O_2

$\underset{R'}{\overset{R}{>}}CH-OH \Longrightarrow \underset{R'}{\overset{R}{>}}C=O$ ------ $NaBH_4$, EtOH or a. $LiAlH_4$; b. H_3O^+

$\underset{H_3C}{\overset{R}{>}}CH-OH \Longrightarrow RCH=CH_2$ ------ 1. $Hg(OAc)_2$, H_2O; 2. $NaBH_4$, NaOH

---

g. 1°, 2° and 3° Amines

$RCH_2NH_2 \Longrightarrow$
- $RCONH_2$ ------ a. $LiAlH_4$, Et_2O; b. H_3O^+
- RCN ------ a. $LiAlH_4$, Et_2O; b. H_3O^+
- $RCHO$ ------ NH_3, $NaBH_3CN$, EtOH (reductive amination)
- RCH_2OH ------ 1. MsCl; 2. NaN_3; 3. Ph_3P, H_2O [k]
- RCH_2N_3 ------ H_2N-NMe_2, $FeCl_3 \cdot 6H_2O$ [l]

$RCH_2NHR' \Longrightarrow$
- $RCONHR'$ ------ a. $LiAlH_4$, Et_2O; b. H_3O^+
- $RCHO$ ------ $R'NH_2$, $NaBH_3CN$, EtOH (reductive amination)

$RCH_2NR'_2 \Longrightarrow$
- $RCONR'_2$ ------ a. $LiAlH_4$, Et_2O; b. H_3O^+
- $RCHO$ ------ R'_2NH, $NaBH_3CN$, EtOH (reductive amination)

---

[k] *Tetrahedron Lett.* **1983**, *24*, 763. [l] *Chem. Lett.* **1998**, 593.

h. Aldehydes and Ketones

$RCH_2CHO \Longrightarrow$
- RCH_2MgX ------ a. Me_2NCHO (DMF); b. H_3O^+
- RCH_2CH_2OH ------ PCC, Swern[m] or Dess-Martin oxidation[n]
- $RC\equiv CH$ ------ a. Sia_2BH; b. NaOH, H_2O_2
- $RCH_2CH=CH_2$ ------ a. O_3, −78 °C; b. Me_2S
- RCH_2COCl ------ H_2, $Pt/BaSO_4$[o] or $LiAlH(Ot\text{-}Bu)_3$[p]
- RCH_2COOH ------ thexylchloroborane
- RCH_2COOR' ------ a. $i\text{-}Bu_2AlH$, −78 °C; b. H_3O^+
- RCH_2CN ------ a. $i\text{-}Bu_2AlH$, −78 °C; b. H_3O^+

$\underset{CH_3}{\overset{R}{>}}C=O \Longrightarrow$
- $RCOCl$ ------ a. $(CH_3)_2CuLi$, −78 °C; b. H_3O^+
- $RCOOH$ ------ a. MeLi (2 eq); b. H_3O^+
- $RC\equiv CH$ ------ $HgSO_4$, H_3O^+

$\underset{R'}{\overset{R}{>}}C=O \Longrightarrow$
- $\underset{R'}{\overset{R}{>}}CH-OH$ ------ Jones, Swern, or NaOCl, HOAc oxidation
- $RCOCl$ ------ organometallic reagents[q]
- $RC(O)N\underset{OCH_3}{\overset{CH_3}{<}}$ ------ a. R'Li or R'MgX; b. H_3O^{+}[r]

$\underset{Ph}{\overset{R}{>}}C=O \Longrightarrow$ benzene ------ $RCOCl$ + $AlCl_3$

---

[m] *Synthesis* **1981**, *165*; *Org. Synth.* **1986**, *64*, 164. [n] *J. Org. Chem.* **1983**, *48*, 4155.
[o] Rosenmund reduction: *Org. React.* **1948**, *4*, 362. [p] *Tetrahedron* **1979**, *35*, 567.
[q] *Tetrahedron* **1999**, *55*, 4177. [r] Weinreb amide: *Tetrahedron Lett.* **1981**, *22*, 3815; see also
J. Org. Chem. **1991**, *56*, 2911.

i. Carboxylic Acids

$RCH_2COOH \Longrightarrow$
- RCH_2CHO ------ Jones reagent or $NaClO_2$, H_2O
- RCH_2CH_2OH ------ PDC, DMF or $Na_2Cr_2O_7$, H_2SO_4
- RCH_2CN ------ H_3O^+ or HO^-, heat
- RCH_2COOR' ------ a. LiOH, THF, H_2O; b. H_3O^+
- $RCH_2COOt\text{-}Bu$ ------ CF_3CO_2H, CH_2Cl_2, 25 °C
- RCH_2Ph ------ RuO_4[s]

$RCOOH \Longrightarrow$
- $RCH=CH_2$ ------ $NaIO_4$, $KMnO_4$, $t\text{-}BuOH$, H_2O
- $RC(O)CH_3$ ------ LiOCl, chlorox[t]

---

[s] *J. Org. Chem.* **1981**, *46*, 3936. [t] *Org. Prep. Proc. Intl.* **1998**, *30*, 230.

(continued)

Table 1.4 Selected Functional Group Interconversions (*continued*)

j. Alkenes

Alkenes $\Longrightarrow$

$\begin{cases} \text{R—X; R—OTs} \text{ ------- } t\text{-BuOK or DBU (E2 elimination)} \\ \text{R—OH} \text{ ------- } \text{KHSO}_4 \text{ or TsOH or H}_3\text{PO}_4 \text{ (dehydration)} \\ \text{RC}\equiv\text{CR}' \begin{cases} \text{Lindlar catalyst + H}_2 \\ \text{Ni(OAc)}_2, \text{NaBH}_4, \text{H}_2\text{N(CH}_2)_2\text{NH}_2 \\ 1.\ \text{Sia}_2\text{BH; } 2.\ \text{AcOH} \end{cases} \begin{matrix} cis \\ \text{alkenes} \end{matrix} \\ \qquad\quad \text{Li}^\circ, \text{liq. NH}_3, t\text{-BuOH} \longrightarrow trans \text{ alkenes} \\ \text{RCHO or R}_2\text{CO} \text{ ------- } \text{Wittig reagent (ylide)} \end{cases}$

k. Alkynes

RC$\equiv$CH $\Longrightarrow$

$\begin{cases} \text{RCH}=\text{CH}_2 \text{ ------- } 1.\ \text{Br}_2, \text{CCl}_4; 2.\ \text{NaNH}_2, \text{NH}_3 \text{ (liq.)} \\ \text{HC}\equiv\text{CH} \text{ ------- } 1.\ n\text{-BuLi; } 2.\ \text{R—Br (1}^\circ \text{ alkyl only)} \\ \text{R–C}(=\text{O})\text{CH}_3 \text{ ------- } \begin{matrix} 1a.\ \text{LDA; b. ClP(O)(OEt)}_2; \\ 2a.\ \text{LDA; b. H}_2\text{O}^u \end{matrix} \\ \text{RCHO} \text{ ------- } \begin{cases} \text{(MeO)}_2\text{P(O)CHN}_2, t\text{-BuOK}^v \\ 1.\ \text{CBr}_4, \text{Ph}_3\text{P, Zn dust; 2a. } n\text{-BuLi;} \\ \text{b. H}_2\text{O}^w \end{cases} \end{cases}$

RC$\equiv$CR$'$ $\Longrightarrow$ RC$\equiv$CH ------- a. n-BuLi; b. R$'$—Br (1° alkyl only)

---

^{u} *Org. Synth.* **1986**, *64*, 44. ^{v} *J. Org. Chem.* **1979**, *44*, 4997. ^{w} Corey-Fuchs procedure; *Tetrahedron Lett.* **1972**, 3769; for examples, see *Helv. Chim. Acta* **1995**, *78*, 242.

Control of Relative Stereochemistry

It is important to use stereoselective and stereospecific reactions (where applicable), such as

- S_N2 displacement reactions; E2 elimination reactions
- Catalytic hydrogenation of alkynes (cis product)
- Metal ammonia reduction of alkynes (trans product)
- Oxidation of alkenes with osmium tetroxide
- Addition of halogens, interhalogens (e.g., BrI) or halogen-like species (e.g., PhSCl, BrOH) to double bonds
- Hydroboration reactions
- Epoxidation of alkenes; ring opening of epoxides
- Cyclopropanation

Control of Enantioselectivity

Control of enantioselectivity will be discussed in the corresponding sections on carbonyl reduction (Chapter 4); alkene hydrogenation, epoxidation, and dihydroxylation (Chapter 5); aldol condensation (Chapter 6); allylation and crotylation (Chapter 7); Claisen rearrangement (Chapter 8); and the Diels-Alder reaction (Chapter 9).

1.4   CHOICE OF SYNTHETIC METHOD

The choice of a method for synthesizing a compound derived from a retrosynthetic analysis should be based on the following criteria:

- *Regiochemistry*, the preferential addition of the reagent in only one of two possible regions or directions, exemplified by the preferential alkylation of 2-methylcyclohexanone by the derived enolate at $C_{(2)}$ and not at $C_{(6)}$

- *Chemoselectivity*, selective reaction of one functional group in the presence of other functional groups, exemplified by the preferential reaction of an aldehyde in the presence of a keto group

- *Stereoselectivity,* the exclusive or predominant formation of one of several possible stereoisomeric products, exemplified by the preferential formation of *cis*-3-methylcyclohexanol on reduction of 3-methylcyclohexanone with lithium aluminum hydride in THF or Et_2O

- *Efficiency*, fewest number of steps

- *High yields* in each step; of paramount concern in any chemical reaction is the yield

- *Availability* and *costs of starting materials*

- *Most environmentally friendly route*. Ideally, the atoms of the substrate and any additional reagents used for the reaction should appear in the final product, called "atom economy"[14] —no by-products are formed, isolation of desired product is facilitated, and waste disposal is minimized (e.g., the Diels-Alder reaction and metal-catalyzed reactions such as the example below[15]):

94% yield
$Z : E = 91 : 9$

- *Simplicity* of selected procedures. Over the years, a large number of reagents have been developed that require special techniques for handling. If possible, one should use procedures that are less demanding in their execution.

- *Isolation and purification of reaction products.*[16] Despite recent advances in methodologies for the synthesis of very complex molecules, one important aspect of synthesis has not changed much over the past decades: isolation and purification. A recent excellent review entitled "Strategy-Level Separations in Organic Synthesis: From Planning to Practice" discusses various techniques for the separation of reaction mixtures.[17] The yield and hence the utility of every reaction is limited by the ability to separate and recover the reaction product from other materials.

- Possibility of a *convergent synthesis* or a *"one-pot process"* (cascade or tandem reactions).

Linear and Convergent Syntheses[3h,3j,18]

The overall yield in a multistep step synthesis is the product of the yields for each separate step. In a *linear synthetic scheme,* the hypothetical TM is assembled in a stepwise manner. For the seven-step synthesis of the hypothetical TM below, if the yield of the intermediate at each step is 80%, the overall yield will be 21% ($0.8^7 \times 100$); for a 70% yield at each step, the overall yield would be only 8%.

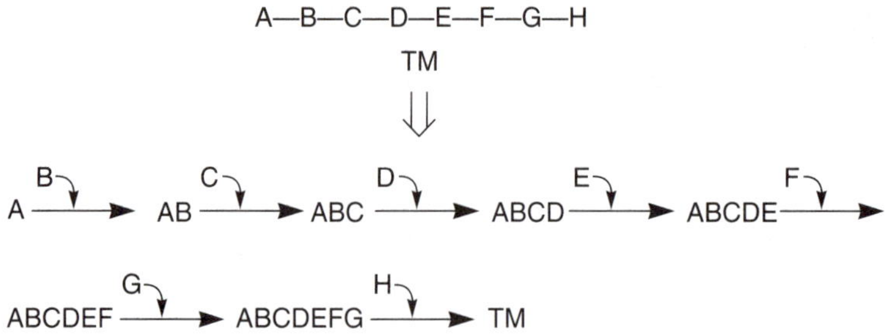

$$A—B—C—D—E—F—G—H$$
$$TM$$
$$\Downarrow$$

$$A \xrightarrow{B} AB \xrightarrow{C} ABC \xrightarrow{D} ABCD \xrightarrow{E} ABCDE \xrightarrow{F}$$

$$ABCDEF \xrightarrow{G} ABCDEFG \xrightarrow{H} TM$$

Since the overall yield of the TM decreases as the number of individual steps increases, a *convergent synthesis* should be considered in which two or more fragments of the TM are prepared separately and then joined at the latest-possible stage of the synthesis. The overall yield in a convergent synthesis is the product of yields of the longest linear sequence. For the synthesis of the above TM, only three stages are involved in the convergent strategy shown below, with an overall yield of 51% ($0.8^3 \times 100$).

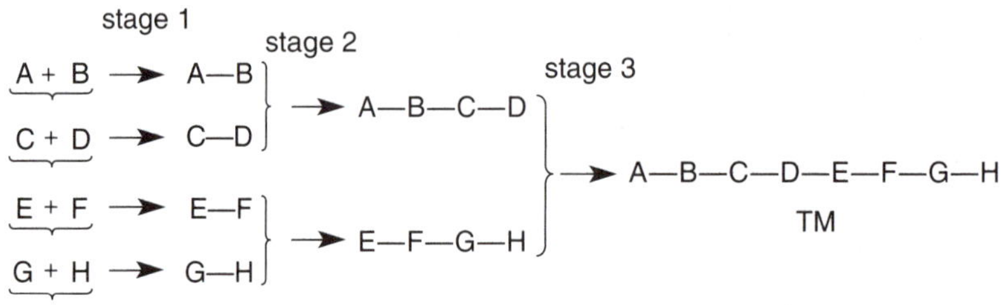

stage 1
stage 2
stage 3

$$A + B \rightarrow A—B$$
$$C + D \rightarrow C—D$$
$$\Big\} \rightarrow A—B—C—D$$

$$E + F \rightarrow E—F$$
$$G + H \rightarrow G—H$$
$$\Big\} \rightarrow E—F—G—H$$

$$\Big\} \rightarrow A—B—C—D—E—F—G—H$$
$$TM$$

It should be noted, however, that the simple overall yield calculation is somewhat misleading since it is computed on one starting material, whereas several are used and the number of reactions is the same! Nevertheless, the increased efficiency of a convergent synthesis compared to the linear approach is derived from the fact that the preparation of a certain amount of a product can be carried out on a smaller scale.

Another important consideration in choosing a convergent protocol is that failure of a single step in a multistep synthesis does not nullify the chosen synthetic approach as a whole, whereas failure of a single step in a linear scheme may require a revision of the whole plan. An example of a triply convergent protocol is the synthesis of the prostaglandin PGE_2 derivative shown below, where the three fragments were prepared separately. The two side chains were then coupled sequentially with the cyclopentenone.[19] Introduction of the first fragment involved conjugate addition of the nucleophilic vinylic organocopper reagent to the enone, followed by trapping of the resulting enolate with the electrophilic side chain.

Analysis

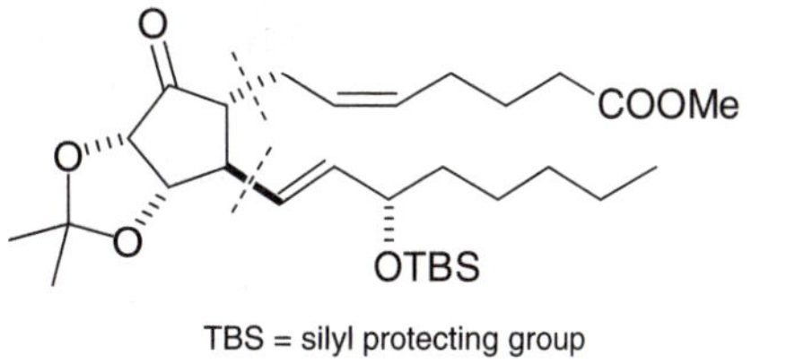

TBS = silyl protecting group

three principal fragments for a convergent coupling

Synthesis

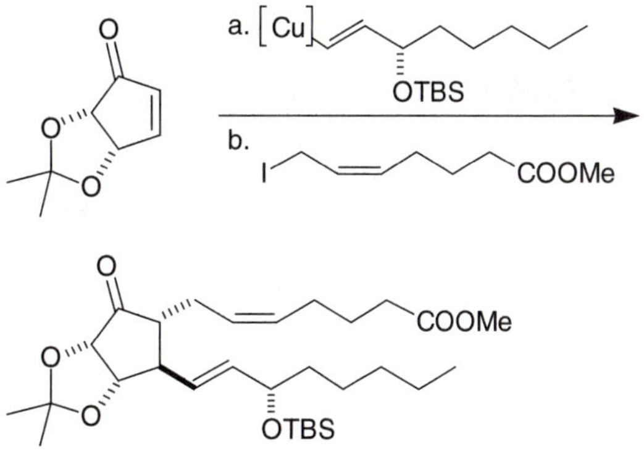

Convergent syntheses involve *consecutive reactions*, where the reagents or catalysts are added sequentially into "one pot," as illustrated in the example below.[20]

Analysis

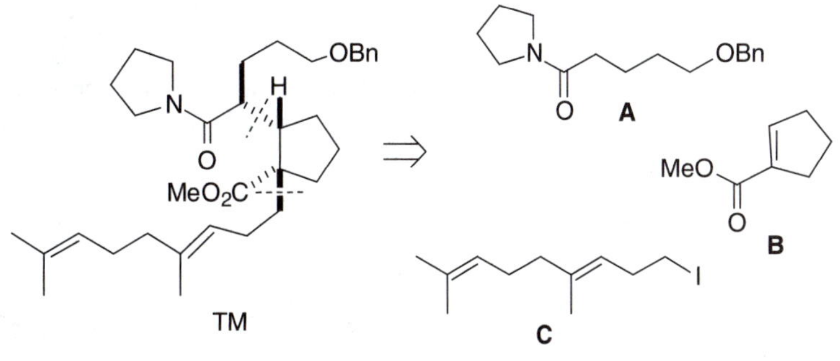

Synthesis

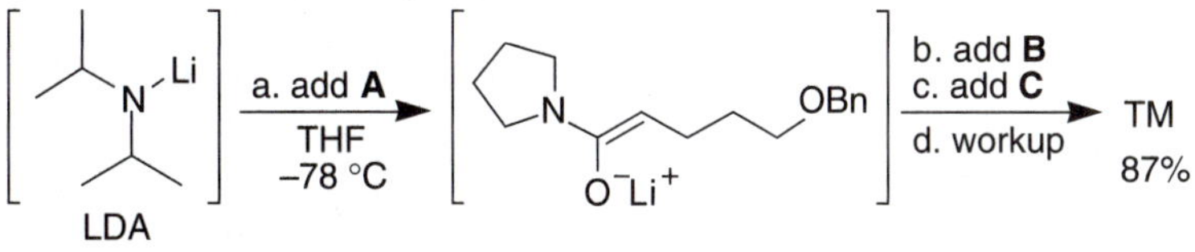

1.5 DOMINO REACTIONS (ALSO CALLED CASCADE OR TANDEM REACTIONS)[21]

Domino-type reactions involve careful design of a multistep reaction in a one-pot sequence in which the first step creates the functionality to trigger the second reaction and so on, making this approach economical and environmentally friendly. A classical example of a tandem reaction is the *Robinson annulation* (a Michael reaction followed by aldol condensation and dehydration).

1.6 COMPUTER-ASSISTED RETROSYNTHETIC ANALYSIS

Computer programs are available that suggest possible disconnections and retrosynthetic pathways.[22] Such programs utilize the type of systematic analysis outlined above to identify key bonds[2] for disconnection and plausible functional group interconversions. In doing so, "retrosynthetic trees" of possible pathways that connect a synthetic target to simple (and/or commercially available) starting materials are generated. The strength of such programs is their thoroughness — in principle, all possible disconnections for any target molecule can be considered. For any molecule of even moderate complexity, however, this process would lead to a plethora of possible synthetic routes too large for any synthetic chemist to analyze in a reasonable amount of time. Fortunately, synthesis programs generally also include routines that rank the synthetic pathways they produce based on well-defined criteria such as fewest number of synthetic steps (efficiency), thus allowing chemists to focus their energy on evaluating the viability and aesthetic appeal of key disconnections. Still, each program is limited by the synthetic strategies (transforms and FGI) contained in its library of possible reactions. Synthetic programs are unlikely to ever replace creative chemists, but this is generally not the intent of those who have created them.

PROBLEMS

The more challenging problems are identified by an asterisk (*).

1. **Functional Group Interconversion.** Show how each of the following compounds can be prepared from the given starting material.

   a. $Cl(CH_2)_6OH \longrightarrow EtO_2C(CH_2)_6CN$

   b.

c.

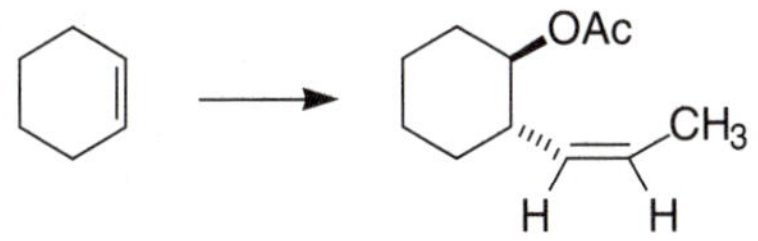

d.

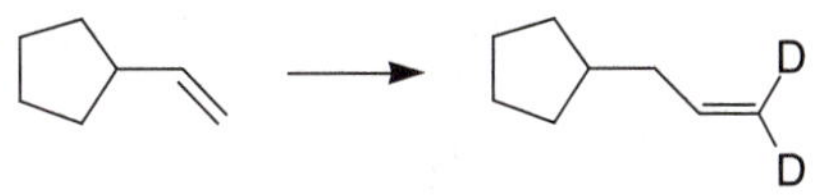

***e.**

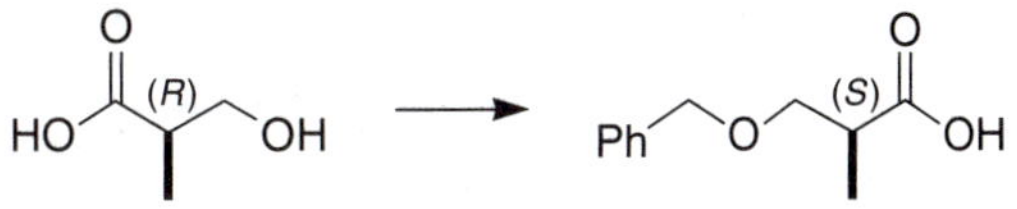

***f.**

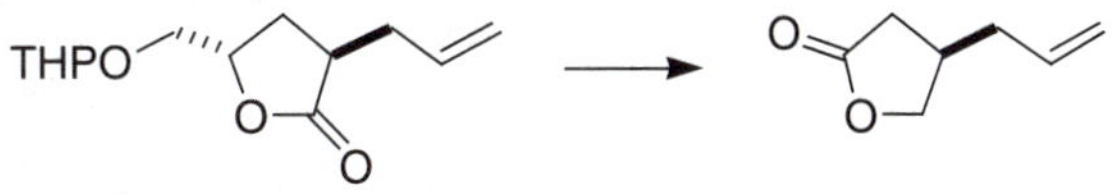

2. **Umpolung.** Show how each of the following compounds can be prepared from the given starting material using either a formyl or an acyl anion equivalent in the synthetic scheme.

a.

$I(CH_2)_5Cl \longrightarrow$

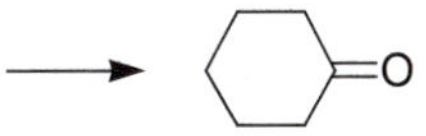

b.

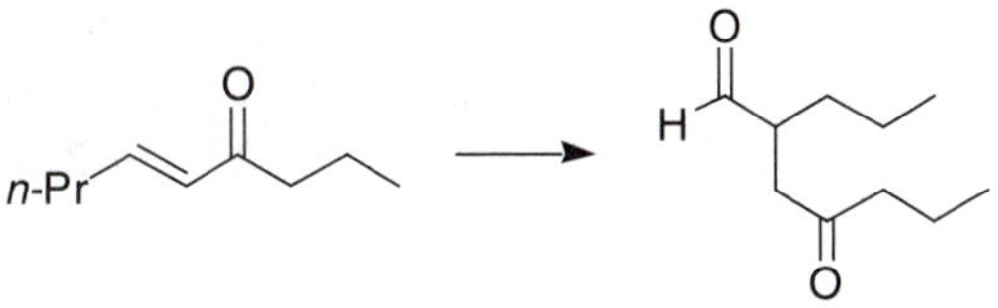

c.

$ClCH_2(CH_2)_2-C{\equiv}C-(CH_2)_4CH_3 \longrightarrow CH_3(CH_2)_9\overset{O}{\overset{\parallel}{C}}(CH_2)_3CH{=}CH(CH_2)_4CH_3$
$\qquad\qquad\qquad\qquad\qquad\qquad\qquad\qquad\qquad\qquad cis$

***d.**

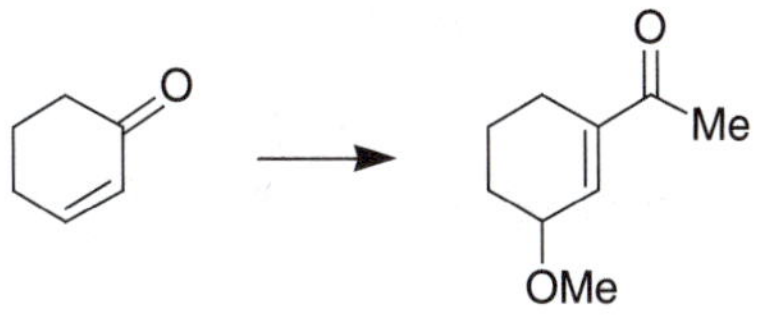

***e.**

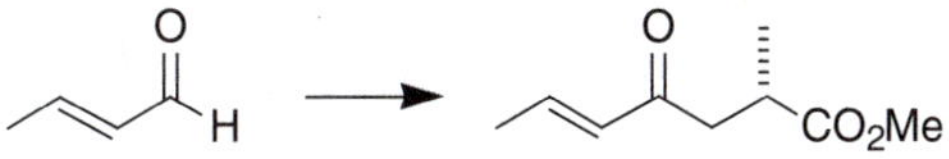

3. **Retrosynthetic Analysis — *One-Step Disconnections.*** For each of the following compounds, suggest a one-step disconnection. Use FGIs as needed. Show charge patterns, the synthons, and the corresponding synthetic equivalents.

a.

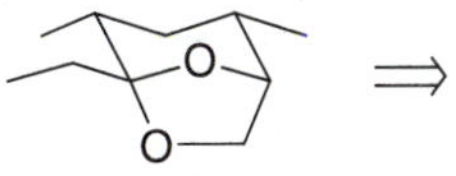

(±)-multistriatin

b.

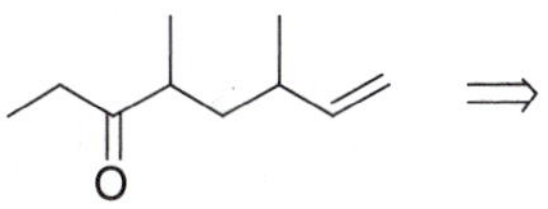

c.

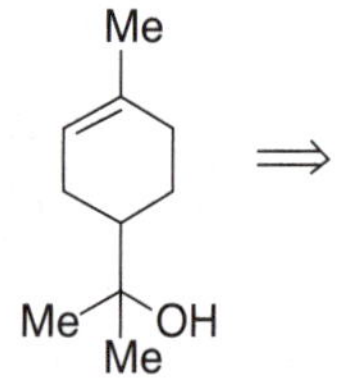

(±)-terpineol

d.

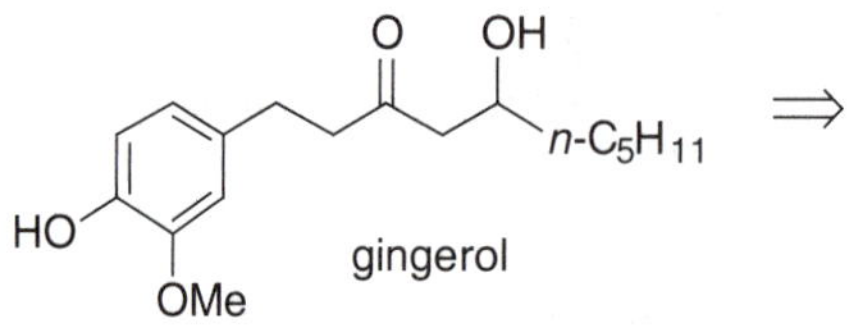

gingerol

e.

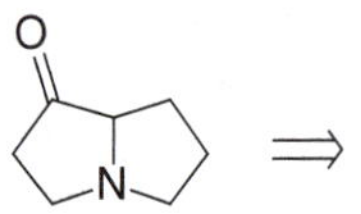

f.

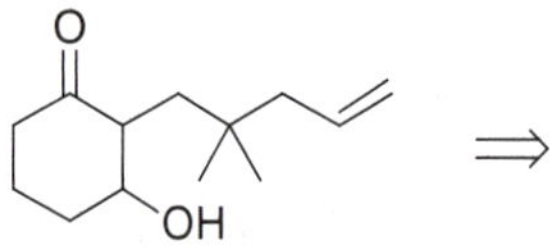

4. **Synthesis.** Outline a retrosynthetic scheme for each of the following target molecules using the indicated starting material. Show (1) the *analysis* (including FGI, synthons, synthetic equivalents) and (2) the *synthesis* of each TM.

a.

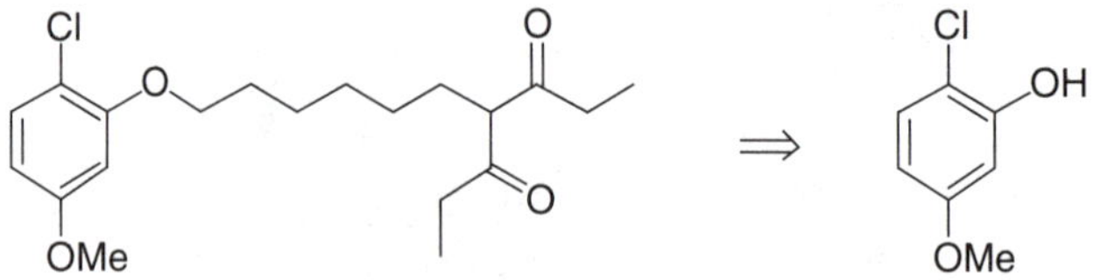

b.

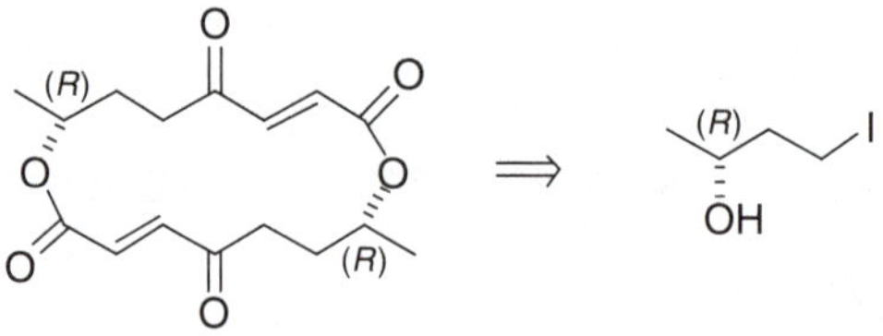

(−)-pyrenophorin
(antifungal compound)

c.

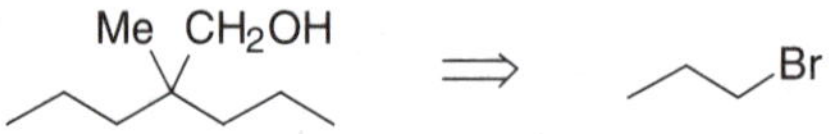

d.

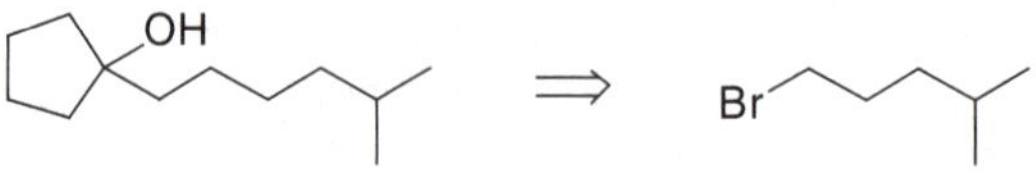

e.

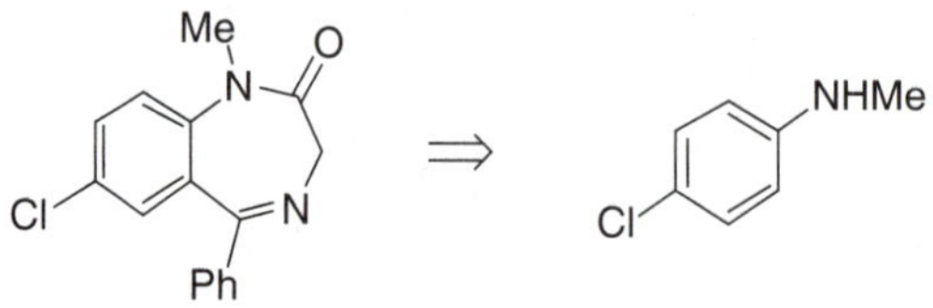

Valium
(tranquilizer)

f. OH O OH ⟹ acetal protecting group for 1,2-diols; cleaved by H^+, H_2O

HO / OH — O — CO$_2$Et

g. OH (CH$_2$)$_4$CH$_3$ ⟹ CHO

*h. HO / HO ⟹ O — COOH / HO—H / H—OH / COOH

*i. O ⟹ O

REFERENCES

1. Several syntheses of pumiliotoxin C have been reported. See, for example, (a) Ibuka, T., Mori, Y., Inubushi, Y. *Tetrahedron Lett.* **1976**, *17*, 3169. (b) Overman, L. E., Jessup, P. J. *J. Am. Chem. Soc.* **1978**, *100*, 5179. (c) Mehta, G., Praveen, M. *J. Org. Chem.* **1995**, *60*, 279.

2. Corey, E. J. *Pure & Appl. Chemistry* **1967**, *14*, 19.

3. (a) Warren, S. *Organic Synthesis: The Disconnection Approach*, Wiley: New York, 1982. (b) Corey, E. J., Cheng, X.-M. *The Logic of Chemical Synthesis*, Wiley: New York, 1989. (c) Mackie, R. K., Smith, D.M., Aitken, R. A. *Guidebook to Organic Synthesis*, 3rd ed., Longman: Harlow, UK, 1999. (d) Ho, T.-L. *Tactics of Organic Synthesis*, Wiley: New York, 1994. (e) Smith, M. B. *Organic Synthesis*, 2nd ed., McGraw-Hill: Boston, 2002. (f) Laszlo, P. *Organic Reactions — Simplicity & Logic*, Wiley: New York, 1995. (g) Willis, C. L., Wills, M. *Organic Synthesis*, Oxford University Press: Oxford, 1995. (h) Smit, W. A., Bochkov, A. F., Caple, R. *Organic Synthesis. The Science Behind the Art*, Royal Society of Chemistry: Cambridge, UK, 1998. (i) Boger, D. L. *Modern Organic Synthesis*, TSTI Press: La Jolla, 1999. (j) Furhop, J. H., Li, G. *Organic Synthesis Concepts and Methods*, 3rd ed., Wiley-VCH: Weinheim, 2003.

4. (a) Evans, D. A. *Acc. Chem. Res.* **1974**, *7*, 147. (b) Serratosa, F. *Organic Chemistry in Action,* 2nd ed., Elsevier: Amsterdam, 1996. (c) Ho, T.-L. *Polarity Control of Synthesis*, Wiley: New York, 1991.

5. (a) Corey, E. J., Seebach, D. *Angew. Chem., Int. Ed.* **1965**, *4*, 1075. (b) Gröbel, B.-T., Seebach, D. *Synthesis* **1977**, 357. (c) Seebach, D. *Angew. Chem. Int. Ed.* **1979**, *18*, 239. (d) Hase, T. A. *Umpoled Synthons: A Survey of Sources and Uses in Synthesis*, Wiley: New York, 1987. (e) Hassner, A., Lokanatha

Rai, K. M. *Comp. Org. Synthesis*. In Trost, B. M., Fleming, I., Eds., Pergamon Press: Oxford, UK, 1991, Vol. 1, p 541. (f) Smith, A. B. III, Adams, C. M. *Acc. Chem. Res.* **2004**, *37*, 365.

6. (a) Bernardi, F., Csizmadia, I. G., Mangini, A., Schlegel, H. B., Whangbo, M.-H., Wolfe, S. *J. Am. Chem. Soc.* **1975**, *97*, 2209. (b) Whitham, G. H. *Organosulfur Chemistry*, Oxford University Press: Oxford, UK, 1995.

7. Brown, C. A., Yamaichi, A. *Chem. Commun.* **1979**, 100.

8. Seebach, D., Seuring, B., Kalinowski, H-O., Lubosch, W., Renger, B. *Angew. Chem., Int. Ed.* **1977**, *16*, 264.

9. (a) Dubs, P., Stüssi, R. *Helv. Chim. Acta* **1978**, *61*, 990. (b) Galobardes, M. R., Pinnick, H. W. *Tetrahedron Lett.* **1981**, *22*, 5235. (c) Pinnick, H. W. *Org. React.* **1990**, *38*, 655.

10. (a) Stork, G., Maldonado, L. *J. Am. Chem. Soc.* **1971**, *93*, 5286. (b) Deuchert, K., Hertenstein, U., Hünig, S., Wehner, G. *Chem. Ber.* **1979**, *112*, 2045. (c) Stetter, H. *Angew. Chem., Int. Ed.* **1976**, *15*, 639. (d) Stetter, H., Kuhlmann, H. *Org. React.* **1981**, *40*, 407.

11. Recent acidity measurements of nitriles have revealed higher pK_a values than originally reported; see Richard, J. P., Williams, G., Gao, J. *J. Am. Chem. Soc.* **1999**, *121*, 715.

12. (a) Schöllkopf, U., Hänssle, P. *Ann.* **1972**, *763*, 208. (b) Baldwin, J. E., Lever, O. W., Tzodikov, N. R. *J. Org. Chem.* **1976**, *41*, 2312. (c) Gould, S. J., Remillard, B. D. *Tetrahedron Lett.* **1978**, *19*, 4353. (d) Soderquist, J. A., Hsu, G. J.-H. *Organometallics* **1982**, *1*, 830.

13. (a) Meakins, G. D. *Functional Groups: Characteristics and Interconversions*, Oxford University Press: Oxford, UK, 1996.

(b) Larock, R. C. *Comprehensive Organic Transformations: A Guide to Functional Group Preparations*, 2nd ed., Wiley-VCH: New York, 1999. *Note*: This book is a "must" for synthetic chemists.

14. Trost, B. M. *Angew. Chem., Int. Ed.* **1995**, *34*, 259.

15. Trost, B. M., Oi, S. *J. Am. Chem. Soc.* **2001**, *123*, 1230.

16. Ho, T.-L. *Distinctive Techniques for Organic Synthesis: A Practical Guide*, World Scientific: Singapore, 1998.

17. Curran, D. P. *Angew. Chem., Int. Ed.* **1998**, *37*, 1174.

18. Hendrickson, J. B. *J. Am. Chem. Soc.* **1977**, *99*, 5439.

19. Johnson, C. R., Penning, T. D. *J. Am. Chem. Soc.* **1988**, *110*, 4726.

20. Heathcock, C. H. *Angew. Chem., Int. Ed.* **1992**, *31*, 665.

21. (a) Ho, T.-L. *Tandem Organic Reactions*, Wiley-Interscience: New York, 1992. (b) Buce, R. A. *Tetrahedron* **1995**, *51*, 13103. (c) Tietze, L. *Chem. Rev.* **1996**, *96*, 115.

22. (a) *Computer Assisted Organic Synthesis*, Wipke, W. T., Howe, W. J., Eds., ACS Symp. Ser. No. 61, 1977. (b) Corey, E. J., Long, A. K., Rubenstein, S. D. *Science* **1985**, *228*, 408. (c) Hendrickson, J. B. *Acc. Chem. Res.* **1986**, *19*, 274. (d) Hendrickson, J. B. *Chemtech* **1998**, *28*, 35.

CHAPTER 2

Stereochemical Considerations in Planning Syntheses

Chemical synthesis always has some element of planning in it.
But the planning should never be too rigid.
R. B. Woodward

2.1 CONFORMATIONAL ANALYSIS[1]

Molecules that differ from each other by rotation about single bonds are called *conformational isomers* or *conformers*. Derek H. R. Barton (1918–98; Nobel Prize, jointly with Odd Hassel, in 1969) showed that the chemical and physical properties of complicated molecules can be interpreted in terms of their specific or preferred rotational arrangements and that a knowledge of the conformations of molecules is crucial to understanding the stereochemical basis of many reactions.[2]

Acyclic Systems[3]

Ethane

The eclipsed conformation of ethane is ~3 kcal/mol less stable than the staggered conformation (~1 kcal/mol for each eclipsed H/H pair).[*] Any conformation between staggered and eclipsed is referred to as a skew conformation.

eclipsed

dihedral angle: H_1—C—C—H_2	0°
H_1—H_2 distance:	2.29 Å

staggered

dihedral angle: H_1—C—C—H_2	60°
H_1—H_2 distance:	2.44 Å

The *instability* of the eclipsed form of ethane was originally postulated to result from repulsion of filled hydrogen orbitals. However, state-of-the-art quantum chemical calculations now indicate that *two* main factors contribute to the preference for the staggered conformation of ethane.[4] First, the eclipsed form is selectively *destabilized* by unfavorable four-electron interactions between the filled C–H bonding orbitals of

---

[*]1 kcal/mol = 4.184 kJ/mol.

each pair of eclipsed bonds. Second, the staggered conformer is selectively *stabilized* by favorable orbital interactions between filled C–H bonding orbitals and unfilled C–H antibonding orbitals of antiperiplanar C–H bonds (hyperconjugation).

The energy required to rotate the ethane molecule about the C–C bond is called its *torsional energy*. *Torsional strain* is the repulsion between neighboring bonds (electron clouds) that are in an eclipsed relationship.

Propane

The CH_3–H eclipsed interaction imposes 1.4 kcal/mol of strain on top of the 2.0 kcal/mol H–H torsional strains in the eclipsed conformation of propane. The 0.4 kcal/mol of additional strain is referred to as *steric strain*, the repulsion between *nonbonded atoms* or *groups*.

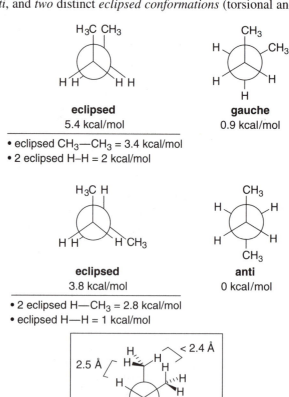

3.4 kcal/mol   0 kcal/mol

Types of interactions:

• eclipsed H—CH_3 = 1.4 kcal/mol
• 2 eclipsed H—H = 2 kcal/mol

Butane

A potential energy plot for rotation about the C_2–C_3 bond in butane shows unique maxima and minima. There are *two* kinds of staggered conformations, *gauche* (steric strain) and *anti*, and *two* distinct *eclipsed conformations* (torsional and steric strain).

eclipsed   **gauche**
5.4 kcal/mol   0.9 kcal/mol

• eclipsed CH_3—CH_3 = 3.4 kcal/mol
• 2 eclipsed H–H = 2 kcal/mol

eclipsed   **anti**
3.8 kcal/mol   0 kcal/mol

• 2 eclipsed H—CH_3 = 2.8 kcal/mol
• eclipsed H—H = 1 kcal/mol

2.5 Å   < 2.4 Å

gauche conformation
steric strain

At room temperature, *n*-butane is a mixture of 70% anti and 30% gauche conformations. To separate these two species, one would have to slow down the interconversion by working at –230°C!

Ring Systems

In addition to torsional strain (eclipsing interaction) and steric strain (nonbonded interaction), the compression of internal bond angles in ring systems leads to an additional type of strain: *angle strain*.

Cyclopropane

Compression of the ideal 109.5° Csp^3 bond angle to an internal bond angle of 60° in cyclopropane results in considerable angle strain, which is manifested by "bent bonds." Six pairs of eclipsed hydrogens adds an additional ~6 kcal/mol of torsional strain. Total strain is ~27 kcal/mol.

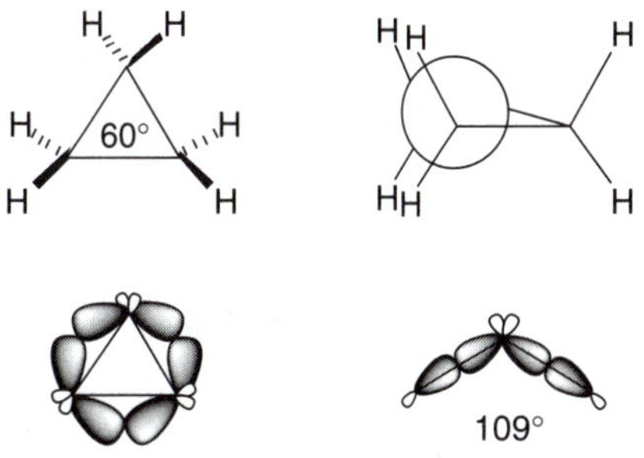

bent bonds
poor overlap = weaker bonds   typical sp³ bond

The poor orbital overlap in cyclopropane allows for C–C bond cleavage under conditions where typical Csp^3–Csp^3 bonds are stable. For example, cyclopropanes undergo hydrogenolysis.[5] Activated cyclopropanes (i.e., bonded to an electron-withdrawing group) are cleaved by nucleophiles, as illustrated below.[6]

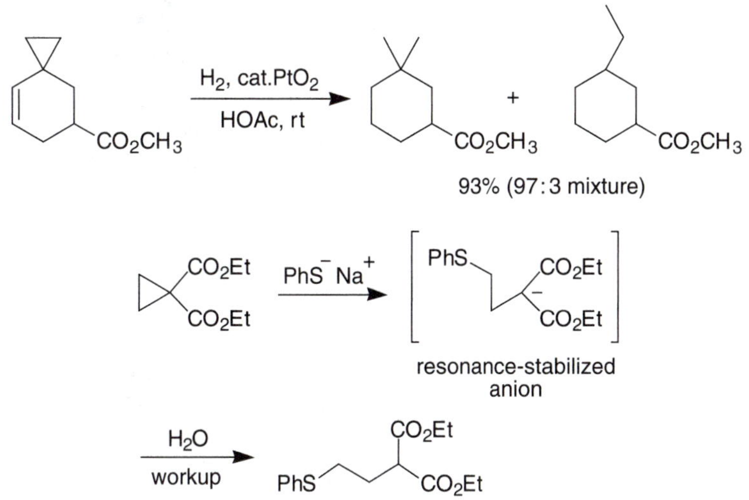

93% (97:3 mixture)

resonance-stabilized anion

Cyclobutane

Spectroscopic measurements indicate that cyclobutane is slightly bent (puckered), so that one carbon atom lies about 25° above the plane of the other three. This slightly increases the angle strain (~88° internal bond angle) but decreases the torsional strain (< 8 kcal/mol) until a minimum energy balance between the two opposing effects is reached. Total strain is ~26 kcal/mol.

Cyclopentane

There would be no angle strain in cyclopentane *if it were planar* (108° internal bond angle); however, *if planar*, there would be ~10 kcal/mol torsional strain. Consequently, cyclopentane adopts a puckered, out-of-plane conformation that strikes a balance between increased angle strain and decreased torsional strain. Four of the carbons are in approximately the same plane, and the hydrogens of the out-of-plane methylene group are nearly staggered with respect to their neighbors. In this conformation, the hydrogens at three of the five carbons can adopt quasi-equatorial or quasi-axial positions. The cyclopentane ring is not static but is in constant motion in such a way that each carbon alternates as the point of the envelope. Total strain, after puckering, is 6 kcal/mol.

envelope conformations of cyclopentane

Cyclohexane

Cyclohexane is the most important of the carbocycles; its structural unit is widely encountered in various natural products. It can adopt a *chair conformation* that is essentially strain free. The chair form of cyclohexane has two distinct types of hydrogens: *equatorial* and *axial*.

Cyclohexane is a dynamic structure, and the chair conformations rapidly flip. Its room temperature ^{1}H-NMR spectrum displays a broad singlet at δ 1.43 ppm (spin averaging), which resolves at −106°C into absorptions at δ 1.20 (axial H's) and at δ 1.66 (equatorial H's) ppm. The interconversion of the two conformations has an enthalpy of activation of 10.8 kcal/mol.

chair
0 kcal/mol

half chair
10.8 kcal/mol

twist boat
5.5 kcal/mol

boat
7.0 kcal/mol

chair

half chair

twist boat

The boat form is an alternate conformation of cyclohexane. Actually, by a slight twist, the nonbonded interactions in the boat form can be reduced (twist boat conformation).

1.83Å

twist boat
5.5 kcal/mol

boat
~7 kcal/mol

twist boat
5.5 kcal/mol

Although the chair form of cyclohexane is the preferred conformation, other conformations are known and in some systems are required.

$$\xrightarrow[-H_2O]{heat}$$

twist boat
(a lactone)

The lactone moeity acts as a stereochemical bridge, maintaining the hydroxyl and carboxylic acid groups in a cis relationship.

2.2 EVALUATION OF NONBONDED INTERACTIONS

Monosubstituted Cyclohexanes

Stereoanalysis of monosubstituted cyclohexanes involves two distinct stages[7]:

1. Determination of the topology of the molecule
2. Assessment of the topology and its effects on the course of a reaction

Because of the 5.5 kcal/mol difference between the chair and the higher-energy twist form of cyclohexane, the vast majority of compounds containing a six-member ring exist almost entirely in the chair form. If a six-member ring system can be said to be in one chair conformation, then for the purpose of synthetic planning, stereochemical predictions can, in many instances, be made with considerable confidence. The factor that contributes to the instability of a *monosubstituted* cyclohexane is the presence of an *axial* substituent. The destabilization caused by an *axial* substituent (e.g., CH_3) is due to its 1,3-diaxial interaction with the two hydrogens on the ring (*n*-butane gauche-type interactions).

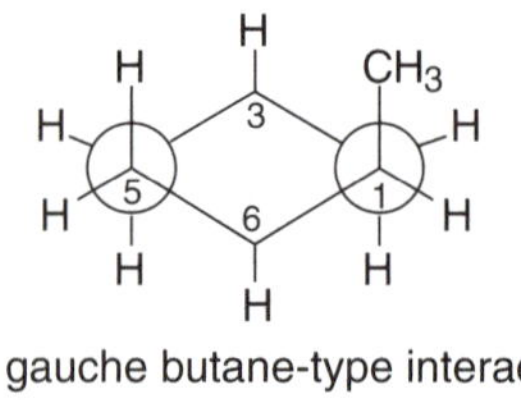

two gauche butane-type interactions:

$C_{(1)}-CH_3$ ⅄ $C_{(3)}-H_{ax}$ and $C_{(1)}-CH_3$ ⅄ $C_{(5)}-H_{ax}$

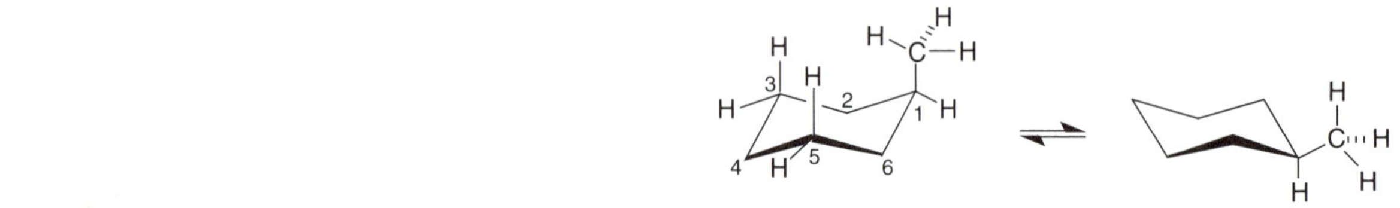

$\Delta G° = -(2 \times 0.9) = -1.8$ kcal/mol

The equilibrium population of any conformer is given by

$$\Delta G° = -RT \ln K_{eq}$$

Therefore, we can compute the equilibrium composition of the two methylcyclohexane conformers as shown below:

$$\Delta G° = -RT \ln K_{eq} \text{ (axial} \rightleftharpoons \text{equatorial)}$$

At 25 °C (298 Kelvin): $\Delta G° = -(1800 \text{ cal/mol}) = -(1.987 \text{ cal/deg•mol}) (298 \text{ K}) \ln K_{eq}$

$$\ln K_{eq} = \frac{1800}{1.987 \times 298} = 3.04 \qquad K_{eq} = 21 = \frac{[\text{equatorial conformer}]}{[\text{axial conformer}]}$$

$$[\text{axial conformer}] + [\text{equatorial conformer}] = 1$$

$$\% \text{equatorial conformer} = \frac{21}{22} (100) = 95\%$$

Table 2.1 shows the population dependence of the favored conformation on K_{eq} and $-\Delta G°$.

Often interaction energy values of conformations are reported as potential energies E by assuming that the difference in free energy $\Delta G°$ between isomers is equal to E and that ΔE may be equated with $\Delta H°$ and $\Delta S° \sim 0$, which is probably true for methylcyclohexane.

$$\Delta G° = \Delta H° - T\Delta S°$$

It should be noted that ΔS may not be equal to zero in some di- and polysubstituted cyclohexanes. However, since the entropy term $\Delta S°$ (in cal) will be relatively small compared to $\Delta H°$ (in kcal), we assume that $\Delta S° \sim 0$ and hence $\Delta G° \sim \Delta H°$.

Conformational energies for axial-equatorial interconversion for a number of monosubstituted cyclohexanes have been reported. These are often referred to as *A values* and allow us to estimate steric effects in reactions.

Table 2.1	Population of the Favored Conformation at 25 °C				
K_{eq} (25 °C)	1	2	4	10	100
$-\Delta G°$ (kcal/mol)	0	0.41	0.82	1.4	2.7
Population %	50	67	80	91	99

The strong preference of the *tert*-butyl group to occupy an equatorial position makes it a highly useful group to investigate conformational equilibria. Thus, a tert butyl group will ensure that the equilibrium lies on the side in which the *tert*-butyl group occupies the equatorial position. Note, however, that a *tert*-butyl group does not "lock" a system in a single conformation; conformational inversion still takes place, although with a high energy barrier.

Disubstituted Cyclohexanes

Depending on the substitution pattern, three principal interactions dictate the conformational equilibrium:

1. The presence of a single axial substituent (butane-type gauche interaction)
2. The interaction of a pair of 1,2-diequatorial substituents (butane-type gauche interaction)
3. The interaction of a pair of *cis*-1,3-diaxial substituents (1,3-diaxial interaction; Table 2.2)

trans-1,2-Dimethylcyclohexane

four 1,3-diaxial CH₃—H interactions
= 4 x 0.9 = 3.6 kcal/mol

one gauche CH₃—CH₃ interaction
= 1 x 0.9 = 0.9 kcal/mol

$$\Delta G° = -(3.6 - 0.9) = -2.7 \text{ kcal/mol}$$

cis-1,2-Dimethylcyclohexane

two 1,3-diaxial CH₃—H interactions
one gauche CH₃—CH₃ interaction
= 1.8 + 0.9 = 2.7 kcal/mol

two 1,3-diaxial CH₃—H interactions
one gauche CH₃—CH₃ interaction
= 1.8 + 0.9 = 2.7 kcal/mol

$$\Delta G° = 0$$

Large steric interactions result when two groups are situated *syn* axially, as in *cis*-1,3-dimethylcyclohexane:

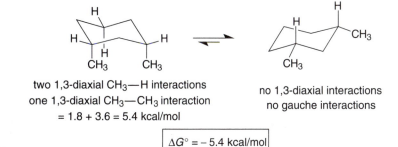

two 1,3-diaxial CH₃—H interactions
one 1,3-diaxial CH₃—CH₃ interaction
= 1.8 + 3.6 = 5.4 kcal/mol

no 1,3-diaxial interactions
no gauche interactions

$$\Delta G° = -5.4 \text{ kcal/mol}$$

Table 2.2　1,2-Diequatorial and 1,3-Diaxial Interaction Energies (kcal/mol)[7]

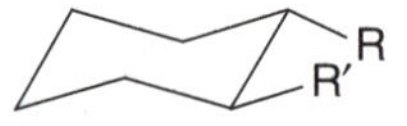

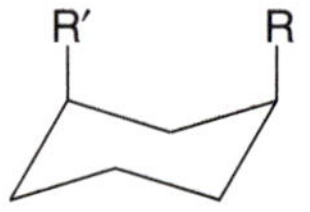

R	R′	1,2-Diequatorial	1,3-Diaxial
–Cl	–Cl	0.7–1.5	5.5
–CH$_3$	–CH$_3$	0.75–0.80	3.6
–OH	–OH	0.35	1.9
–CH$_3$	–OH	0.38	1.9–2.7
–CH$_3$	CO$_2$Et	0.2	2.8–3.2

Evaluation of Destabilization Energies (E$_D$)

Corey and Feiner[7] have developed a computer program (LHASA) for conformational analysis and for determining the destabilization energies (E$_D$) in substituted cyclohexane derivatives. In the following discussion, we will adopt their A, G and U designations and use the corresponding E$_D$ values for evaluating steric interactions.

Each substituent R has associated with it three appendage interaction values (Table 2.3):

1.　An A$_R$ value for R–H 1,3-diaxial interactions
2.　A G$_R$ value for gauche R–R′ 1,2-diequatorial interactions
3.　A U$_R$ value for R-R′ 1,3-diaxial interactions

Table 2.3　A, G, and U Values of Cyclohexane Derivatives (kcal/mol)[a,7]

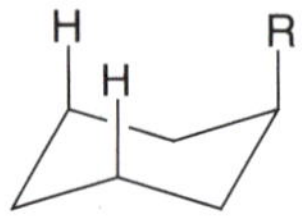

	E$_D$ = A$_R$	E$_D$ = G$_R$ + G$_{R'}$	E$_D$ = U$_R$ + U$_{R'}$ $+ \frac{1}{2}[A_R + A_{R'}]$
R or R′	**A**	**G**	**U**
F	0.2	0	0
Cl	0.4	0.5	0.4
Br	0.4	0.8	0.4
I	0.4	1.0	0.4
OH, OR	0.9	0.2	0.9
NHR	1.3	0.3	1.3
NR$_2$	2.1	0.5	2.1
N=	0.5	0.1	0.5
N≡	0.2	0.1	1.2
C=	1.3	0.2	0.9
Aryl	3.0	1.2	1.1
C≡	0.2	0	1.2
CO$_2$H, CO$_2$R	1.2	0.5	1.2
CHO	0.8	0.3	0.8
CH$_2$R	1.8	0.4	1.8
CHR$_2$	2.1	0.8	2.1
CR$_3$	6.0	2.5	6.0

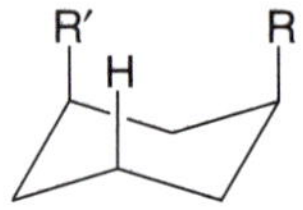

[a] No values are given for *cis*-1,2-axial-equatorial interactions because inversion leads again to a 1,2-equatorial-axial interaction. This is not the case for the *trans*-1,2-diequatorial isomer, since inversion affords a diaxial conformer. For conversion of kcal to kilojoules, use 1 kcal/mol = 4.184 kJ/mol.

It is assumed that conformational effects are *additive*, that is, the destabilizing interactions operate independently of each other. However, this is not always the case. The magnitude of the E_D values for polar substituents may be affected by the *polarity of the solvent*, by *hydrogen bonding*, and by *dipole-dipole* interactions.

A: $E_D = U_{Me} + U_{OH} + G_{OMe} + G_{Cl} + \frac{1}{2}(A_{Me} + A_{OH})$
$ = 1.8 + 0.9 + 0.2 + 0.5 + 0.9 + 0.45$
$ = 4.75 \text{ kcal/mol}$

B: $E_D = A_{OMe} + A_{Cl}$
$ = 0.9 + 0.4$
$ = 1.3 \text{ kcal/mol}$

At 25 °C, $\Delta E_D = 4.75 - 1.3 = 3.45 \text{ kcal/mol}$

Assuming $\Delta E_D \sim \Delta G°$, then $K = 339$ and **B** = 99.7%

Atypical Disubstituted Cyclohexanes

Dipole-Dipole Interactions

The preferred conformations of 2-bromo- and 2-chlorocyclohexanones depend on the polarity of the solvent.[8] In the diequatorial conformer there is considerable electrostatic repulsion. Note that parallel dipoles are destabilizing in a nonpolar solvent.

favored conformation in methanol
(polar solvent,
solvation of Cl and C=O)

favored conformation in octane
(nonpolar solvent,
smaller dipole moment)

Hydrogen Bonding

Intramolecular hydrogen bonding between 1,3-diaxial OH groups in nonpolar solvents confers appreciable stability to a conformer. In polar solvents, however, the solvent competes for *inter*molecular H-bond formation, resulting in normal steric effects dominating the equilibrium.

2.3   SIX-MEMBER HETEROCYCLIC SYSTEMS

Tetrahydropyrans

The steric interaction between any axial substituent and a β-situated heteroatom is counted as zero.[7] Thus, in 3-substituted tetrahydropyrans, the destabilization due to an axial R substituent is computed as half of the A_R.

$$E_D = 1/2\ A_R$$

Because of the shorter C–O bond distance (C–O 1.43 Å), the 1,3 R · · · H interaction is expected to increase. This is especially evident when two heteroatoms are present in the cyclohexane ring, as in 1,3-dioxanes. For example, the conformational equilibrium in the following example favors the axial *t*-butyl group! It should be noted that interference by bonded atoms or groups is more severe than by nonbonding (lone pairs) electrons.

2 x CH₃—H ~3.5 kcal/mol
(because of the shorter C—O bond)

C—O 1.43 Å
C—C 1.54 Å

preferred conformation
t-Bu (ax) ~1.4 kcal/mol

ΔE_D ~ 2.1 kcal/mol

Anomeric Effect

The anomeric effect refers to the tendency of a group X at C(1) of a pyranose ring to assume the axial rather than the equatorial orientation.[9] This phenomenon is important in carbohydrate chemistry since it influences the composition of isomeric mixtures and hence their reactivities. It has been suggested that the effect is caused mainly by a stabilizing interaction between the axial lone pair of electrons on the ring oxygen atom and the antiperiplanar, antibonding σ* orbital of the C–X bond.[10] This leads to a shortening of the bond between the ring oxygen and the anomeric carbon and a lengthening of the C–X bond.

Alternatively, electrostatic repulsive forces between the dipoles due to the ring oxygen lone pairs and the exocyclic oxygen or halogen may account in part for the observed axial preferences.[11]

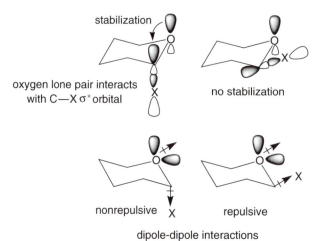

stabilization

oxygen lone pair interacts
with C—X σ* orbital

no stabilization

nonrepulsive

repulsive

dipole-dipole interactions

The anomeric effect is solvent and substituent dependent and decreases in the following order: Cl > OAc > OMe > OH, as exemplified by the equilibrium concentrations of the α and β anomers of substituted D-glucose in various protic solvents at 25 °C.[12]

A (α anomer)   oxocarbenium ion intermediate   **B** (β anomer)

X	R	% A	% B	Solvent (25 °C)
OH	H	36	64	H_2O
OMe	H	67	33	MeOH
OAc	Ac	86	14	AcOH
Cl	Ac	94	6	MeC≡N

2.4   POLYCYCLIC RING SYSTEMS

Hydrindane–Bicyclo[4.3.0]nonane

Due to a deformation of the cyclohexane ring, *cis*-hydrindane (the numbers in [ ] refer to carbon atoms between the bridgehead carbons) is only slightly less stable than the *trans* isomer (~1 kcal/mol). 8-Methylhydrindane is of interest since it occurs as part structure of the C/D rings of steroids (as a *trans* fused ring system). Introduction of an angular CH_3 group makes the *cis* isomer more stable than the trans because the *cis* isomer has fewer 1,3-diaxial interactions.

cis-8-methylhydrindane

trans-8-methylhydrindane

Decalin–Bicyclo[4.4.0]decane

The decalin structural feature is contained in steroids, many of which have useful biological activity. Decalin can exist in two *isomeric* forms depending on whether the rings are *trans* fused or *cis* fused. *Trans*-decalin is constrained to a rigid conformation, whereas *cis*-decalin is conformationally mobile.

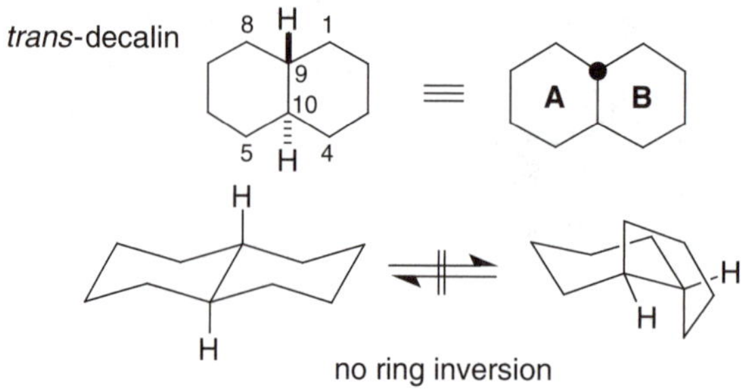

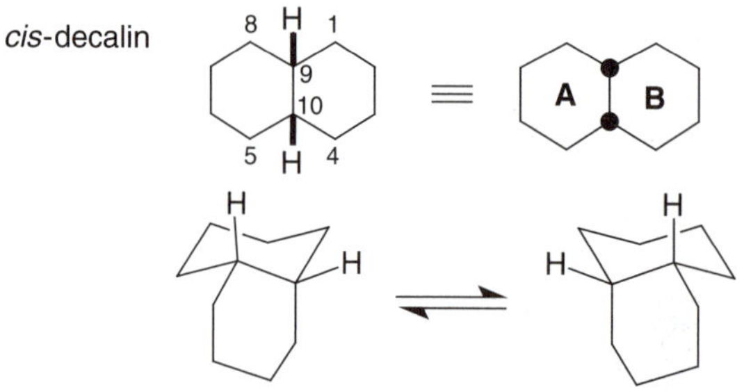

trans-decalin

no ring inversion

cis-decalin

Problems involving the flexible conformations of *cis*-decalin are more easily solved with the following chair-chair conformations. The *descriptors* α and β denote that the substituent is below the molecular plane or above the molecular plane, irrespective of axial versus equatorial.

β-substituents positioned *above* the plane of the ring

α-substituents positioned *below* the plane of the ring

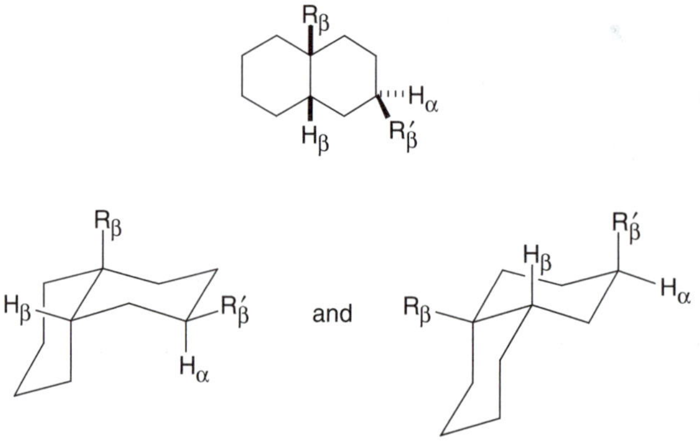

and

Note that *trans*- and *cis*-decalin are stereoisomers and *not* conformational isomers! *cis*-Decalin is less stable than *trans*-decalin by ~2.7 kcal/mol.

trans-decalin

methylene substituents are equatorial
and hence antiperiplanar

cis-decalin

1,3-diaxial interactions

Ring A: **CH** at C(5) with CH at C(1) ⎤
    CH at C(5) with CH at C(3) ⎬ $E_D = 3 \times 0.9 = 2.7$ kcal/mol
Ring B: **CH** at C(7) with CH at C(1) ⎦
    CH at C(5) with CH at C(1) --- already counted

Evaluation of ΔE_D for 2β-CH$_3$ *cis*-decalin

$E_D = 2.7 + 1.8 = 4.5$ kcal/mol $E_D = 2.7$ kcal/mol

$$\Delta E_D = 4.5 - 2.7 = 1.8 \text{ kcal/mol}$$

Both conformations have the same *cis*-decalin gauche interactions (2.7 kcal/mol).

Evaluation of E_D for 2β-OH, 5α-CH$_3$ *trans*-decalin

$E_D = A_{OH} + G_{CH_3} + G_{CH_2 (C_4)}$
   $= 0.9 + 0.4 + 0.4 = 1.7$ kcal/mol

Bridged Bicyclic Systems The bridged bicyclic systems bicyclo[2.2.1]heptane, also called *norbornane*, and bicyclo[2.2.2]octane contain "locked" boat cyclohexane rings. Because of their rigidity, these systems have played an important role in the development of theories of structure-reactivity relationships.

bicyclo[2.2.1]heptane bicyclo[2.2.2]octane

Bredt's rule (Julius Bredt, 1924)[13] is a qualitative generalization that bridged, bicyclic ring systems cannot have a double bond at the bridgehead. This rule does not apply to fused or large-member ring systems.

Fused

Bridged

Bredt's rule violation

Double bond to the
bridgehead is stable
(independent of *n*).

Double bond to the
bridgehead is highly
strained when *n* < 3.

Tricyclic Systems

For multiple, fused cyclohexyl ring systems, several *anti* and *syn* configurations are possible, as exemplified below.

trans–anti–trans

(5-10) (10-9) (9-8)

cis–syn–trans

2.5 CYCLOHEXYL SYSTEMS WITH sp²-HYBRIDIZED ATOMS

Cyclohexanones

In the absence of *one* 1,3-diaxial R–H interaction in 3-substituted cyclohexanones, one might anticipate that E_D should be reduced by half. However, conformational studies have revealed varied results.[1b,14] Corey and Feiner[7] propose to use the values of two-thirds A_R.

$\Delta E_D = 0.2$ kcal/mol

$E_D = 1.8$ kcal/mol

$E_D = \frac{2}{3}(3.0) = 2.0$ kcal/mol

Cyclohexenes

The conformational equilibria for monosubstituted cyclohexenes are known to be less biased against the axial conformer than is the case with the saturated analog.[15] This may be attributed to the absence of one 1,3-diaxial hydrogen interaction for either an axial or a pseudoaxial substituent in the half-chair conformations:

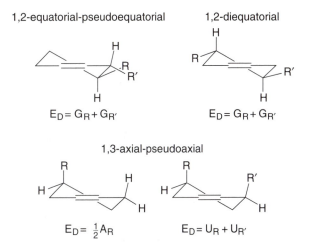

ax = axial
ax′ = pseudoaxial
eq = equatorial
eq′ = pseudoequatorial

half-chair form

Evaluation of Destabilization Interactions in Cyclohexene Systems

For half-chair cyclohexenes, there are three types of destabilizing interactions: (1) 1,3-axial-pseudoaxial, (2) 1,2-equatorial-pseudoequatorial, and (3) 1,2-diequatorial. Unfortunately, there are few values published for such interactions. Therefore, we adopt the computational A, G, and U values of Corey and Feiner.[7]

1,2-equatorial-pseudoequatorial

$E_D = G_R + G_{R'}$

1,2-diequatorial

$E_D = G_R + G_{R'}$

1,3-axial-pseudoaxial

$E_D = \frac{1}{2} A_R$

$E_D = U_R + U_{R'}$

Allylic Strain

Another bond type with a marked conformational preference is the vinylic bond. Eclipsing interactions between vinyl and allylic substituents result in *allylic* 1,2-strain ($A^{1,2}$ strain) and *allylic* 1,3-strain ($A^{1,3}$ strain).[16] Both the $A^{1,2}$ and the $A^{1,3}$ strain play an important role in determining the stereochemical outcome of many important reactions involving acyclic and cyclic systems (e.g., alkylation of enamines, alkylation and protonation of enolates).

$A^{1,2}$ strain in cyclohexene systems

$A^{1,2}$ strain   at 25 °C   ~ 30 : 70 mixture

$A^{1,3}$ strain (exocyclic)

$A^{1,3}$ strain   favored

$A^{1,3}$ strain in acyclic systems

| rel. energies (kcal/mol) | 0.73 | 0 | > 2 kcal/mol |

| rel. energies (kcal/mol) | 3.4 | 0 | > 4 kcal/mol |

2.6   SIGNIFICANT ENERGY DIFFERENCE[17]

The stereochemical course of reactions at three-, four-, and five-member rings can be reliably predicted by assuming the relative congestion of the two faces. As the ring size increases above six so does the conformational mobility and hence the uncertainty of the stereochemical outcome. Even with seven-member rings, predictions are generally difficult.

Intermediate in complexity are six-member ring systems that prefer either the chair or, much less commonly, the boat and twist boat conformations. As pointed out by Barton, knowledge of the conformation of a six-member ring system, and hence its axial/equatorial substitution pattern, is crucial to an understanding of the stereochemical basis for many reactions in ring systems.[2]

A significant energy difference for a conformer is considered to be *1.8 kcal/mol*. This corresponds to roughly 95% preponderance of one conformer at room temperature. Thus conformational homogeneity would be predicted for 2-methylcyclohexanone (ΔE_D = 1.8 kcal/mol) but not for *cis*-2-methyl-5-phenylcyclohexanone (ΔE_D = 0.2 kcal/mol) (see page 44).

ΔE_D = 1.8 kcal/mol

2.7   COMPUTER-ASSISTED MOLECULAR MODELING

Theoretical and computational chemistry contribute greatly to our understanding of conformational preferences of both stable molecules and transition states.[18] Molecular mechanics methods (classical or force field)[18] help define the conformational preferences of reactants and products. These methods are empirical, having been parameterized to reproduce experimental structures and energies and/or data provided by high-level quantum mechanical calculations. Different force fields have been parameterized to perform best for different classes of molecules (e.g., Allinger's MM2-MM4[19] for typical organic molecules, Kollman's AMBER[20], and Karplus's CHARMM[21] for biological macromolecules), so care is necessary in choosing which method to apply to a particular problem.

In contrast, quantum mechanical calculations are more time consuming but are not dependent on empirical parameterization (i.e., they are ab initio). These methods have long been used to deduce and rationalize the structures and relative energies not only of reactants and products but also of transition states, thus providing a comprehensive view of reactivity and selectivity.[22]

2.8   REACTIVITY AND PRODUCT DETERMINATION AS A FUNCTION OF CONFORMATION

The effect of conformation on reactivity is intimately associated with the details of the mechanism of a reaction.[1a,1b,23] Two extreme situations[24] can arise:

1.   As a warning against predicting product stereochemistry based on reactant conformation, the *Curtin-Hammett principle* states that the rate of reaction of a molecule is a function not only of the concentration of any reacting conformation but also of its transition state energy. As illustrated in the figure below, the conformational barrier between reactant conformations A and B is substantially lower than the reaction barriers to products P_A and P_B. While conformation A is more populated, product P_B will be formed preferentially. Thus, when a reaction mechanism demands a particular conformation, the product distribution depends not only on the population of reactive starting conformers, but also on the transition state energies leading to the respective products. The best way to encourage a reaction from a single conformation is to suppress the concentrations of all other conformers by introducing appropriately placed substituents and/or lowering the reaction temperature.

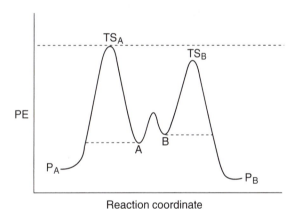

Reaction coordinate

2.   The conformational barrier (A <=> B) is substantially higher than the reaction barriers TS_A and TS_B. This case is known as the *conformational equilibrium control*, where the ratio of products is equal to the ratio of the population of the starting states.

Hammond Postulate   *Early transition states* are frequently typical of fast, exothermic reactions in which there is no substantial bond making or breaking. Hence, early transition states resemble starting materials in structure. *Late transition states* are frequently typical of relatively slow, endothermic transformations. There is substantial bond formation in late transition states; consequently, late transition states resemble products.

Stereoelectronic Effects   Certain reactions have distinct stereoelectronic requirements that must be met if the reaction is to proceed efficiently.[11,25] The effect of conformation on reactivity is conveniently divided into (1) steric effects—close approach of two groups, and (2) stereoelectronic effects—spatial disposition of a particular electron pair, either bonded or nonbonded, as in S_N2 and E2 reactions.

Terminology of Selectivity[26]

- *Stereoisomers* have the same molecular formula, but their atoms have a different spatial arrangement (*configuration*).

- A *stereoselective* reaction leads to the exclusive or predominant formation of one stereoisomer over another in a chemical reaction.

- In a *stereospecific* reaction, two stereochemically different substrates react to give stereochemically different products. All stereospecific reactions are necessarily stereoselective.

- A *chemoselective* reagent reacts selectively with *one functional group* in the presence of other functional groups.

- In a *regioselective* reaction, the reagent reacts at only one of two or more possible sites.

- In an *enantioselective* reaction, an achiral substrate is converted by the action of a chiral reagent or auxiliary to one of two enantiomers.

- In a *diastereoselective* reaction, one diastereomer is formed preferentially.

- *Thermodynamic control* leads to a major product based on its relative stability (meaning the product predominates at equilibrium), whereas *kinetic control* leads to a major product based on reaction rate (the product that is formed the most rapidly is the kinetic product).

Esterification and Saponification

The rate of esterification and the rate of saponification will depend on $\Delta G^{\ddagger}$ for the rate determining step (RDS). In the examples below, the steric requirements of OH and CO_2Et are greater in the tetrahedral intermediate than in the ground state. Hence, $\Delta G^{\ddagger}_{axial/equat}$ is enhanced over $\Delta G^{\circ}_{ax/eq}$ for the starting materials. Thus, the equatorial isomers B and D will react faster than the corresponding axial isomers A and C. Note that the conformational energies of sp^3 groups are generally greater than those of sp^2 groups.

Esterification

$\Delta G^{\circ} = 0.9$ kcal/mol $\Delta G > 0.9$ kcal/mol

Saponification

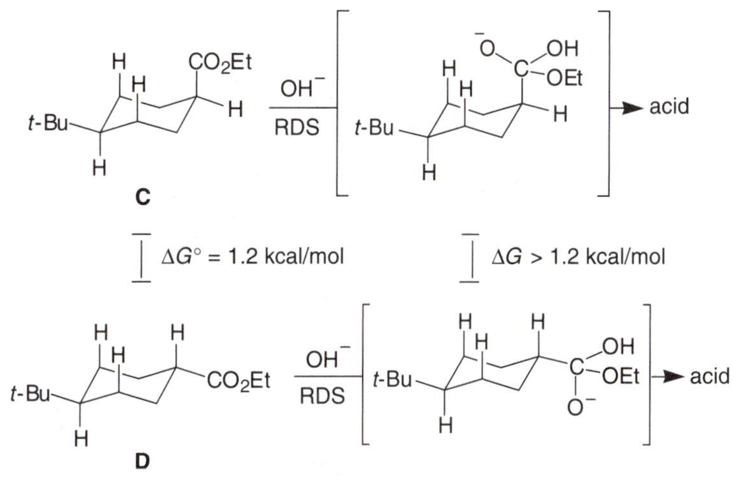

SₙN2-Type Reactions

Whereas cyclopentyl compounds undergo S_N2 reactions at rates comparable to acyclic systems, cyclohexyl compounds react rather slowly in S_N2 reactions. The TS of an S_N2 reaction involves a pentacoordinated carbon and proceeds with inversion of configuration. Steric hindrance to the approaching nucleophile plays an important role in these reactions. For example, the S_N2 reaction of thiophenolate (PhS⁻) with the axial *cis*-1-bromo-4-*tert*-butylcyclohexane A proceeds ~60 times faster than with the equatorial *trans* isomer B. Inspection of the transition states A‡ and B‡ reveals that backside attack by the nucleophile (PhS⁻) on the equatorial isomer B is hindered by the two axial hydrogens, destabilizing TS B‡ relative to TS A‡. Moreover, because of the higher ground state energy of the axial bromide (0.4 kcal/mol), the activation energy is further reduced.

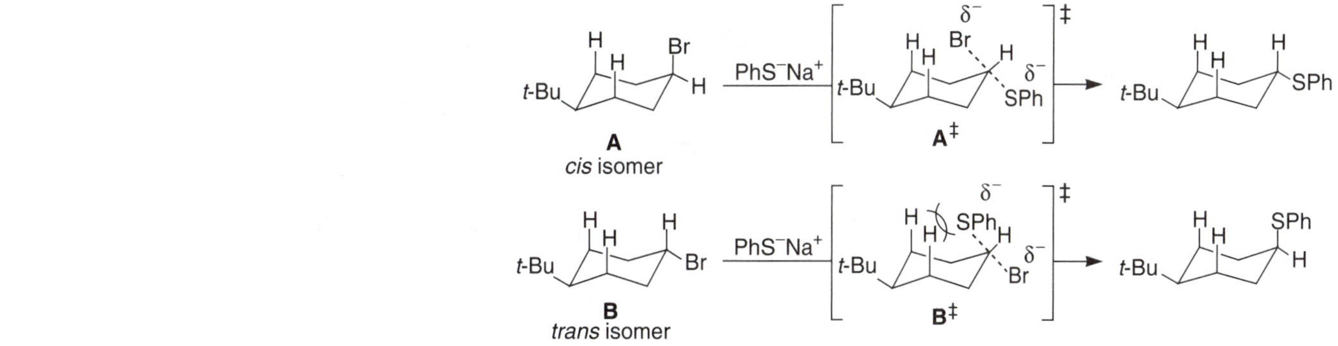

Nucleophilic substitution at an allylic substrate under S_N2 conditions may proceed via nucleophilic attack at the γ-carbon, especially when substitution at the α-carbon sterically impedes the normal S_N2 reaction. These S_N2' reactions with cyclohexenyl systems generally proceed via an anti addition of the nucleophile to the double bond, as depicted below (best overlap of participating orbitals).[16d,27]

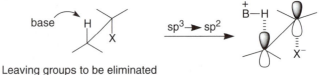

Michael-Type Additions

The stereochemistry of nucleophilic 1,4-additions to enones (Michael-type additions) is controlled by stereoelectronic factors. In the absence of compelling steric effects, the nucleophile approaches the β-carbon of the enone *antiparallel* to the neighboring (γ) pseudoaxial substituent (circled H in the example below).

E2 Elimination Reaction

For stereoelectronic reasons (overlap of reacting orbitals), the two reacting groups, H and X, must be either *antiperiplanar* or *synperiplanar*. During an electron transition there should be the least change in the position of the atoms involved (Frank-Condon effect). Strong bases with a *large steric requirement* will suppress the S_N2 mode of reaction.

Leaving groups to be eliminated
should be antiperiplanar.

Most E2 elimination reactions of cyclohexyl systems proceed through an *anti* conformation except when the *synperiplanar* arrangement is readily attainable. For example, with cyclopentane derivatives and the rigid bicyclo[2.2.1]heptanes[28] and bicyclo[2.2.2]octanes, where the H and X can adopt a *synperiplanar* arrangement, *syn* eliminations are observed.

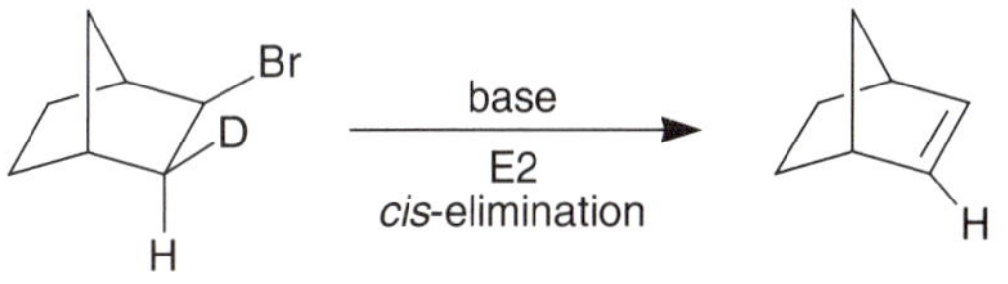
Br
D
H
base
E2
cis-elimination
H

The following ("non–Curtin-Hammett") example demonstrates the reactivity and product differentiating effects in E2 reactions. Generally, the overall rate of a reaction involving a reactant in various conformers will depend on the rate constants for the individual conformers *and their populations*.

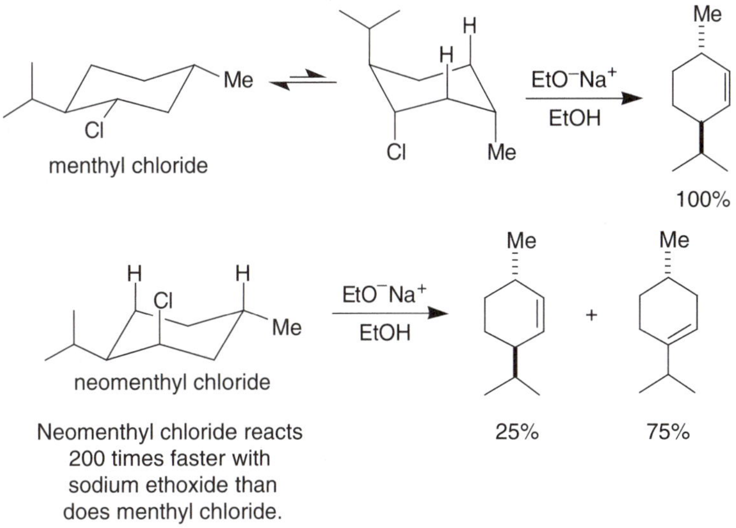
Me
Cl
menthyl chloride
H
H
Cl
Me
EtO⁻Na⁺
EtOH
Me
100%
H
H
Cl
Me
neomenthyl chloride
EtO⁻Na⁺
EtOH
Me
Me
+
25%
75%
Neomenthyl chloride reacts 200 times faster with sodium ethoxide than does menthyl chloride.

Addition Reactions to Double Bonds

Addition of Electrophiles to Cyclic Compounds

Addition of bromine to cyclohexene is subject to stereoelectronic control and proceeds via formation of a bromonium ion intermediate. This is followed by spontaneous *antiperiplanar* (S_N2 type), *antiparallel* opening by the bromide anion via a chairlike TS. Thus, under kinetic control, the predominant or exclusive product formed is via path b resulting from diaxial addition of Br_2 to the double bond.

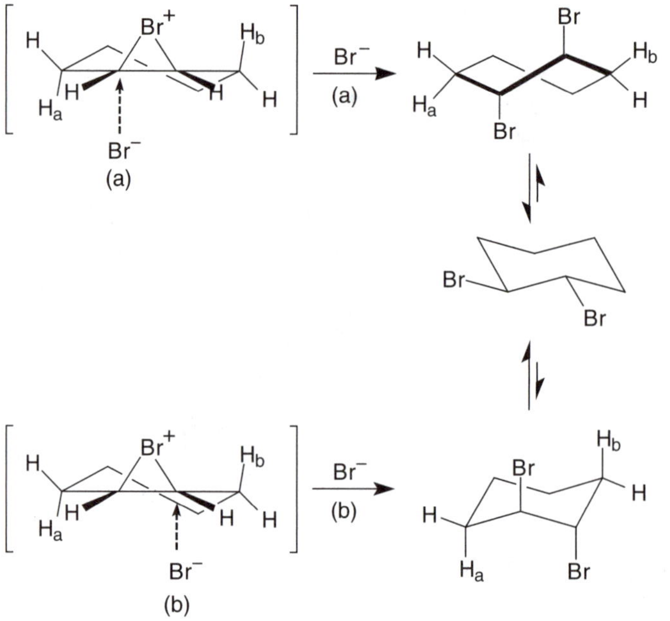

This is evidenced by the nearly exclusive formation of the diaxial dibromide on treatment of 4-*tert*-butylcyclohexene with bromine.

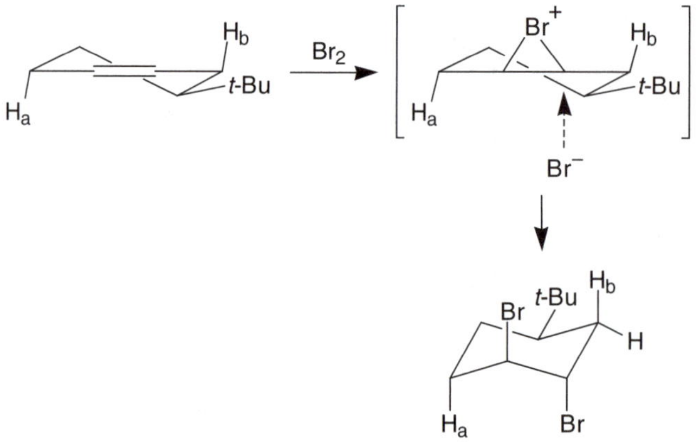

Preferential formation of diaxial products is further illustrated by the reaction of the octalin (octahydronaphthalene) depicted below with an aqueous solution of bromine. Addition of the electrophilic "Br⁺" to the less hindered α-face of the double bond and *antiperiplanar, antiparallel* opening of the resultant bromonium ion intermediate by HO⁻ furnishes the diaxial bromohydrin.

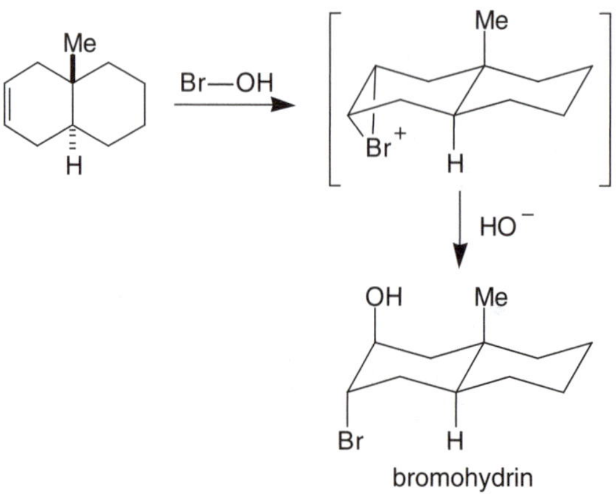

bromohydrin

Epoxidation of the octalin below with *m*-chloroperoxybenzoic acid (*m*CPBA) occurs preferentially from the less hindered side to produce the α-epoxide A, which on treatment with CH₃ONa (or with aqueous acid) gives the *diaxial* product B. Attack of CH₃O⁻ at carbon C(3) of epoxide A would lead to the diequatorial product D via the higher energy twist boat conformation C in order to maintain the stereoelectronically required diaxial arrangement of the incoming nucleophile and the departing oxyanion.

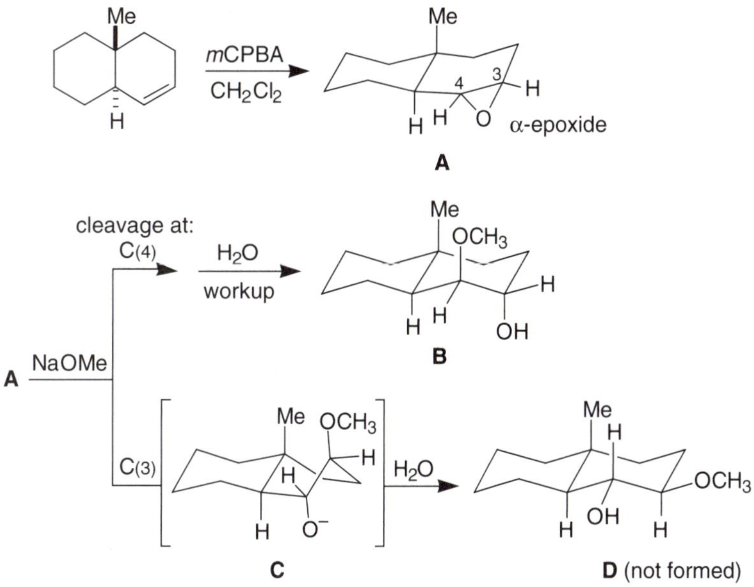

Addition of Electrophiles to Acyclic Compounds

To achieve diastereoselectivity in electrophilic additions to a double bond in acyclic compounds, there must be a facial preference for attack. An $A^{1,3}$ strain provides such an element for conformational control, as exemplified by hydroboration of the alkene shown below.[29] The hydration of a double bond via hydroboration involves (1) anti-Markovnikov addition of the B–H bond, (2) *cis* addition of the B–H bond, (3) addition of the B–H bond from the *less hindered side* of the double bond, and (4) oxidation with retention of configuration.

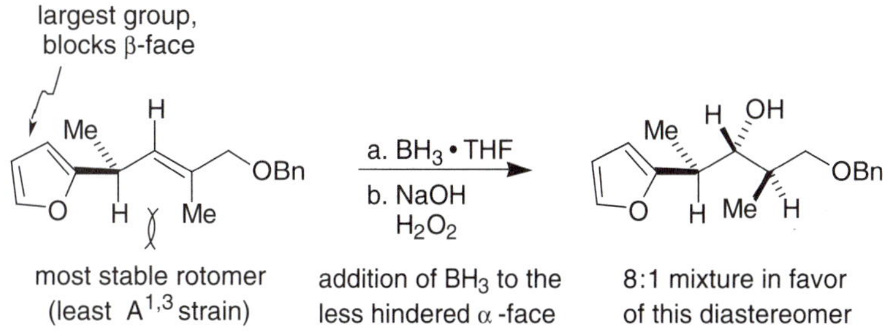

Oxidation of Alcohols

The oxidation of a secondary alcohol to a ketone with chromium (VI) is a complex reaction. With unhindered alcohols, oxidation proceeds via initial rapid formation of the chromate ester followed by a rate-determining E2-type elimination of $HCrO_3^-$ as the leaving group.

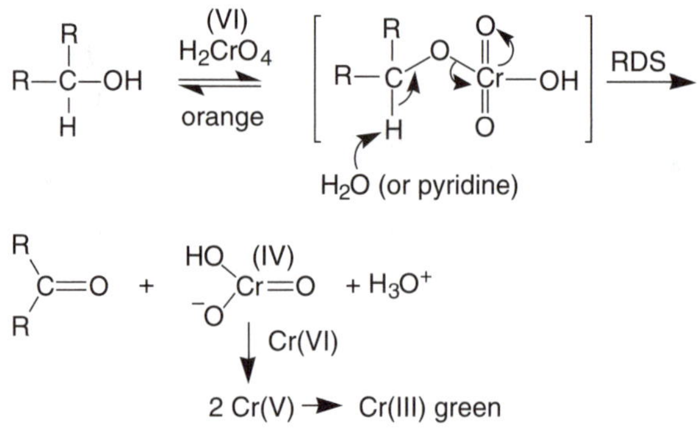

Axial cyclohexanols are more reactive than the corresponding equatorial alcohols toward chromic acid oxidation. To understand the basis for this effect, we must consider the free energies for these reactions. Consider oxidation of the *trans*- and *cis*-3,3,5-trimethylcyclohexanols. A large portion of the conformational free energy of the diaxial interactions of the sp^3 hybridized chromate ester in the *trans* isomer is relieved as the reaction proceeds toward the sp^2 hybridized keto group. Thus, the more strained *trans* alcohol is more reactive (relief of steric strain).[30] Alternatively, the enhanced reactivity of the *trans* chromate ester may be rationalized in terms of the greater accessibility of the equatorial hydrogen for removal by the base in the rate-determining step.

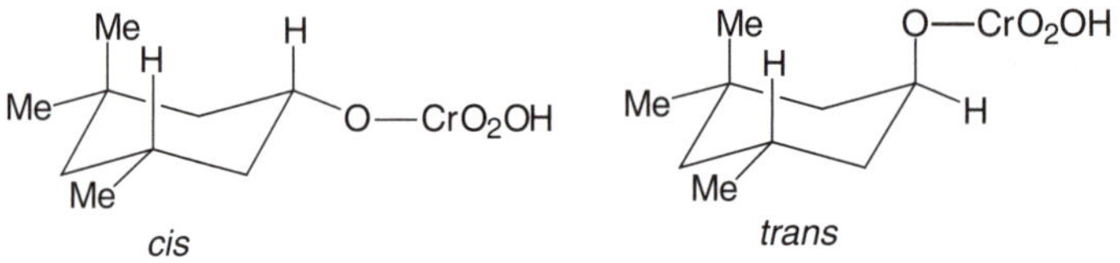

	cis	*trans*
relative rates:	1	34

PROBLEMS

1. Draw the chair or the half-chair conformations (where applicable) for each of the molecules shown below and determine the corresponding E_D and ΔE_D values. Use the A, G, and U values from Table 2.3 and assume 0.7 kcal/mol for Me/H $A^{1,2}$ strain.

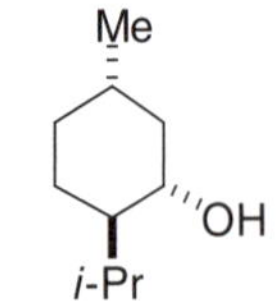

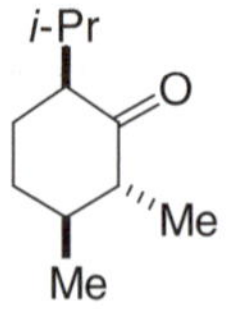

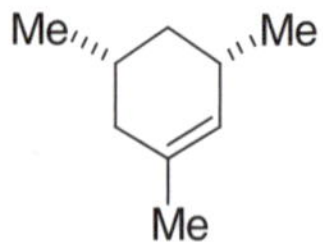

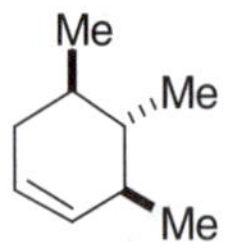

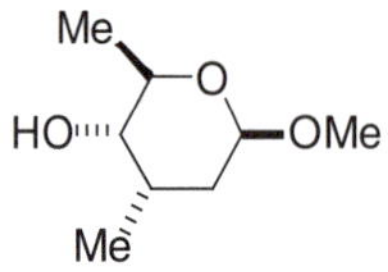

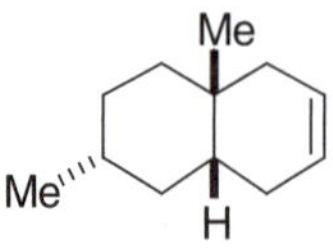

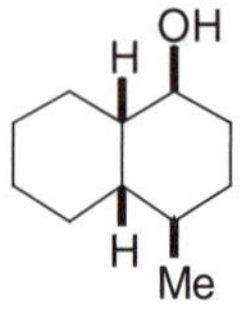

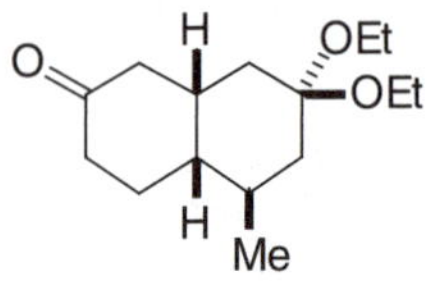

2. Draw the most stable conformation for each of the compounds shown below. You do *not* need to compute the E_D and ΔE_D values.

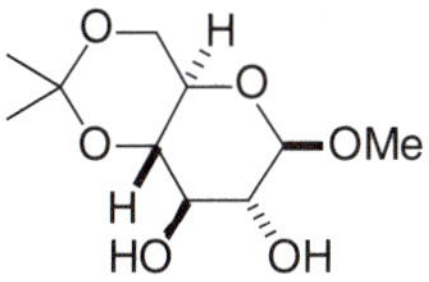

A 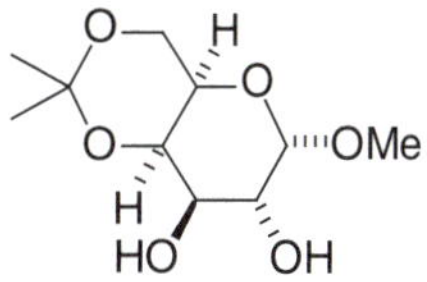   B 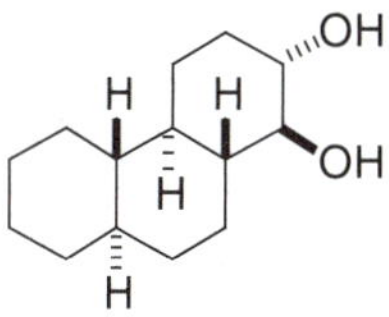   C 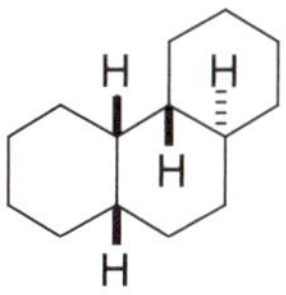   D

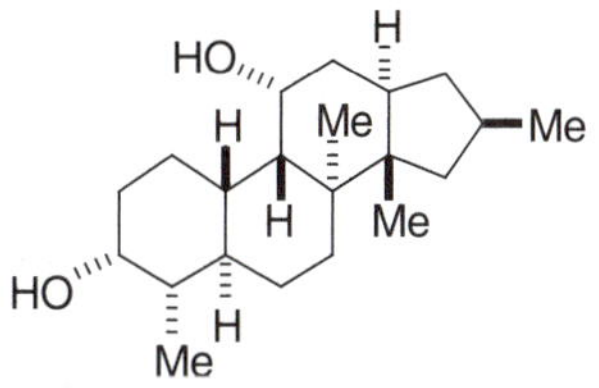

E 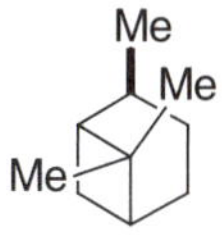   F

3. For each molecule shown below, calculate the percentage (%) of the more stable conformation at the temperature indicated.

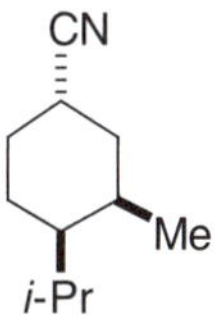

A 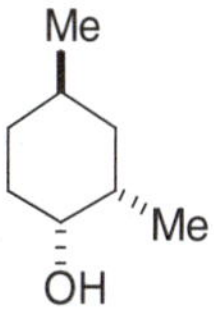   B 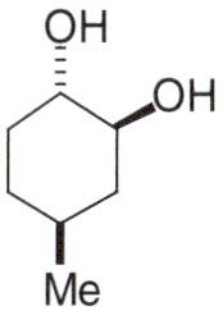   C

at: 25 °C,   50 °C   25 °C
      100 °C

4. Show the conformation of each of the following alcohols and arrange them in order of decreasing ease of esterification with *p*-nitrobenzoyl chloride.

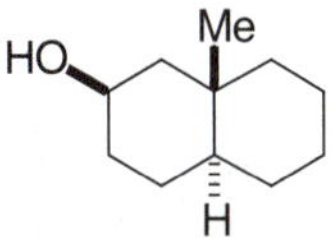

A 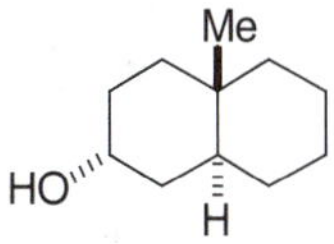   B 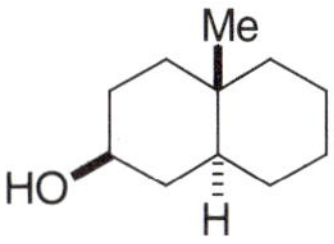   C

5. Given below are the observed α : β epoxide ratios from epoxidations of the octalins **A** and **B** with *m*-chloroperbenzoic acid (*m*CPBA) in CHCl₃. How do you explain the differences in stereoselectivity?

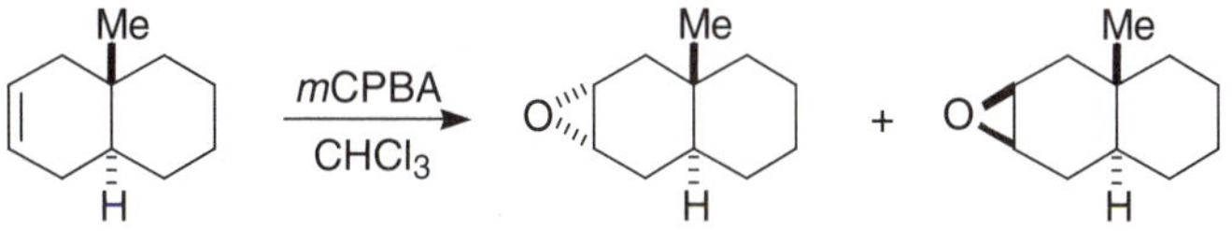

A $\xrightarrow[\text{CHCl}_3]{m\text{CPBA}}$ 95   +   5

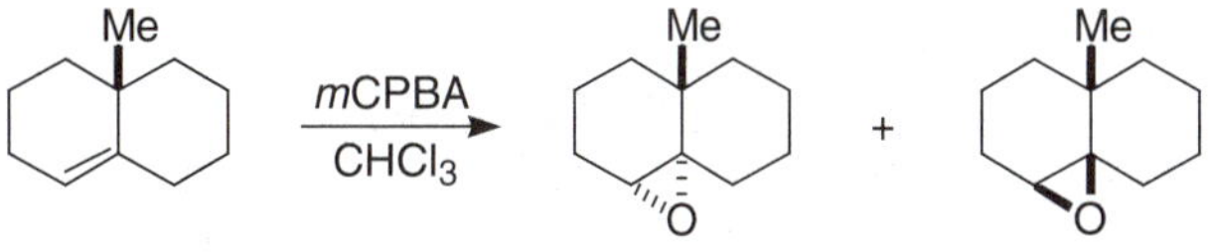

B $\xrightarrow[\text{CHCl}_3]{m\text{CPBA}}$ 60   +   40

6. **Reagents**. Show the major product formed for each of the following reactions.

a.

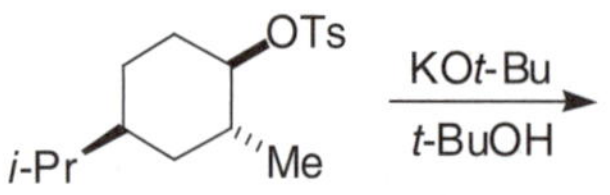

$$\xrightarrow[\text{\textit{t}-BuOH}]{\text{KO\textit{t}-Bu}}$$

b.

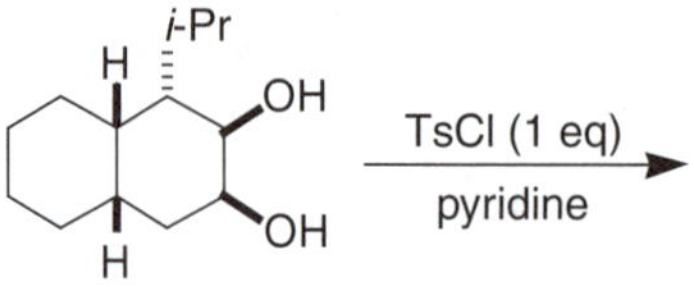

$$\xrightarrow[\text{pyridine}]{\text{TsCl (1 eq)}}$$

c.

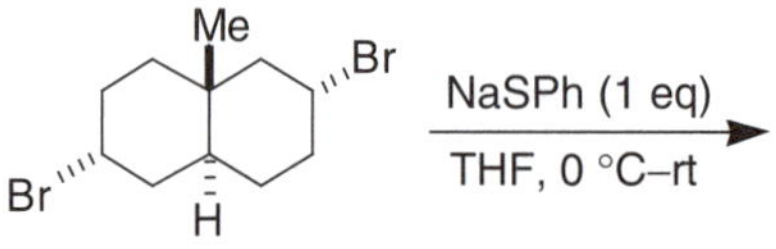

$$\xrightarrow[\text{THF, 0 °C–rt}]{\text{NaSPh (1 eq)}}$$

d.

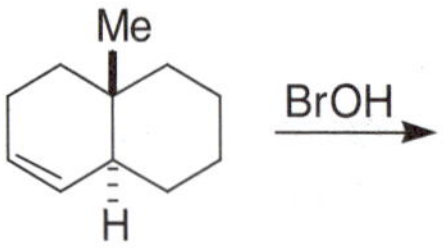

$$\xrightarrow{\text{BrOH}}$$

e.

$$\xrightarrow[\text{EtOH}]{\text{NaOEt}}$$

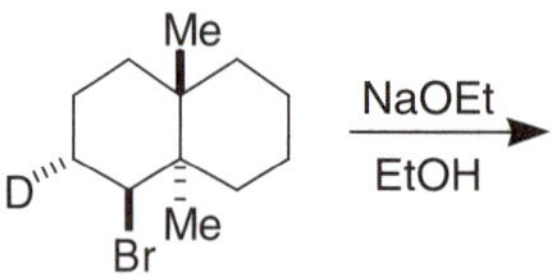

*7. Consider the conformational equilibrium of the ketone below. Explain why conformer **A** predominates in DMSO (100% **A**), whereas **B** is the major conformer in isooctane (22% **A** : 78% **B**).

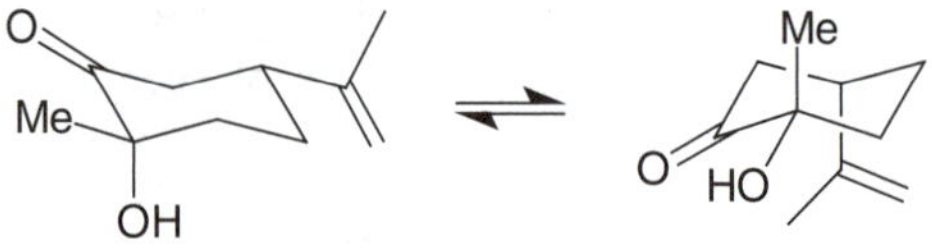

*8. Suggest a reason why the equilibrium below favors the conformation on the right.

*9. Why does the adamantyl compound shown below behave more like a ketone than an amide? (*Hint*: Draw the corresponding resonance hybrid.)

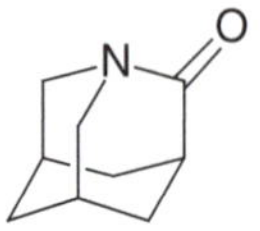

REFERENCES

1. (a) Nasipuri, D. *Stereochemistry of Organic Compounds: Principles and Applications*, Wiley: New York, 1991. (b) Eliel, E. L., Wilen, S. H., Mander, L. N. *Stereochemistry of Organic Compounds*, Wiley: New York, 1994. (c) Juaristi, E. *Conformational Behavior of Six-Membered Rings: Analysis, Dynamics, and Stereoelectronic Effects*, VCH: New York, 1995. (d) Lightner, D. A., Gurst, E. *Organic Conformational Analysis and Stereochemistry from Circular Dichroism Spectroscopy*, Wiley-VCH: New York, 2000.

2. Barton, D. H. R. *Experientia* **1950**, *6*, 316.

3. For nomenclature regarding conformations of linear hydrocarbons, see Michl, J., West, R. *Acc. Chem. Res.* **2000**, *33*, 821.

4. (a) Pophristic, V. T.; Goodman, L. *Nature* **2001**, *411*, 565. (b) Bickelhaupt, F. M., Baerends, E. J. *Angew. Chem., Int. Ed.* **2003**, *42*, 4183. (c) Weinhold, F. *Angew. Chem., Int. Ed.* **2003**, *42*, 4188. (d) Mo, Y., Wu, W., Song, L., Lin, M., Zhang, Q., Gao, J. *Angew. Chem., Int. Ed.* **2004**, *43*, 1986.

5. Zutterman, F., Krief, A. *J. Org. Chem.* **1983**, *48*, 1135.

6. Penelle, J., Xie, T. *Macromolecules* **2000**, *33*, 4667.

7. Corey, E. J., Feiner, N. F. *J. Org. Chem.* **1980**, *45*, 765.

8. Basso, E. A., Kaiser, C., Rittner, R.; Lambert, J. B. *J. Org. Chem.* **1993**, *58*, 7865.

9. (a) Kirby, A. J. *The Anomeric Effect and Related Stereoelectronic Effects at Oxygen*, Springer-Verlag: New York, 1983. (b) Cramer, C. J. *J. Org. Chem.* **1992**, *57*, 7034. (c) Collins, P., Ferrier, R. *Monosaccharides*, Wiley: Chichester, UK, 1994.

10. (a) Salzner, U. *J. Org. Chem.* **1995**, *60*, 986. (b) Perrin, C. L. *Tetrahedron* **1995**, *51*, 11901.

11. (a) Perrin, C. L., Armstrong, K. B., Fabian, M. A. *J. Am. Chem. Soc.* **1994**, *116*, 715. (b) Wiberg, K. B.; Marquez, M. *J. Am. Chem. Soc.* **1994**, *116*, 2197.

12. Kirby, A. J. *Stereoelectronic Effects*, Oxford University Press: New York, 1996.

13. Wiseman, J. R., Pletcher, W. A. *J. Am. Chem. Soc.* **1970**, *92*, 956. (b) Buchanan, G. L. *Chem. Soc. Rev.* **1974**, *3*, 41.

14. Allinger, N. L., Chen, K., Rahman, M., Pathiaseril, A. *J. Am. Chem. Soc.* **1991**, *113*, 4505.

15. Rabideau, P. W. *Conformational Analysis of Cyclohexenes, Cyclohexadienes and Related Hydrocarbon Compounds*, VCH: New York, 1989.

16. (a) Johnson, F. *Chem. Rev.* **1968**, *68*, 375. (b) Wiberg, K. B., Martin, E. *J. Am. Chem. Soc.* **1985**, *107*, 5035. (c) Hoffmann, R. W. *Chem. Rev.* **1989**, *89*, 1841. (d) Atkinson, R. S. *Stereoselective Synthesis*, Wiley: Chichester, UK, 1995.

17. Corey, E. J., Feiner, N. F. *J. Org. Chem.* **1980**, *45*, 757.

18. (a) Leach, A. R. *Molecular Modelling: Principles and Applications*, 2nd ed., Prentice Hall: Harlow, UK, 2001. (b) Cramer, C. J. *Essentials of Computational Chemistry: Theories and Models*, Wiley: New York, 2002.

19. Bowen, J. P., Allinger, N. L. *Molecular Mechanics: The Art and Science of Parameterisation*. In *Reviews in Computational Chemistry*, Lipkowitz, K. B., Boyd, D. B. Eds., VCH: New York, 1991, Vol. 2, pp. 81–97.

20. Cornell, W. D., Cieplak, P., Bayly, C. I., Gould, I. R., Merz, K. M., Jr., Ferguson, D. M., Spellmeyer, D. C., Fox, T., Caldwell, J. W., Kollmann, P. A. *J. Am. Chem. Soc.* **1995**, *117*, 5179.

21. MacKerell, A. D., Bashford, D., Bellott, M., Dunbrack, R. L., Evanseck, J. D., Field, M. J., Gao, J., Guo, H., Ha, S., Joseph-McCarthy, D., Kuchnir, L., Kuczera, K., Lau, T. F. K., Mattos, C., Michnick, S., Nago, T., Nguyen, D. T., Prodhom, B., Reiher, W. E., Roux, B., Schlenkrich, M., Smith, J. C., Stote, R., Straub, J., Watanabe, M., Wiorkievicz-Kuczera, J., Yin, D., Karplus, M. *J. Phys. Chem. B* **1998**, *102*, 3585.

22. K. N. Houk, a pioneer in this field, developed many models of selectivity used by synthetic organic chemists; for a recent representative example of the use of quantum mechanics in the development of models of stereoselectivity, see Bahmanyar, S., Houk, K. N. *J. Am. Chem. Soc.* **2001**, *123*, 12911.

23. Nógrádi, M. *Stereoselective Synthesis: A Practical Approach*, VCH: Weinheim, Germany, 1995.

24. Zefirov, N. S. *Tetrahedron* **1977**, *33*, 2719.

25. Deslongchamps, P. *Stereoelectronic Effects in Organic Chemistry*, Pergamon: Oxford, UK, 1983.

26. Ward, R. S. *Selectivity in Organic Synthesis*, Wiley: Chichester, UK, 1999.

27. Goering, H. L., Kantner, S. S. *J. Org. Chem.* **1984**, *49*, 422.

28. Kwart, H., Takeshita, T., Nyce, J. L. *J. Am. Chem. Soc.* **1964**, *86*, 2606.

29. Schmid, G., Fukuyama, T., Akasaka, K., Kishi, Y. *J. Am. Chem. Soc.* **1979**, *101*, 259.

30. Eliel, E. L., Schroeter, S. H., Brett, T. J., Biros, F. J., Richer, J.-C. *J. Am. Chem. Soc.* **1966**, *88*, 3327.

CHAPTER 3

The Concept of Protecting Functional Groups

There is no denying that the sheer sense of challenge posed by a complex molecular target serves to stimulate the creative impulses of the synthetic chemist.

S. J. Danishefsky

2-*O*-Me-α-D-mannopyranose

The presence of several functional groups in a molecule can greatly complicate the design of a synthesis, especially if a reagent reacts indiscriminately with the various functional groups present (for example, sugars). When a chemical reaction is to be carried out selectively at one reactive site in a multifunctional compound, other reactive sites must be temporarily blocked. Thus, selection of protective groups plays an important part in planning a synthesis.[1]

A protecting group must fulfill a number of requirements:

1. The protecting group reagent must react *selectively* (kinetic chemoselectivity) *in good yield* to give a protected substrate that is stable to the projected reactions.

2. The protecting group must be *selectively* removed in good yield by readily available reagents.

3. The protecting group should not have additional functionality that might provide additional sites of reaction.

Introduction of a protective group adds additional steps to a synthetic scheme. Hence, one should strive to keep the use of protecting groups to a *minimum* and avoid them if possible.

3.1 PROTECTION OF NH GROUPS[2]

Primary and secondary amines are prone to oxidation, and N–H bonds undergo metallation on exposure to organolithium and Grignard reagents. Moreover, the amino group possesses a lone electron pair, which can be protonated or reacted with electrophiles (R–X). To render the lone pair less reactive, the amine can be converted into an amide via acylation. Protection of the amino group in amino acids plays a crucial role in peptide synthesis.

N-Benzylamines

$$\text{N}-\text{CH}_2\text{Ph} \equiv \text{N}-\text{Bn}$$

N-Benzyl groups (*N*-Bn) are especially useful for replacing the N–H protons in primary and secondary amines when exposed to organometallic reagents or metal hydrides. Depending on the reaction conditions, primary amines can form mono- and/or dibenzylated products. Hydrogenolysis of benzylamines with Pd catalysts and H_2 in the presence of an acid regenerates the amine.[1] Generally, benzylamines are not cleaved by Lewis acids.

The nonpyrophoric $Pd(OH)_2/C$ (*Pearlman's catalyst*[3]) catalyzes the selective hydrogenolysis of benzylamines in the presence of benzyl ethers.[4]

Amides

Acylation of primary and secondary amines with acetic anhydride or acid chlorides furnishes the corresponding amides in which the basicity of the nitrogen is reduced, making them less susceptible to attack by electrophilic reagents.

Benzamides (*N*-Bz) are formed by the reaction of amines with benzoyl chloride in pyridine or trimethylamine. The group is stable to pH 1–14, nucleophiles, organometallics (except organolithium reagents), catalytic hydrogenation, and oxidation. It is cleaved by strong acids (6N HCl, HBr) or diisobutylaluminum hydride.[5]

Carbamates

Treatment of primary and secondary amines with methyl or ethyl chloroformate in the presence of a *tertiary*-amine furnishes the corresponding methyl and ethyl carbamates, respectively. The protected amines behave like amides; hence, they no longer act as nucleophiles. They are stable to oxidizing agents and aqueous bases but may react with reducing agents. Iodotrimethylsilane is often the reagent of choice for the removal of the *N*-methoxycarbonyl group.[6]

The *benzyloxycarbonyl* group (abbreviated as Cbz or Z) is one of the most important nitrogen-protecting groups in organic synthesis, especially in peptide synthesis. It is introduced by reacting the amine with benzyloxycarbonyl chloride in the presence of a *tertiary*-amine. The protected amine is stable to both aqueous base and aqueous acid, but can be removed by dissolving metal reduction (Li°, liq. NH_3), catalytic hydrogenation (Pd/C, H_2), or acid hydrolysis (HBr).

Benzyl esters and carbamates in the presence of other easily reducible groups such as aryl bromides, cyclopropanes, and alkenes are selectively cleaved with triethylsilane and palladium chloride.[7]

The *t-butoxycarbonyl* group (Boc) is another widely used protecting group for primary and secondary amines.[8] It is inert to hydrogenolysis and resistant to bases and nucleophilic reagents but is more prone to cleavage by acids than the Cbz group. Deprotection of the *N*-Boc group is conveniently carried out with CF_3CO_2H, neat or in CH_2Cl_2 solution.[1] Selective cleavage of the *N*-Boc group in the presence of other protecting groups is possible when using $AlCl_3$.[9]

3.2   PROTECTION OF OH GROUPS OF ALCOHOLS

The most important protecting groups for alcohols are ethers and mixed acetals. The proper choice of the protecting group is crucial if chemoselectivity is desired.

Reactivity of alcohols: 1° > 2° > 3° ROH

The stability of ethers and mixed acetals as protecting groups for alcohols varies from the very stable methyl ether to the highly acid-labile trityl ether. *However, all ethers are stable to basic reaction conditions.* Hence, ether or mixed acetal protecting groups specifically tolerate

- RMgX and RLi reagents
- Nucleophilic reducing reagents such as $LiAlH_4$ and $NaBH_4$
- Oxidizing agents such as $CrO_3 \cdot 2$ pyridine, pyridinium chlorochromate (PCC), and MnO_2
- Wittig reagents
- Strong bases such as LDA

Alkyl Ethers

Methyl Ethers $\boxed{RO-CH_3}$

Methyl ethers are readily accessible via the Williamson ether synthesis, but harsh conditions are required to deprotect them. For hindered alcohols, the methylation should be carried out in the presence of KOH/DMSO.[10]

Reagents for cleaving methyl ethers include Me_3SiI (or $Me_3SiCl + NaI$) in CH_2Cl_2[6b] and BBr_3 (or the solid $BBr_3 \cdot SMe_2$ complex) in CH_2Cl_2.[11] BBr_3 is especially effective for cleaving $PhOCH_3$.[12]

$$ROH \xrightarrow[\text{THF, 0 °C}]{\text{NaH}} \left[RO^- Na^+ \right] \xrightarrow{CH_3I} ROCH_3 \xrightarrow[CH_2Cl_2]{\text{TMSI}} ROH$$
$$\text{(aq. acid workup)}$$

Methylation of *sec*-OH groups in sugars with methyl iodide and silver oxide is often the method of choice.

$$ROH \xrightarrow[\text{DMF}]{CH_3I, Ag_2O} ROCH_3$$

tert-Butyl Ethers $\boxed{RO-CMe_3}$

t-Butyl ethers are readily prepared and are stable to nucleophiles, hydrolysis under basic conditions, organometallic reagents, metal hydrides, and mild oxidations. However, they are cleaved by dilute acids (S_N1 reaction).

$$ROH \xrightarrow[\substack{\text{conc. } H_2SO_4 \\ \text{or } BF_3 \cdot OEt_2}]{t\text{-BuOH or } Me_2C=CH_2} ROt\text{-Bu} \xrightarrow{\text{4 N HCl}} ROH$$

Benzylic Ethers

Benzyl Ethers $\boxed{RO-Bn}$

Benzyl ethers are quite stable under both acidic and basic conditions and toward a wide variety of oxidizing and reducing reagents. Hence, they are frequently used in organic syntheses as protecting groups. It should be noted, however, that *n*-BuLi may deprotonate a benzylic hydrogen, especially in the presence of TMEDA (tetramethylethylenediamine) or HMPA (hexamethylphosphoramide).

Formation: Methods for cleavage:

$$ROH \xrightarrow[\text{b. PhCH}_2\text{Br}]{\text{a. NaH, THF}} ROCH_2Ph \xrightarrow[\substack{\text{or} \\ \text{Ra-Ni, EtOH} \\ \text{or} \\ \text{Na}^\circ,\ \text{NH}_3\ (l),\ \text{EtOH}}]{\text{Pd/C, H}_2,\ \text{EtOH}} ROH\ +\ H_3CPh$$

Catalytic hydrogenolysis offers the mildest method for deprotecting benzyl ethers. Hydrogenolysis of 2°- and 3°-benzyl ethers may be sluggish. Protection of alcohols using (benzyloxy)methyl chloride produces the corresponding (benzyloxy)methyl ethers (RO-BOM), which are cleaved more readily than the corresponding ROBn ethers.[13]

$$ROH \xrightarrow[\text{(i-Pr)}_2\text{NEt, CH}_2\text{Cl}_2]{\text{PhCH}_2\text{OCH}_2\text{Cl}} \underbrace{RO-CH_2-OBn}_{\text{BOM group}} \xrightarrow[\text{H}_2,\ \text{EtOH}]{\text{cat.Pd(OH)}_2} ROH$$

Acid-Catalyzed Benzylation. Benzyl trichloroacetimidate, $Cl_3CC(=NH)OBn$, reacts with hydroxyl groups under acid catalysis to give the corresponding benzyl ethers in good yield.[14] The method is particularly useful for the protection of base-sensitive substrates (i.e., alkoxide-sensitive), such as hydroxy esters[15] or hydroxy lactones, as exemplified below.[16]

benzyl
trichloroacetimidate   active benzylating agent   TfOH

protected
alcohol

97%

78%

p-Methoxybenzyl Ethers $\boxed{\text{RO}-\text{PMB}}$

The PMB ether, also referred to as an MPM ether [(4-methoxyphenyl)methyl], is less stable to acids than a benzyl ether. Its utility as a protecting group stems from the fact that it can be removed oxidatively with DDQ (2,3-dichloro-5,6-dicyano-1,4-benzo-quinone) under conditions that do not affect protecting groups such as acetals, RO-Bn (or RO-BOM), RO-MOM, RO-MEM, RO-THP, RO-TBS, benzoyl, tosyl, or acetate groups, nor do they affect epoxides or ketones.[17] Alternatively, RO-PMB ethers can be cleaved with $(NH_4)_2Ce(NO_3)_6$.[18]

DMAP [(4-(*N*,*N*-dimethylamino)pyridine,[19] pK_a DMAP-H$^+$ 9.70] is a versatile hypernucleophilic catalyst for the preparation of ethers and esters.[20] Toxicological data of DMAP show that the compound should be handled with care. DABCO (1,4-diazabicyclo[2.2.2]octane)[21] appears to be a less hazardous substitute for the widely used DMAP.

Trityl Ethers

Triphenylmethyl Ethers   RO—CPh₃ ≡ RO—Tr

Trityl ethers have played an important role in the *selective* protection and manipulation of –CH₂OH groups in carbohydrate chemistry. In recent years, however, silyl ethers are increasingly used in place of the trityl ether protective group.

Trityl ethers are stable to bases and nucleophiles but are readily cleaved by acids or by hydrogenolysis (Pd, H₂). The trityl group may be selectively cleaved in the presence of *tert*-butyldimethylsilyl, triethylsilyl, or benzoyl (Bz) groups.[22]

Silyl Ethers

The popularity of silicon protecting groups stems from the fact that they are readily introduced and removed under mild conditions.[23] Moreover, a wide variety of silylating agents are available for tailor-made protection of ROH groups. The chemoselectivity of silylating agents for alcohols and the stability of the resultant silyl ethers toward acid and base hydrolysis, organometallic reagents, and oxidizing and reducing agents increases with increased steric size of the groups attached to silicon. Generally, the sterically least-hindered alcohol is the most readily silylated but is also the most labile to acid or base hydrolysis.

Bases generally employed for the preparation of silyl ethers include R₃N, imidazole, DMAP, and DBU (1,8-diazabicyclo[5.4.0]undec-7-ene). Hindered ROH groups are best converted to the corresponding alkoxides with NaH, MeLi, or *n*-BuLi prior to silylation.

R₃SiOTf = trialkylsilyl trifluoromethanesulfonate

2,6-lutidine =   DBU =

Depending on the structure of silyl ethers, they can be deprotected by H_2O, aqueous acids, and fluoride salts. Since silicon has a strong affinity for fluoride ion (bond energy, kcal/mol: Si–F, 143; Si–O, 111), the O–SiR_3 bond is especially prone to cleavage by fluoride salts, such as $n\text{-}Bu_4N^+F^-$, which is soluble in organic solvents such as THF and CH_2Cl_2.

Trimethylsilyl Ethers $\boxed{\text{RO—TMS}}$

$$ROH + Me_3SiCl \xrightarrow{\text{Et}_3\text{N, THF}} RO\text{—}SiMe_3$$

Unfortunately, trimethylsilyl ethers are very susceptible to solvolysis in protic media, either in the presence of acids or bases. Cleavage of RO-TMS occurs on treatment with citric acid in CH_3OH at 20 °C (10 min), or K_2CO_3 in CH_3OH at 0 °C, or $n\text{-}Bu_4N^+F^-$ in THF at 0 °C (within seconds).

Hexamethyldisilazane (HMDS, $Me_3SiNHSiMe_3$), is a convenient silylating agent for ROH, RCO_2H, phenols, and enolizable ketones.[24]

Triethylsilyl Ethers $\boxed{\text{RO—TES}}$

Triethylsilyl ethers have been used as protective groups in Grignard additions, Swern and Dess-Martin oxidations, Wittig reactions, metallations with R_2NLi reagents, and cleavage of RO-PMB ethers with DDQ.

t-Butyldimethylsilyl Ethers $\boxed{\text{RO—TBS}}$

The *t*-butyldimethylsilyl group is the most widely used of the silicon protecting groups.[25] The rate of silylation of alcohols with TBSCl follows the trend: 1° ROH > 2° ROH > 3° ROH. The large difference in rate of silylation between primary and secondary OH groups makes the TBSCl reagent well suited for the selective protection of the –CH_2OH group in methyl glycosides.

For protecting a primary OH group in the presence of a secondary OH group, one should use TBS-Cl and Et_3N with DMAP as a catalyst. Hindered 2° and 3° alcohols can be silylated with $t\text{-}BuMe_2SiOTf$ and 2,6-lutidine as a base. The $t\text{-}BuMe_2SiOTf$ reagent is prepared from $t\text{-}BuMe_2SiCl$ and triflic acid.[26]

The TBS ether protecting group is more stable to hydrolysis than the trimethylsilyl ether by a factor of 10^4 but is still readily cleaved by exposure to either n-Bu$_4$NF in THF, HF-pyridine, CsF in DMF, or H$_2$SiF$_6$.[27] Diisobutylaluminum hydride is another reagent for the deprotection of t-butyldimethylsilyl ethers under mild conditions.[28]

The example shown below illustrates how a compound containing three different, often-employed protecting groups can be selectively and sequentially deprotected, starting with TBS, followed by THP, and finally with the MEM ether.

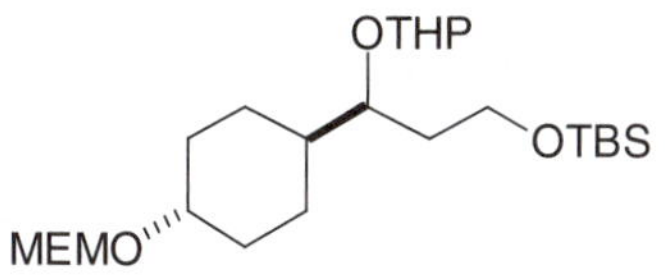

	Conditions		Group cleaved
1. n-Bu$_4$N$^+$F$^-$		→	TBS
2. AcOH, H$_2$O, THF		→	THP
3. TiCl$_4$ (or ZnBr$_2$)		→	MEM

Thexyldimethylsilyl Ethers RO—TDS

Thexyldimethylsilyl chloride[29] is a liquid and is less expensive then TBSCl. Furthermore, ROTDS ethers react more slowly under hydrolytic conditions than do RO-TBS ethers.

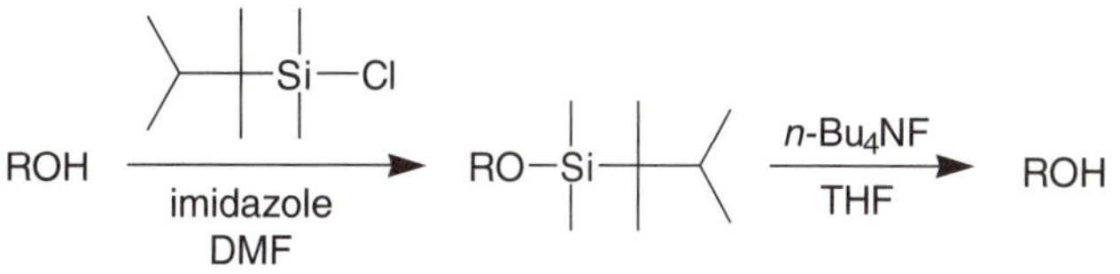

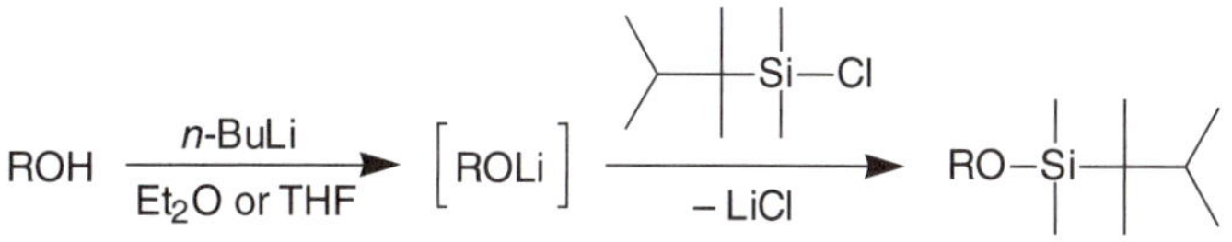

Triisopropylsilyl Ethers RO—TIPS

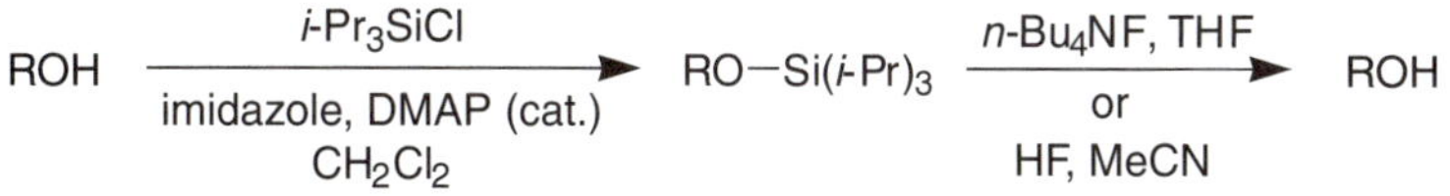

Triisopropylsilyl chloride (TIPS-Cl) is an excellent reagent for the selective protection of a primary OH in the presence of a secondary OH group.[30] A simple and efficient method for silylation of alcohols and phenols is using TIPS-Cl and imidazole under microwave irradiation.[31] The TIPS group is stable under a wide range of reaction conditions, such as acid and basic hydrolysis, and toward powerful nucleophiles.

t-Butyldiphenylsilyl Ethers RO—BPS

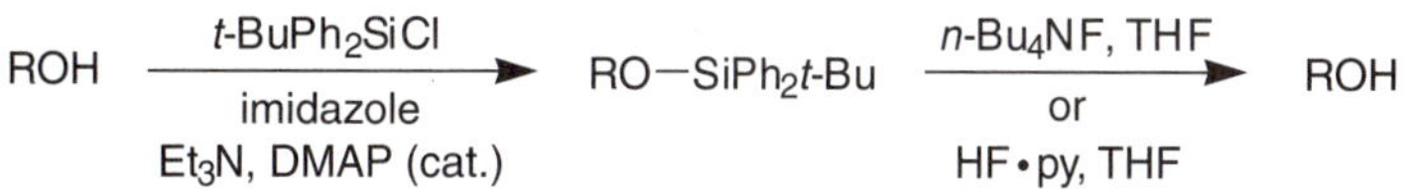

Treatment of primary or secondary alcohols with BPS-Cl in CH$_2$Cl$_2$ in the presence of imidazole or DMAP affords the corresponding t-butyldiphenylsilyl ethers. Tertiary alcohols are not silylated.

The BPS group is more stable toward acidic hydrolysis than TBS or the TIPS groups. It survives DIBAL (diisobutylaluminum hydride) reductions, cleavage of THP with aqueous AcOH, and cleavage of isopropylidene and benzylidene acetals with aqueous CF_3CO_2H. BPS-Cl has been used to selectively protect primary OH groups in inositol (hexahydroxycyclohexane).[32]

Acetals

Tetrahydropyranyl Ethers   RO—THP

The THP group is a widely used protecting group; it is readily introduced by reaction of the enol ether dihydropyran with an alcohol in the presence of an acid catalyst, such as TsOH, $BF_3 \cdot OEt_2$, or $POCl_3$. For sensitive alcohols such as allylic alcohols, PPTS (pyridinium p-toluenesulfonate) is used as a catalyst for tetrahydropyranylation.[33] As an acetal, the THP group is readily hydrolyzed under aqueous acidic conditions with AcOH-THF, TsOH, PPTS-EtOH, or Dowex-H (cation exchange resin).

dihydropyran, an enol ether

tetrahydropyranyl ether, an acetal, ~100%

PPTS

Protection of chiral alcohols as THP ethers leads to mixtures of diastereoisomers. To avoid this problem, use 5,6-dihydro-4-methoxy-2H-pyran.[34]

single diastereomer

Methoxymethyl Ethers   RO—CH_2OCH_3 ≡ RO—MOM

α-Halo ethers are often used for the protection of alcohols.[35] The high reactivity of α-halo ethers in nucleophilic displacement reactions by alkoxides permits the protection of alcohols under mild conditions. Moreover, as acetals the alkoxy-substituted methyl ethers are cleaved with a variety of reagents.

The reaction of chloromethyl methyl ether (MOM-Cl, a *carcinogen*) with an alkoxide or with an alcohol in the presence of i-Pr$_2$NEt (Hünig's base) furnishes the corresponding formaldehyde acetal.[36] Alkylation of 3°-alcohols requires the more reactive MOM-I, derived from MOM-Cl and NaI in the presence of i-Pr$_2$NEt.[37]

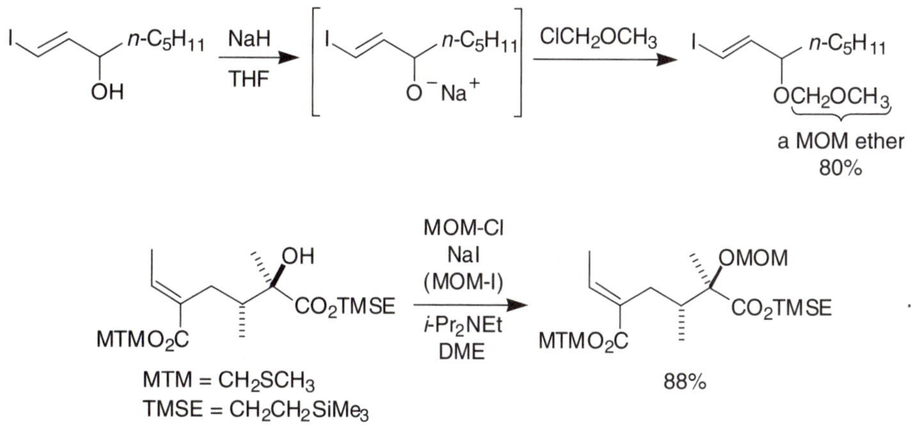

a MOM ether
80%

MTM = CH$_2$SCH$_3$
TMSE = CH$_2$CH$_2$SiMe$_3$

88%

Cleavage of the MOM group with dilute acids or with PPTS in *t*-BuOH regenerates the alcohol.[38] A mild and selective reagent for removing the MOM group in the presence of methyl or benzyl ethers, –SiPh$_2$*t*-Bu ethers, or esters is bromotrimethylsilane. However, TMSBr will also cleave acetals and trityl ethers.[39]

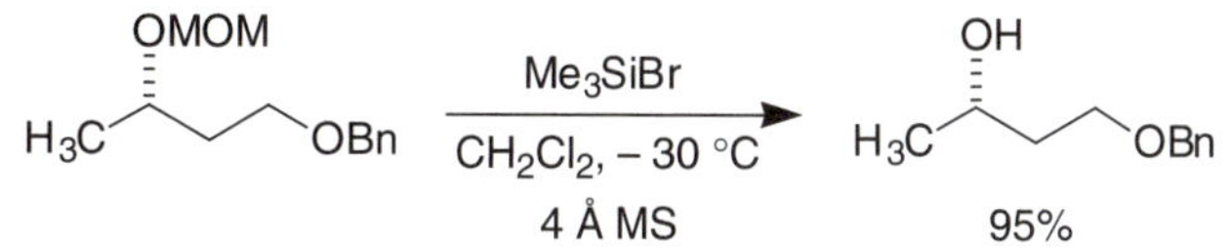

2-Methoxyethoxymethyl Ethers   RO–CH$_2$OCH$_2$CH$_2$OCH$_3$ ≡ RO–MEM

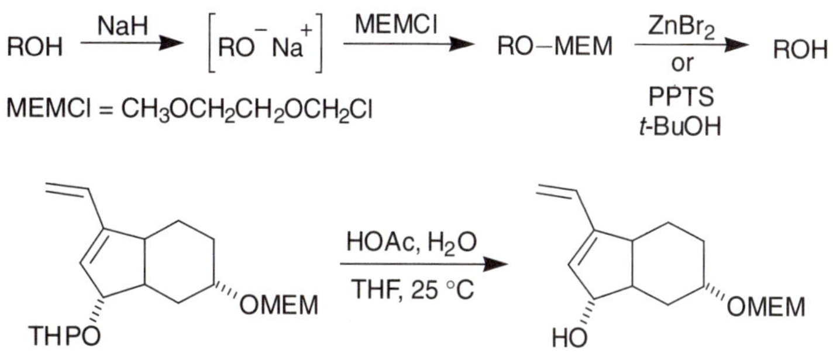

MEMCl = CH$_3$OCH$_2$CH$_2$OCH$_2$Cl

MEM ethers are excellent protecting groups for 1°, 2°, and 3° alcohols and even 3° allylic alcohols. They are stable toward strong bases, organometallic reagents, and many oxidizing agents and are more stable to acidic conditions than THP ethers.[40] The cleavage of MEM ethers to regenerate the corresponding alcohols is usually carried out with anhydrous zinc bromide in methylene chloride. PPTS is a mild reagent for removing MEM ethers of allylic alcohols.[38]

1-Methoxyisopropyl Ethers   RO–C(CH$_3$)$_2$OMe ≡ RO–MIP

The MIP protective group (formerly named MME) is a useful protecting group for alcohols that are sensitive to strongly acidic conditions. The use of this group avoids the formation of diasteromers.

PPTS has been used as a catalyst for the preparation of MIP ethers[41] as well as for their cleavage.[42]

Esters

Carboxylic Acid Esters

The use of carboxylic acid esters as protective groups for alcohols is limited since they may undergo acyl substitution, hydrolysis or reduction. Reagents used for the preparation of esters in the presence of Et_3N or pyridine are Ac_2O, Ac_2O-DMAP (note that DMAP increases the rate of acylation of alcohols with Ac_2O by a factor of 10^4), PhCOCl, $(PhCO)_2O$, and t-BuCOCl (pivaloyl chloride).[43] Deprotection of esters is usually done under basic conditions.[44]

A simple, convenient method for the selective acylation of a primary OH in the presence of a secondary OH group is its conversion into the t-butanoyl ester (also known as a pivalate ester, OPv). Moreover, the steric bulk of the t-butyl group makes these esters resistant to nucleophilic attack, including hydrolysis under mild basic conditions. The pivalate ester can be cleaved using metal hydride reagents.

Selective esterification of a primary hydroxyl group in the presence of secondary hydroxyl groups also has been achieved with methyl α-D-glucopyranoside.[45]

Relative reactivities of esters toward hydrolysis and nucleophilic reagents follows the order $RCOOCH_3 > RCOOPh > RCOOt$-Bu.

p-Toluenesulfonate Esters

$RO-SO_2C_6H_4CH_3 \equiv RO-Ts$

p-Toluenesulfonate esters are often used for the regioselective protection of OH groups in carbohydrates[46] and nucleic acids. Primary hydroxyl groups react faster with *p*-toluenesulfonyl chloride (TsCl)[47] than do secondary hydroxyl groups.[48] Extraction of the tosylate with a solution of $CuSO_4$ removes traces of residual pyridine. Methods for the selective deprotection of tosyl groups include reductive cleavage with Na(Hg) amalgam-methanol[49] and irradiation by UV light in the presence of an electron-rich aromatic compound.[50]

$$ROH \ + \ TsCl \quad \xrightarrow[\substack{\text{b. H}^+\text{, H}_2\text{O} \\ \text{c. CuSO}_4\text{, H}_2\text{O} \\ (\text{b + c = workup})}]{\text{a. pyridine}} \quad ROTs \quad \xrightarrow[\text{CH}_3\text{OH, H}_2\text{O}]{\text{Na(Hg) (4\%)}} \quad ROH$$

Reductive cleavage of sulfonates with reducing agents may proceed either via C–O or O–S cleavage leading to alkanes and alcohols, respectively. For example, reduction of primary mesylates or tosylates with LAH furnishes preferentially the corresponding alkanes.[51]

For the more hindered secondary tosylates, hydride attack may occur at sulfur, regenerating the original alcohol.[52]

Reduction of primary as well as secondary tosylates with $LiEt_3BH$ (Superhydride) provides a convenient method for the deoxygenation of alcohols to the corresponding alkanes.[53]

3.3 PROTECTION OF DIOLS AS ACETALS

Acetalization of 1,2- and 1,3-diols plays an important role in manipulating the reactivity of cyclic and acyclic polyhydroxy compounds. Acetals derived from 1,2- and 1,3-diols are readily accessible via their reactions with ketones or aldehydes in the presence of an acid catalyst. Once they are formed, acetals are very stable to basic conditions but are labile toward acids.

Notable features of diol acetalization include

- Diols react with aldehydes and ketones in the presence of an acid catalyst to yield acetals in a reversible reaction.

- Acetal formation allows the selective blocking of pairs of HO groups in polyhydroxy compounds.
- Five- and six-member rings are formed preferentially.

The equilibrium of diol acetalization is shifted to the acetal side by removing the H_2O. This may be accomplished by

- Azeotropic distillation (via a Dean-Stark or Merlic trap)
- Addition of molecular sieves (4 Å), anhydrous $CuSO_4$, or Drierite
- Transacetalization: acetal exchange with acetals, orthoesters, or enol ethers, avoiding the formation of water

1,2-Diols

Only vicinal *cis*-OH groups of cyclic 1,2-diols readily form acetals.[54] Acetal exchange is the most common method for preparing isopropylidene acetals (1,3-dioxolanes).

The strong tendency for acetalization of *cis*-1,2-diols is exemplified by the reaction of glucopyranose A, which on treatment with acetone in the presence of a catalytic amount of H_2SO_4 furnishes the 1,2:5,6-di-*O*-isopropylidene-α-D-glucofuranose B (thermodynamic product). However, kinetic acetalization with 2-methoxypropene in DMF in the presence of TsOH as a catalyst at 0 °C occurs without rearrangement to give the 4,6-isopropylidene glucopyranose C.[55]

Selective protection of *trans*-diequatorial 1,2-diols in carbohydrate systems has been reported via formation of di-spiroacetals.[56]

Isopropylidene acetal formation in the acyclic 1,2,4-butanetriol D again favors the five-member 1,3-dioxolane ring even if one of the OH groups is tertiary.[57]

Acetalization of D-mannitol (E) with acetone leads to the preferential blocking of the two terminal 1,2-diol moieties.[58]

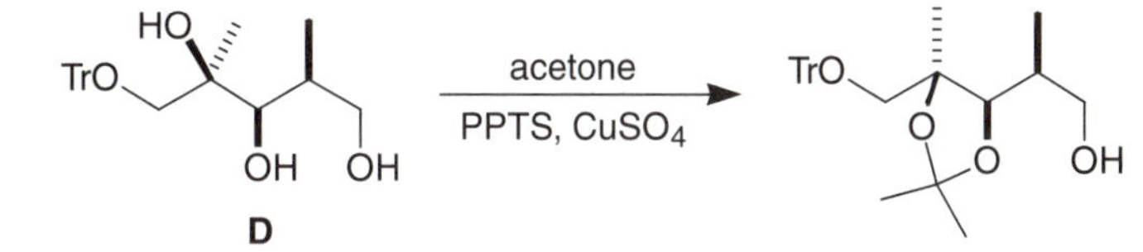

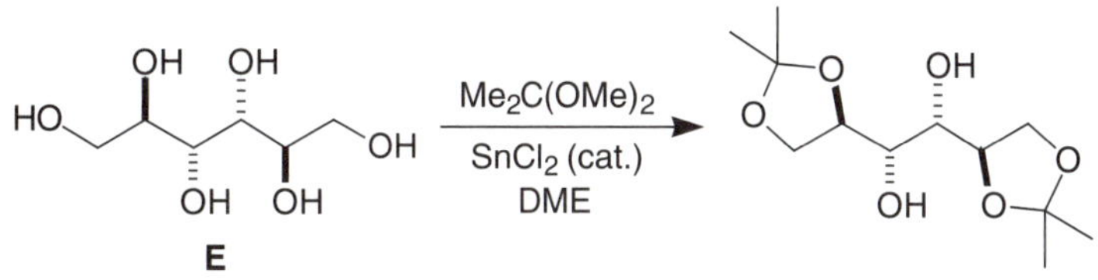

1,3-Diols

Both *cis*- and *trans*-1,3-diols form cyclic acetals with aldehydes in the presence of an acid catalyst to furnish the corresponding benzylidene and ethylidene derivatives, respectively.

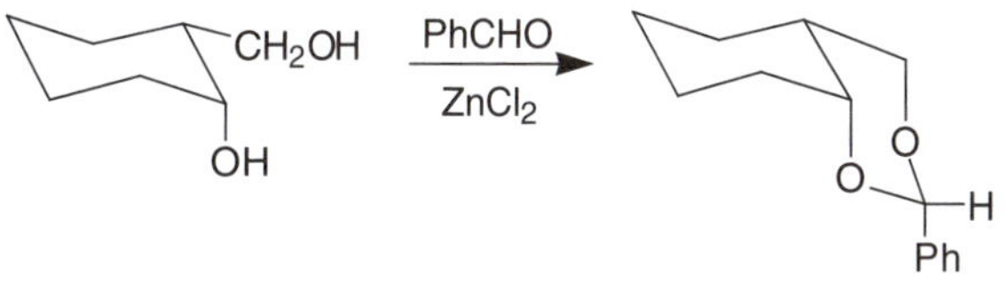

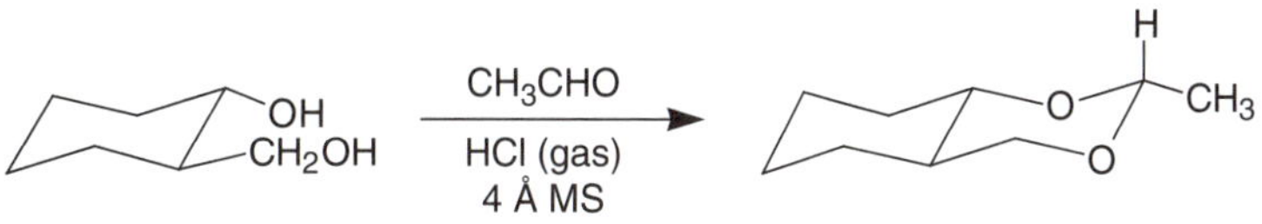

Treatment of methyl α-D-glucopyranoside with benzaldehyde dimethyl acetal in the presence of camphorsulfonic acid (CSA) gives methyl 4,6-*O*-benzylidene-α-D-glucopyranoside.[59]

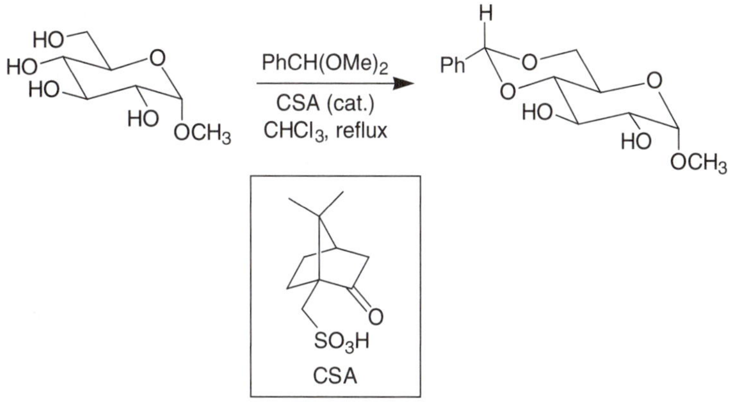

3.4 PROTECTION OF CARBONYL GROUPS IN ALDEHYDES AND KETONES

Via O,O-Acetals

Acyclic and cyclic acetals are the most important carbonyl protecting groups of aldehydes and ketones, and also serve as efficient chiral auxiliaries for the synthesis of enantiomerically pure compounds.[60]

The acetal protective group is introduced by treating the carbonyl compound with an alcohol, an orthoester, or a diol in the presence of a Lewis acid catalyst. In recent years, several transition metal catalysts such as $TiCl_4$ have been shown to offer major advantage over general Brønsted acid catalysts.[61]

Notable features of carbonyl acetalization are

- The general order of reactivity of various carbonyl groups (steric effects may cause a reversal of the reactivity order) is as follows:

- 1,3-Dioxanes (six-member ring acetals) derived from ketones hydrolyze faster than the corresponding 1,3-dioxolanes (five-member ring acetals)
- The selective protection of a reactive carbonyl group in the presence of a less reactive one is possible
- Acetals are stable to
  - Strong aqueous bases
  - Nucleophilic reducing agents
  - Organometallic reagents (RLi; RMgX, etc.)
  - Oxidations under nonacidic conditions
  - $Na°$ or $Li°/NH_3$ reductions
- Acetals are cleaved by
  - Acid-catalyzed hydrolysis—this is the most common method for deprotecting acetals
  - Selective deprotection[62] of acetals—determined by the relative rate of hydrolysis[63] as influenced by steric, inductive, and stereoelectronic effects

Acetalization with Alcohols[64]

Acetalization with Trialkyl Orthoformates. In an acetal exchange reaction, trialkyl orthoformates will convert carbonyl groups to their corresponding acetal derivatives without concomitant formation of water. Weak acids such as NH_4NO_3 or amberlyst-15 (a sulfonic acid resin) catalyze the acetalization.[65]

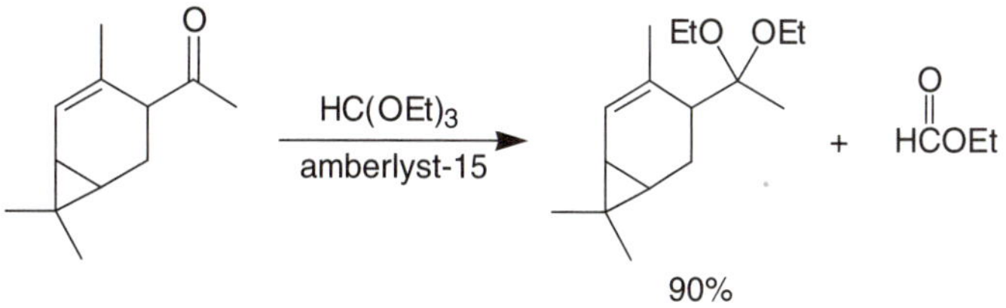

90%

Acetalization with Diols. 1,3-Dioxolane (five-member ring acetal) is the most widely used C=O protecting group. The formation of acetals with diols provides an entropic advantage over the use of two equivalents of an alcohol. The water formed is removed by azeotropic distillation.

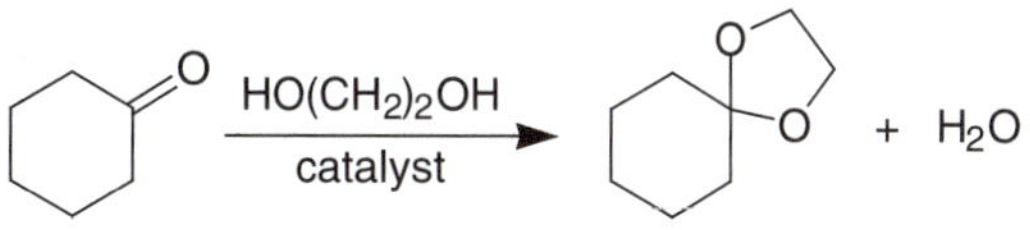

a 1,3-dioxolane

catalysts for acetalization:
PPTS, $BF_3 \cdot OEt_2$, TsOH, or amberlyst-15

cleavage of 1,3-dioxolanes:
TsOH and H_2O, or 5% HCl in THF,
or amberlyst-15 in acetone and H_2O

Utilization of orthoformate esters[66] and Me_3SiCl[67] are standard procedures for water removal in acetalization. In the latter case, water is removed as hexamethyldisiloxane.

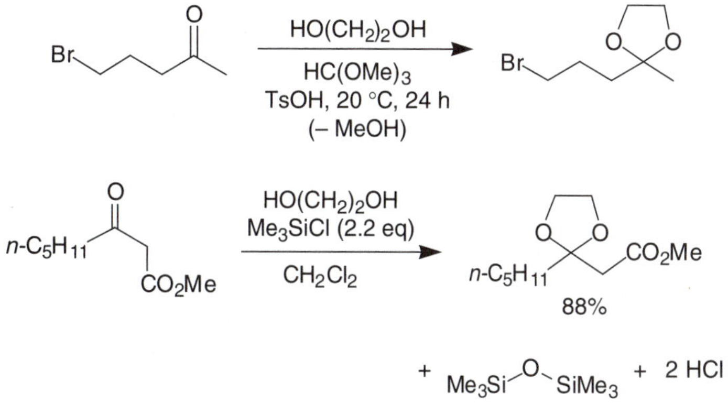

Acid-catalyzed acetalization of α,β-unsaturated ketones may result in double bond migration. The extent of migration of the double bond of enones depends on the strength of the acid catalyst used.[1d,68]

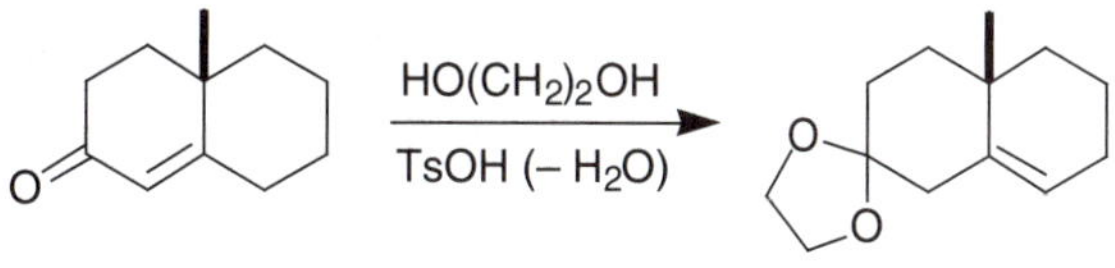

R. Noyori developed a procedure that avoids migration of the double bond during acetalization of α,β-unsaturated ketones (enones).[69] Treatment of the enone with the silylated diol 1,2-bis[(trimethylsilyl)oxy]ethane in the presence of a catalytic amount of trimethylsilyl trifluoromethanesulfonate (TMSOTf[70]) affords the acetal *without* double bond migration (kinetic control). The stability of the Me₃Si–O–SiMe₃ formed (instead of H₂O) drives the reaction to completion. Modifications of Noyori's procedure in which preparation of the silylated diol is circumvented have been reported.[71]

Noyori's acetalization procedure and its variants have been used for the preparation of chiral cyclic acetals. The rigid 1,3-dioxolanes formed play an important role as temporary chiral auxiliaries in stereoselective reactions and as electrophilic reactants.[72]

Chemoselectivity of Acetalization. Electronic and/or steric effects may influence the chemoselectivity of 1,3-dioxolane formation in compounds containing more than one carbonyl group.

Steric hindrance at one carbonyl group may direct acetalization to a less hindered one.

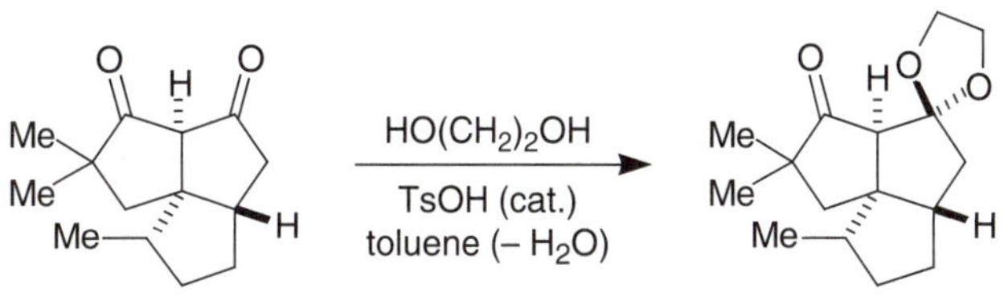

The presence of a double bond in enones increases the electron density at the carbonyl carbon, thereby reducing its reactivity toward acetalization.

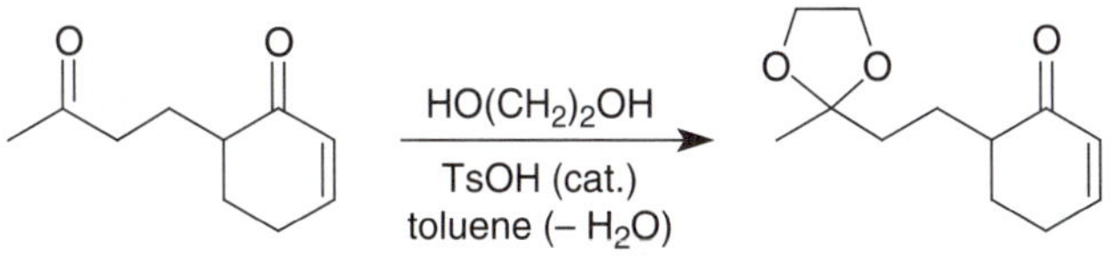

Interestingly, Noyori's conditions lead to chemoselective acetalization of the conjugated keto group.[73]

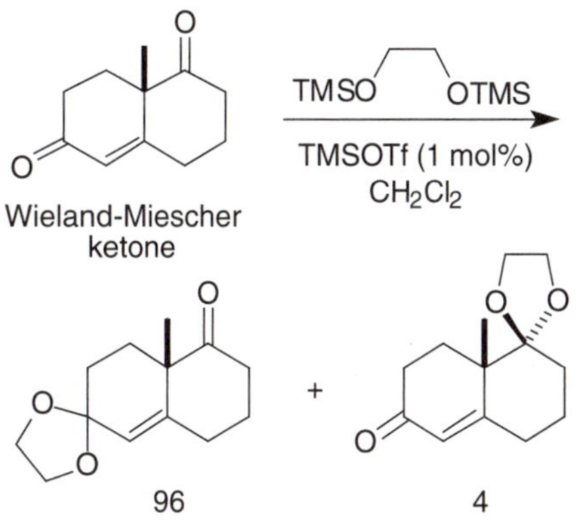

Chemoselective acetalization of an α,β-unsaturated ketone moiety in the presence of a saturated keto group can also be achieved by using 2,4,6-collidinium p-toluenesulfonate (CPTS) as a catalyst.[74]

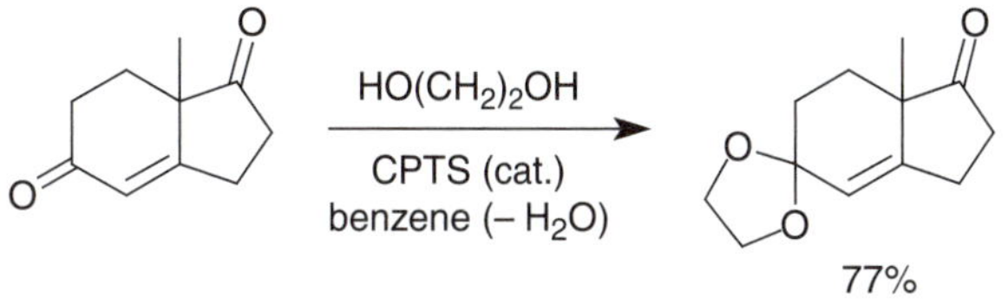

Generally, the more reactive aldehyde group undergoes selective acetalization in the presence of a keto group. However, there are few methods for the selective acetalization of a keto group in the presence of an aldehyde group. For example, sequential treatment of the dicarbonyl compound shown below with TMSOTf in the presence of Me_2S, followed by addition of $TMSOCH_2CH_2OTMS$ and removal of the 1-silyloxysulfonium salt, resulted in the chemoselective acetalization of the keto group.[75]

Remarkable chemoselectivities in acetalization of carbonyl groups promoted by microwave irradiation have been reported. For example, cyclic ketones can be selectively converted to their corresponding acetals, while acyclic ones remain unchanged under the same experimental conditions. Moreover, α,β-unsaturated aldehydes and ketones react faster then the corresponding saturated compounds.[76]

Via S,S-Acetals[77]   Unlike with the corresponding O,O-acetals, there is less selectivity between cyclic and acyclic thioacetals in their ease of formation and cleavage. Moreover, since cleavage of thioacetals often requires Hg(II) salts, the reaction is environmentally less attractive when compared to O,O-acetalization.

The preparation of thioacetals involves treatment of the carbonyl substrate with a dithiol in the presence of an acid catalyst, usually TsOH or $BF_3 \cdot OEt_2$. Since thioacetals are quite stable toward hydrolysis, there is no special need to remove the H_2O formed during the reaction. Also, since it is more difficult to equilibrate thioacetals than acetals via protonation, double bond migration in thioacetalization of enones is usually not observed.

The lower basicity of RS–H (pK_a 10) as compared to RO–H (pK_a 16) makes thionium ion formation more difficult and thus renders thioacetals more resistant to hydrolytic cleavage. Hence, an O,O-acetal moiety can be selectively deprotected in the presence of a thioacetal protecting group.[78]

The method of choice for deprotection of S,S-acetals takes advantage of the high affinity of sulfur for heavy metal ions, such as Hg^{2+}. Neutralization of the acid generated during hydrolysis with HgO or $CaCO_3$ allows for the presence of a wide variety of functional groups.

The alkylation of sulfur with reactive alkylating agents, such as MeI, Me_3OBF_4, Et_3OBF_4, or $MeOSO_2CH_3$, results in *S,S*-acetal deprotection without using Hg^{2+}.[79] Reductive removal of the sulfur of thioacetals with Raney-nickel (Ra-Ni)[81] provides a frequently used procedure for deoxygenation of aldehydes or ketones.

Several other procedures are available for the deoxygenation of the carbonyl groups of aldehydes and ketones to the corresponding methyl and methylene groups, respectively,[80] as outlined below.

Deoxygenation of Aldehydes and Ketones

Reduction of Thioacetals

Ra-Ni. A mild procedure for deoxygenation of aldehydes or ketones is via desulfurization of their thioacetal derivatives. For example, reduction of thioacetals with Ra-Ni, derived by treatment of an Ni-Al alloy with NaOH, produces the corresponding alkane moieties.[82] The hydrogen atoms in the deoxygenated product come from the hydrogen gas adsorbed on the Ra-Ni surface during its preparation.

$$NiAl_2 + 6\ NaOH + H_2O \longrightarrow Ni\ (ppt) + 3\ Na_3AlO_3 + 3\ H_2$$

There are problems associated with the use of Ra-Ni. Its reactivity depends on the mode of preparation, and it is pyrophoric in the dry state. Furthermore, the reagent is used in large excess, and it may cause the reduction or hydrogenolysis of other functional groups present in the molecule.

Ni-Boride. To circumvent the shortcomings of Ra-Ni, thioacetals may be reduced in the presence of a Ni-reagent, such as Ni-boride[83] generated in situ from $NiCl_2$ or $Ni(OAc)_2$ and $NaBH_4$. Ni-boride is more conveniently prepared and handled than Ra-Ni and reduces thioacetals without concomitant formation of olefin by-products.[84]

$$NiCl_2\ (1\ eq) + NaBH_4\ (2\ eq) \longrightarrow Ni\text{-boride} + thioacetal \longrightarrow reduction\ product$$

90%

***n*-Bu₃SnH.** Another effective reagent for the desulfurization of thioacetals is tri-*n*-butyltin hydride in the presence of 2,2′-azobis(isobutyronitrile) (AIBN), which proceeds by a radical mechanism.[85]

Clemmensen Reduction[86]

The direct deoxygenation of ketones via the Clemmensen reduction requires strong acidic conditions and high temperatures; hence it is not suitable for acid-sensitive compounds.

Wolff-Kishner Reduction

Deoxygenation of the carbonyl group of aldehydes and ketones via the intermediacy of their hydrazone derivatives, known as the Wolff-Kishner reduction,[87] offers an alternative to the thioacetal desulfurization method. The Wolff-Kishner reduction in the presence of hydrazine and NaOH (or KOH) has been replaced largely by the *Huang-Minlon method*,[88] where the deoxygenation is carried out with hydrazine in refluxing ethylene glycol.

A milder approach for the deoxygenation of aldehydes and ketones involves treatment of the preformed hydrazone with *t*-BuO⁻K⁺ in DMSO at room temperature.[89] Alternatively, conversion of the carbonyl group of aldehydes and ketones into the corresponding tosylhydrazone and reduction of these with NaBH₃CN[90] or with $(RCO_2)_2BH$[91] produces the desired methylene compounds in good yields.

3.5   PROTECTION OF THE CARBOXYL GROUP

Protecting groups for carboxylic acids are used to avoid reaction of the acidic COOH hydrogen with bases and nucleophiles or to prevent nucleophilic additions at the carbonyl carbon. Below are depicted several procedures for the protection of the carboxyl group by its conversion to the ester group.[43,44]

Alkyl Esters   Classical methods for ester formation include the following approaches:

$$RCOOH + R'OH \text{ (excess)} \xrightarrow{\text{strong acid catalyst}} RCOOR'$$

$$RCOCl + R'OH \xrightarrow[\text{or Et}_3\text{N, DMAP}]{\text{py or Et}_3\text{N}} RCOOR'$$

$$RCOONa + R'Br \text{ or } R'I \xrightarrow{S_N2} RCOOR' \text{ (primary halides only)}$$

A mild method for the specific preparation of methyl esters is the reaction of carboxylic acids with diazomethane.[92] Since diazomethane is very toxic and explosive, it must be handled with care, and is best suited for small-scale preparations.

diazomethane

An operationally simple method for preparing methyl esters involves treatment of a carboxylic acid in methanol with of two equivalents of Me_3SiCl.[93] The esterification proceeds via initial formation of the trimethylsilyl ester followed by displacement of the silanol by methanol.

Other methods of esterification:

- $RCOOH + BF_3 \cdot Et_2O + MeOH$. This procedure has been used for the preparation of unsaturated methyl esters.[94]

- $RCOOH + R'X$ in the presence of DBU (1,8-diazabicyclo[5.4.0]undec-7-ene) as a catalyst.

- $RCOOH$ with 1°, 2° and 3° alcohols in the presence of DCC (dicyclohexylcarbodiimide) and DMAP gives the corresponding esters in good yield under mild conditions (Steglich esterification).[95]

N,N'-dicyclohexylurea

- The *Mitsunobu* esterification of carboxylic acids with alcohols in the presence of Ph_3P and DEAD (diethyl azodicarboxylate) occurs under neutral conditions and provides the corresponding esters in high yields.[96]

85%

- A very mild esterification method involves reaction of mixed carboxylic-carbonic anhydrides derived from carboxylic acids and alkyl chloroformates in the presence of a catalytic amount of DMAP.[97]

95%

t-Butyl Esters[98]

An economical procedure for the formation of t-butyl esters is the reaction of a carboxylic acid with 2-methyl propene in the presence of an acid catalyst. For a laboratory-scale preparation, formation of the mixed anhydride using MsCl in the presence of t-BuOH gives the ester in good yield.[98a]

$$ RCO_2H + H_2C{=}CMe_2 \xrightarrow[\text{Et}_2\text{O, 25 °C}]{\text{conc. H}_2\text{SO}_4 \text{ (cat.)}} RCO_2CMe_3 $$

72%

The t-BuO group provides steric shielding of the carbonyl carbon, thereby lowering its susceptibility to attack by nucleophilic reagents. t-Butyl esters are rapidly cleaved with CF_3COOH or HCOOH or with TsOH (catalyst) in refluxing benzene.

Benzyl Esters

Benzyl esters are versatile protecting groups for carboxylic acids. They are usually prepared from acid chlorides and benzyl alcohol in the presence of pyridine or from carboxylic acids and benzyl chloroformate in the presence of Et_3N and a catalytic amount of DMAP. A very useful feature of benzyl esters is that they are readily deprotected by hydrogenolysis.[44,99]

$$ RCO_2CH_2Ph \xrightarrow[\text{25 °C}]{\text{Pd/C, H}_2} RCO_2H + H_3CPh $$

Aryl Esters   2,6-Di-*tert*-butyl-4-methylphenyl esters (BHT esters) and 2,6-di-*tert*-butyl-4-methoxyphenyl esters (BHA esters) provide steric suppression of the carboxyl reactivity so they do not react with RMgX or RLi reagents.[100] BHT and BHA esters derived from α,β-unsaturated acids undergo 1,4-addition with a variety of organolithium reagents rather than the usually observed 1,2-additions.[101] Although, these esters are very resistant to hydrolysis, oxidative cleavage of the esters with $(NH_4)_2Ce(NO_3)_6$ regenerates the acids.

Silyl Esters

An alternative, operationally simple method for protecting the carbonyl group of RCO_2H against nucleophilic attack by RMgX or RLi reagents is its conversion to a TIPS ester.[30,102] Generally, silyl esters are sensitive toward acid-induced hydrolytic cleavage. In the case of TIPS esters, desilylation has been achieved by treatment with HF•pyridine.

Oxazolines

1,3-Oxazolines protect *both* the carbonyl and hydroxyl group of a carboxyl group. The starting material, 2-amino-2-methylpropanol, is readily available.[103]

The oxazoline moiety serves as a protecting group toward RMgX and $LiAlH_4$, but not for RLi because the protons at C_α may be deprotonated.[104]

Ortho Esters

Ortho esters are stable toward base but are readily hydrolyzed on exposure to mild acids.[105] The cyclic ortho ester of 1-alkyl-4-methyl-2,6,7-trioxabicyclo[2.2.2]octanes

(OBO)[106] is frequently used for protecting carboxyl groups.[107] Lewis acid–mediated rearrangements of carboxylic esters derived from 3-hydroxymethyl-3-methyloxetane cleanly afford the corresponding OBO ortho esters.

3.6 PROTECTION OF DOUBLE BONDS

There are few reported efficient methods available for protecting double bonds. Both halogenation-dehalogenation and epoxidation-deoxygenation are procedures that have been used for protection-deprotection of double bonds. However, these procedures are limited in their application, especially in the presence of other functional groups. Selective protection of double bonds has been achieved using cyclopentadienyl iron dicarbonyl tetrafluoroborate as a protecting group.[108]

3.7 PROTECTION OF TRIPLE BONDS

Masking the potentially acidic proton of 1-alkynes (pK_a 25) is readily achieved by their conversion to the corresponding 1-silyl-1-alkynes.[30]

Depending on the nature of the trialkylsilyl group, it can be removed under a variety of conditions (MeONa in MeOH, n-Bu$_4$NF in THF, or AgNO$_3$ in EtOH followed by NaCN[109]).

Protection of an internal triple bond (or an internal triple bond in the presence of a double bond)[110] can be done by converting the former to the dicobaltoctacarbonyl complex. The following alkene hydroboration-oxidation example illustrates this approach.[110a]

PROBLEMS

1. **Reagents.** Give the structures of the major products (**A**–**G**) expected from the following reactions. Assume standard aqueous workup conditions are used for product isolation.

a.

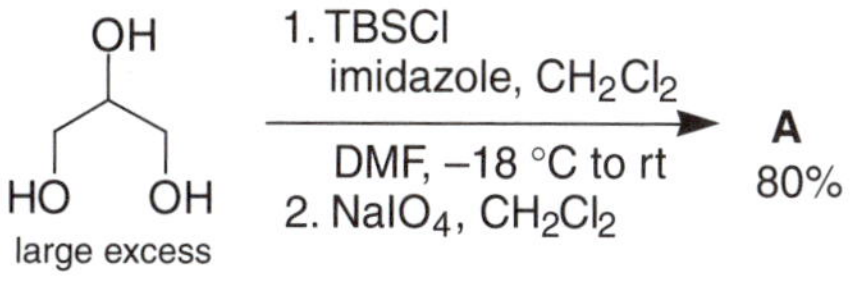
OH
HO OH
large excess
1. TBSCl
imidazole, CH2Cl2
DMF, –18 °C to rt
2. NaIO4, CH2Cl2
A
80%

b.

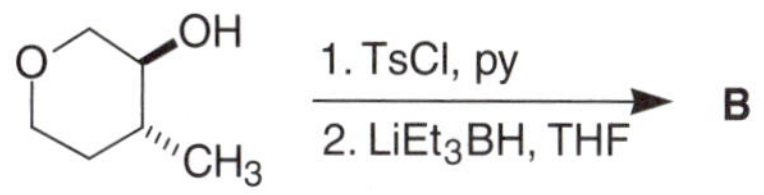
O
OH
CH3
1. TsCl, py
2. LiEt3BH, THF
B

c.

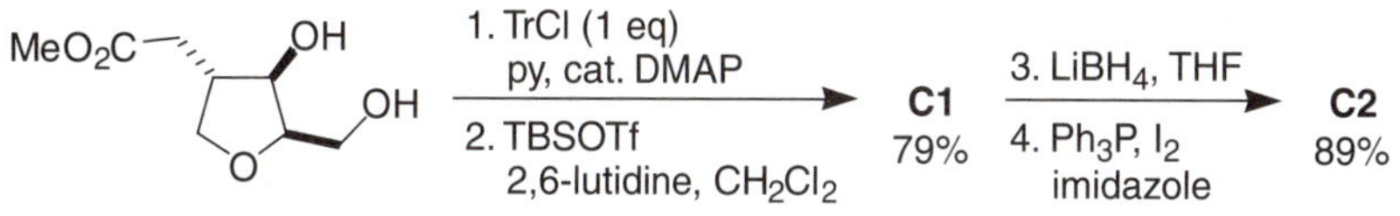
MeO2C
OH
OH
O
1. TrCl (1 eq)
py, cat. DMAP
2. TBSOTf
2,6-lutidine, CH2Cl2
C1
79%
3. LiBH4, THF
4. Ph3P, I2
imidazole
C2
89%

d.

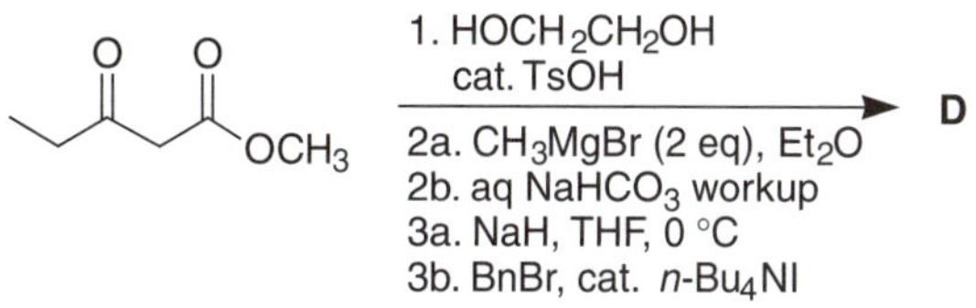
O O
OCH3
1. HOCH2CH2OH
cat. TsOH
2a. CH3MgBr (2 eq), Et2O
2b. aq NaHCO3 workup
3a. NaH, THF, 0 °C
3b. BnBr, cat. n-Bu4NI
D

e.

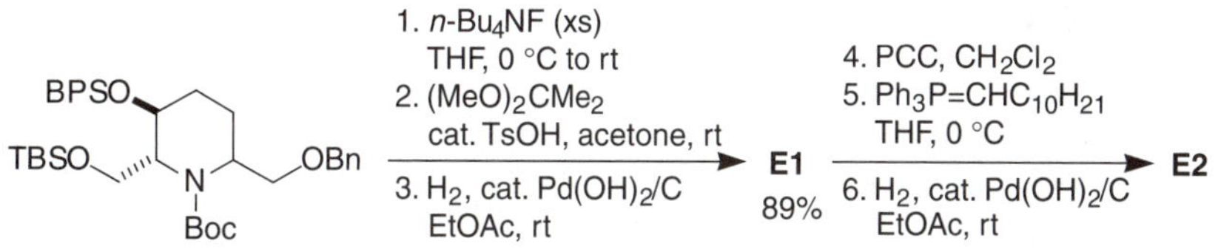
BPSO
TBSO
N
Boc
OBn
1. n-Bu4NF (xs)
THF, 0 °C to rt
2. (MeO)2CMe2
cat. TsOH, acetone, rt
3. H2, cat. Pd(OH)2/C
EtOAc, rt
E1
89%
4. PCC, CH2Cl2
5. Ph3P=CHC10H21
THF, 0 °C
6. H2, cat. Pd(OH)2/C
EtOAc, rt
E2

f.

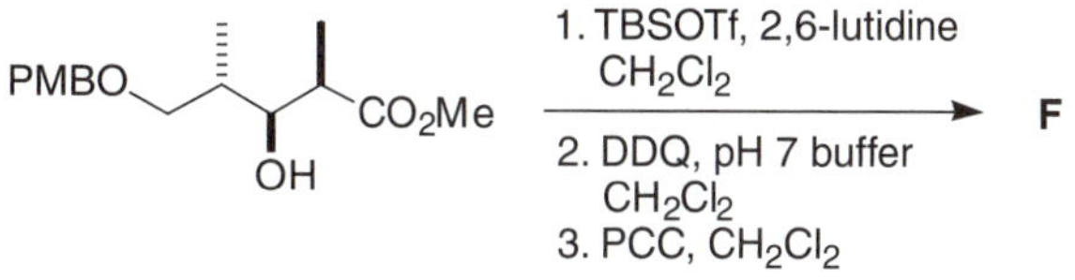
PMBO
OH
CO2Me
1. TBSOTf, 2,6-lutidine
CH2Cl2
2. DDQ, pH 7 buffer
CH2Cl2
3. PCC, CH2Cl2
F

g.

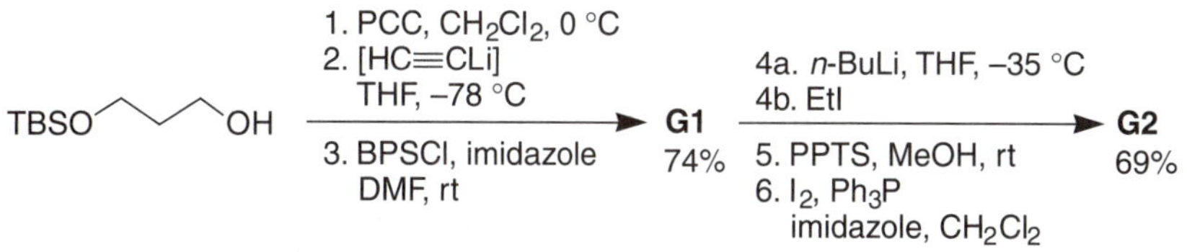
TBSO
OH
1. PCC, CH2Cl2, 0 °C
2. [HC≡CLi]
THF, –78 °C
3. BPSCl, imidazole
DMF, rt
G1
74%
4a. n-BuLi, THF, –35 °C
4b. EtI
5. PPTS, MeOH, rt
6. I2, Ph3P
imidazole, CH2Cl2
G2
69%

2. **Selectivity.** Show the product(s) expected for the following transformations.

a.

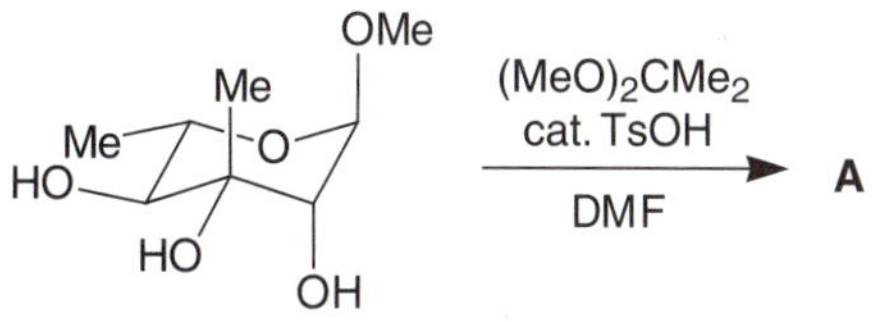
OMe
Me
Me
O
HO
HO
OH
(MeO)2CMe2
cat. TsOH
DMF
A

b.

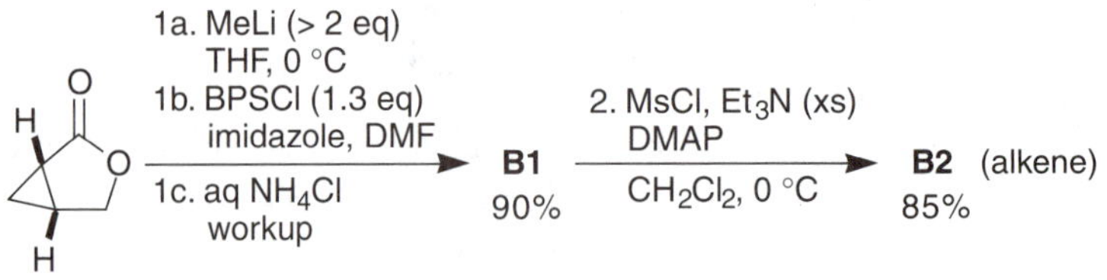

1a. MeLi (> 2 eq)
THF, 0 °C
1b. BPSCl (1.3 eq)
imidazole, DMF → **B1** 90%
1c. aq NH$_4$Cl
workup

2. MsCl, Et$_3$N (xs)
DMAP
CH$_2$Cl$_2$, 0 °C → **B2** (alkene) 85%

c.

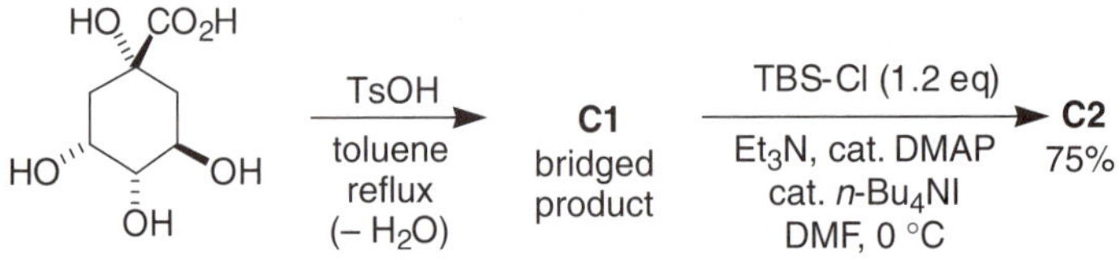

TsOH
toluene
reflux
(− H$_2$O) → **C1** bridged product

TBS-Cl (1.2 eq)
Et$_3$N, cat. DMAP
cat. n-Bu$_4$NI
DMF, 0 °C → **C2** 75%

d.

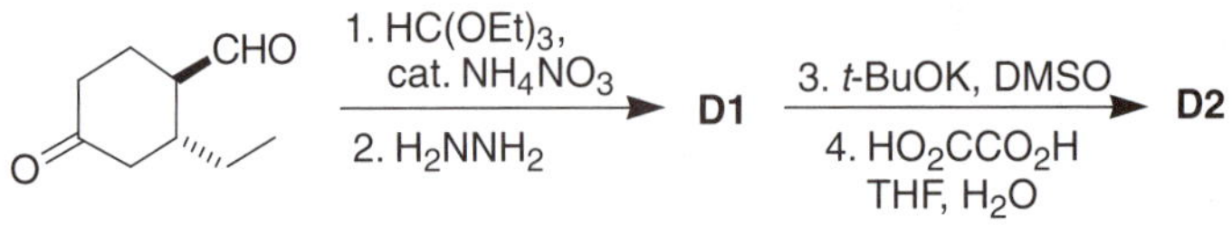

1. HC(OEt)$_3$,
cat. NH$_4$NO$_3$
2. H$_2$NNH$_2$ → **D1**

3. t-BuOK, DMSO → **D2**
4. HO$_2$CCO$_2$H
THF, H$_2$O

e.

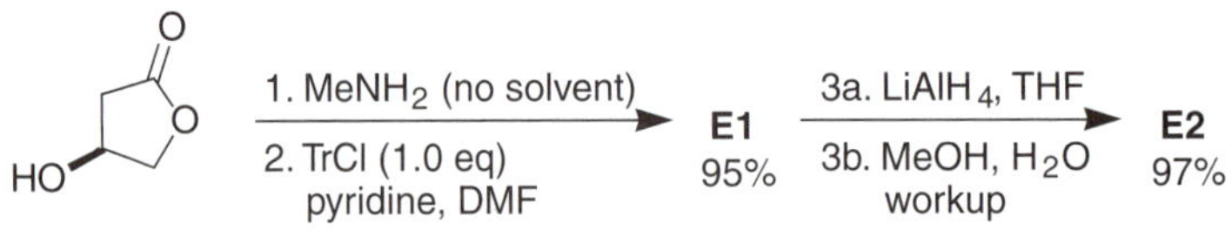

1. MeNH$_2$ (no solvent)
2. TrCl (1.0 eq)
pyridine, DMF → **E1** 95%

3a. LiAlH$_4$, THF
3b. MeOH, H$_2$O
workup → **E2** 97%

f.

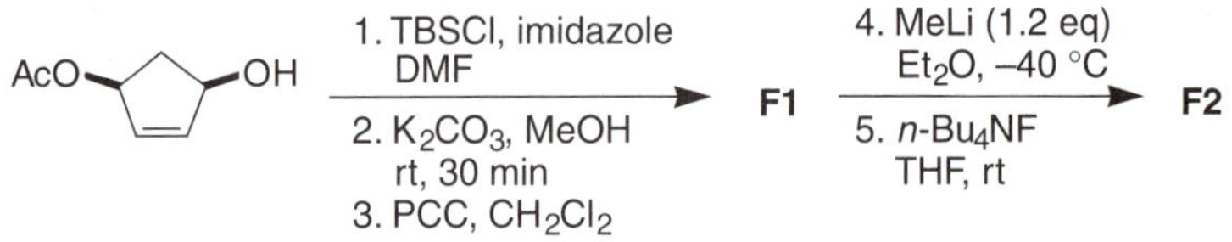

AcO — OH

1. TBSCl, imidazole
DMF
2. K$_2$CO$_3$, MeOH
rt, 30 min
3. PCC, CH$_2$Cl$_2$ → **F1**

4. MeLi (1.2 eq)
Et$_2$O, −40 °C
5. n-Bu$_4$NF
THF, rt → **F2**

g.

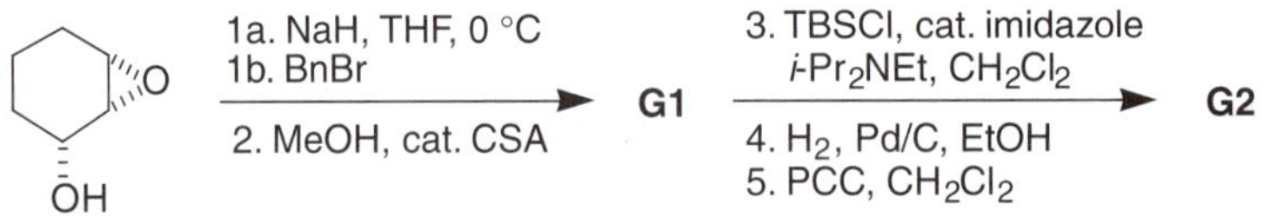

1a. NaH, THF, 0 °C
1b. BnBr
2. MeOH, cat. CSA → **G1**

3. TBSCl, cat. imidazole
i-Pr$_2$NEt, CH$_2$Cl$_2$
4. H$_2$, Pd/C, EtOH
5. PCC, CH$_2$Cl$_2$ → **G2**

3. **Retrosynthetic analysis.** Outline a retrosynthetic scheme for each of the following target molecules. Show (1) the *analysis* (including FGI, synthons, synthetic equivalents) and (2) the *synthesis* of each TM. You may only use compounds with five or fewer carbons as starting materials.

a H$_3$C— ... ⟹

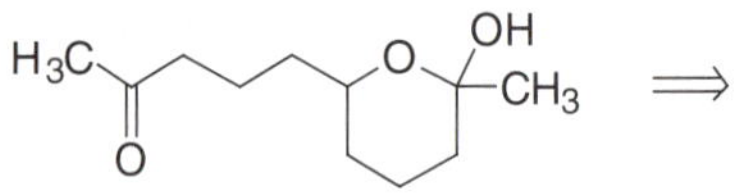

b. Br$^-$ Ph$_3$P$^+$ ... ⟹

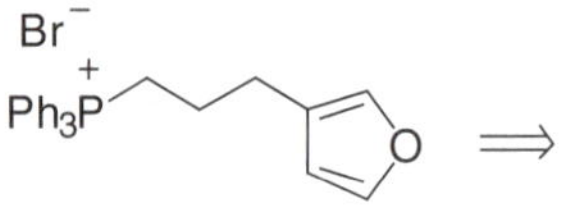

4. **Synthesis.** Supply the missing reagents required to accomplish each of the following transformations. Be sure to control the relative stereochemistry.

a.

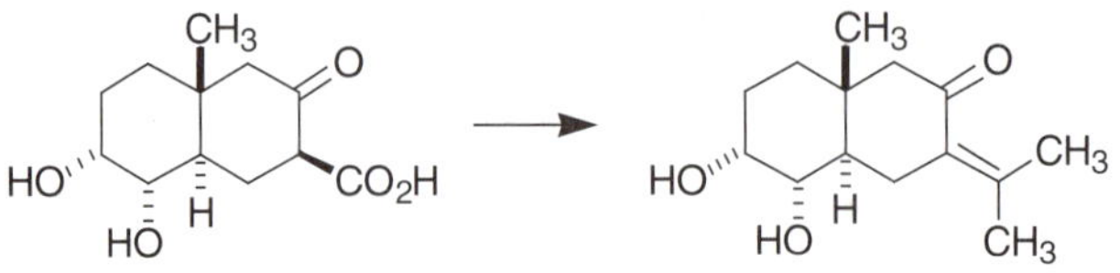

b.

c.

d.

e.

f.

*5. Explain the regio- and stereochemical outcome of the following sequence of reactions by showing the structures of the intermediates obtained after each step.

1. PhCHO, H⁺
2. LiAlH₄ • AlCl₃ (1:1)
3. PhCH(OMe)₂, H⁺

REFERENCES

1. (a) Schelhaas, M., Waldmann, H. *Angew. Chem., Int. Ed.* **1996**, *35*, 2057. (b) Greene, T. W., Wuts, P. G. M. *Protective Groups in Organic Synthesis*, 3rd ed., Wiley: New York, 1999. (c) Smith, M. B. *Organic Synthesis*, 2nd ed., McGraw-Hill: Boston, 2002. (d) Kocienski, P. J. *Protecting Groups*, 3rd ed., Georg Thieme: Stuttgart, 2004.

2. Theodoridis, G. *Tetrahedron* **2000**, *56*, 2339.

3. Pearlman, W. M. *Tetrahedron Lett.* **1967**, 1663.

4. Bernotas, R. C., Cube, R. V. *Synth. Commun.* **1990**, *20*, 1209.

5. Gutzwiller, J., Uskokovic, M. *J. Am. Chem. Soc.* **1970**, *92*, 204.

6. (a) Lott, R. S., Chauhan, V. S., Stammer, C. H. *J. Chem. Soc., Chem. Comm.* **1979**, 495. (b) Olah, G. A., Narang, S. C., Gupta, B. G. B., Malhotra, R. *J. Org. Chem.* **1979**, *44*, 1247.

7. Coleman, R. S., Shah, J. A. *Synthesis* **1999**, 1399.

8. (a) Keller, O., Keller, W. E., van Look, G., Wersin, G. *Org. Synth., Col., Vol. VII*, **1990**, 70. (b) Paleveda, W. J., Holly, F. W., Veber, D. F. *Org. Synth., Col., Vol. VII*, **1990**, 75.

9. Bose, D. S., Lakshminarayana, V. *Synthesis* **1999**, 66.

10. (a) Johnstone, R. A. W., Rose, M. E. *Tetrahedron* **1979**, *35*, 2169. (b) Sánchez, A. J., Konopelski, J. P. *J. Org. Chem.* **1994**, *59*, 5445.

11. Williard, P. G., Fryhle, C. P. *Tetrahedron Lett.* **1980,** *21*, 3731.

12. McOmie, J. F., West, D. E. *Org. Synth., Col., Vol. V,* **1973**, 412.

13. Tanner, D., Somfai, P. *Tetrahedron* **1987**, *43*, 4395.

14. Iversen, T., Bundle, D. R. *J. Chem. Soc., Chem. Comm.* **1981**, 1240.

15. Widmer, U. *Synthesis* **1987**, 568.

16. (a) White, J. D., Reddy, G. N., Spessard, G. O. *J. Am. Chem. Soc.* **1988**, *110*, 1624. (b) Balasubramaniam, R. P., Moss, D. K., Wyatt, J. K., Spence, J. D., Gee, A., Nantz, M. H. *Tetrahedron* **1997**, *53*, 7429.

17. Horita, K., Yoshioka, T., Tanaka, T., Oikawa, Y., Yonemitsu, O. *Tetrahedron* **1986**, *42*, 3021.

18. Lizarzaburu, M. E., Kurth, M. J., Nantz, M. H. *Tetrahedron Lett.* **1999**, *40*, 8985.

19. Murugan, R., Scriven, E. F. V. *Aldrichimica Acta* **2003**, *36*, 21.

20. Scriven, E. F. V. *Chem. Soc. Rev.* **1983**, *12*, 129.

21. Hartung, J., Hünig, S., Kneuer, R., Schwarz, M., Wenner, H. *Synthesis* **1997**, 1433.

22. Jones, G. B., Hynd, G., Wright, J. M., Sharma, A. *J. Org. Chem.* **2000**, *65*, 263.

23. (a) Lalonde, M., Chan, T. H. *Synthesis* **1985**, 817.
(b) Nelson, T. D., Crouch, R. D. *Synthesis* **1996**, 1031.

24. Bruynes, C. A., Jurriens, T. K. *J. Org. Chem.* **1982**, *47*, 3966.

25. Corey, E. J., Venkateswarlu, A. K. *J. Am. Chem. Soc.* **1972**, *94*, 6190.

26. Corey, E. J., Cho, H., Rücker, C., Hua, D. H. *Tetrahedron Lett.* **1981**, *22*, 3455.

27. Pilcher, A. S., DeShong, P. *J. Org. Chem.* **1993**, *58*, 5130.

28. Corey, E. J., Jones, G. B. *J. Org. Chem.* **1992**, *57*, 1028.

29. Wetter, H., Oertle, K. *Tetrahedron Lett.* **1985**, *26*, 5515.

30. Rücker, C. *Chem. Rev.* **1995**, *95*, 1009, and references cited therein.

31. Khalafi-Nezhad, A., Alamdari, R. F., Zekri, N. *Tetrahedron* **2000**, *56*, 7503.

32. Bruzik, K. S., Tsai, M.-D. *J. Am. Chem. Soc.* **1992**, *114*, 6361.

33. Miyashita, M., Yoshikoshi, A., Grieco, P. A. *J. Org. Chem.* **1977**, *42*, 3772.

34. van Boom, J. H., Herschied, J. D. M. *Synthesis* **1973**, 169.

35. Benneche, T. *Synthesis* **1995**, 20.

36. Kluge, A. F., Untch, K. G., Fried, J. H. *J. Am. Chem. Soc.* **1972**, *94*, 7827.

37. Narasaka, K., Sakakura, T., Uchimaru, T., Guédin-Vuong, D. *J. Am. Chem. Soc.* **1984**, *106*, 2954.

38. Monti, H., Léandri, G., Klos-Ringuet, M., Corriol, C. *Synth. Comm.* **1983**, *13*, 1021.

39. Hanessian, S., Delorme, D., Dufresne, Y. *Tetrahedron Lett.* **1984**, *25*, 2515.

40. Corey, E. J., Gras, J.-L., Ulrich, P. *Tetrahedron Lett.* **1976**, *17*, 809.

41. Suzuki, K., Tomooka, K., Katayama, E., Matsumoto, T., Tsuchihashi, G. *J. Am. Chem. Soc.* **1986**, *108*, 5221.

42. Just, G., Luthe, C., Viet, M. T. P. *Can. J. Chem.* **1983**, *61*, 712.

43. Otera, J. *Esterification: Methods, Reactions and Applications*, Wiley-VCH: Weinheim, Germany, 2003.

44. Salomon, C. J., Mata, E. G., Mascaretti, O. A. *Tetrahedron* **1993**, *49*, 3691.

45. Ishihara, K., Kurihara, H., Yamamoto, H. *J. Org. Chem.* **1993**, *58*, 3791.

46. Aspinall, G. O., Zweifel, G. *J. Chem. Soc.* **1957**, 2271.

47. Fieser, L. F., Fieser, M. *Reagents for Organic Synthesis*, Vol. 1, 1967, p. 1179.

48. Johnson, W. S., Collins, J. C., Jr., Pappo, R., Rubin, M. B., Kropp, P. J., Johns, W. F., Pike, J. E., Bartmann, W. *J. Am. Chem. Soc.* **1963**, *85*, 1409.

49. Tipson, R. S. *Methods in Carbohydrate Chem.*, Vol. II, 1963, p. 250.

50. Nishida, A., Hamada, T., Yonemitsu, O. *J. Org. Chem.* **1988**, *53*, 3386.

51. Rossi, R., Salvadori, P. A. *Synthesis* **1979**, 209.

52. Seyden-Penne, J. *Reductions of the Alumino- and Borohydrides in Organic Synthesis*, VCH/Lavoisier-Tec & Doc: Paris, 1991.

53. Krishnamurthy, S., Brown, H. C. *J. Org. Chem.* **1976**, *41*, 3064.

54. Ley, S. V., Baeschlin, D. K., Dixon, D. J., Foster, A. C., Ince, S. J., Priepke, H. W. M., Reynolds, D. J. *Chem. Rev.* **2001**, *101*, 53.

55. (a) Gelas, J., Horton, D. *Heterocycles* 1981, *16*, **1587**.
(b) Manzo, E., Barone, G., Michelangelo, P. *Synlett* **2000**, 887.

56. (a) Ley, S. V., Leslie, R., Tiffin, P. D., Woods, M. *Tetrahedron Lett.* **1992**, *33*, 4767. (b) Ziegler, T. *Angew. Chem., Int. Ed.* **1994**, *33*, 2272.

57. Nakata, T., Fukui, M., Oishi, T. *Tetrahedron Lett.* **1988**, *29*, 2219.

58. Schmid, C. R., Bryant, J. D. *Org. Synth.* **1995**, *72*, 6.

59. Ferro, V., Mocerino, M., Stick, R. V., Tilbrook, D. M. G. *Aust. J. Chem.* **1988**, *41*, 813.

60. (a) Whitesell, J. K. *Chem. Rev.* **1989**, *89*, 1581. (b) Alexakis, A., Mangeney, P. *Tetrahedron: Asymmetry* **1990**, *1*, 477.

61. Clerici, A., Pastori, N., Porta, O. *Tetrahedron* **2001**, *57*, 217.

62. (a) Majumdar, S., Bhattachariya, A. *J. Org. Chem.* **1999**, *64*, 5682. (b) Eash, K. J., Pulia, M. S., Wieland, L. C., Mohan, R. S. *J. Org. Chem.* **2000**, *65*, 8399.

63. Deslongchamps, P., Dory, Y. L., Li, S. *Tetrahedron* **2000**, *56*, 3533.

64. Gemal, A. L., Luche, J.-L. *J. Org. Chem.* **1979**, *44*, 4187.

65. Patwardhan, S. A., Dev, S. *Synthesis* **1974**, 348.

66. Bhushan, B., Sharma, A. K., Kaushik, N. K. *Bull. Soc. Chim. France* **1983**, 297.

67. Chan, T. H., Brook, M. A., Chaly, T. *Synthesis* **1983**, 203.

68. DeLeeuw, J. W., De Waard, E. R., Beetz, T., Huisman, H. D. *Rec. Trav. Chim. Pays-Bas* **1973**, *92*, 1047.

69. Tsunoda, T., Suzuki, M., Noyori, R. *Tetrahedron Lett.* **1980**, *21*, 1357.

70. Stang, P. J., White, M. R. *Aldrichimica Acta* **1983**, *16*, 15.

71. Kurihara, M., Miyata, N. *Chem. Lett.* **1995**, 263.

72. Alexakis, A., Mangeney, P. *Tetrahedron: Asymmetry* **1990**, *1*, 477.

73. Hwu, J. R., Wetzel, J. M. *J. Org. Chem.* **1985**, *50*, 3946.

74. Nitz, T. J., Paquette, L. A. *Tetrahedron Lett.* **1984**, *25*, 3047.

75. Kim, S., Kim, Y. G., Kim, D. *Tetrahedron Lett.* **1992**, *33*, 2565.

76. Kalita, D. J., Borah, R., Sarma, J. C. *Tetrahedron Lett.* **1998**, *39*, 4573.

77. Greene, T. W., Wuts, P. G. M. *Protective Groups in Organic Synthesis*, 3rd ed., Wiley: New York, 1999, pp. 329–344.

78. Ellison, R. A., Lukenbach, E. R., Chiu, C. *Tetrahedron Lett.* **1975**, *16*, 499.

79. (a) Oishi, T., Kamemoto, K., Ban Y. *Tetrahedron Lett.* **1972**, *13*, 1085. (b) Takano, S., Hatakeyama, S., Ogasawara, K. *J. Chem. Soc., Chem. Comm.* **1977**, 68.

80. Caubère, P., Coutrot, P. In *Comp. Org. Synthesis*, Trost, B. M., Fleming, I., Eds., Pergamon Press: Oxford, UK, 1991, Vol. 8, p. 835.

81. (a) Pettit, G. R., van Tamelen, E. E. *Org. React.* **1962**, *12*, 356. (b) Enders, D., Pieter, R., Renger, B., Seebach, D. *Org. Synth., Col. Vol. VI*, **1988**, 542.

82. Sondheimer, F., Wolfe, S. *Can. J. Chem.* **1959**, *37*, 1870.

83. Boar, R. B., Hawkins, D. W., McGhie, J. F., Barton, D. H. R. *J. Chem. Soc., Perkin Trans I* **1973**, 654.

84. Zaman, S. S., Sarmah, P., Barua, N. C. *Chem. Ind.* **1989**, 806.

85. Gutierrez, C. G., Stringham, R. A., Nitasaka, T., Glasscock, K. G. *J. Org. Chem.* **1980**, *45*, 3393.

86. Vedejs, E. *Org. React.* **1975**, *22*, 401.

87. Todd, D. *Org. React.* **1948**, *4*, 378.

88. (a) Huang-Minlon. *J. Am. Chem. Soc.* **1946**, *68*, 2487. (b) For a recent application of the method, see Gunatilaka, A. A. L., Nanayakkara, N. P. D. *Tetrahedron* **1984**, *40*, 805.

89. Cram, D. J., Sahyun, M. R., Knox, G. R. *J. Am. Chem. Soc.* **1962**, *84*, 1734.

90. (a) Caglioti, L. *Tetrahedron* **1966**, *22*, 487. (b) Hutchins, R. O., Milewski, C. A., Maryanoff, B. E. *J. Am. Chem. Soc.* **1973**, *95*, 3662. (c) Hutchins, R. O., Karcher, M., Rua, L. *J. Org. Chem.* **1975**, *40*, 923.

91. Kabalka, G. W., Summers, S. T. *J. Org. Chem.* **1981**, *46*, 1217.

92. (a) Hecht, S. M., Kazarich, J. W. *Tetrahedron Lett.* **1973**, *14*, 1397. (b) Hudlicky, M. *J. Org. Chem.* **1980**, *45*, 5377. (c) Shioiri, T., Aoyama, T., Mori, S. *Org. Synth.* **1989**, *68*, 1.

93. (a) Brook, M. A., Chan, T. H. *Synthesis* **1983**, 201. (b) Gerspracher, M., Rapoport, H. *J. Org. Chem.* **1991**, *56*, 3700.

94. Kadaba, P. K. *Synthesis* **1971**, 316.

95. (a) Neises, B., Steglich, W. *Angew. Chem., Int. Ed.* **1978**, *17*, 522. (b) Boden, E. P., Keck, G. E. *J. Org. Chem.* **1985**, *50*, 2394.

96. Mitsunobu, O. *Synthesis* **1981**, 1.

97. Kim, S., Lee, J. I., Kim, Y. C. *J. Org. Chem.* **1985**, *50*, 560.

98. (a) Chandrasekaran, S., Turner, J. V. *Synth. Comm.* **1982**, *12*, 727. (b) Ohta, S., Shimabayashi, A., Aono, M., Okamoto, M. *Synthesis* **1982**, 833.

99. Hartung, W. H., Simonoff, R. *Org. React.* **1953**, *7*, 263.

100. (a) Heathcock, C. H., Pirrung, M. C., Montgomery, S. M., Lampe, J. *Tetrahedron* **1981**, *37*, 4087. (b) Häner, R., Laube, T., Seebach, D. *J. Am. Chem. Soc.* **1985**, *107*, 5396.

101. Cooke, M. P., Jr. *J. Org. Chem.* **1986**, *51*, 1637.

102. Evans, D. A., Trotter, B. W., Côfé, B., Coleman, P. I., Dias, L. C., Tyler, A. N. *Angew. Chem., Int. Ed.* **1997**, *36*, 2744.

103. Meyers, A. I., Temple, D. L., Jr. *J. Am. Chem. Soc.* **1970**, *92*, 6644.

104. Meyers, A. I., Temple, D. L., Haidukewych, D., Mihelich, E. D. *J. Org. Chem.* **1974**, *39*, 2787.

105. Deslongchamps, P., Dory, Y. L., Shigui, L. *Tetrahedron* **2000**, *56*, 3533.

106. Corey, E. J., Raju, N. *Tetrahedron Lett.* **1983**, *24*, 5571.

107. Blaskovich, M. A., Evindar, G., Rose, N. G. W., Wilkinson, S., Luo Y., Lajoie, G. A. *J. Org. Chem.* **1998**, *63*, 3631.

108. Nicholas, K. M. *J. Am. Chem. Soc.* **1975**, *97*, 3254, and references cited therein.

109. Corey, E. J., Kirst, H. A. *Tetrahedron Lett.* **1968**, 5041.

110. (a) Seyferth, D., Nestle, M. O., Wehman, A. T. *J. Am. Chem. Soc.* **1975**, *97*, 7417. (b) Ganesh, P., Nicholas, K. M. *J. Org. Chem.* **1993**, *58*, 5587. (c) Magnus, P., Pitterna, T. *J. Chem. Soc., Chem. Commun.* **1991**, 541.

CHAPTER 4

Functional Group Transformations: Oxidation and Reduction

Discovery consists of seeing what everybody has seen and thinking what nobody has thought.

Albert Szent-Györgyi

4.1   OXIDATION OF ALCOHOLS TO ALDEHYDES AND KETONES

The classical procedure for oxidizing primary alcohols to aldehydes and secondary alcohols to ketones involves treatment of the appropriate alcohol with a chromium(VI) reagent.[1] Oxidation of primary alcohols to aldehydes requires anhydrous conditions. In the presence of water, the resultant aldehyde can form the hydrate, which may be further oxidized to the carboxylic acid.

Recently, procedures for oxidation of alcohols to aldehydes and ketones have been developed that obviate the toxicity associated with the use of chromium reagents. Because of the greater stability of ketones to most oxidizing conditions, the conversion of secondary alcohols to ketones can be accomplished with a wide variety of reagents and conditions (Table 4.1).

The E2-like process depicted for the general oxidation mechanism in Table 4.1 is supported by the observation that deuterium substitution of the α-H in isopropanol slows the rate of chromic acid oxidation by sevenfold.[2] Deuterium replacement at the methyl positions does not diminish the oxidation rate. Since C–D bonds are broken more slowly than C–H bonds, these results suggest that the α-H is removed in a slow step.

Relative rate of oxidation with H_2CrO_4	1.0	0.16	1.0

| Table 4.1 | Methods for Alcohol Oxidation |

Name	Reagents	Z
Jones	CrO$_3$, H$_2$SO$_4$, H$_2$O	(Cr–OH structure)
Swern	DMSO, oxalyl chloride	$-\overset{+}{\text{S}}\text{Me}_2$
Dess-Martin	(AcO–I structure)	(AcO–I structure)
TEMPO	(H$_3$C / N–O structure) NaOCl, NaBr	(H$_3$C / N–O structure)

REAGENTS AND PROCEDURES FOR ALCOHOL OXIDATION

Jones Reagent

$$CrO_3 + H_2SO_4 + H_2O + \text{acetone}$$

$$\text{or} \quad Na_2Cr_2O_7 + H_2SO_4 + H_2O + \text{acetone}$$

The Jones reagent is an excellent reagent for the oxidation of secondary alcohols that do not contain acid-sensitive groups such as acetals.[3] Oxidation of primary alcohols with Jones reagent may result in the conversion of the aldehydes initially formed to the corresponding carboxylic acids. The reagent is added to the alcohol contained in acetone at 0–25 °C, and the excess Cr(VI) is destroyed in the reaction workup by adding some isopropyl alcohol (color change from orange to blue green).

Chromic acid oxidation may also be performed in the presence of water-immiscible solvents. Addition of a stoichiometric amount of aqueous sodium dichromate and sulfuric acid to a solution of the secondary alcohol in diethyl ether at 25 °C affords the corresponding ketone.[4]

Collins-Ratcliff Reagent

$CrO_3 \cdot 2C_5H_5N$ is a mild reagent for the oxidation of alcohols that contain acid-sensitive groups.[5] The reagent is prepared by adding CrO_3 (carcinogenic) to a mixture of pyridine-CH_2Cl_2, and the RCH_2OH or R_2CHOH is then added to the oxidant in solution. Unfortunately, to achieve rapid and complete oxidation, a large excess of the reagent is required.

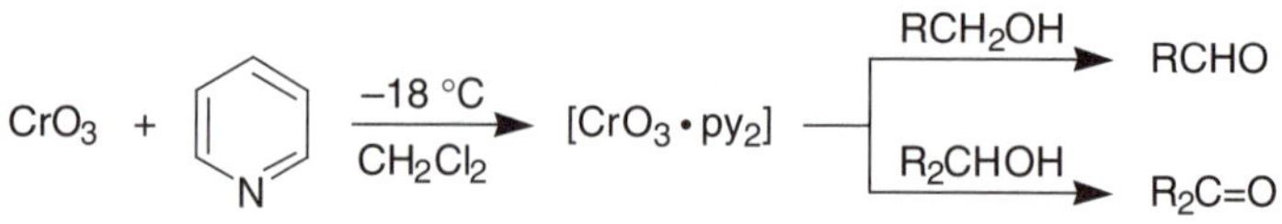

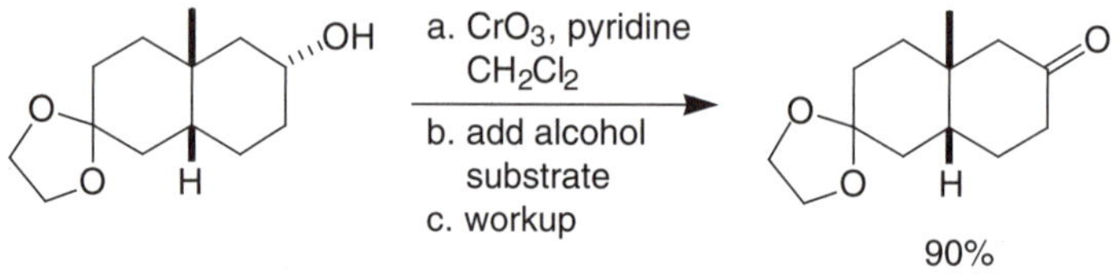

Pyridinium Chlorochromate (PCC)

$$RCH_2OH + PCC \xrightarrow{2e^-} RCHO + py \cdot HCl + CrO_2 + H_2O$$

Primary and secondary alcohols are readily oxidized in CH_2Cl_2 utilizing 1 to 1.5 equivalents of PCC.[6,7] Since PCC is slightly acidic, oxidations of compounds containing acid-sensitive groups should be carried out in the presence of powdered NaOAc.[6a]

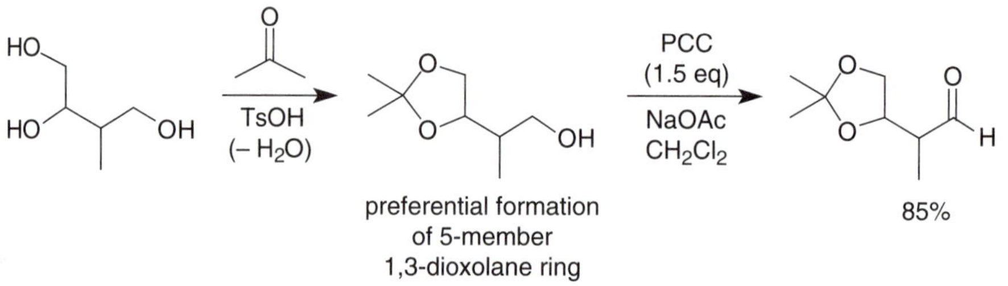

Oxidation of primary alcohols with PCC in the presence of molecular sieves (3 Å or 4 Å) results in higher yields of aldehydes.[8] Ultrasound also has been used to facilitate PCC oxidations.[9]

Pyridinium Dichromate (PDC)

PDC is soluble in H_2O, DMF, and DMSO but sparingly soluble in CH_2Cl_2 or $CHCl_3$. The reagent is less acidic than PCC. Hence, oxidations in CH_2Cl_2 can be carried out under nearly neutral conditions. This permits the conversion of primary alcohols containing acid-sensitive groups into the corresponding aldehydes or ketones,[10] as illustrated below.

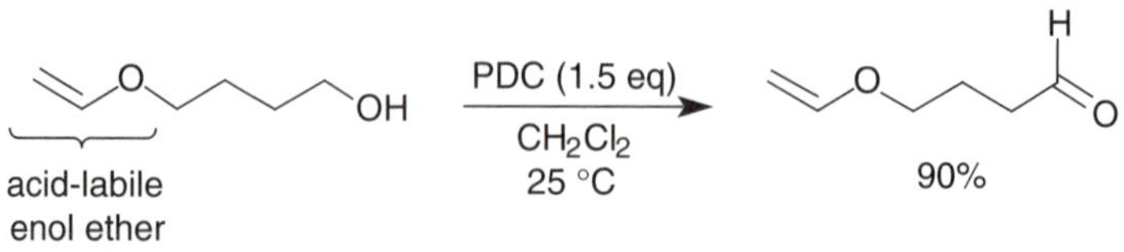

In DMF solution, however, PDC oxidizes nonconjugated primary alcohols to the corresponding carboxylic acids.[10] Oxidations of 2° alcohols by PDC in DMF proceed to give the corresponding ketones.

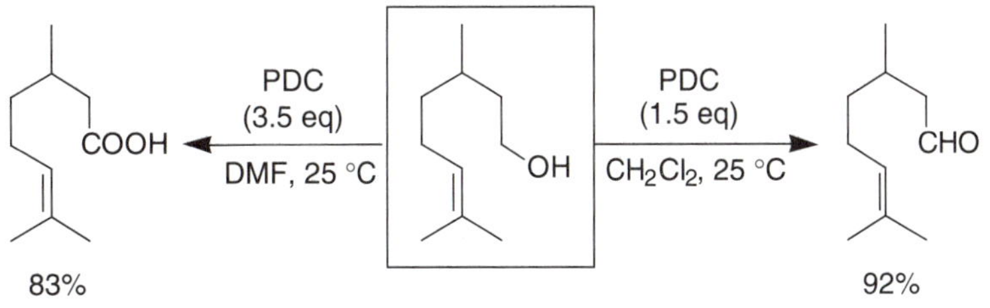

83%   92%

Swern Oxidation

Activation of dimethylsulfoxide (DMSO) by oxalyl chloride produces intermediate A (below) that decomposes rapidly at –78 °C to furnish chlorodimethylsulfonium chloride (B) along with CO and CO_2. Reaction of B with RCH_2OH (or R_2CHOH) leads to intermediate C, which upon addition of Et_3N and warming affords the corresponding carbonyl compound, Me_2S and $Et_3NH^+Cl^-$.[11]

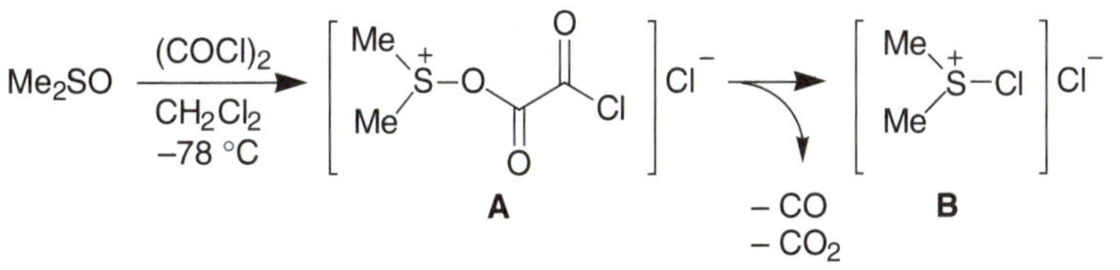

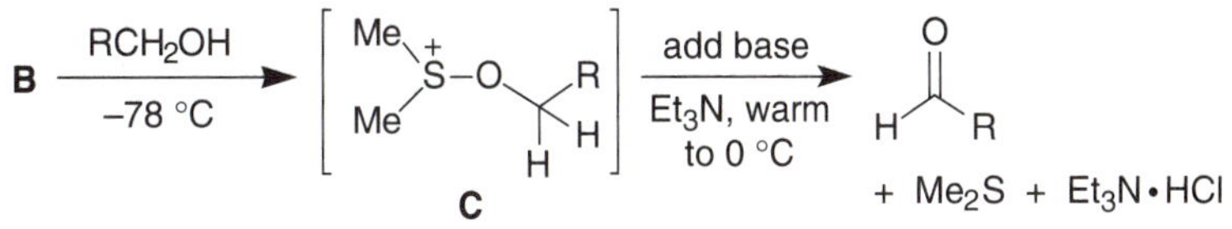

The Swern oxidation proceeds rapidly at low temperatures and thus can be employed for the preparation of α-keto aldehydes and acylsilanes, which are hyperactive carbonyl compounds and prone to hydration, polymerization, and air oxidation.[12]

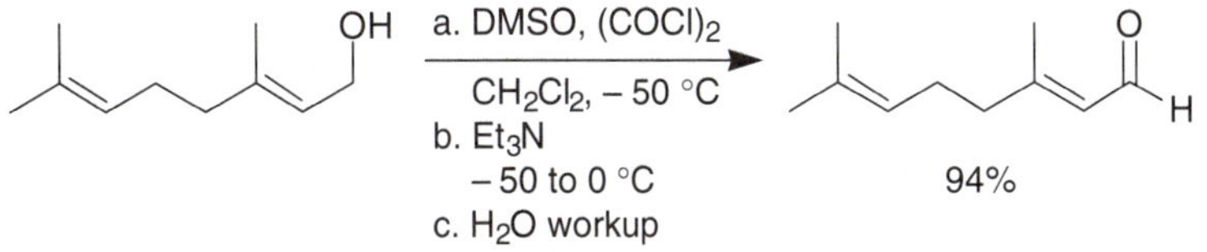

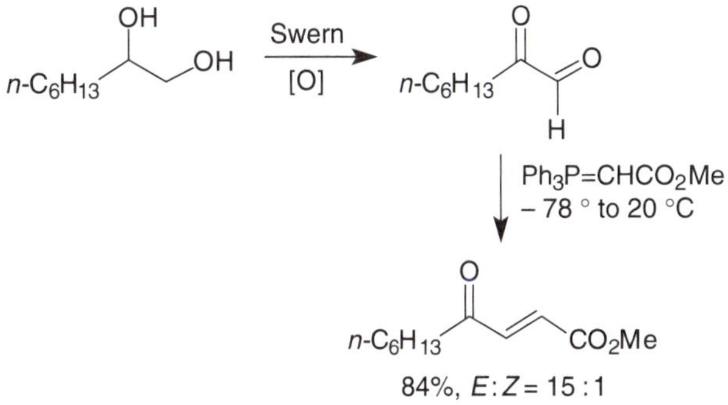

If formation of chlorinated side products is a problem, the Swern oxidation can be performed with DMSO, P_2O_5, and Et_3N.[13]

Dess-Martin Periodinane (DMP) Oxidation[14]

Oxidation of 2-iodobenzoic acid with Oxone (2 $KHSO_5$–$KHSO_4$–K_2SO_4) furnishes the oxidizing agent o-iodooxybenzoic acid, IBX,[15] a periodinane. (Note: IBX is reported to be explosive when heated > 130 °C.) Acetylation of IBX with Ac_2O in the presence of a catalytic amount of TsOH produces the Dess-Martin periodinane, DMP, 1,1,1-triacetoxy-1,1-dihydro-1,2-benziodoxol-3(1H)-one, in high yield.[16]

The Dess-Martin oxidation of alcohols has proven to be an efficient method for the conversion of primary and secondary alcohol to aldehydes and ketones, respectively. The rate of oxidation is markedly accelerated in the presence of water.[16b] The oxidation proceeds under mild reaction conditions and is especially suitable for multifunctional substrates containing acid-sensitive groups, as exemplified below.[17]

The DMP reagent has several advantages over Cr(VI)- and DMSO-based oxidizing reagents. These include the use of DMP in a near 1 to 1 stoichiometry, relative ease of preparation, shorter reaction times, simplified workups, and lower toxicity.

IBX, the precursor of DMP, is a valuable oxidant of functionalized alcohols and nitrogen- and sulfur-containing substrates when dissolved in DMSO.[18] In contrast to the oxidative cleavage observed with 1,2-diols by the DMP reagent, IBX converts glycols to α-ketols or α-diketones[19a] and 1,4-diols to lactols.[19b]

Tetrapropylammonium Perruthenate (TPAP)

TPAP ($Pr_4N^+RuO_4^-$) is an air-stable oxidant for primary and secondary alcohols.[20] It is commercially available and environmentally friendly since it is used in catalytic amounts in the presence of a co-oxidant such as N-methylmorpholine-N-oxide (NMO).

It is important that the H_2O produced in the reaction or any H_2O derived from hydrated NMO be removed with molecular sieves. Oxidations can be performed on small or large scales. For large-scale reactions it is important to moderate the reaction by slow addition of NMO while cooling.

TPAP tolerates a wide variety of functional groups, including double bonds, enones, halides, epoxides, esters, and lactones. Protecting groups, such as MEM, trityl, silyl and benzyl ethers, THP, and acetals, are not affected.

4.3 CHEMOSELECTIVE AGENTS FOR OXIDIZING ALCOHOLS

Activated Manganese Dioxide

$$MnCl_2 + KMnO_4 + H_2O \longrightarrow MnO_2 \text{ (ppt)}$$

MnO_2 is a highly chemoselective oxidant—allylic, benzylic, and propargylic alcohols are oxidized faster than saturated alcohols.[21] The oxidation takes place under mild conditions in H_2O, acetone, or $CHCl_3$.

The disadvantages of oxidations with manganese dioxide include the large excess of MnO_2 that is typically required for complete conversion of the starting material (it is not unusual to use > 40 equivalents), the long reaction times, and the difficulty associated with obtaining highly activated MnO_2.

Barium Manganate[22]

$$KMnO_4 + BaCl_2 + NaOH + KI \longrightarrow Ba[MnO_4]_2 \text{ (ppt)}$$

$Ba[MnO_4]_2$ possesses similar chemoselectivities as MnO_2 in oxidations of alcohols, but it is more readily available and does not require special treatment for its activation.[23]

Silver Carbonate on Celite (Fetizon's Reagent)

Silver carbonate is especially useful for small-scale oxidations since the products usually are recovered in high purity by simply filtering the Ag° and evaporating the solvent.[24] The ease of alcohol oxidation follows the trend: allylic, benzylic–OH > 2° ROH > 1° ROH.[24a] Highly hindered –OH groups are not oxidized. Oxidation of allylic α, ω-diols gives high yields of five- or six-member ring lactones.[24b]

Silver carbonate has been used to oxidize lactols (hemiacetals) to lactones.[25]

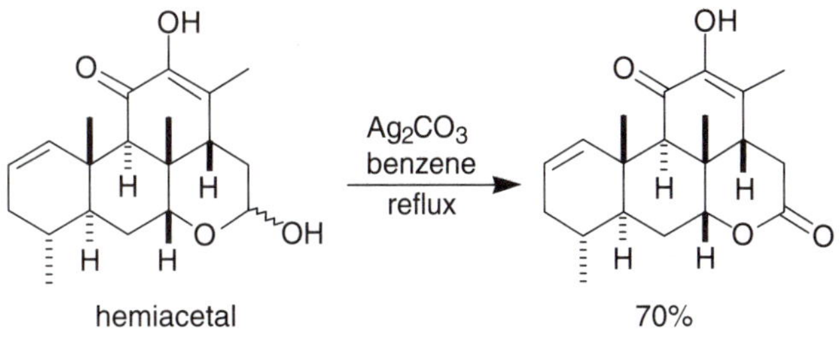

hemiacetal   70%

2,2,6,6-Tetramethyl-1-piperidinyloxy (TEMPO)

TEMPO is a commercially available nitroxyl radical-containing reagent that catalyzes the oxidation of primary and secondary alcohols in conjunction with co-oxidants (oxygen, hypochlorite, bromite, hypervalent iodine, or peroxy acids).[26] The catalyst is particularly useful for the oxidation of optically active α-alkoxy- or α-amino alcohols to the corresponding aldehydes without loss of enantiomeric purity.[27]

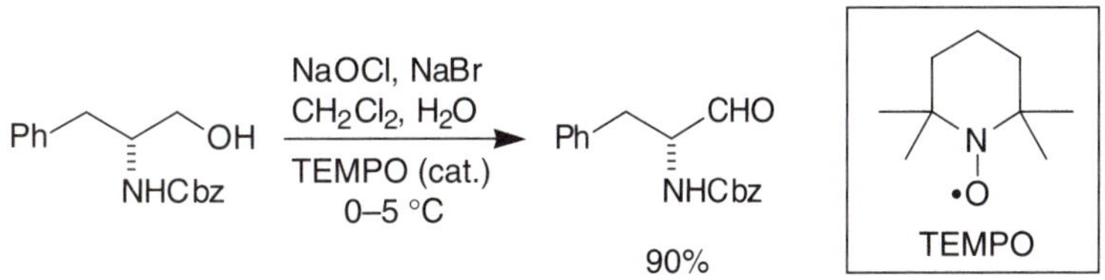

TEMPO-catalyzed oxidations are chemoselective for primary alcohols.[28] Secondary alcohols are oxidized slowly, as illustrated below.[28b]

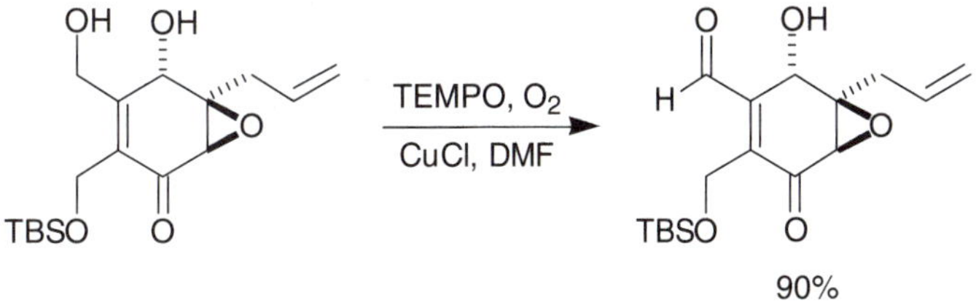

Polymer-supported TEMPO is a metal-free catalyst for chemoselective oxidations. It is readily removed from the reaction mixture and can be reused several times.[29]

Ceric Ammonium Nitrate

Chemoselective oxidation of a secondary OH group in the presence of a primary OH group has been achieved with $(NH_4)_2Ce(NO_3)_6$, $NaBrO_3$.[30] Note, however, that the reagent does not tolerate the presence of double bonds.

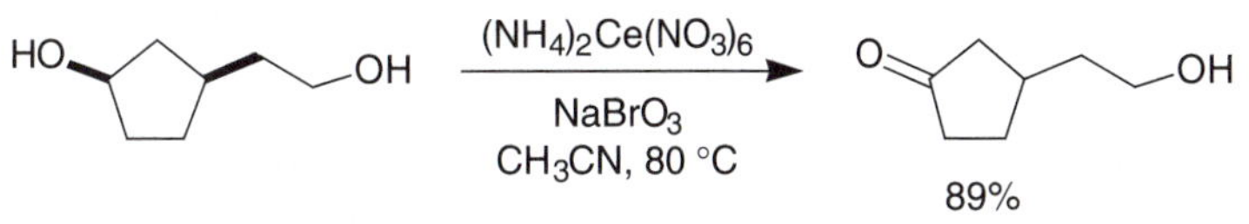

Triphenylcarbenium Tetrafluoroborate

Triphenylcarbenium salts ($Ph_3C^+X^-$) selectively oxidize secondary t-butyl or triphenylmethyl (trityl) ethers derived from alcohols.[31] The oxidation proceeds via initial hydride abstraction followed by loss of the group on oxygen.[32] The secondary-over-primary selectivity results from preferential formation of an oxocarbenium ion intermediate at the secondary center (R_2^+C–OTr is formed faster than RH^+C–OTr).

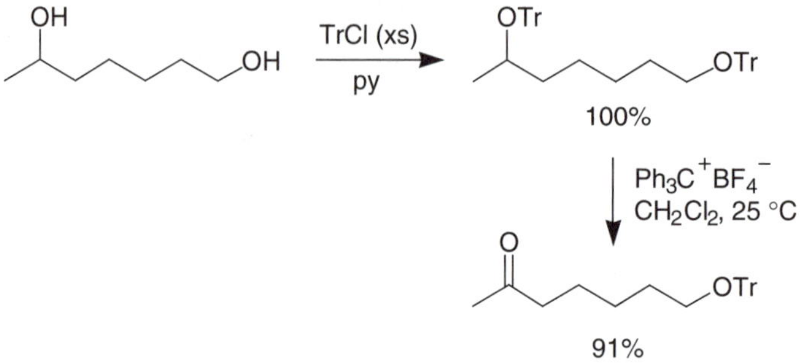

100%

Ph₃C⁺BF₄⁻
CH₂Cl₂, 25 °C

91%

Sodium Hypochlorite

Chemoselective oxidation of a secondary OH group in the presence of a primary OH group is possible with NaOCl in aqueous acetic acid.[33]

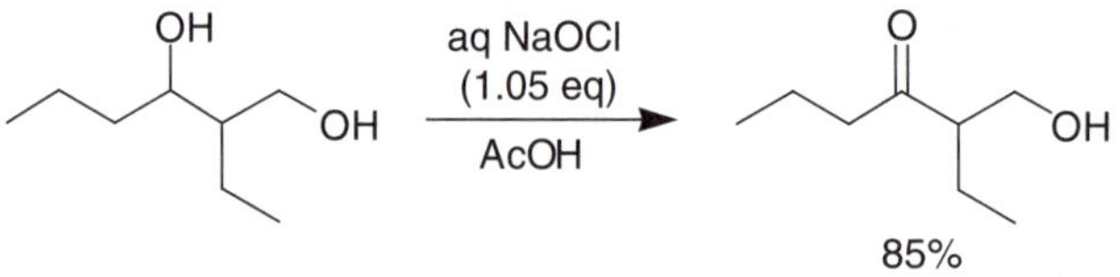

aq NaOCl
(1.05 eq)

AcOH

85%

4.4   OXIDATION OF ACYLOINS

Many oxidants transform acyloins (α-hydroxy ketones) into α-diketones. Copper and bismuth oxidants are commonly selected for this transformation.

Cupric Acetate

Cu(II) salts (e.g., Cu(OAc)₂ or CuSO₄) in stoichiometric amounts convert acyloins into diketones.[34] Catalytic amounts of cupric acetate can be used in conjunction with ammonium nitrate.

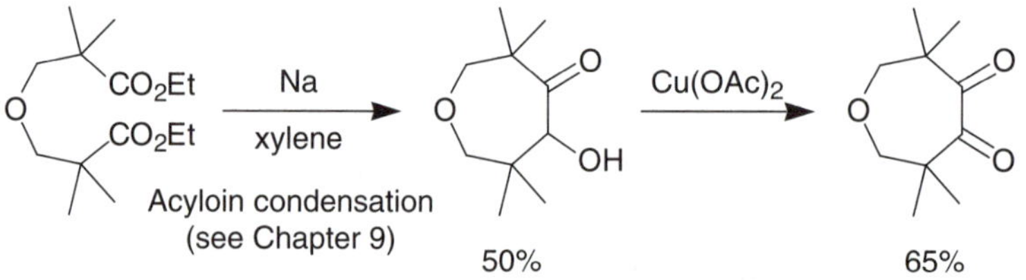

Na
xylene

Acyloin condensation
(see Chapter 9)

Cu(OAc)₂

50%   65%

Bismuth Sesquioxide

Commercially available Bi₂O₃, in the presence of acetic acid, oxidizes acyloins to α-diketones in good yields.[35]

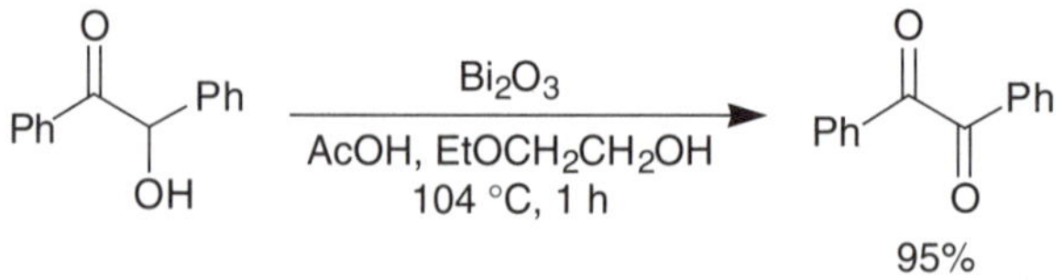

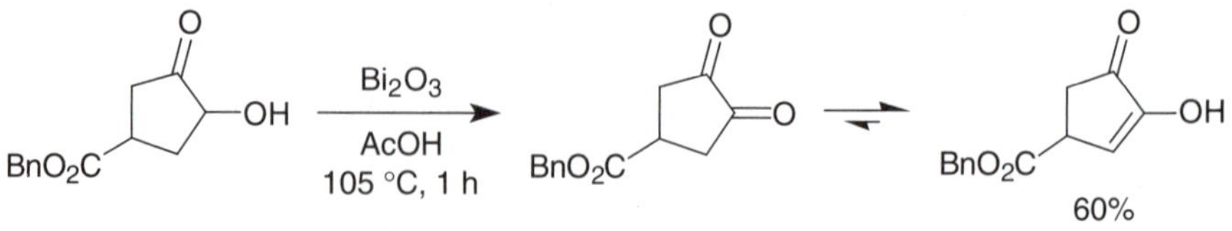

Bi₂O₃

AcOH, EtOCH₂CH₂OH
104 °C, 1 h

95%

Bi₂O₃

AcOH
105 °C, 1 h

60%

The required α-hydroxy ketones are accessible via a variety of synthetic methods, including the acyloin condensation (see Chapter 9), oxidation of ketone enolates,[36a] oxidation of enol ethers,[36b] and oxidation of α, β-unsaturated ketones,[37] as depicted below.

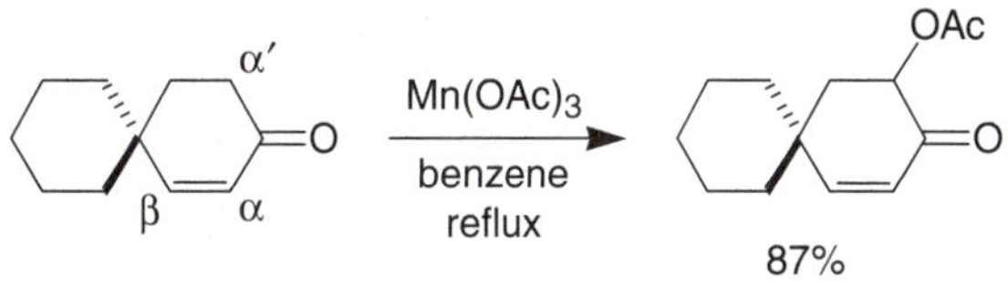

4.5   OXIDATION OF TERTIARY ALLYLIC ALCOHOLS (THE BABLER OXIDATION)[38]

A *carbonyl transposition* can be effected via the addition of a vinyl or an alkyl Grignard reagent to an α,β-unsaturated ketone. Acid-catalyzed rearrangement of the resultant allylic alcohol during oxidation with PCC affords the transposed α,β-unsaturated carbonyl substrate. This reaction represents a useful alternative when Wittig olefination of the ketone is problematic.

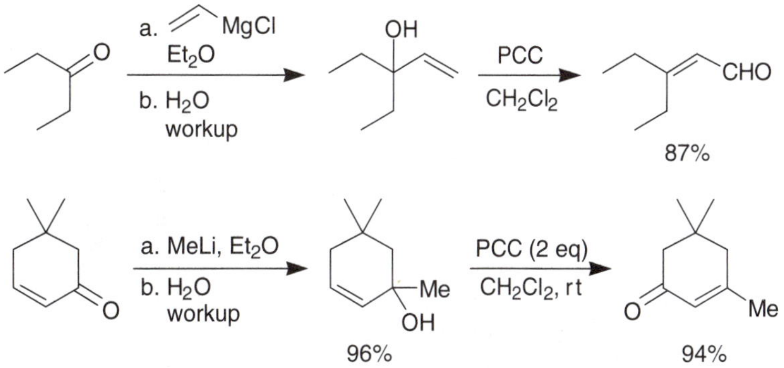

Tertiary *bis(allylic)* alcohols are oxidized by PCC or PDC to the carbonyl transposed dienones.[39]

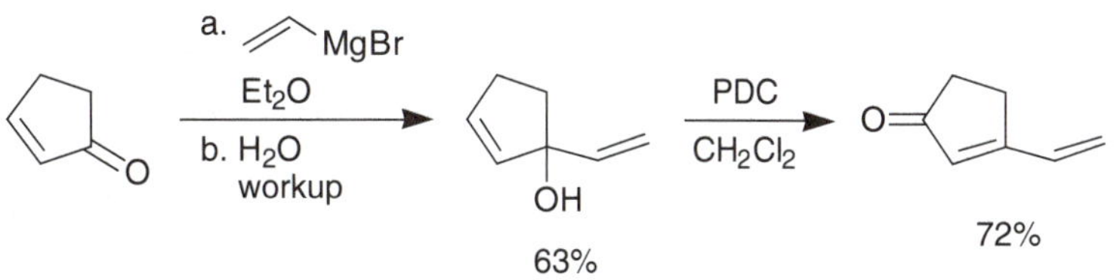

Addition of silica gel (SiO$_2$) to the PCC reaction greatly facilitates the workup, and application of ultrasound enhances the rate of the reaction and the yield of the product.[40]

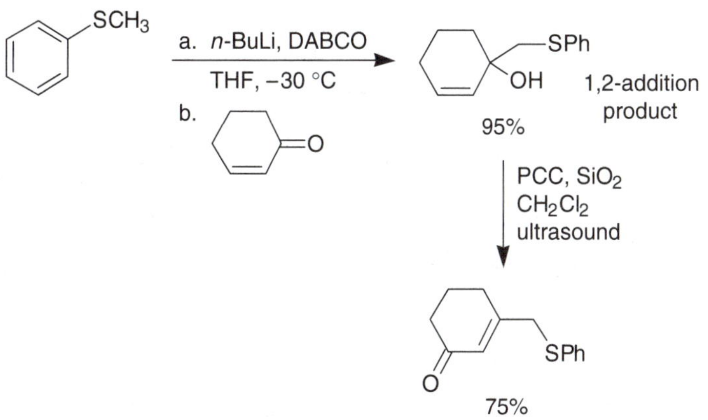

4.6   OXIDATIVE PROCEDURES TO CARBOXYLIC ACIDS

Oxidation of Aldehydes to Carboxylic Acids

Many of the oxidants employed to prepare aldehydes from primary alcohols may be used to further oxidize the aldehyde initially formed to the corresponding carboxylic acid. The most common oxidants for this purpose include $KMnO_4$,[41] chromic acid, sodium chlorite,[42] silver oxide,[43] and PDC in DMF.[10]

Oxidation of Aldehydes to Carboxylic Acid Esters[44]

α,β-*Unsaturated* aldehydes are converted directly to carboxylate esters by MnO_2 and NaCN in an alcohol solvent.[44a] Sodium cyanide catalyzes the oxidation by forming a cyanohydrin that is susceptible to MnO_2 oxidation. Methanolysis of the acyl cyanide intermediate in the example below gives the methyl ester in excellent yield.

The oxidation of allylic alcohols directly to methyl carboxylates has been reported using MnO_2 and sodium cyanide in methanol.[44a]

Aldehydes dissolved in alcohols react with 1 to 2 equivalents of aqueous $NaOCl$[33] or Oxone (a potassium triple-salt that contains potassium peroxymonosulfate, $KHSO_5$)[44b] to furnish the corresponding esters.

Oxidation of Terminal Alkynes

Since 1-alkynes are relatively resistant to oxidation, strong oxidants are required to effect oxidative cleavage of terminal alkynes to carboxylic acids with loss of one carbon. Iodosylbenzene in combination with Ru catalysts[45a] or potassium permanganate[45b] cleaves 1-alkynes to carboxylic acids. Phase transfer agents (quaternary ammonium salts[46]) are used in the $KMnO_4$ oxidations to overcome problems associated with the low solubility of permanganate in nonpolar solvents.

Conversion of 1-alkynes into substituted acetic acids *without the loss of one carbon* is accomplished via hydroboration-oxidation, as exemplified below.[47]

Ruthenium- or permanganate-mediated oxidations of internal alkynes are highly dependent on solvent conditions and generally afford the corresponding α-diketones.[48]

4.7   ALLYLIC OXIDATION OF ALKENES

Selenium Dioxide

Alkenes possessing allylic C–H bonds are oxidized by SeO_2 either to allylic alcohols or esters or to α,β-unsaturated aldehydes or ketones, depending on the experimental conditions.[49] The reaction involves an *ene*-type reaction (A) followed by a *sigmatropic* [2,3]-shift (B) to give the selenium ester (C), which is converted to the corresponding allylic alcohol (D) on solvolysis.[50]

If the alkene possesses a methyl substituent, oxidative cleavage of C (elimination of a selenium atom and water) furnishes the corresponding α,β-unsaturated aldehyde.[50b] Lower yields of products were obtained when using stoichiometric amounts of SeO_2. *t*-Butyl hydroperoxide is used to reoxidize selenium. Besides giving the desired carbonyl compounds, the reaction may also produce allylic alcohols as side products.

Chromic Anhydride[51] Reaction of chromic anhydride (CrO_3) with *t*-butanol yields *t*-butyl hydrogen chromate, a powerful oxidant suitable for allylic oxidation of electron-deficient alkenes.[51b] Oxidations using t-BuOCrO$_3$H in CCl_4 are highly exothermic and should be performed with caution.

PCC and CrO_3 pyridine complexes oxidize allylic and benzylic methylene groups of *electron-rich* π systems to the carbonyl groups.[52]

t-Butyl Peroxybenzoate[53]   Copper(I) salts catalyze the allylic oxidation of alkenes in the presence of peresters, such as *tert*-BuO$_2$COPh, to afford the corresponding allylic benzoate esters.[53b] In the case of terminal alkenes, internal allylic esters are formed in preference over the terminal isomers. The mechanism is believed to involve addition of an allylic radical to copper(II) benzoate.[54] Rearrangement of the copper(III) intermediate then produces the product and regenerates the copper(I) catalyst.

Asymmetric allylic oxidation of alkenes using peresters is possible when the ligand L$_n$ of the Cu(III) intermediate is chiral. Copper complexes of chiral bis(pyridine)- and bis(oxazoline)-type ligands have been used with *tert*-butyl perbenzoate to obtain optically active allylic benzoates.[55]

4.8 TERMINOLOGY FOR REDUCTION OF CARBONYL COMPOUNDS

Most reductions of carbonyl groups are now done with reagents that transfer a hydride ion from a Group III atom. Over the years, a large number of reducing agents have become available that provide chemo-, stereo-, and enantioselectivity in reductions.[56] Since formation of a new stereogenic center during carbonyl group reduction may require either relative or absolute stereocontrol, the use of chiral reducing agents has become a popular means for controlling the stereochemical outcome of reductions. Below is a brief review of terminology associated with regio- and stereochemical issues of carbonyl reduction.[57]

- A *chemoselective reagent* reacts selectively with one functional group in the presence of other functional groups.

- In a *regioselective reaction*, the reagent adds at only one of two possible regions (directions).

- A *stereoselective reaction* leads to the exclusive or predominant formation of one of several possible stereoisomeric products. Thus, one reaction pathway from a given substrate is favored over the other (as in nucleophilic additions to cyclic ketones or alkylations of enolate ions).

trans-isomer
(as major product)

- In a *stereospecific reaction,* a given substrate isomer leads to one product while another stereoisomer leads to the opposite stereoisomeric product. Thus, the starting material specifies the stereochemical outcome of the reaction (as in S_N2 reactions and epoxidation of *cis*- and *trans*-alkenes). All stereospecific reactions are necessarily stereoselective.

meso-isomer
(as sole dibromide)

- A *prochiral center* is a trigonal carbon of C=O and C=C that is not a stereogenic center but can be made chiral by addition reactions.

- A *stereogenic carbon atom* (chiral center, chiral atom, asymmetric atom) is bound to four unlike groups and thus generates chirality. Note that a molecule may possess a molecular chirality without having a stereogenic center.
- *Stereoisomers* have the same molecular formula, but their atoms have a different spatial arrangement — a different *configuration*. Stereoisomers are classified as

-*Enantiomers* when the isomers *are* mirror images of each other and cannot be superimposed

-*Diastereoisomers* when the isomers *are not* mirror images of each other (this includes alkene *E, Z* isomers). Diastereoisomers possess different physical and chemical properties.

- *Asymmetric induction* is the preferential formation of one enantiomer or diastereomer over another due to the influence of a chiral center (or chiral element) either in the substrate, reagent, catalyst, or solvent of the reaction.

- In an *enantioselective reaction,* an achiral substrate is converted selectively to one of two enantiomers by a chiral reagent or catalyst. This process is an example of asymmetric induction.

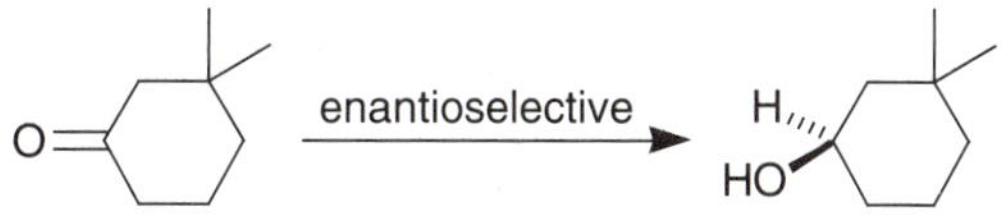

- *Enantiomeric excess* or % ee (or % optical purity) = [observed specific rotation] divided by [specific rotation of pure enantiomer] $\times$ 100. Note that there are examples where the linear relationship between enantiomeric excess and optical rotation fails.[58] A percent enantiomeric excess (% ee) of less than 100% indicates that the compound is *"contaminated"* with the other enantiomer. The *ratio* of enantiomers in a sample of known (measured) optical purity may be calculated as follows: fraction of the major isomer = [(% ee) + 0.5 (100 – % ee)]. Thus, 86% ee represents a sample consisting of a 93:7 ratio of enantiomers [major enantiomer = 86 + 0.5 (14)].

- *Diastereomeric excess* or % de = [% of major diastereomer – % of minor diastereomer].

- A *racemate,* or racemic mixture, is an equimolar mixture of two enantiomers.[59]

- Objects and molecules are said to be *homochiral* when they possess the same sense of chirality.[60] For example, L-alanine and its methyl ester derivative shown below are said to be homochiral. *This term should not be used to signify that a compound is enantiomerically pure.*

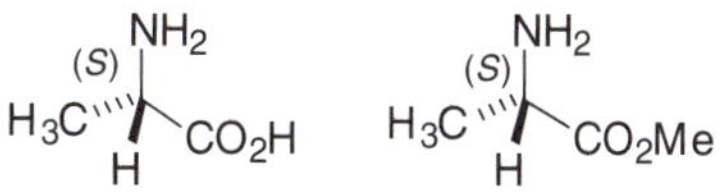

4.9 NUCLEOPHILIC REDUCING AGENTS

The majority of reductions of carbonyl compounds and nitriles with nucleophilic reducing agents, such as $M[AlH_4]$ and $M[BH_4]$, proceed via nucleophilic transfer of a hydrogen atom with two electrons called a "hydride" from the reducing agent to the carbonyl or cyano carbon. The rate of reduction and the chemoselectivity of a reducing agent toward a given substrate depends on factors such as

- The nature of the metal cation (Li^+, Na^+, Zn^{2+}), which serves as a Lewis acid to activate the carbonyl or cyano moiety toward hydride transfer

- Substitution of the reducing agent hydrogens by alkyl, –OR, or –CN groups

- The reaction medium (Et_2O, THF, ROH, H_2O)

- The reactivity order of substrates is: $RCHO > R_2CO > RCO_2R' > RCONR_2 > RCO_2H$

Aluminum Hydrides

Lithium Aluminum Hydride–LiAlH₄

Lithium aluminum hydride (LAH) is a powerful reducing agent but is not very chemoselective. It must be used in *nonprotic* solvents such as Et_2O or THF; and one is advised to determine its concentration by hydrolysis.[61] The reagent is usually used in excess, with only three hydrides of $LiAlH_4$ being utilized. To decompose any excess Al–H, first add ethyl acetate, followed by methanol and then H_2O.[62] If feasible, use acidic conditions for workup to bring the resultant precipitate $(Al(OH)_3)$ into solution. Also, addition of Na-K tartrate (*Rochelle's salt*)[62c] or $N(CH_2CH_2OH)_3$[62d] during workup may be helpful.

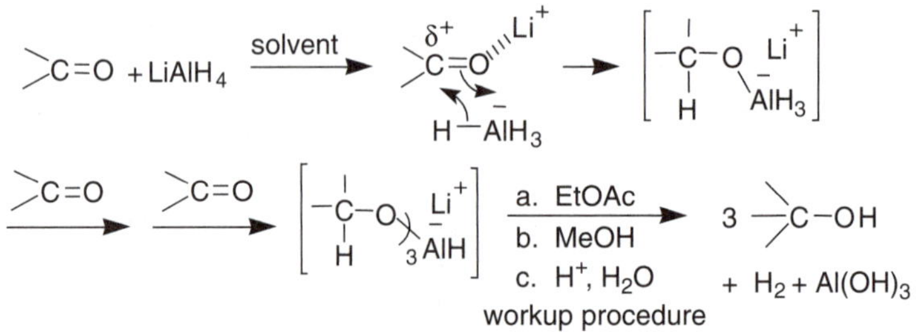

As depicted above, the mechanism is believed to involve a hydride transfer from the aluminate species onto the carbonyl carbon. The resultant alkoxide ion coordinates the remaining aluminum hydride to form an alkoxytrihydroaluminate ion, which is capable of reducing the next carbonyl group.[63]

The reduction of esters to primary alcohols and the reduction of amides to amines requires two hydrides, whereas reduction of carboxylic acids to primary alcohols consumes three hydrides (Table 4.2).

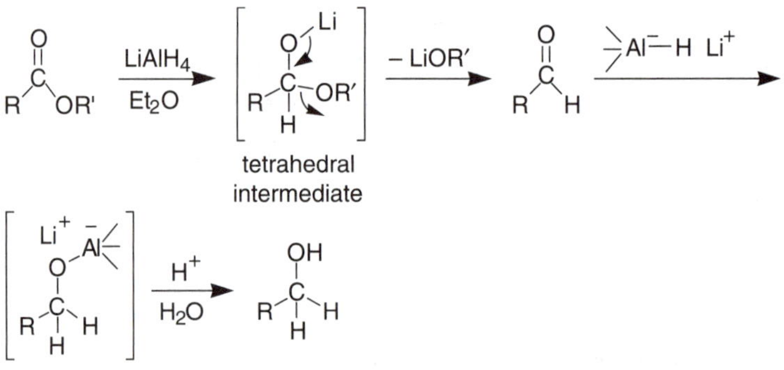

Table 4.2	Hydride Consumption in $LiAlH_4$ Reductions	
Compound	**Hydrides consumed**	**Product**
RCHO	1	RCH_2OH
$R_2C=O$	1	R_2CHOH
RCO_2H	3 ($-H_2$)	RCH_2OH
RCO_2R'	2	$RCH_2OH + R'OH$
$RCONR'_2$	2	$RCH_2NR'_2$
$RC{\equiv}N$	2	RCH_2NH_2
Lactone	2	α,ω-Diol
Epoxide	1	Alcohol
C=C (isolated)	No reaction	

The reactivity of LAH may be tempered by the addition of dialkyl amines. For example, both aliphatic and aromatic carboxylic esters are reduced to the corresponding aldehydes by LAH in the presence of diethylamine at room temperature.[64]

$$\text{CO}_2\text{Et} \xrightarrow[\text{THF}]{\begin{array}{c}\text{LiAlH}_4 \text{ (1 eq)}\\ \text{Et}_2\text{NH (2 eq)}\end{array}} \text{CHO}$$

93%

Lithium Trialkoxyaluminum Hydride—Li[AlH(OR)₃][65,66]

The reactivity and selectivity of LAH can be modified by replacing three of its hydrides with *t*-butoxy or ethoxy groups. The resulting reagents are *less* reactive but more selective than LAH and are best prepared just prior to use in situ.

$$\text{LiAlH}_4 + 3\ t\text{-BuOH} \xrightarrow{\text{THF}} \text{Li[AlH(O}t\text{-Bu)}_3] + 3\ \text{H}_2$$

lithium tri-*t*-butoxyaluminum hydride

$$\text{LiAlH}_4 + 3\ \text{EtOH} \xrightarrow[\text{or Et}_2\text{O}]{\text{THF}} \text{Li[AlH(OEt)}_3] + 3\ \text{H}_2$$

lithium triethoxyaluminum hydride

Lithium tri-t-butoxyaluminum hydride readily reduces aldehydes and ketones to the corresponding alcohols and reduces acid chlorides to aldehydes. Epoxides, esters, carboxylic acids, *tert*-amides, and nitriles are not, or only slowly, reduced. Thus, the reagent may be used for chemoselective reductions.[65]

Note that before lithium trialkoxyaluminum hydrides became available, acid chlorides were converted to aldehydes using the more tedious *Rosenmund* procedure (Pd/BaSO₄, H₂, quinoline, sulfur).[67]

Lithium triethoxyaluminum hydride is a more powerful reagent which reduces *tert*-amides and nitriles to the corresponding aldehydes.[65]

Sodium Bis(2-methoxyethoxy)aluminum Hydride—Na[AlH₂(OCH₂CH₂OCH₃)₂][66]

Sodium bis(2-methoxyethoxy)aluminum hydride, or Red-Al, is a versatile, commercially available reducing agent. It is thermally more stable than LAH and may be used in aromatic hydrocarbon as well as in ether solvents. Overall, the reducing properties of Red-Al are similar to those of LAH (reductions of aldehydes, ketones, esters, etc.).

$$CH_2=CH(CH_2)_8CO_2Et \xrightarrow[\text{b. } H^+, H_2O \text{ workup}]{\substack{\text{a. Red-Al (1.1 eq)} \\ \text{toluene}}} CH_2=CH(CH_2)_8CH_2OH$$

$$> 95\%$$

A notable difference in Red-Al selectivity relative to LAH involves the reduction of nitriles. Aliphatic nitriles do not react with Red-Al.

Borohydrides [56,65] The nucleophilic borohydrides—sodium borohydride [$NaBH_4$], lithium borohydride [$LiBH_4$], zinc borohydride [$Zn(BH_4)_2$], lithium- and potassium trialkylborohydrides [Li-, K-R_3BH], and sodium cyanoborohydride [$NaBH_3CN$]—exhibit, depending on the metal cation and the ligands, characteristic reducing properties toward functional groups. All borohydrides reduce aldehydes and ketones; reduction of esters and carboxylic acids to primary alcohols requires specific reagents.

Sodium Borohydride

Sodium borohydride is a mild, selective reducing agent, and its handling does not require special precautions. The reagent is insoluble in Et_2O, very slightly soluble in THF and DME, but soluble in diglyme and hydroxylic solvents. EtOH is usually the solvent of choice. $NaBH_4$ reduces RCHO and R_2CO in EtOH or aqueous solutions rapidly at 25 °C to the corresponding alcohols. Esters are slowly reduced to RCH_2OH.[68]

71%

$NaBH_4$ in the presence of I_2 reduces carboxylic acids to the corresponding primary alcohols.[69] This approach is especially attractive for the conversion of amino acids to amino alcohols, which are important ligands in asymmetric synthesis.[70]

73%

Lithium Borohydride

Lithium borohydride is a commercially available, solid reagent that rapidly decomposes when exposed to moist air. It is conveniently prepared from $NaBH_4$ and LiBr in Et_2O or in THF.[71] The NaBr precipitates as it is formed, and the clear supernatant solution is used as such for reductions after determination of its hydride concentration.[61]

$$NaBH_4 + LiBr \xrightarrow{Et_2O} LiBH_4 + NaBr \text{ (ppt)}$$

The Li^+ cation is a stronger Lewis acid than the Na^+ cation. Li^+ coordination with the carbonyl group enhances the electrophilicity of the carbonyl carbon, thereby facilitating hydride transfer. Lithium borohydride is a more powerful reducing agent than sodium borohydride: it reduces esters to primary alcohols but is unreactive towards amides.[72]

$$n\text{-}C_{15}H_{31}\underset{\underset{\displaystyle OH}{|}}{CH}\text{---}\underset{\underset{\displaystyle NHCOCH_3}{|}}{CH}\text{---}CO_2Et \quad \xrightarrow[\text{THF}]{\text{LiBH}_4} \quad n\text{-}C_{15}H_{31}\underset{\underset{\displaystyle OH}{|}}{CH}\text{---}\underset{\underset{\displaystyle NHCOCH_3}{|}}{CH}\text{---}CH_2OH$$

Zinc Borohydride

Zinc borohydride is not commercially available; it is prepared from anhydrous $ZnCl_2$ and $NaBH_4$ solutions in ether solvents. The chloride-free supernatant solutions then are used as required.

$$2\,NaBH_4 + ZnCl_2 \quad \xrightarrow{\text{THF}} \quad Zn(BH_4)_2 + 2\,NaCl\ (ppt)$$

Zinc borohydride has some interesting properties: it is less basic than $NaBH_4$ and thus it is especially suitable for the reduction of base-sensitive compounds. Also, the zinc cation has a better coordinating ability than either Na^+ or Li^+, making $Zn(BH_4)_2$ often the reagent of choice for chelation-controlled, stereoselective reductions of acyclic ketones (see Section 4.12).

Besides aldehydes and ketones, $Zn[BH_4]_2$ reduces aliphatic esters and carboxylic acids.[73] Moreover, it is an effective *hydroborating agent* for alkenes, dienes, and alkynes.[73a]

L-Leucine → L-Leucinol 85%

90%

Lithium or Potassium Trialkylborohydride[56c,e]

The presence of three alkyl groups in lithium trialkylborohydrides imparts increasing nucleophilicity to the hydride, making them more powerful reducing agents than lithium borohydride itself. Aldehydes, ketones, and esters are rapidly and quantitatively reduced to the corresponding alcohols even at $-78\ °C$. The most frequently used trialkylborohydrides are lithium triethylborohydride (Superhydride) and lithium and potassium tri-*sec*-butylborohydride (L- and K-Selectride). These reagents are commercially available as solutions in ether solvents and must be handled in the absence of air. The trialkylboron species formed from these reagents should be oxidized on completion of the reduction, prior to workup, to the correponding alcohols with alkaline hydrogen peroxide.

Lithium triethylborohydride ($LiEt_3BH$) is a super-nucleophile that reduces primary alkyl bromides[74] and tosylates[75] more effectively to the corresponding hydrocarbons than does $LiAlH_4$. Epoxides are readily cleaved to give alcohols by attack of the hydride at the less substituted carbon.[76]

The sterically encumbered L- and K-Selectrides also reduce aldehydes and ketones rapidly and quantitatively to the corresponding alcohols. One of the remarkable features of these reagents is their unusual ability to introduce major steric control in the reduction of cyclic ketones. Their applications as highly selective reducing agents will be discussed later in this chapter.

Sodium Cyanoborohydride—NaBH$_3$CN

Because of the presence of the electron withdrawing cyano group, NaBH$_3$CN is less nucleophilic and hence is more selective than NaBH$_4$.[77]

The utility of NaBH$_3$CN as a reducing agent is greatly enhanced by its stability toward low pH (stable to pH 3). Thus, the reagent permits reductions under conditions that would rapidly hydrolyze NaBH$_4$. NaBH$_3$CN is soluble in a variety of solvents: H$_2$O, ROH, THF, but it is insoluble in Et$_2$O and hydrocarbon solvents. Under neutral conditions in H$_2$O or MeOH, reduction of RCHO and R$_2$CO is negligible. However, in acidic solution, carbonyl reduction does occur (protonated carbonyl group).

$$\text{C=O} + \text{NaBH}_3\text{CN} \xrightarrow[\text{HCl}]{\text{MeOH}} \text{H--C--OH} + \text{B(OMe)}_3 + \text{HCN (toxic)} + \text{NaCl}$$

NaBH$_3$CN is a chemoselective reducing agent. For example, it is possible to selectively reduce an aldehyde group in the presence of a keto group or a keto group in the presence of an ester group using NaBH$_3$CN. Even under diverse reaction conditions, functional groups such as RCONH$_2$, –C≡N, and –NO$_2$ are inert toward the reagent.

Reductive Amination with NaBH₃C≡N[78]

Since the reduction of an iminium salt by NaBH₃CN occurs more readily than the reduction of a carbonyl group, NaBH₃CN is the reagent of choice for the reductive amination of aldehydes and ketones.[79] The reaction entails condensation of the carbonyl compound with NH₃, RNH₂, or R₂NH at pH 5 to 8 to give the corresponding iminium salts. These are reduced in situ with NaBH₃CN to furnish primary, secondary, or tertiary amines, respectively.

Sodium triacetoxyborohydride Na[BH(OAc)₃][80] and hydrogenation (H₂, Pd/C)[81] are used as alternatives to sodium cyanoborohydride for the reductive amination of carbonyl compounds. Also, Zn[BH₄]₂ is a particularly effective agent for the reductive amination of α,β-unsaturated aldehydes and ketones.[82]

Deoxygenation via Tosyl Hydrazone Reduction

In situ reduction of tosylhydrazones by NaBH₃CN provides an efficient method for the deoxygenation of carbonyl compounds to furnish the corresponding hydrocarbons (see also Section 3.4).[83] In the case of tosylhydrazones derived from α,β-unsaturated carbonyl compounds, the reduction leads to a stereoselective migration of the double bond to give the corresponding *trans*-alkene.

a tosyl hydrazone

70%
stereoselective
for *trans*-alkene

4.10 ELECTROPHILIC REDUCING AGENTS

Whereas nucleophilic reducing agents react fast with electron-deficient carbonyl groups, the reactivity of electrophilic reducing agents such as R₂AlH and BH₃, characterized by their coordination with the carbonyl oxygen prior to hydride transfer, favor reductions of electron-rich carbonyl groups.

Diisobutylaluminum Hydride (DIBAL-H)

Diisobutylaluminum hydride (DIBAL-H) can be obtained commercially neat or in hydrocarbon solvents. Reductions with DIBAL-H must be carried out in the absence of air and moisture. DIBAL-H is a very versatile reagent for the selective reduction of appropriately substituted esters or nitriles to the corresponding aldehydes and for the reduction of lactones to lactols.[84]

Reduction of Esters to Aldehydes

In contrast to LAH reductions of esters, nitriles and lactones, where two hydrides are utilized, reductions of these compounds by DIBAL-H at low temperature can be stopped after the transfer of one hydride to the carbonyl carbon. Hydrolytic workup of the tetrahedral intermediates furnishes the corresponding aldehydes.

The reduction of esters to aldehydes generally works best when alkoxy or amino functionality is in close proximity to the ester group, as in α- or β-alkoxy esters. A neighboring alkoxy group will stabilize the tetrahedral intermediate through chelation and prevent overreduction. DIBAL-H-mediated mono-reduction of lactones delivers the corresponding lactols (hemiacetals).[85]

Reduction of Nitriles to Aldehydes

Treatment of nitriles with one equivalent of DIBAL-H at −78 °C followed by hydrolytic workup affords the corresponding aldehydes. Reduction of nitriles with two equivalents of DIBAL-H produces, after hydrolytic workup, the corresponding primary amines. In the example below, note that the double bond is not reduced (hydroaluminated) under the reaction condition employed.

Reduction of Vinyl Esters to Allylic Alcohols

DIBAL-H is the reducing agent of choice for regioselective reduction of α,β-unsaturated esters to allylic alcohols.[86] Horner-Wadsworth-Emmons olefination (see Section 8.3b) followed by DIBAL-H reduction provides a tandem synthesis of (E)-allylic alcohols.[87]

Borane • Tetrahydrofuran and Borane • Dimethylsulfide[56a,c,e]

Diborane, B_2H_6, is a gas and is difficult to handle. However, borane complexed with donors such as THF or dimethylsulfide are commercially available and have become valuable reagents for the reduction of various functional groups. BH_3•SMe_2 is soluble in and unreactive toward a wide variety of aprotic solvents such as THF, Et_2O, CH_2Cl_2, and hydrocarbons.

An attractive feature of BH_3•THF and BH_3•SMe_2 is the facile reduction of carboxylic acids to primary alcohols.[88] The reduction involves initial formation of a triacylborate with concomitant evolution of 3 H_2, followed by fast hydride transfer to the carbonyl carbon to furnish, after workup, the corresponding primary alcohol.

Both BH_3•THF and BH_3•SMe_2 allow the selective reduction of a –COOH group in the presence of other functional groups.

A summary of reactivities of BH_3•THF toward various functional groups is shown in Table 4.3.[56a]

Reduction of saturated carboxylic acids with the borane derivative, thexylchloroborane, provides a *direct* route to aldehydes without their prior conversion to carboxylic acid derivatives.[89] The aldehydes initially formed were isolated either as their bisulfite adducts or as their hydrazones, from which the aldehydes were regenerated.

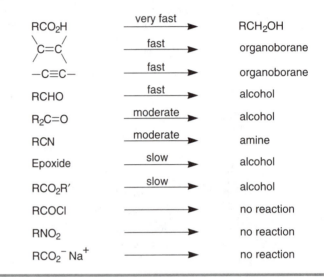

Br(CH₂)₅COOH $\xrightarrow[\text{CH}_2\text{Cl}_2]{}$ Br(CH₂)₅CHO   76%

Table 4.3	Selectivity in BH₃ • THF Reductions

RCO₂H	very fast →	RCH₂OH
>C=C<	fast →	organoborane
—C≡C—	fast →	organoborane
RCHO	fast →	alcohol
R₂C=O	moderate →	alcohol
RCN	moderate →	amine
Epoxide	slow →	alcohol
RCO₂R′	slow →	alcohol
RCOCl	→	no reaction
RNO₂	→	no reaction
RCO₂⁻ Na⁺	→	no reaction

4.11   REGIO- AND CHEMOSELECTIVE REDUCTIONS

Reductions of α,β-Unsaturated Aldehydes and Ketones

Hydride reductions of α,β-unsaturated aldehydes and ketones may proceed via 1,2- or 1,4-additions to furnish the corresponding allylic alcohols or saturated aldehydes and ketones, respectively.

i-Bu₂AlH,[84] Zn(BH₄)₂,[90] (i-PrO)₂TiBH₄,[91] 9-BBN,[92] and CeCl₃-NaBH₄ (Luche reagent)[93] favor 1,2-additions. The latter reagent is especially suited for selective reductions of α,β-unsaturated aldehydes and ketones to the corresponding allylic alcohols.

NaBH₄, LiAlH₄, and Li[AlH(OR)₃] effect both 1,2 and 1,4-additions (Table 4.4). The regioselectivity of enone reductions can be strongly influenced by the nature of the reagent, the presence of substituents on the substrate, and the reaction conditions.

Table 4.4   Reduction of 2-Cyclopentenone[94]

				Reference
LiAlH$_4$, THF, 0 °C	14	2	84	94
LiAlH(OMe)$_3$, THF, 0 °C	91	0	9	94
LiAlH(Ot-Bu)$_3$, THF, 0 °C	0	11	89	94
NaBH$_4$, EtOH, 0 °C	0	0	100	94
NaBH$_4$, CeCl$_3$, MeOH	97	0	3	93
i-Bu$_2$AlH, hexane-C$_6$H$_6$, 0 °C	99	0	1	84d
9-BBN^a, THF, 0 °C	99	0	1	92

^{a} 9-BBN: 9-borabicyclo[3.3.1]nonane.

For example, K-Selectride reduces β-*unsubstituted* cyclohexenones to cyclohexanones (1,4-addition) and β-substituted cyclohexenones to the corresponding allylic alcohols (1,2-addition).[95]

Reduction of Aldehydes in the Presence of Ketones[96,97]

A number of chemoselective reductions of the more reactive R–CHO group in the presence of an R$_2$CO group have been reported using 9-BBN•pyridine[97] or K[BH(OAc)$_3$], as exemplified below.[96b]

Zinc borohydride is a mild reducing agent permitting the reduction of aldehydes in the presence of ketones.[98] Moreover, it selectively reduces a nonconjugated keto group in the presence of a conjugated keto group.

Reduction of Ketones in the Presence of Aldehydes

There are few methods for effecting chemoselective reductions of ketones in the presence of aldehydes. $NaBH_4$–$CeCl_3$–$EtOH$-$HC(OCH_3)_3$ reduces chemoselectively a keto group in the presence of an aldehyde and an ester group.[99] The reason for the selective reduction of the keto group in the presence of the reactive aldehyde is that the CHO group undergoes preferential acetalization with $CeCl_3$ and the trimethyl orthoformate.

Reduction of Ketones in the Presence of Esters

Selective reduction of a keto group in the presence of an ester can be accomplished using either $BH_3 \bullet SMe_2$ in THF or $NaBH_4$ in ethanol and H_2O.[100]

Reduction of Carboxylic Acids in the Presence of Ketones or Esters

The facile reduction of the –COOH group by $BH_3 \bullet THF$ or $BH_3 \bullet SMe_2$ has been employed for chemoselective reductions of the carboxyl group in the presence of ester[101] or lactone[102] functionalities using a stoichiometric quantity of the borane. The carbonyl group in triacylboranes resembles the reactivity of an aldehyde or a ketone more than of an ester (ester resonance) due to electron delocalization from the acyl oxygen into the p orbital of boron.

Reduction of Esters in the Presence of Amides or Nitriles

Treatment of DIBAL-H with one equivalent of *n*-BuLi generates Li[(*i*-Bu)$_2$(*n*-Bu)AlH], a reductant that selectively reduces esters in the presence of amides or nitriles.[103]

Et$_2$N⟶CO$_2$CH$_3$

a. "Li[*i*-Bu$_2$BuAlH]" (2 eq)
   THF, hexane
b. NaBH$_4$, EtOH
c. H$^+$, H$_2$O

Et$_2$N⟶OH
83%

NC⟶CO$_2$CH$_3$

a. "Li[*i*-Bu$_2$BuAlH]" (2 eq)
   THF, hexane
b. NaBH$_4$, EtOH
c. H$^+$, H$_2$O

NC⟶OH
80%

4.12   DIASTEREOSELECTIVE REDUCTIONS OF CYCLIC KETONES[104]

Generally, the diastereoselectivity in reductions of cyclic ketones with nucleophilic hydrides is determined not only by the steric congestion of the ketone but also by the nature of the nucleophilic reducing agent, as shown by the reduction of 3-methylcyclohexanone with various nucleophilic reducing agents (Table 4.5).[105] The first entry in Table 4.5, the *Meerwein-Ponndorf-Verley* (MPV) reduction, involves treatment of a ketone with aluminum triisopropoxide [Al(OCHMe$_2$)$_3$].[106] This is an equilibration process favoring the *more stable* stereoisomer, which in the case of an alkyl-substituted cyclohexanone is the equatorial alcohol. A mild variant of the MPV reduction uses SmI$_2$.[107]

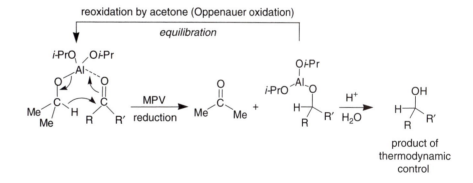

Table 4.5   Reduction of 3-Methylcyclohexanone[105]

Conditions	cis (%)	trans (%)
MPV^a (Al(O*i*-Pr)$_3$, *i*-PrOH)	70	30
LiAlH$_4$, THF	76	24
LiAlH(O*t*-Bu)$_3$, THF	90	10
NaBH$_4$, MeOH	77	23
L-Selectride (LiBH(*sec*-Bu)$_3$), THF	5	95
LTSBH,^{b} THF	< 1	> 99

^{a} Meerwein-Ponndorf-Verley reduction.
^{b} Li-tri-1,2-dimethylpropylborohydride (Li-trisiamylborohydride).

Rationalization of the stereochemical results obtained from reduction of substituted cyclohexanones with hydrides has resulted in controversy.[108] Reductions of ketones with metal hydrides are *exothermic* and according to the Hammond postulate, the transition states should be *reactant-like.*

The trajectory of nucleophilic additions to the carbonyl group brings the nucleophile over the smallest group in a nonperpendicular attack. Although perpendicular attack would result in maximum overlap between the HOMO of the nucleophile and the p-orbital at the carbonyl carbon that makes up part of the LUMO π^*, there is a significant antibonding overlap with the other p-orbital on oxygen.[109] Therefore, the best compromise is an angle of attack of ~107° (Bürgi-Dunitz trajectory).[110]

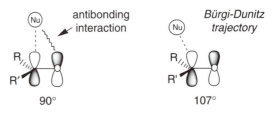

There are two possible modes of delivery of the "hydride" to the carbonyl carbon of cyclohexanone: (1) *axial* attack with formation of the *equatorial* alcohol and (2) *equatorial* attack with formation of the *axial* alcohol. Two factors are competing with each other: (1) *steric interaction* of the incoming "hydride" with the 3,5-diaxial hydrogens in the axial attack and (2) *torsional strain* of the incoming "hydride" with the 2,6-diaxial hydrogens in the equatorial attack.

axial attack (LiAlH$_4$, NaBH$_4$) ····▶ equatorial alcohol
steric interaction with the axial H$_3$ and H$_5$

equatorial attack (Selectrides, LTSBH) ··▶ axial alcohol
torsional strain with the axial H$_2$ and H$_6$

For sterically unhindered reducing agents (LiAlH$_4$, Li[AlH(OR)$_3$], NaBH$_4$), the effect of torsional strain prevails over the steric effect, leading to predominant axial attack and formation of the equatorial alcohol. On the other hand, both L-Selectride and LTSBH (Li-tri-1,2-dimethylpropylborohydride) have large steric requirements, resulting in the equatorial delivery of hydride to the carbonyl carbon. However, if there are bulky axial substituents at C(3) or and at C(5) (as in 3,3,5-trimethylcyclohexanone, Table 4.6) or if the carbonyl group is embedded in a rigid system, steric effects prevail, resulting in preferential transfer of the "hydride" from the less hindered side (Table 4.7).[56,111]

Table 4.6	Hydride Reduction of 3,3,5-Trimethylcyclohexanone	

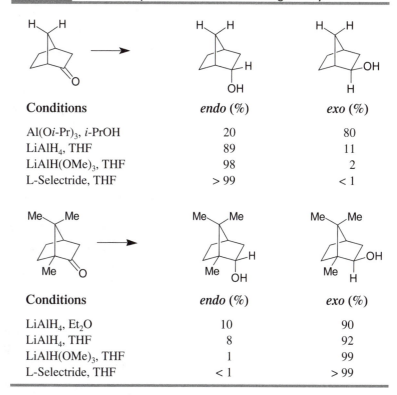

Conditions	*cis* (%)	*trans* (%)
NaBH$_4$, EtOH	33	67
LiAlH$_4$, THF	26	74
LiAlH(OMe)$_3$, THF	8	92
L-Selectride	< 1	> 99

Table 4.7	Selected Hydride Reductions of Bridged Bicyclic Ketones	

Conditions	*endo* (%)	*exo* (%)
Al(O*i*-Pr)$_3$, *i*-PrOH	20	80
LiAlH$_4$, THF	89	11
LiAlH(OMe)$_3$, THF	98	2
L-Selectride, THF	> 99	< 1

Conditions	*endo* (%)	*exo* (%)
LiAlH$_4$, Et$_2$O	10	90
LiAlH$_4$, THF	8	92
LiAlH(OMe)$_3$, THF	1	99
L-Selectride, THF	< 1	> 99

4.13   INVERSION OF SECONDARY ALCOHOL STEREOCHEMISTRY (THE MITSUNOBU REACTION)[112]

The availability of a stereodefined alcohol from reduction of a cyclic ketone raises the question of how one can obtain the corresponding stereoisomer.

A solution to this problem is the Mitsunobu reaction, which provides a powerful tool for inverting the configuration of a given alcohol to its stereoisomer.[113] The reaction involves conversion of the alcohol into a good leaving group capable of being displaced by a relatively weak nucleophile, generally a carboxylate ion (RCOO⁻). The product of stereochemical inversion is an ester, which on saponification leads to the alcohol of opposite configuration. The mechanism of the Mitsunobu reaction is proposed to involve an alkoxyphosphonium intermediate that undergoes S_N2 inversion.[114]

85%

4.14   DIASTEREOFACIAL SELECTIVITY IN ACYCLIC SYSTEMS

Diastereofacial selection in nucleophilic additions to acyclic aldehydes and ketones is of major importance in synthetic organic chemistry.[115] The greater conformational freedom of acyclic systems makes predictions as to the stereochemical outcome of such reactions more difficult than when the carbonyl group is contained within a cyclic framework.

We may distinguish between *enantiotopic* or *diastereotopic* faces in trigonal moieties (> C=O and > C=C <). The faces of prochiral moieties can be differentiated as the *Re* or *Si* face by applying (*R*) or (*S*) sequence rules to the three substituents in the plane as viewed from each face.[116]

Enantiotopicity   The two faces of the carbonyl moiety in acetaldehyde or in butanone, for example, are enantiotopic and are designated as *Re* face if the priority of the substituents is in a clockwise order ($=O > R_L > R_S$) and *Si* face if the priority of the substituents are ordered in an anticlockwise manner. Attack at the *Re* and *Si* enantiotopic faces by a *chiral* reactant (Nu*) gives diastereomers, which may not be formed in equal amounts. The product ratio reflects the facial bias of the chiral reactant. When a single chirality element directs the selective formation of one stereoisomer over another, the process is known as *single asymmetric induction*.

Diastereotopicity—Asymmetric Induction

The two π-faces of an aldehyde or of a ketone *with at least one stereogenic center* are diastereotopic. As a result, the *Re* and *Si* attack by an achiral nucleophile (i.e., LiAlH$_4$, EtMgBr, PhLi) or an achiral enolate ion differ in energy, so unequal amounts of products are formed (A : B ≠ 1). The ratio of products reflects the bias of the chiral substrate to undergo preferential addition on one diastereotopic face.

When two of the reaction components are chiral (any combination of substrate, reagent, catalyst, or solvent), the chirality elements of each reactant will operate either in concert (matched pair) or in opposition (mismatched pair) and together influence the stereochemical outcome of the reaction.[117] In this case, the reaction is subject to *double asymmetric induction*. Unless the diastereofacial selectivities of both chiral reactants are in opposition *and* identical in magnitude, the ratio C : D ≠ 1. Applications of double asymmetric induction in synthesis will be discussed in Chapter 5.

double asymmetric induction

**Prediction of *Re* vs. *Si*
Addition: Cram's Rule**

In 1952, Donald J. Cram (1919–2001, Nobel Prize in Chemistry, 1987) and his coworker Fathy A. Abd Elhafez presented an experimental rule for the diastereoface-differentiating reaction of ketones having a *chiral* center at the vicinal carbon with metal hydride complexes or with organometallic reagents.[118] Cram's rule states that ketones (A; L, M, and S are large, medium, and small groups, respectively) with a chiral center at the vicinal α-carbon probably coordinate with metal hydride complexes or with organometallic compounds, making the carbonyl group sterically more encumbered and more electrophilic. In such cases, the conformation in which the coordinated carbonyl group bisects the least bulky groups (S, M) on the adjacent stereocenter should predominate (as in rotomer B). The nucleophile of the reagent Y⁻–M⁺ then attacks the carbonyl carbon at the less hindered diastereoface, furnishing an excess of product C.

C
major product
Cram product

D
minor product
anti-Cram product

The Cram rule as originally formulated is only valid when there is no chelating group attached to the substrate and so neglects any dipolar interactions with the nucleophile. Moreover, there is considerable torsional strain between the L and the R groups. Several subsequent models[119] have addressed these shortcomings, the Felkin-Anh model being the most popular.

Felkin-Anh Model[120]

In this model, nucleophilic additions to ketones occur from a conformation that places the entering group (Nu) in an *antiperiplanar* arrangement with the largest group L at the adjacent chiral center. In E (favorable, S // R interaction), the trajectory of nucleophilic addition to the carbonyl group brings the nucleophile over the *smallest* group in a nonperpendicular attack with an angle of ~107° (see the discussion in Section 4.12).[112] In F (unfavorable, M // R interaction), attack by the nucleophile also occurs in a nonperpendicular manner, but the nucleophile interacts with the medium-size group M. Thus, based on the Felkin-Anh model, the Cram products should dominate.

In reductions of ketones with complex metal hydrides, diastereoselection is dependent on both the nature of the reducing agent and the substrate (Table 4.8).[121] The degree of Cram and Felkin-Anh selectivity depends on a number of other factors. For example, replacement of Grignard or lithium reagents with trialkoxy titanates gives superior results.[122] Generally, aldehydes react with lower stereoselectivity than ketones with the same reagent.

Table 4.8   Examples of 1,2-Asymmetric Induction in Carbonyl Addition Reactions

Conditions	R	Cram	anti-Cram
$LiAlH_4$, Et_2O	Me	71	29
Li[sec-Bu_3BH]	Me	96	4
$LiAlH_4$, Et_2O	Et	76	24
$LiAlH_4$, Et_2O	t-Bu	98	2

Conditions	Cram	anti-Cram
CH_3Li	65	35
CH_3MgBr	66	34
$CH_3Ti(Oi$-$Pr)_3$, THF, 0 °C	88	12
$CH_3Ti(OPh)_3$, Et_2O, –50 °C	93	7

Chelation-Controlled Addition Reactions

Cram discussed a cyclic model for nucleophilic additions to chiral carbonyl compounds containing an α-alkoxy, α-hydroxy, and α-amino group capable of forming a *chelate* with the organometallic reagent.[123] Incorporation of chelate organization into the design of stereoselective processes is an important control element in diastereoselective[124] and enantioselective[125] carbonyl additions.

α-*Chelation* not only increases the electrophilic character of the carbonyl carbon but also prevents free rotation about the C_α-C(=O) bond, thus directing the nucleophile to the less hindered *Si* diastereoface. As exemplified below, these *1,2-asymmetric induction*s in such rigid chelated systems provide a high degree of diastereoselectivity. A prerequisite for achieving high diastereoselectivities in chelation-controlled nucleophilic additions to carbonyl compounds is that the rate of reaction of the chelated substrate should be greater than for the nonchelated substrate.[115, 126]

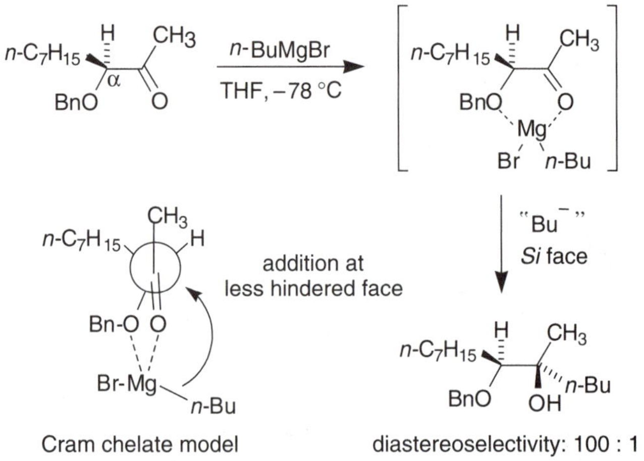

Cram chelate model   diastereoselectivity: 100 : 1

High diastereoselectivities in *β-chelation-controlled* reactions have also been observed with aldehydes and ketones in the presence of a Lewis acid where the chiral center is β to the carbonyl group, resulting in *1,3-asymmetric induction*.[127] The high diastereoselectivity observed in the reaction shown below, using LiI as the Lewis acid arises from β-chelation of both the ketone and the ether oxygen with Li^+. This locks the conformation of the β-alkoxy ketone chain, and the hydride attacks from the less hindered side.[128]

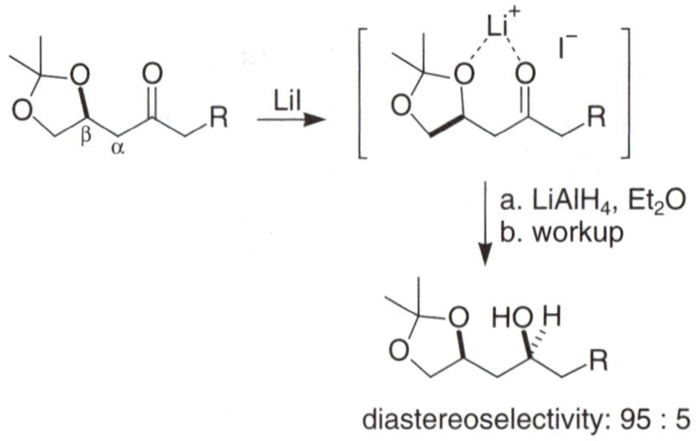

diastereoselectivity: 95 : 5

In general, five-member ring chelates are formed in preference over six-member ring chelates. Thus, chelation also can be used to control the regiochemistry of addition reactions, as exemplified below.[129] No reduction of the C(4) ester was observed because of the less favorable β-chelation by $MgBr_2$ (six-member ring chelate between –OBn to C(4) carbonyl).

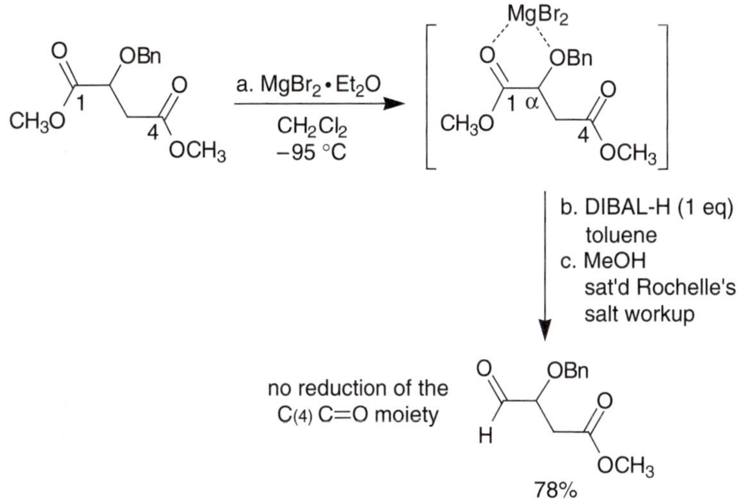

no reduction of the
C(4) C=O moiety

78%

Hydroxyl-Directed Reduction of β-Hydroxy Ketones

Alkoxydialkylboranes ($R'OBR_2$) react with β-hydroxy ketones to form boron chelate intermediates that on subsequent reduction give the 1,3-*syn* diols.[130] Methoxydiethylborane and $NaBH_4$ are the reagents of choice for this transformation. The chelate is reduced via *intermolecular* hydride delivery at the face opposite from the β-R group, as shown below.

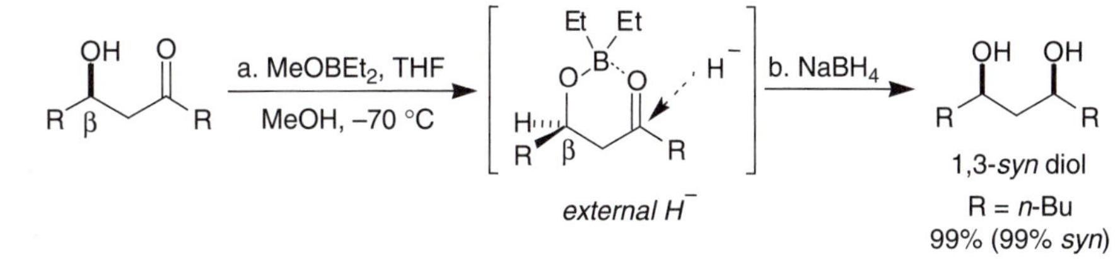

external H^-

1,3-*syn* diol
R = *n*-Bu
99% (99% *syn*)

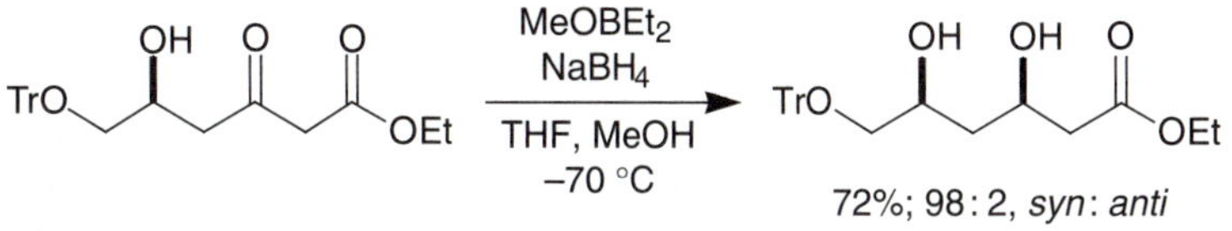

72%; 98 : 2, *syn* : *anti*

Trialkylboranes are also effective chelation agents in stereoselective $NaBH_4$ reductions of β-hydroxy ketones to 1,3-*syn* diols.[131]

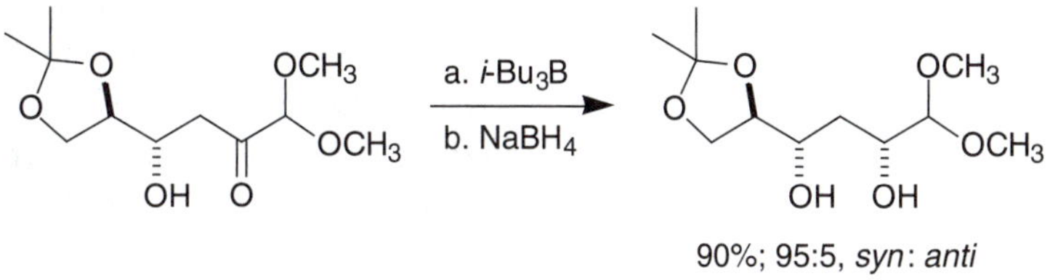

90%; 95:5, *syn* : *anti*

Treatment of β-hydroxy ketones with tetramethylammonium triacetoxyborohydride [$Me_4NHB(OAc)_3$] complements the chelation approach described above by affording the corresponding 1,3-*trans* diols.[132,133] Acetic acid-promoted ligand exchange provides an alkoxydiacetoxyborohydride intermediate in which the proximal ketone is stereoselectively reduced by *intramolecular* hydride transfer. Alkyl substituents in the α-position do not diminish the 1,3-asymmetric induction.

internal H⁻

1,3-*anti* diol

intramolecular hydride delivery

X = Me, Y = H

Me₄N[BH(OAc)₃]
CH₃CN, HOAc
18 h, –20 °C

84%; 98 : 2, *anti* : *syn*

X = H, Y = Me | Me₄N[BH(OAc)₃]
CH₃CN, HOAc
18 h, –20 °C

92%; 98 : 2, *anti* : *syn*

4.15   ENANTIOSELECTIVE REDUCTIONS

Several approaches are available for the synthesis of enantiomerically enriched alcohols from ketones.[134] The three main strategies to obtain enantiomerically enriched (*nonracemic*) material are listed below.[135]

- *Optical resolution* of a racemic mixture: Not very economical, since 50% of the product is lost!

- *Derivatization*: Start with a chiral compound and manipulate it in such a way as to maintain chirality throughout the reactions (S_N2 reactions or chirality transfer in pericyclic reactions).

- *Asymmetric synthesis*: (1) Use a *chiral auxiliary* (chiral acetal—the synthetic equivalent of an aldehyde; chiral hydrazone—the synthetic equivalent of a ketone) covalently attached to an achiral substrate to control subsequent bond formations. The auxiliary is later disconnected and recovered, if possible. (2) Use a *chiral reagent* to distinguish between enantiotopic faces or groups (asymmetric induction) to mediate formation of a chiral product. The substrate and reagent combine to form diastereomeric transition states. (3) Use a *chiral catalyst* to discriminate enantiotopic groups or faces in diastereomeric transition states but only using catalytic amounts of a chiral species.

Most often, asymmetry is created on conversion of a *prochiral* trigonal carbon of carbonyl, enol, imine, enamine, and olefin groups to a tetrahedral center. One of the easiest methods for the preparation of optically active alcohols is the reduction of prochiral ketones. This transformation is achieved using chiral reductants in which chiral organic moieties are ligated to boron or to a metal hydride (Table 4.9).

Table 4.9   Reagents for the Reduction of Prochiral Aldehydes and Ketones

Name	Structure	Source	Reference
Ipc-9-BBN [*R*- and *S*-Alpine-Borane]		(+)- or (−)-α-pinene + 9-BBN	136
Ipc₂BCl [(+)- and (−)-DIP Cl]		(+)- or (−)-Ipc₂BH + HCl in Et₂O	139
trans-2,5-Dimethylborolane [(*R*,*R*)- and (*S*,*S*)-DMB]		 (optical resolution required)	142
2,5-Oxazaborolidines (CBS reagents)		D- or L-Proline + PhMgBr, then BH₃	144
BINAL-H [(*R*)- and (*S*)-BINAL-H]		(+)- or (−)-binaphthol + LiAlH₄, then MeOH or EtOH	147

Alpine-Borane[136]   Alpine-Borane, prepared by hydroboration of α-pinene with 9-borabicyclo[3.3.1]nonane (9-BBN), reduces aldehydes, α-keto esters and acetylenic ketones with excellent enantioselectivity.[137] The reduction proceeds via a cyclic process similar to the MPV reaction.

Saturated ketones are not readily reduced by Alpine-Borane. However, the intramolecular version of this reduction using (Ipc)$_2$BH proceeds with good yield and facial selectivity.[138]

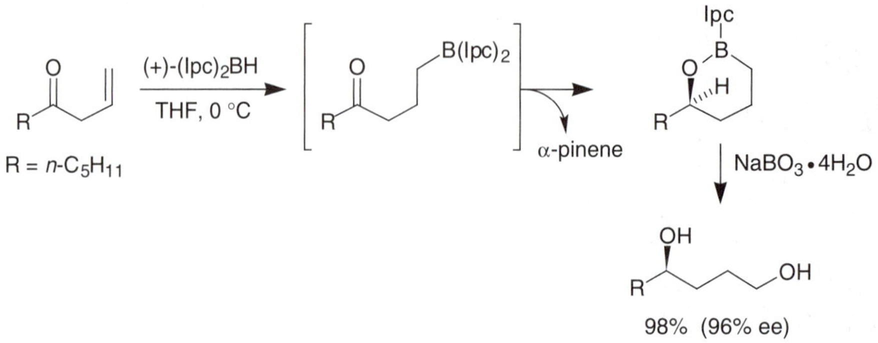

98% (96% ee)

B-Chlorodiisopinocampheyl-borane[139]

(−)-(Ipc)$_2$BCl [from (+)-α-pinene] and (+)-(Ipc)$_2$BCl [from (−)-α-pinene] are excellent reagents for the enantioselective reduction of aryl- and alkyl ketones, cyclic ketones, α-keto esters[140], and α,β-unsaturated ketones.[141]

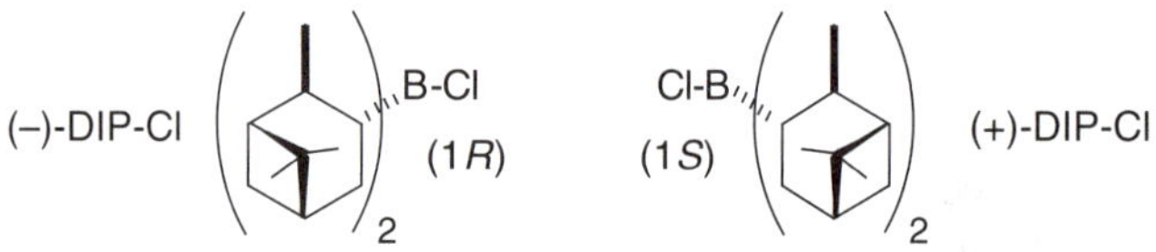

(−)-DIP-Cl   (1R)   (1S)   (+)-DIP-Cl

The transfer of the hydride from (Ipc)$_2$BCl to the keto group likely proceeds via a pathway similar to that depicted above for Alpine-Borane, where the 9-BBN moiety is replaced by Cl. Diethanolamine is usually added in the workup procedure to remove the boron components.

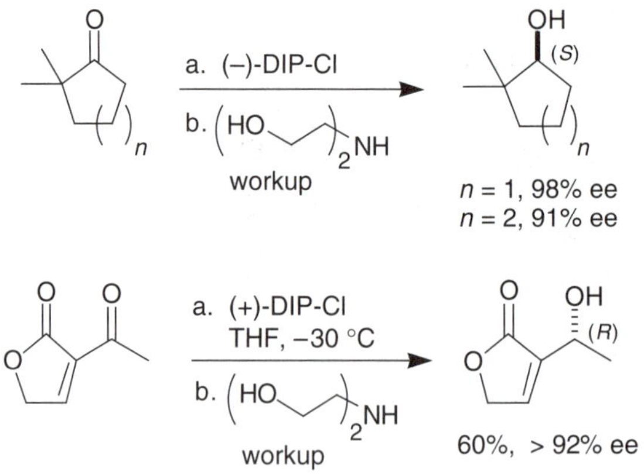

n = 1, 98% ee
n = 2, 91% ee

60%, > 92% ee

---

trans-2,5-Dimethylborolane[142]

Asymmetric reductions of prochiral ketones with (R,R)- or (S,S)-2,5-dimethyl-borolanes proceed with high enantioselectivity. The C(2) symmetry, which makes both faces of the borolane ring equivalent, is an important feature of these reagents. In the majority of cases, the presence of a C(2) symmetry axis within the chiral auxiliary can serve the very important function of dramatically reducing the number of competing diastereomeric transition states.

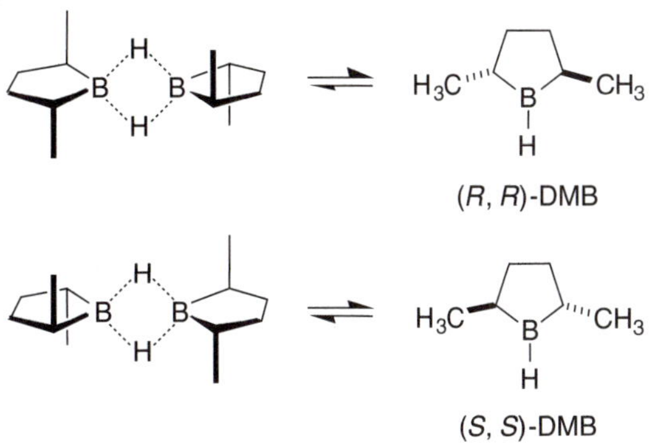
H
B
B
H
H_3C
B
CH_3
H
(R, R)-DMB
H
B
B
H
H_3C
B
CH_3
H
(S, S)-DMB

The DMB reagents are used in conjunction with a catalytic amount of the homochiral 2,5-dimethylborolanyl mesylate (DMB-OSO_2CH_3 = DMB-OMs).[143] Hydride transfer occurs at the face of the mesylate-ketone coordination complex that exhibits the least steric interactions. The availability of both forms of enantiopure DMB allows the preparation of enantioenriched (*R*)- or (*S*)-alcohols in a predictable and controlled manner.

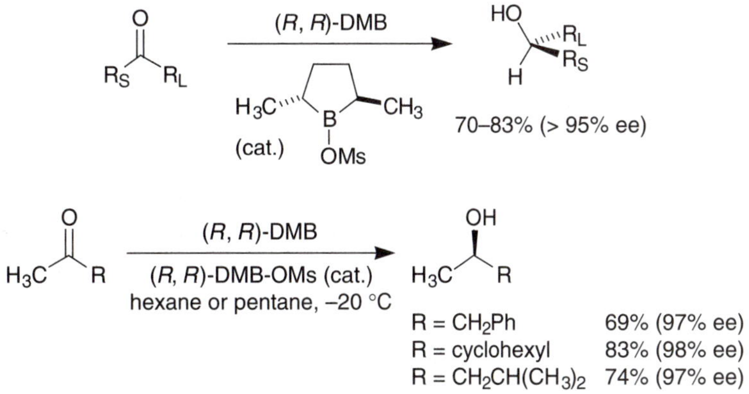
O
R_S
R_L
(R, R)-DMB
H_3C
B
CH_3
(cat.)
OMs
HO
R_L
R_S
H
70–83% (> 95% ee)
O
H_3C
R
(R, R)-DMB
(R, R)-DMB-OMs (cat.)
hexane or pentane, –20 °C
OH
H_3C
R
R = CH_2Ph 69% (97% ee)
R = cyclohexyl 83% (98% ee)
R = $CH_2CH(CH_3)_2$ 74% (97% ee)

Oxazaborolidines— CBS Reduction[144]

Oxazaborolidines have emerged as important reagents for the enantioselective reduction of a variety of prochiral ketones. CBS reduction (chiral oxazaborolidine-catalyzed reduction)[145] of unsymmetrical ketones with diphenyl oxazaborolidine in the presence of BH_3 proceeds *catalytically* to provide alcohols of predicable absolute stereochemistry in high enantiomeric excess.

The reduction presumably involves complexation of BH_3 with the ring nitrogen, followed by coordination of the ketone oxygen to the ring boron and finally transfer of the hydride to the carbonyl group via a six-member chair-like transition state.[146]

BINAL-H[147]   BINAL-H is prepared by adding binaphthol to a solution of LiAlH$_4$ in THF followed by one equivalent of either methanol or ethanol. The reagent generally is used in three-fold excess to achieve optimal selectivities. It reduces acetylenic ketones,[147a] aromatic ketones,[147b] and α,β-unsaturated ketones with high optical purities.[147c]

(S)-BINAL-H   (R)-BINAL-H

The (R)-BINAL-H reagent selectively adds the hydride to the *Si* face of the substrate, whereas the (S)-reagent adds it to the *Re* face.

R = n-C$_4$H$_9$ 91%, 91% ee

R = H   87%, 85% ee
R = n-C$_4$H$_9$ 85%, 90% ee

Enzymatic Reductions of Keto Groups

Enzymes provide an alternative to chemical methods for the enantioselective reduction of prochiral ketones.[148] These reductions are usually carrried out in water or buffered aqueous suspensions with sugars as nutrients.[149]

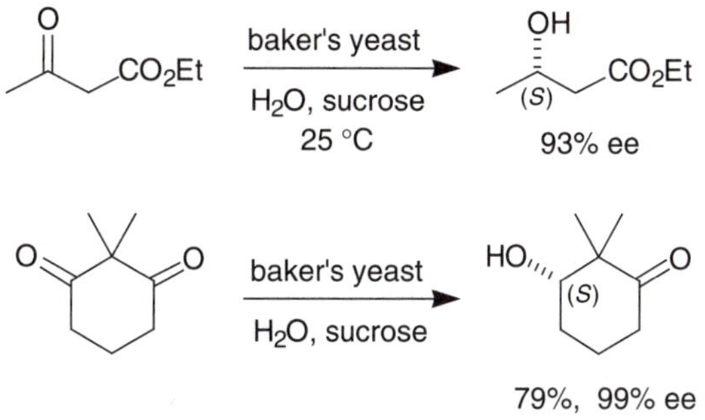

O
CO2Et
baker's yeast
H2O, sucrose
25 °C
OH
(S)
CO2Et
93% ee
O
O
baker's yeast
H2O, sucrose
HO
(S)
O
79%, 99% ee

The use of ester *hydrolase* preparations such as pig liver esterase (*PLE*)[150] or pig liver acetone powder (*PLAP*)[151] complements reduction approaches for the synthesis of enantio-enriched alcohols.[152]

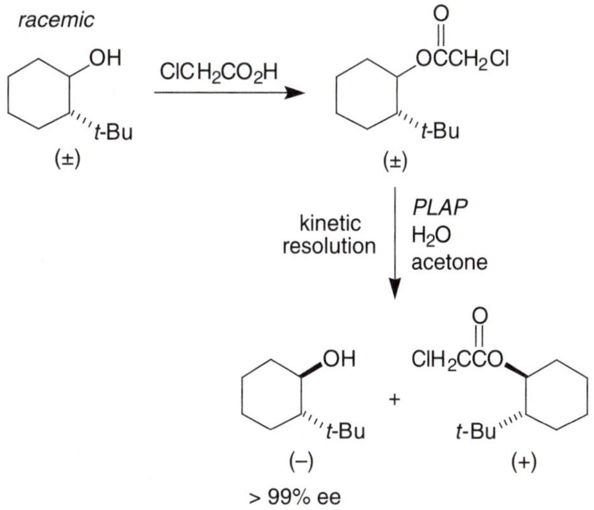

racemic
OH
t-Bu
(±)
ClCH2CO2H
O
OCCH2Cl
t-Bu
(±)
kinetic
resolution
PLAP
H2O
acetone
OH
t-Bu
(−)
> 99% ee
+
O
ClH2CCO
t-Bu
(+)

PROBLEMS

1. **Reagents.** Give the structures of the major products (**A–H**) expected from the following reactions. Be sure to indicate product stereochemistries. Assume that standard aqueous workup conditions are used for product isolation.

a.

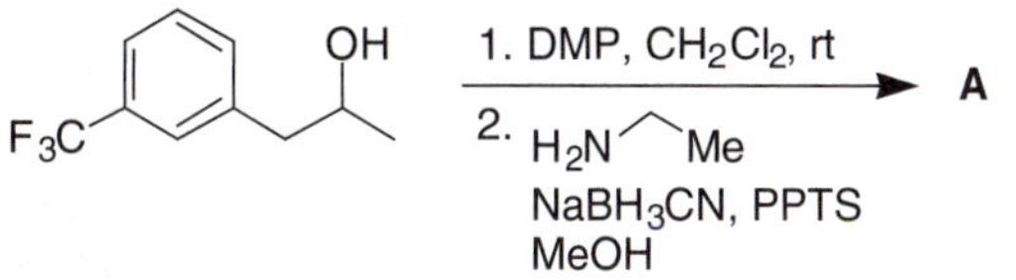

OH
F3C
1. DMP, CH2Cl2, rt
2. H2N Me
NaBH3CN, PPTS
MeOH
A

b.

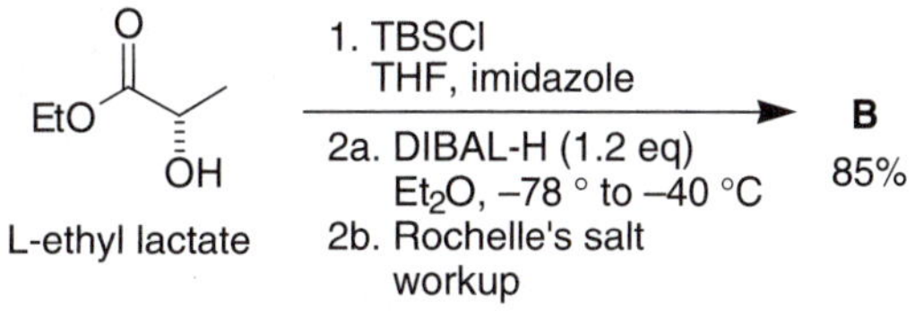

O
EtO
OH
L-ethyl lactate
1. TBSCl
THF, imidazole
2a. DIBAL-H (1.2 eq)
Et2O, −78 ° to −40 °C
2b. Rochelle's salt
workup
B
85%

c.

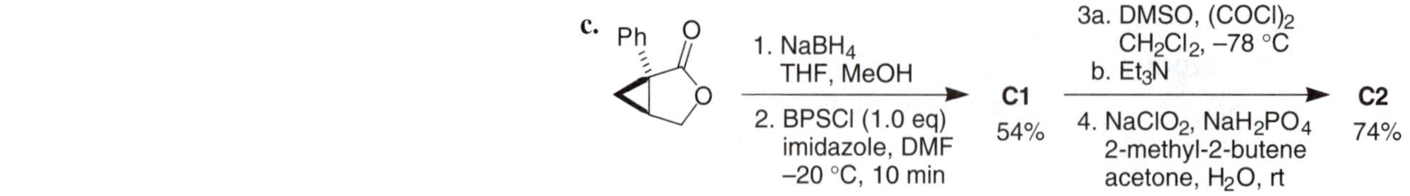

1. NaBH$_4$
   THF, MeOH
2. BPSCl (1.0 eq)
   imidazole, DMF
   −20 °C, 10 min

C1
54%

3a. DMSO, (COCl)$_2$
    CH$_2$Cl$_2$, −78 °C
 b. Et$_3$N

4. NaClO$_2$, NaH$_2$PO$_4$
   2-methyl-2-butene
   acetone, H$_2$O, rt

C2
74%

d.

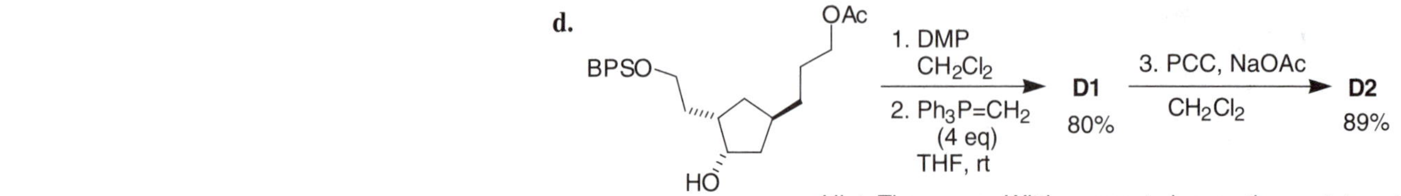

1. DMP
   CH$_2$Cl$_2$

2. Ph$_3$P=CH$_2$
   (4 eq)
   THF, rt

D1
80%

3. PCC, NaOAc

CH$_2$Cl$_2$

D2
89%

Hint: The excess Wittig reagent cleaves the acetate ester.

e.

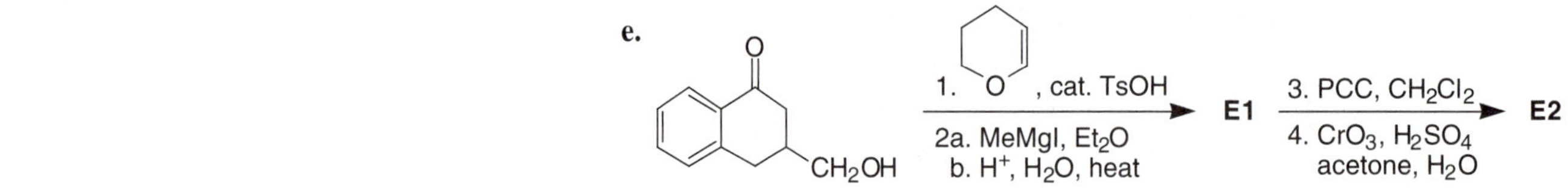

1. [pyran], cat. TsOH

2a. MeMgI, Et$_2$O
 b. H$^+$, H$_2$O, heat

E1

3. PCC, CH$_2$Cl$_2$
4. CrO$_3$, H$_2$SO$_4$
   acetone, H$_2$O

E2

f.

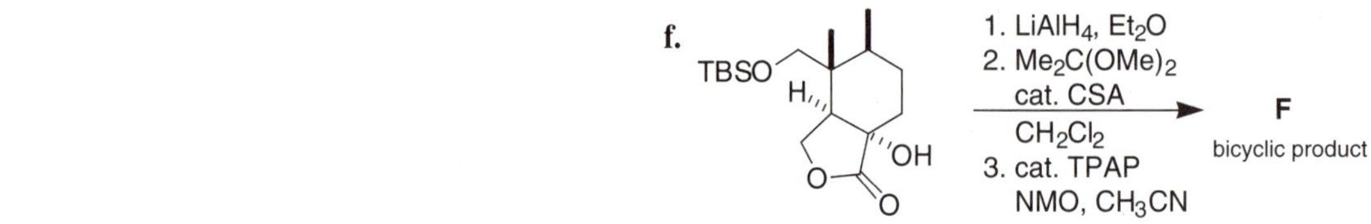

1. LiAlH$_4$, Et$_2$O
2. Me$_2$C(OMe)$_2$
   cat. CSA
   CH$_2$Cl$_2$
3. cat. TPAP
   NMO, CH$_3$CN

F
bicyclic product

g.

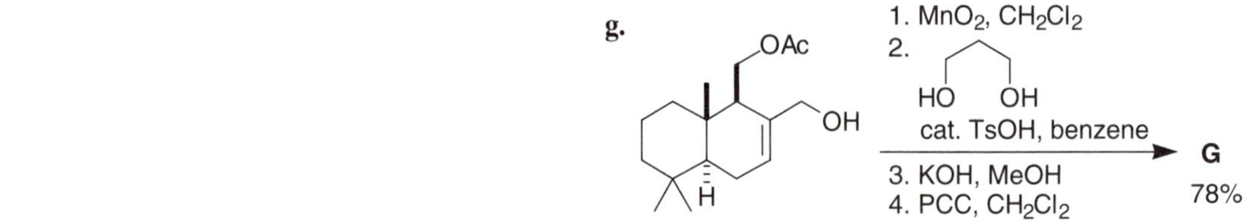

1. MnO$_2$, CH$_2$Cl$_2$
2. [HO OH]
   HO   OH
   cat. TsOH, benzene
3. KOH, MeOH
4. PCC, CH$_2$Cl$_2$

G
78%

*h.

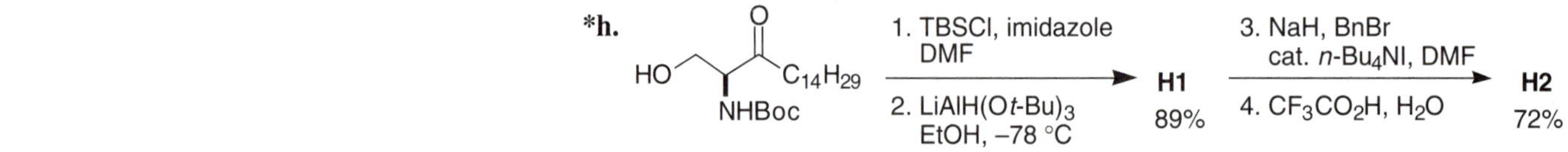

1. TBSCl, imidazole
   DMF

2. LiAlH(Ot-Bu)$_3$
   EtOH, −78 °C

H1
89%

3. NaH, BnBr
   cat. n-Bu$_4$NI, DMF

4. CF$_3$CO$_2$H, H$_2$O

H2
72%

2. Selectivity. Show the product(s) obtained or the appropriate reagent(s) to be used for the following transformations.

a.

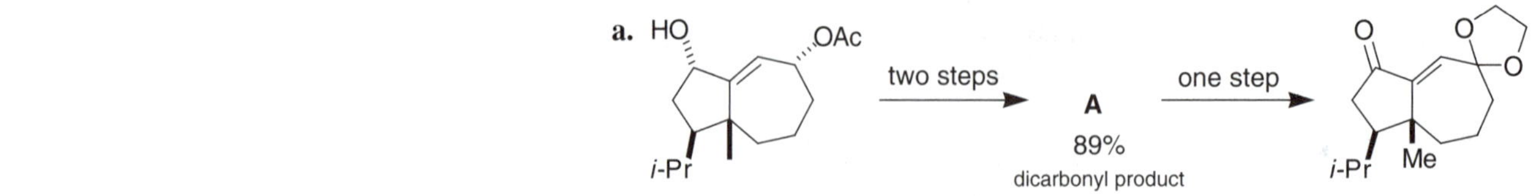

two steps

A
89%
dicarbonyl product

one step

b.

1. NaBH$_4$
   EtOH, 0 °C

2. cat. TsOH,
   [dihydropyran]

B1

3. LiAlH$_4$, THF
4. CBr$_4$, PPh$_3$

B2

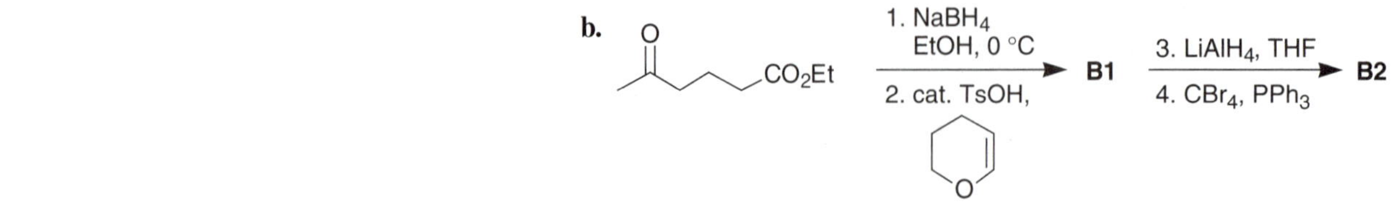

c.

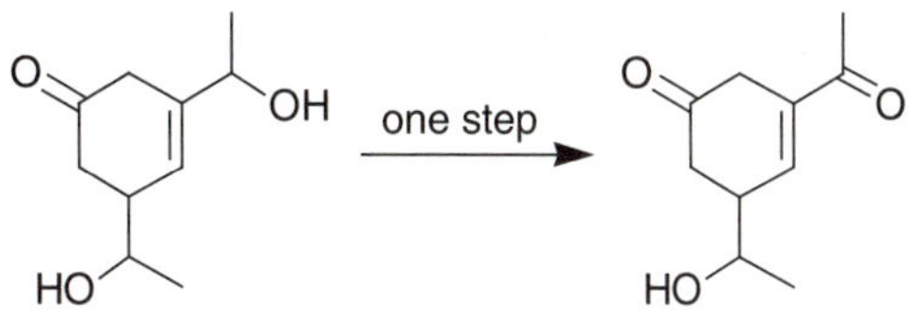

one step

d.

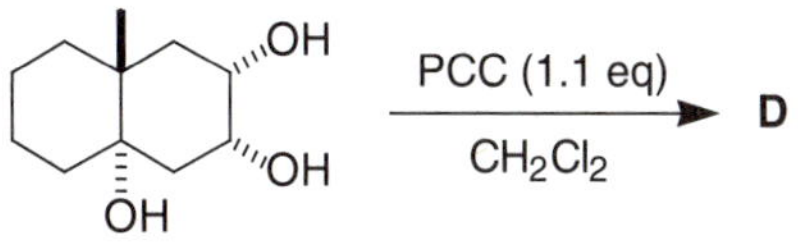

$$\frac{PCC\ (1.1\ eq)}{CH_2Cl_2}$$ **D**

e.

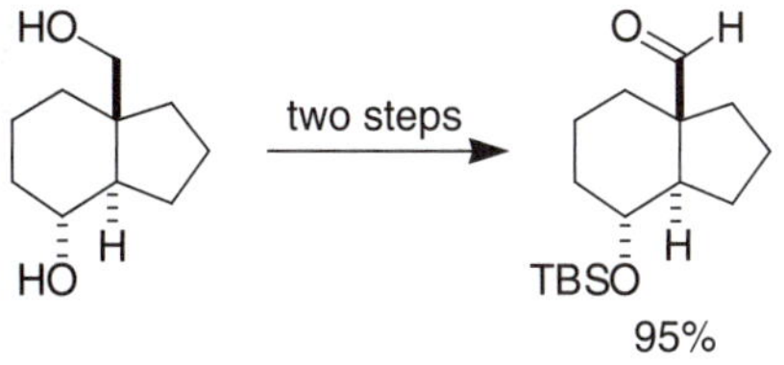

two steps

95%

*f.

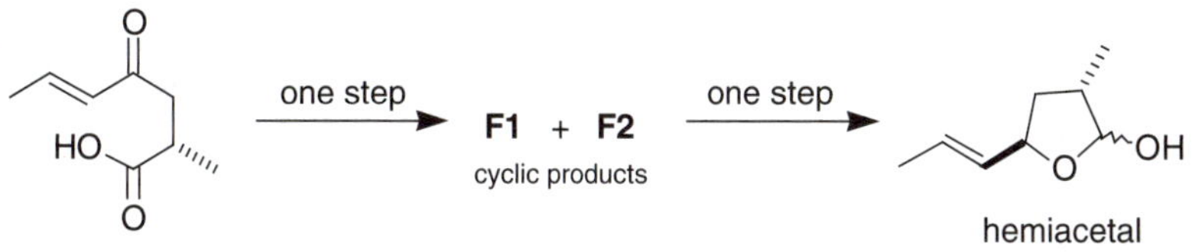

one step

F1 + **F2**

cyclic products

one step

hemiacetal

*g.

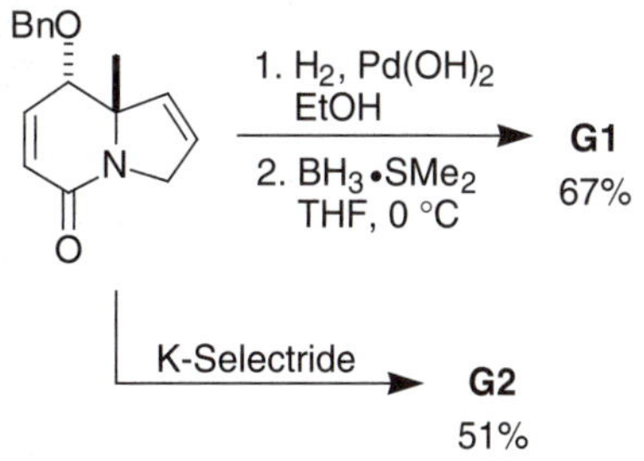

1. H$_2$, Pd(OH)$_2$
   EtOH

G1

2. BH$_3$•SMe$_2$
   THF, 0 °C

67%

K-Selectride **G2**

51%

3. **Stereochemistry.** Predict the stereochemistries of the major products formed
   (**A–H**) in the following reactions. Explain your choices.

a.

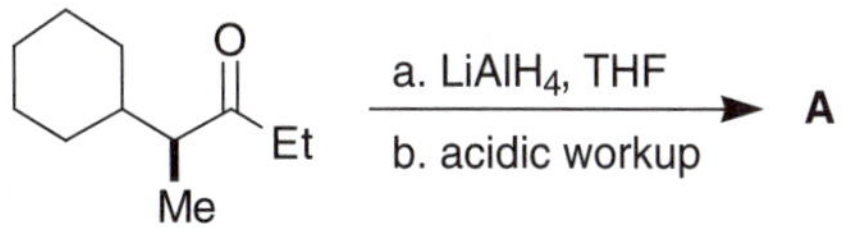

a. LiAlH$_4$, THF

b. acidic workup

A

b.

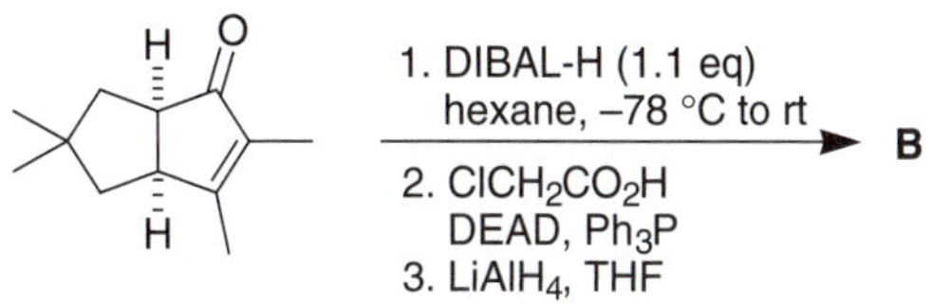

1. DIBAL-H (1.1 eq)
   hexane, –78 °C to rt

2. ClCH$_2$CO$_2$H
   DEAD, Ph$_3$P

3. LiAlH$_4$, THF

B

c.

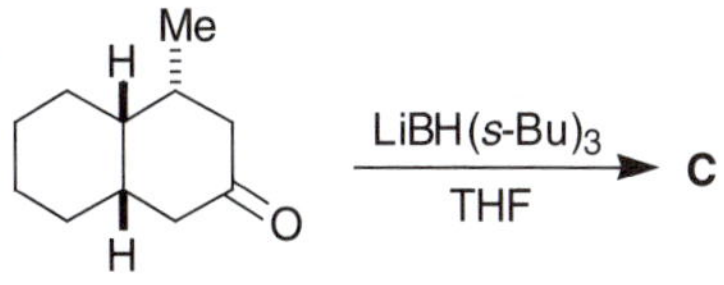

$$\frac{LiBH(s\text{-}Bu)_3}{THF}$$ **C**

d.

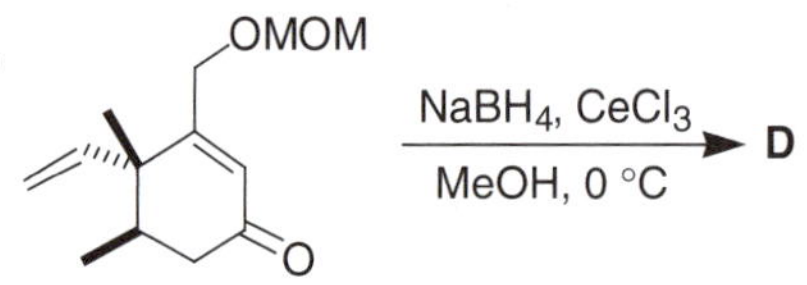

$$\xrightarrow[\text{MeOH, 0 °C}]{\text{NaBH}_4, \text{CeCl}_3} \textbf{D}$$

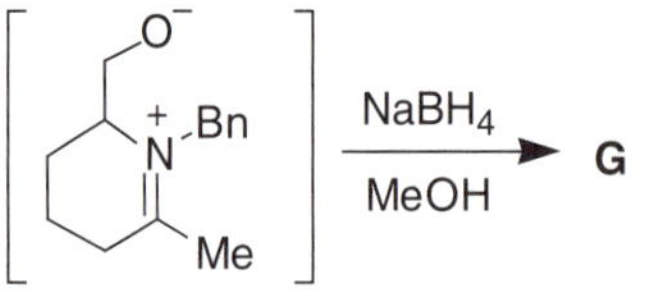

e.

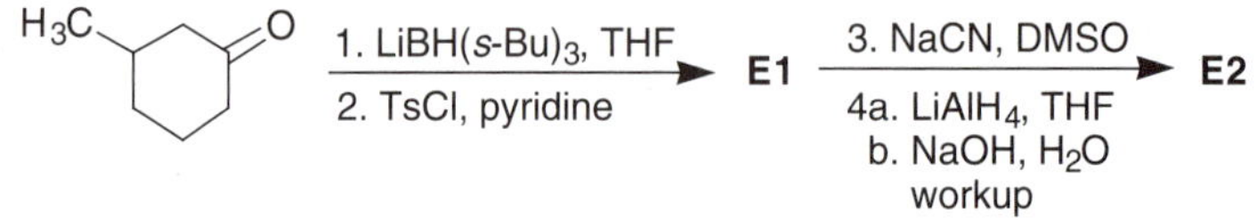

$$\xrightarrow[\text{2. TsCl, pyridine}]{\text{1. LiBH(s-Bu)}_3, \text{THF}} \textbf{E1} \xrightarrow[\substack{\text{4a. LiAlH}_4, \text{THF} \\ \text{b. NaOH, H}_2\text{O} \\ \text{workup}}]{\text{3. NaCN, DMSO}} \textbf{E2}$$

*f.

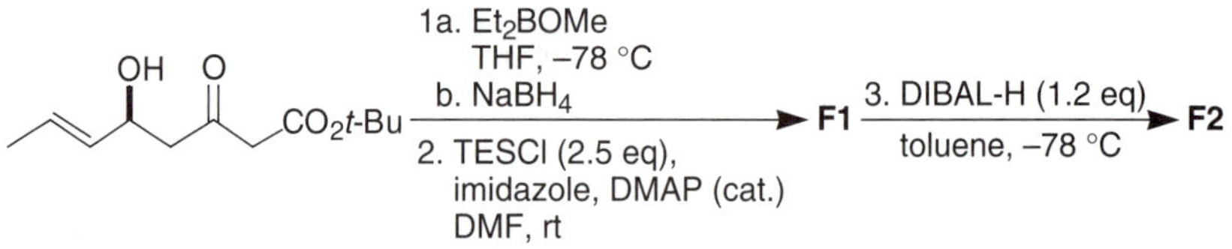

$$\xrightarrow[\substack{\text{2. TESCl (2.5 eq),} \\ \text{imidazole, DMAP (cat.)} \\ \text{DMF, rt}}]{\substack{\text{1a. Et}_2\text{BOMe} \\ \text{THF, −78 °C} \\ \text{b. NaBH}_4}} \textbf{F1} \xrightarrow[\text{toluene, −78 °C}]{\text{3. DIBAL-H (1.2 eq)}} \textbf{F2}$$

*g.

$$\xrightarrow[\text{MeOH}]{\text{NaBH}_4} \textbf{G}$$

*h.

$$\xrightarrow[\text{12 h}]{\text{neat}} \textbf{H}$$

(−)-DIPCl

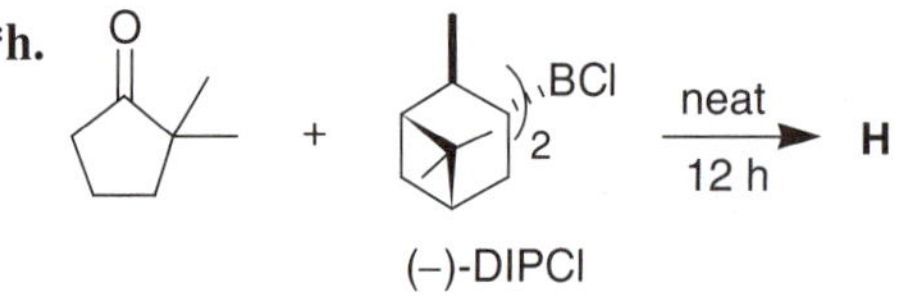

*i. Propose a method to accomplish the following stereochemical inversion.

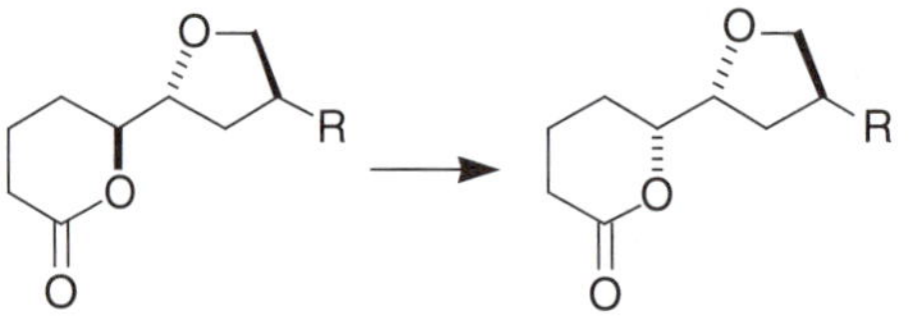

4. Reactivity. Explain the regioselectivity and stereochemistry observed in the each of transformations shown below.

a.

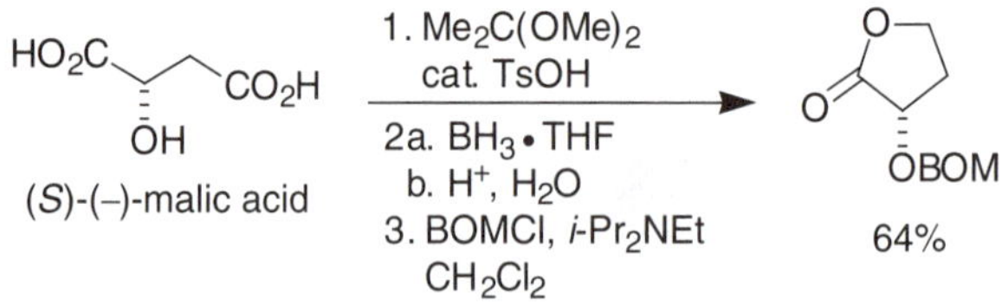

(S)-(−)-malic acid

$$\xrightarrow[\substack{\text{2a. BH}_3 \cdot \text{THF} \\ \text{b. H}^+, \text{H}_2\text{O} \\ \text{3. BOMCl, i-Pr}_2\text{NEt} \\ \text{CH}_2\text{Cl}_2}]{\substack{\text{1. Me}_2\text{C(OMe)}_2 \\ \text{cat. TsOH}}}$$

64%

b.

$$\xrightarrow[\text{EtOH, H}_2\text{O}]{\text{NaBH}_4, \text{CeCl}_3 \cdot 7 \text{ H}_2\text{O}}$$

*c.

$$\xrightarrow[\substack{\text{CF}_3\text{CO}_2\text{H} \\ \text{toluene} \\ 72\%}]{\text{PhCHO}}$$

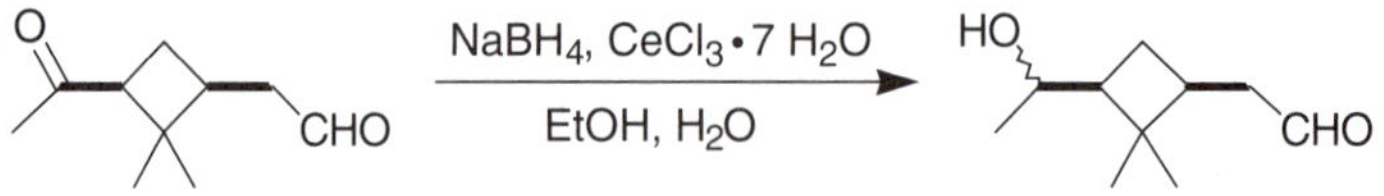

5 : 1

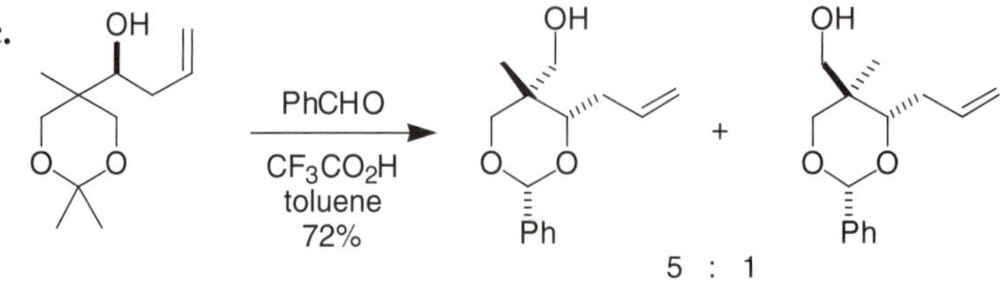

*d.

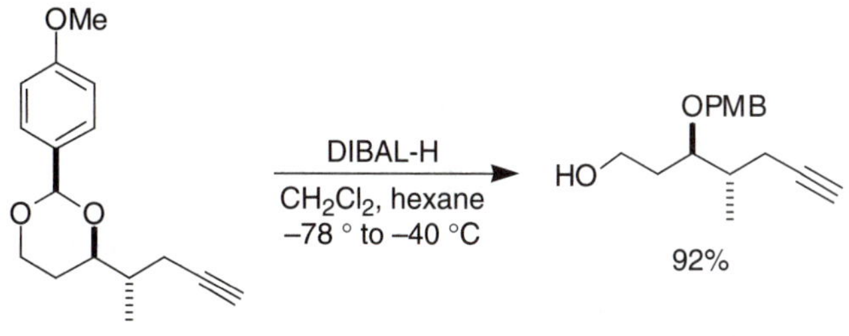

DIBAL-H
CH₂Cl₂, hexane
−78 ° to −40 °C

92%

5. **Synthesis.** Supply the missing reagents required to accomplish each of the following syntheses. Be sure to control the relative stereochemistry.

a.

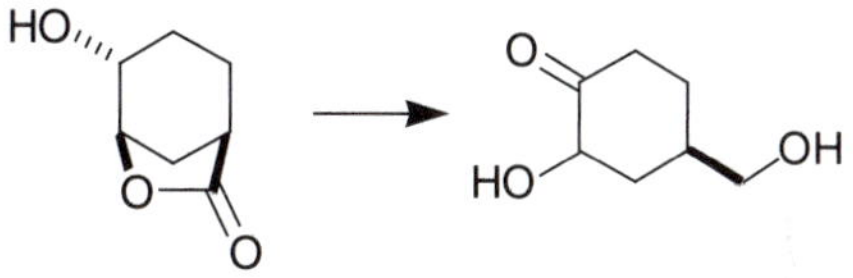

b.

c.

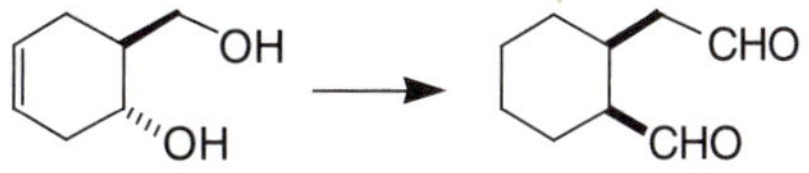

d.

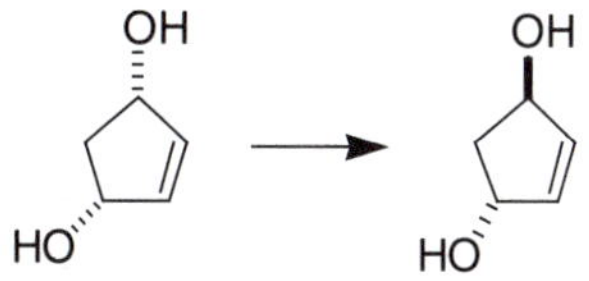

e.

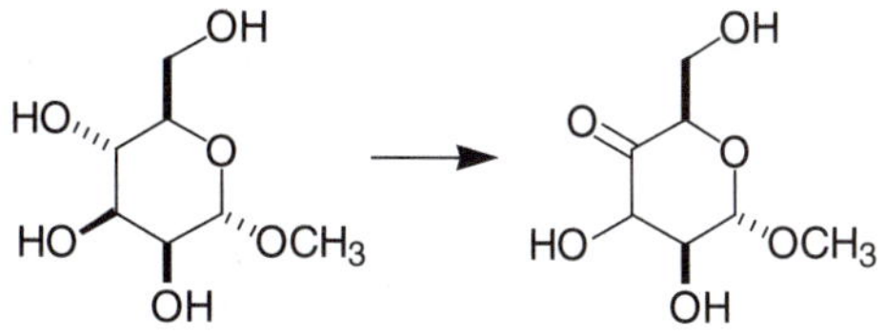

*f.

*g.

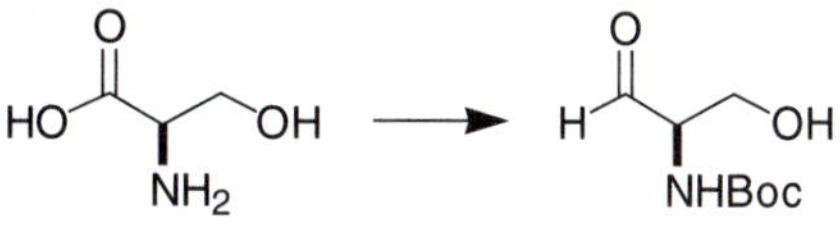

*h.

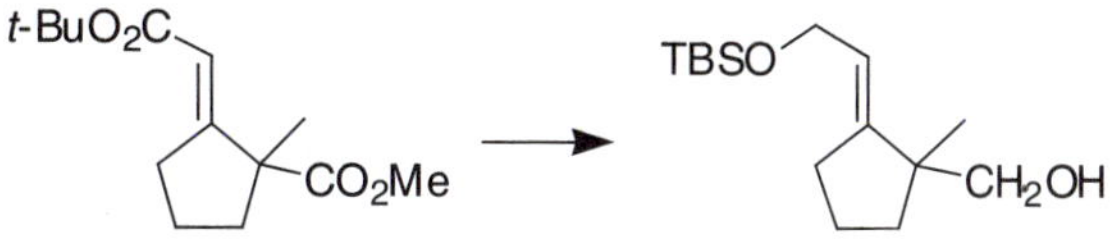

*i.

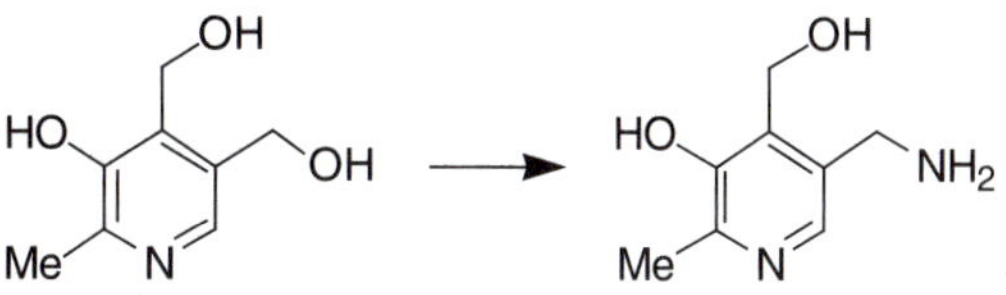

REFERENCES

1. (a) Cainelli, G. *Chromium Oxidations in Organic Chemistry*, Springer-Verlag: New York, 1984. (b) Haines, A. H. *Methods for Oxidation of Organic Compounds*, Academic Press: London, 1988. (c) Hudlicky, M. *Oxidations in Organic Chemistry*, Am. Chem. Soc. Monograph 186: Washington, DC, 1990. (d) Luzzio, F. A. *Org. React.* **1998**, *53*, 1. (d) Donohoe, T. J. *Oxidation and Reduction in Organic Synthesis*, Oxford University Press: Oxford, UK, 2000.

2. Roberts, J. D., Caserio, M. C. *Basic Principles of Organic Chemistry*, 2nd ed., Benjamin: Menlo Park, CA, 1977, p. 641.

3. Bowden, K., Heilbron, I. M., Jones, E. R. H., Weedon, B. C. L. *J. Chem. Soc.* **1946**, 39.

4. Brown, H. C., Garg, C. P., Liu, K.-T. *J. Org. Chem.* **1971**, *36*, 387.

5. (a) Ratcliffe, R., Rodehorst, R. *J. Org. Chem.* **1970**, *35*, 4000. (b) Collins, J. C., Hess, W. W. *Org. Synth.* **1972**, 52, 5. (c) Ratcliffe, R. *Org. Synth.* **1975**, *55*, 84.

6. (a) Corey, E. J., Suggs, J. W. *Tetrahedron Lett.* **1975**, 2647. (b) Piancatelli, G., Scettri, A., D'Auria, M. *Synthesis* **1982**, 245.

7. Brown, H. C., Rao, C. G., Kulkarni, S. U. *J. Org. Chem.* **1979**, *44*, 2809.

8. (a) Herscovici, J., Antonakis, K. *J. Chem. Soc., Chem. Comm.* **1980**, 561. (b) Corey, E. J., Pan, B.-C., Hua, D. H., Deardorff, D. R. *J. Am. Chem. Soc.* **1982**, *104*, 6816.

9. (a) Luche, J.-L. *Synthetic Organic Sonochemistry*, Plenum Press: New York, **1998**. (b) Mason, T. J., Lorimer, J. P. *Applied Sonochemistry*, Wiley-VCH: Weinheim, Germany, 2002.

10. Corey, E. J., Schmidt, G. *Tetrahedron Lett.* **1979**, 399.

11. (a) Mancuso, A. J., Swern, D. *Synthesis* **1981**, 165. (b) Marx, M., Tidwell, T. T. *J. Org. Chem.* **1984**, *49*, 788. (c) Leopold, E. J. *Org. Synth.* **1985**, *64*, 164. (d) Tidwell, T. T. *Org. React.* **1990**, *39*, 297.

12. Ireland, R. E., Norbeck, D. W. *J. Org. Chem.* **1985**, *50*, 2198.

13. Taber, D. F., Amedio, J. C., Jr., Jung, K.-Y. *J. Org. Chem.* **1987**, *52*, 5621.

14. (a) Dess, D. B., Martin, J. C. *J. Am. Chem. Soc.* **1991**, *113*, 7277. (b) Boeckman, R. K., Jr., Shao, P., Mullins, J. J. *Org. Synth.* **2000**, *77*, 141.

15. Frigerio, M., Santagostino, M., Sputore, S. *J. Org. Chem.* **1999**, *64*, 4537.

16. (a) Ireland, R. E., Lui, L. *J. Org. Chem.* **1993**, *58*, 2899. (b) Meyer, S. D., Schreiber, S. L. *J. Org. Chem.* **1994**, *59*, 7549.

17. Nicolaou, K. C., Yue, E. W., Naniwa, Y., De Riccardis, F., Nadin, A., Leresche, J. E., La Greca, S., Yang, Z. *Angew. Chem., Int. Ed.* **1994**, *33*, 2184.

18. Nicolaou, K. C., Mathison, C. J. N., Montagnon, T. *Angew. Chem., Int. Ed.* **2003**, *42*, 4077.

19. (a) Frigerio, M., Santagostino, M. *Tetrahedron Lett.* **1994**, *35*, 8019. (b) Corey, E. J., Palani, A. *Tetrahedron Lett.* **1995**, *36*, 3485.

20. Ley, S. V., Norman, J., Griffith, W. P., Marsden, S. P. *Synthesis* **1994**, 639.

21. (a) Fatiadi, A. J. *Synthesis* **1976**, 65. (b) Fatiadi, A. J. *Synthesis* **1976**, 133.

22. Firouzabadi, H., Ghaderi, E. *Tetrahedron Lett.* **1978**, 839.

23. (a) Garigipati, R. S., Freyer, A. J., Whittle, R. R., Weinreb, S. M. *J. Am. Chem. Soc.* **1984**, *106*, 7861. (b) Hollinshead, D. M., Howell, S. C., Ley, S. V., Mahon, M., Ratcliffe, N. M., Worthington, P. A. *J. Chem. Soc., Perkin Trans. 1*, **1983**, 1579.

24. (a) Tronchet, J. M. J., Tronchet, J., Birkhäuser, A. *Helv.* **1970**, *53*, 1489. (b) Fetizon, M., Golfier, M., Louis, J.-M. *Tetrahedron* **1975**, *31*, 171. (c) McKillop, A., Young, D. W. *Synthesis* **1979**, 401.

25. Grieco, P. A., Lis, R., Ferrino, S., Jaw, J. Y. *J. Org. Chem.* **1982**, *47*, 601.

26. (a) Cella, J. A., Kelley, J. A., Kenehan, E. F. *J. Org. Chem.* **1975**, *40*, 860. (b) Anelli, P. L., Banfi, S., Montanari, F., Quici, S. *J. Org. Chem.* **1989**, *54*, 2970. (c) Inokuchi, T., Atsumoto, S. *J. Org. Chem.* **1990**, *31*, 2177. (d) De Mico, A., Margarita, R., Parlanti, L., Vescovi, A., Piancatelli, G. *J. Org. Chem.* **1997**, *62*, 6974.

27. Pierce, M. E., Harris, G. D., Islam, Q., Radesca, L. A., Storace, L., Waltermire, R. E., Wat, E., Jadhav, P. K., Emmett, G. C. *J. Org. Chem.* **1996**, *61*, 444.

28. (a) Semmelhack, M. F., Schmid, C. R., Cortés, D. A., Chou, C. S. *J. Am. Chem. Soc.* **1984**, *106*, 3374. (b) Mehta, G., Pan, S. C. *Org. Lett.* **2004**, *6*, 3985.

29. Pozzi, G., Quici, S., Benaglia, M., Dell'Anna, G. *Org. Lett.* **2004**, *6*, 441.

30. (a) Ho, T. L. *Synthesis* **1978**, 936. (b) Tomioka, H., Oshima, K., Nozaki, H. *Tetrahedron Lett.* **1982**, *23*, 539.

31. Jung, M. E., Speltz, L. M. *J. Am. Chem. Soc.* **1976**, *98*, 7882.

32. Olah, G. A., Gupta, B. G. B., Fung, A. P. *Synthesis* **1980**, 897.

33. Stevens, R. V., Chapman, K. T., Stubbs, C. A., Tam, W. W., Albizati, K. F. *Tetrahedron Lett.* **1982**, *23*, 4647.

34. Krebs, A., Burgdörfer, G. *Tetrahedron Lett.* **1973**, *23*, 2063.

35. Rigby, W. *J. Chem. Soc.* **1951**, 793.

36. (a) Vedejs, E., Larsen, S. *Org. Synth.* **1985**, *64*, 127. (b) Rubottom, G. M., Gruber, J. M., Juve, H. D., Jr. *Org. Synth.* **1985**, *64*, 118.

37. (a) Williams, G. J., Hunter, N. R. *Can. J. Chem.* **1976**, *54*, 3830. (b) Dunlap, N. K.; Sabol, M. R., Watt, D. S. *Tetrahedron Lett.* **1984**, *25*, 5839.

38. (a) Babler, J. H., Coghlan, M. J. *Synth. Commun.* **1976**, *6*, 469. (b) Dauben, W. G., Michno, D. M. *J. Org. Chem.* **1977**, *42*, 682. (c) For a recent modification, see Shibuya, M., Ito, S., Takahashi, M., Iwabuchi, Y. *Org. Lett.* **2004**, *6*, 4303.

39. Majetich, G., Condon, S., Hull, K., Ahmad, S. *Tetrahedron Lett.* **1989**, *30*, 1033.

40. Luzzio, F. A., Moore, W. J. *J. Org. Chem.* **1993**, *58*, 2966.

41. (a) Abiko, A., Roberts, J. C., Takemasa, T., Masamune, S. *Tetrahedron Lett.* **1986**, *27*, 4537. (b) Roush, W. R., Michaelides, M. R., Tai, D. F., Lesur, B. M., Chong, W. K. M., Harris, D. J. *J. Am. Chem. Soc.* **1989**, *111*, 2984.

42. (a) Lindgren, B. O., Hilsson, T. *Acta Chem. Scand.* **1973**, *27*, 888. (b) Kraus, G. A., Roth, B. *J. Org. Chem.* **1980**, *45*, 4825. (c) Bal, B. S., Childers, W. E., Jr., Pinnick, H. W. *Tetrahedron* **1981**, *37*, 2091.

43. (a) Campaigne, E., LeSuer, W. M. *Org. Synth., Col. Vol. IV*, **1963**, 919. (b) Colombo, L., Gennari, C., Resnati, G., Scolastico, C. *Synthesis* **1981**, 74.

44. (a) Corey, E. J., Gilman, N. W., Ganem, B. E. *J. Am. Chem. Soc.* **1968**, *90*, 5616. (b) Travis, B. R., Sivakumar, M., Hollist, G. O., Borhan, B. *Org. Lett.* **2003**, *5*, 1031.

45. (a) Müller, P., Godoy, J. *Helv. Chim. Acta* **1981**, *64*, 2531. (b) Krapcho, A. P., Larson, J. R., Eldridge, J. M. *J. Org. Chem.* **1977**, *42*, 3749.

46. (a) Makosza, M. *Pure Appl. Chem.* **1975**, *43*, 439. (b) Makosza, M., Fedorynski, M. In *Handbook of Phase Transfer Catalysis*, Sasson, Y., Neumann, R., Eds., Chapman Hill: London, 1997.

47. (a) Zweifel, G., Backlund, S. J. *J. Am. Chem. Soc.* **1977**, *99*, 3184. (b) Midland, M. M., Lee, P. E. *J. Org. Chem.* **1981**, *46*, 3933.

48. (a) Gopal, H., Gordon, A. J. *Tetrahedron Lett.* **1971**, 2941. (b) Lee, D. G., Chang, V. S. *J. Org. Chem.* **1979**, *44*, 2726.

49. (a) Rabjohn, N. *Org. React.* **1976**, *24*, 261. (b) Carruthers, W. *Some Modern Methods of Organic Synthesis*, 3rd ed., Cambridge University Press: Cambridge, UK, 1986, p. 348.

50. (a) Umbreit, M. A., Sharpless, K. B. *J. Am. Chem. Soc.* **1977**, *99*, 5526. (b) Brown, H. C., Dhokte, U. P. *J. Org. Chem.* **1994**, *59*, 2025.

51. (a) Ginsburg, D., Pappo, R. *J. Chem. Soc.* **1951**, 516. (b) Mori, K. *Tetrahedron* **1978**, *34*, 915.

52. Parish, E. J., Chitrakorn, S., Wei, T.-Y. *Synth. Commun.* **1986**, *16*, 1371.

53. (a) Kharasch, M. S., Sosnovsky, G., Yang, N. C. *J. Am. Chem. Soc.* **1959**, *81*, 5819. (b) Pedersen, K., Jakobson, P., Lawesson, S.-O. *Org. Synth., Col. Vol. V*, **1973**, 70.

54. Kochi, J. K., Bemis, A. *Tetrahedron* **1968**, *24*, 5099.

55. (a) Andrus, M. B., Chen, X. *Tetrahedron* **1997**, *53*, 16229. (b) Sekar, G., DattaGupta, A., Singh, V. K. *J. Org. Chem.* **1998**, *63*, 2961. (c) Malkov, A. V., Bella, M., Langer, V., Kocovsky, P. *Org. Lett.* **2000**, *2*, 3047.

56. (a) Brown, H. C. *Boranes in Organic Chemistry*, Cornell University Press: Ithaca, NY, 1972. (b) Hajos, A. *Complex Hydrides and Related Reducing Agents in Organic Synthesis*, Elsevier: Amsterdam, 1979. (c) Pelter, A., Smith, K., Brown, H. C. *Borane Reagents*, Academic Press: London, 1988. (d) Hudlicky, M. *Reductions in Organic Chemistry*, 2nd ed., Am. Chem. Soc. Monograph 188, Washington, DC, 1996. (e) Brown, H. C., Ramachandran, P. V. In *Reductions in Organic Synthesis*, Abdel-Magid, A. F., Ed., Am. Chem. Soc. Symposium Series 641, Washington, DC, 1996. (f) Smith, M. B., March, J. *March's Advanced Organic Chemistry*, 5th ed., Wiley: New York, 2001, p. 1544.

57. (a) Eliel, E. L., Wilen, S. H., Mander, L. N. *Stereochemistry of Organic Compounds*, Wiley: New York, 1994. (b) Ward, R. S. *Selectivity in Organic Synthesis*, Wiley: Chichester, UK, 1999.

58. Horeau, A. *Tetrahedron Lett.* **1969**, 3121.

59. (a) Eliel, E. L., Wilen, S. H. *Chem. Eng. News* **1991** (July 22), 3. (b) Brewster, J. H. *Chem. Eng. News* **1992** (May 18), 3.

60. Ruch, E. *Acc. Chem. Res.* **1972**, *5*, 49.

61. Brown, H. C., Kramer, G. W., Levy, A. B., Midland, M. M. *Organic Syntheses via Boranes*, Wiley: New York, 1975, p. 244.

62. (a) Fieser, L. F., Fieser, M. *Reagents for Organic Synthesis*, Wiley: New York, Vol. 1, 1967, p. 583. (b) Burke, S. D., Murtiashaw, C. W., Dike, M. S., Strickland, S. M. S., Saunders, J. O. *J. Org. Chem.* **1981**, *46*, 2400. (c) Burgstahler, A. W., Nordin, I. C. *J. Am. Chem. Soc.* **1961**, *83*, 198 (experimental procedure given on page 205). (d) Powell, J., James, N., Smith, S. J. *Synthesis* **1986**, 338.

63. (a) Seyden-Penne, J. *Reductions by the Alumino- and Borohydrides in Organic Synthesis*, VCH/Lavoisier-Tec & Doc: Paris, 1991. (b) Ashby, E. C., Boone, J. R. *J. Am. Chem. Soc.* **1976**, *98*, 5524.

64. Cha, J. S., Kwon, S. S. *J. Org. Chem.* **1987**, *52*, 5486.

65. Brown, H. C., Krishnamurthy, S. *Tetrahedron* **1979**, *35*, 567.

66. (a) Málek, J. *Org. React.* **1985**, *34*, 1. (b) Málek, J. *Org. React.* **1988**, *36*, 249.

67. Mosettig, E., Mozingo, R. *Org. React.* **1948**, *4*, 362.

68. Ward, D. E., Rhee, C. *Can. J. Chem.* **1989**, *67*, 1206.

69. Kanth, J. V. B., Periasamy, M. *J. Org. Chem.* **1991**, *56*, 5964.

70. McKennon, M. J., Meyers, A. I., Drauz, K., Schwarm, M. *J. Org. Chem.* **1993**, *58*, 3568.

71. Brown, H. C., Choi, Y. M., Narasimhan, S. *Inorg. Chem.* **1981**, *20*, 4454.

72. Sallay, I., Dutka, F., Fodor, G. *Helv. Chim. Acta* **1954**, *37*, 778.

73. (a) Narasimhan, S., Madhavan, S., Prasad, K. G. *J. Org. Chem.* **1995**, *60*, 5314. (b) Narasimhan, S., Balakumar, R. *Aldrichimica Acta* **1998**, *31*, 19.

74. Brown, H. C., Krishnamurthy, S. *J. Am. Chem. Soc.* **1973**, *95*, 1669.

75. Brown, H. C., Kim, S. C., Krishnamurthy, S. *J. Org. Chem.* **1980**, *45*, 1.

76. Krishnamurthy, S., Schubert, R. M., Brown, H. C. *J. Am. Chem. Soc.* **1973**, *95*, 8486.

77. (a) Lane, C. F. *Synthesis* **1975**, 135. (b) Hutchins, R. O., Natale, N. R. *Org. Prep. Proced. Int.* **1979**, *11*, 201.

78. Baxter, E. W., Reitz, A. B. *Org. React.* **2002**, 59, 1.

79. Borch, R. F., Bernstein, M. D., Durst, H. D. *J. Am. Chem. Soc.* **1971**, *93*, 2897.

80. (a) Abdel-Magid, A. F., Maryanoff, C. A. In *Reductions in Organic Synthesis*, Abdel-Magid, A. F., Ed., Am. Chem. Soc. Symposium Series 641, Washington, D C, 1996, p. 201. (b) Abdel-Magid, A. F., Carson, K. G., Harris, B. D., Maryanoff, C. A., Shah, R. D. *J. Org. Chem.* **1996**, *61*, 3849.

81. Emerson, W. S. *Org. React.* **1948**, *4*, 174.

82. Ranu, B. C., Majee, A., Sarkar, A. *J. Org. Chem.* **1998**, *63*, 370.

83. Shapiro, R. H. *Org. React.* **1976**, *23*, 405.

84. (a) Winterfeldt, E. *Synthesis* **1975**, 617. (b) Zweifel, G. In *Comprehensive Organic Chemistry*, Vol. 3, Barton, D. H. R. Ollis, W. D., Eds., Pergamon Press: Oxford, UK, 1979, p. 1013. (c) Yoon, N. M., Gyoung, Y. S. *J. Org. Chem.* **1985**, *50*, 2443. (d) Wilson, K. E., Seidner, R. T., Masamune, S. *J. Chem. Soc., Chem. Commun.* **1970**, 213.

85. (a) Baran, J. S. *J. Org. Chem.* **1965**, *30*, 3564. (b) Corey, E. J., Weinshenker, N. M., Schaaf, T. K., Huber, W. *J. Am. Chem. Soc.* **1969**, *91*, 5675.

86. Smith, A. B., III, Branca, S. J., Pilla, N. N., Guaciaro, M. A. *J. Org. Chem.* **1982**, *47*, 1855.

87. Taber, D. F., Silverberg, L. J., Robinson, E. D. *J. Am. Chem. Soc.* **1991**, *113*, 6639.

88. Yoon, N. M., Pak, C. S., Brown, H. C., Krishnamurthy, S., Stocky, T. P. *J. Org. Chem.* **1973**, *38*, 2786.

89. Brown, H. C., Cha, J. S., Yoon, N. M., Nazer, B. *J. Org. Chem.* **1987**, *52*, 5400.

90. Yoon, N, M., Lee, H. J., Kang, J., Chung, J. S. *J. Korean Chem. Soc.* **1975**, *19*, 468.

91. Ravikumar, K. S., Baskaran, S., Chandrasekaran, S. *J. Org. Chem.* **1993**, *58*, 5981.

92. Krishnamurthy, S., Brown, H. C. *J. Org. Chem.* **1977**, *42*, 1197.

93. Gemal, A. L., Luche, J.-L. *J. Am. Chem. Soc.* **1981**, *103*, 5454.

94. Brown, H. C., Hess, H. M. *J. Org. Chem.* **1969**, *34*, 2206.

95. Ganem, B. *J. Org. Chem.* **1975**, *40*, 146.

96. (a) Gribble, G. W., Ferguson, D. C. *J. Chem. Soc., Chem. Comm.* **1975**, 535. (b) Tolstikov, G. A., Odinokov, V. N., Galeeva, R. I., Bakeeva, R. S., Akhunova, V. R. *Tetrahedron Lett.* **1979**, 4851. (c) Gribble, G. W., Nutaitis, C. F. *Org. Prep. Proced. Int.* **1985**, *17*, 317.

97. Brown, H. C., Kulkarni, S. U. *J. Org. Chem.* **1977**, *42*, 4169.

98. Ranu, B. C. *Synlett* **1993**, 885, and references cited therein.

99. (a) Luche, J.-L., Gemal, A. L. *J. Am. Chem. Soc.* **1979**, *101*, 5848. (b) Gemal, A. L., Luche, J.-L. *J. Org. Chem.* **1979**, *44*, 4187.

100. Reed, L. J., Niu, C.-I. *J. Am. Chem. Soc.* **1955**, *77*, 416.

101. Kende, A. S., Fludzinski, P. *Org. Synth.* **1985**, *64*, 104.

102. Cohen, N., Lopresti, R. J., Saucy, G. *J. Am. Chem. Soc.* **1979**, *101*, 6710.

103. Kim, S., Ahn, K. H. *J. Org. Chem.* **1984**, *49*, 1717.

104. (a) Wu, Y.-D., Houk, K. N. *J. Am. Chem. Soc.* **1987**, *109*, 908. (b) Wu, Y.-D., Houk, K. N., Trost, B. M. *J. Am. Chem. Soc.* **1987**, *109*, 5560. (c) For a rationalization based on stereoelectronic factors, see Cieplak, A. S., Tait, B. D., Johnson, C. R. *J. Am. Chem. Soc.* **1989**, *111*, 8447.

105. (a) Brown, H. C., Krishnamurthy, S. *J. Am. Chem. Soc.* **1972**, *94*, 7159. (b) Krishnamurthy, S. *Aldrichimica Acta* **1974**, *7*, 55. (c) Krishnamurthy, S., Brown, H. C. *J. Am. Chem. Soc.* **1976**, *98*, 3383.

106. Wilds, A. L. *Org. React.* **1944**, *2*, 178.

107. (a) Collin, J., Namy, J.-L., Kagan, H. B. *Nouv. J. Chim.* **1986**, *10*, 229. (b) Evans, D. A., Kaldor, S. W., Jones, T. K., Clardy, J., Stout, T. J. *J. Am. Chem. Soc.* **1990**, *112*, 7001.

108. (a) Gung, B. W. *Tetrahedron* **1996**, *52*, 5263. (b) Tomoda, S., Senju, T. *Tetrahedron* **1999**, *55*, 3871. (c) Yadav, V. K., Jeyaraj, D. A., Balamurugan, R. *Tetrahedron* **2000**, *56*, 7581.

109. Procter, G. *Stereoselectivity in Organic Synthesis*, Oxford University Press: New York, 1998.

110. (a) Bürgi, H. B., Dunitz, J. D., Schefter, E. *J. Am. Chem. Soc.* **1973**, *95*, 5065. (b) Bürgi, H. B., Dunitz, J. D., Lehn, J. M., Wipff, G. *Tetrahedron* **1974**, *30*, 1563. (c) Bürgi, H. B., Dunitz, J. D. *Acc. Chem. Res.* **1983**, *16*, 153.

111. (a) Haubenstock, H., Eliel, E. L. *J. Am Chem. Soc.* **1962**, *84*, 2363. (b) Rei, M.-H. *J. Org. Chem.* **1983**, *48*, 5386. (c) Weissenberg, M., Levisalles, J. *Tetrahedron* **1995**, *51*, 5711.

112. (a) Mitsunobu, O. *Synthesis* 1981, 1. (b) Hughes, D. L. *Org. React.* **1992**, *42*, 335. (c) Dodge, J. A., Nissen, J. S., Presnell, M. *Org. Synth.* **1995**, *73*, 110.

113. Meyers, A. I., Bienz, S. *J. Org. Chem.* **1990**, *55*, 791.

114. For an interpretation of the mechanism, see Varasi, M., Walker, K. A. M., Maddox, M. L. *J. Org. Chem.* **1987**, *52*, 4235.

115. (a) Lodge, E. P., Heathcock, C. H. *J. Am. Chem. Soc.* **1987**, *109*, 2819. (b) Nogradi, M. *Stereoselective Synthesis: A Practical Approach*, VCH: Weinheim, Germany, 1995. (c) Nishio, M.; Hirota, M. *Tetrahedron* **1989**, *45*, 7201. (d) Ho, T.-L. *Stereoselectivity in Synthesis*, Wiley: New York, 1999.

116. Hanson, K. R. *J. Am. Chem. Soc.* **1966**, *88*, 2731.

117. Masamune, S., Choy, W., Petersen, J. S., Sita, L. R. *Angew. Chem., Int. Ed.* **1985**, *24*, 1.

118. Cram, D. J., Elhafez, F. A. A. *J. Am. Chem. Soc.* **1952**, *74*, 5828.

119. (a) Cornforth, J. W., Cornforth, R. H., Matthew, K. K. *J. Chem. Soc.* **1959**, 112. (b) Karabatsos, G. J. *J. Am. Chem. Soc.* **1967**, *89*, 1367.

120. (a) Chérest, M., Felkin, H., Prudent, N. *Tetrahedron Lett.* **1968**, 2199. (b) Anh, N. T., Eisenstein, O. *Nouv. J. Chem.* **1977**, *1*, 61. (c) Anh, N. T. *Top. Curr. Chem.* **1980**, *88*, 145.

121. Yamamoto, Y., Matsuoka, K., Nemoto, H. *J. Am. Chem. Soc.* **1988**, *110*, 4475.

122. (a) Reetz, M. *Topi. Curr. Chem.* **1982**, *106*, 1. (b) Weidmannn, B., Seebach, D. *Angew. Chem., Int. Ed.* **1983**, *22*, 31.

123. Cram, D. J., Kopecky, K. R. *J. Am. Chem. Soc.* **1959**, *81*, 2748.

124. Reetz, M. T. *Acc. Chem. Res.* **1993**, *26*, 462.

125. (a) Evans, D. A., Kozlowski, M. C., Murry, J. A., Burgey, C. S., Campos, K. R., Connell, B. T., Staples, R. J. *J. Am. Chem. Soc.* **1999**, *121*, 669. (b) Evans, D. A., Burgey, C. S., Kozlowski, M. C., Tregay, S. W. *J. Am. Chem. Soc.* **1999**, *121*, 686.

126. Atkinson, R, S. *Stereoselective Synthesis*, Wiley: New York, 1995, p. 304.

127. (a) Reetz, M. T., Jung, A. *J. Am. Chem. Soc.* **1983**, *105*, 4833. (b) Evans, D. A., Allison, B. D., Yang, M. G. *Tetrahedron Lett.* **1999**, *40*, 4457.

128. Mori, Y., Kuhara, M., Takeuchi, A., Suzuki, M. *Tetrahedron Lett.* **1988**, *29*, 5419.

129. Keck, G. E., Andrus, M. B., Romer, D. R. *J. Org. Chem.* **1991**, 56, 417.

130. Chen, K.-M., Hardtmann, G. E., Prasad, K., Repic, O., Shapiro, M. J. *Tetrahedron Lett.* **1987**, *28*, 155.

131. (a) Narasaka, K., Pai, F.-C. *Tetrahedron* **1984**, *40*, 2233. (b) For chelation control using Li[BH(n-Bu)$_3$], see Faucher, A.-M., Brochu, C., Landry, S. R., Duchesne, I., Hantos, S., Roy, A., Myles, A., Legault, C. *Tetrahedron Lett.* **1998**, *39*, 8425.

132. Saksena, A. K., Mangiaracina, P. *Tetrahedron Lett.* **1983**, 24, 273.

133. (a) Evans, D. A., Chapman, K. T. *Tetrahedron Lett.* **1986**, *27*, 5939. (b) Evans, D. A., Chapman, K. T., Carreira, E. M. *J. Am. Chem. Soc.* 1988, *110*, 3560.

134. (a) Kruger, D., Sopchik, A. E., Kingsbury, C. A. *J. Org. Chem.* **1984**, *49*, 778. (b) Brown, H. C., Park, W. S., Cho, B. T., Ramachandran, P. V. *J. Org. Chem.* **1987**, *52*, 5406. (c) Singh, V. K. *Synthesis* **1992**, 605. (d) Deloux, L., Srebnik, M. *Chem. Rev.* **1993**, *93*, 763. (e) Seyden-Penne, J. *Chiral Auxiliaries and Ligands in Asymmetric Synthesis*, Wiley: New York, 1995. (f) Itsuno, S. *Org. React.* **1998**, *52*, 395.

135. Corey, E. J. Lecture, Robert Welch Foundation Conferences on Chemical Research, XXVII, 1993.

136. Midland, M. M. *Chem. Rev.* **1989**, *89*, 1553.

137. (a) Midland, M. M., Greer, S., Tramontano, A., Zderic, S. A. *J. Am. Chem. Soc.* **1979**, *101*, 2352. (b) Midland, M. M., McDowell, D. C., Hatch, R. L., Tramontano, A. *J. Am. Chem. Soc.* **1980**, *102*, 867. (c) Midland, M. M., Graham, R. S. *Org. Synth.* **1985**, *63*, 57.

138. Molander, G. A., Bobbitt, K. L. *J. Org. Chem.* **1994**, *59*, 2676.

139. Ramachandran, P. V., Brown, H. C. *Recent Advances in Asymmetric Reductions with B-Chlorodiisopinocampheyl borane*, Am. Chem. Soc. Monograph 641, Washington, DC, 1996.

140. (a) Brown, H. C., Chandrasekharan, J., Ramachandran, P. V. *J. Am. Chem. Soc.* **1988**, *110*, 1539. (b) Ramachandran, P. V., Teodorovic, A. V., Rangaishenvi, M. V., Brown, H. C. *J. Org. Chem.* **1992**, *57*, 2379. (c) Rogic, M. M. *J. Org. Chem.* **2000**, *65*, 6868.

141. Horne, D. A., Fugmann, B., Yakushijin, K., Büchi, G. *J. Org. Chem.* **1993**, *58*, 62.

142. Masamune, S., Kim, B. M., Petersen, J. S., Sato, T., Veenstra, J. S., Imai, T. *J. Am. Chem. Soc.* **1985**, *107*, 4549.

143. (a) Imai, T., Tamura, T., Yamamuro, A., Sato, T., Wollmann, T. A., Kennedy, R. M., Masamune, S. *J. Am. Chem. Soc.* **1986**, *108*, 7402. (b) Masamune, S., Kennedy, R. M., Petersen, J. S., Houk, K. N., Wu, Y. *J. Am. Chem. Soc.* **1986**, *108*, 7404.

144. (a) Corey, E. J., Bakshi, R. K., Shibata, S. *J. Am. Chem. Soc.* **1987**, *109*, 5551. (b) Corey, E. J., Shibata, S., Bakshi, R. K. *J. Org. Chem.* **1988**, *53*, 2861. (c) Lohrag, B. B., Bhushan, V. *Oxazaborolidines*, Wiley-VCH: New York, 1998.

145. Corey, E. J., Bakshi, R. K., Shibata, S., Chen, C.-P., Singh, V. K. *J. Am. Chem. Soc.* **1987**, *109*, 7925.

146. Jones, D. K., Liotta, D. C., Shinkai, I., Mathre, D. J. *J. Org. Chem.* **1993**, *58*, 799.

147. (a) Nishizawa, M., Yamada, M., Noyori, R. *Tetrahedron Lett.* 1981, *22*, 247. (b) Noyori, R., Tomino, I., Tamimoto, Y., Nishizawa, M. *J. Am. Chem. Soc.* **1984**, *106*, 6709. (c) Noyori, R., Tomino, I., Yamada, M., Nishizawa, M. *J. Am. Chem. Soc.* **1984**, *106*, 6717.

148. (a) Wong, C.-H., Whitesides, G. M. *Enzymes in Synthetic Organic Chemistry*, Pergamon Press: Oxford, UK, 1994. (b) Smith, M. B. *Organic Synthesis*, 2nd ed., McGraw-Hill: Boston, 2002, pp. 415–420.

149. (a) Seebach, D., Sutter, M. A., Weber, R. H., Züger, M. F. *Org. Synth.* **1984**, *63*, 1. (b) Mori, K., Mori, H. *Tetrahedron* **1987**, *43*, 4097.

150. Huang, F.-C., Lee, L. F. H., Mittal, R. S. D., Ravikumar, P. R., Chan, J. A., Sih, C. J., Caspi, E., Eck, C. R. *J. Am. Chem. Soc.* **1975**, *97*, 4144.

151. Jones, J. B. *Pure Appl. Chem.* **1990**, *62*, 1445.

152. Esser, P., Buschmann, H., Meyer-Stork, M., Scharf, H.-D. *Angew. Chem., Int. Ed.* **1992**, *31*, 1190.

Functional Group Transformations: The Chemistry of Carbon-Carbon π-Bonds and Related Reactions

The goal is always finding something new, hopefully unimagined and, better still, hitherto unimaginable.

K. Barry Sharpless

5.1 REACTIONS OF CARBON-CARBON DOUBLE BONDS

Hydrogenation of Carbon-Carbon Double Bonds

Heterogeneous Catalytic Hydrogenation[1]

Hydrogenation of carbon-carbon double bonds is frequently carried out in the presence of a heterogenous metal catalyst and generally proceeds under mild conditions. Selective reduction of a double bond in the presence of other unsaturated groups is usually possible, except when the compound contains triple bonds, nitro groups, or an acyl halide.

The mechanism of heterogenous hydrogenation involves (1) dissociative chemisorption of H_2 on the catalyst, (2) coordination of the alkene to the surface of the catalyst, and (3) addition of the two hydrogen atoms to the activated π-bond in a *syn*-manner.

Catalyst Selection. For low-pressure hydrogenations (1–30 atm), Pt, Pd, Rh, and Ru are used. The reactivity of a given catalyst decreases in the following order: Pt > Pd > Rh ~ Ru > Ni. For high-pressure hydrogenations (100–300 atm), Ni is usually the metal of choice.

Platinum, prepared by reduction of PtO_2 (Adams catalyst) with H_2, is pyrophoric. Usually 0.1–1% of the catalyst is employed. A more convenient procedure for the preparation of Pt is by reduction of chloroplatinic acid with $NaBH_4$ in ethanol.[2]

$$H_2PtCl_6 \cdot 6\,H_2O + EtOH + NaBH_4 \longrightarrow Pt\text{-catalyst}$$

Palladium is available as a metal deposit on the surface of an inert support such as carbon.

Nickel is used for high-pressure hydrogenations. Supported-Ni catalysts such as *Raney-Ni* and *Ni-Boride* are employed for hydrogenolysis of the C–S bonds in thioacetals.[3] Raney-Ni is prepared by treating nickel-aluminum alloy with NaOH. Freshly prepared Raney-Ni absorbs much of the H_2 produced in the reaction.

$$NiAl_2 + 6\ NaOH + H_2O \longrightarrow Ni\ (ppt) + 2Na_3AlO_3 + 3H_2$$

Ni-Boride, readily prepared by reduction of $Ni(OAc)_2$ with $NaBH_4$, is more reactive than Raney-Ni.[4] Treatment of $NiCl_2$ with $NaBH_4$ in methanol followed by heating yields a stable suspension of Ni_2B, which catalyzes the regioselective 1,4-reduction of α,β-unsaturated aldehydes and ketones.[5]

$$NiCl_2 \cdot 6H_2O + NaBH_4 \text{ in MeOH} \longrightarrow Ni_2B \text{ suspension}$$
(store up to six months at 25 °C)

78%

93%

Solvent. The activity of a given catalyst generally is increased by changing from a neutral to a polar to an acid solvent. EtOAc, EtOH, and HOAc are the most frequently used solvents for low-pressure hydrogenations.

Substrate Reactivity. The reactivity of unsaturated substrates decreases in the following order: RCOCl > RNO_2 > RC=CR > RCH≡CHR > RCHO > RC≡N > RCOR > benzene. Both Pt and Pd catalysts fail to reduce RCOOR′, RCOOH, and $RCONH_2$ groups. Pd is usually more selective than Pt.

The ease of reduction of an olefin *decreases* with increasing substitution of the double bond. Conjugation of a double bond with a carbonyl group can markedly increase the rate of hydrogenation of the double bond.

98%

Stereochemistry. In general, hydrogenation takes place by a *syn*-addition of hydrogen to the *less hindered* side of the double bond.

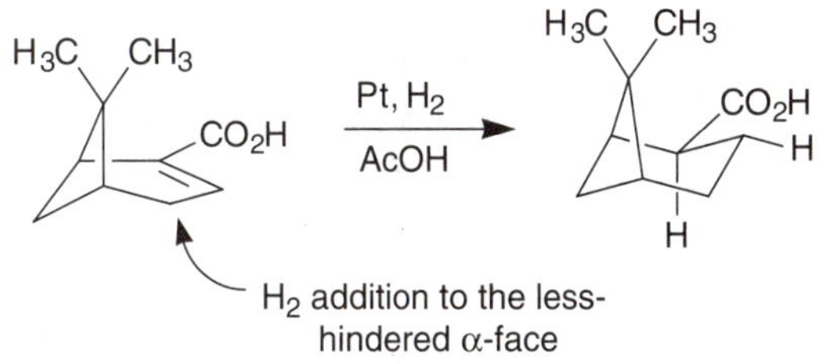

H₂ addition to the less-hindered α-face

Hydrogenolysis. With Pt and Pd catalysts, hydrogenation of *allylic and benzylic* alcohols, ethers, esters, amines, and halides is often accompanied by hydrogenolysis of the C–X bond where X = OH, OR, OAc, NR_2, or halide, respectively. Rh catalysts are particularly useful for hydrogenations when concomitant hydrogenolysis of an oxygen function is to be avoided. Divalent sulfur, Hg, and, to a lesser degree, amines poison hydrogenation catalysts.

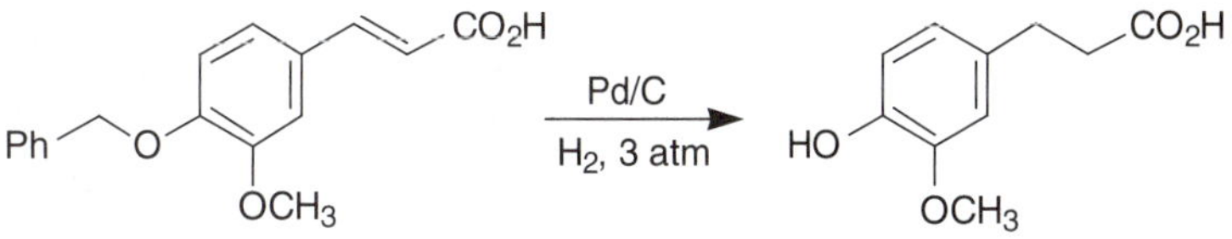

Isomerization. With Pd catalysts, and, to a lesser extent, with Pt catalysts, a mixture of isomeric products may be obtained due to positional isomerization of double bonds during hydrogenation.[6] As illustrated below, *syn*-addition of H_2 to either face of the double bond in alkene A furnishes *cis*-decalin C. However, *syn*-addition of H_2 to the isomerized alkene B can produce the *cis*-decalin C and/or *trans*-decalin D, depending on which face of the double bond undergoes addition by H_2. In fact, hydrogenation of A in the presence of Pt furnishes 80% of the thermodynamically more stable *trans*-decalin and only 20% of *cis*-decalin.

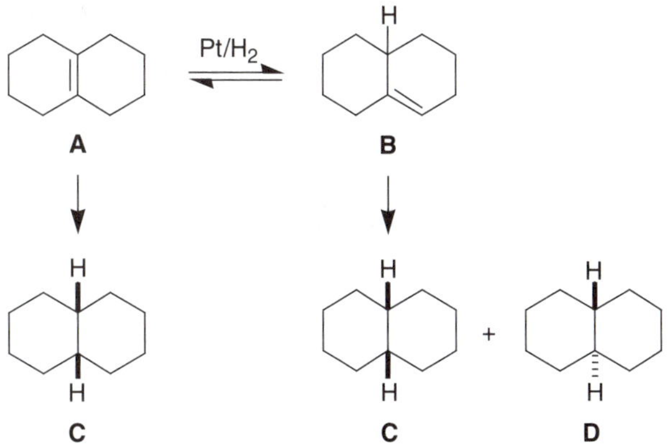

Homogeneous Catalytic Hydrogenation[7]

Chlorotris(triphenylphosphine)rhodium (Wilkinson's catalyst) is among the most efficient catalysts and permits hydrogenation in homogeneous solution.[8] The Rh complex is readily prepared by heating rhodium chloride with excess triphenylphosphine in ethanol.

$$RhCl_3 \cdot H_2O + Ph_3P \text{ (excess)} \longrightarrow (Ph_3P)_3RhCl \qquad \textit{Wilkinson's catalyst}$$

Characteristic features of this Rh-catalyst include (1) hydrogenation of double bonds via *syn*-addition of H_2, (2) *cis*–double bonds are hydrogenated faster than *trans*–double bonds, (3) terminal double bonds are hydrogenated more rapidly than more substituted double bonds,[9] (4) less isomerization of double bonds, (5) little hydrogenolysis of allylic or benzylic ethers and amines, and (6) R–C≡N, R–NO₂, R–Cl, RCOOH, RCOOR′, and R₂C=O are not reduced.

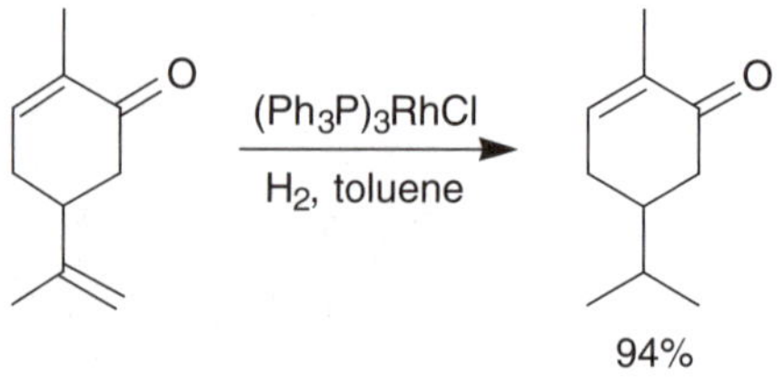

94%

Directed Hydrogenation[10]

Intramolecular H-bonding or chelation by an adjacent functionality, such as a hydroxyl group, can direct the approach of a metal catalyst to favor hydrogenation of one diastereotopic π-face over another. The most effective catalysts for directed hydrogenation are the coordinatively unsaturated Crabtree's catalyst[11] and the 2,5-norbornadiene-Rh(I) catalyst shown below.[12]

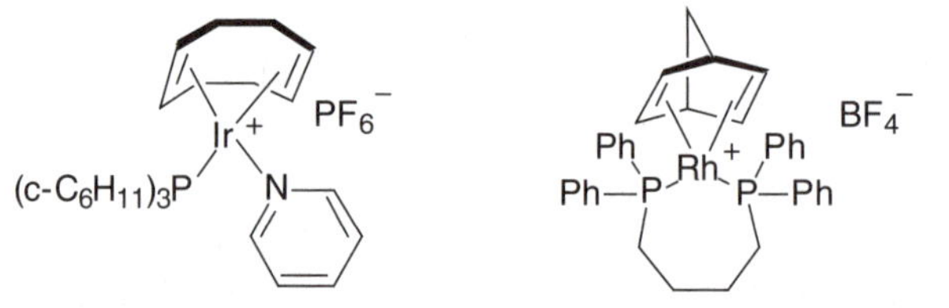

[Ir(cod)(PChx₃)(py)]PF₆
Crabtree's catalyst

Rh(Diphos-4)⁺
2,5-norbornadiene Rh catalyst

For example, in the case of cyclic unsaturated alcohols, *face-selective* hydrogenation occurs when the hydroxyl group binds to the Ir during hydrogenation of the double bond.[11,13]

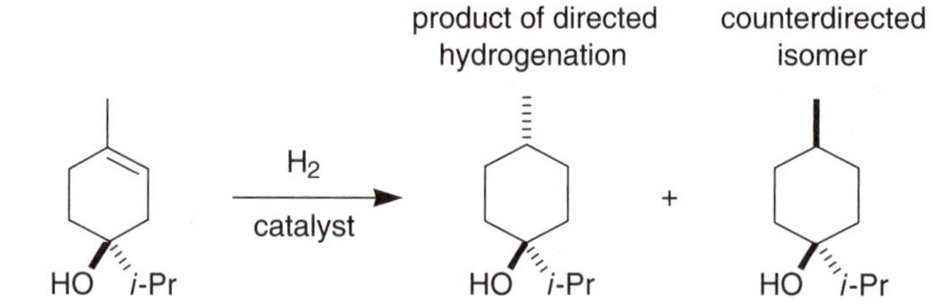

	product of directed hydrogenation	counterdirected isomer
5% Pd/C, EtOH	20	80
5% Pd/C, CH₂Cl₂	53	47
[Ir(cod)(PChx₃)(py)]PF₆	> 99.9	< 0.01

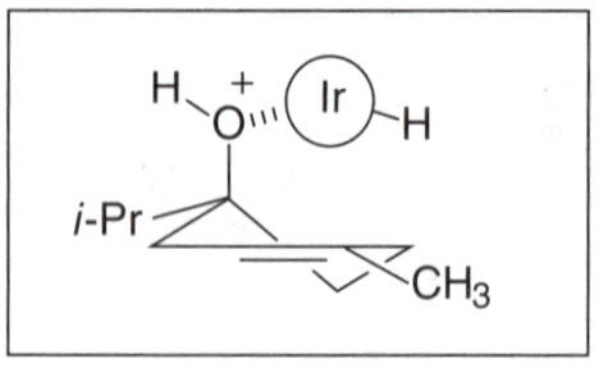

Asymmetric Hydrogenation[14]

Many chiral phosphorus-based auxiliary ligands are available for transition metals in asymmetric, catalytic, homogeneous reductions of alkenes. Particularly noteworthy are the diphosphine ligands DIPAMP, developed by William S. Knowles (Nobel Prize, 2001), and BINAP, developed by Ryoji Noyori (Nobel Prize, 2001).

DIPAMP   (+)-BINAP

BINAP is available as either the (+)- or (−)-enantiomer and displays broad utility in rhodium- and ruthenium-catalyzed asymmetric hydrogenations of β-keto esters and alkenes.[15]

92% yield, 97% ee
Naproxen
anti-inflammatory analgesic

Dissolving Metal Reductions

Solutions of Li, Na, or K in liquid ammonia (bp −33 °C) contain solvated metal cations and electrons. These solutions are able to reduce α, β-unsaturated ketones and aromatic substrates.

$$[NH_3]_{\text{liquid}} \ + \ M° \ \longrightarrow \ M^+ \ [NH_3 \text{---} e^- \text{---} NH_3]$$

solvated electron
blue solution

Reduction of Enones[16]

Reductions of α,β-unsaturated ketones with solutions of Li, Na, or K in liquid NH_3 are chemoselective, resulting in the exclusive reduction of the carbon-carbon double bonds. The reaction involves addition of the substrate dissolved in Et_2O or THF to a well-stirred solution of the metal in liquid NH_3. Addition of *one* equivalent of *t*-BuOH as a proton donor is beneficial for driving the reduction to completion. However, excess alcohol, especially of the more acidic EtOH, causes further reduction of the saturated ketone initially formed to the corresponding alcohol.

Transfer of an electron to the conjugated π-system of the enone furnishes a radical anion, which on protonation (*t*-BuOH) followed by transfer of a second electron affords an enolate ion. Its protonation on workup gives the saturated ketone, or it may be alkylated prior to workup to form a new C–C bond. The *regiospecific* generation of enolate ions from α,β-unsaturated ketones is an important tool in carbon-carbon-bond-forming reactions.

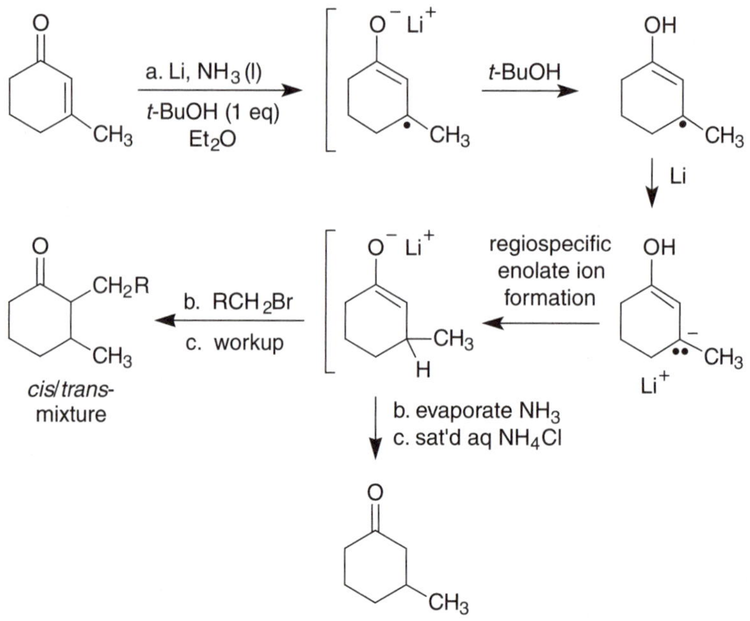

Catalytic hydrogenation of an enone would not be chemoselective if an isolated double bond were also present in the molecule. However, isolated double bonds are inert to dissolving metal reduction. On the other hand, a variety of functional groups are reduced with alkali metals in liquid ammonia. These include alkynes, conjugated dienes, allylic, or benzylic halides and ethers.

It is noteworthy that reduction of the α,β-unsaturated decalone shown below with lithium in liquid ammonia furnishes the *trans*-decalone as the major product in spite of the 1,3-diaxial interaction between the CH$_3$–OCH$_3$ substituents.[17]

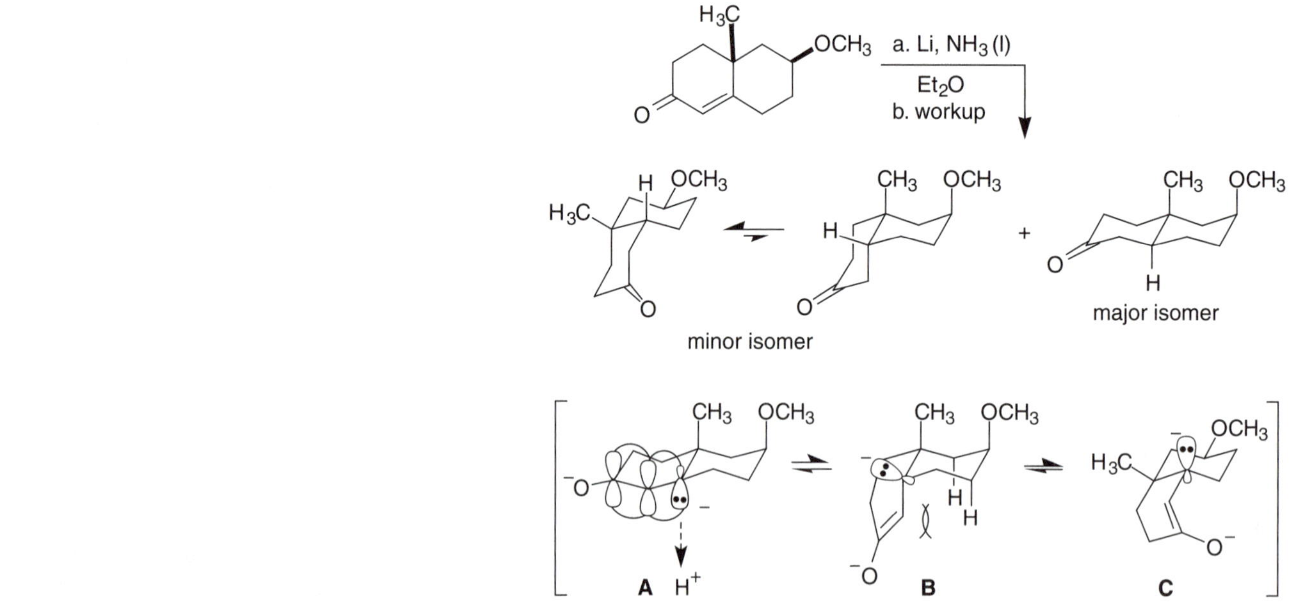

An explanation of the observed isomer distributions is that sp^3-π overlap is possible in conformations A and B, but not in C. Conformation A is favored because the *cis*-decalin-type conformation B experiences additional 1,3-destabilizing interactions at the concave face. Thus, the enolate ion intermediates formed during dissolving metal reduction of bicyclic enones react with electrophiles (e.g., MeO$_2$C–CN) to afford predominantly *trans*-decalone products.[18]

84%

Nonconjugated ketones can be reduced in the dissolving metal medium to the corresponding saturated alcohol in the presence of excess alcohol prior to workup. In the case of cyclic ketones, the thermodynamically more stable alcohol predominates. For example, 4-*t*-butylcyclohexanone on reduction with Na in liquid NH$_3$–Et$_2$O and excess *t*-BuOH furnishes the *trans*-alcohol in greater than 98% isomeric purity, while reduction of the same ketone with LAH in ether provides the corresponding *trans*-alcohol in 89% isomeric purity.

Extension of the dissolving metal reduction of enones to α,β-unsaturated carboxylic acid esters converts the ester moiety to an amide. However, α,β-unsaturated esters undergo double bond reduction on treatment with magnesium in methanol.[19]

96%

Reduction of Aromatic Compounds—The Birch Reduction[20]

Alkali metals in liquid ammonia in the presence of an alcohol reduce aromatic systems to 1,4-cyclohexadienes. These can be further elaborated into a host of derivatives. The availability of a wide variety of substituted aromatic compounds, either commercial or via synthesis, makes the Birch reduction an important tool in organic synthesis.

The reduction is initiated by addition of a solvated electron to the aromatic system to generate a radical anion, which is then protonated by an alcohol cosolvent to furnish a pentadienyl radical. Addition of another electron leads to a pentadienyl anion, which can be protonated either at the *ortho-* or at the *para*-position (with respect to the sp³ carbon). Kinetically controlled protonation of both the initial radical anion and the pentadienyl anion occurs at the *para*-position, which possesses the highest electron density, according to MO calculations.[21] Selectivity for the 1,4-diene product is also predicted by the "principle of least motion";[22] that is, protonation at the *para*-position produces the least change in nuclear position.[23]

The function of the alcohol in the metal –NH₃ reduction is to provide a proton source that is more acidic than ammonia to ensure efficient quenching of the radical anion and pentadienyl anion species. Furthermore, the presence of alcohol represses the formation of the amide ion NH_2^-, which is more basic than RO^-M^+ and is capable of isomerizing the 1,4-cyclohexadiene product to the thermodynamically more stable conjugated 1,3-cyclohexadiene.

Reaction Conditions. A typical procedure for the reduction of *o*-xylene to 1,2-dimethyl-1,4-cyclohexadiene is as follows: sodium metal (or Li wire) is cut into small pieces and slowly added to a solution of the aromatic substrate in a solvent mixture of liquid NH₃, Et₂O (or THF), and EtOH (or *t*-BuOH).[20b] The alcohol does not react with the metal at −33 °C (bp of liquid ammonia). Relative rates of benzene reduction are Li = 360, Na = 2, and K = 1.

Birch reduction of substrates containing methoxy or *N,N*-dimethylamino groups may be contaminated with appreciable amounts of conjugated products. In these cases, it is conceivable that the isomerization occurs during workup.[24]

Allylic and benzylic heteroatom substituents such as –OR, –SR, and halogens undergo concomitant hydrogenolysis during Birch reduction. However, benzylic –OH groups are converted to alkoxides, and the resultant electron-rich –CH$_2$O$^-$ moiety resists further reduction.

Regiochemistry. Birch reduction of monosubstituted benzenes could furnish either of two possible 1,4-cyclohexadienes, A or B below.

The regiochemical course of the reduction of substituted benzenes is determined by the site of initial protonation of the radical anion species. Generally, *electron-donating groups* (D) retard electron transfer and remain on *unsaturated* carbons.

The reason why groups such as –C(O)R, –CHO, and –CO$_2$R *behave as electron-donating groups* in this reaction is that they are reduced to –CH$_2$O$^-$ before reduction of the aromatic system occurs. Isolated double bonds are generally stable under Birch reduction. Conjugated dienes and alkynes are reduced, the latter to *trans*-alkenes.

The deactivating effects of alkyl and alkoxy substituents on the regiochemistry of reduction of substituted naphthalenes are exemplified below.[25]

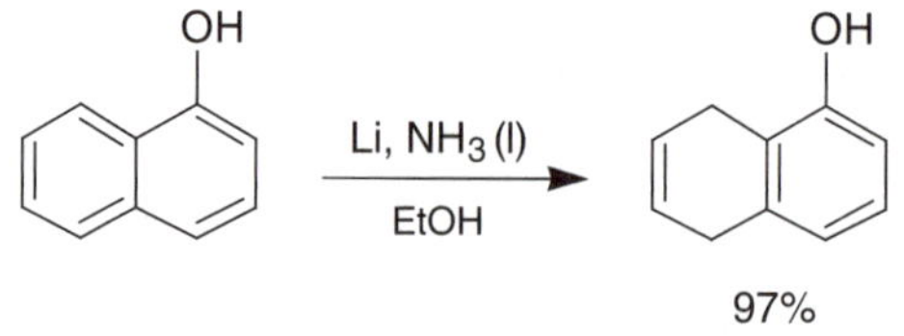

97%

Electron-withdrawing groups (EWG) facilitate electron transfer and reside on *saturated* carbons. As with all Birch reductions, the saturated (sp^3) carbons are para to each another.

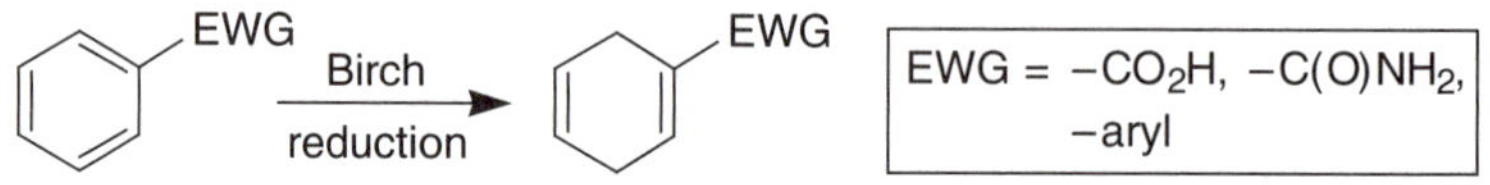

EWG = $-CO_2H$, $-C(O)NH_2$, $-aryl$

Birch reduction of benzoic acids in the presence of an alcohol (proton donor) furnishes 1,4-dihydrobenzoic acids. The carboxylate salt ($-CO_2^-M^+$) formed during reduction of benzoic acid derivatives is sufficiently electron rich that it is not reduced.

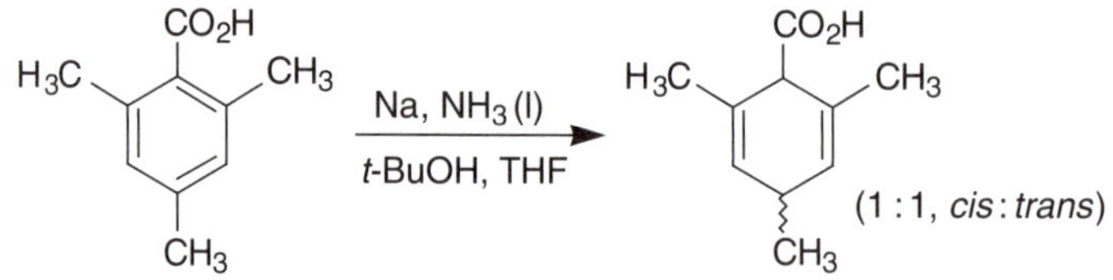

(1:1, *cis*:*trans*)

The carboxy group generally *dominates* the regiochemistry of the reduction when other substituents are present.

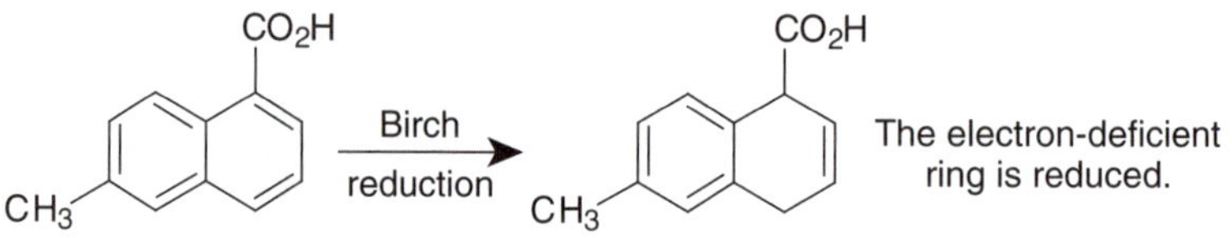

The electron-deficient ring is reduced.

The strong activation effect by the carboxyl group allows reduction to occur when only *one* equivalent of alcohol is present or even without an alcohol. In these cases, the intermediate dianion persists in solution and can be trapped with electrophilic reagents to generate a *quaternary* carbon center.[26]

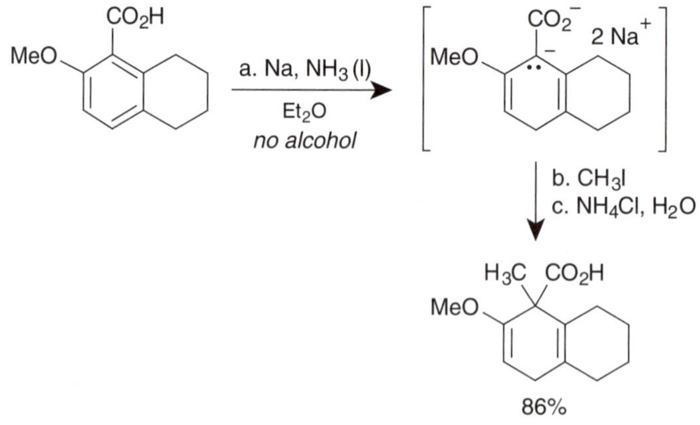

86%

While ester groups are reduced competitively with the aromatic ring under the usual Birch conditions, addition of one or two equivalents of H_2O or *t*-BuOH to NH_3 before metal addition preserves the ester moiety.[27]

64%

Synthetic Applications of the Birch Reaction

Target molecules containing a 1,4-cyclodhexadiene unit are probably best prepared via the Birch reaction. These primary reduction synthons can be further elaborated into a variety of synthetically useful compounds, as exemplified below.

Formation of 1,3-Cyclohexadienes. Isomerizations of 1,4-cyclohexadienes with KOt-Bu in DMSO furnish the thermodynamically more stable conjugated 1,3-cyclohexadienes.

Formation of Cyclohexenones. Hydrolysis of the initial enol ether (vinyl ether) formed from Birch reduction of anisole or substituted anisoles under mild acidic conditions leads to β,γ-unsaturated cyclohexenones. Under more drastic acidic conditions, these isomerize to the conjugated α,β-cyclohexenones. Birch reduction of anisoles followed by hydrolytic workup is one of the best methods available for preparing substituted cyclohexenones.[28]

It should be noted that Birch reduction of 4-substituted anisoles followed by acidic workup (aq HCl, THF) produces mixtures of isomeric cyclohexenones containing an appreciable amount of the β,γ-unsaturated product.[29]

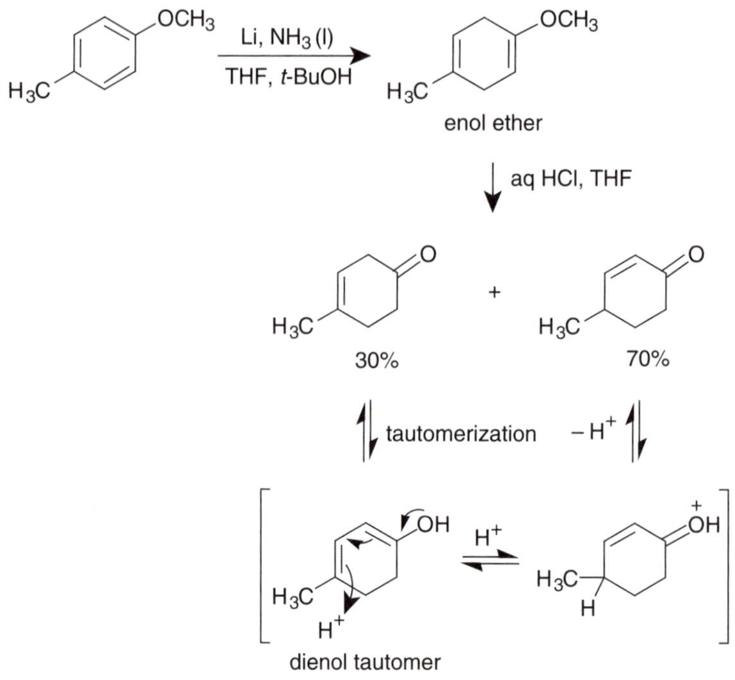

enol ether

aq HCl, THF

30%   +   70%

tautomerization   − H⁺

dienol tautomer

Formation of Acyclic Compounds. Selective cleavage of the more nucleophilic double bond of anisole-derived 1,4-cyclohexadienes by ozone provides highly functionalized acyclic compounds containing a stereodefined double bond.

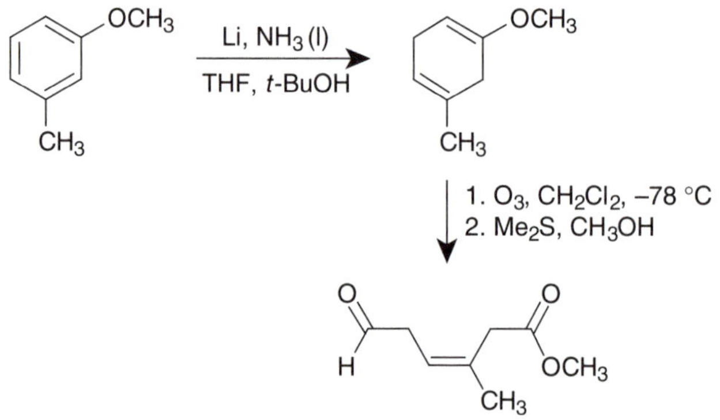

1. O₃, CH₂Cl₂, −78 °C
2. Me₂S, CH₃OH

Formation of Chiral Quaternary Carbon. Birch reduction-alkylation of benzoic acids and esters establishes quaternary carbon centers. Neighboring stereocenters will influence the stereochemical outcome of the tandem reaction sequence. The following example illustrates how a chiral auxiliary (derived from prolinol)[30] controls the stereoselection in the Birch reduction-alkylation step.[31, 32]

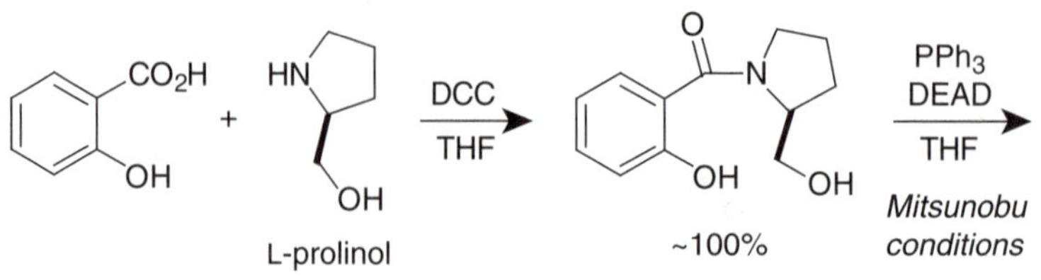

L-prolinol

DCC
THF

~100%

PPh₃
DEAD
THF

Mitsunobu conditions

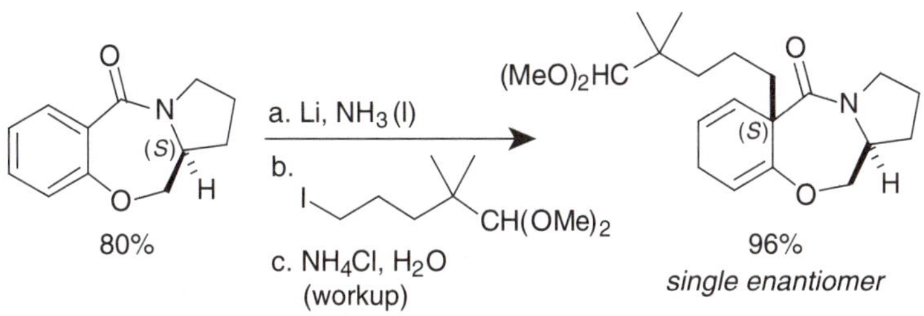

80%   96%
single enantiomer

a. Li, NH₃ (l)
b.
c. NH₄Cl, H₂O (workup)

Synthesis of Aromatic Compounds for the Birch Reduction

Electrophilic Aromatic Substitution Reactions. Friedel-Crafts alkylation, acylation, and the Vilsmeier-Haack formylation,[33] shown below, are excellent reactions for the synthesis of substituted aromatic compounds.

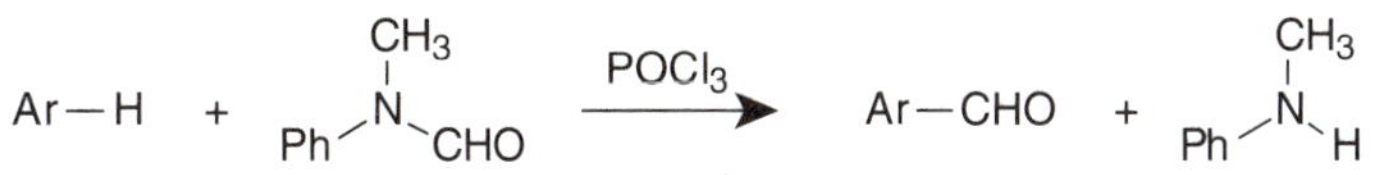

Reactions of Aromatic Lithium Compounds. See Chapter 7, Section 7.1.

Hydration of Alkenes

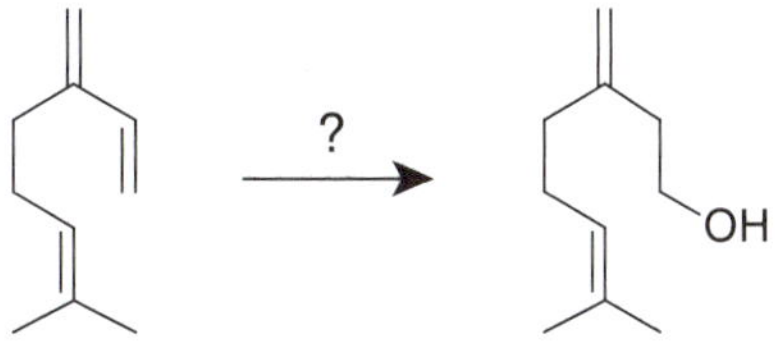

Hydroboration-Oxidation[34]

The hydration of alkenes via hydroboration-oxidation, developed by Herbert C. Brown (1912–2004; Nobel Prize, 1979) and coworkers, provides a valuable tool for the synthesis of a wide variety of alcohols of predictable regio- and stereochemistry.

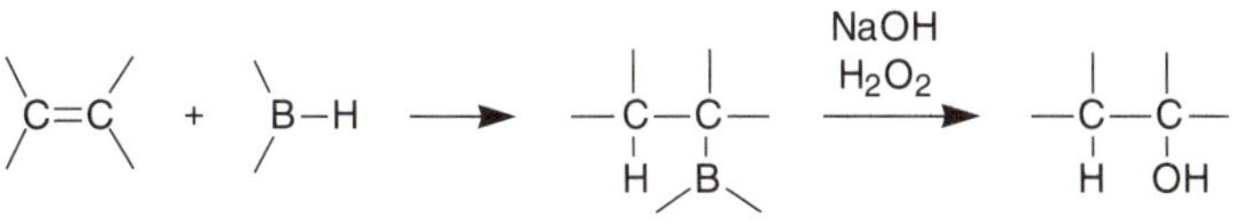

Preparation of Organoboranes.[35] Borane-tetrahydrofuran (BH_3•THF) and borane-dimethyl sulfide (BH_3•SMe_2) are the reagents of choice for hydroboration of alkenes. They are commercially available — BH_3•THF as a 1–2 M solution in THF and BH_3•SMe_2 neat, 10.0–10.2 M in borane. The latter reagent is especially suitable as a hydroborating agent since it can be stored at 0 °C for long periods and can be used in a variety of aprotic solvents.

    Except for a few very hindered double bonds, virtually all alkenes undergo hydroboration. Extreme crowding may cause isomerization of the initial organoborane.[36] The presence of a functional group may interfere with hydroboration (see Table 4.3). In such cases, it is often necessary to use a sterically hindered organoborane. For example, dicyclohexylborane (Chx_2BH) is an excellent reagent for the selective hydroboration of 1-alkenes in the presence of RCHO and $R_2C{=}O$ functionality.[37]

Scope. Borane is a trifunctional molecule and reacts with alkenes to give three types of organoboranes: R_3B, trialkylboranes; R_2BH, dialkylboranes; and RBH_2, monoalkylboranes. Whereas R_3B reagents are monomeric, both R_2BH and RBH_2 reagents are *dimeric*. For convenience they are referred to as monomers in the following discussion.

Unhindered alkenes such as monosubstituted alkenes react rapidly with BH_3 to produce R_3B.

Moderately hindered alkenes and certain cycloalkenes also can form R_3B. However, control of the stoichiometry and reaction conditions permits the preparation of R_2BH.

For tetrasubstituted alkenes and cycloalkenes, hydroboration to the trialkylborane stage is difficult. In these cases, it is possible to produce cleanly the monoalkylboranes RBH_2.

Selective Hydroborating Agents. The mono- and dialkylboranes are themselves useful hydroborating agents for sterically less hindered alkenes.[38] Some important mono- and dialkylboranes are given below.

- *Thexylborane* ($ThxBH_2$, 1,1,2-trimethylpropylborane) is readily prepared by treating $BH_3 \cdot THF$ or $BH_3 \cdot SMe_2$ in THF with 2,3-dimethyl-2-butene in a 1:1 ratio.[39] Although this monoalkylborane is crowded, it will hydroborate two additional alkenes provided they are not too hindered.

- *Thexylchloroborane*, prepared by the reaction of thexylborane with an equimolar amount of hydrogen chloride in ether[40] or by hydroboration of 2,3-dimethyl-2-butene with $BH_2Cl \bullet SMe_2$,[41] is a valuable reagent for the synthesis of mixed thexyl-*n*-dialkylboranes and for the reduction of carboxylic acids to the corresponding aldehydes (see Section 4.10).

- *Disiamylborane* (Sia_2BH, 1,2-dimethylpropylborane), a dialkylborane with a large steric requirement, is prepared by hydroboration of 2-methyl-2-butene with either $H_3B \bullet THF$ or $H_3B \bullet SMe_2$. The reagent should be used immediately after preparation. Disiamylborane provides for chemo-, regio-, and stereoselective hydroborations.

- *Dicyclohexylborane* (Chx_2BH) may be made using either $H_3B \bullet THF$ or $H_3B \bullet SMe_2$. It is a white crystalline solid, sparingly soluble in Et_2O and THF, and is used immediately after its preparation. Dicyclohexylborane is more stable than disiamylborane and exhibits similar reaction characteristics.

- *9-Borabicyclo[3.3.1.]nonane* (9-BBN) is available by hydroboration of 1,5-cyclooctadiene with one equivalent of BH_3,[42] or commercially either as a crystalline dimer or as a THF solution. The reagent is thermally more stable than dicyclohexylborane. It is frequently used as an *anchor* group in organoborane reactions, allowing an efficient utilization of valuable alkenes.

- *Diisopinocampheylborane* [Ipc_2BH, bis(pinan-3β-yl)borane] is a chiral reagent obtained by hydroboration of (+)- or (−)-α-pinene. Hydroboration of *cis*-alkenes with (+) or (−)-Ipc_2BH creates new chiral centers.[43]

(+)-α-pinene

Regiochemistry. An important feature of the hydroboration reaction is the regio-selectivity observed with unsymmetrical alkenes. The direction of B–H addition to the double bond is influenced by inductive, resonance, and steric effects of the substrate and by the nature of the hydroborating agent. Although the mechanism of hydroboration is complex, the reaction can be depicted to involve interaction of the vacant p-orbital of boron with the π-orbital of the alkene. This leads to a partial negative charge on boron and a partial positive charge on the carbon. Transfer of a hydrogen with its electron pair from boron to the carbon that can best accommodate the partial positive charge proceeds via a four-center-type transition state.

Generally, boron attacks positions of highest electron density and lowest steric congestion. This working hypothesis rationalizes the regioselectivity and stereo-selectivity of most hydroboration reactions.[44] Because of the larger steric requirement of a dialkylborane, R_2BH reagents exhibit greater regioselectivity for addition to the less hindered carbon of a double bond in comparison to $BH_3 \cdot THF$ or $BH_3 \cdot SMe_2$ (Table 5.1). It is important to note that the presence of a functional group may influence not only the regioselectivity of the alkene hydroboration, but also the stability of the resulting organoborane.[45]

Table 5.1 Regiochemistry of Alkene Hydroboration[a]

	$RCH=CH_2$		i-$PrCH=CHCH_3$		$(CH_3)_2C=CHCH_3$		$PhCH=CH_2$	
	↑	↑	↑	↑	↑	↑	↑	↑
$BH_3 \cdot THF$	6	94	43	57	2	98	19	81
Sia_2BH	1	99	3	97			2	98
9-BBN	1	99	1	99			2	98

[a] Determined from the ratios of the alcohols obtained after oxidation of the organoboranes with alkaline hydrogen peroxide.

Chemoselectivity. The rates of hydroboration of alkenes with dialkylboranes vary over a wide range. Thus, it is possible to selectively react one double bond in dienes, as shown in Table 5.2.[46] Relative rates of hydroboration with dialkylboranes are as follows:

9-BBN: $R_2C=CH_2$ > $RCH=CH_2$ > cyclopentene > (E) RCH=CHR >
  $R_2C=CHR$ > (Z) RCH=CHR > cyclohexene > $R_2C=CR_2$

Sia_2BH: $RCH=CH_2$ > $R_2C=CH_2$ > (Z) RCH=CHR, cyclopentene >
  (E) RCH=CHR > cyclohexene > $R_2C=CHR$

Table 5.2	Chemoselectivity of Diene Hydroboration

Stereochemistry. The four-center transition state for hydroboration of alkenes discussed above implies that addition of the B–H bond to a double bond proceeds in a *syn*-manner. Moreover, hydroboration of a bicyclic alkene, such as α-pinene, results in the addition of B–H from the less hindered face of the molecule to give Ipc_2BH.

Hydroboration involves:

- *Regioselective* addition of B–H
- *cis*-Addition of B–H
- Addition of B–H from the less-hindered face

Oxidation of Organoboranes. The facile reactions of olefins and dienes with various hydroborating agents makes a variety of organoboranes readily available. Organoboranes tolerate many functional groups and are formed in a stereospecific manner. The boron atom in these organoboranes can be readily substituted with a variety of functional groups.[47] For example, hydroboration followed by in situ oxidation by alkaline hydrogen peroxide provides for the *anti-Markovnikov hydration* of double bonds.[48]

$$R_3B + 3 H_2O_2 + NaOH \longrightarrow 3 ROH + NaB(OH)_4$$

As an alternative, organoboranes can be oxidized with Na-perborate ($NaBO_3 \cdot 4 H_2O$)[49] or with trimethylamine N-oxide (Me_3NO).[50] Oxidation of trialkylboranes with pyridinium chlorochromate (PCC) leads directly to the corresponding carbonyl compounds.[51]

Oxidation of the C–B bond occurs with *complete retention of configuration*, allowing the synthesis of alcohols of predictable stereochemistry.[52]

R = *trans*-2-methylcyclohexyl

100%

Asymmetric Synthesis of Alcohols from Alkenes[53]

Diisopinocampheylborane. (–)-Ipc$_2$BH and (+)-Ipc$_2$BH are enantiomerically pure hydroborating agents that are readily accessible via hydroboration of (+)-α-pinene and (–)-α-pinene, respectively.[41,54]

(+)-α-pinene   (–)-Ipc$_2$BH

(–)-α-pinene   (+)-Ipc$_2$BH

Hydroboration of *cis*-alkenes with these reagents creates new chiral centers. Oxidation of the resultant organoboranes with retention furnishes alcohols of predictable *absolute* stereochemistry. Ipc$_2$BH is perhaps one of the most versatile chiral reagents for laboratory use. A rationalization of the observed stereoselectivities has been proposed.[55] Under similar experimental conditions but using (+)-Ipc$_2$BH, hydroboration of *cis*-2-butene produces the (+)-(S) alcohol.

cis-2-butene   (–)-*R*, 99% ee

Monoisopinocampheylborane. Although asymmetric hydroboration using Ipc$_2$BH is exceptionally effective with *cis*-disubstituted alkenes, it is less than satisfactory with *trans*-disubstituted alkenes and trisubstituted alkenes. However, monoisopinocampheylborane (IpcBH$_2$) achieves asymmetric hydroboration of *trans*-alkenes and of trisubstituted alkenes with good asymmetric induction.[56]

R = CH$_3$, 67% ee
R = Ph, 100% ee

2,5-Dimethylborolane. Excellent results in asymmetric hydroborations are obtained with Masamune's *trans*-2,5-dimethylborolane reagents.[57] The C(2)-symmetry of the reagent ensures that both faces of the borolane ring are equivalent, which reduces the number of competing *anti*-Markovnikov transition states.[58] Unfortunately, the reagent is not readily available.

(*R,R*)-2,5-dimethylborolane

α-face approach   β-face approach

favored   *vs.*   *disfavored*

NaOH
H$_2$O$_2$

~98% ee

Diastereoselective Hydroboration. Hydroborations of *chiral acyclic* alkenes with *achiral* boranes often furnish, after oxidative workup, the corresponding alcohols with good diastereoselectivities. In these cases one uses a stereocenter present in the molecule to introduce the selective formation of a new chiral center (an example of *single asymmetric induction*).[59] It is important that the stereocenter is adjacent to the double bond (allylic position) undergoing hydroboration. The preferred conformations for such alkenes are those in which the allylic A^{1,3} strains are minimized with the smallest group of the stereocenter eclipsing the double bond. In the absence of strong electronic or chelating effects, the hydroborating agent attacks the double bond from the face opposite to the larger group, as exemplified below.[60]

The hydroboration is not only stereoselective but also regioselective, placing boron at the less hindered, mono-substituted carbon of the double bond.

In addition to the 1,3-allylic strain concept, Houk has employed a model for π-facial stereoselection of electrophilic additions to chiral alkenes, such as hydroboration, epoxidation, and dihydroxylation, with similar predictive success.[55]

Oxymercuration – Demercuration[61]

Acid catalyzed hydration of alkenes is not well suited for laboratory preparation of alcohols. Since the reaction proceeds via carbocation intermediates, mixtures of alcohols may be formed. However, oxymercuration-demercuration of alkenes provides a simple tool for regioselective hydration of alkenes whereby rearrangements are seldom observed.

Treatment of an alkene with mercuric acetate in aqueous THF results in the electrophilic addition of mercuric ion to the double bond to form an intermediate mercurium ion. Nucleophilic attack by H_2O at the *more substituted* carbon yields a stable organomercury compound, which upon addition of $NaBH_4$ undergoes reduction. Replacement of the carbon-mercury bond by a carbon-hydrogen bond during the reduction step proceeds via a radical process. The overall reaction represents *Markovnikov hydration of a double bond,* which contrasts with the hydroboration-oxidation process.

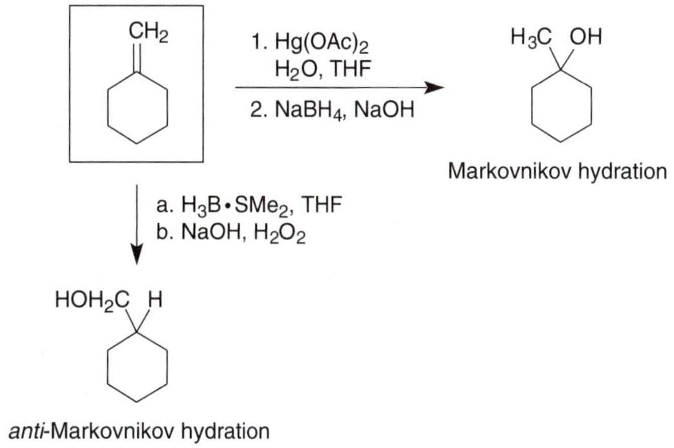

Markovnikov hydration

anti-Markovnikov hydration

Solvomercuration-Demercuration

Mercuration of 1-alkenes in the presence of nucleophilic solvents such as alcohols, amines, and nitriles or in the presence of sodium azide provides convenient access to the corresponding ethers,[62] amines,[63] amides,[64] and azides.[65]

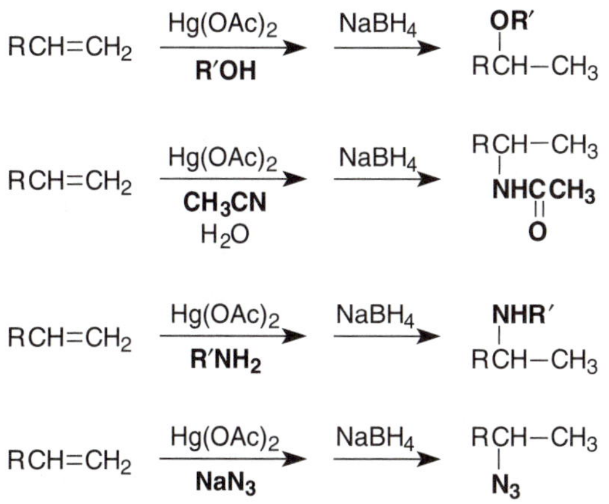

Oxidation of RHgX

The *intramolecular* version of the oxymercuration reaction affords cyclic ethers. Furthermore, treatment of organomercury compounds (R-HgX) with $NaBH_4$ in DMF in the presence of O_2 replaces the carbon-mercury bond by a carbon-oxygen bond and yields the corresponding alcohol (R-OH).[66]

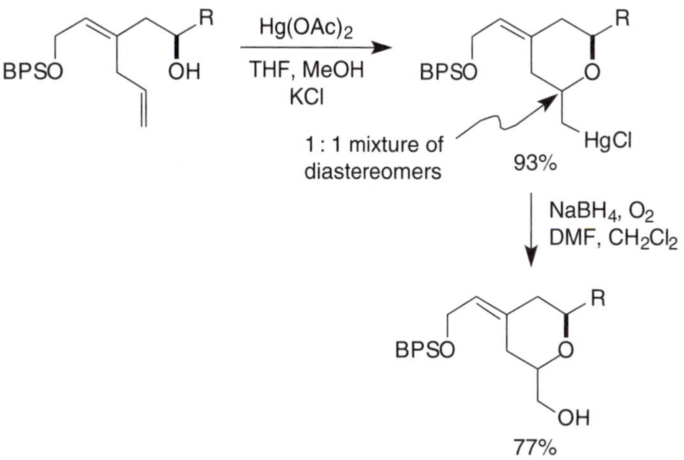

1 : 1 mixture of diastereomers   93%

77%

Epoxidation of Alkenes[67]

Epoxides (oxiranes) are widely used as versatile synthetic intermediates because regio- and stereoselective methods exist both for their construction and subsequent reactions. Reactions of epoxides are dominated by the electrophilic nature of the strained three-membered ring, which is susceptible to attack by a variety of nucleophiles. The 1,2-bifunctional pattern makes epoxides the charge affinity counterpart of enol derivatives, which are nucleophilic at the α-position.[68]

Epoxidation of Alkenes Using Peroxy Acids

The reaction of alkenes with peroxy acids provides for convenient and selective oxidation of double bonds. The peroxy acids most commonly used in the laboratory are m-chloroperoxybenzoic acid (mCPBA), monoperoxyphthalic acid magnesium salt (MMPP),[69] peroxybenzoic acid, peroxyformic acid, peroxyacetic acid, trifluoroperoxyacetic acid, and t-butyl hydroperoxide [$(CH_3)_3COOH$]. The order of reactivity for epoxidation of alkenes follows the trend $CF_3CO_3H > mCPBA \sim HCO_3H > CH_3CO_3H \gg H_2O_2 > t\text{-BuOOH}$.

mCPBA MMPP

Stereoselectivity. Epoxidation involves an electrophilic *syn*-addition of the oxygen moiety of the peroxy acid to the double bond. The concerted formation of two new C–O bonds ensures that the reaction is stereospecific: *cis*-alkenes furnish the corresponding *cis*-epoxides and *trans*-alkenes the corresponding *trans*-isomers (racemic).

With conformationally rigid cyclic alkenes, the reagent preferentially approaches the double bond from the less hindered side.

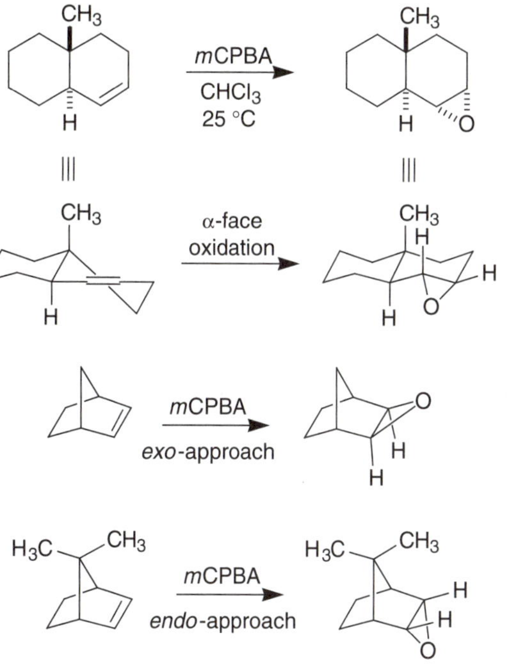

mCPBA
CHCl₃
25 °C

α-face
oxidation

mCPBA
exo-approach

mCPBA
endo-approach

Chemoselectivity. The rate of epoxidation increases with the number of electron-donating substituents on the double bond. The order of alkene reactivity with peroxy acids is as follows:

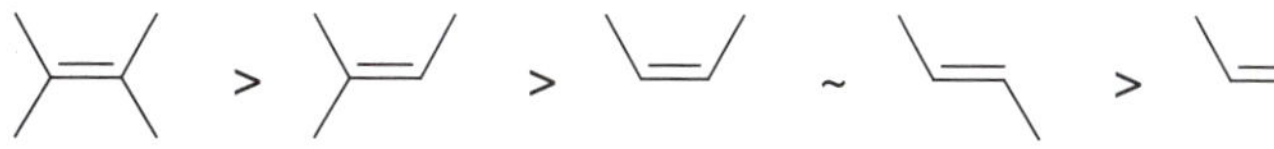

Hence, the more nucleophilic (more substituted) double bonds of the diene and triene depicted below undergo preferential epoxidation when treated with an equimolar amount of the peroxy acid.

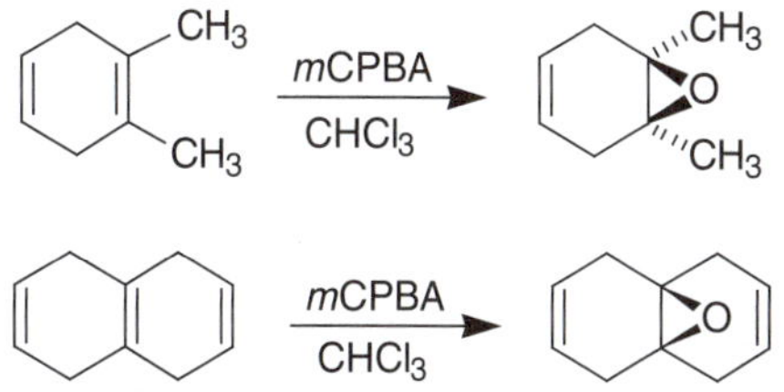

mCPBA
CHCl₃

mCPBA
CHCl₃

Conjugation of the alkene double bond with an electron-withdrawing group reduces the rate of epoxidation. Thus, α,β-unsaturated carboxylic acids and esters require a stronger oxidant, such as trifluoroperoxyacetic acid, for oxidation.

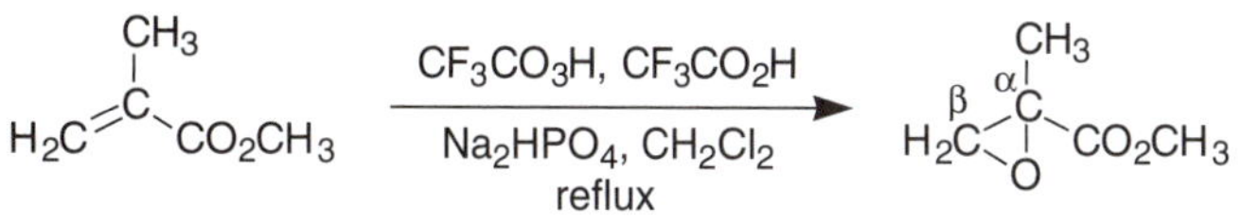

CF_3CO_3H, CF_3CO_2H
Na_2HPO_4, CH_2Cl_2
reflux

Alternatively, epoxy esters may be prepared via the Darzens reaction, which involves base-mediated condensation of α-chloro esters with aldehydes or ketones.[70]

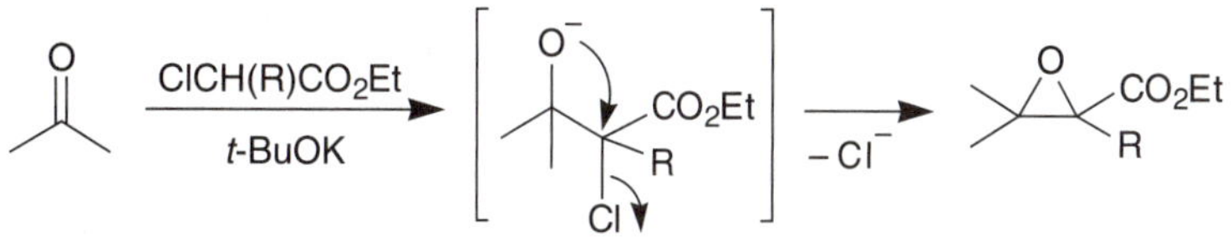

$ClCH(R)CO_2Et$
t-BuOK

$- Cl^-$

Epoxidation of the electron-deficient double bond in α,β-unsaturated ketones may be complicated by the *Baeyer-Villiger* reaction, an oxidation involving the carbonyl group.

However, if the double bond and the carbonyl group are not conjugated, the former generally reacts faster with peroxy acids than the carbonyl group.

Metal-catalyzed epoxidations of alkenes with H_2O_2 provides an economical alternative to oxidations using peroxyacids.[71]

Baeyer-Villiger Reaction.[72] Adolf von Baeyer (Nobel Prize, 1905) in collaboration with Victor Villiger showed that treatment of cyclic ketones with monoperoxysulfuric acid (HO_2SO_3H) produced lactones.[73] The great synthetic utility of the reaction is derived from its stereospecificity and often high degree of regioselectivity. Moreover, the reaction proceeds under mild conditions and can be applied to acyclic, cyclic, and aromatic ketones.

The Baeyer-Villiger oxidation involves an initial attack of a peroxy acid at the carbonyl carbon, which is followed by migration of an adjacent group from the carbonyl carbon to the electron-deficient oxygen of the peroxy acid moiety. The rearrangement proceeds in a concerted manner and is *stereospecific*. Thus, a chiral migrating group maintains its chiral integrity in the product. The overall reaction represents an insertion of an oxygen between the carbonyl carbon and the migrating group. The Baeyer-Villiger reaction applied to acyclic ketones provides esters, whereas cyclic ketones furnish lactones.

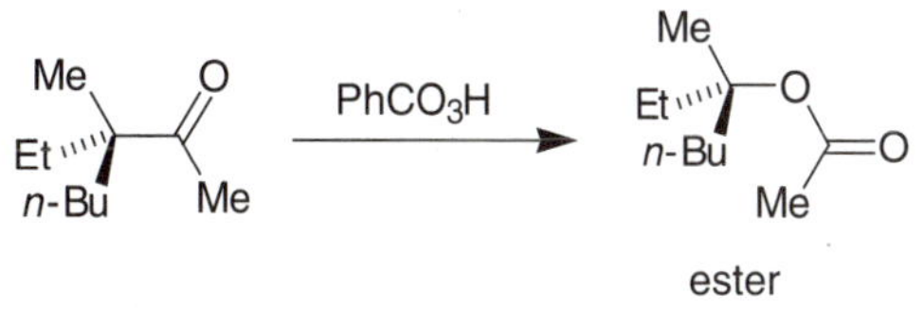

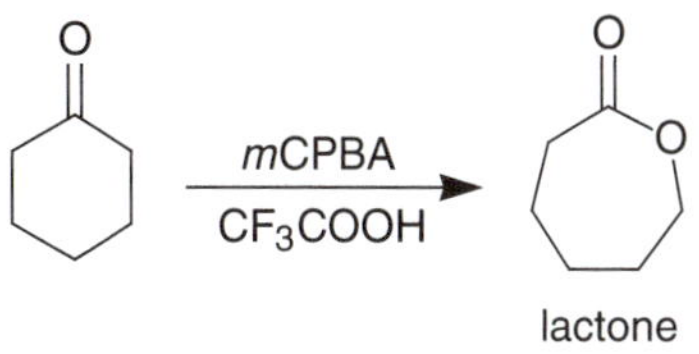

The observed relative ease of migration, *tert*-alkyl > *sec*-alkyl > phenyl > *n*-alkyl > methyl, reflects the ability of the migrating group to accommodate a partial positive charge at the transition state. In addition to electronic factors, steric and conformational constraints as well as reaction conditions may influence the ease of migration. However, the regiochemistry can usually be controlled by proper choice of migrating group. For example, oxidation of methyl ketones results almost exclusively in the formation of acetates. Thus, the Baeyer-Villiger oxidation is not only *stereospecific* but frequently *regioselective*.

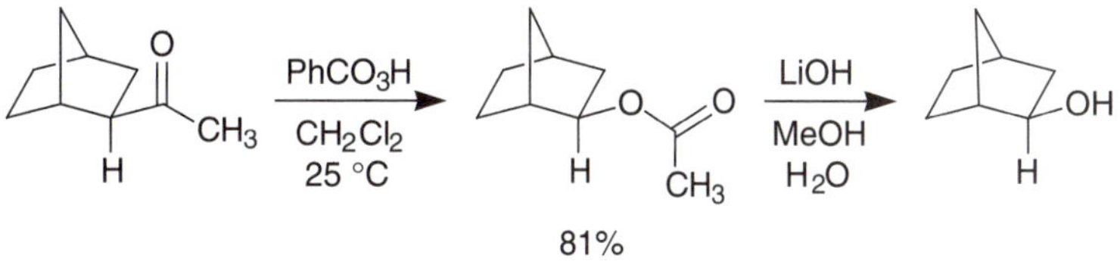

By controlling reaction conditions and by proper choice of the peroxy acid, it is often possible to favor the Baeyer-Villiger reaction over epoxidation. An illustrative example of the usefulness of the Baeyer-Villiger reaction is the *stereospecific*, and *regio*- and *chemoselective* conversion of the unsaturated bicyclic ketone shown below to a cyclopentene containing three consecutive stereogenic centers.[74]

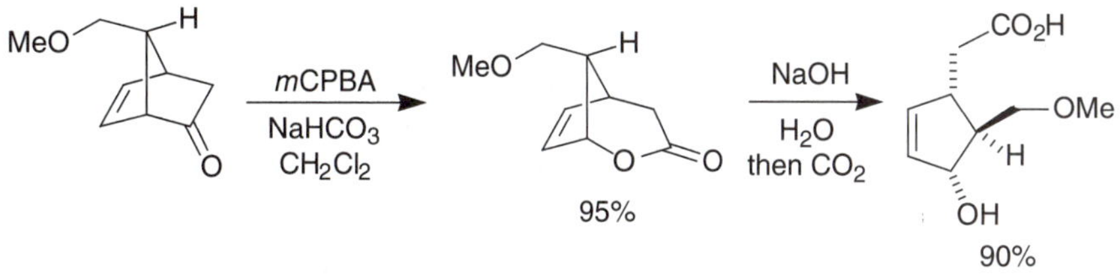

Epoxidation of α,β-Unsaturated Ketones Using Alkaline Hydrogen Peroxide

Oxidation of α,β-unsaturated ketones with alkaline hydrogen peroxide produces the corresponding keto epoxides in good yields. This nucleophilic epoxidation proceeds via an initial Michael-type addition of the hydroperoxide anion to the enone system, which is then followed by elimination of HO^-. It should be noted that the inductive electron-withdrawing effect of the neighboring oxygen in HOO^- makes it a weaker base than HO^- but a better nucleophile than hydroxide.

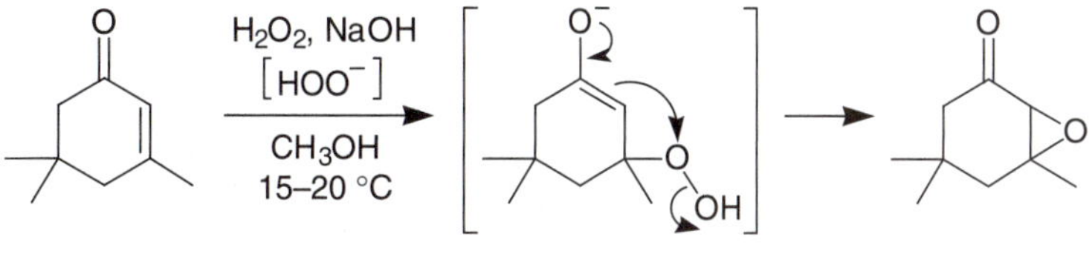

In contrast to the *stereospecific* epoxidation of acyclic alkenes with peroxy acids, oxidation of acyclic α,β-unsaturated ketones with alkaline hydrogen peroxide is *stereoselective* in that only one stereoisomer is formed from *cis-* and *trans-*enones.[75]

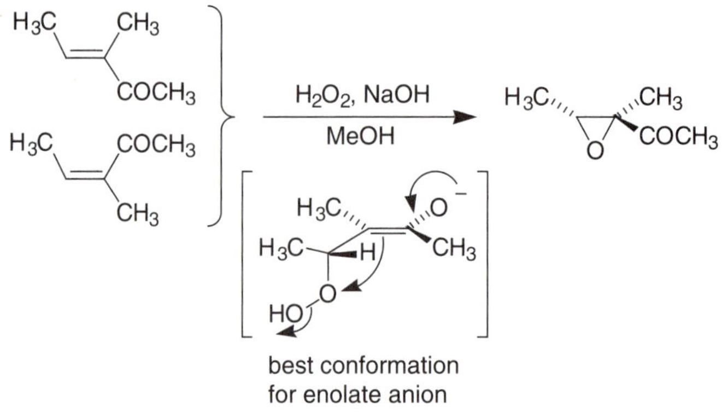

best conformation
for enolate anion

Treatment of enones with basic *tert*-butyl hydroperoxide provides an alternative route for epoxidation of enones when the alkaline hydrogen peroxide procedure fails. For example, the enone shown below did not react with alkaline hydrogen peroxide but underwent chemo- and stereoselective oxidation with *tert*-butyl hydroperoxide in the presence of the base trimethylbenzylammonium hydroxide (Triton-B). To avoid a 1,3-diaxial interaction with the angular methoxymethyl substituent, the hydroperoxide anion attacked the enone from the α-face of the molecule.[76]

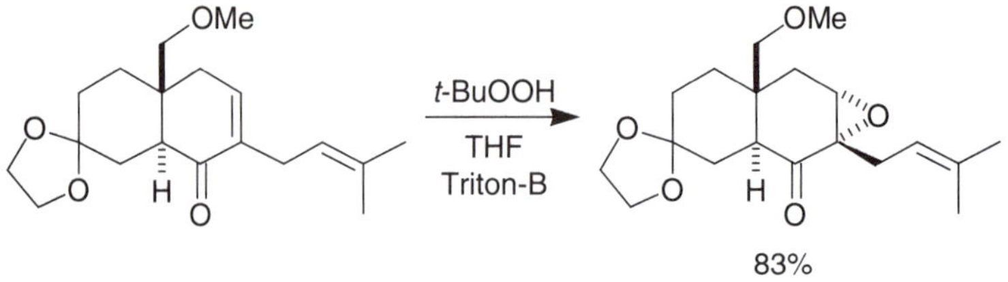

83%

Epoxidation of Alkenes Using Dimethyldioxirane[77]

Dimethyldioxirane (DMDO) is a mild reagent for epoxidation under neutral conditions of electron-rich as well as of electron-deficient alkenes.[78] Moreover, dimethyldioxirane is often the oxidant of choice for the preparation of labile epoxides.[79] The reagent is prepared by oxidation of acetone with potassium caroate $KHSO_5$ (Oxone) and is stable in acetone solution at $-20\ °C$ for several days.[80]

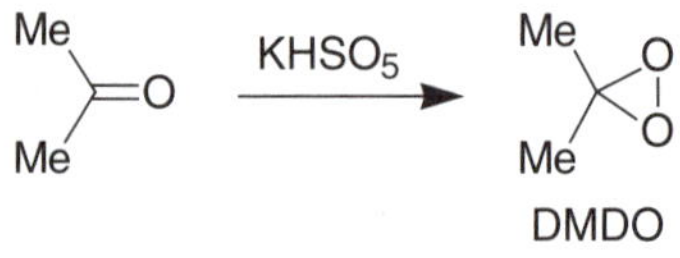

DMDO

Use of dimethyldioxirane obviates the problems associated with epoxidations of conjugated acids and esters with peroxy acids and of enones with alkaline hydrogen peroxide. The only by-product of the reaction is acetone, making this procedure an environmentally friendly one.

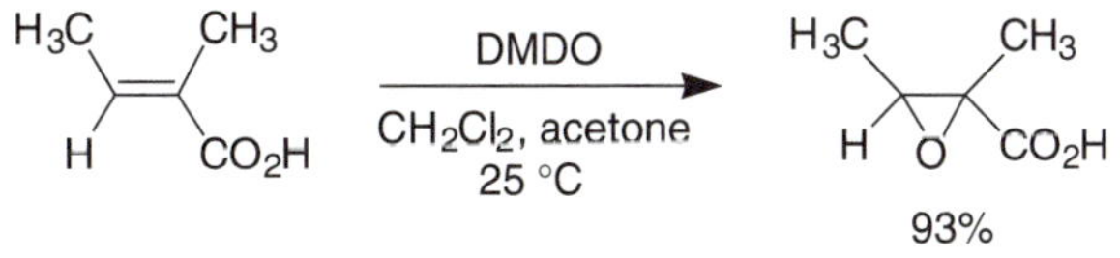

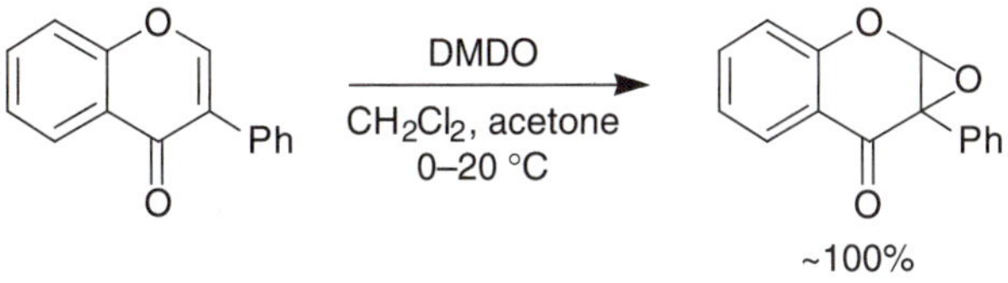

93%

~100%

Methyl(trifluoromethyl)dioxirane (TFDO), prepared from 1,1,1-trifluoroacetone and KHSO$_5$,[81] is more reactive than DMDO by a factor of ~600.[82] In addition to facile epoxidation of alkenes, TFDO can be used to regioselectively oxidize tertiary over secondary C–H bonds via an "oxenoid" (butterfly) mechanism.[79]

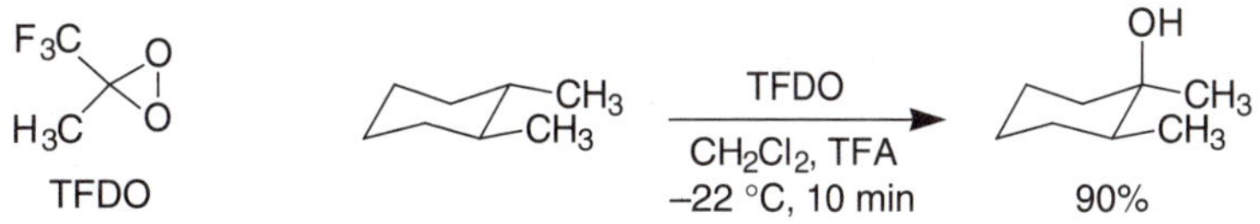

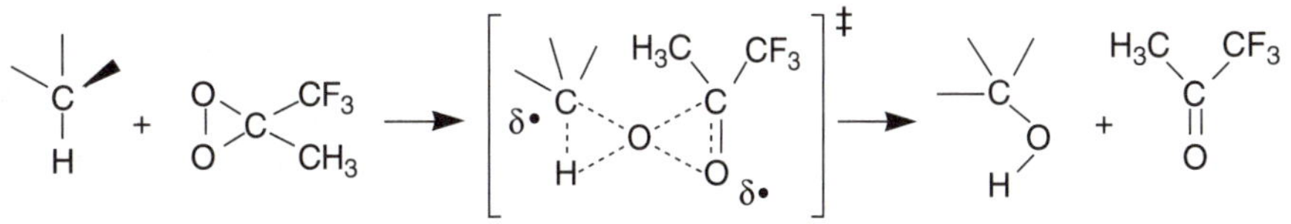

90%

Preparation of Epoxides from Halohydrins[83]

The reaction of chloro- or bromohydrins with bases provides an economical route for the preparation of epoxides. Halohydrins are readily accessible by treatment of an alkene with either hypochlorous acid (Cl$_2$ + H$_2$O → HOCl), hypochlorite bleach solution (NaOCl), or hypobromous acid (NBS + H$_2$O → HOBr). These reactions involve the initial formation of a halohydrin via *anti*-addition of X$^+$ and HO$^-$, followed by internal "S$_N$2" displacement of the halide by the oxyanion.[83b] The regiochemistry of $^\delta$HO–X$^{\delta+}$ addition is highly dependent on the halide and on steric and electronic factors, as shown below.[83c]

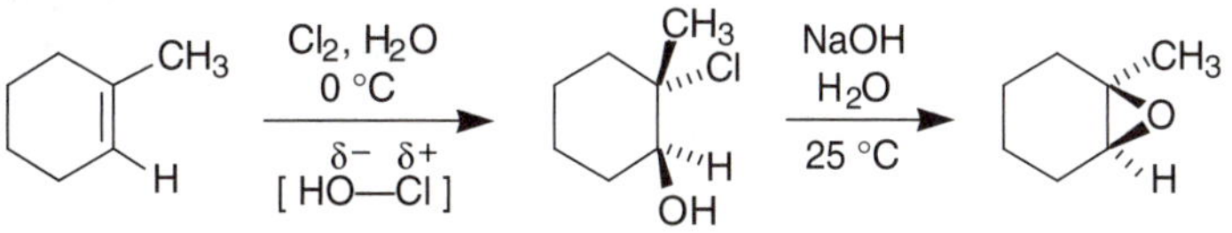

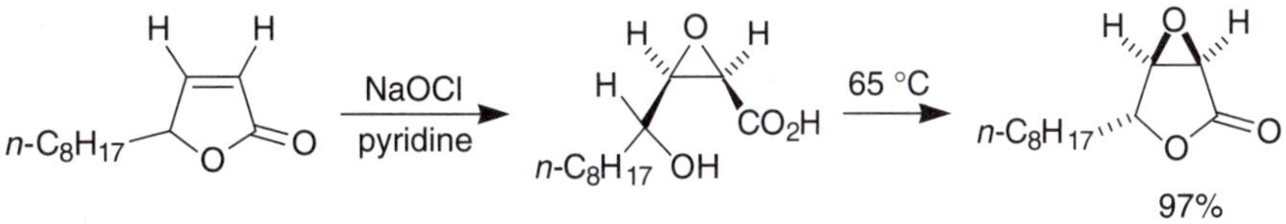

97%

A notable feature of the halohydrin route is that it makes available epoxides with the stereochemistry opposite of that obtained using peroxy acids.

Preparation of Epoxides from Ketones

An alternative approach to epoxides utilizes aldehydes or ketones instead of alkenes as precursors for epoxide preparation.

Dimethylsulfonium Methylide.[84] Methylation of dimethylsulfide with methyl iodide produces trimethylsulfonium iodide. The positive charge on sulfur enhances the acidity of the methyl protons so that treatment of the sulfonium salt with a base converts it to dimethylsulfonium methylide. This "unstabilized" ylide should be used immediately after its preparation.

Reaction of the sulfur ylide with the carbonyl group of aldehydes, ketones, or enones forms a betaine intermediate, which decomposes by intramolecular displacement of Me$_2$S by the oxyanion to yield the corresponding epoxide.

Dimethyloxosulfonium Methylide.[84] Deprotonation of trimethylsulfoxonium iodide forms a sulfur ylide that is significantly more stable than dimethylsulfonium methylide and may be prepared and used at room temperature.

This "stabilized" ylide reacts with aldehydes and ketones to furnish epoxides. The difference in reactivity between dimethylsulfonium methylide and dimethyloxosulfonium methylide is apparent when considering their reactions with α,β-unsaturated ketones. Whereas the "nonstabilized" ylide yields the epoxide, the "stabilized" ylide affords a cyclopropane via conjugate addition followed by ring closure and loss of dimethyl sulfoxide.

89%   81%

There is striking difference in the mode of reaction of sulfur ylides and phosphorus ylides (Wittig reagent) with aldehydes and ketones. The former ylides lead to epoxides, whereas the Wittig reagent furnishes alkenes, pointing to the low affinity of sulfur toward oxygen compared to that of phosphorus.

Chloroiodomethane. Chloroiodomethane on treatment with methyllithium-lithium bromide or *n*-butyllithium at low temperature undergoes lithium iodide exchange to form a lithium chlorocarbenoid species [LiCH₂Cl]. This highly reactive "carbanion" intermediate can be intercepted by the carbonyl group of aldehydes or ketones before it undergoes α-elimination to generate carbene and LiCl. Displacement of chloride from the initially formed carbonyl adducts furnishes the corresponding epoxides in high yields.[85]

Reactions of Epoxides[86]

The inherent strain (~27 kcal/mol) of epoxides makes them prone to (1) ring opening by a wide range of nucleophiles, (2) base-induced rearrangement, and (3) acid-catalyzed isomerization.

Nucleophilic Ring-Opening Reactions.[87] Generally, nucleophilic opening of an unsymmetrically substituted epoxide is *regioselective*, placing the nucleophile at the less substituted carbon of the epoxide (S_N2 type reaction) as shown below.

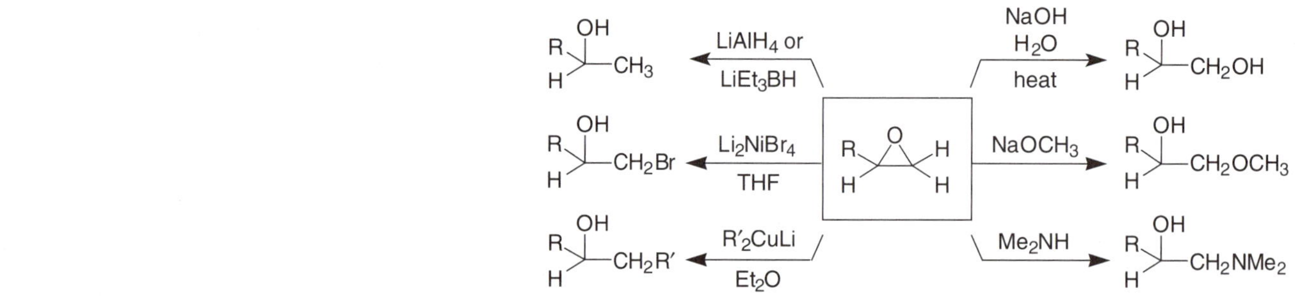

Reactions of substituted epoxides with Grignard reagents are often accompanied by rearrangement. Thus, organocuprates are better suited for carbon-carbon bond forming reactions with epoxides. For a detailed discussion of the preparation and reactions of organocuprates, see Section 7.5.

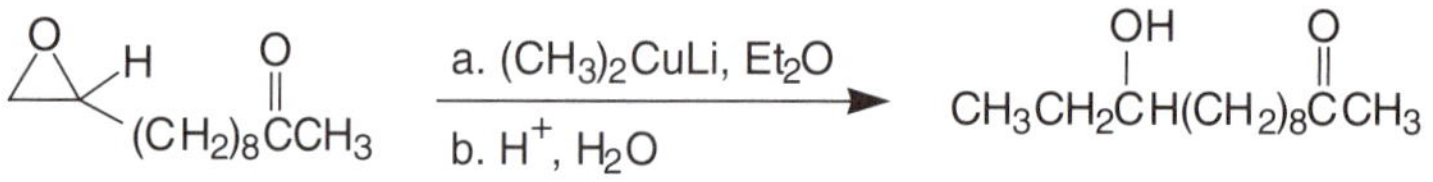

Lithium enolates of ketones, esters, and amides do not react directly with epoxides but require a promoter such as Et_2AlCl, $LiClO_4$, or $LiCl$.[88] However, alkylations of epoxides do occur with $Li[CH_2CO_2C(CH_3)_3]$, $Na[HC(CO_2Et)_2]$, $Li[CR_2CO_2Li]$, or $Li[CR_2CO_2Na]$.[89]

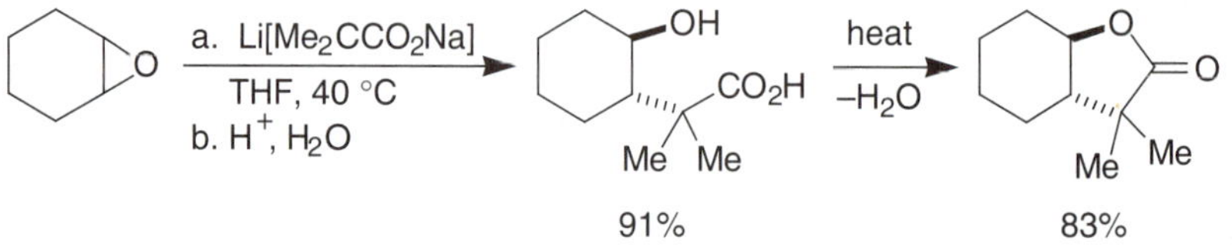

Stereochemistry. In cyclohexane derivatives, opening of the epoxide ring with nucleophilic reagents proceeds stereospecifically in the majority of cases via an S_N2 reaction, placing the oxygen of the epoxide and the attacking nucleophile in a *trans-* and *diaxial*-relationship. Thus, a single diastereomer of an epoxide gives upon ring opening a single diastereomer of the product.

In the case below, where R = H, epoxide opening via an *antiparallel* approach of the nucleophile at $C_{(4)}$ leads to a chairlike transition state preferentially, furnishing the *trans-diaxial*-product. A parallel approach of the nucleophile at $C_{(3)}$ would produce, via a higher-energy twisted boat transition state, the diequatorial product. However, if R = CH_3, the *syn*-axial interaction between the CH_3 group and the incoming nucleophile influences the regiochemistry of epoxide cleavage and, depending on the nucleophile size, may result in preferential attack of the nucleophile at $C_{(3)}$.[90]

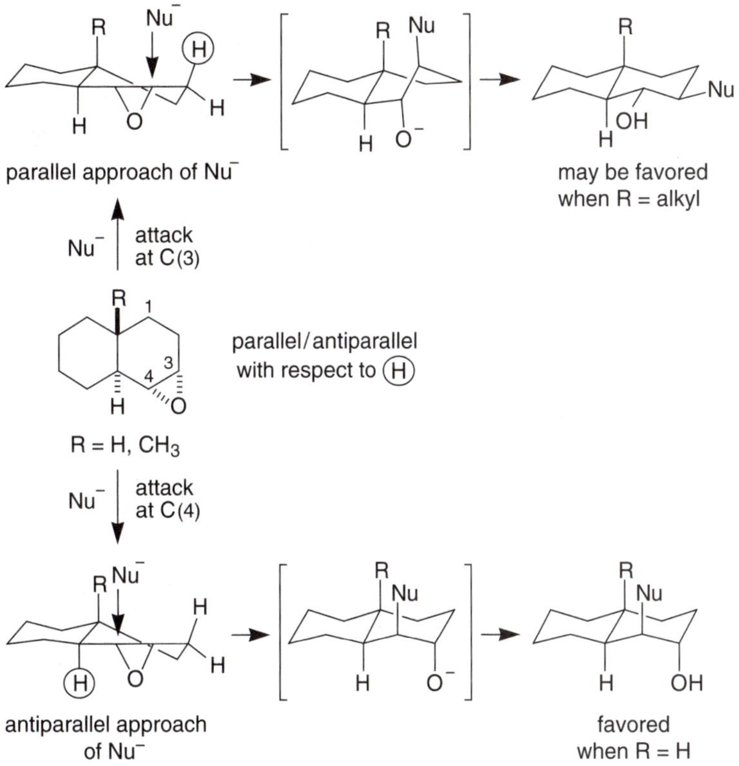

parallel approach of Nu⁻   may be favored when R = alkyl

Nu⁻ attack at C(3)

R = H, CH₃   parallel/antiparallel with respect to (H)

Nu⁻ attack at C(4)

antiparallel approach of Nu⁻   favored when R = H

Intramolecular nucleophilic opening reactions of epoxides lead to the formation of cyclic systems.[91] The process is stereospecific with inversion of configuration at the epoxide carbon. The nucleophile must approach the epoxide C–O bond from a conformation that allows for a *collinear S_N2-like displacement.* Of the two possible modes of intramolecular epoxide ring opening, formation of the smaller ring is generally favored, as illustrated below.[92] This is because it is difficult to attain the necessary coplanar arrangement at the transition state of (1) the attacking group, (2) the epoxide carbon, and (3) the oxyanion leaving group. The preference for ring size can be influenced by the selection of base.[93]

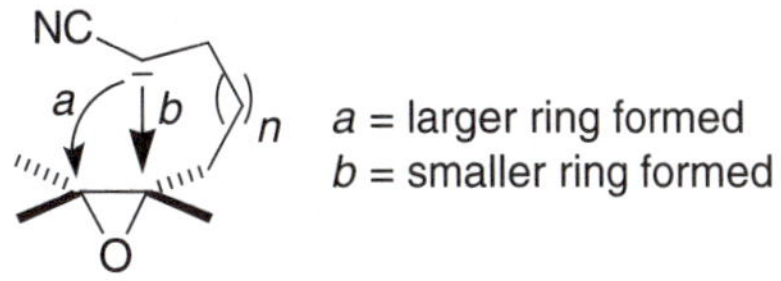

a = larger ring formed
b = smaller ring formed

4 ring vs. 5 ring:

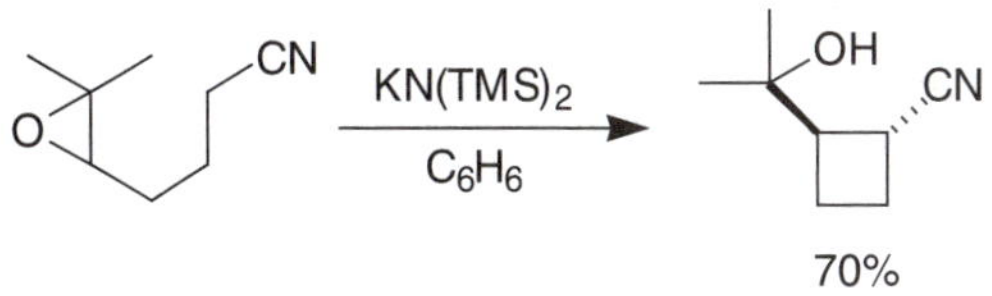

$$\xrightarrow[C_6H_6]{KN(TMS)_2}$$

70%

5 ring vs. 6 ring:

$$\xrightarrow[NH_3\,(l),\ glyme]{KNH_2}$$

77%

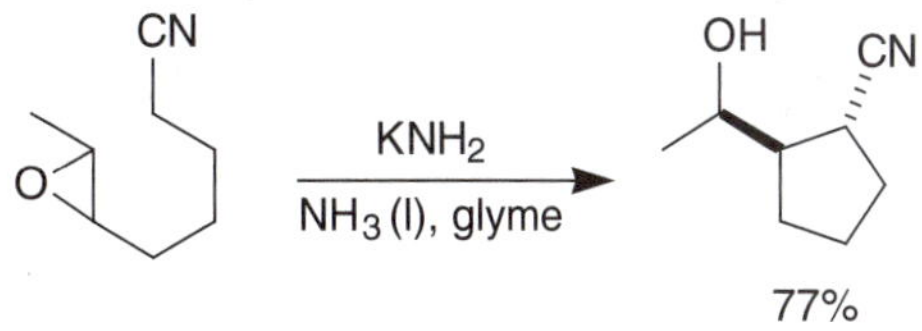

In cases involving cyclohexene oxides, the requirement for opening via an *antiparallel* approach dictates the ring size, as illustrated below.[92a]

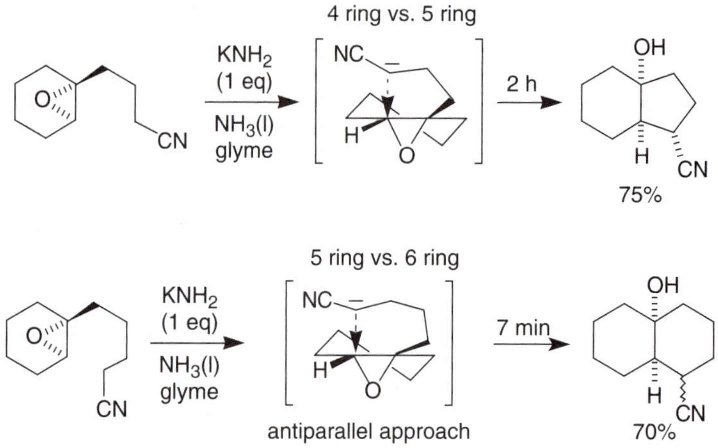

Inversion of Olefin Stereochemistry.[94] The preparation of alkenes via inversion of the double bond geometry is an important synthetic transformation. For example, interconversion of the (Z)-alkene to the (E)-isomer depicted below involves treatment of the (Z)-epoxide with the nucleophilic LiPPh₂[95] followed by phosphorus alkylation to furnish the betaine,[96] which undergoes *syn*-elimination to produce the (E)-alkene. The alkene inversion works for di-, tri-, and tetra-substituted olefins.

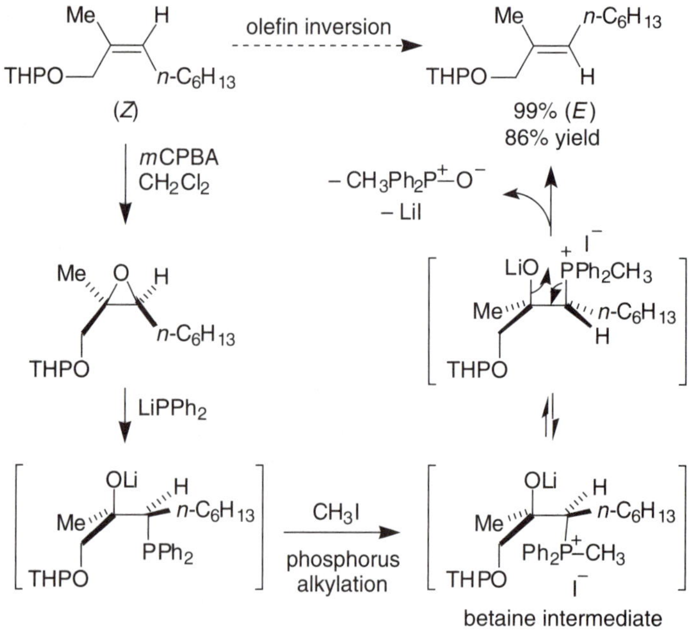

Base-Induced Elimination of Epoxides.[97] Epoxides may serve as precursors for stereoselective syntheses of allylic alcohols. Thus, starting with an alkene and converting it to an epoxide followed by its treatment with a strong base produces the corresponding allylic alcohol.[98]

The overall reaction amounts to a transposition of the initial alkene double bond. In contrast to ring opening of epoxides by nucleophiles, the base-induced elimination requires a non-nucleophilic strong base, such as a lithium dialkylamide (e.g., LDA). The reaction is believed to proceed via a *syn*-elimination involving a boatlike transition state.[99]

Silyl-protected allylic alcohols can be obtained by treatment of epoxides with a non-nucleophilic base in the presence of stoichiometric amounts of TMSOTf or TBSOTf (Tf = SO_2CF_3).[100]

Acid-Catalyzed Ring Opening Reactions. The acid-catalyzed opening of epoxides with H_3O^+ proceeds faster than the base-induced reaction with OH^-. In acidic medium, protonation of the ring oxygen places the ensuing partial positive charge on the more substituted epoxide ring carbon. Thus, backside displacement of the neutral leaving group (–OH) by a nucleophile (generally a nucleophilic solvent) occurs preferentially at the more substituted carbon of the epoxy ring. The general rule that ring opening of cyclohexene-derived epoxides positions the –OH and –Nu in a *trans*- and *diaxial*-relationship is also applicable under the acid-catalyzed conditions.

In contrast to the reductive cleavage of 1-methylcyclohexene epoxide with LiAlH₄ or, better, with LiEt₃BH[101] to produce 1-methylcyclohexanol, reduction of the epoxide with sodium cyanoborohydride in the presence of boron trifluoride etherate furnishes *cis*-2-methylcyclohexanol.[102] In this case, complexation of the epoxide oxygen with the Lewis acid BF₃ now directs "hydride" addition to the more substituted carbon, which can better sustain the induced partial positive charge.

Acid-catalyzed opening of saturated hydroxy epoxides via intramolecular participation of the hydroxyl group leads to regio- and stereoselective formation of tetrahydrofuran and tetrahydropyran systems.[103]

Interestingly, acid-catalyzed intramolecular opening of allylic epoxides by hydroxyl groups proceeds regioselectively via the π-stabilized carbenium ion intermediate to form the tetrahydropyran ring.[103]

Lewis Acid–Mediated Rearrangement of Epoxides.[104] This procedure is useful for the conversion of alkenes into carbonyl compounds via epoxides. Boron trifluoride etherate is the Lewis acid generally employed for this purpose. The reaction proceeds via a carbenium ion intermediate, and the migratory aptitude of groups follows that observed in carbenium ion rearrangements. The usefulness of the reaction depends on structural and conformational factors leading to selective rearrangements. The formation of the epoxide and its opening may be carried out in one vessel by using a mixture of peroxytrifluoracetic acid and boron trifluoride etherate.[105]

Epoxidation of Allylic Alcohols

Hydroxyl-Directed Epoxidation with Peroxy Acids[106]

Although epoxidation of cyclic alkenes occurs preferentially from the less hindered side, the presence of a polar substituent near the double bond may reverse the facial direction of attack by the peroxide. For example, introduction of an allylic OH group onto the substrate provides a means for interaction with the peroxy acid by hydrogen bonding and thereby promotes stereoselective epoxidation. Cycloalkenes with an OH group at the allylic position are attacked by peroxy acids *syn* to the double bond to give the corresponding *cis*-epoxy alcohol diastereomers, even if that face is sterically more hindered. It is assumed that the directive effect of the hydroxyl group is due to hydrogen bonding between the hydroxyl group and the peroxy acid, stabilizing the transition state for epoxidation.[106b] If the OH group is replaced by an acetate group, a nearly equimolar mixture of diastereomeric epoxides is observed.

An instructive example of a π-facial, homoallylic OH-directed stereoselective epoxidation in an acyclic system, used for the construction of the natural product monensin, is depicted below.[107] As expected, the more electron-rich trisubstituted double bond in A would be more susceptible to epoxidation than the terminal double bond. To minimize allylic 1,3-strain between the ethyl group and the CH$_2$CH=CH$_2$ appendage, A should preferentially adopt the conformation B, in which the smallest substituent H (hydrogen) is now in the same plane as the ethyl group. This places the hydroxymethyl moiety (CH$_2$OH) in proximity to the β-face of the double bond, leading, after treatment with mCPBA, to the formation of epoxide diastereomer C.

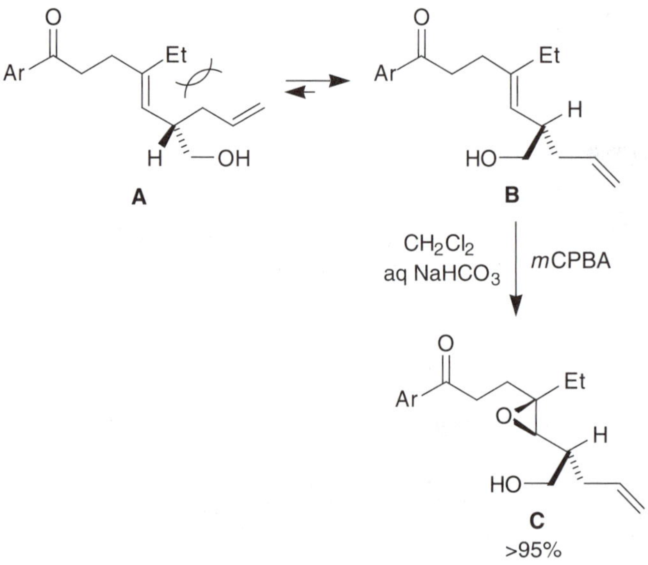

The hydroxyl-directed epoxidation of alkenes using peroxy acids is sensitive to the nature of the solvent. In the example below, alkene epoxidation occurs on the more accessible α-face when Et$_2$O is used as solvent, whereas the use of CH$_2$Cl$_2$ promotes hydroxyl-directed epoxidation on the β-face of the alkene.[108]

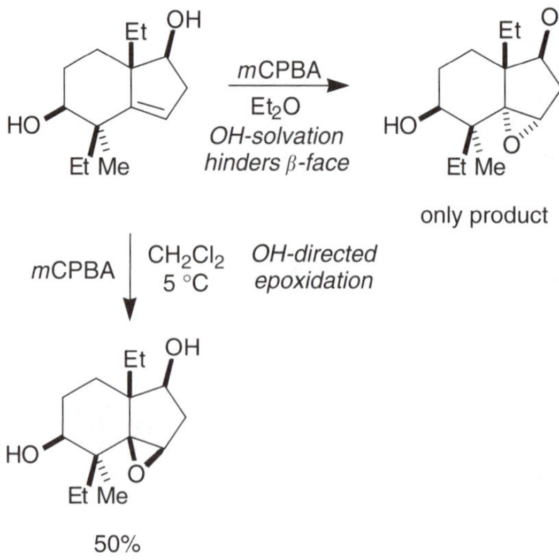

Transition-Metal-Catalyzed Epoxidation of Allylic Alcohols

Straight-chain allylic alcohols have been crucial to the development of stereoselective syntheses of complex natural products since they can be converted stereoselectively into epoxides and via further elaboration of these into a host of functionalized compounds of predictable stereochemistry.

Control of Chemoselectivity. The reaction of allylic alcohols with *t*-butyl hydroperoxide and vanadium acetylacetonate or titanium tetraisopropoxide provides a highly chemoselective method for the preparation of epoxides, as exemplified below.[109] Catalysis by vanadium is envisioned to involve a complex of *t*-BuOOH, the OH group of the allylic alcohol, and the metal.

*m*CPBA:	33%	67%
cat. VO(acac)$_2$, *t*-C$_4$H$_9$OOH:	99%	1%

Control of Stereochemistry. There are two principal ways to control the stereochemistry of a reaction:[110]

Substrate-control strategy (internal control): If the chiral carbon is in proximity to the double bond, it is often possible to achieve *diasteroselective* epoxidation. The new stereocenters formed bear a specific relationship to the stereocenter already in the molecule (relative asymmetric induction). Unfortunately, for a given substrate often only one of the two possible stereoisomers is available.

The selectivity is determined by the substrate since an achiral epoxidizing agent is used. To obtain the other diastereomer, one must exert *reagent control* to override the substrate's preference.

Reagent-control strategy (external control): Powerful enantiomerically pure catalysts or auxiliaries are used for constructing chiral molecules in a diastereo- and enantioselective manner. Using this strategy, it is often possible to enhance or reverse the modest diastereofacial preference exhibited by a chiral substrate. The Sharpless asymmetric epoxidation (*SAE*) is an important example of a reagent-control strategy in which the *diastereoselective* oxidation is made *enantioselective* by using a chiral auxiliary or a chiral ligand.

The Sharpless Asymmetric Epoxidation (SAE) Reaction[111]

The Sharpless asymmetric epoxidation reaction is often used as a key step in synthetic protocols involving the synthesis of natural products such as terpenes, carbohydrates, insect pheromones, and pharmaceutical products. The SAE reaction is characterized by its simplicity and reliability. The epoxides are obtained with predictable absolute configuration and in high enantiomeric excess (ee). Moreover, 2,3-epoxy alcohols serve as versatile intermediates for a host of stereospecific transformations.

Enantioselectivity. In 1980, T. Katsuki and K. B. Sharpless (Nobel Prize, 2001) reported a method whereby *prochiral* allylic alcohols are epoxidized in the presence of *t*-BuOOH, Ti(*i*-OPr)$_4$, and (+)-or (–)-diethyl tartrate (DET) with high regio- and stereoselectivity to produce the corresponding optically active epoxides.[112]

Diethyl tartrate, the allylic alcohol, and the oxidant *t*-BuOOH displace the isopropoxide groups on titanium to form the active Ti-catalyst in a complex ligand exchange pathway.[110d] From structural and kinetic studies,[113] Sharpless proposed that oxygen transfer occurs from a dimeric complex that has one tartaric ester moiety per titanium atom.

dimeric intermediate complex

The absolute configuration of the epoxy alcohol is predictable using the mnemonic model depicted below in which the *CH₂OH group is positioned at the lower right.* As a rule, the transfer of the epoxide oxygen occurs from the upper face of the allyl alcohol when (+)-(*R, R*)-tartaric esters are used, while the lower face is attacked when the (–)-(*S, S*)-tartrates are used.[114]

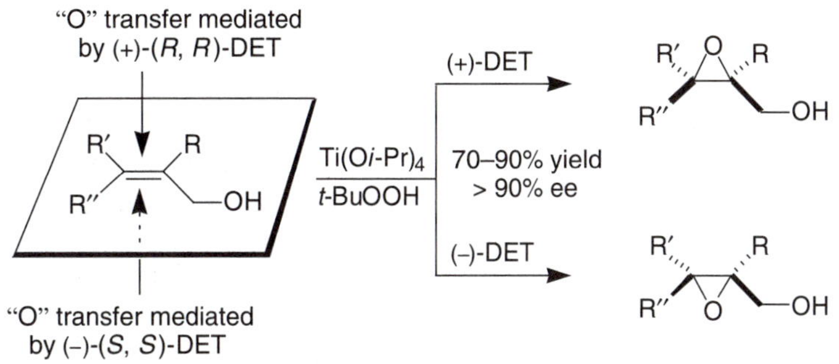

The SAE proceeds with *catalytic* amounts of the Ti-(DET) complex provided that 3 Å or 4 Å molecular sieves (MS) are added. Usually a 5–7.5 mol % of Ti-(DET) is sufficient to obtain high ee and yields.[115]

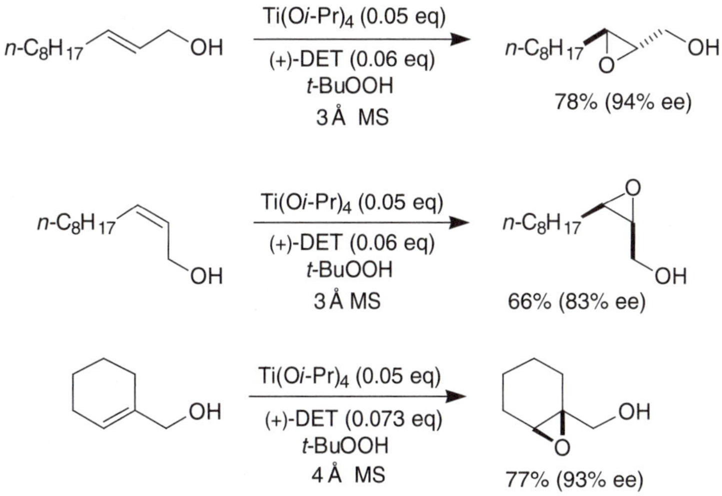

Diastereoselectivity.[116,117] In applying the *reagent-control* strategy to SAE reactions, consider the following reported experiments:

1. Epoxidation of the *chiral* allylic alcohol A with the *achiral* titanium alkoxide catalyst gives a 2.3 : 1 mixture of the α- and β-epoxides B and C, indicating the chiral substrate's preference for α-attack.

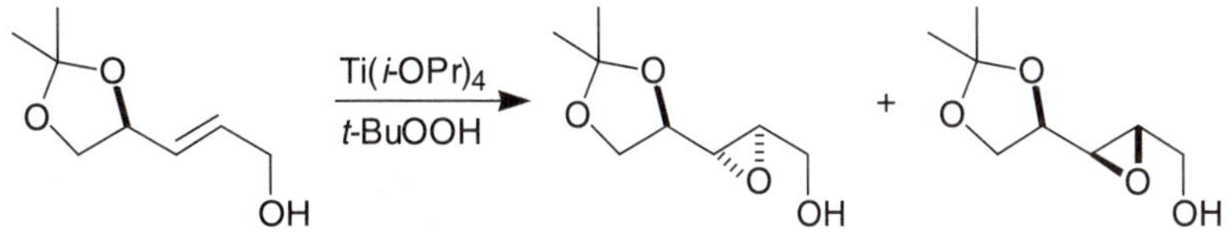

2. Reaction of the *achiral* substrate D with the *chiral* Ti-(+)-DET catalyst results in a 99 : 1 preference for β-attack to furnish the epoxy alcohol F.

3. In the mismatched reaction of the *chiral* substrate A with the *chiral* Ti-(+)-DET catalyst, the two preferences are opposed. Since the *reagent preference* [Ti-(+)-DET] is much stronger than the *substrate preference* [alcohol A], the reagent preference prevails and the epoxide H is formed with good diastereoselectivity (22 : 1).

4. Employing the *chiral* substrate A and the *chiral* Ti-(–)-DET catalyst reverses the reagent's preference to α-attack. The substrate and the reagent preferences are now *matched* and the epoxide I is formed with high diastereoselectivity (90 : 1).

The conclusion that emerges from these experiments is that for the development of an effective reagent-control strategy, it is important to use an asymmetric reagent with a much larger diastereofacial preference than the chiral substrate so that the former prevails over the latter.

Applications of the SAE Reaction[118]

The utility of the SAE reaction stems from the fact that chiral epoxy alcohols possess reactive sites at $C_{(1)}$, $C_{(2)}$, and $C_{(3)}$ that can be transformed into a large number of other functionalities while maintaining their preexisting chiral centers. Moreover, either of the two enantiomers may be obtained.

Nucleophilic Opening at $C_{(1)}$. The latent reactivity at $C_{(1)}$ can be activated via the *Payne rearrangement*[119] by treatment of 2,3-epoxy alcohol A with aqueous NaOH in the presence of a cosolvent. This results in equilibration of A with the isomeric 1,2-epoxy 3-ol B. Even if epoxide A is preferred at equilibrium, $C_{(1)}$ in isomer B is sterically less hindered and hence should react faster with the nucleophile in an S_N2 manner. Once B is formed, it will react selectively and irreversibly with the nucleophile to furnish product C. The success of epoxide opening (B → C) by nucleophiles depends on whether the reagent is compatible with the alkaline aqueous medium required for the Payne rearrangement.

Nucleophilic Opening at C(2) and C(3). This opening depends on steric and electronic factors. For example, in the presence of camphorsulphonic acid, nucleophilic attack by methanol occurs at the more substituted carbon. However, with epoxy alcohols having the same number of substituents at C(2) and C(3), epoxide opening with nucleophiles occurs preferentially at C(3) because the presence of the electron-withdrawing OH group at C(1) retards S_N2 substitution at C(2).

Titanium isopropoxide (a Lewis acid) induces ring opening of 2,3-epoxy alcohols and 2,3-epoxy carboxylic acids with a variety of nucleophiles under mild experimental conditions with fair to excellent C(3) regio- and stereoselectivity.[120] The 2,3-epoxy carboxylic acids are readily available by RuO_4 oxidation of the corresponding epoxy alcohols.[121]

Reduction of 2,3-epoxy alcohols with Red-Al [sodium bis(2-methoxyethoxy)aluminum hydride] cleaves the C(2)–O bond to furnish a 1,3-diol. The observed regioselectivity may result by an intramolecular delivery of hydride from the aluminate formed on reaction of Red-Al with the –CH$_2$OH group.[122]

Lewis Acid-Catalyzed Rearrangement of 2,3-Epoxy-1-ols. Epoxides with adjacent hydroxyl groups, as in 2,3-epoxy-1-ols, undergo rearrangement on treatment with BF$_3$•Et$_2$O to afford aldol-like products (e.g., β-hydroxy-aldehydes).[123]

Resolution of Racemic Mixtures of Allylic Alcohols.[124] An important application of the SAE reaction is the kinetic resolution of *racemic* mixtures of secondary allylic alcohols. In this case, the chiral catalyst reacts faster with one enantiomer than with the other since the two transition states are diastereomeric. Thus, using 0.5 mole of *t*-BuOOH for each mole of racemic allylic alcohol, the faster-reacting enantiomer will consume the *t*-BuOOH to furnish the epoxide. This leaves behind the unreacted slower-reacting allylic alcohol in high enantiomeric excess, which is then separated from the epoxide via chromatography.

Jacobsen Epoxidation[125]

The rigid, chiral salen complexes of Mn(III) shown below catalyze the asymmetric epoxidation of alkenes when treated with commercial bleach (NaOCl). This synthesis of enantio-enriched epoxides is particularly powerful since the method is applicable to unfunctionalized olefins. In general, (Z)-1,2-disubstituted alkenes afford higher enantioselectivities than do the (E)-isomers or trisubstituted alkenes.[126] The reaction mechanism is complex[127] and proceeds via the formation of a Mn(III,IV) dinuclear species.[128]

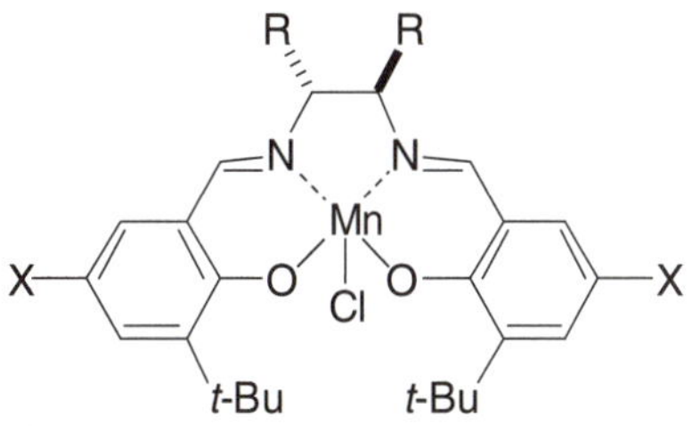

chiral (salen)Mn(III) complexes

X = H, alkyl, OMe
R = Ph or –(CH$_2$)$_4$–

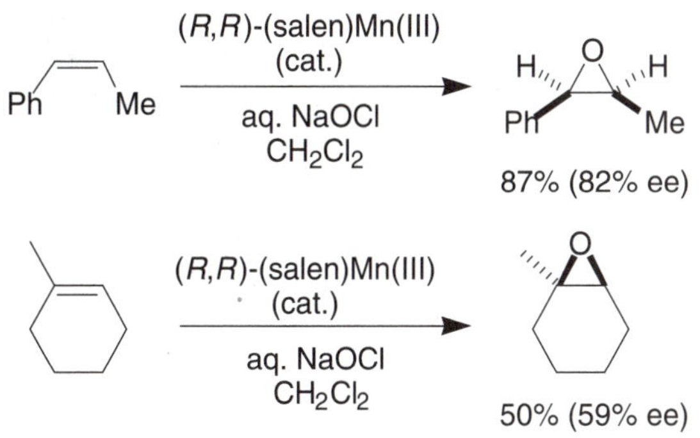

87% (82% ee)

50% (59% ee)

Dihydroxylation of Alkenes—Formation of 1,2-Diols

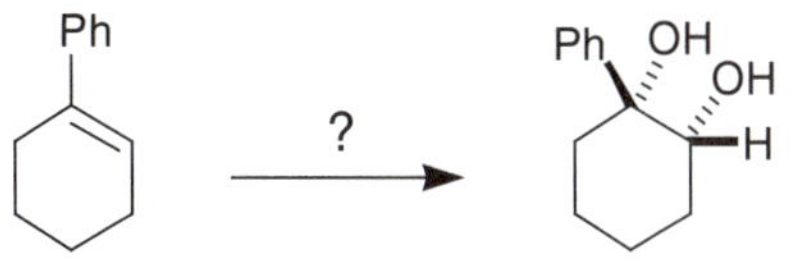

(1R, 2R)-diol

Treatment of alkenes either with osmium tetroxide or with alkaline potassium permanganate results in *syn*-dihydroxylation of the double bond.

Osmium Tetroxide[129]

The dihydroxylation of olefins with OsO$_4$ (a very toxic and volatile reagent that must be handled with care) provides a reliable method for the preparation of *cis*-1,2-diols. Although osmium tetroxide is expensive, it is the reagent of choice for *syn*-dihydroxylation because yields of diols obtained are usually high.

The reaction can be depicted as a concerted *syn*-addition of the reagent to the double bond, forming the cyclic osmate ester, which upon hydrolysis or reduction (H$_2$S, NaHSO$_3$, or Na$_2$SO$_3$) produces the *cis*-1,2-diol.

The reaction is stereospecific in that (E)-alkenes furnish the *syn*-1,2-diols, whereas (Z)-alkenes give the *anti*-1,2-diols.

The disadvantages associated with the high cost and safety in handling of OsO_4 when used in stoichiometric amounts are minimized using procedures that require only catalytic amounts of the reagent. In this case, a stoichiometric amount of a co-oxidant is added to oxidatively hydrolyze the intermediate osmate ester with regeneration of OsO_4. The most widely used co-oxidant is *N*-methylmorpholine *N*-oxide (NMO).[130] Other possible co-oxidants are (1) trimethylamine *N*-oxide (Me_3NO, H_2O)[131], (2) *tert*-butyl hydroperoxide (*t*-BuOOH)[132], and (3) potassium ferricyanide/potassium carbonate ($K_3Fe(CN)_6$, K_2CO_3).[133]

Dihydroxylation usually occurs from the less hindered side of the double bond, but appears to be less susceptible to steric hindrance than epoxidation with peroxy acids.[134]

Osmium-tetroxide-catalyzed dihydroxylation of sterically hindered olefins proceeds more efficiently with trimethylamine *N*-oxide in the presence of pyridine.[131] The base appears to catalyze not only formation of the osmate ester, but also its hydrolysis.[135]

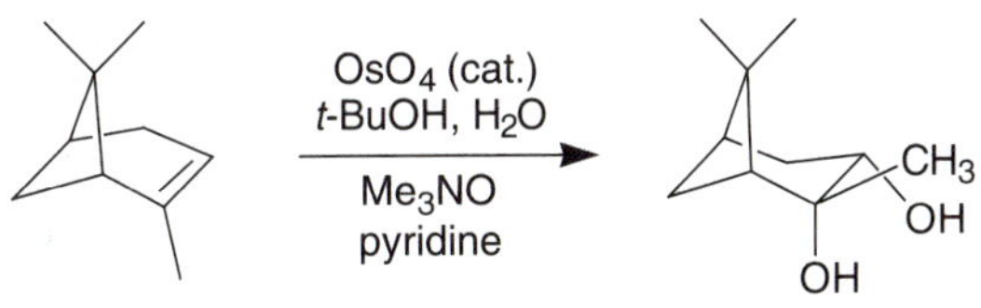

Under controlled conditions, the osmium-tetroxide-mediated dihydroxylation is also chemoselective, reacting preferentially with the more nucleophilic double bond in the presence of a triple bond.[136]

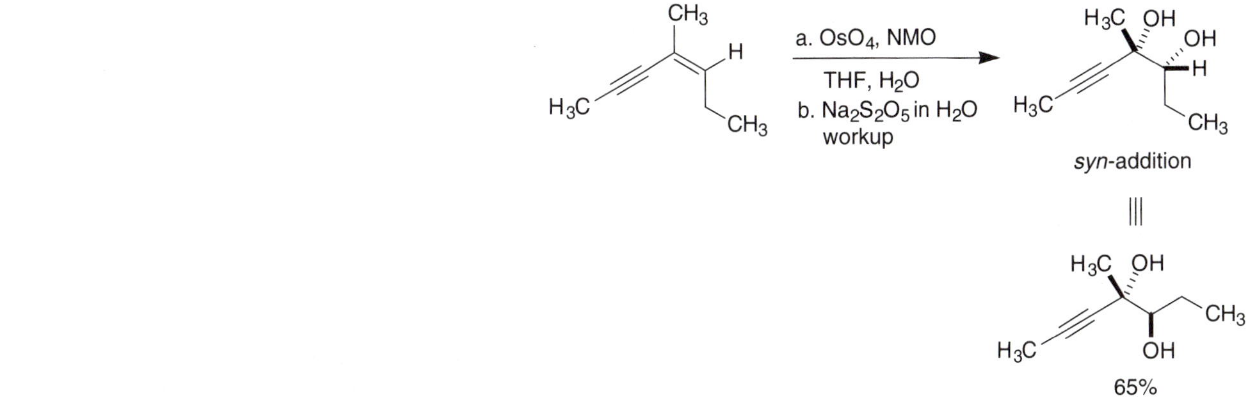

syn-Dihydroxylation on the *more hindered side* of cycloalkenes can be achieved with iodine and silver acetate in moist acetic acid (Woodward procedure)[137] or more economically with iodine and potassium iodate in acetic acid.[138]

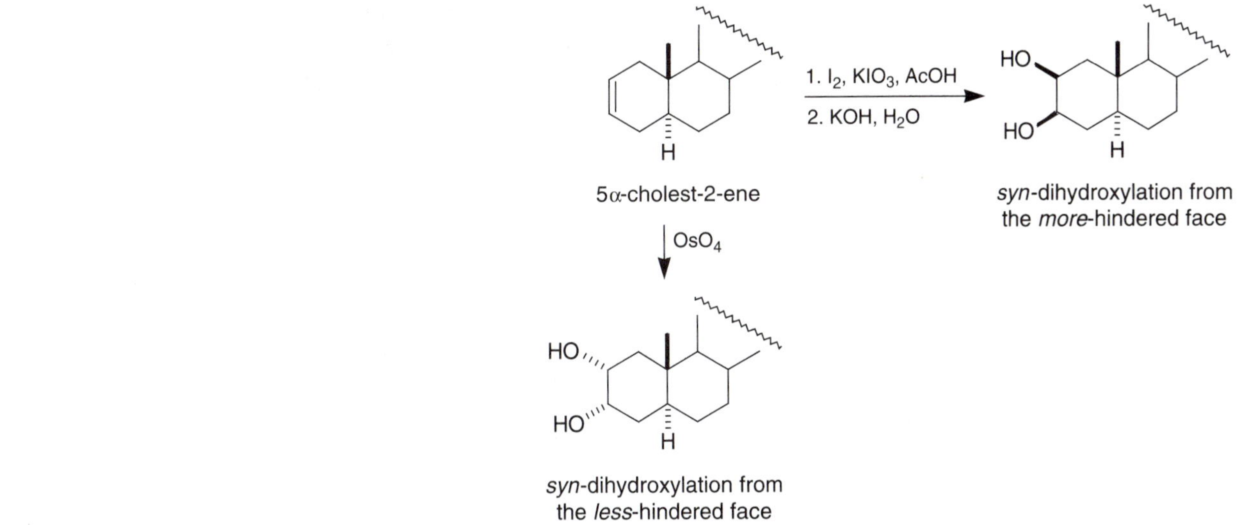

It should also be noted that epoxidation of olefins followed by ring opening with OH⁻ provides 1,2 diols with different stereochemistry from that observed with osmium-tetroxide-mediated *syn*-dihydroxylation. Therefore, even when only one diastereomeric olefin is available, it is possible to prepare both diastereomeric 1,2-diols, as shown below.

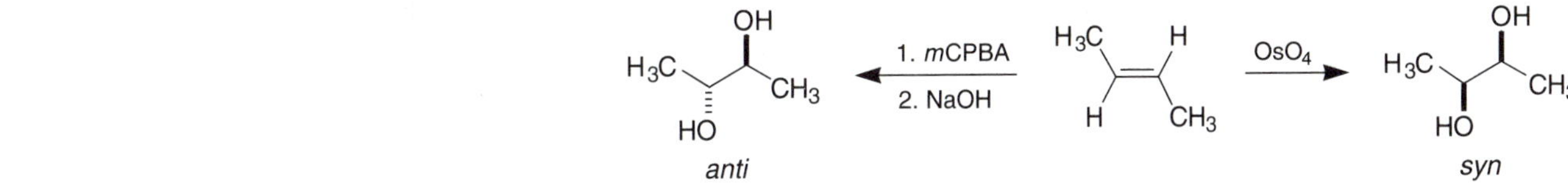

Potassium Permanganate[139]

The reaction of alkenes with alkaline potassium permanganate proceeds rapidly via formation of a cyclic manganese ester, which is hydrolyzed to the 1,2-diol. To avoid overoxidation to an acyloin (α-ketol), the pH of the reaction medium has to be monitored. Although the yields of *cis*-diols obtained are usually modest (~50%), the procedure is less hazardous and much less expensive than using osmium tetroxide and thus is well suited for large-scale preparations. Improved yields of diols are obtained when the oxidation is carried out in water in the presence of a phase transfer agent such as benzyltriethylammonium chloride.[140]

Asymmetric Dihydroxylation[141]

Reagent-controlled asymmetric dihydroxylation (AD) of prochiral alkenes is feasible using chiral auxiliaries. Sharpless and coworkers showed that treatment of prochiral alkenes with catalytic amounts of the solid, nonvolatile dipotassium osmate dihydrate ($K_2[Os^{VI}O_2(OH)_4]$), potassium ferricyanide ($K_3Fe(CN)_6$, a stoichiometric oxidant, and a chiral ligand results in enantioselective *syn*-dihydroxylation. The required enantiomerically pure ligands are readily available from the cinchona alkaloids dihydroquinine (DHQ) and dehydroquinidine (DHQD). Since use of the phthalazine spacer (PHAL) outperformed the monomeric alkaloid ligands, the (DHQ)$_2$PHAL and (DHQD)$_2$PHAL analogs have become the first-choice ligands for most AD reactions. Mechanistic studies of the origin of enantioselectivity for the cinchona alkaloid-catalyzed AD of alkenes have been reported.[142]

[DHQ]$_2$PHAL in AD-mix-α

[DHQD]$_2$PHAL in AD-mix-β

(DHQ)$_2$PHAL and (DHQD)$_2$PHAL are commercially available—packaged as a mixture with dipotassium osmate dihydrate, potassium ferricyanide, and K_2CO_3—and are known by the trade names *AD-mix-α* and *AD-mix-β*, respectively. With the exception of (Z)-alkenes, excellent enantioselectivites with most olefinic substrates have been achieved using the $(DHQ)_2PHAL$ and $(DHQD)_2PHAL$ ligands.[143]

A mnemonic for predicting the absolute configurations of the diols derived from AD reactions has been proposed. To use this model, one evaluates which of the substituents on the alkene is the largest and places it at the lower left corner of the model as shown below. AD-mix-α dihydroxylates the bottom (α)-face of the alkene, while AD-mix-β induces dihydroxylation from the top (β)-face. For alkenes with substituents of similar size, predictions as to the stereochemical outcome may be difficult.

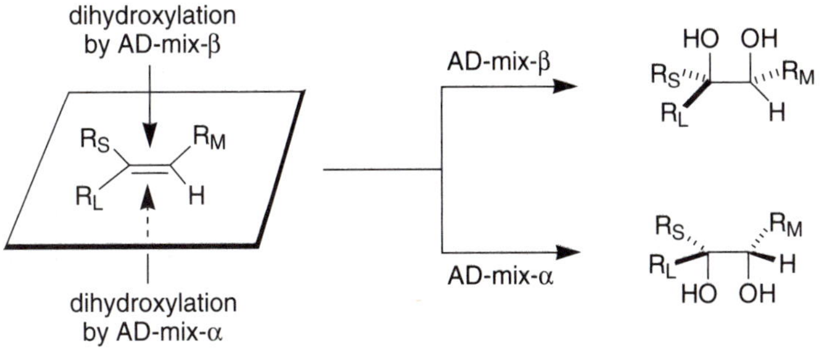

To speed up hydrolysis of the osmium(VI) glycolate, the reaction may be carried out in the presence of one equivalent of $MeSO_2NH_2$, except in cases of terminal mono- and terminal disubstituted alkenes.[142b]

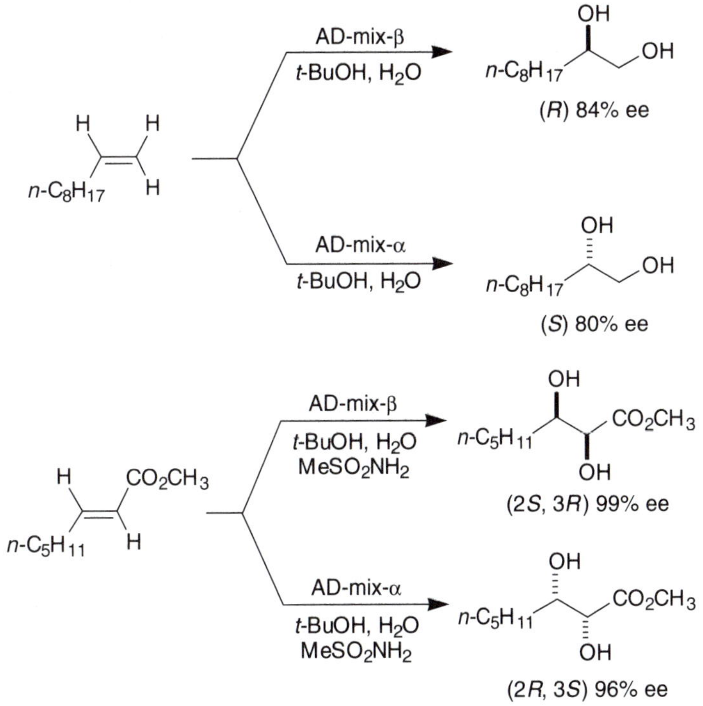

Halolactonization[144]

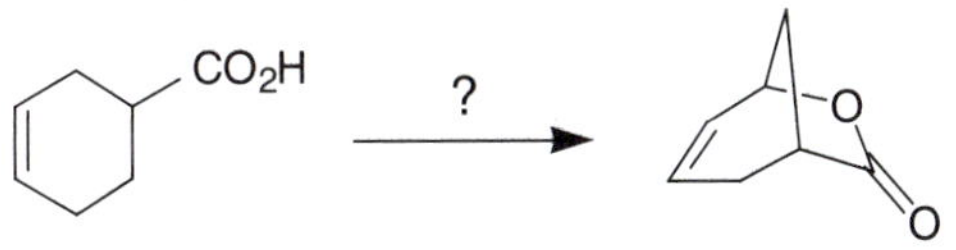

Halolactonization of unsaturated acids is a powerful tool for introducing new functional stereocenters onto a double bond in both cyclic and acyclic systems. The reaction is reminiscent of the acid-catalyzed opening of an epoxide by a neighboring hydroxyl group. In both cases, the ring oxonium ion and the bromonium (or iodonium) ion intermediates, respectively, are attacked intramolecularly by the proximal nucleophile.

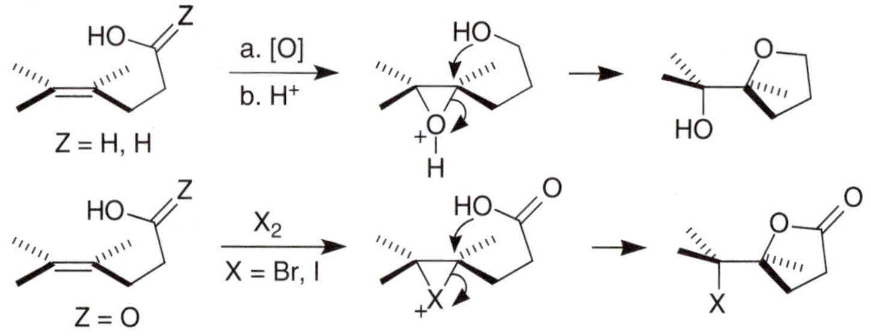

Cycloalkenes tethered with a γ,δ- or δ,ε-unsaturated acid side chains react with Br_2 or I_2 to furnish the corresponding halolactones. Iodolactonization is more commonly used than bromolactonization since iodine is easier to handle (solid) and is more chemoselective (less reactive) than bromine. Halolactonization with aqueous base is *kinetically* controlled and proceeds via addition of a Br^+ or I^+ atom to the double bond to form a transient halonium ion. In the absence of strong directing steric effects, formation of the halonium ion may occur at either diastereoface of the double bond. However, only the halonium ion intermediate which allows *trans-diaxial* S_N2 opening by the neighboring carboxylate nucleophile leads, if the *intramolecular* reaction is sterically favorable, to the lactone.

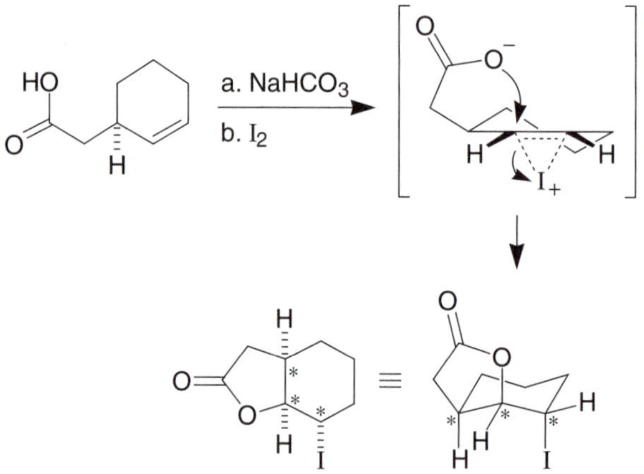

In the example shown above, halolactonization controls the relative stereochemistries of three contiguous stereocenters (∗): the two newly created stereocenters (C–O and C–I) are *trans* to each other and the lactone is by necessity part of a *cis*-fused ring system.

Reductive removal of the halogen atom with either Raney-Ni or with tributyltin hydride in the presence of AIBN (azobisisobutyronitrile) as a free-radical initiator furnishes the halide-free lactone. Halolactonization followed by base-induced *anti*-elimination of H–I with DBU (1,8-diazabicyclo[5.4.0]undec-7-ene) produces the unsaturated lactone.

Lactone Ring Size. For stereoelectronic reasons, formation of five-member ring lactones is favored over six-member ring lactones.

A superb example of the synthetic utility of the iodolactonization reaction in organic synthesis is the conversion of the bicyclo[2.2.1]heptenone below into the Corey lactone. The latter compound possesses *four contiguous stereogenic* centers and is an important precursor for the synthesis of prostaglandins.[145]

Corey lactone

The iodolactonization of *acyclic unsaturated acids* in the presence of aqueous NaHCO$_3$ provides a method for *acyclic stereocontrol.* Again, five- and six-member ring lactones are the favored products. For example, iodolactonization of *trans*-3-pentenoic acid introduces two additional stereocenters at C(3) and C(4) with high diastereoselectivity resulting from stereospecific *trans*-addition to the (E)-alkene.[146] Assuming that the conformational preference is reflected in the transition state, the stereocenter at C(4) arises from *trans*-addition involving a conformation in which the destabilizing 1,3-allylic strain is minimized.

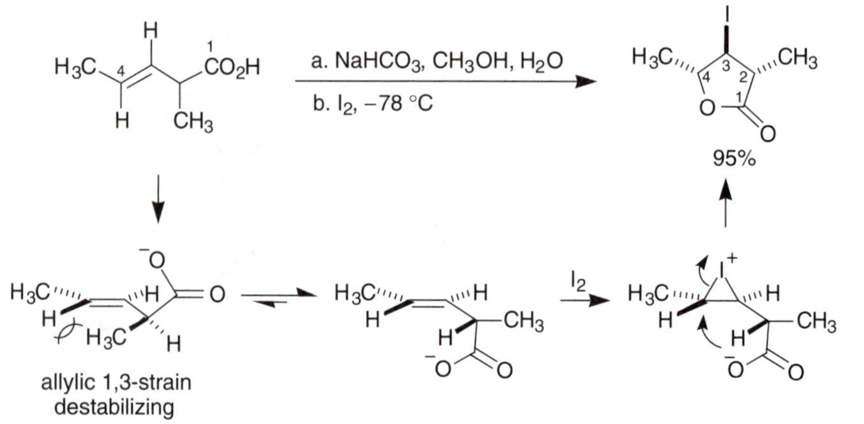

allylic 1,3-strain
destabilizing

In the absence of a base and in an *aprotic neutral solvent* such as acetonitrile, *cis-trans* equilibration occurs via the protonated lactone to furnish with high stereoselectivity the thermodynamically more stable *trans*-isomer.[147]

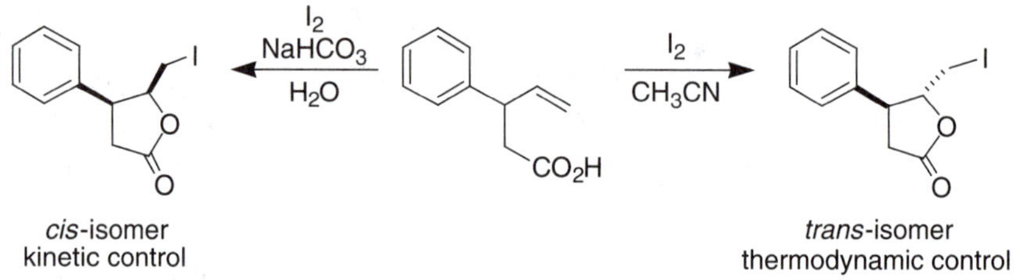

cis-isomer
kinetic control

trans-isomer
thermodynamic control

Cleavage of Carbon-Carbon Double Bonds

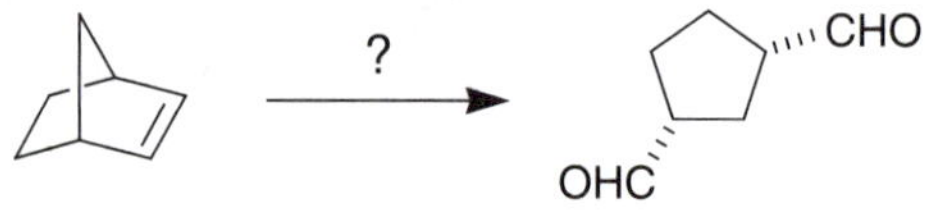

Ozonolysis[148]

Ozonolysis is widely used both in degradation work to locate the position of double bonds and in synthesis for the preparation of aldehydes, ketones, and carboxylic acids. Ozone is a 1,3-dipole and undergoes 1,3-dipolar cycloadditions with alkenes.

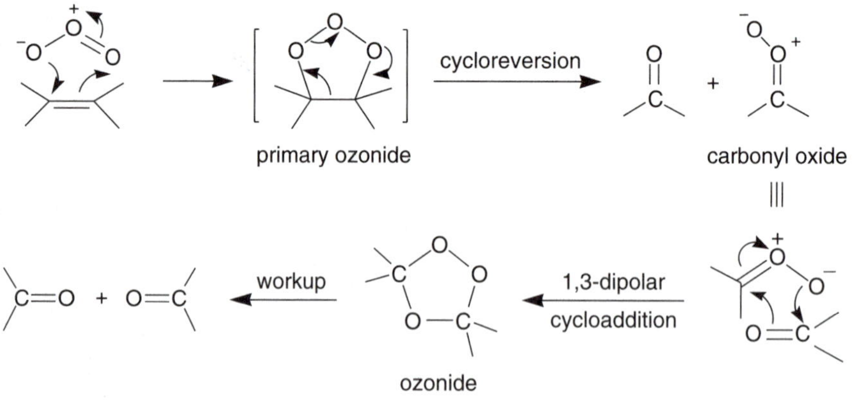

primary ozonide

carbonyl oxide

ozonide

There are two modes used to decompose ozonides:

1. Reductive: Me$_2$S, Zn–AcOH, H$_2$/Pd, KI–CH$_3$OH, metal hydrides
2. Oxidative: H$_2$O$_2$

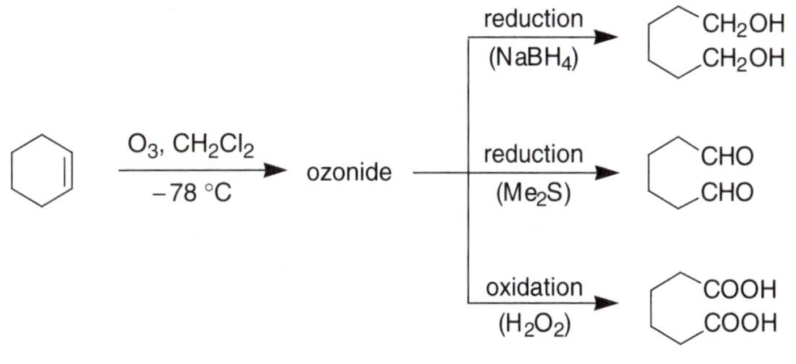

Ozonolysis produces peroxide intermediates that pose a potential explosion hazard. Formation of insoluble substances on the wall of the reaction vessel during ozonolysis points to solid peroxides and special care must be taken during workup.[149] The solvent chosen should be able to dissolve not only the starting material, but also the ozonide and any peroxide substance formed.

In the preparation of aldehydes by ozonolysis of alkenes, it is important to add the correct amount of ozone to the solution because an excess of O_3 can lead to side reactions. Ozonolysis in alcoholic solvents traps the carbonyl oxide as a hydroperoxide. Dimethyl sulfide reduces hydroperoxides under very mild conditions and generates the corresponding aldehydes in excellent yields. This workup procedure is recommended when the aldehyde is the desired reaction product.[149,150]

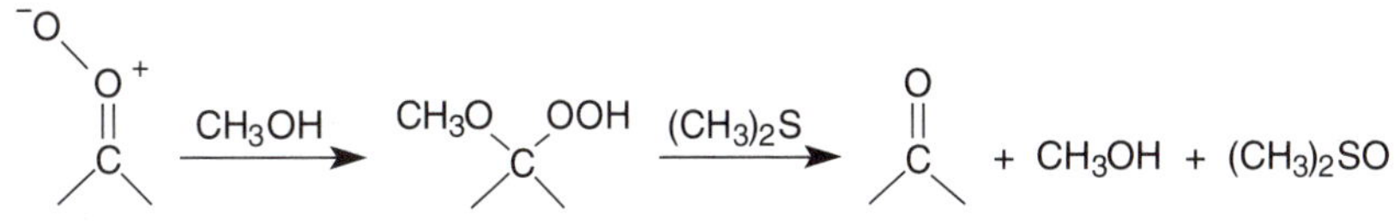

Since the electrophilic ozone preferentially attacks a more nucleophilic double, it is possible to achieve chemoselective cleavage in compounds containing two or more double bonds by limiting the amount of ozone. The relative reactivity of double bonds toward ozone decreases in the following order:

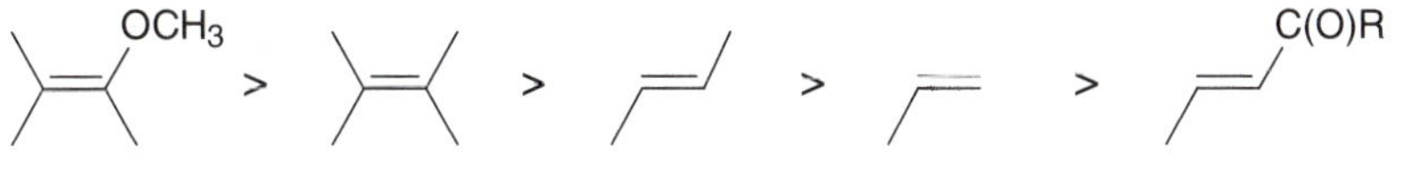

The example depicted below indicates preferential cleavage of the electron-rich enol ether double bond over the trisubstituted one by the electrophilic ozone. Thus, Birch reduction of methoxy-substituted benzenes followed by ozonolysis of the resultant enol ethers provides a powerful route to functionally substituted (Z) alkenes.[151]

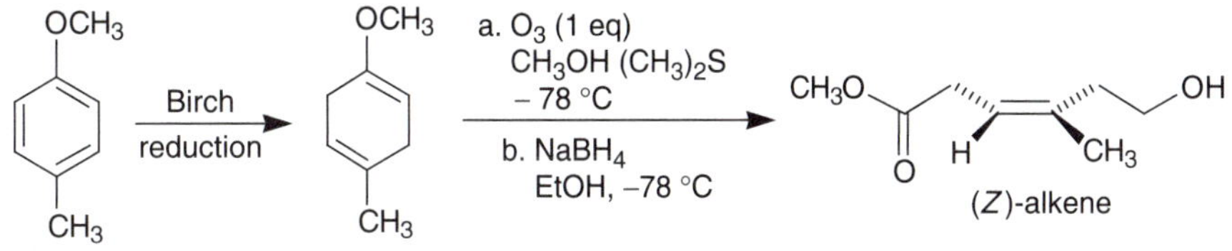

In competition of a more nucleophilic trisubstituted double bond with a terminal disubstituted one, ozone exhibits a high preference for the former.[152]

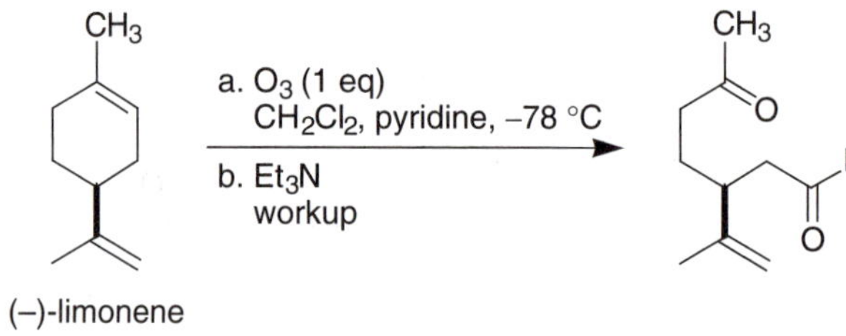

(−)-limonene

Ozonolysis of alkenes in pyridine leads directly to the corresponding carbonyl compounds.[153]

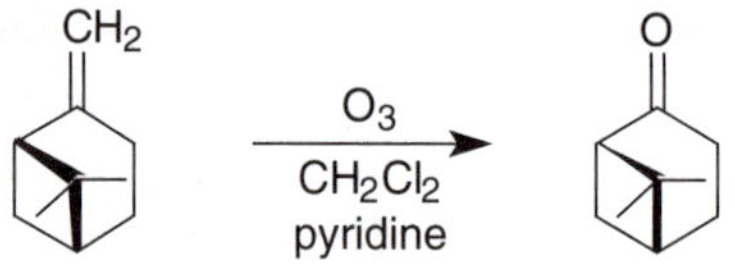

Cleavage of symmetrically substituted double bonds by ozone under different reaction conditions generates two carbons that are chemodifferentiated.[154]

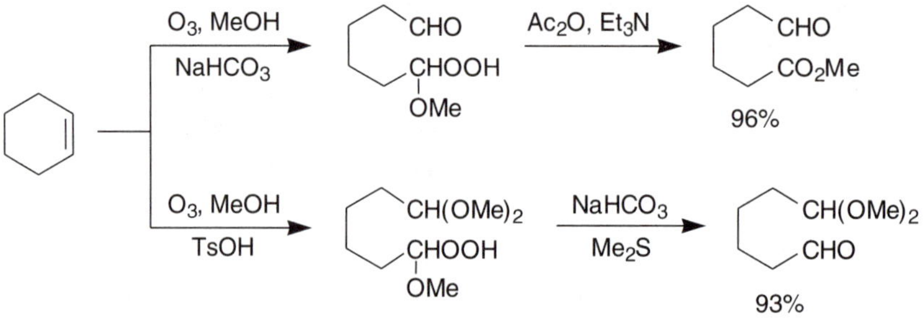

Reduction of the ozonide with LiAlH₄ or with NaBH₄ produces the corresponding alcohols.[151,155]

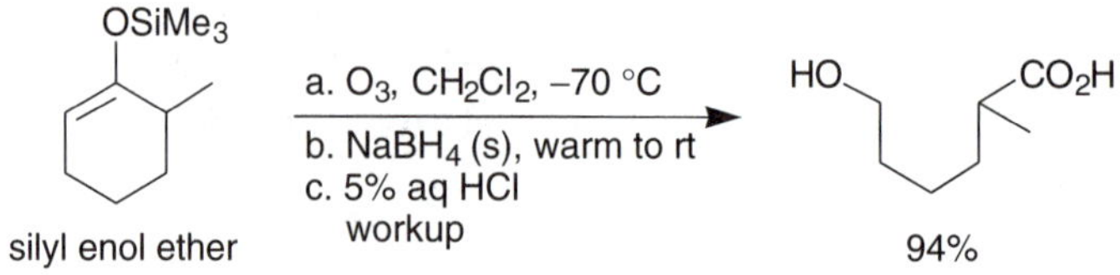

silyl enol ether

Glycol Formation and Cleavage[156]

Osmium Tetroxide and Co-oxidant. A useful alternative to ozonolysis, especially when carried out on a large scale, is the oxidation of alkenes with a catalytic amount of OsO₄ (1–5 mol %) in the presence of NaIO₄ or KIO₄ as co-oxidants (*Lemieux-Johnson oxidation*).[157] The periodate has a dual function: cleavage of the 1,2-diol (glycol) formed by the oxidizing agent and re-oxidation of the reduced osmium dioxide to osmium tetroxide. Periodate oxidation of the diol does not proceed beyond the aldehyde stage. Moreover, triple bonds are not affected by OsO₄ or by periodate.[158]

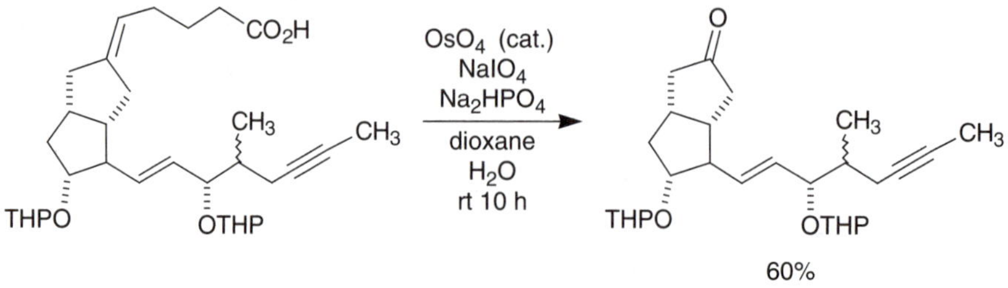

60%

An isolated double bond will be cleaved selectively in competition with an electron-deficient conjugated double bond of an enone or of an α,β-unsaturated ester.[159]

Formation of an $OsO_4 \cdot (pyridine)_2$ complex enhances the chemoselectivity of attack at the electron-rich, easily accessible double bond in compounds with multiple double bonds.[160]

Potassium Permanganate and Co-oxidant. The less volatile and less expensive potassium permanganate may also be used for the cleavage of alkenes in the presence of periodate. However, if the cleavage of the resulting diol by periodate leads to an aldehyde fragment, it will be further oxidized by potassium permanganate to the corresponding carboxylic acids. Ketones are stable under potassium permanganate catalyzed oxidation.

Cleavage of glycols derived from OsO_4 or $KMnO_4$ oxidation of alkenes by periodates or by periodic acid ($HIO_4 \cdot 2H_2O$) results in two carbonyl fragments whose structures depend on the degree of substitution of the double bond. The mechanism for periodate cleavage of glycols involves a cyclic periodate ester, which undergoes decomposition into two carbonyl fragments. Both *cis-* and *trans-*1,2-dihydroxycyclohexanes are cleaved but the *trans-*isomer at slower rate.

Below are summarized some cleavage possibilities of 1,2-diols, an α-hydroxyketone, and a 1,2-ketone with sodium periodate in $EtOH–H_2O$.[161]

Lead Tetraacetate. $Pb(OAc)_4$ may be used either in toluene or in THF for oxidation of glycols that are only slightly soluble in aqueous media. It accomplishes the same types of reactions effected by $NaIO_4$ with H_2O-soluble compounds. Being a more powerful oxidizing agent, $Pb(OAc)_4$ will also cleave 1,2-diketones, α-hydroxy ketones, and α-hydroxy acids.[162]

via a hemiacetal-glycol   75%

Ruthenium Tetroxide. RuO_4 is a more powerful oxidizing agent than OsO_4 or $KMnO_4$ and converts alkenes directly to ketones or carboxylic acids, depending on the substitution of the double bond. Whereas ketones are stable under the reaction conditions, aldehydes are oxidized to the carboxylic acids. In the catalytic reaction, $NaIO_4$ or KIO_4 serves to reoxidize the ruthenium dioxide formed after double bond cleavage back to ruthenium tetroxide. Since ruthenium tetroxide attacks solvents such as ether, tetrahydrofuran, benzene, and pyridine, the oxidation is carried out in CCl_4, $CHCl_3$, CH_2Cl_2, or aqueous acetone.[163] In the following example, the initial cleavage of the enone produces a 1,2-dicarbonyl compound, which is then further oxidized by ruthenium tetroxide with loss of one carbon atom to a carboxyl group.[163a]

80%

$$RuO_2 + IO_4^- \longrightarrow RuO_4 + IO_3^-$$

Besides ruthenium tetroxide, other ruthenium salts, such as ruthenium trichloride hydrate, may be used for oxidation of carbon-carbon double bonds. Addition of acetonitrile as a cosolvent to the carbon tetrachloride-water biphase system markedly improves the effectiveness and reliability of ruthenium-catalyzed oxidations. For example, $RuCl_3 \cdot H_2O$ in conjunction with $NaIO_4$ in acetonitrile-CCl_4-H_2O oxidizes (*E*)-5-decene to pentanoic acid in 88% yield.[121,164] Ruthenium salts may also be employed for oxidations of primary alcohols to carboxylic acids, secondary alcohols to ketones, and 1,2-diols to carboxylic acids under mild conditions at room temperature, as exemplified below. However, in the absence of such readily oxidized functional groups, even aromatic rings are oxidized.

75%

96% ee   92% yield, 94% ee

5.2   REACTIONS OF CARBON-CARBON TRIPLE BONDS

The ethynyl moiety is a versatile group for numerous chemical transformations. It is readily introduced onto a variety of organic substrates via ethynylation or coupling reactions (see Chapter 7) and provides a reaction site for manifold further modifications. The following discussion will be centered on the conversion of alkynes into stereodefined alkenes and into carbonyl compounds.

Catalytic Semireduction of Alkynes[165]

The partial reduction of substrates containing triple bonds is of considerable importance not only in research, but also commercially for stereoselectively introducing (Z)–double bonds into molecular frameworks of perfumes, carotenoids, and many natural products. As with catalytic hydrogenation of alkenes, the two hydrogen atoms add *syn* from the catalyst to the triple bond. The high selectivity for alkene formation is due to the strong absorption of the alkyne on the surface of the catalyst, which displaces the alkene and blocks its re-adsorption. The two principal metals used as catalysts to accomplish semireduction of alkynes are palladium and nickel.

Preparation of (Z)-Alkenes Using Lindlar-Type Catalysts

The classical route for the chemo- and stereoselective semihydrogenation of internal alkynes to *cis*-alkenes uses Lindlar's catalyst, Pd deposited on $CaCO_3$ or $BaCO_3$ and treated with $Pb(OAc)_2$. In the presence of quinoline, the procedure usually affords the corresponding (Z)-alkenes in high yields and isomeric purities.[166] Quinoline deactivates the palladium catalyst, reducing the chance of overreduction to saturated compounds. The catalyst is pyrophoric in hydrocarbon solvents. Since (Z)-alkenes are thermodynamically less stable than (E)-alkenes, isomerization may be a problem. For example, when the reduction is carried out at room temperature, the (Z)-alkene formed is usually contaminated with 1–3% of the (E)-isomer. At –10 to –30 °C, however, very little isomerization is observed.

The ease of reduction is 1-alkynes > disubstituted alkynes > 1-alkenes, indicating that selective semihydrogenation of triple bonds can be achieved in molecules containing double bonds. For example, nonconjugated enynes can be hydrogenated regio- and stereospecifically in high yields.

96% yield, 98% *E,Z*

Reduction of nonconjugated diynes with Lindlar's catalyst produces the corresponding (Z, Z)-dienes in good yields. An acetylenic trimethylsilyl group can be used as a blocking group to slow semihydrogenation of the silyl-bearing alkyne relative to other internal ethynyl moieties, as illustrated below.[167,168]

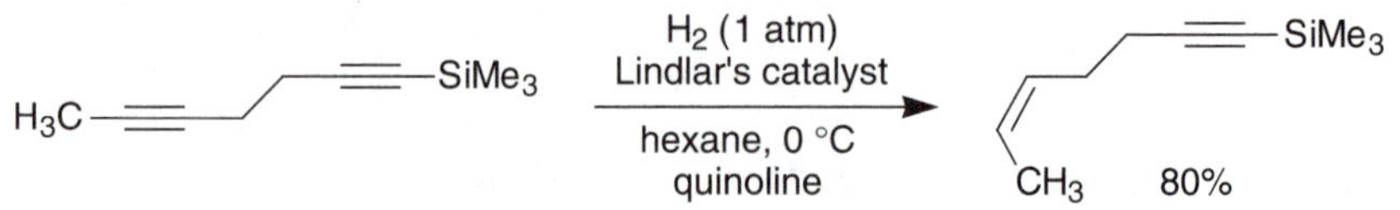

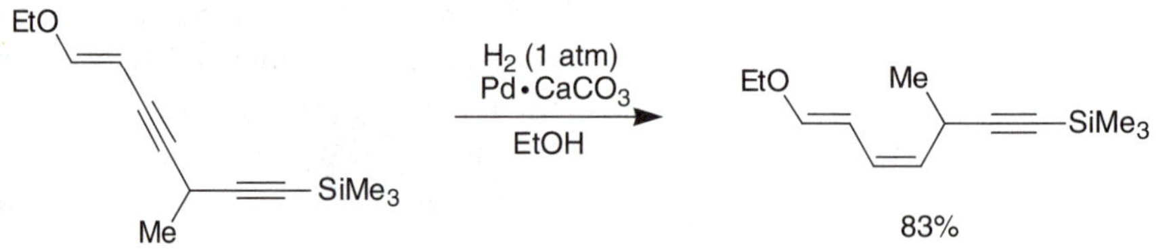

Many modifications of the original Lindlar's catalyst, particularly regarding solid supports, bases, and solvents, have been reported. Palladium precipitated on $BaSO_4$ in the presence of quinoline, but without added $Pb(OAc)_2$, has been suggested to be superior to Lindlar's catalyst in reproducibility and ease of preparation.[169] The triple bond can be hydrogenated selectively in compounds containing other potentially reducible groups, as exemplified below.[170]

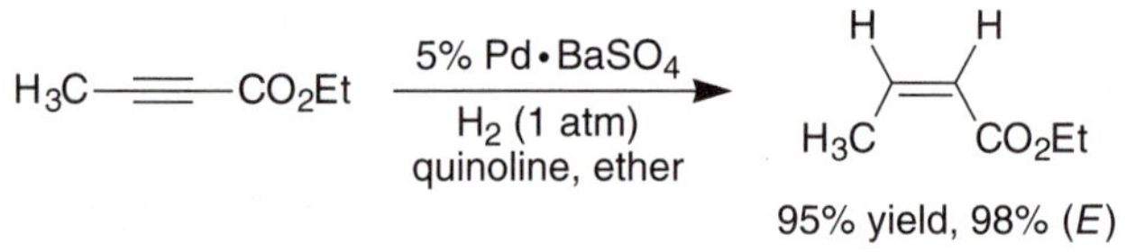

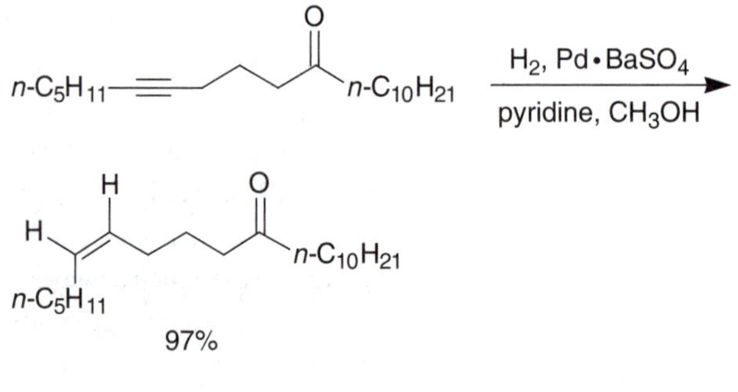

Preparation of (Z)-Alkenes Using Nickel Boride

Highly chemoselective and stereoselective semihydrogenations of internal alkynes to *cis*-alkenes have been achieved with Brown's P-2 Nickel catalyst in the presence of ethylenediamine.[171] The catalyst is a black suspension of Ni boride prepared by addition of a solution of $NaBH_4$ in 95% ethanol to $Ni(OAc)_2 \cdot 4\ H_2O$. The hydrogen is generated in situ from addition of aqueous sodium borohydride to a reservoir of acetic acid (using an apparatus known as a *Brown^2 gasimeter*).[172] Advantages of the P-2 Ni catalyst for semihydrogenation of alkynes include its ease of preparation from readily available starting materials and its nonpyrophoric nature.

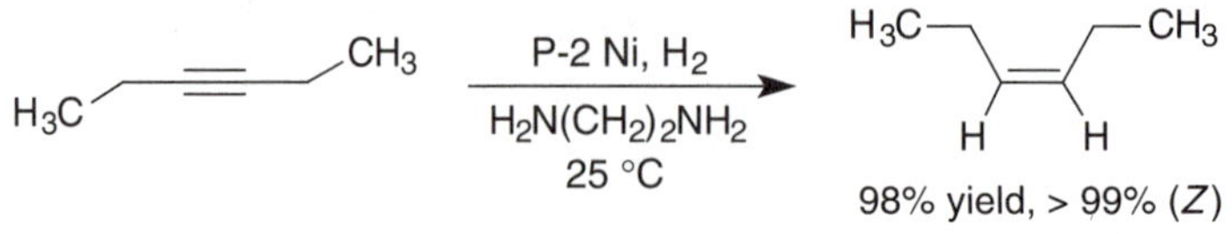

94% yield, > 99% (Z)

Semireductions of enynes and diynes with P-2 Ni produce the corresponding dienes, as exemplified below.[173]

92%

Nickel-boride-type catalysts are also employed for desulfurization, dehalogenation, hydrogenolysis, and reduction of nitro and other functional groups.[174]

Preparation of (Z)-Alkenes Using Zn Powder

The highly activated Rieke zinc, prepared in situ from $ZnBr_2$ and $K°$,[175] selectively reduces alkynes to *cis*-alkenes, enynes to dienes, diynes to enynes or dienes, and propargylic alcohols to *cis*-allylic alcohols.[176]

84% yield, > 95% (Z)-reduction

Other activated zinc reagents reported for the selective semireduction of triple bonds in the presence of double bonds are Zn/KCN[177] and $Zn/Cu/Ag$.[178]

90%

Reduction of Alkynes via Protonolysis of Alkenylboranes[179]

Hydroborations of alkynes directed toward the synthesis of alkenylboranes furnishes versatile intermediates for a wide array of chemical transformations. For example, protonolysis of alkenylboranes provides a noncatalytic method for the semireduction of alkynes.

Hydroboration of Alkynes and Enynes

Hydroboration of a 1-alkyne with borane-THF in a 3 : 1 ratio produces, via *bis*-addition of the B–H bond to the triple bond, dihydroboration products along with the starting alkyne.

However, hydroboration of internal alkynes with a stoichiometric amount of borane results predominantly in monoaddition of the B–H bond to the triple bond to furnish the corresponding (Z)-trialkenylborane.

With sterically hindered dialkylboranes, such as dicyclohexylborane (Chx$_2$BH) and disiamylborane (Sia$_2$BH), the stoichiometric hydroborations of both terminal and internal alkynes stops at the monoaddition stage. However, the stoichiometric reaction of 1-alkynes with 9-BBN affords mainly the 1,1-diborylalkanes.[180]

The monohydroboration of alkynes is stereoselective, involving a *cis*-addition of the B–H bond to the triple bond. The reaction of 1-alkynes with dialkylboranes is not only stereoselective, but also regioselective, placing the boron at the terminal carbon of the triple bond.

Hydroborations of unsymmetrically disubstituted alkynes furnish mixtures of regioisomers except when the substituents are of markedly different steric size. In these cases the dialkylboranes add preferentially to the sterically less-hindered carbon of the triple bond.[181]

The relative reactivity of disiamylborane toward unsaturated carbon-carbon bonds is[182]

$$RC≡CH > RC≡CR > RCH=CH_2 > RCH=CHR$$

Thus, chemoselective hydroborations of the triple bond in various enynes have been achieved. A similar chemoselectivity is predicted for hydroborations with the slightly less hindered dicyclohexylborane.[183]

A reversal of the chemoselectivity is observed when enynes are hydroborated using 9-BBN.[183]

Protonolysis of Vinylic Boranes

Treatment of alkenylboranes with carboxylic acids at room temperature or, in the case of more hindered alkenylboranes, with hot propanoic acid results in the selective cleavage of the vinylic B–C bond with *retention of configuration.*[179]

For regiodefined alkenylboranes, cleavage of the vinylic carbon-boron bond with deuterioacetic acid provides a simple, stereospecific preparation of deuterium-substituted alkenes.[181]

The cleavage of alkenylboranes may be carried out under non-acidic conditions by converting the alkenylborane into the *ate* complex with *n*-BuLi prior to hydrolysis with aqueous NaOH.[184] Also, alkenylboranes derived from internal alkynes may be protonated under neutral conditions using catalytic amounts of $Pd(OAc)_2$ in THF or acetone[185] or with aqueous silver ammonium nitrate $[Ag(NH_3)_2 NO_3]$.[186]

The mildness and selectivity of the hydroboration-protonolysis procedure allows for the presence of functional groups and its ready adaptability to the synthesis of natural products.[187]

93% (> 98% isomeric purity)

Functionally substituted alkynes such as 1-halo-1-alkynes[188] and ethyl 2-alkynoates[189] undergo regio- and stereoselective monohydroboration to produce, after protonolysis, the synthetically useful (Z)-1-halo-1-alkenes and cis-α,β-unsaturated esters, respectively.

Preparation of *trans*-Alkenes

Reduction with Li or Na in Liquid NH₃[190]

Lithium or sodium in liquid ammonia reduces disubstituted alkynes to *trans*-alkenes.[191] The reaction is carried out by addition of the alkyne in ether to a mixture of Na in NH$_3$ (l), and the alkene produced is the (E)-isomer. Overreduction and isomerization of the alkene are suppressed by addition of t-BuOH.[192]

The reduction proceeds by addition of one electron from the metal to the triple bond to form a linear radical anion, which picks up a proton from the NH$_3$ solvent or from an added proton source, usually t-butanol, to give a vinyl radical. Subsequent transfer of another electron from the metal leads to the vinyl anion having the more stable *trans*-configuration. Protonation of the vinyl anion furnishes the *trans*-alkene.

The electron-rich acetylide anion produced from deprotonation of 1-alkynes with Na in NH_3 (liq.) is reluctant to accept an electron, allowing the selective reduction of an internal triple bond in the presence of a terminal one. Reduction to the corresponding 1-alkenes can be achieved in the presence of ammonium sulfate.[191b]

Indium selectively reduces propargyl ethers, amines or esters to the corresponding terminal alkenes. A number of readily reducible functional groups, such as RCHO, $R_2C=O$, OBn, OTs, and Cl, are inert to these reaction conditions.[193] The mechanism of the reduction is not yet clear.

Reduction of Propargylic Alcohols with LiAlH₄ or with Red-Al

The reduction of propargylic alcohols with $LiAlH_4$ is a standard procedure for preparing *trans*-allylic alcohols.[194] The propargylic alcohols themselves are readily available by the reaction of acetylide anions with either paraformaldehyde,[195] aldehydes or ketones.[196] There is a striking solvent dependence on the stereochemical course of the reduction. The amount of *trans*-reduction increases with the Lewis basicity of the solvent. THF is the solvent of choice for *trans* reduction[197] whereas diisopropyl ether gives principally *cis*-reduction. Homopropargylic alcohols react slowly with $LiAlH_4$.[198]

Evidence that the reaction proceeds via an alkenylaluminum intermediate resulting from *trans*-addition of the Al–H bond to the triple bond stems from the fact that it exhibits characteristics similar to those observed in reactions with typical alkenylaluminum compounds.[194]

Under milder conditions, treatment of propargylic alcohols with *n*-BuLi followed by diisobutylaluminum hydride at –78 °C also affords *trans*-allylic alcohols with excellent stereoselectivity.[199] More recently, Red-Al [Na(AlH$_2$(OCH$_2$CH$_2$OCH$_3$)$_2$)], sodium *bis*(2-methoxyethoxy)aluminum hydride] is the reagent of choice for the reduction of acetylenic alcohols. The reaction proceeds cleanly with high *trans*-selectivity.[200]

Hydration of Alkynes

Hydroboration-Oxidation

Monohydroboration of 1-alkynes followed by oxidation gives the corresponding aldehydes in high yields.[201] Oxidation of the vinyl carbon-boron bond produces the enol, which then tautomerizes to the carbonyl group. To minimize aldol condensation of the aldehyde formed during oxidation, the reaction should be carried out at pH ~8 or in buffered medium.[202]

Hydroboration-oxidation of an alkenylborane derived from a *symmetrically* substituted alkyne yields a single ketone, whereas *unsymmetrically* disubstituted alkynes furnish mixtures of ketones.

Hydroboration of alkynylsilanes with dicyclohexylborane proceeds in a stereo- and regioselective manner, placing the boron at the silicon-bearing carbon. Oxidation of the resultant alkenylborane with excess alkaline hydrogen peroxide produces, via an acyl silane intermediate (see Section 7.9), the carboxylic acids in greater than 80% yields. The alkynylsilanes may be generated in situ as illustrated by the one-pot transformation below.[203]

Oxymercuration—Preparation of Methyl Ketones[204]

The mercuric ion-catalyzed hydration of alkynes probably proceeds in a similar manner to the oxymercuration of alkenes (see Section 5.1). Electrophilic addition of Hg^{2+} to the triple bond leads to a vinylic cation, which is trapped by water to give an vinylic organomercury intermediate. Unlike the alkene oxymercuration, which requires reductive removal of the mercury by $NaBH_4$, the vinylic mercury intermediate is cleaved under the acidic reaction conditions to give the enol, which tautomerizes to the ketone. Hydration of terminal alkynes follows the Markovnikov rule to furnish methyl ketones.[205]

Methyl ketones also may be prepared from

- *Carboxylic acids*, by addition of two equivalents CH_3Li[206] (see Section 7.1)
- *Amides*, using "CH_3CeCl_2" [207]

- *Acid chlorides*, via $(CH_3)_2CuLi$ addition (see Section 7.5)
- *Terminal alkenes*, using the *Wacker* process[208]

Symmetrically disubstituted alkynes give a single ketone, whereas unsymmetrical alkynes produce a mixture of ketones.

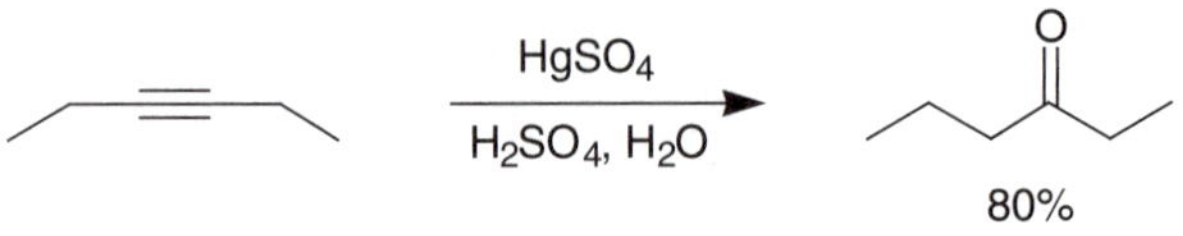
$HgSO_4$
H_2SO_4, H_2O
O
80%

PROBLEMS

1. **Reagents**. Give the structures of the major product(s) expected from each of the following reactions. Indicate product stereochemistry where applicable.

a.

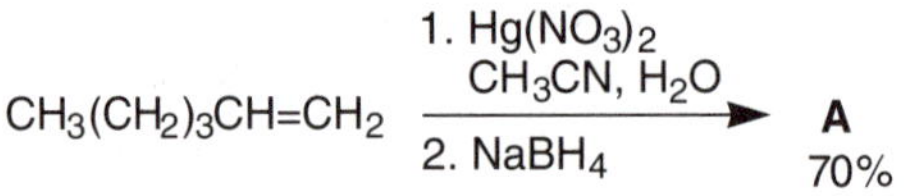
$CH_3(CH_2)_3CH{=}CH_2$
1. $Hg(NO_3)_2$
CH_3CN, H_2O
2. $NaBH_4$
A
70%

b.

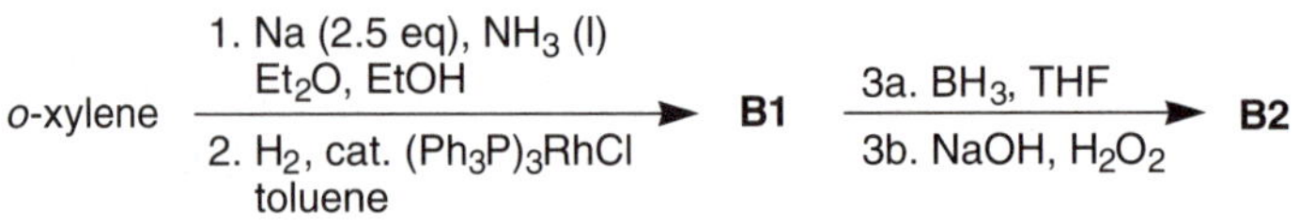
o-xylene
1. Na (2.5 eq), NH_3 (l)
Et_2O, EtOH
2. H_2, cat. $(Ph_3P)_3RhCl$
toluene
B1
3a. BH_3, THF
3b. NaOH, H_2O_2
B2

c.

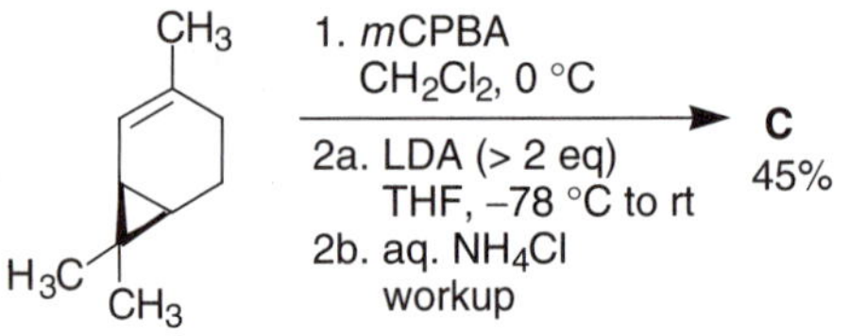
CH_3
H_3C
CH_3
1. mCPBA
CH_2Cl_2, 0 °C
2a. LDA (> 2 eq)
THF, –78 °C to rt
2b. aq. NH_4Cl
workup
C
45%

d.

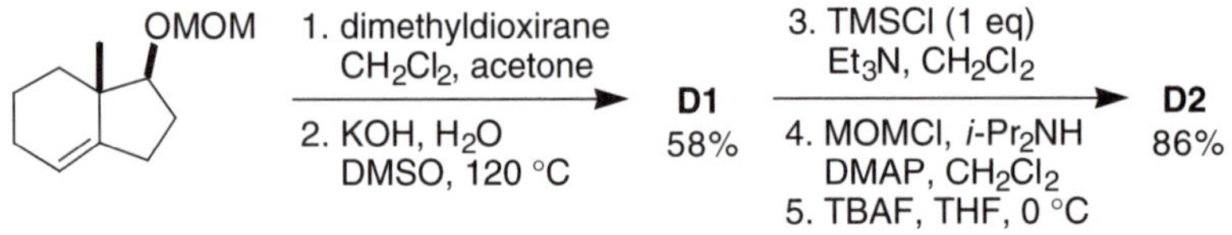
OMOM
1. dimethyldioxirane
CH_2Cl_2, acetone
2. KOH, H_2O
DMSO, 120 °C
D1
58%
3. TMSCl (1 eq)
Et_3N, CH_2Cl_2
4. MOMCl, i-Pr_2NH
DMAP, CH_2Cl_2
5. TBAF, THF, 0 °C
D2
86%

e.

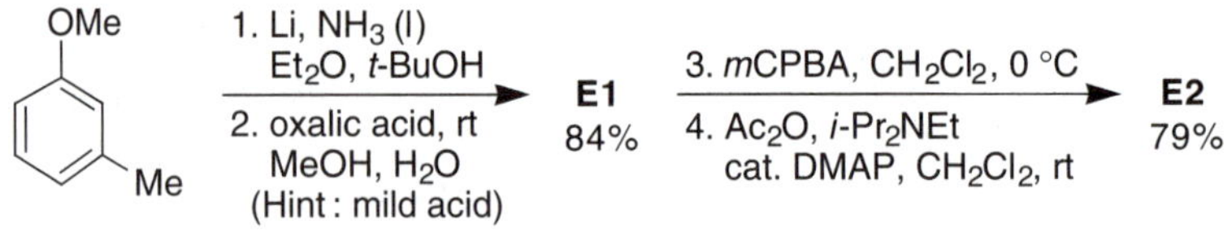
OMe
Me
1. Li, NH_3 (l)
Et_2O, t-BuOH
2. oxalic acid, rt
MeOH, H_2O
(Hint: mild acid)
E1
84%
3. mCPBA, CH_2Cl_2, 0 °C
4. Ac_2O, i-Pr_2NEt
cat. DMAP, CH_2Cl_2, rt
E2
79%

f.

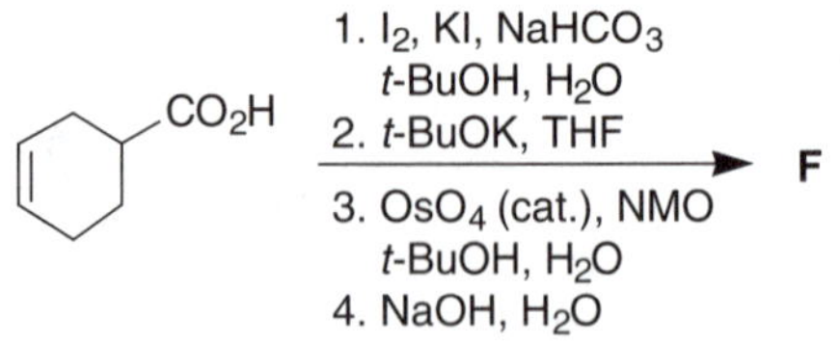
CO_2H
1. I_2, KI, $NaHCO_3$
t-BuOH, H_2O
2. t-BuOK, THF
3. OsO_4 (cat.), NMO
t-BuOH, H_2O
4. NaOH, H_2O
F

g.

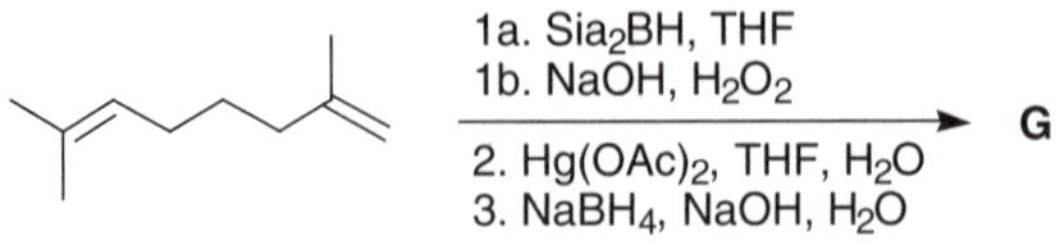
1a. Sia_2BH, THF
1b. NaOH, H_2O_2
2. $Hg(OAc)_2$, THF, H_2O
3. $NaBH_4$, NaOH, H_2O
G

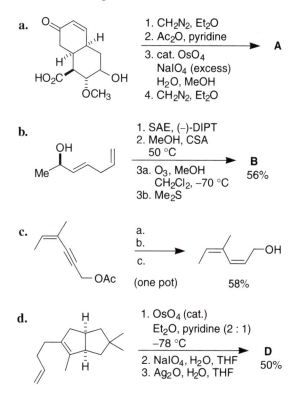

h.

t-Bu —≡— CH₃ → 1a. Na, NH₃ (l)
1b. aq. NH₄Cl
work-up
2. mCPBA, CH₂Cl₂ **H1** → 3a. [Li-C≡CH]
THF
3b. H₂O workup **H2**

i. 1. mCPBA
cyclohexane
rt, 7 days
2. K₂CO₃ (1.2 eq)
MeOH
3. PDC, CH₂Cl₂ **I**
90%

j. 1a. Na, NH₃ (l)
Et₂O
1b. Br⟍⟍⟋
1c. aq. NH₄Cl
workup **J**
91%

*k. 1. ⟍⟋MgCl / Me
2. DMSO
(CF₃CO)₂O
3. DIBAL-H, Et₂O **K**
83%

Step 2 is a modified Swern oxidation procedure; see *J. Org. Chem.* **1998**, *63*, 8522.

2. Selectivity. Show the product(s) obtained or the appropriate reagent(s) to be used for the following transformations.

a. 1. CH₂N₂, Et₂O
2. Ac₂O, pyridine
3. cat. OsO₄
NaIO₄ (excess)
H₂O, MeOH
4. CH₂N₂, Et₂O **A**

b. 1. SAE, (–)-DIPT
2. MeOH, CSA
50 °C
3a. O₃, MeOH
CH₂Cl₂, –70 °C
3b. Me₂S **B**
56%

c. a.
b.
c. → (one pot) 58%

d. 1. OsO₄ (cat.)
Et₂O, pyridine (2 : 1)
–78 °C
2. NaIO₄, H₂O, THF
3. Ag₂O, H₂O, THF **D**
50%

e.

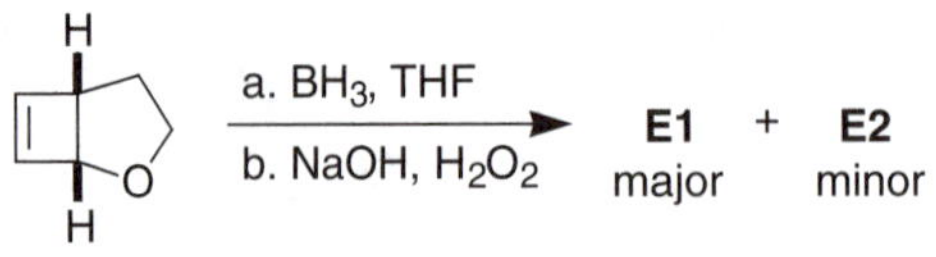

$$\text{a. BH}_3\text{, THF}$$
$$\text{b. NaOH, H}_2\text{O}_2$$

E1   +   **E2**
major   minor

*f.

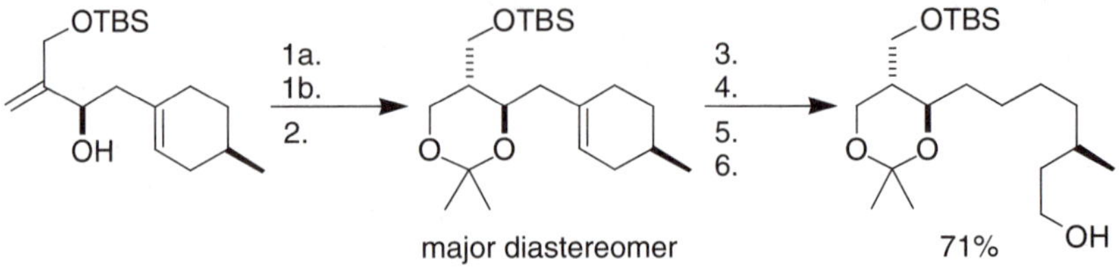

major diastereomer   71%

3. **Stereochemistry**. Give the structure and predict the stereochemistry of the major product formed in each of the following reactions. Give an explanation for your choice of stereochemistry.

a.

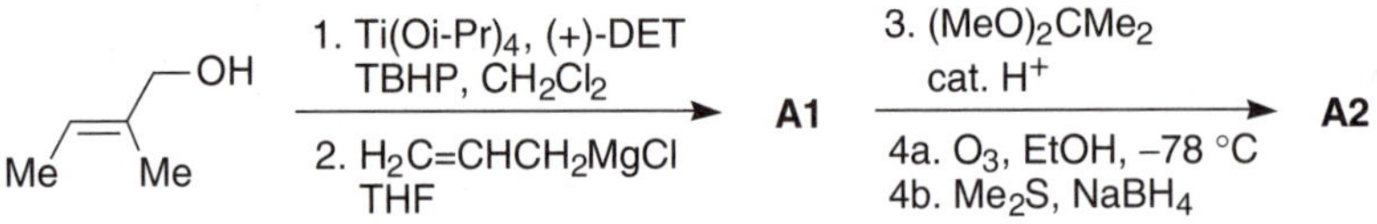

1. Ti(Oi-Pr)$_4$, (+)-DET
   TBHP, CH$_2$Cl$_2$

2. H$_2$C=CHCH$_2$MgCl
   THF

A1

3. (MeO)$_2$CMe$_2$
   cat. H$^+$

4a. O$_3$, EtOH, -78 °C
4b. Me$_2$S, NaBH$_4$

A2

b.

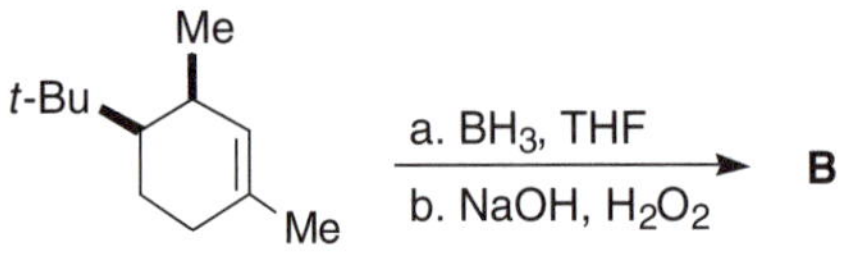

a. BH$_3$, THF

b. NaOH, H$_2$O$_2$

B

c.

1. Ti(Oi-Pr)$_4$
   ($-$)-DET, t-BuOOH
   CH$_2$Cl$_2$, -20 °C

2. Red-Al, THF, rt

C

d.

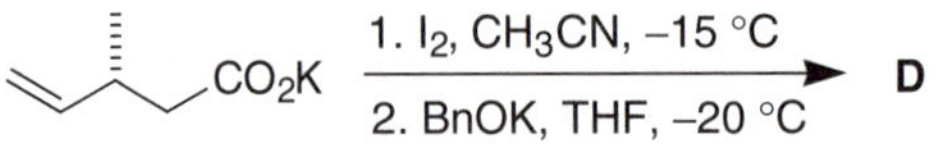

1. I$_2$, CH$_3$CN, -15 °C

2. BnOK, THF, -20 °C

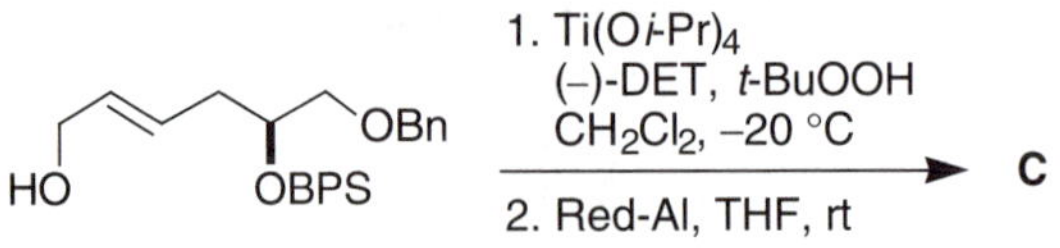

D

e. Select the appropriate reducing agent from the list shown below. Explain your choice.

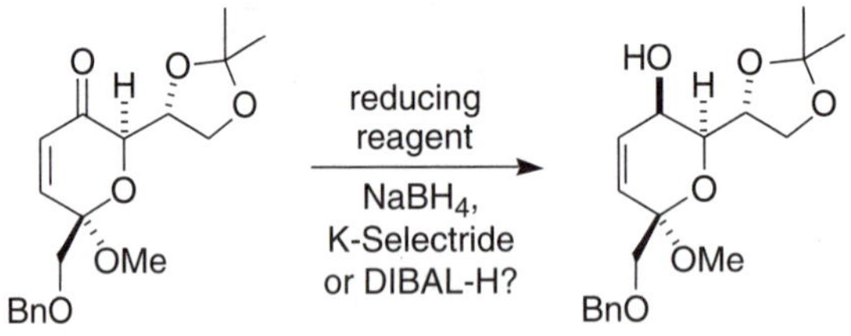

reducing
reagent

NaBH$_4$,
K-Selectride
or DIBAL-H?

4. **Reactivity**. Explain the regioselectivity and stereochemistry observed in each of the following transformations.

a.

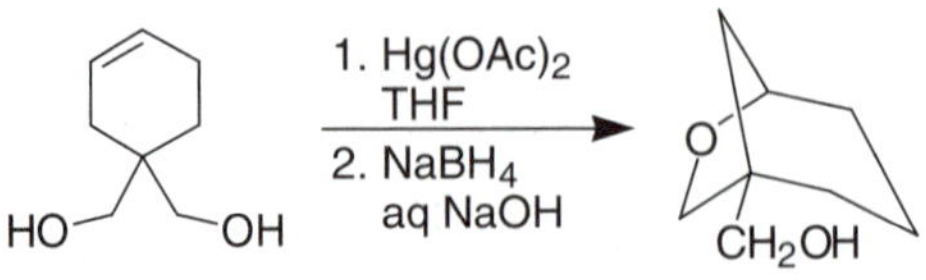

1. Hg(OAc)$_2$
   THF

2. NaBH$_4$
   aq NaOH

b.

a. cat. MgBr$_2$
   PhCH$_2$CH$_2$MgBr

b. aq NH$_4$Cl

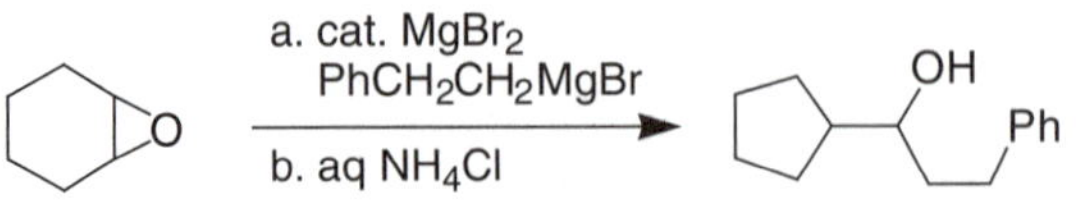

c.

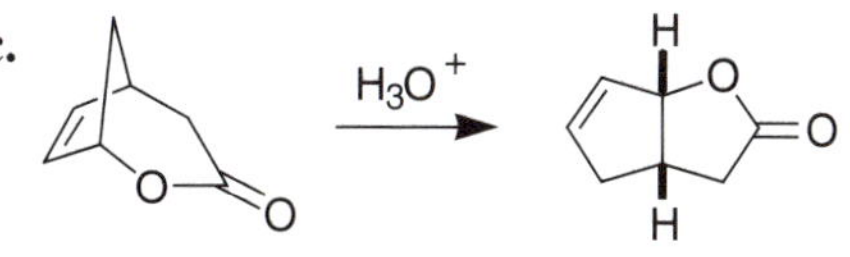
H₃O⁺

d. Why is the β-epoxide not formed on *m*CPBA epoxidation although the β-epoxide is predicted based on OH-directed epoxidation? Propose a strategy to synthesize the β-epoxide.

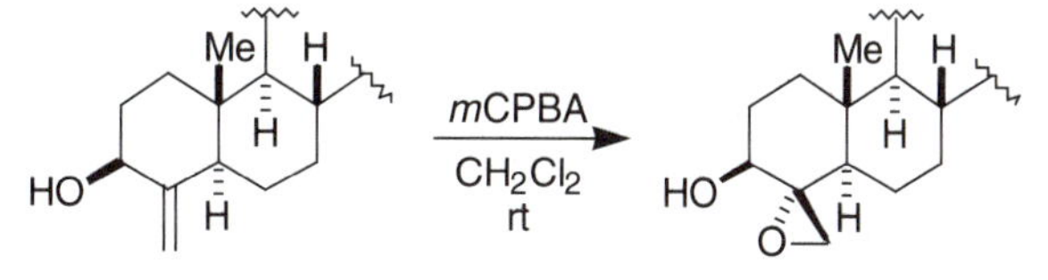
mCPBA
CH₂Cl₂
rt

85%

e.

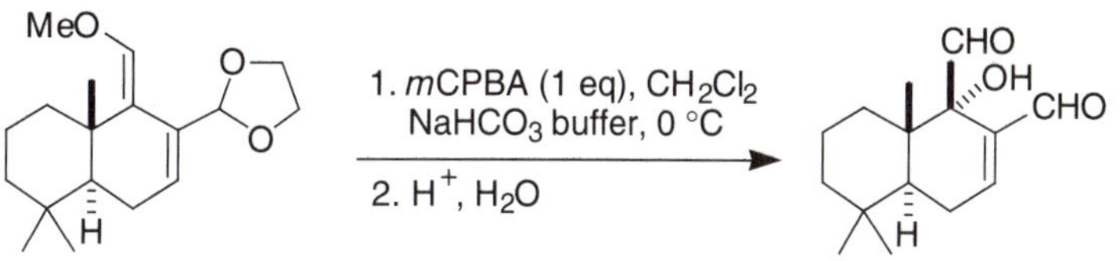
1. mCPBA (1 eq), CH₂Cl₂
NaHCO₃ buffer, 0 °C
2. H⁺, H₂O

80%, warburganal

***f.** Provide a mechanistic interpretation for the formation of the aldehydes shown below.

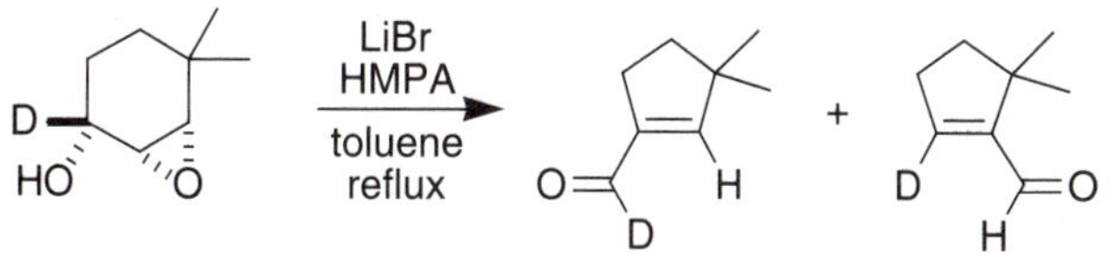
LiBr
HMPA
toluene
reflux

F1 (19%) **F2** (70%)

5. **Synthesis.** Supply the reagents required to accomplish each of the following syntheses. Indicate the relative stereochemistry, where applicable, of the products obtained at each step.

a.

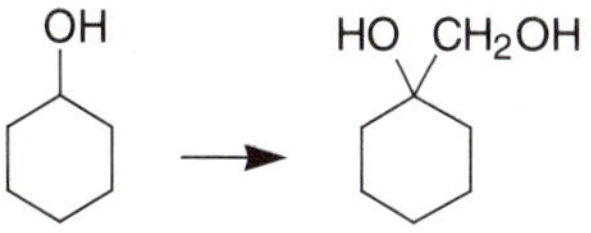

b.

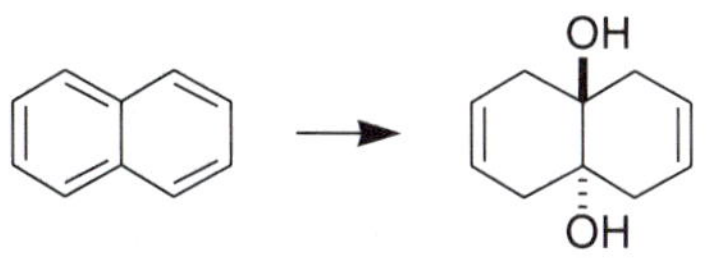

c.

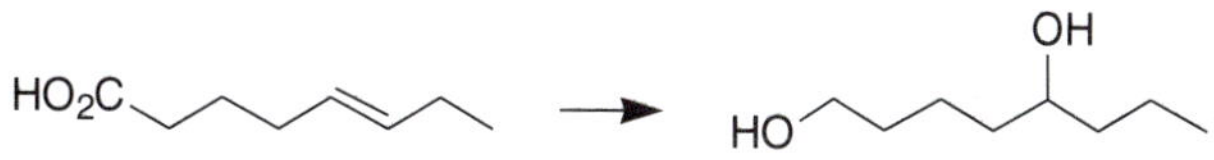

d.

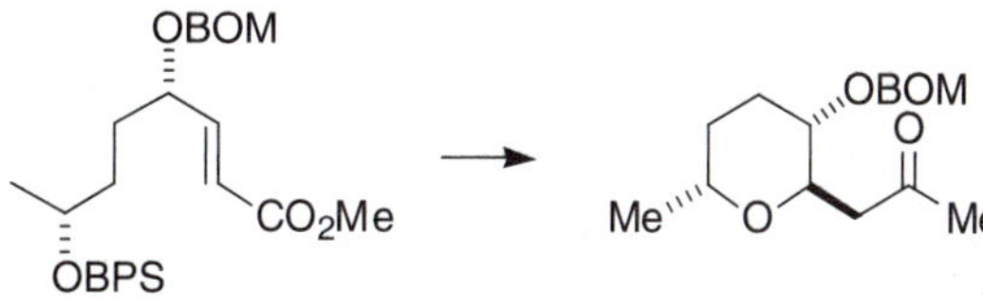

e.

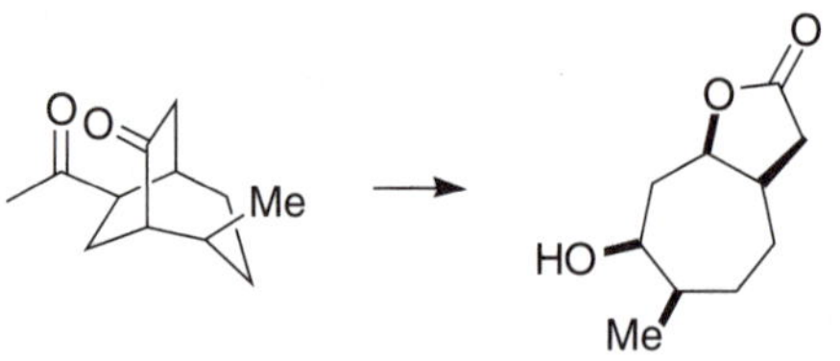

f.

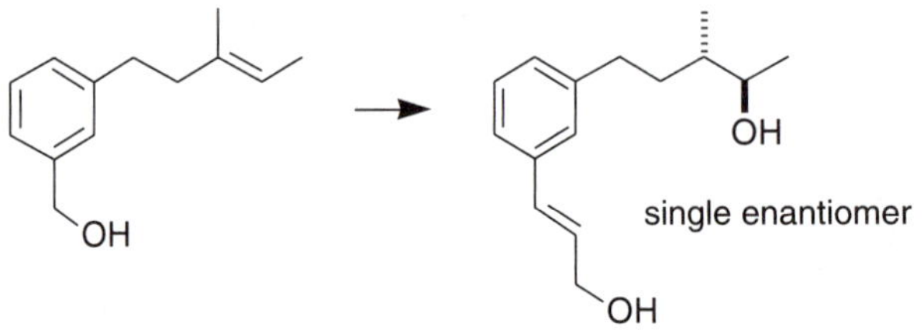

single enantiomer

*g.

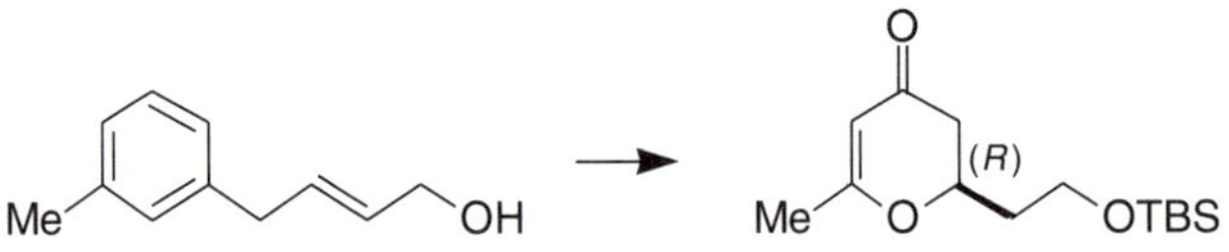

*h.

chiral methyl group,
single enantiomer

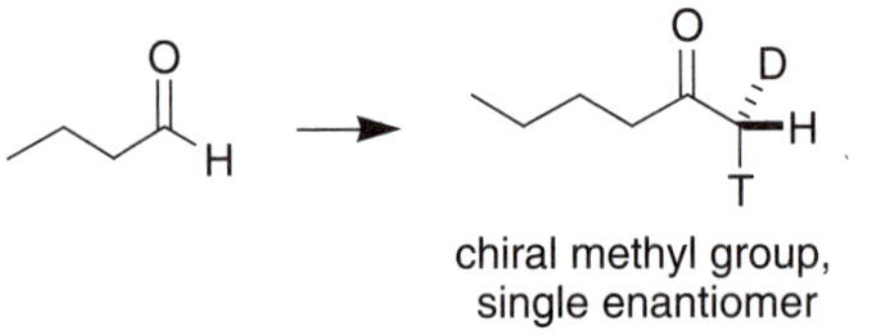

Hint: No SAE; however, the
sequence does involve
an allylic alcohol intermediate.

6. Consider the reactions **A-F**. Assume that the Sharpless epoxidations proceed with *complete π-facial selectivity* regardless of substrate. Select the best answer among the following choices regarding the stereochemical outcome of each of the reactions.

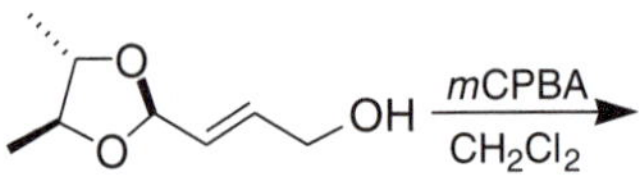

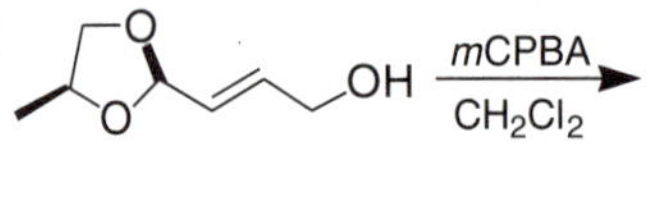

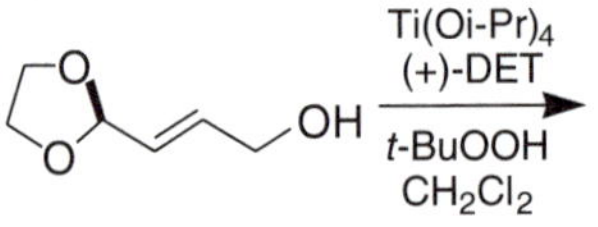

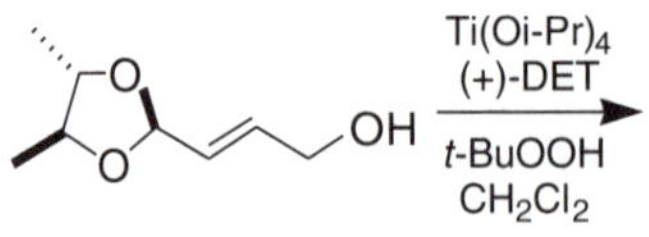

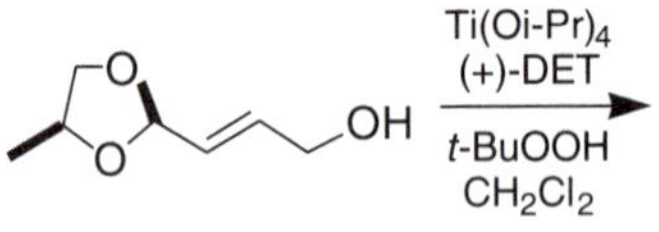

a. Assuming that the stereocenters, if any, in the starting allylic alcohols are racemic, which of the above reactions lead(s) to a mixture of *enantiomers*?

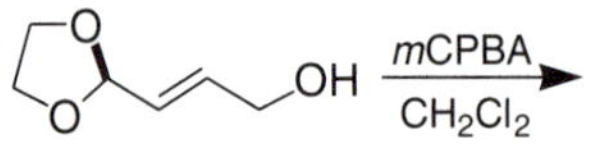

(i) reaction **A** only (iv) reactions **A** and **B**

(ii) reactions **B** and **E** (v) reactions **A**, **B**, and **C**

(iii) reactions **D** and **F**

b. Assuming that the starting allylic alcohols, if chiral, are enantiomerically pure, which of the above reactions lead(s) to a mixture of *diastereomers*?

    (i) reactions **E** and **F**   **(iv)** reaction **E** only

    (ii) reaction **B** only   **(v)** all of them would except reactions **A** and **D**

    (iii) reactions **B** and **C**

*7. Retrosynthetic Analysis. For each of the following syntheses show (1) your retrosynthetic analysis and (2) all reagents and reaction conditions required to transform a commercially available starting material to the target molecule.

a. Propose a synthesis of (+)-*trans*-3-hydroxypipecolic acid using the Sharpless asymmetric dihydroxylation procedure to establish the absolute stereochemistry.

(+)-*trans*-3-hydroxypipecolic acid

b. Propose a synthesis of the following lactone using the Sharpless asymmetric epoxidation procedure to establish the absolute stereochemistry.

REFERENCES

1. (a) Kieboon, A. P. G., van Rautwijk, F. *Hydrogenation and Hydrogenolysis in Synthetic Organic Chemistry*, Delft University, 1977. (b) Freifelder, M. *Catalytic Hydrogenation in Organic Synthesis: Procedures and Commentary*, Wiley: New York, 1978. (c) Rylander, P. N. Catalytic *Hydrogenation in Organic Synthesis*, Academic Press: New York, 1997.

2. Brown, C. A. *J. Am. Chem. Soc.* **1969**, *91*, 5901.

3. Pettit, G. R., van Tamelen, E. E. *Org. React.* **1962**, *12*, 356.

4. (a) Brown, C. A. *J. Org. Chem.* **1970**, *35*, 1900. (b) Brown, C. A., Ahuja, V. K. *J. Org. Chem.* **1973**, *38*, 2226.

5. Belisle, C. M., Young, Y. M., Singaram, B. *Tetrahedron Lett.* **1994**, *35*, 5595.

6. Sauvage, J.-F., Baker, R. H., Hussey, A. S. *J. Am. Chem. Soc.* **1961**, *83*, 3874.

7. Birch, A. J., Williamson, D. H. *Org. React.* **1976**, *24*, 1.

8. (a) Osborn, J. A., Jardine, F. H., Young, J. F., Wilkinson, G. *J. Chem. Soc. A* **1966**, 1711. (b) Harmon, R. E., Gupta, S. K., Brown, D. J. *Chem. Rev.* **1973**, *73*, 21.

9. Ireland, R. E., Bey, P. *Org. Synth.* **1973**, *53*, 63.

10. Brown, J. M. *Angew. Chem., Int. Ed.* **1987**, *26*, 190.

11. Crabtree, R. H., Davis, M. W. *J. Org. Chem.* **1986**, *51*, 2655.

12. Evans, D. A., Morrissey, M. M., Dow, R. L. *Tetrahedron Lett.* **1985**, *26*, 6005.

13. Brown, J. M., Hall, S. A. *Tetrahedron Lett.* **1984**, *25*, 1393.

14. (a) Kagan, H. B. In *Comprehensive Organometallic Chemistry*, Wilkinson, G., Stone, F. G. A., Abel, E. W., Eds., Pergamon: Oxford, UK, 1982, Vol. 8, p. 463. (b) Ojima, I., Clos, N., Bastos, C. *Tetrahedron* **1989**, *45*, 6901. (c) Gridner, I. D., Imamoto, T. *Acc. Chem. Res.* **2004**, *37*, 633.

15. (a) Noyori, R., Takaya, H. *Acc. Chem. Res.* **1990**, *23*, 345. (b) Taber, D. F., Silverberg, L. J. *Tetrahedron Lett.* **1991**, *32*, 4227.

16. Caine, D. *Org. React.* **1976**, *23*, 1.

17. Stork, G., Darling, S. D. *J. Am. Chem. Soc.* **1964**, *86*, 1761.

18. Crabtree, S. R., Mander, L. N., Sethi, S. P. *Org. Synth.* **1991**, *70*, 256.

19. Youn, I. K., Yon, G. H., Pak, C. S. *Tetrahedron Lett.* **1986**, *27*, 2409.

20. (a) Birch, A. J. *J. Chem. Soc.* **1944**, 430. (b) Rabideau, P. W., Marcinow, Z. *Org. React.* **1992**, *42*, 1.

21. Fleming, I. *Frontier Orbitals and Organic Chemical Reactions*, Wiley: London, **1976**, pp. 200–202.

22. Hine, J. *Adv. Phys. Org. Chem.* **1977**, *15*, 1.

23. Rabideau, P. W. *Tetrahedron* **1989**, *45*, 1579.

24. Birch, A. J., Hutchinson, E. G., Rao, G. S. *J. Chem. Soc., Chem. Commun.* **1970**, 657.

25. Gutsche, C. D., Peter, H. H. *Org. Synth., Col. Vol. IV* **1963**, 887.

26. Bachi, M. D., Epstein, J. W., Herzberg-Minzly, Y., Loewenthal, H. J. E. *J. Org. Chem.* **1969**, *34*, 126.

27. Rabideau, P. W., Wetzel, D. M., Young, D. M. *J. Org. Chem.* **1984**, *49*, 1544.

28. Braude, E. A., Webb, A. A., Sultanbawa, M. U. S. *J. Chem. Soc.* **1958**, 3328.

29. House, H. O. *Modern Synthetic Reactions*, W. A. Benjamin: Menlo Park, CA, **1972**, p. 644.

30. Schultz, A. G., Sundararaman, P. *Tetrahedron Lett.* **1984**, *25*, 4591.

31. Schultz, A. G., Puig, S. *J. Org. Chem.* **1985**, *50*, 915.

32. Schultz, A. G., Dittami, J. P. *Tetrahedron Lett.* **1983**, *24*, 1369. (b) Schultz, A. G., Dittami, J. P., Lavieri, F. P., Salowey, C., Sundararaman, P., Szymula, B. *J. Org. Chem.* **1984**, *49*, 4429.

33. (a) Vilsmeier, A., Haack, A. *Ber.* **1927**, *60*, 119. (b) Jones, G., Stanforths, S. P. *Org. React.* **1997**, *49*, 1.

34. (a) Brown, H. C. *Hydroboration*, W. A. Benjamin: New York, 1962. (b) Brown, H. C. *Boranes in Organic Chemistry*, Cornell University Press: Ithaca, NY, 1972.

35. (a) Zweifel, G., Brown, H. C. *Org. React.* **1963**, *13*, 1. (b) Brown, H. C., Kramer, G. W., Levy, A. B., Midland, M. M. *Organic Syntheses via Boranes*, Wiley: New York, 1975. (c) Suzuki, A., Dhillon, R. S. In *Topics in Current Chemistry*, Springer-Verlag: Berlin, 1986, Vol. 130, p. 23. (d) Pelter, A., Smith, K., Brown, H. C. *Borane Reagents*, Academic Press: London, 1988.

36. Brown, H. C., Racherla, U. S., Taniguchi, H. *J. Org. Chem.* **1981**, *46*, 4313.

37. Kabalka, G. W., Yu, S., Li, N.-S. *Tetrahedron Lett.* **1997**, *38*, 5455.

38. Brown, H. C., Mandal, A. K., Kulkarni, S. U. *J. Org. Chem.* **1977**, *42*, 1392.

39. Negishi, E., Brown, H. C. *Synthesis* **1974**, 77.

40. Zweifel, G., Pearson, N. R. *J. Am. Chem. Soc.* **1980**, *102*, 5919.

41. Kulkarni, S. U., Lee, H. D., Brown, H. C. *J. Org. Chem.* **1980**, *45*, 4542.

42. Soderquist, J. A., Negron, A. *Org. Synth.* **1991**, *70*, 169.

43. (a) Brown, H. C., Ayyangar, N. R., Zweifel, G. *J. Am. Chem. Soc.* **1964**, *86*, 397. (b) Brown, H. C., Singaram, B. *J. Org. Chem.* **1984**, *49*, 945.

44. (a) Brown, H. C., Zweifel, G. *J. Am. Chem. Soc.* **1960**, *82*, 4708. (b) Nelson, D. J., Brown, H. C. *J. Am. Chem. Soc.* **1982**, *104*, 4907.

45. Pelter, A., Smith, K., Brown, H. C. In *Borane Reagents*, Academic: London, 1988, pp. 168–170.

46. Brown, H. C., Liotta, R., Scouten, C. G. *J. Am. Chem. Soc.* **1976**, *98*, 5297.

47. Brown, H. C., Singaram, B. *Pure Appl. Chem.* **1987**, *59*, 879.

48. Brown, H. C., Synder, C., Subba Rao, B. C., Zweifel, G. *Tetrahedron* **1986**, 42, 5505.

49. (a) Kabalka, G. W., Maddox, J. T., Shoup, T., Bowers, K. R. *Org. Synth.* **1996**, *73*, 116. (b) McKillop, A., Sanderson, W. R. *Tetrahedron* **1995**, *51*, 6145. (c) Muzart, J. *Synthesis* **1995**, 1325.

50. (a) Köster, R., Morita, Y. *Angew. Chem., Int. Ed.* **1966**, *5*, 580. (b) Hassner, A., Soderquist, J. J. *Organomet. Chem.* **1977**, *131*, C1. (c) Miller, J. A., Zweifel, G. *Synthesis* **1981**, 288.

51. Brown, H. C., Kulkarni, S. U., Rao, C. G. *Synthesis* **1980**, 151.

52. (a) Brown, H. C., Zweifel, G. *J. Am. Chem. Soc.* **1961**, *83*, 2544. (b) Brener, L., Brown, H. C. *J. Org. Chem.* **1977**, *42*, 2702.

53. (a) Brown, H. C., Jadhav, P. K. In *Asymmetric Synthesis*, Morrison, J. D., Ed., Academic: New York, 1983, Vol. 2, p. 1. (b) Brown, H. C., Jadhav, P. K., Singaram, B. In *Modern Synthetic Methods*, Scheffold, R., Ed., Springer-Verlag: New York, 1986, Vol. 4, p. 307.

54. Brown, H. C., Desai, M. C., Jadhav, P. K. *J. Org. Chem.* **1982**, *47*, 5065.

55. Houk, K. N., Rondan, N. G., Wu, Y.-D., Metz, J. T., Paddon-Row, M. N. *Tetrahedron* **1984**, *40*, 2257.

56. Brown, H. C., Jadhav, P. K., Mandal, A. K. *J. Org. Chem.* **1982**, *47*, 5074.

57. Masamune, S., Kim, B. M., Petersen, J. S., Sato, T., Veenstra, J. S., Imai, T. *J. Am. Chem. Soc.* **1985**, *107*, 4549.

58. Whitesell, J. K. *Chem. Rev.* **1989**, *89*, 1581.

59. Bartlett, P. A. *Tetrahedron* **1980**, *36*, 2.

60. Schmid, G., Fukuyama, T., Akasaka, K., Kishi, Y. *J. Am. Chem. Soc.* **1979**, *101*, 259.

61. (a) Larock, R. C. *Organomercury Compounds in Organic Synthesis*, Springer-Verlag: Berlin, 1985. (b) Larock, R. C. *Solvomercuration/Demercuration Reactions in Organic Synthesis*, Springer-Verlag: New York, 1986. (c) Brown, H. C., Geoghegan, P. J., Jr. *J. Org. Chem.* **1970**, *35*, 1844. (d) Brown, H. C., Lynch, G. J., Hammar, W. J., Liu, L. C. *J. Org. Chem.* **1979**, *44*, 1910.

62. Brown, H. C., Rei, M.-H. *J. Am. Chem. Soc.* **1969**, *91*, 5646.

63. Aranda, V. G., Barluenga, J., Yus, M., Asensio, G. *Synthesis* **1974**, 806.

64. Brown, H. C., Kurek, J. T. *J. Am. Chem. Soc.* **1969**, *91*, 5647.

65. Heathcock, C. H. *Angew. Chem., Int. Ed.* **1969**, *8*, 134.

66. Kageyama, M., Tamura, T., Nantz, M. H., Roberts, J. C., Somfai, P., Whritenour, D. C., Masamune, S. *J. Am. Chem. Soc.* **1990**, *112*, 7407.

67. (a) Rao, A. S., Paknikar, S. K., Kirtane, J. G. *Tetrahedron* **1983**, *39*, 2323. (b) Smith, M. B. *Organic Synthesis*, McGraw-Hill: Boston, 2002, pp. 226–245. (c) Procter, G. *Stereoselectivity in Organic Synthesis*, Oxford Science Publications: Oxford, UK, 1998, pp. 71–76.

68. Seebach, D. *Angew. Chem., Int. Ed.* **1979**, *18*, 239.

69. Brougham, P., Cooper, M. S., Cummerson, D. A., Heaney, H., Thompson, N. *Synthesis* **1987**, 1015.

70. Newman, M. S., Magerlein, B. J. *Org. React.* **1949**, *5*, 413.

71. Lane, B. S., Burgesse, K. *Chem. Rev.* **2003**, *103*, 2457.

72. (a) Hassall, C. H. *Org. React.* **1957**, *9*, 73. (b) House, H. O. *Modern Synthetic Reactions*, 2nd ed., W. A. Benjamin: Menlo Park, CA, 1972, p. 323. (c) Krow, G. R. In *Comprehensive Organic Synthesis*, Trost, B. M., Fleming, I., Eds., Pergamon: Oxford, UK 1991, Vol. 7, p. 671. (d) Krow, G. R. *Org. React.* **1993**, *43*, 251. (e) Strukul, G. *Angew. Chem., Int. Ed.* **1998**, *37*, 1199.

73. Baeyer, A., Villiger, V. *Chem. Ber.* **1899**, *32*, 3625.

74. Corey, E. J., Weinshenker, N. M., Schaaf, T. K., Huber, W. *J. Am. Chem. Soc.* **1969**, *91*, 5675.

75. House, H. O., Ro, R. S. *J. Am. Chem. Soc.* **1958**, 80, 2428.

76. Grieco, P. A., Nishizawa, M., Oguri, T., Burke, S. D., Marinovic, N. *J. Am. Chem. Soc.* **1977**, *99*, 5773.

77. Adam, W., Saha-Möller, C. R., Zhao, C.-G. *Org. React.* **2002**, *61*, 219.

78. (a) Murray, R. W. *Chem. Rev.* **1989**, *89*, 1187. (b) Curci, R., Dinoi, A., Rubino, M. F. *Pure Appl. Chem.* **1995**, *67*, 811. (c) Ferraz, H. M. C., Muzzi, R. M., Vieira, T. O., Viertler, H. *Tetrahedron Lett.* **2000**, *41*, 5021.

79. (a) Adam, W., Hadjiarapoglou, L., Nestler, B. *Tetrahedron Lett.* **1990**, *31*, 331. (b) Adam, W., Hadjiarapoglou, L., Levai, A. *Synthesis* **1992**, 436.

80. Adam, W., Chan, Y.-Y., Cremer, D., Gauss, J., Scheutzow, D., Schindler, M. *J. Org. Chem.* **1987**, *52*, 2800.

81. Mello, R., Fiorentino, M., Sciacovelli, O., Curci, R. *J. Org. Chem.* **1988**, *53*, 3890.

82. Mello, R., Fiorentino, M., Fusco, C., Curci, R. *J. Am. Chem. Soc.* **1989**, *111*, 6749.

83. (a) Osterberg, A. E. *Org. Synth., Col., Vol. 1* **1944**, 185. (b) Dalton, D. R., Dutta, V. P. *J. Chem. Soc. B* **1971**, 85. (c) Jakubowski, A. A., Guziec, F. S., Tishler, M. *Tetrahedron Lett.* **1977**, 2399.

84. (a) Corey, E. J., Chaykovsky, M. *J. Am. Soc.* **1965**, *87*, 1353. (b) Corey, E. J., Chaykovsky, M. *Org. Synth., Col. Vol. V* **1973**, 755.

85. Sadhu, K. M., Matteson, D. S. *Tetrahedron Lett.* **1986**, 795.

86. Smith, J. G. *Synthesis* **1984**, 629.

87. Ager, D. J., East, M. B. *Asymmetric Synthetic Methodology*, CRC: Boca Raton, FL, 1996, p. 255 (Table 10.1).

88. Taylor, S. K. *Tetrahedron* **2000**, *56*, 1149.

89. (a) Creger, P. L. *J. Org. Chem.* **1972**, *37*, 1907. (b) Petragnani, N., Yonashiro, M. *Synthesis* **1982**, 521.

90. Nasipuri, D. *Stereochemistry of Organic Compounds*, Wiley: New York, 1991, p. 358.

91. (a) Tenud, L., Farooq, S., Seibl, J., Eschenmoser, A. *Helv. Chim. Acta* **1970**, *53*, 2059. (b) Babler, J., Tortorello, A. *J. Org. Chem.* **1976**, *41*, 885.

92. (a) Stork, G., Cama, L. D., Coulson, D. R. *J. Am. Chem. Soc.* **1974**, *96*, 5268. (b) Stork, G., Cohen, J. F. *J. Am. Chem. Soc.* **1974**, *96*, 5270. (c) Lallemand, J. Y., Onanga, M. *Tetrahedron Lett.* **1975**, 585.

93. (a) Corbel, B., Decesare, J. M., Durst, T. *Can. J. Chem.* **1978**, *56*, 505. (b) Decesare, J. M., Corbel, B., Durst, T., Blount, J. F. *Can. J. Chem.* **1981**, *59*, 1415.

94. Sonnet, P. E. *Tetrahedron* **1980**, *36*, 557.

95. (a) Martinez, A. G., Ruiz, M. O. *Synthesis* **1983**, 663. (b) Aguiar, A. M., Beisler, J., Mills, A. *J. Org. Chem.* **1962**, *27*, 1001.

96. (a) Vedejs, E., Fuchs, P. J. *J. Am. Chem. Soc.* **1973**, *95*, 822. (b) Vedejs, E., Snoble, K. A. J., Fuchs, P. L. *J. Org. Chem.* **1973**, *38*, 1178.

97. (a) Apparu, M., Barrelle, M. *Tetrahedron* **1978**, *34*, 1541. (b) Crandall, J. K., Apparu, M. *Org. React.* **1983**, *29*, 345.

98. Crandall, J. K., Lin, L.-H. C. *J. Org. Chem.* **1968**, *33*, 2375.

99. Thummel, R. P., Rickborn, B. *J. Am. Chem. Soc.* **1970**, *92*, 2064.

100. (a) Murata, S., Suzuki, M., Noyori, R. *J. Am. Chem. Soc.* **1979**, *101*, 2738. (b) Nantz, M. H., Fuchs, P. L. *J. Org. Chem.* **1987**, *52*, 5298.

101. Krishnamurty, S., Schubert, R. M., Brown, H. C. *J. Am. Chem. Soc.* **1973**, *95*, 8486.

102. Hutchins, R. O., Taffer, I. M., Burgoyne, W. *J. Org. Chem.* **1981**, *46*, 5214.

103. Nicolaou, K. C., Prasad, C. V. C., Somers, P. K., Hwang, C.-K. *J. Am. Chem. Soc.* **1989**, 111, 5330.

104. (a) Coxon, J. N., Hartshorn, M. P., Rae, W. J. *Tetrahedron* **1970**, *26*, 1091. (b) House, H. O. *Modern Synthetic Reactions*, W. A. Benjamin: Menlo Park, CA, 1972, pp. 318–320.

105. Hart, H., Lerner, L. R. *J. Org. Chem.* **1967**, *32*, 2669.

106. (a) Henbest, H. B., Wilson, R. A. I. *J. Chem. Soc.* **1957**, **1958**. (b) Hoveyda, A. H., Evans, D. A., Fu, G. C. *Chem. Rev.* **1993**, *93*, 1307.

107. (a) Fukuyama, T., Wang, C.-L. J., Kishi, Y. *J. Am. Chem. Soc.* **1979**, *101*, 260. (b) Nicolaou, K. C., Sorensen, E. J. *Classics*

in Total Synthesis, VCH: Weinheim, Germany, 1996, pp. 199–200.

108. Zurflüh, R., Wall, E. N., Siddall, J. B., Edwards, J. A. *J. Am. Chem. Soc.* **1968**, *90*, 6224.

109. Sharpless, K. B., Verhoeven, T. R. *Aldrichimica Acta* **1979**, *12*, 63.

110. (a) Masamune, S., Choy, W., Petersen, J. S., Sita, L. R. *Angew. Chem., Int. Ed.* **1985**, *24*, 1. (b) Nicolaou, K. C., Sorensen, E. J. *Classics in Total Synthesis*, VCH: Weinheim, Germany, 1996, pp. 293–315.

111. (a) Pfenninger, A. *Synthesis* **1986**, 89. (b) Ager, D. J., East, M. B. *Asymmetric Synthetic Methodology*, CRC: Boca Raton, FL, 1996, p. 269. (c) Johnson, R. A., Sharpless, K. B. In *Catalytic Asymmetric Synthesis*, Ojima, I., Ed., 2nd ed., Wiley-VCH: New York, 2000, p. 231. (d) Katsuki, T., Martin, V. S. *Org. React.* **1996**, *48*, 1.

112. (a) Katsuki, T., Sharpless, K. B. *J. Am. Chem. Soc.* **1980**, *102*, 5974. (b) Hill, J. G., Sharpless, K. B. *Org. Synth.* **1984**, *63*, 66.

113. (a) Woodard, S. S., Finn, M. G., Sharpless, K. B. *J. Am. Chem. Soc.* **1991**, *113*, 106. (b) Finn, M. G., Sharpless, K. B. *J. Am. Chem. Soc.* **1991**, *113*, 113.

114. (a) Seebach, D., Weidmann, B., Widler, L. In *Modern Synthetic Methods*, Scheffold, R., Ed., Verlag Sauerländer: Aarau, Germany, 1983, Vol. 3, pp. 324–340. (b) Schinzer, D. In *Organic Synthesis Highlights II*, Waldmann, H., Ed., VCH: Weinheim, Germany, 1995, p. 3.

115. Gao, Y., Hanson, R. M., Klunder, J. M., Ko, S. Y., Masamune, H., Sharpless, K. B. *J. Am. Chem. Soc.* **1987**, *109*, 5765.

116. Masamune, S., *Heterocycles* **1984**, *21*, 107. (b) Johnson, R. A., Sharpless, K. B. In *Comprehensive Organic Synthesis*, Trost, B. M., Fleming, I., Eds., Pergamon: Oxford, UK, 1991, Vol. 7, p. 389.

117. Nicolaou, K. C., Sorensen, E. J. *Classics in Total Synthesis*, VCH: Weinheim, Germany, 1996, pp. 293–315.

118. (a) Katsuki, T., Martin, V. S. *Org. React.* **1996**, *48*, 1. (b) Ager, D. J., East, M. B. *Asymmetric Synthetic Methodology*, CRC: Boca Raton, FL, 1996, p. 280.

119. (a) Payne, G. B. *J. Org. Chem.* **1962**, *27*, 3819. (b) Hanson, R. M. *Org. React.* **2002**, *60*, 1.

120. (a) Caron, M., Sharpless, K. B. *J. Org. Chem.* **1985**, *50*, 1557. (b) Chong, J. M., Sharpless, K. B. *J. Org. Chem.* **1985**, *50*, 1560.

121. Carlsen, P. H. J., Katsuki, T., Martin, V. S., Sharpless, K. B. *J. Org. Chem.* **1981**, *46*, 3936.

122. Katsuki, T., Martin, V. S. *Org. React.* **1996**, *48*, 1.

123. Jung, M. E., D'Amico, D. C. *J. Am. Chem. Soc.* **1993**, *115*, 12208.

124. Martin, V. S., Woodard, S. S., Katsuki, T., Yamada, Y., Ikeda, M., Sharpless, K. B. *J. Am. Chem. Soc.* **1981**, *103*, 6237.

125. (a) Zhang, W., Loebach, J. L., Wilson, S. R., Jacobsen, E. N. *J. Am. Chem. Soc.* **1990**, *112*, 2801. (b) Jacobsen, E. N., Zhang, W., Güler, M. L. *J. Am. Chem. Soc.* **1991**, *113*, 6703.

126. (a) Zhang, W., Jacobsen, E. N. *J. Org. Chem.* **1991**, *56*, 2296. (b) Irie, R., Noda, K., Ito, Y., Katsuki, T. *Tetrahedron Lett.* **1991**, *32*, 1055.

127. (a) Palucki, M., Finney, N. S., Pospisil, P. J., Güler, M. L., Ishida, T., Jacobsen, E. N. *J. Am. Chem. Soc.* **1998**, *120*, 948. (b) For a discussion of the application of vicinal diamines as ligands in asymmetric synthesis, see Lucet, D., Le Gall, T., Mioskowski, C. *Angew. Chem., Int. Ed.* **1998**, *37*, 2580.

128. (a) Campbell, K. A., Lashley, M. R., Wyatt, J. K., Nantz, M. H., Britt, R. D. *J. Am. Chem. Soc.* **2001**, *123*, 5710. (b) Schmitt, H., Lomoth, R., Magnuson, A., Park, J., Fryxelius, J., Kritikos, M., Mårtensson, J., Hammarström, L., Sun, L., Åkermark, B. *Chem.— Eur. J.* **2002**, *8*, 3757.

129. (a) Schröder, M. *Chem. Rev.* **1980**, *80*, 187. (b) Haines, A. H. In *Comprehensive Organic Synthesis*, Trost, B. M., Fleminng, I., Eds., Pergamon: Oxford, UK, **1991**, Vol. 7, p. 437.

130. VanRheenen, V., Kelly, R. C., Cha, D. Y. *Tetrahedron Lett.* **1976**, *25*, 1973.

131. Ray, R., Matteson, D. S. *Tetrahedron Lett.* **1980**, *21*, 449.

132. Sharpless, K. B., Akashi, K. *J. Am. Chem. Soc.* **1976**, *98*, 1986.

133. Minato, M., Yamamoto, K., Tsuji, J. *J. Org. Chem.* **1990**, *55*, 766.

134. Goldsmith, D. J., Sakano, I. *J. Org. Chem.* **1976**, *41*, 2095.

135. Fieser, L. F., Fieser, M. *Reagents for Organic Synthesis*, Wiley: New York, 1967, Vol. 1, p. 759.

136. (a) Corey, E. J., Hopkins, P. B., Kim, S., Yoo, S., Nambiar, K. P., Falck, J. R. *J. Am. Chem. Soc.* **1979**, *101*, 7131. (b) Vorbrüggen, H., Bennua, B. *Synthesis* **1985**, 925.

137. Woodward, R. B., Brutcher, F. V., Jr. *J. Am. Chem. Soc.* **1958**, *80*, 209.

138. Mangoni, L., Adinolfi, M., Barone, G., Parrilli, M. *Tetrahedron Lett.* **1973**, 4485.

139. Fatiadi, A. J. *Synthesis* **1987**, 85.

140. Ho, T.-L. *Distinctive Techniques for Organic Synthesis: A Practical Approach*, World Scientific: Singapore, 1988, pp. 1–51.

141. (a) Waldmannn, H. In *Organic Synthesis Highlights II*, Waldmann, H., Ed.,VCH: Weinheim, Germany, 1995, p. 9. (b) Nicolaou, K. C., Sorensen, E. J. *Classics in Total Synthesis*, VCH: Weinheim, Germany, 1996, pp. 675–691. (c) Procter, G. *Stereoselectivity in Organic Synthesis*, Oxford Science Publications: Oxford, UK, 1998, pp. 76–78. (d) Johnson, R. A., Sharpless, K. B. In *Catalytic Asymmetric Synthesis*, Ojima, I., Ed., Wiley-VCH: New York, 2000, p. 357.

142. (a) Corey, E. J., Noe, M. C. *J. Am. Chem. Soc.* **1996**, *118*, 11038. (b) Nelson, D. W., Gypser, A., Ho, P. T., Kolb, H. C., Kondo, T., Kwong, H.-L., McGrath, D. V., Rubin, A. E., Norrby, P.-O., Gable, K. P., Sharpless, K. B. *J. Am. Chem. Soc.* **1997**, *119*, 1840.

143. For an excellent review of the AD reaction, see Kolb, H. C., Van Nieuwenhze, M. S., Sharpless, K. B. *Chem. Rev.* **1994**, *94*, 2483.

144. (a) Dowle, M. D., Davies, D. I. *Chem. Soc. Rev.* **1979**, *8*, 171. (b) Mulzer, J. In *Organic Synthesis Highlights*, Mulzer, J., Altenbach, H.-J., Braun, M., Krohn, K., Reissig, H.-U., Eds., VCH: Weinheim, Germany, 1991, p. 158.

145. (a) Corey, E. J., Weinshenker, N. M., Schaaf, T. K., Huber, W. *J. Am. Chem. Soc.* **1969**, *91*, 5675. (b) Corey, E. J., Schaaf, T. K., Huber, W., Koelliker, U., Weinshenker, N. M. *J. Am. Chem. Soc.* **1970**, *92*, 397.

146. Corey, E. J., Hase, T. *Tetrahedron Lett.* **1979**, 335.

147. (a) Bartlett, P. A., Myerson, J. *J. Am. Chem. Soc.* **1978**, *100*, 3950. (b) Gonzàlez, F. B., Bartlett, P. A. *Org. Synth.* **1985**, *64*, 175.

148. (a) Bailey, P. S. *Ozonation in Organic Chemistry*, Academic: New York, 1978, Vol 1. (b) Lee, D. G. In *Comprehensive Organic Synthesis*, Trost, B. M., Flemming, I., Eds., Pergamon: Oxford, UK, 1991, Vol. 7, pp. 541–558. (c) Smith, M. B. *Organic Synthesis*, 2nd ed., McGraw-Hill: Boston, 2002, p. 267.

149. Lavallée, P., Bouthillier, G. *J. Org. Chem.* **1986**, *51*, 1362 (see footnote 27).

150. (a) Pappas, J. J., Keaveney, W. P., Gancher, E., Berger, M. *Tetrahedron Lett.* **1966**, 4273. (b) Keaveney, W. P., Berger, M. G., Pappas, J. J. *J. Org. Chem.* **1967**, *32*, 1537.

151. Corey, E. J., Katzenellenbogen, J. A., Gilman, N. W., Roman, S. A., Erickson, B. W. *J. Am. Chem. Soc.* **1968**, *90*, 5618.

152. Knöll, W., Tamm, C. *Helv. Chim. Acta* **1975**, *58*, 1162.

153. Conia, J.-M., Leriverend, P. *Comptes Rendues, France,* **1960**, *250*, 1078.

154. (a) Schreiber, S. L., Claus, R. E., Reagan, J. *Tetrahedron Lett.* **1982**, *23*, 3867. (b) Claus, R. E., Schreiber, S. L. *Org. Synth.* **1986**, *64*, 150.

155. (a) Greenwood, F. L. *J. Org. Chem.* **1955**, *20*, 803. (b) Clark, R. D., Heathcock, C. H. *J. Org. Chem.* **1976**, *41*, 1396.

156. Fatiadi, A. J. *Synthesis* **1974**, 229.

157. Pappo, R., Allen, D. S., Jr., Lemieux, R. U., Johnson, W. S. *J. Org. Chem.* **1956**, *21*, 478.

158. Vorbrüggen, H., Bennua, B. *Synthesis* **1985**, 925.

159. Baggiolini, E. G., Iacobelli, J. A., Hennessy, B. M., Batcho, A. D., Sereno, J. F., Uskokovic, M. R. *J. Org. Chem.* **1986**, *51*, 3098.

160. (a) Slomp, G., Jr., Johnson, J. L. *J. Am. Chem. Soc.* **1958**, *80*, 915. (b) Corey, E. J., Ohno, M., Mitra, R. B., Vatakencherry, P. A. *J. Am. Chem. Soc.* **1964**, *86*, 478. (c) Haag, T., Luu, B., Hetru, C. *J. Chem. Soc., Perkin Trans.,* **1988**, 2353.

161. (a) Jackson, E. L. *Org. React.* **1944**, *2*, 341. (b) Fatiadi, A. J. In *Synthetic Reagents*, Pizey, J. S., Ed., Wiley: New York, 1981, Vol. 4, p. 147.

162. (a) Baer, E. *J. Am. Chem. Soc.* **1942**, *64*, 1416. (b) Butler, R. N. In *Synthetic Reagents*, Pizey, J. S., Ed., Wiley: New York, 1977, Vol. 3, p. 277.

163. (a) Piatak, D. M., Bhat, H. B., Caspi, E. *J. Org. Chem.* **1969**, *34*, 112. (b) Lee, D. G., van den Engh, M. In *Oxidation in Organic Chemistry*, Trahanovsky, W. S., Ed., Academic: New York, 1973, Part B.

164. For recent developments in alkene oxidative cleavage, see Griffith, W. P., Kwong, E. *Synth. Commun.* **2003**, *17*, 2945, and references therein.

165. (a) Marvell, E. N., Li, T. *Synthesis* **1973**, 457. (b) Kieboom, A. P. G., van Rantwijk, F. *Hydrogenation and Hydrogenolysis in Synthetic Organic Chemistry*, Delft University Press, 1977. (c) Rylander, P. N. *Hydrogenation Methods*, Academic: London, 1985. (d) Augustine, R. L. *Homogeneous Catalysis for the Synthetic Chemist*, Dekker,: New York, 1966, p. 387. (e) Siegel, S. In *Comprehensive Organic Synthesis*, Trost, B. M., Flemming, I., Eds., Pergamon: Oxford, UK, 1991, Vol. 8, p. 430.

166. (a) Lindlar, H. *Helv. Chim. Acta* **1952**, *35*, 446. (b) Lindlar, H., Dubuis, R. *Org. Synth.* **1966**, *46*, 89. (c) Henrick, C. A. *Tetrahedron* **1977**, *33*, 1845.

167. Henrick, C. A. *Tetrahedron* **1977**, *33*, 1845.

168. Shakhovskoi, B. G., Stadnichuk, M. D., Petrov, A. A. *Zh. Obshch. Khim.* **1964**, *34*, 2625.

169. Cram, D. J., Allinger, N. L. *J. Am. Chem. Soc.* **1956**, *78*, 2518.

170. (a) Kocienski, P. J., Cernigliaro, G. J. *J. Org. Chem.* **1976**, *41*, 2927. (b) Taschner, M. J., Rosen, T., Heathcock, C. H. *Org. Synth.* **1985**, *64*, 108.

171. (a) Brown, C. A., Ahuja, V. K. *J. Chem. Soc., Chem. Commun.* **1973**, 553. (b) Brown, C. A., Ahuja, V. K. *J. Org. Chem.* **1973**, *38*, 2226.

172. Brown, C. A., Brown, H. C. *J. Am. Chem. Soc.* **1962**, *84*, 2829.

173. (a) Jain, S. C., Dussourd, D. E., Conner, W. E., Eisner, T., Guerrero, A., Meinwald, J. *J. Org. Chem.* **1983**, *48*, 2266. (b) Durand, S., Parrain, J.-L., Santelli, M. *Synthesis* **1998**, 1015.

174. Ganem, B., Osby, J. O. *Chem. Rev.* **1986**, *86*, 763.

175. Rieke, R. D., Uhm, S. J. *Synthesis* **1975**, 452.

176. (a) Winter, M., Näf, F., Furrer, A., Pickenhagen, W., Giersch, W., Meister, A., Willhalm, B., Thommen, W., Ohloff, G. *Helv. Chim. Acta* **1979**, *62*, 135. (b) Chou, W.-N., Clark, D. L., White, J. B. *Tetrahedron Lett.* **1991**, *32*, 299.

177. (a) Näf, F., Decorzant, R., Thommen, W., Willhalm, B., Ohloff, G. *Helv. Chim. Acta* **1975**, *58*, 1016. (b) Tellier, F., Descoins, C. *Tetrahedron Lett.* **1990**, *31*, 2295.

178. (a) Boland, W., Schroer, N., Sieler, C., Feigel, M. *Helv. Chim. Acta* **1987**, *70*, 1025. (b) Auignon-Tropis, M., Pougny, J. R. *Tetrahedron Lett.* **1989**, *30*, 4951.

179. (a) Brown, H. C., Zweifel, G. *J. Am. Chem. Soc.* **1961**, *83*, 3834. (b) Pelter, A., Smith, K., Brown, H. C. *Borane Reagents*, Academic: London, 1988.

180. Brown, H. C., Scouten, C. G., Liotta, R. *J. Am. Chem. Soc.* **1979**, *101*, 96.

181. Zweifel, G., Clark, G. M., Polston, N. L. *J. Am. Chem. Soc.* **1971**, *93*, 3395.

182. Brown, H. C., Moerikofer, A. W. *J. Am. Chem. Soc.* **1963**, *85*, 2063.

183. Brown, C. C., Coleman, R. A. *J. Org. Chem.* **1979**, *44*, 2328.

184. Negishi, E., Chiu, K.-W. *J. Org. Chem.* **1976**, *41*, 3484.

185. Yatagai, H., Yamamoto, Y., Maruyama, K. *J. Chem. Soc., Chem. Commun.* **1978**, 702.

186. Corey, E. J., Ravindranathan, T. *J. Am. Chem. Soc.* **1972**, *94*, 4013.

187. Negishi, E., Abramovitch, A. *Tetrahedron Lett.* **1977**, 411.

188. (a) Zweifel, G., Arzoumanian, H. *J. Am. Chem. Soc.* **1967**, *89*, 5086. (b) Brown, H. C., Blue, C. D., Nelson, D. J., Bhat, N. G. *J. Org. Chem.* **1989**, *54*, 6064.

189. Plamondon, J., Snow, J. T., Zweifel, G. *Organometal. Chem. Syn.* **1971**, *1*, 249.

190. (a) House, H. O. *Modern Synthetic Reactions*, Benjamin: Menlo Park, CA, 1972, pp. 205–209. (b) Warthen, J. D., Jr., Jacobson, M. *Synthesis* **1973**, 616.

191. (a) Campbell, K. N., Eby, L. T. *J. Am. Chem. Soc.* **1941**, *63*, 216. (b) Henne, A. L., Greenlee, K. W. *J. Am. Chem. Soc.* **1943**, *65*, 2020.

192. Boland, W., Hansen, V., Jaenicke, L. *Synthesis* **1979**, 114.

193. Ranu, B. C., Dutta, J., Guchhait, S. K. *J. Org. Chem.* **2001**, *66*, 5624.

194. Grant, B., Djerassi, C. *J. Org. Chem.* **1974**, *39*, 968.

195. Zwierzak, A., Tomassy, B. *Synth. Commun.* **1996**, *26*, 3593.

196. Joung, M. J., Ahn, J. H., Yoon, N. M. *J. Org. Chem.* **1996**, *61*, 4472.

197. (a) Normant, J. F., Commercon, A., Villieras, J. *Tetrahedron Lett.* **1975**, 1465. (b) Trost, B. M., Lee, D. C. *J. Org. Chem.* **1989**, *54*, 2271.

198. Bahlman, F., Enkelmann, R., Plettner, W. *Chem. Ber.* **1964**, *97*, 2118.

199. Doolittle, R. E. *Synthesis* **1984**, 730.

200. (a) Jones, T. K., Denmark, S. E. *Org. Synth.* **1985**, *64*, 182. (b) Hoveyda, A. H., Evans, D. A., Fu, G. C. *Chem. Rev.* **1993**, *93*, 1307 (see also p. 1350). (c) Uwai, K., Oshima, Y., Sugihara, T., Ohta, T. *Tetrahedron* **1999**, *55*, 9469. (d) Schreiber, S. L., Schreiber, T. S., Smith, D. B. *J. Am. Chem. Soc.* **1987**, *109*, 1525.

201. (a) Brown, H. C., Zweifel, G. *J. Am. Chem. Soc.* **1961**, *83*, 3834. (b) Brown, H. C., Snyder, C., Subba Rao, B. C., Zweifel, G. *Tetrahedron* **1986**, *42*, 5505.

202. (a) Lewis, R. G., Gustafson, D. H., Erman, W. F. *Tetrahedron Lett.* **1967**, 401. (b) Corey, E. J., Wess, G., Xiang, Y. B., Singh, A. K. *J. Am. Chem. Soc.* **1987**, *109*, 4717.

203. Zweifel, G., Backlund, S. J. *J. Am. Chem. Soc.* **1977**, *99*, 3184.

204. Larock, R. C. *Solvomercuration-Demercuration Reactions in Organic Synthesis*, Springer-Verlag: Berlin, 1986.

205. Stacy, G. W., Mikulec, R. A. *Org. Synth., Col. Vol. IV* **1963**, 13.

206. Jorgenson, M. *J. Org. React.* **1970**, *18*, 2.

207. Kurosu, M., Kishi, Y. *Tetrahedron Lett.* **1998**, *39*, 4793.

208. Tsuji, J., Nagashima, H., Nemoto, H. *Org. Synth., Col. Vol VII*, **1990**, 137.

Formation of Carbon–Carbon Single Bonds via Enolate Anions

CHAPTER 6

> *Nature, it seems, is an organic chemist having some predilection
> for the aldol and related condensations.*
>
> John W. Cornforth

ydrogen atoms alpha to a carbonyl, nitrile, or sulfonyl group are relatively
acidic (Table 6.1). The acidity of the C–H bond in these compounds is due to
a combination of the inductive electron-withdrawing effect of the neighboring functionality and resonance stabilization of the anion that is formed by removal of
the α-proton. Since these α-carbon anions possess carbanionic reactivity, they undergo a host of carbon-carbon bond-forming reactions by alkylation with R–X, addition
to RR′CO, or acylation with RC(O)X.[1]

6.1   1,3-DICARBONYL AND RELATED COMPOUNDS

The acidities of 1,3-dicarbonyl compounds are sufficiently high that they can be substantially converted to their conjugate bases by oxyanions such as hydroxide and

Table 6.1   Approximate pK_a Values for the α-Protons of Carbonyl, Nitrile, and Sulfonyl Compounds[2]

Ketones	pK_a	Esters	pK_a	Amides and nitriles	pK_a	Sulfones	pK_a
cyclopentanone	(26)	t-BuO / CH$_3$ ester	24.5	NC–CH$_3$	29	PhO$_2$S–CH$_3$	(29)
H$_3$C–CO–CH$_3$	20	EtO–CO–CH$_2$–N=Ph	(20)	Me$_2$N–CO–CH$_2$–Ph	(27)	PhO$_2$S allyl	(23)
Ph–CO–CH$_3$	19	EtO–CO–CH$_2$–CO–OEt	13	pyrrolidine–CO–CH$_2$–CN	(17)	MeO$_2$S–CH$_2$–SO$_2$Me	12
H$_3$C–CO–CH$_2$–CO–CH$_3$	9	EtO–CO–CH$_2$–CO–CH$_3$	11	NC–CH$_2$–CN	11	PhO$_2$S–CH$_2$–CO–CH$_3$	(11)

Note: Values are relative to H$_2$O or dimethyl sulfoxide (DMSO values in parentheses).

alkoxides (case A below). Treatment of 1,3-dicarbonyl compounds with amines will form the corresponding enolate species in equilibrium with the dicarbonyl compound (case B). Deprotonation with sodium hydride (usually in THF or DME as solvent) transforms 1,3-dicarbonyl compounds to enolate species irreversibly by loss of H_2 (case C). The selection of base and solvent for alkylation, acylation, or condensation reactions of 1,3-dicarbonyl compounds must take into account whether the overall reaction requires the presence of a conjugate acid for participation in an equilibrium process (e.g., as in the Knoevenagel condensation, discussed below).

Case A—deprotonation by a conjugate base of a weaker acid

Case B—deprotonation by a conjugate base of an acid with a comparable pK_a

Case C—irreversible deprotonation

Malonic Acid Esters

Alkylation

Mono- and dialkylations of malonic acid esters generally are performed in an alcoholic solution of a metal alkoxide.[3] Alkylation of a monoalkylated malonic ester requires the presence of another equivalent of alkoxide and the appropriate alkyl halide. The alkylation works well with RCH_2X (X=I, Br, OTs), $PhCH_2X$ (X=Cl, Br) and even with unhindered *sec* alkyl bromides.[4] Subsequent hydrolysis of the diester under acidic or basic conditions followed by heat-induced decarboxylation yields the α-alkylated carboxylic acid. Thus, dialkyl malonates are the synthetic equivalents (SE) of acetate enolate anions and can be used to obtain mono- or disubstituted carboxylic acids.

A potential side reaction in malonate alkylation is E2-elimination, which mainly depends on the structure of the alkyl halide. Also, dialkylation may occur if the malonate contains two active hydrogens.[5] Dialkylation may be circumvented by carrying out the deprotonation with one equivalent of NaH.

Decarboxylation
Decarboxylation of mono- or dialkylated malonic acid esters can be effected by heating in the presence of aqueous acids (e.g., 48% aq HBr, reflux).

Saponification of the diester followed by acidification gives the corresponding malonic acid which is then decarboxylated. The decarboxylation step may require high temperatures (150–250 °C). Treatment of malonic acids with a catalytic amount of Cu(I) oxide in acetonitrile is reported to accelerate the decarboxylation step and affords the monoacid products in good yield under milder conditions.[6]

3:1 *cis* : *trans*   *cis*-isomer only

Conjugate Addition–Michael-Type Reactions
The condensation of enolates derived from malonic esters and other active methylene compounds with α,β-unsaturated aldehydes, ketones, esters, or nitriles proceeds exclusively by 1,4-addition. The conjugate addition to α,β-unsaturated compounds, often called Michael acceptors, is promoted by treatment of the active methylene species with either an excess of a weak base (e.g., Et$_3$N or piperidine) or using a stronger base in catalytic amounts (e.g., 0.1–0.3 equivalents NaH, NaOEt, or *t*-BuOK).

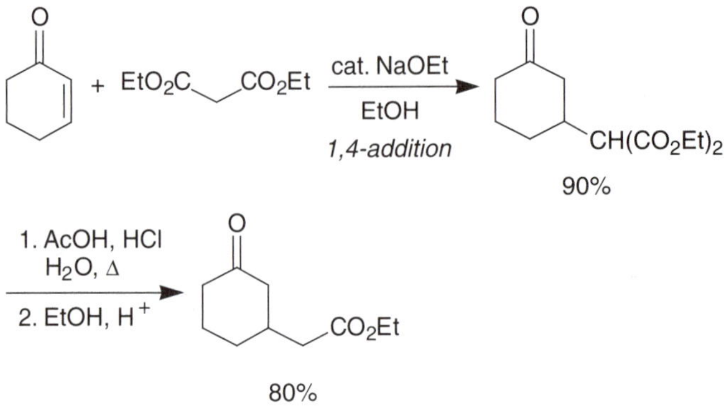

90%

80%

The Knoevenagel Condensation[7]

Diesters of malonic acid, as well as other active methylene compounds such as ace-toacetic esters and cyanoacetic esters, condense with aldehydes and ketones in the presence of catalytic amounts of primary or secondary amines or ammonium salts.

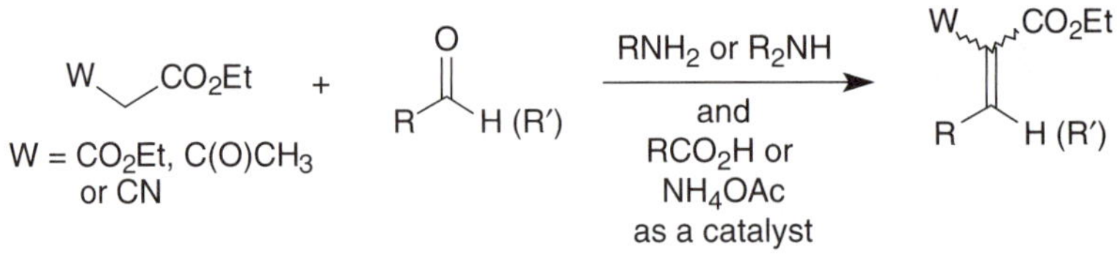

W = CO_2Et, $C(O)CH_3$
or CN

In some cases the condensation is stereoselective.[8]

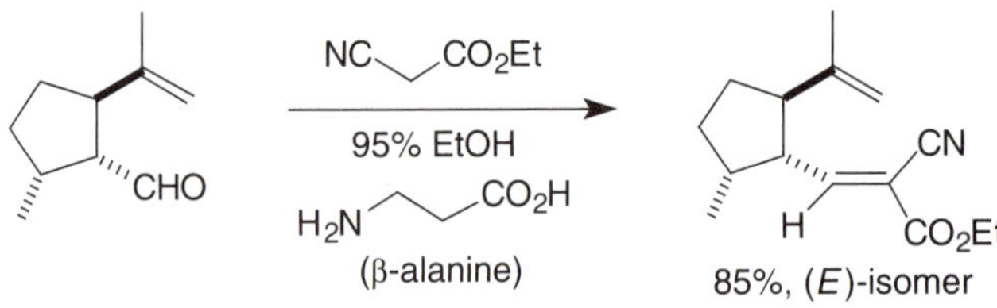

85%, (*E*)-isomer

The *Doebner condensation* uses malonic acid, malonic acid monoesters, or cyano-acetic acid instead of the corresponding dialkyl malonates. Usually the product stereoisomer with the carboxyl group *trans* to the larger substituent predominates.

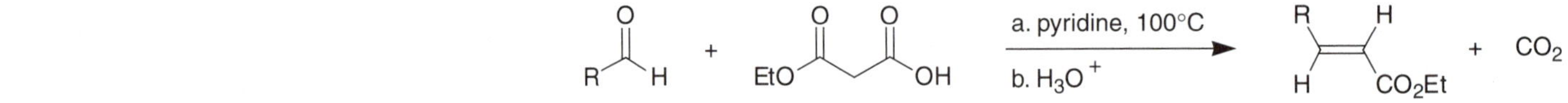

The required malonic acid monoesters are readily prepared by heating the strong-ly acidic isopropylidene malonate (Meldrum's acid,[9] pK_a 7.3) in an alcohol. Acetone is liberated in the process.[10]

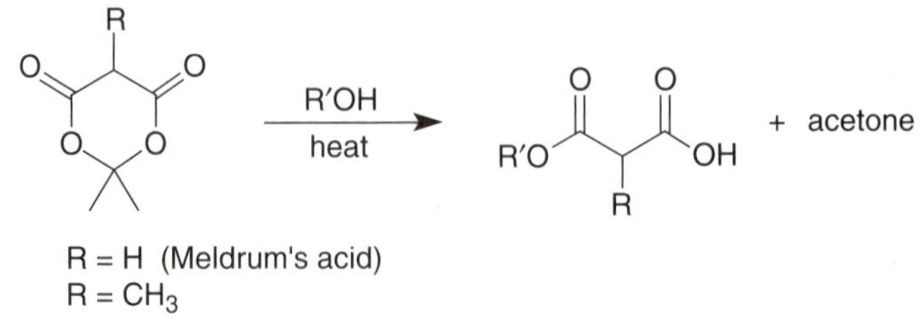

R = H (Meldrum's acid)
R = CH_3

β-Keto Esters

Preparation via the Claisen Condensation[11]

In the Claisen condensation, a nucleophilic ester enolate donor is added to the carbonyl group of a second ester molecule. Loss of alkoxide from the resultant intermediate — the tetrahedral adduct — forms a β-keto ester, which is much more acidic than the starting ester. Hence, deprotonation of the initial product by alkoxide drives the overall reaction to completion and protects the β-keto ester from further carbonyl addition reactions. Thus, the starting ester must have *at least two* α-hydrogens.

Mixed Claisen condensations are feasible when one partner does not have acidic α-hydrogens, as in EtO-CHO (a formylating agent), $PhCO_2Et$, EtO_2CCO_2Et, and $(EtO)_2C=O$ (an ethoxycarbonylating agent). Formate esters are the most reactive since they are "part aldehyde."

Preparation via the Dieckmann Condensation[12]

The Dieckmann condensation is usually defined to include only intramolecular condensations of diesters.[13]

The intramolecular condensation of ester enolates provides efficient access to 5- and 6-member ring β-keto esters. Similar to the Claisen condensation, the Dieckmann condensation is driven to completion by deprotonation of the initially formed β-keto ester. Thus, at least one of the ester groups must have two α-hydrogens for the reaction to proceed.

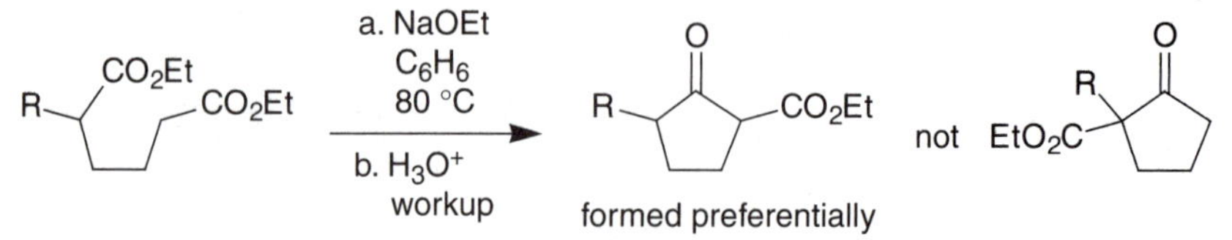

A variety of electron-withdrawing groups such as those in ketones, nitriles, and sulfones are also suitable for the Dieckmann reaction.[14]

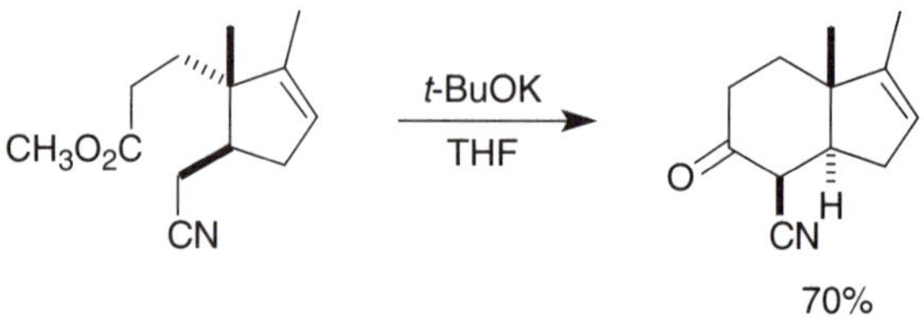

The Dieckmann condensation in tandem with the Claisen condensation or Michael reaction can be used to assemble highly functionalized carbocycles, including the 1,2-, 1,3-, and 1,4-dione species depicted below.[15]

1,2-Dione:

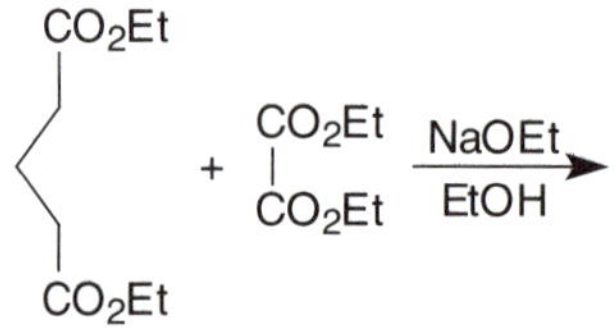

1,3-Dione:

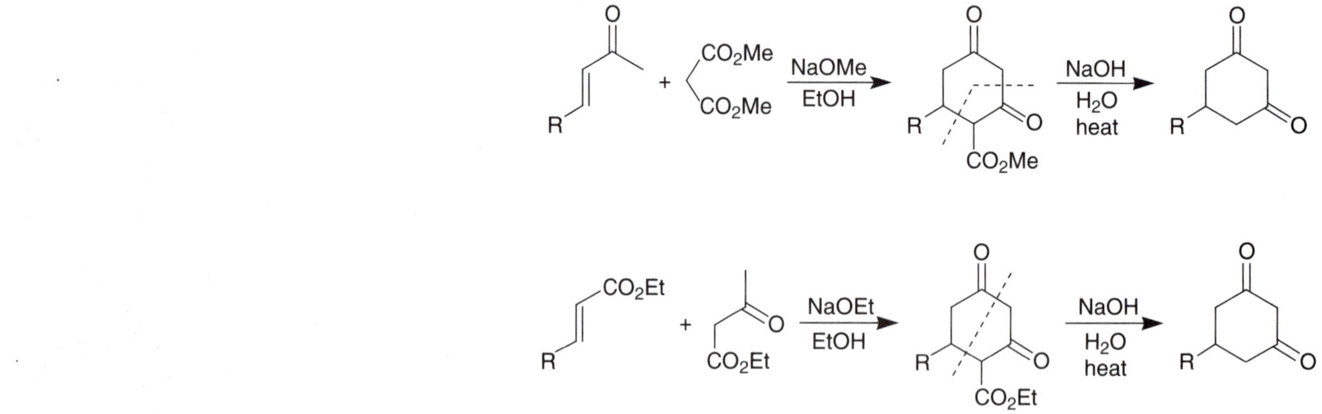

1,4-Dione:

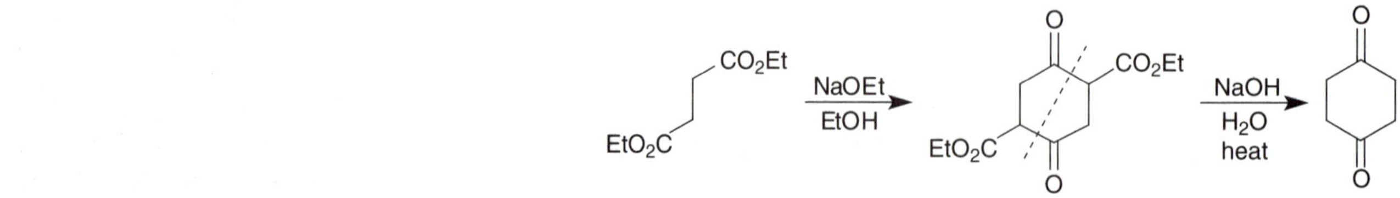

Dicarboxylic esters in the presence of Na in ether or in benzene (carcinogen) cyclize to furnish not β-keto esters but instead α-hydroxy ketones (acyloins). This *acyloin condensation* involves reductive dimerization of a ketyl radical anion (see Chapter 9).[16]

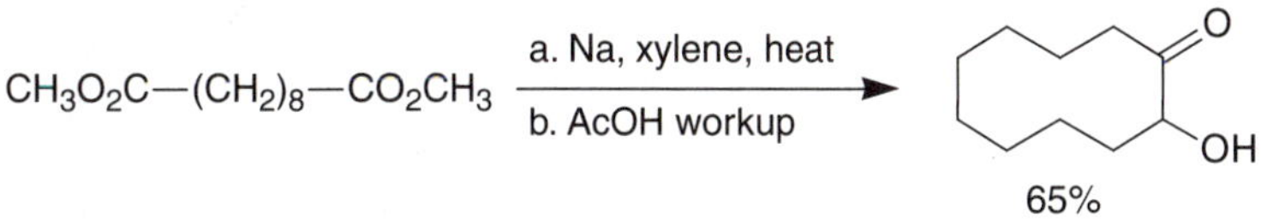

$$CH_3O_2C-(CH_2)_8-CO_2CH_3 \xrightarrow[\text{b. AcOH workup}]{\text{a. Na, xylene, heat}}$$

65%

Preparation of Acetoacetic Esters

Treatment of alcohols with diketene in the presence of tertiary amines is a simple and efficient method for the preparation of β-keto esters. As illustrated below, this method is specific for the formation of acetoacetic esters.[17]

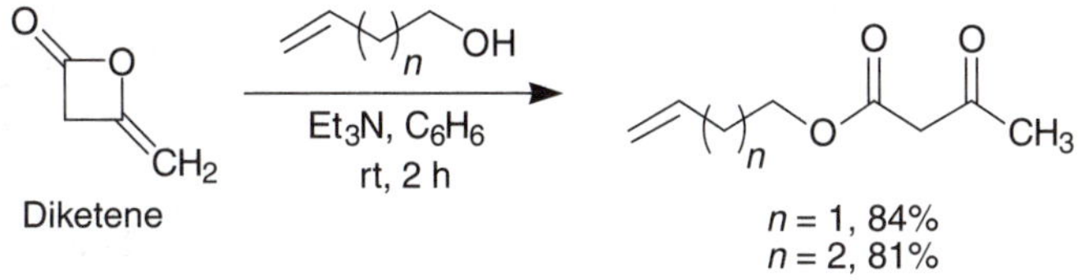

Diketene

$$\xrightarrow[\substack{\text{Et}_3\text{N, C}_6\text{H}_6 \\ \text{rt, 2 h}}]{}$$

n = 1, 84%
n = 2, 81%

Preparation Via C-Acylation of Ketones or Esters

β-Ketoesters may be prepared via acylation of enolates derived from the corresponding ketones. *Acylation usually takes place at the less substituted carbon* because of the formation of a delocalized β-keto ester anion.

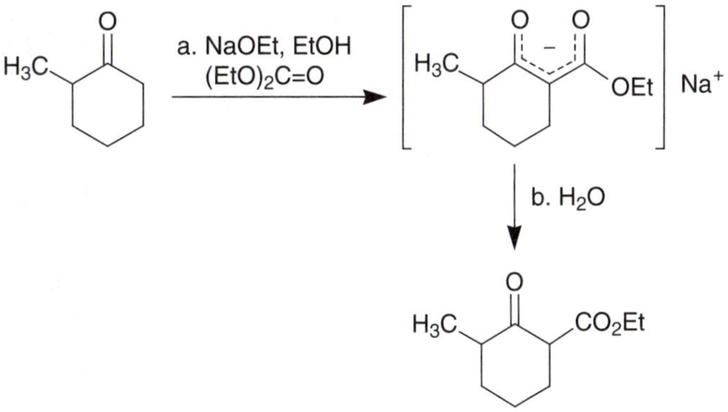

$$\xrightarrow[\text{(EtO)}_2\text{C=O}]{\text{a. NaOEt, EtOH}}$$

Na⁺

b. H₂O

As an alternative to using dialkyl carbonates [$(RO)_2C=O$] for α-acylation of ketones,[18] the reaction of ketone enolates with alkyl cyanoformates ($NC–CO_2R$, *Mander's reagent*) provides β-keto esters in good yields.[19] Reaction of lithium enolates with acyl halides or anhydrides, however, usually leads to mixtures of *O*- and *C*-acylated products.[20]

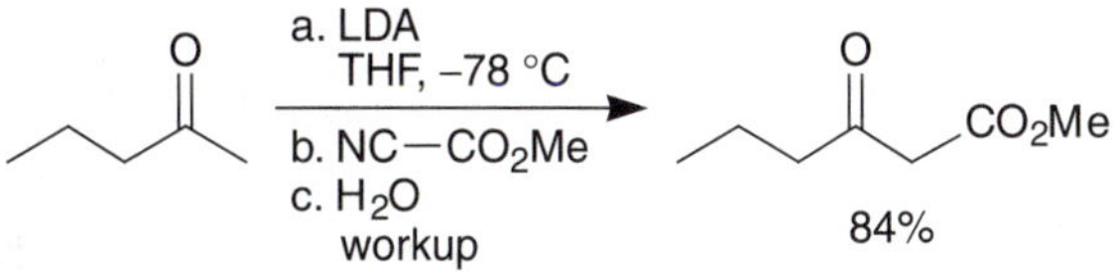

$$\xrightarrow[\substack{\text{b. NC-CO}_2\text{Me} \\ \text{c. H}_2\text{O} \\ \text{workup}}]{\substack{\text{a. LDA} \\ \text{THF, -78 °C}}}$$

84%

An operationally simple, one-pot *C*-acylation procedure for the conversion of carboxylic acids to β-keto esters under nearly neutral conditions is the successive addition of carbonyldiimidazole (CDI) and the monomethyl malonate[21] magnesium salt to the carboxylic acid in THF.[22,23]

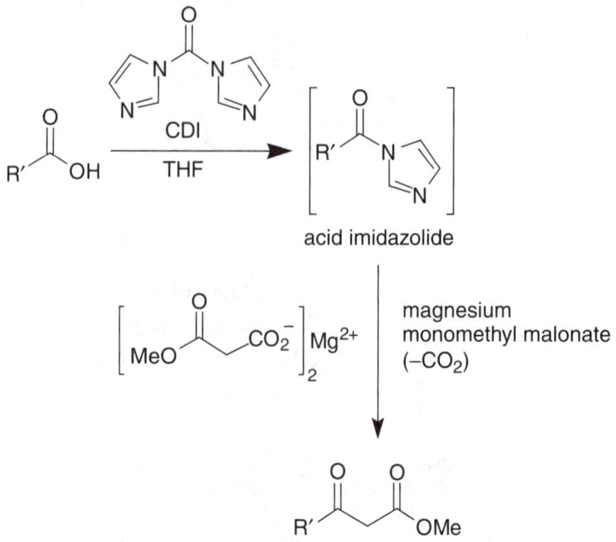

acid imidazolide

The *C*-acylation reactions of magnesium monomethyl malonate proceed in the presence of acid- and base-sensitive functionality,[24] as illustrated in the reaction sequence shown below.[25]

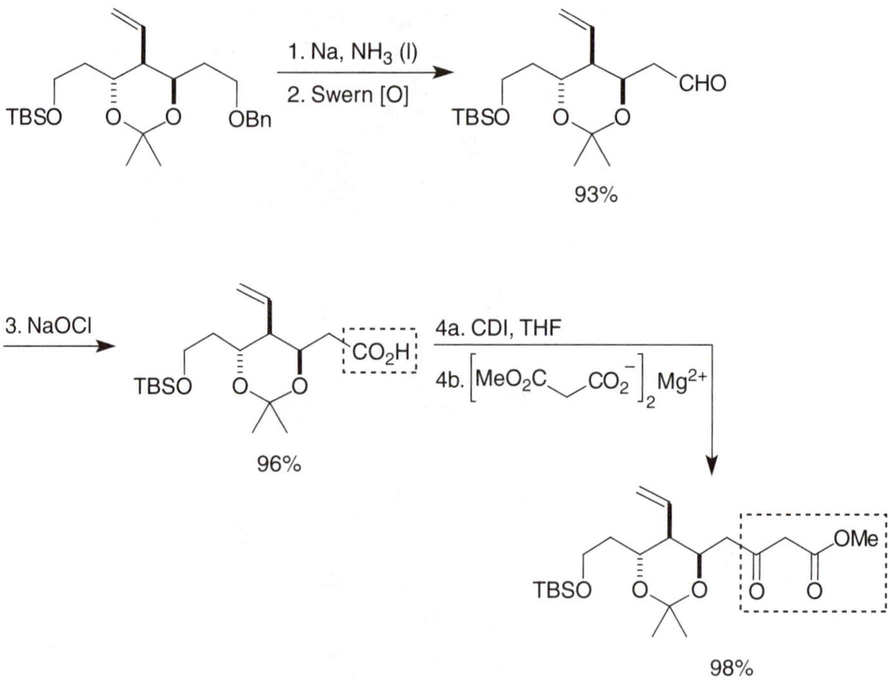

93%

98%

Alkylation of β-Keto Esters

The alkylation of β-keto ester enolates followed by decarboxylation affords substituted ketones (*acetoacetic ester synthesis*). The ester group acts as a temporary activating group. Retro-Claisen condensation can be a serious problem during hydrolysis of the ester, particularly in basic solution if the product has no protons between the carbonyl groups. In these cases, the hydrolysis should be carried out under acidic conditions or using one of the methods of decarbalkoxylation described in the next section.

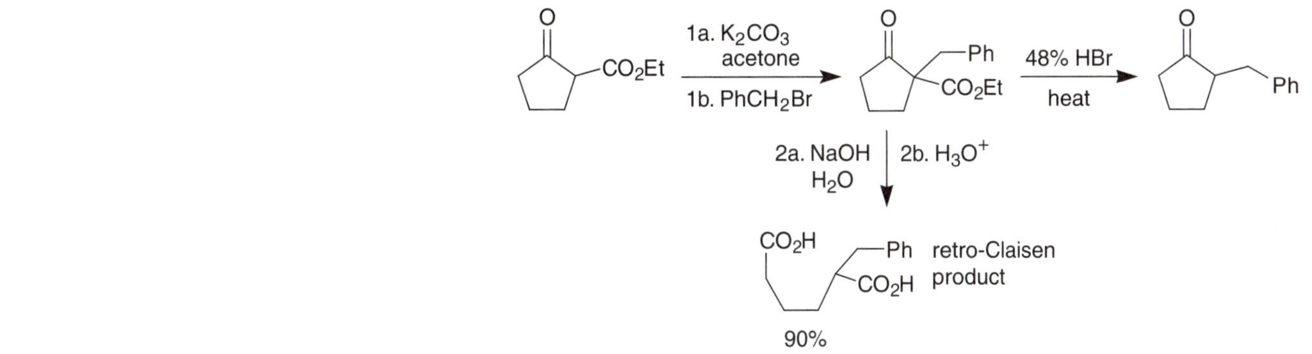

Using two equivalents of LDA provides for regioselective alkylation at the carbon having the higher electron density.[26]

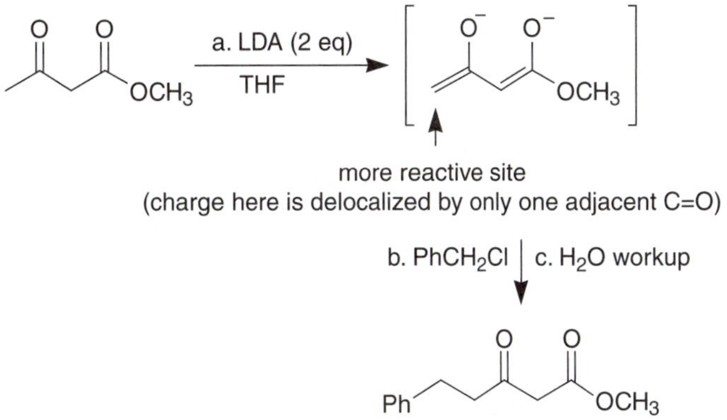

more reactive site
(charge here is delocalized by only one adjacent C=O)

An alternative procedure for β-keto ester dianion formation involves deprotonation of $H_{α'}$ using one equivalent of NaH followed by deprotonation of an $H_α$ using either MeLi or n-BuLi.[27] This method is especially useful in cases where the α-position is too hindered for deprotonation by LDA.[28]

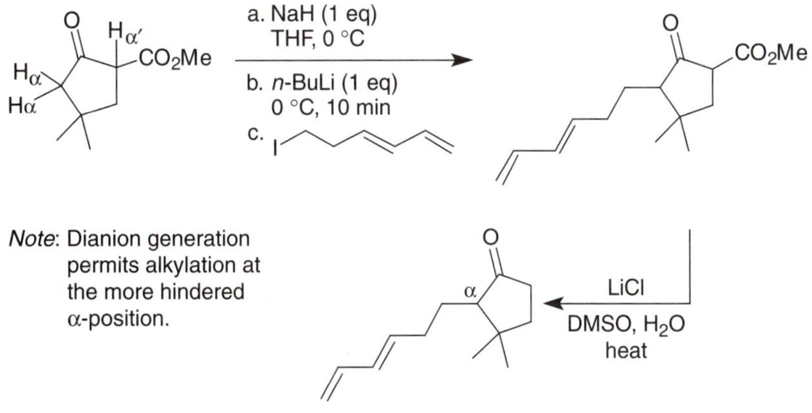

Note: Dianion generation permits alkylation at the more hindered α-position.

Decarboxylation of β-Keto Esters

In addition to the classic sequence of saponification-acidification followed by heating, the decarboxylation of β-keto esters can be effected by more direct methods such as *decarbalkoxylation*. For example, heating β-keto esters in the presence of lithium halides in H_2O-DMSO[29] or in H_2O-collidine produces ketone products in one-pot transformations, as shown below.[30]

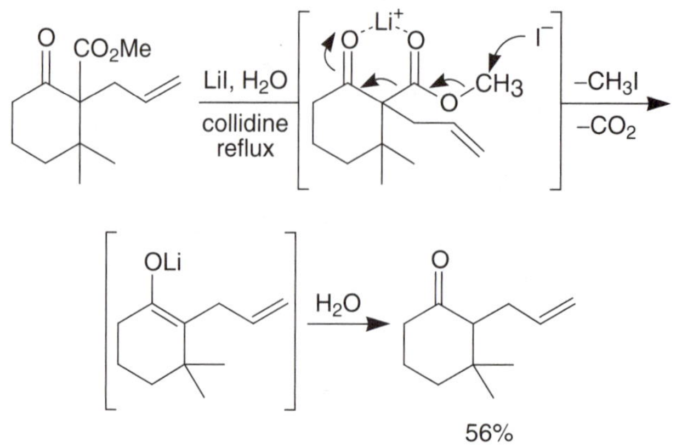

56%

Other procedures for decarbalkoxylation include heating β-keto esters in the presence of boric acid[31] or 4-(*N,N*-dimethylamino)pyridine (DMAP) in H_2O-toluene.[32] This latter approach is also useful for the decarboxylation of α,α-*disubstituted* β-keto esters, which are prone to undergo cleavage via retro-Claisen condensation.

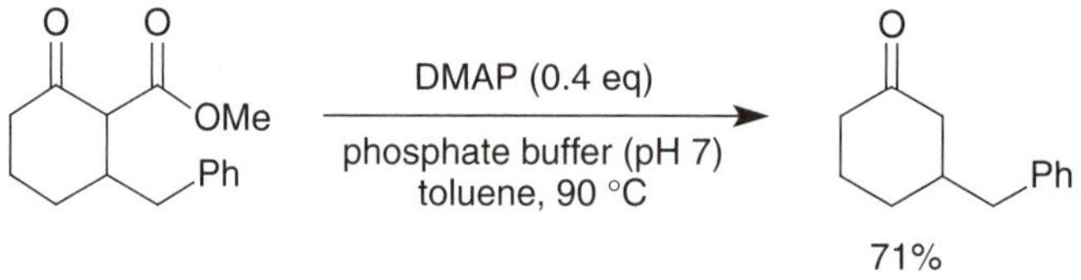

71%

1,3-Diketones   Direct alkylation of β-diketones may be accomplished with reactive alkylating reagents such as primary alkyl iodides and allylic and benzylic halides.

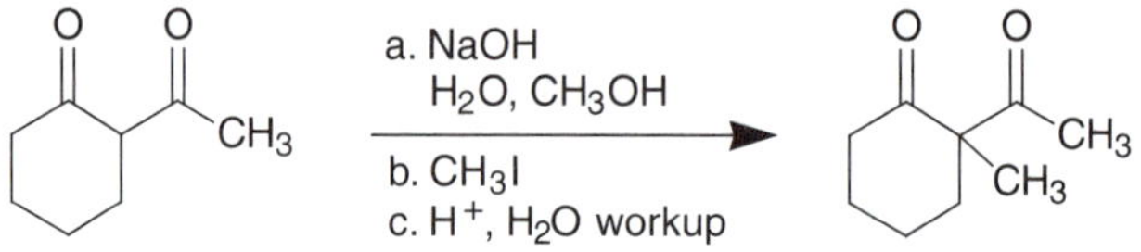

However, since enolate anions are ambident nucleophiles with the distribution of charge between the α-carbon and oxygen conferring reactivity to both sites, alkylation may result at either site.

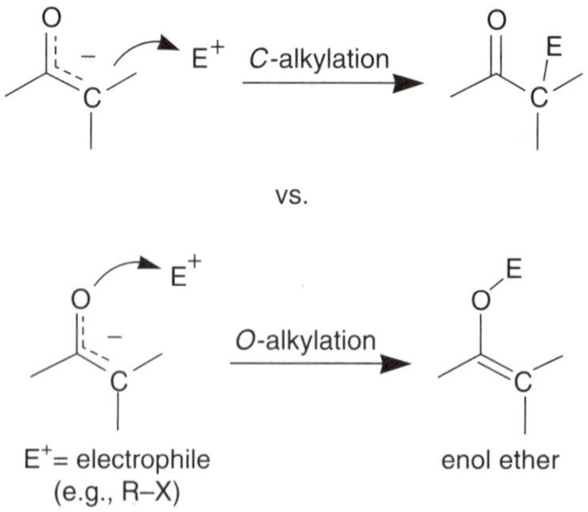

In general, there is significant competition between *C*- and *O*-alkylation when the equilibrium concentration of the enol tautomer is relatively high, as in the case of 1,3-cyclohexanedione.[33]

minor + major

To obviate *O*-alkylation, the readily accessible 1,5-dimethoxy-1,4-cyclohexa-diene surrogate derived from Birch reduction of 1,3-dimethoxybenzene can be used. Enol ether hydrolysis affords the alkylated 1,3-cyclohexadiones in excellent yields.[34]

R = $-CH_2CH_2CH = CH_2$ 95%
R = $-CH_2CH_2Ph$ 93%

O-Alkylation may occur when the enolate anion and its counteranion are dissociated. Shielding of the enolate by the cation represses *O*-alkylation. The tendency of cations to dissociate from oxygen follows the trend $R_4N^+ > K^+ > Na^+ > Li^+$. The selection of the proper solvent also plays an important role in determining the site of alkylation. Polar aprotic solvents such as DMSO and hexamethylphosphoramide (HMPA, a potential carcinogen) favor *O*-alkylation. Therefore, these solvents should be avoided when *C*-alkylation is desired. Toluene in conjunction with a phase transfer reagent such as *n*-Bu$_4$NBr provides a method for the high-yield preparation of both mono- or dialkylated 1,3-diones.[35]

40 °C   74%

6.2   DIRECT ALKYLATION OF SIMPLE ENOLATES

Ester Enolates[36]   Esters (pK_a ~25) do not form high concentration of enolates when treated with alkoxides (alcohol pK_a 16–18). Therefore, strong, non-nucleophilic bases, such as LDA (*i*-Pr$_2$NH pK_a = 36), are required to rapidly and quantitatively deprotonate them at –78 °C in THF. The ester is added to a solution of LDA in Et$_2$O or in THF (*inverse* addition). After the ester enolate is completely formed *under these kinetic conditions,* it is treated with a reactive alkyl halide. LDA is particularly well suited as a base for these alkylations because it reacts very slowly even with reactive alkyl halides.[37] The products of direct ester enolate alkylation are also accessible by dialkylation of malonic esters described in Section 6.1.

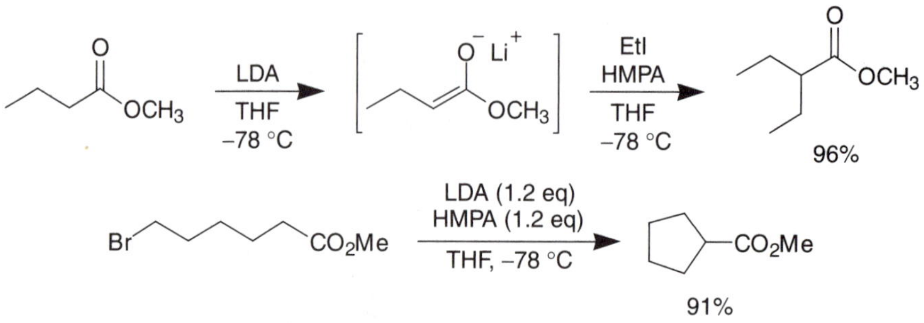

When competing Claisen condensation of the ester is a problem, the use of the sterically hindered *t*-butyl esters is recommended.[38] Unlike with ketone enolates, the *O*-alkylation of ester enolates generally is not a problem. Consequently, HMPA may be added to ester enolate alkylations to improve yields. Many S_N2 reactions proceed more readily in HMPA than in THF, DME, or DMSO. A solvent for replacing the carcinogenic HMPA in a variety of alkylation reactions is 1,3-dimethyl-3,4,5,6-tetrahydro-2(1H)pyrimidinone (*N,N'*-dimethylpropyleneurea, DMPU), which also has a strong dipole to facilitate metal counterion coordination.[39]

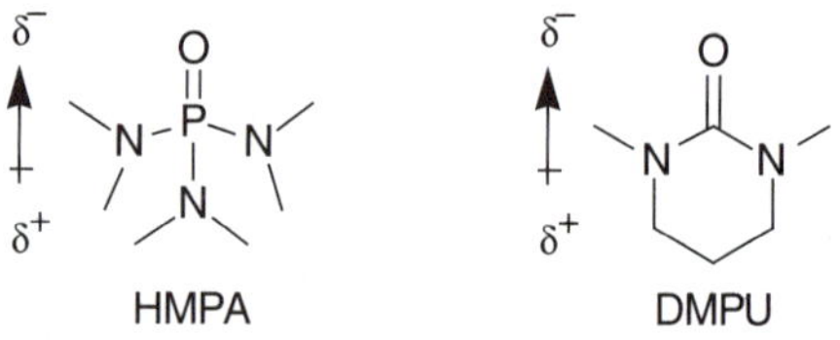

If the ester possesses a β-stereocenter and the β-substituents are of widely different steric size, then alkylation of the enolate derived from such an ester usually leads to high diastereoselectivity. In the example shown below, to minimize the $A^{1,3}$ strain the enolate adopts a conformation in which the smallest group (H) is nearly eclipsing with the double bond. The electrophile, MeI, then approaches the enolate from the side opposite the larger group (PhMe$_2$Si).

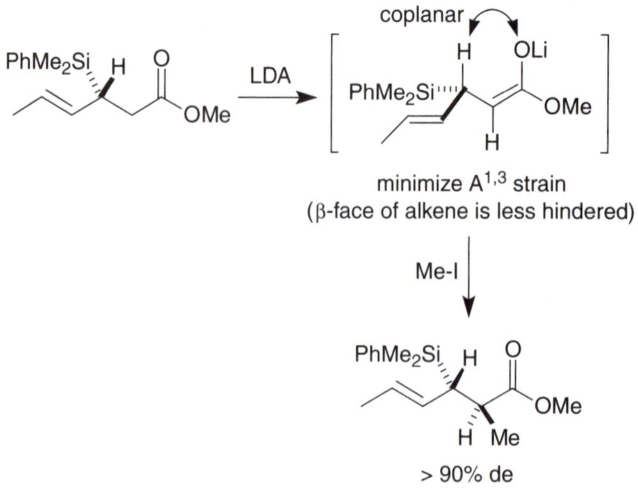

Enolates Derived from Carboxylic Acids,[36] Amides, and Nitriles[40]

The dilithio dianions of carboxylic acids can be prepared in THF-HMPA solution using LDA as the base. These dianions are more stable than α-ester anions but can still be readily C-alkylated.[26,41]

N,N-disubstituted amides are efficiently α-alkylated by treatment with a strong base followed by reaction with an alkyl halide.

Deprotonation of alkylnitriles with LDA or lithium hexamethyldisilazide (LHMDS[42]) and treatment of the resultant ambient α-nitrile anions with 1° and 2°-alkyl halides affords C-alkylated products in good yield. However, the α-anions of highly substituted nitriles may undergo N-alkylation to give amides on aqueous workup.[43]

Acylation of alkylnitriles is best accomplished by low-temperature reaction with dialkyl carbonates or alkyl chloroformates [ClC(O)OR] in the presence of *excess* LDA.[44]

Ketone Enolates[1a,1c,1g,45]

Ketones require a stronger base than NaOEt or NaOCH$_3$ to convert them into enolate anions in high enough concentration to be useful for subsequent alkylation.

Common base and solvent conditions used to generate ketone enolates for alkylation include (1) *t*-BuOK in DMSO, *t*-BuOH, THF, or DME; (2) NaH in THF or DME; and (3) NaNH$_2$ in Et$_2$O, THF, or DME. One of the most efficient methods for generating enolates from ketones is the use of lithium dialkylamides (R$_2$NLi) in

Et$_2$O, THF, or DME. These are powerful bases yet weak nucleophiles.[46] Although alkyllithiums are even stronger bases than R$_2$NLi, their use as bases for deprotonating ketones is limited due to competing nucleophilic addition to the carbonyl group.

Regioselective Enolate Formation via Deprotonation[1c]

The deprotonation of an unsymmetrically substituted ketones having both α and α′-hydrogens furnishes two regioisomeric enolates. Much effort has been devoted to uncover methods to control the regiochemistry of enolate formation from such ketones.

Thermodynamic enolates are generated at room temperature or reflux by conducting the deprotonation in the presence of a small amount of a weak acid (usually a 1–2% excess of the ketone starting material). Lithiated bases are preferred for thermodynamic enolate generation because lithium enolates have a fairly covalent oxygen-lithium bond. Consequently, the most stable enolates will be those that have the most highly substituted double bond.

The ratio of enolate regioisomers can be determined by reacting the enolate mixture with R$_3$SiCl and isolating the resultant stable silyl enol ethers. Silyl halide reagents such as Me$_3$SiCl are *oxophilic* and react nearly exclusively via *O*-silylation, forming an oxygen-silicon bond (142 kcal/mol) rather than a carbon-silicon bond (85 kcal/mol). Silyl enol ether mixtures can be separated by distillation or using chromatographic methods.

Regioselective formation of thermodynamic enolates (or their corresponding silyl enol ethers) can be accomplished by treatment of unsymmetrical ketones with KH, Et_3B[47] or with KH, t-$BuMe_2SiCl$ in the presence of HMPA.[48]

Kinetic enolates are obtained by slow addition of the ketone (1.00 eq) to an excess of a hindered strong base (1.05 eq) at low temperature in an aprotic solvent (nonequilibrating conditions). It should be noted that deprotonation of acyclic ketones may furnish (*E*)/(*Z*)-mixtures of enolates. Since the stereochemistry of enolates plays a pivotal role in controlling the product stereochemistry in aldol reactions, methodologies used for selectively preparing either one of the isomers are discussed in the following subsections.

temp (°C)		
−78	96	4
0	84	16

Deprotonation of α,β-unsaturated ketones by LDA proceeds kinetically via α'-proton abstraction to give cross-conjugated enolates. Thus, if alkylation is carried out with 1.05 equivalents of LDA at low temperature, α'-substitution prevails.[49] However, γ-deprotonation can be effected by reaction under equilibrating conditions (e.g., using a proton source, higher temperature, and longer reaction time) to give the more stable, conjugated enolate.

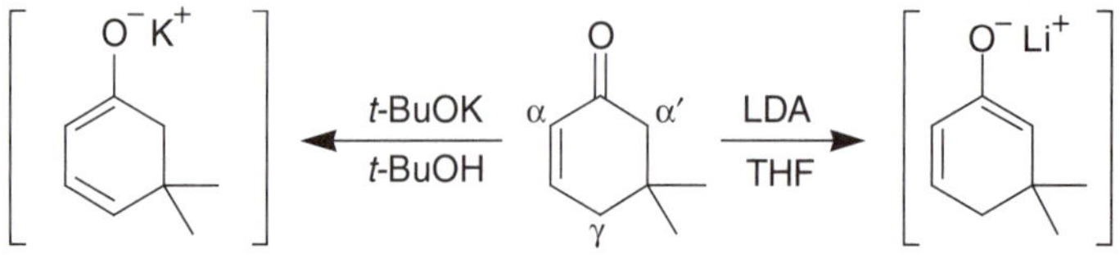

thermodynamic enolate
conjugated or
extended enolate

kinetic enolate
cross-conjugated enolate

Depending on the conditions used to deprotonate α,β-unsaturated ketones, subsequent alkylation gives either α'-alkyl-α,β-unsaturated ketones or α-alkyl-β,γ-unsaturated ketones.[50] In many cases, the initial α-alkylation product may undergo further α-alkylation to give α,α-dialkyl-β,γ-unsaturated ketones.

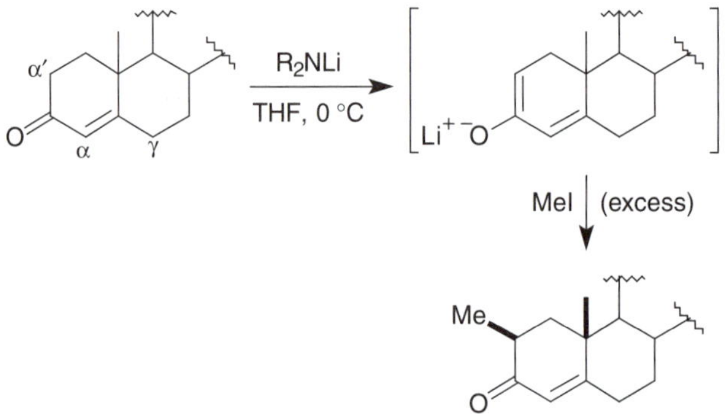

kinetic product, 85%

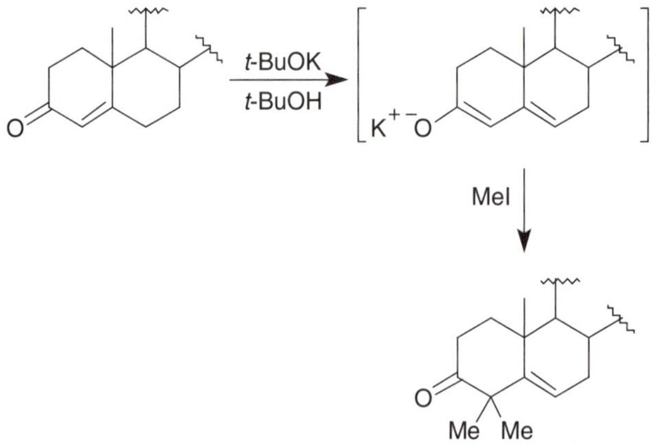

thermodynamic product
(overalkylation)

Lithium hexamethyldisilazide (LHMDS, LiN(TMS)$_2$) and Ph$_3$CLi[51] are useful bases for γ-deprotonation of enones to generate conjugated enolates.

Since conjugated enolates react almost exclusively at the α-position, low temperature protonation of a conjugated enolate provides the β,γ-unsaturated ketone. This deconjugation protocol can be used to functionalize the γ-position of an enone, as shown below.

Via Cleavage of Silyl Enol Ethers[52]

Regioselectively generated silyl enol ethers react with methyllithium to afford regiochemically pure lithium enolates. Treatment of these enolates with reactive electrophiles leads to regiospecifically alkylated ketones.

Similarly, silyl enol ethers may be unmasked by reaction with *anhydrous* fluoride reagents, available as siliconate $[(Et_2N)_3S(Me_3SiF_2)]$[53] or as ammonium salts (benzyltrimethyl- or adamantyltrimethylammonium fluoride).[54]

Via Metal-Ammonia Reduction of Enones[55]

Dissolving metal reduction of α,β-unsaturated ketones regiospecifically produces enolates that, on removal of ammonia, may be reacted with electrophiles.

Via Conjugate Addition Reactions

The copper-mediated 1,4-addition of alkyl groups to α,β-unsaturated ketones affords regiochemically pure enolate anions (see also Section 7.5) which may be trapped at oxygen with silyl halides, acyl halides, or dialkylcarbonates to provide silyl enol ethers, enol acetates, or enol carbonates, respectively. These can be unmasked at a later stage by reaction with MeLi to regenerate the enolate for further elaboration.[56]

Conjugate reduction of enones with K-Selectride (K[*sec*-Bu₃BH]) is a valuable method for regiospecifically generating enolate anions.[57] As illustrated below, the enolate species reacts with an alkyl halide to give the corresponding α-alkylated ketone in good yield.

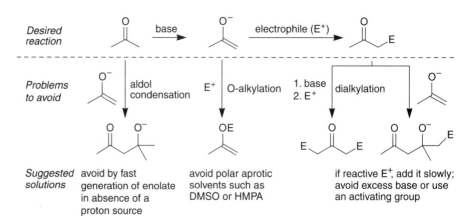

Via Introduction of Activating Groups

An excellent way to control the regiochemistry of deprotonation of ketones is to introduce an activating acyl group (formyl, carboalkoxy), making the proton at the α'-position of the carbonyl group significantly more acidic. Acylation usually takes place at the less substituted carbon.

Intermolecular Alkylation of Ketone Enolates

In summary, several factors must be adjusted to maximize the *C*-alkylation of ketone enolates. These include the following[58]

- Enolate formation must be faster than reaction of the enolate species with the starting ketone (this avoids aldol condensation).

- The selectivity for reaction at carbon rather than at oxygen must be favorable since ambident anions may react at either terminus of the enolate system depending on choice of solvent, counterion, and alkyl halide.

- The electrophile must be added rapidly (an excess if possible) to trap the enolate before it can condense with the product initially formed. Unfortunately, less reactive alkylating agents often allow enolate equilibration to compete with alkylation.

If di- or polyalkylation is a problem, the addition of triethanolamine borate to the reaction mixture will suppress overalkylation.[59] Likewise, enolates formed in the presence of Et_3B react smoothly to form mono-alkylation products, as illustrated in the following examples.[60]

Intramolecular Alkylation–Cyclization Reactions[61]

These reactions are subject to stereoelectronic control. The transition state for S_N2 displacement requires collinear (180°) disposition of the nucleophile (C^-) and the leaving group L. In some cases, for steric reasons, O-alkylation dominates over C-alkylation. See Section 6.3 below for additional details.

6.3   CYCLIZATION REACTIONS–BALDWIN'S RULES FOR RING CLOSURE[62]

Because bifunctional substrates are used in ring formation, *intra*molecular reactions always have to compete with *inter*molecular processes giving rise to oligomers and polymers. Considering entropic effects, stereoelectronic effects, and ring strain, some general rules for ring closure can be given.

- Cyclization reactions to furnish 5–7-member rings proceed faster than the corresponding intermolecular reactions (entropic and strain effects).

- Although entropy effects for the formation of smaller rings (3–4) are favorable because the functional groups involved are held in close proximity, the ring strain in the transition state lowers their rates of formation.

- The most difficult rings to close are those that are of medium size (8–11) because torsional and transannular interactions destabilize the transition state.

- Although larger rings (12 and higher) are almost free of strain, unfavorable entropy (large separation of the reactive centers) slows down the reaction rate and hence increases the chance for intermolecular reactions.

The following facts must be considered in planning intramolecular cyclizations:

- The number of atoms participating in the cyclization;
- Steric effects in the chain connecting the reaction centers;
- Hybridization of the atom that is attacked in the course of the cyclization;
- Geometric constraints imposed by through-space interactions between two orbitals on the same molecule (stereoelectronic effects).

An extensive study of known cyclization reactions led J. E. Baldwin to formulate a set of empirical rules to predict the relative ease of ring-forming processes. These ring closure rules — known as *Baldwin's rules* — apply to nucleophilic, electrophilic, and radical reactions.

In evaluating the feasibility of a proposed ring closure, consider the following three questions:

1. What is the number of atoms in the skeleton of the new ring?

2. Is the breaking bond *exo-* or *endo*cyclic to the smallest-formed ring?

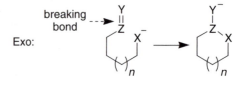

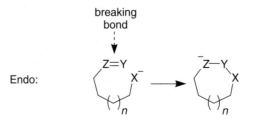

3. What is the hybridization of the carbon being attacked by the nucleophile?

    Attack on sp³: *tet*

    Attack on sp²: *trig*

    Attack on sp: *dig*

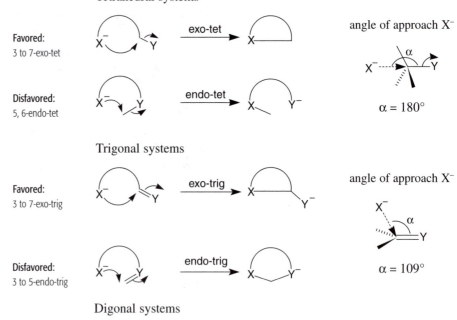

It should be noted that disfavored does not mean that the ring closure will not occur, only that it is more difficult than the corresponding favored ring closure.

Intramolecular Aldol Condensations

For ring closures involving ketone enolates, Baldwin and Lusch modified the original rules as shown below.[62b] Again, these empirical rules are based on stereoelectronic considerations.

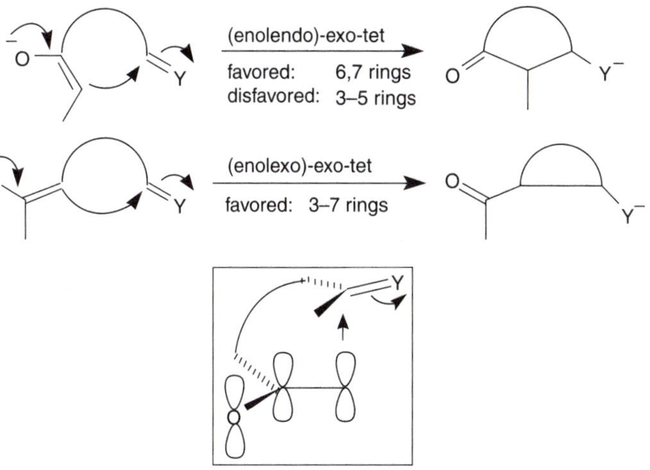

Intramolecular Alkylation of Enolates

Favored ring-closing alkylation reactions include the 6-(enolendo)-exo-tet, 6-(enolexo)-exo-tet, and 5-exo-tet cyclizations shown below.[63] For these transformations, the leaving group may be Cl, Br, I, –OMs or –OTs and the electron-withdrawing group (EWG) generally is a ketone, ester, nitrile, nitro, or sulfone moiety.

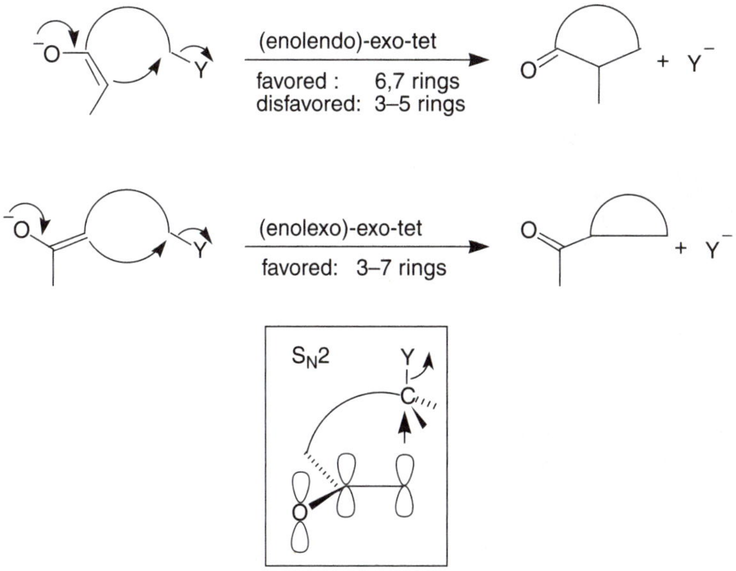

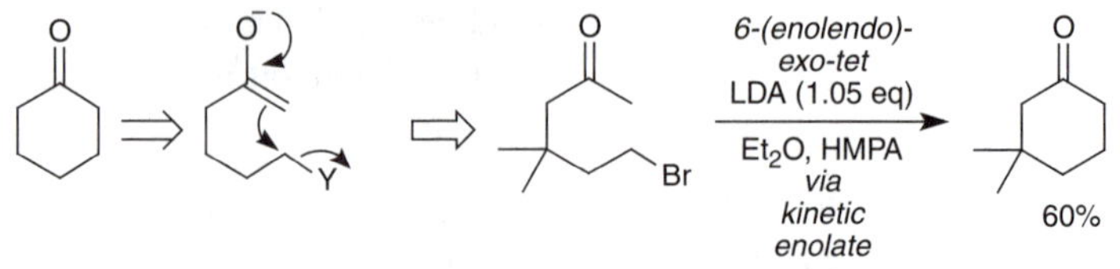

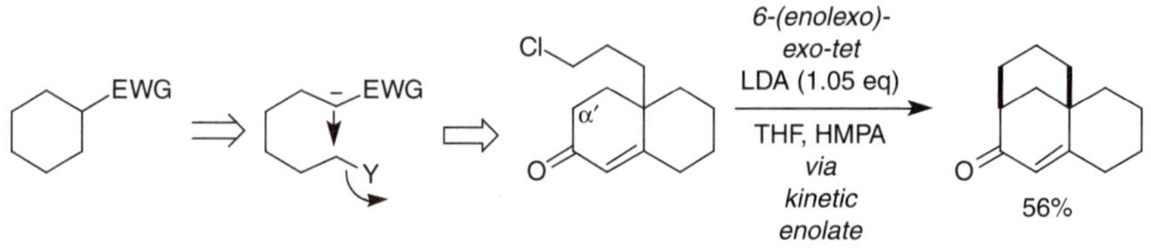

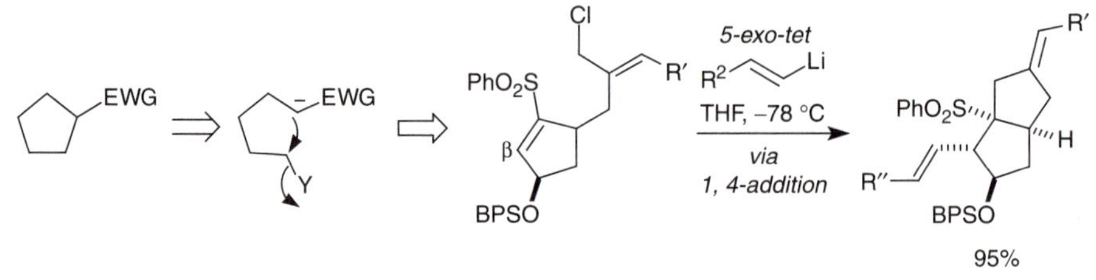

Note that the 5-(enolendo)-exo-tet cyclization shown below is a disfavored ring-closing alkylation.

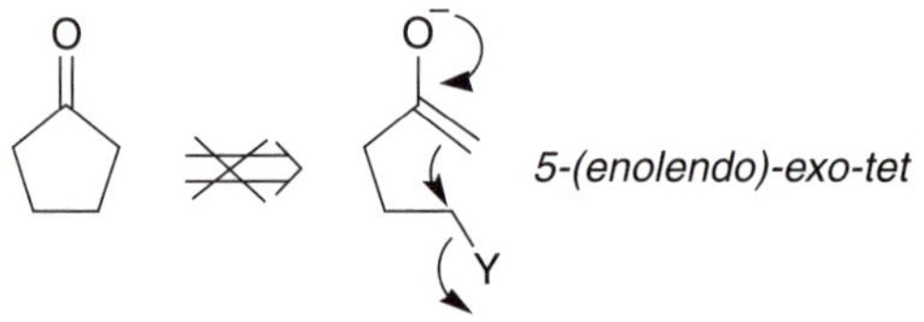

5-(enolendo)-exo-tet

The importance of stereoelectronic effects in intramolecular alkylation-cyclization reactions is exemplified below.

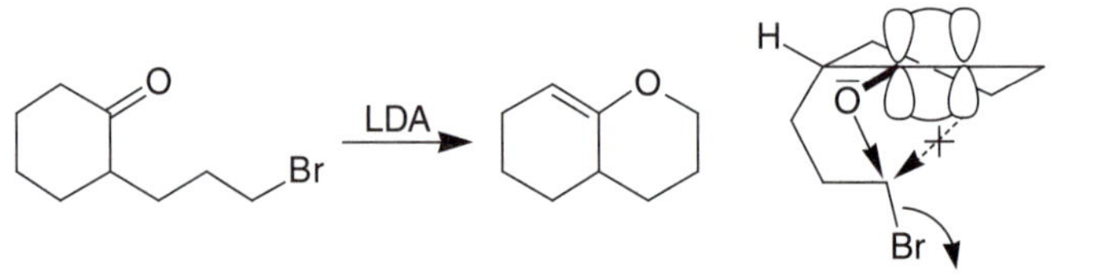

Colinear attack at C–Br by the enolate α-carbon is sterically not possible; thus *O*-alkylation prevails.

However,

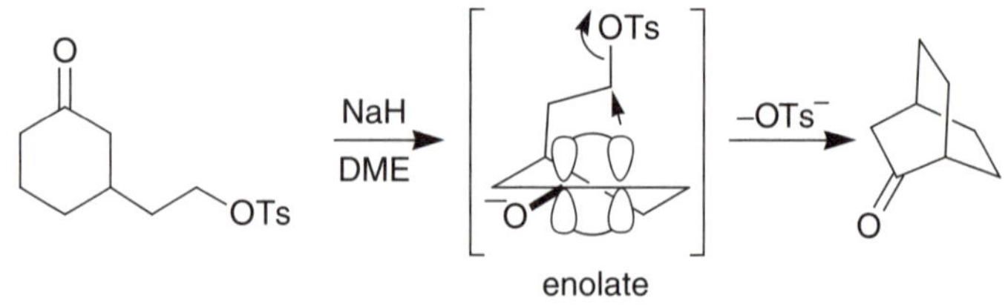

enolate

6.4 STEREOCHEMISTRY OF CYCLIC KETONE ALKYLATION[64]

Enolate reactions involve an early, reactant-like transition state. There is no appreciable bond formation at the transition state; hence, the transition state should resemble the enolate structure. Therefore, there should not be much discrimination between antiparallel and parallel attack. Axial attack via a chair-like TS is usually preferred, although not in a kinetically overwhelming fashion.[65] Equatorial attack via a boat-like TS imparts torsional strain as R passes by the adjacent hydrogen.

ap = antiparallel alkylation
p = parallel alkylation

In conformationally rigid systems, steric factors play an important role in determining the facial selectivity of alkylations of enolates. Whereas alkylation of enolate A proceeds via a chairlike transition state, attack of the electrophile on enolate B in an antiparallel manner is subject to a *synaxial* interaction with the axial Me group. Hence, parallel approach of the electrophile through a twist boat transition state prevails.[66]

Alkylation of the enolate derived from deprotonation of the lactone shown below occurs nearly exclusively from the face opposite from the *t*-butyl group.[67]

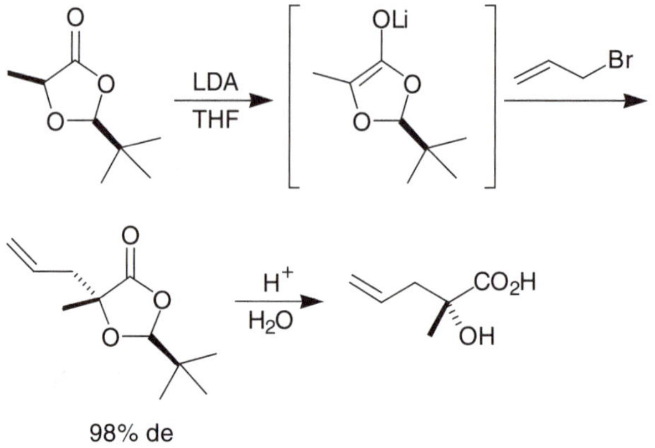

98% de

6.5 IMINE AND HYDRAZONE ANIONS[58a]

Imine anions are superior to the corresponding enolate anions of the parent carbonyl compounds in alkylation reactions because they give only monoalkylated products of predictable regioselectivity. They can be prepared from aldehyde imines as well as from ketone imines by deprotonation with EtMgBr, LDA, or *t*-BuLi.[68]

Imine Alkylation

High yields of alkylated aldehydes and ketones are obtained when using primary and even secondary alkyl halides in the presence of HMPA or DMPU. Little rearrangement is observed with allylic halides.[68b] Moreover, the new alkyl group is generally introduced at the less substituted α-carbon, even with α,β-unsaturated aldehydes or ketones.

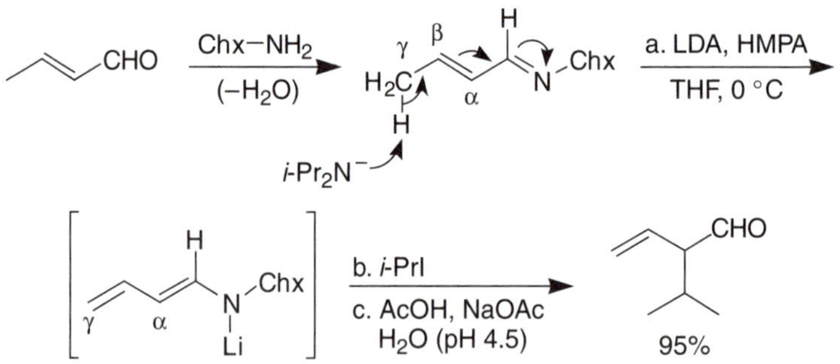

The effect of steric strain in the imine anion assures good regiochemical control for alkylation.[68a]

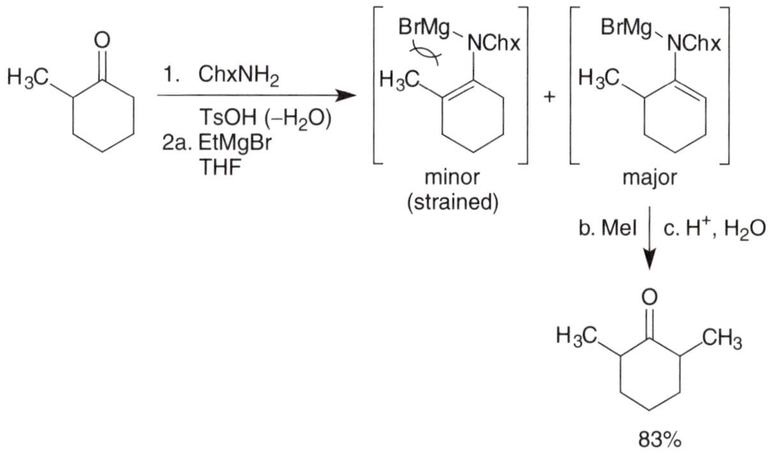

83%

Hydrazone Alkylation

Alkylations of hydrazone anions derived from aldehydes and ketones are closely related to imine anion alkylations. Lithiated hydrazones are particularly powerful nucleophiles for alkylation reactions. The alkylated hydrazones may be cleaved with periodate, as shown below.[69] The choice of solvent and base is critical to the success of hydrazone alkylations; hexane-free medium is essential for complete metalation.[70]

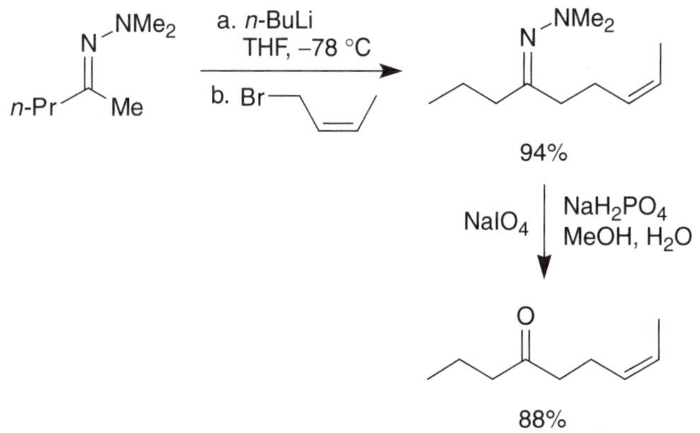

94%

88%

Enantioselective Alkylation of Hydrazones[71]

Reaction of 3-pentanone with the commercially available hydrazine (S)-1-amino-2-methoxymethylpyrrolidine (SAMP) affords the corresponding chiral hydrazone. Deprotonation with LDA followed by alkylation and hydrolysis furnishes (S)-4-methyl-3-heptanone in greater than 99% enantiomeric excess.[71b] Using the corresponding (R)-hydrazine (RAMP) provides (R)-4-methyl-3-heptanone.

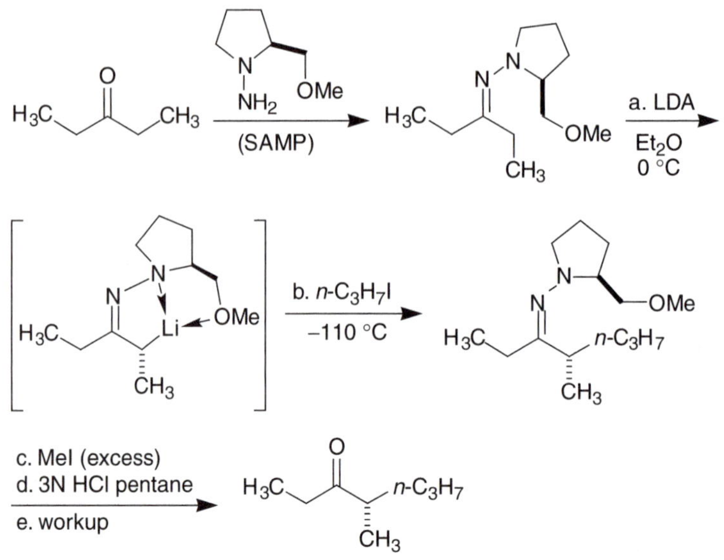

> 99% ee

An alternative to hydrolysis is cleavage of the RAMP- and SAMP-hydrazones by ozonolysis which takes place without epimerization or racemization.[72] This is a reliable method of obtaining the alkylated ketone products with high de and ee selectivity.[73]

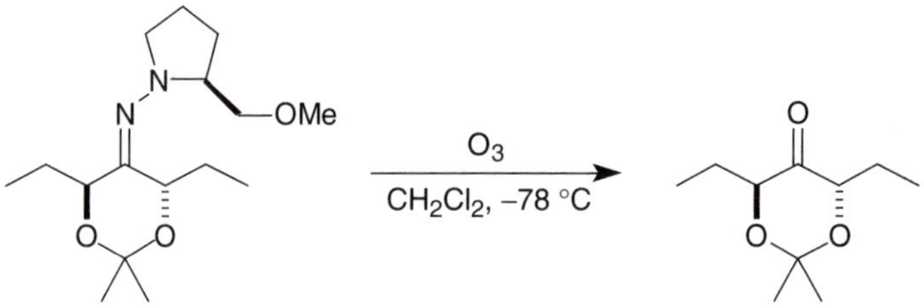

83% (>98% ee)

6.6 ENAMINES[74, 75]

The enamine reaction provides an alternative method for selective alkylation and acylation of *aldehydes* and *ketones*. The enamine group is both a protecting group for carbonyl compounds and a directing or activating group for further elaboration. Note the relationship between enolates and enamines:

$$\mathrm{RCH{=}CH{-}O^{-}} \longleftrightarrow \mathrm{R\bar{C}H{-}CH{=}O} \qquad \mathrm{RCH{=}CH{-}\ddot{N}R_2} \longleftrightarrow \mathrm{R\bar{C}H{-}CH{=}\overset{+}{N}R_2}$$

Preparation of Enamines

Enamines are readily generated by acid-catalyzed condensation of an aldehyde or ketone with a secondary amine such as pyrrolidine, piperidine, morpholine, and di-*tert*-butylamine (for aldehydes).

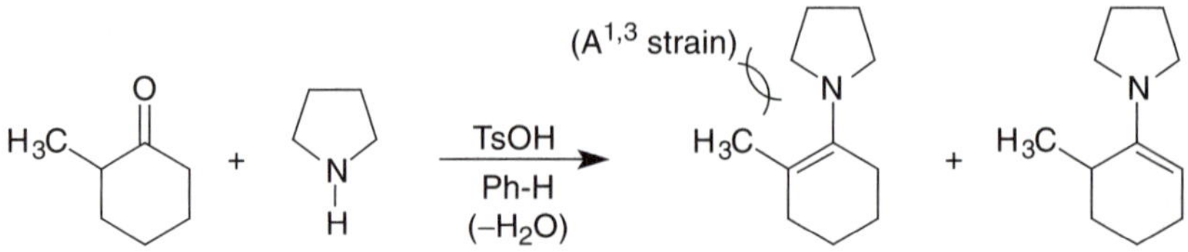

10% 90%

Alkylation of Enamines

Enamines are ambident nucleophiles giving C- and N-alkylated products. Acceptable yields of C-alkylated products are obtained by using reactive alkyl halides such as CH_3I, allylic and benzylic halides, and α-halocarbonyl compounds. The resultant iminium ion intermediates no longer behave as a enolates, thus dialkylation is avoided. The stereochemical course of alkylation of the enamine derived from 2-methylcyclohexanone is depicted below. The reason for the preferred parallel alkylation via a boat-like transition state over antiparallel alkylation via a chair-like transition state is the synaxial RX // CH_3 interaction in the latter case.[74b]

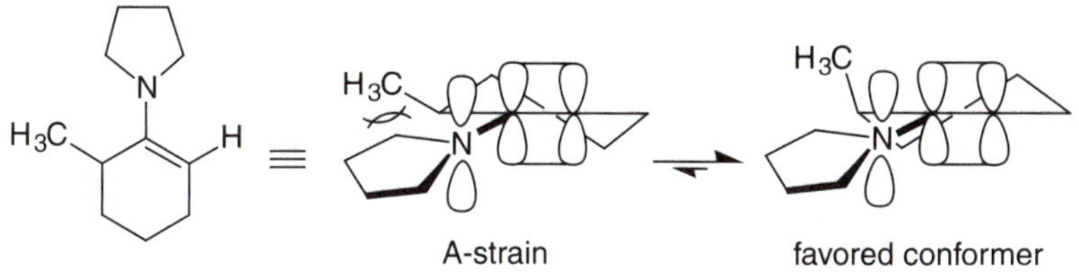

A-strain   favored conformer

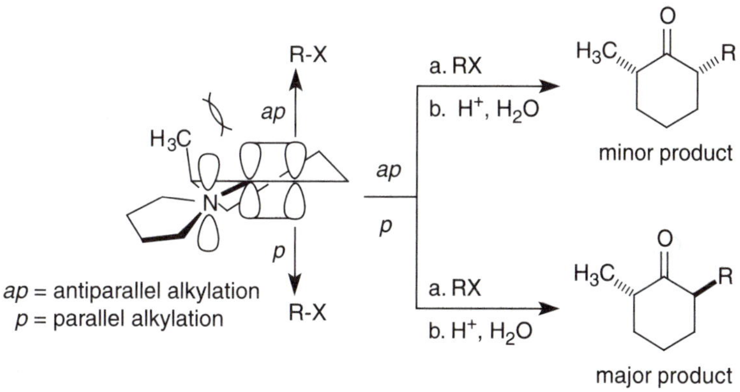

ap = antiparallel alkylation
p = parallel alkylation

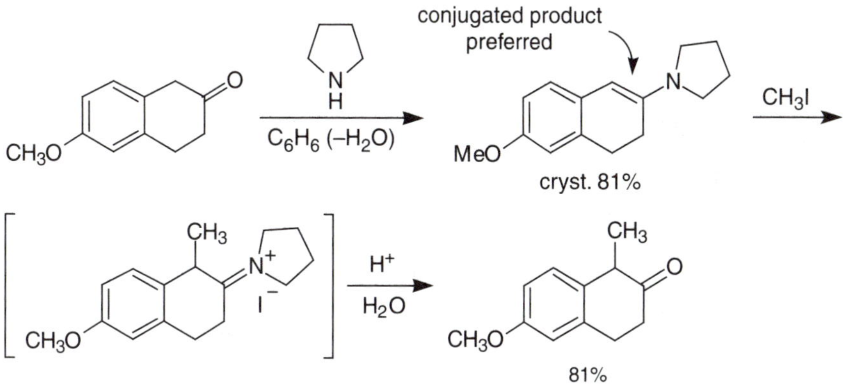

minor product

major product

conjugated product preferred

cryst. 81%

81%

Stereoselective Alkylation of Enamines[76]

Chiral enamines may be prepared by condensation of ketones with enantioenriched *sec*-amines. The C_2-symmetric *trans*-2,5-dimethylpyrrolidine is a frequently used chiral auxiliary for the preparation of enantiomeric enamines.[77] Alkylation of chiral enamines followed by hydrolysis is an effective method for the enantioselective alkylation of ketones.[78]

90% (80% ee)

Acylation

A useful method for preparing 1,3-diketones and β-keto esters from ketones is to first form the enamine and then acylate it with an appropriate acyl chloride derivative. Morpholine, a *sec*-amine, is a good choice for this purpose since the derived enamines have lower reactivity toward acyl halides, allowing for more control of the reaction.

6.7   THE ALDOL REACTION[79]

The aldol reaction is one of the most useful methods for the construction of carbon-carbon bonds. The products of aldol reactions are either β-hydroxy carbonyl compounds or, after dehydration, α,β-unsaturated carbonyl compounds. The aldol reaction is useful not only for making C–C bonds, but also for providing two functional groups, the C=O and a β-OH, which can be further elaborated.

In the strictest sense, the aldol reaction involves the condensation of an enolate derived from an aldehyde or a ketone with another aldehyde or ketone to give a β-hydroxyaldehyde or a β-hydroxyketone, respectively. In a broader sense, the aldol reaction encompasses reactions of enolates—usually derived from ketones, esters, or amides—with an aldehyde or a ketone.

R^1, R^2, R^3, R^4 = H, alkyl, or aryl

Intermolecular Aldol Reactions

The aldol reaction is catalyzed by base or by acid. Both base- and acid-catalyzed condensations are reversible in the 1,2-addition step. The equilibrium constant for the addition step is usually unfavorable for ketones.

Base-catalyzed aldol reaction

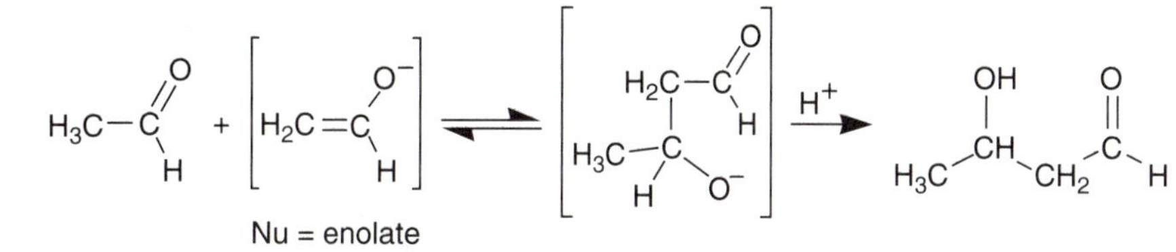

Acid-catalyzed aldol reaction

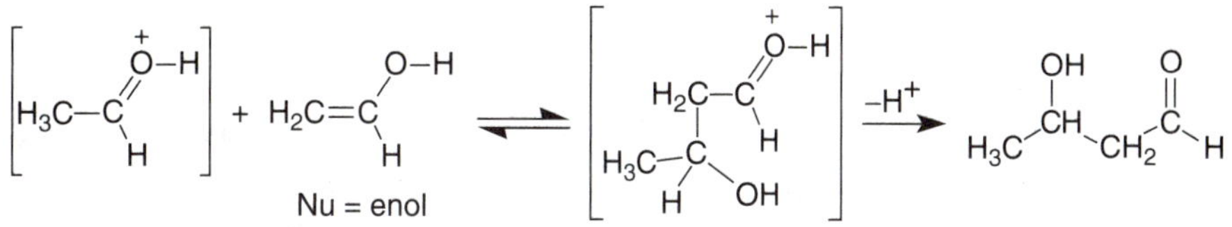

The enol content of simple ketones is quite low under standard acid-catalyzed conditions (for acetone $K_{eq} = 10^{-7}$; cyclohexanone $K_{eq} = 10^{-6}$). The base-catalyzed aldol reaction for aldehydes is slightly exothermic while the reaction for ketones is somewhat endothermic.

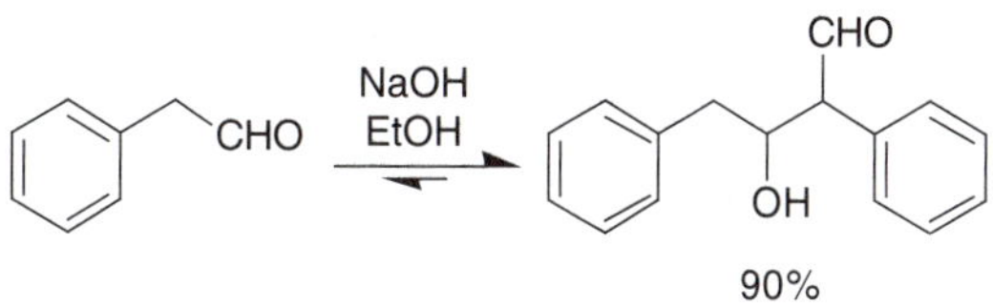

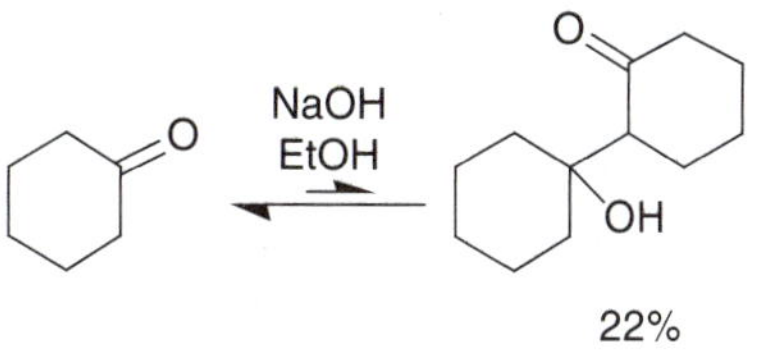

90%

22%

The β-hydroxy aldehydes and β-hydroxy ketones formed in aldol reactions are readily dehydrated to yield conjugated systems. These β-elimination reactions can be effected by either acid or base while heating during the condensation.

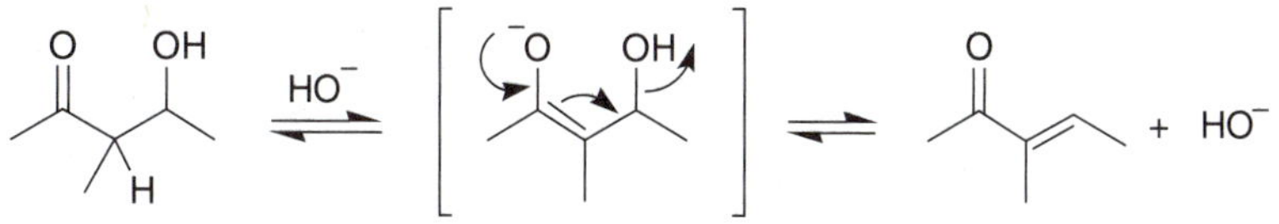

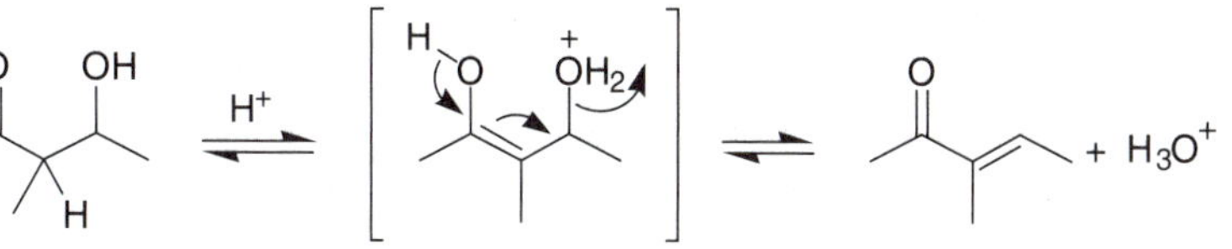

Removal of H_2O from the reaction mixture drives the equilibrium toward product formation. Even though the initial aldol step itself may be unfavorable (as it usually is for ketones), the subsequent dehydration step allows most aldol reactions to be carried out in good yield.

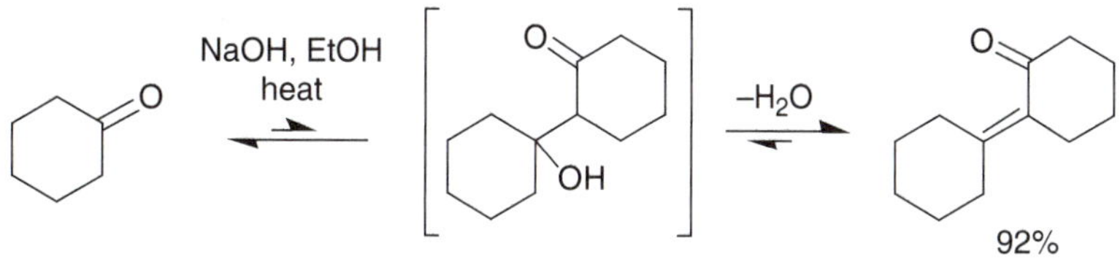

92%

Intramolecular Aldol Reaction[80]

The intramolecular aldol condensation is a powerful tool for obtaining five- and six-member rings. This is an important step in the Robinson annulation reaction (see Section 7.9).

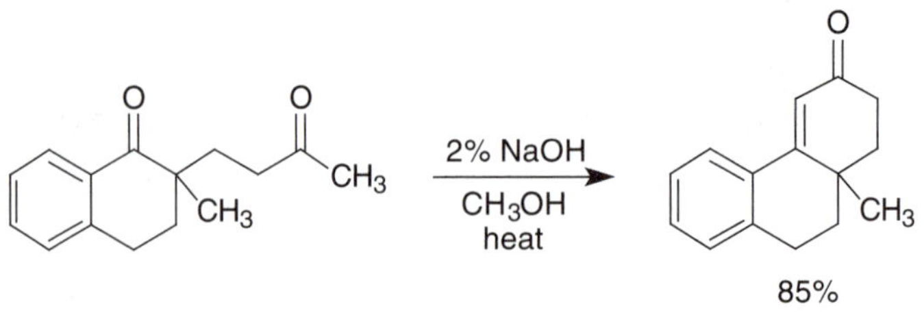

85%

In the following base-catalyzed intramolecular aldol reactions, both enolates are formed reversibly. However, cyclization is faster via attack of the enolate at the less hindered carbonyl group.

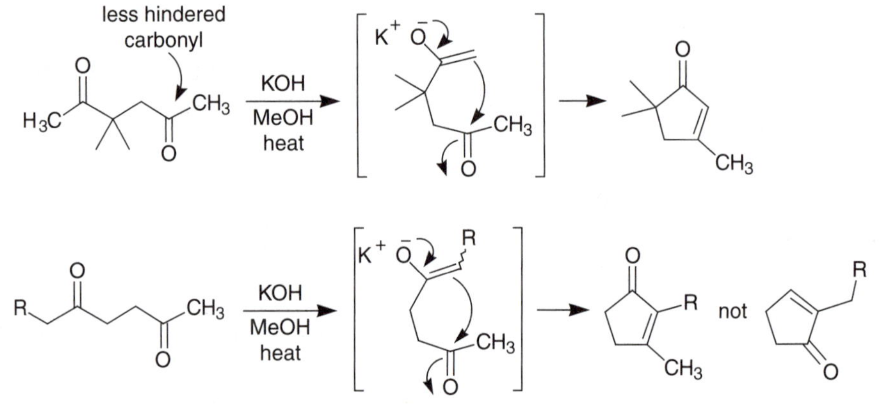

With 1,5-dicarbonyl compounds, two modes of ring closure are often possible. In the example shown below, the more stable (higher-substituted) enone is formed preferentially (thermodynamic control).[81] The observed distribution of products is the result of equilibration via *retro-aldol* reaction.

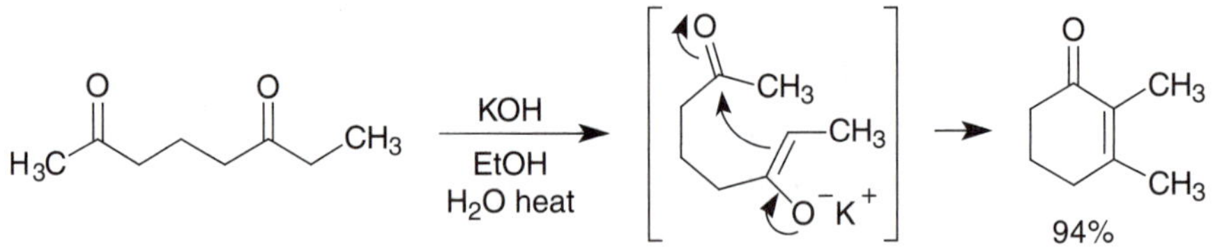

94%

Intramolecular aldol reactions of 1,6-diketones or 1,6-keto-aldehydes afford the corresponding cyclopentenyl carbonyl compounds.[82]

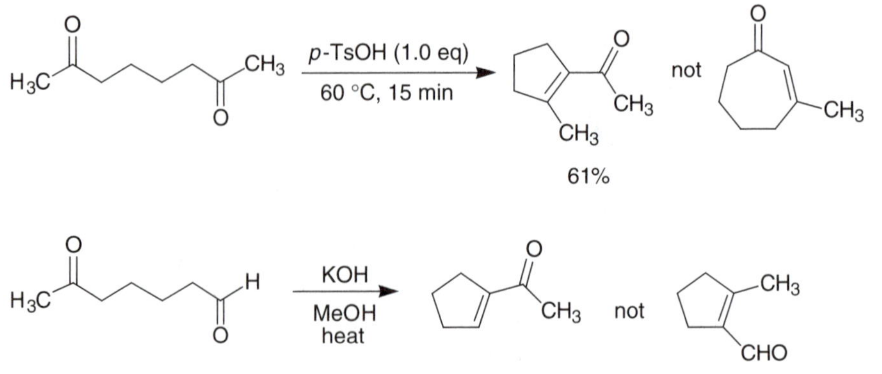

61%

Mixed Aldol Reactions

Aldol reactions between two different carbonyl compounds are called mixed or crossed aldol reactions. With aqueous bases, these reactions are of little synthetic value if both reactants have α-hydrogens because they afford mixtures of products. However, under carefully controlled conditions, it is possible to condense ketones with aldehydes in the presence of dilute sodium hydroxide to furnish β-hydroxyketones, provided that the aldehyde is added slowly to the ketone.

Changing from thermodynamic to kinetic conditions (LDA), it is possible to carry out crossed aldol reactions with good control, as exemplified below.

Claisen–Schmidt Reaction[83]

When one of the carbonyl compounds cannot form an enolate (as with HCHO, PhCHO, or Ph_2CO), crossed aldol condensations are feasible.[84]

The Mukaiyama Reaction[85]—Lewis Acid–Catalyzed Crossed Aldol Reactions

The enol content of simple aldehydes and ketones is low under standard acid-catalyzed conditions. Silyl enol ethers, often available free of regioisomers, are an important source of *enol equivalents* for nucleophilic addition reactions. The reaction of silyl enol ethers with carbonyl compounds in the presence of $BF_3 \cdot Et_2O$, $SnCl_4$, $TiCl_4$, or $InCl_3$[86] proceeds through an open transition state instead of a closed transition state and leads, after hydrolytic workup, to aldol products.

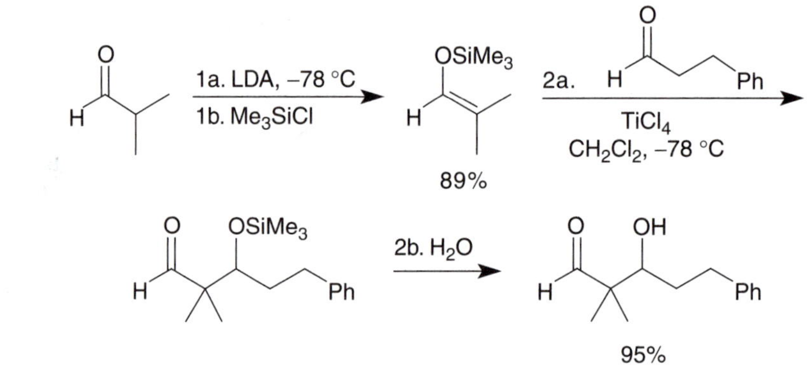

With β-alkoxy-substituted aldehydes, catalysis by Me_2AlCl or $MeAlCl_2$ proceeds via a highly organized chelate and gives aldol products with excellent diastereoselectivity, as shown below.[87]

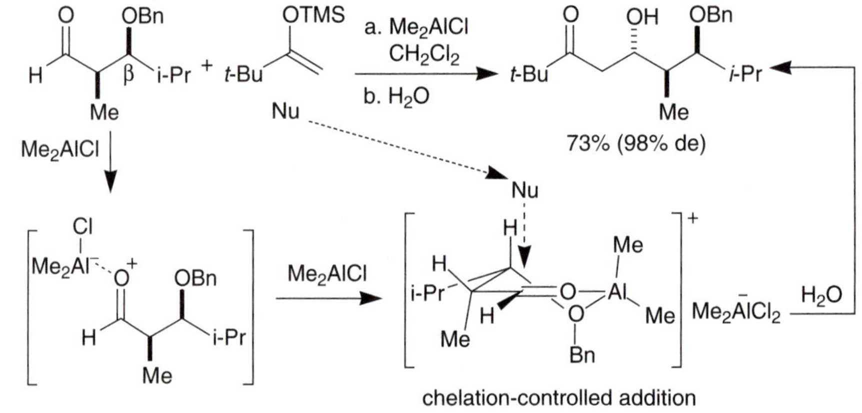

Silyl enol ethers with a chiral auxiliary appendage react with achiral aldehydes to produce, after cleavage of the auxiliary group, enantiomerically enriched β-hydroxy carboxylic acids.[88]

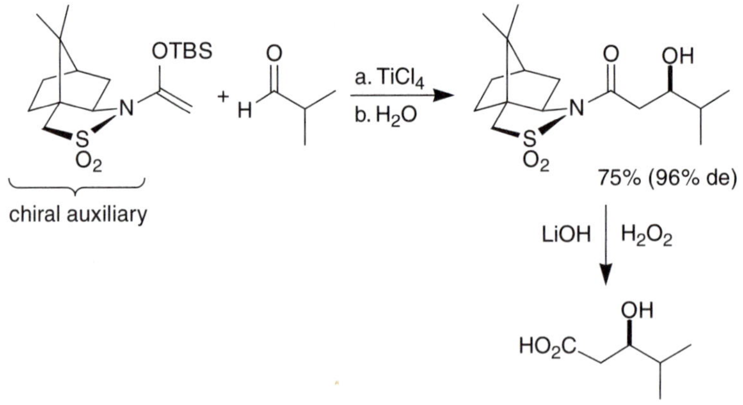

Imine Aldol Reactions[89]

Imine enolates can be prepared without self-condensation, yet they will add rapidly to carbonyl compounds. This circumvents the low reactivity of ketones toward carbanions derived from nonactivated methylene compounds and permits addition of aldehyde enolate equivalents to ketones.

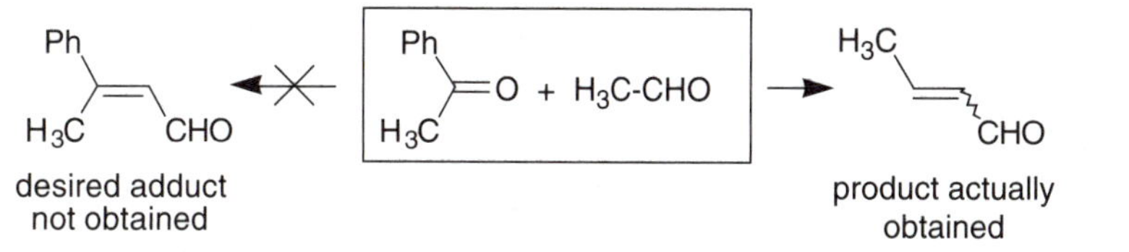

desired adduct
not obtained

product actually
obtained

However,

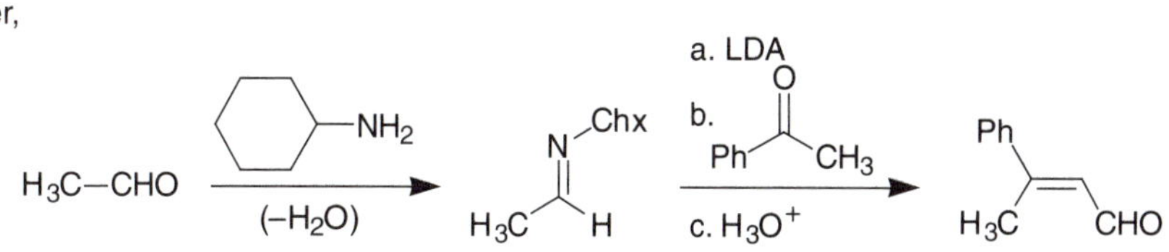

Enamines also undergo condensation with aldehydes to give α,β-unsaturated ketones on hydrolysis.[90]

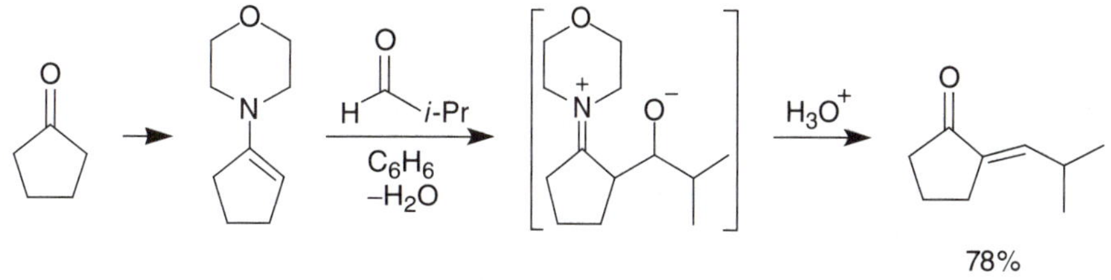

78%

Stereoselective Aldol Reactions[91]

One of the most challenging problems for the synthetic chemist is control of stereochemistry in conformationally nonrigid, open-chain compounds. When planning and executing the synthesis of a complex organic compound, the chemist must cope with the following distinct and interrelated problems:[92]

- Assembly of the requisite molecular framework
- Placement of the necessary functional groups at their proper sites
- Control of the relative chirality at various points of asymmetry in the desired product

Among these problems, effective control of the stereochemistry is often the most challenging, particularly when the elements of asymmetry reside in an acyclic structure. In the evaluation of an aldol reaction as a method for building acyclic molecules containing many stereocenters, there are two types of diastereoselections that must be considered.

The first is *simple diastereoselection*, using achiral substrates. Two newly created chiral centers may be formed at the 2- and 3-positions (numbering is by convention) with either *syn* or *anti* relative configuration.

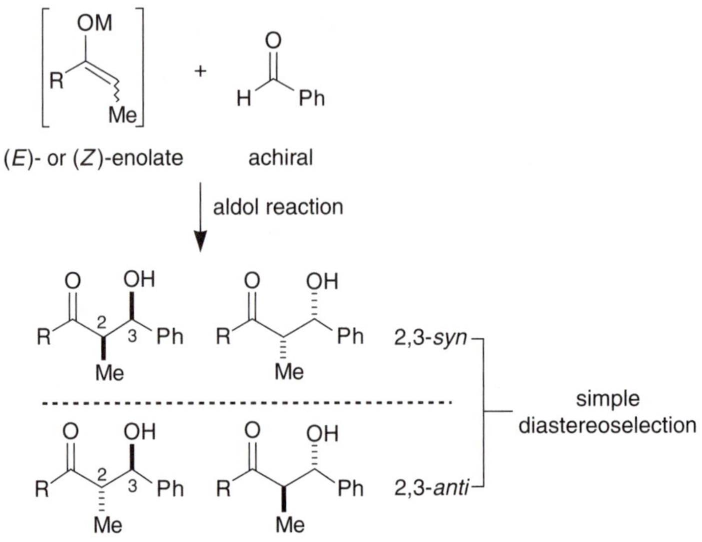

(E)- or (Z)-enolate   achiral

aldol reaction

2,3-*syn*

2,3-*anti*

simple
diastereoselection

The second is referred to as *diastereoface selection*, that is, in many cases one carries out aldol reactions on aldehydes already having one or more chiral centers. The carbonyl faces in these molecules are diastereotopic rather then enantiotopic.

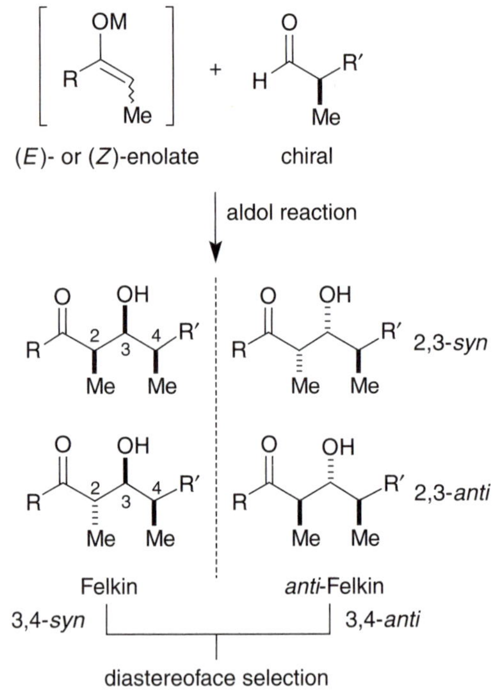

(E)- or (Z)-enolate   chiral

aldol reaction

2,3-*syn*

2,3-*anti*

Felkin   *anti*-Felkin

3,4-*syn*   3,4-*anti*

diastereoface selection

Control of (E)- and (Z)- Enolate Stereochemistry.[93]

Simple diastereoselection from precursors that do not contain a chiral center is controlled by two factors:

1. The configuration of the enolate ion, (E)-(O)- vs. Z-(O)-enolate. Note that the substituents at each end of the double bond having the higher priorities (OM > R) determine the configuration of the double bond.

(E)-(O)-enolate   (Z)-(O)-enolate

2. The orientation of the enolate and aldehyde in the transition state of the aldol
    reaction, "open" transition state vs. the ordered, chelate-controlled transition
    state

Whereas deprotonation of cyclic ketones (4–7-member rings) can only lead to the
(E)-(O)-enolate geometry, control of enolate stereochemistry of acyclic ketones with
lithium amides is rather complicated and depends on the structure of the carbonyl
compound, steric requirements of the base, and reaction conditions.

Ireland's deprotonation model is widely used to rationalize the stereochemistry with
various ethyl ketones and bases.[94] In the absence of additives that solvate the lithium
cation such as HMPA, proton transfer occurs via a chair-like "closed" transition state.
Under these conditions, the (Z)-enolate is *disfavored* because of the 1,3-diaxial interaction between the Me and the *i*-Pr group on nitrogen. As the steric requirement of the
R group increases, so does the $A^{1,3}$ strain between the R and Me groups in forming the
double bond, thus destabilizing the (E)-(O)- relative to the (Z)-(O)-enolate (Table 6.2).

Table 6.2a[94]

R	(E)-(O)-enolate (%)	(Z)-(O)-enolate (%)
OMe	95	5
Et	70	30
i-Pr	44	56
t-Bu	~0	100

HMPA solvates the lithium cation, leading to an "open" transition state, alleviat-
ing the 1,3-diaxial interaction between the Me and *i*-Pr groups but increasing the $A^{1,3}$
strain between the R and Me groups, which favors the (Z)-(O)-enolate.[93]

Table 6.2b[94]

R	(E)-(O)-enolate (%)	(Z)-(O)-enolate (%)
OEt	15	85
Et	8	92

Although stereoselective formation of enolates from acyclic ketones with bases such as LDA is rather difficult, stereodefined boron enolates are more readily accessible.[95] In the Mukaiyama method, an ethyl ketone is treated with a dialkylboron triflate and a tertiary amine, usually i-Pr$_2$NEt. The resultant Z-(O) boron enolates (also known as *enol borinates*) are believed to be formed under kinetic control by deprotonation of the Lewis acid–complexed substrate. Brown and co–workers have shown that E-(O) boron enolates may be prepared by treatment of ethyl ketones with dicyclohexylboron chloride in the presence of Et$_3$N.[96]

In analogous fashion, titanium[97] and tin[98] enolates are formed by the reaction of enolizable ketones with a tertiary amine and TiCl$_4$ or SnOTf$_2$, respectively.[99] The reactions of titanium enolates are highly selective and comparable to boron enolates in aldol condensations.[100]

Simple Diastereoselection – *Syn-Anti* Selectivity.[93b]

When an aldehyde is reacted with a ketone-derived enolate under *equilibrating conditions*, the thermodynamically more stable 2,3-*anti* product predominates regardless of the geometry of the enolate. If, however, the reaction is *kinetically* controlled, the (*Z*)- and (*E*)-enolates furnish 2,3-*syn* and *anti* aldol products, respectively. This behavior has been interpreted in terms of a chair-type transition state known as the *Zimmerman-Traxler* model.[101]

Zimmerman-Traxler transition state model for (*Z*)-(*O*)- and (*E*)-(*O*)-enolates

Note: If the aldol reaction is catalyzed by Lewis acids such as BF$_3$ or TiCl$_4$ the addition reaction will proceed via an acyclic transition state (Mukaiyama aldol).[102]

Although boron enolates are usually more stereoselective in aldol reactions than lithium enolates, the latter are more readily prepared (e.g., using LDA). To obtain synthetically useful levels of aldol stereoselectivities with lithium enolates, the (*E*)-(*O*)- and (*Z*)-(*O*)-enolates must be available with high selectivity (> 95 : 5), and the non-enolized carbonyl group R must be large (Table 6.3).[79]

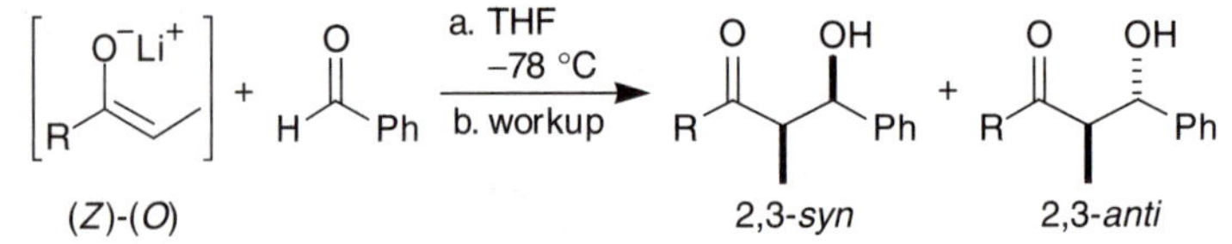

Table 6.3a [79e]

R	2,3-*syn* (%)	2,3-*anti* (%)
t-Bu	99	1
i-Pr	90	10
Et	90	10
H	50	50

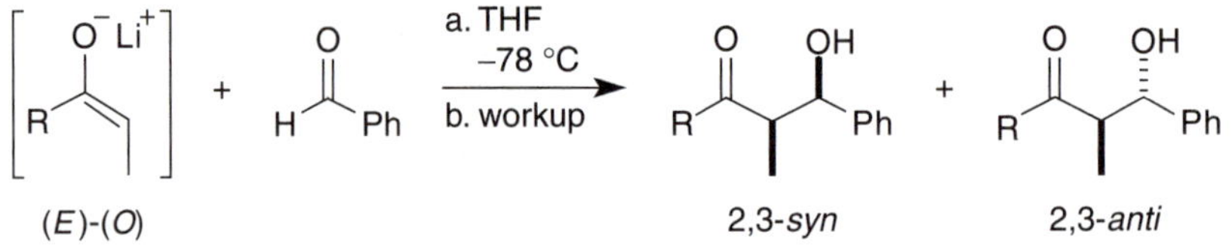

Table 6.3b [79e]

R	2,3-*syn* (%)	2,3-*anti* (%)
i-Pr	50	50
Et	40	60
H	60	40

Aldol reactions of lithium (*Z*)-(*O*)-enolates derived from α-trimethylsilyloxy ketones with aldehydes provide, depending on the workup conditions, either β-hydroxy aldehydes,[103] β-hydroxy acids[104] or β-hydroxy ketones with high levels of 2,3-*syn* selectivity.[105]

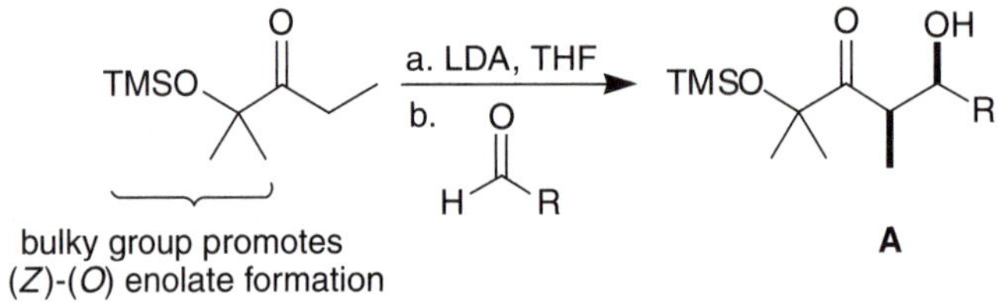

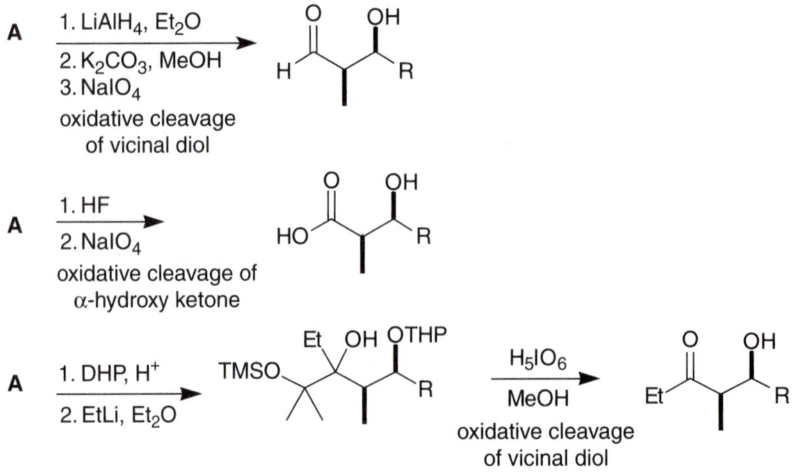

Since the reactions of lithium (E)-(O)-enolates are not stereoselective (see Table 6.3b), formation of 2,3-*anti* aldol products is conveniently accomplished by ketone enolization using dicyclohexylboron chloride. Reaction of the resultant (E)-(O) boron enolates with aldehydes at low temperature affords the 2,3-*anti* products with excellent diastereoselectivity.[106]

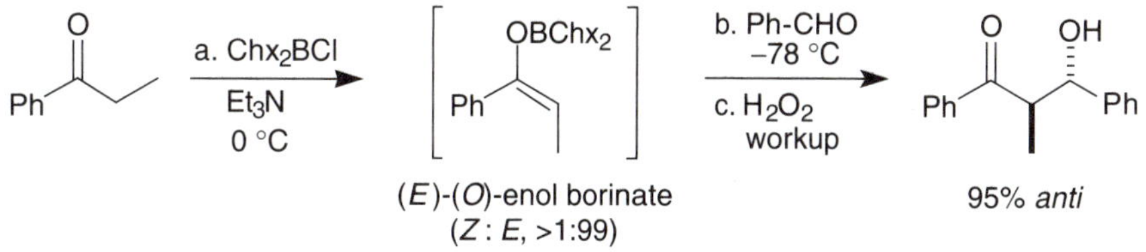

(E)-(O)-enol borinate
$(Z:E, >1:99)$

95% *anti*

Boron enolates are more stereoselective lithium enolates.[107] The reasons for this are (1) B–O bonds are shorter than Li–O bonds and (2) boron enolates are less reactive than lithium enolates, leading to late, product-like transition states where steric repulsions are more important.

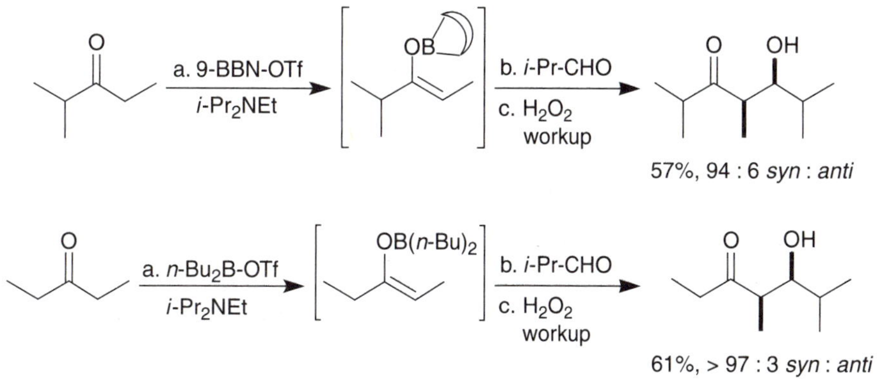

57%, 94 : 6 *syn : anti*

61%, > 97 : 3 *syn : anti*

When treated with dialkylboron triflates, thioesters preferentially form (E)-(O) boron enolates, which react with aldehydes to afford good yields of the 2,3-*anti* aldol products.[108]

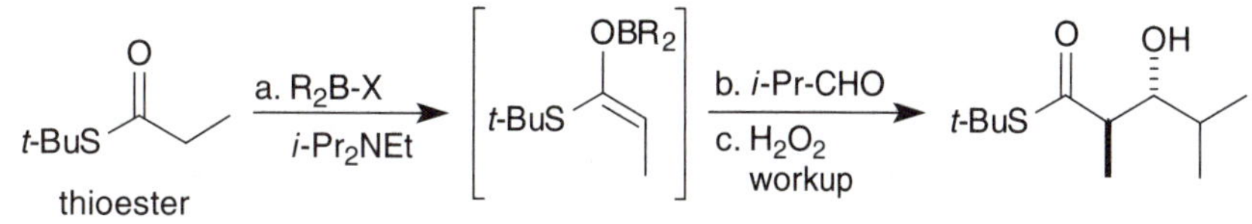

R = cyclopentyl, X = OTf: 79%, > 95 : 5 *anti : syn*
R = (menthyl)CH$_2$–, X = Cl: 61% (98% ee), 96:4 *anti : syn*

The β-hydroxy thioester products may be transformed to 1,3-diols by LiAlH$_4$ reduction. Protection of the hydroxyl group followed by reaction with DIBAL-H leads to the corresponding β-alkoxy aldehydes.[109] Alternatively, treatment with a cuprate reagent produces the corresponding ketone.[110]

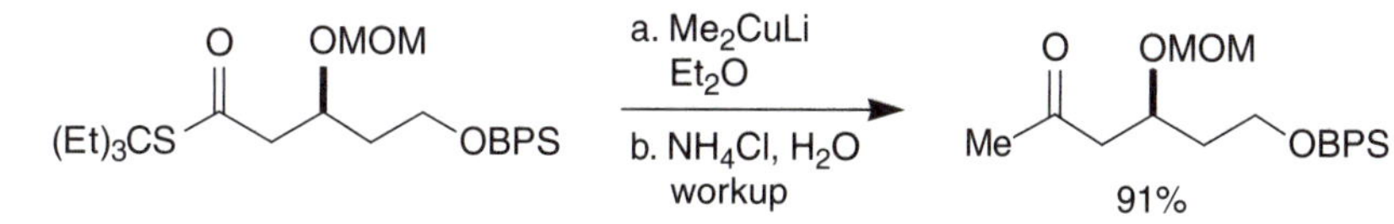

91%

Enantioselective Aldol Reactions[111]

Development of diastereoselective and enantioselective aldol reactions has had a profound impact on the synthesis of two important classes of natural products—the macrolide antibiotics and the polyether ionophores.[112] The aldehyde and the enolate involved in aldol reactions can be chiral, but we shall discuss only the case of chiral enolates.

α-Silyloxyketones

There are several chiral enolate precursors, such as L-t-butylglycine[113] or (R)- or (S)-mandelic acid,[114] which can be used in asymmetric aldol reactions, as depicted below.

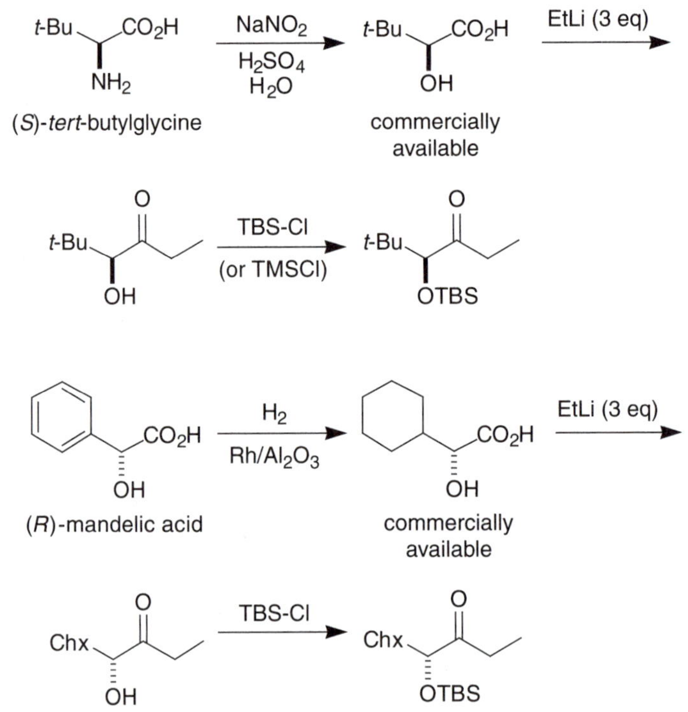

Deprotonation of α-silyloxy ketones with LDA furnishes (Z)-lithium enolates, whereas treatment of ketones with n-Bu₂BOTf in the presence of i-Pr₂EtN gives the corresponding (Z)-(O)-boron enolates. Interestingly, reaction of the Li-enolates with i-PrCHO proceeds with opposite facial preference to that of the boron enolates. Thus, the Si face of the Li-enolate adds to the Si face of the aldehyde and the Si face of the boron enolate adds to the Re face of the aldehyde to furnish the chiral β-hydroxy ketone enantiomers shown below. The reason for the different face selectivity between the lithium enolate and the boron enolate is that lithium can coordinate with three oxygens in the aldol Zimmerman-Traxler transition state, whereas boron has only two coordination sites for oxygen.

It should be noted that these chiral enolate condensations are stoichiometric and *self-immolative;* that is, the chirality of the enolate is sacrificed in its conversion to the β-hydroxy carboxylic acid. This problem can be circumvented by using a chiral auxiliary, such as an N-acyloxazolidinone.

N-Acyloxazolidinones[115]

When employed in stoichiometric amounts, the oxazolidinone auxiliaries can be recovered and reused. Moreover, they are readily prepared by reduction of α–amino acids followed by conversion of the resulting 1,2-amino alcohols to the N-acyloxazolidinones. Both enantiomers of a given oxazolidinone are accessible.[116]

Treatment of N-acyloxazolidinones with di-*n*-butylboron triflate in the presence of Et$_3$N furnishes the (Z)-(O) boron enolates. These on treatment with aldehydes give the corresponding 2,3-*syn* aldol products (the ratio of *syn*- to *anti*- isomers is typically ≥ 99 : 1!). On hydrolysis they produce chiral α-methyl-β-hydroxy carboxylic acids, as exemplified below.[117] The facial selectivity of the chiral boron enolate is attributed to the favored rotomeric orientation of the oxazolidinone carbonyl group, where its dipole is opposed to the enolate oxygen dipole. At the Zimmerman-Traxler transition state, the aldehyde approaches the oxazolidinone appendage from the face of the hydrogen rather than from the benzyl substituent.

89%   90%   84%

The oxazolidinone auxiliary group can also be used to direct the stereochemical outcome to favor *anti*-selective aldol reactions by diverting the reaction to an open transition state using Lewis acid conditions (MgCl$_2$, TMSCl, and Et$_3$N).[118]

A lithium hydroperoxide (LiOH + H$_2$O$_2$) hydrolysis protocol is the method of choice for deacylation of the auxiliary oxazolidinone without racemization to obtain the β-hydroxy carboxylic acid products.[119] The chiral auxiliary can be recovered and recycled. Direct substitution of the oxazolidinone auxiliary with (MeO)MeNH gives the corresponding Weinreb amide.[120] Its subsequent reaction with DIBAL-H or with Grignard reagents converts the amide to an aldehyde or ketone, respectively.[121] Furthermore, reduction of the N-acyloxazolidinone moiety with LiBH$_4$ affords the corresponding alcohol in good yield.[122]

Chiral Ligand-Mediated Aldol Condensations[112,123]

Boron reagents such as (+)- or (−)-(Ipc)$_2$BOTf are chiral promoters in aldol condensations.[124] Enolization of an achiral ketone with (Ipc)$_2$BOTf forms a chiral enolate and thus imparts diastereofacial selectivity (DS) for condensation with a chiral aldehyde. If the ketone is chiral, the DS of the reagent may be *matched* or *mismatched* with the intrinsic DS of the ketone (see Section 5.1 for a discussion of reagent vs. substrate control). Therefore, selection of the proper enantiomer can increase the formation of a desired product diastereomer. This strategy is an example of *double asymmetric induction*[125] (See Section 4.14) and is exemplified below.[126]

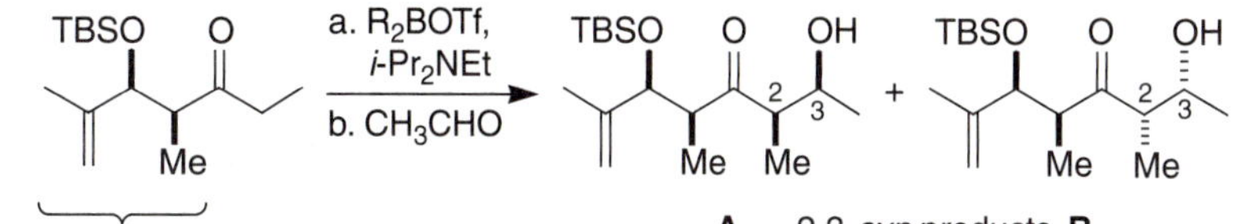

Although *not* a chiral auxiliary, this portion of the molecule acts similarly by imparting diastereofacial selectivity (DS) in the reaction. Its magnitude can be determined using an achiral reagent such as 9-BBN-OTf.

a. R$_2$BOTf, *i*-Pr$_2$NEt
b. CH$_3$CHO

A   2,3-*syn* products B

R$_2$BOTf	Diastereoselectivity (A : B)	
9-BBN-OTf	12 : 1	substrate DS
(+)-(Ipc)$_2$BOTf	47 : 1	matched case
(−)-(Ipc)$_2$BOTf	3 : 1	mismatched case

The stereochemical outcome of an aldol reaction involving more than one chiral component is consistent with the rule of approximate multiplicativity of diastereofacial selectivities intrinsic to the chiral reactants. For a matched case, the diastereoselectivity *approximates* (substrate DS) $\times$ (reagent DS). For a mismatched case, the diastereoselectivity is (substrate DS) $\div$ (reagent DS). Double asymmetric induction also can be used to enforce the inherent facial selectivity of a chiral aldehyde, as shown below.[127]

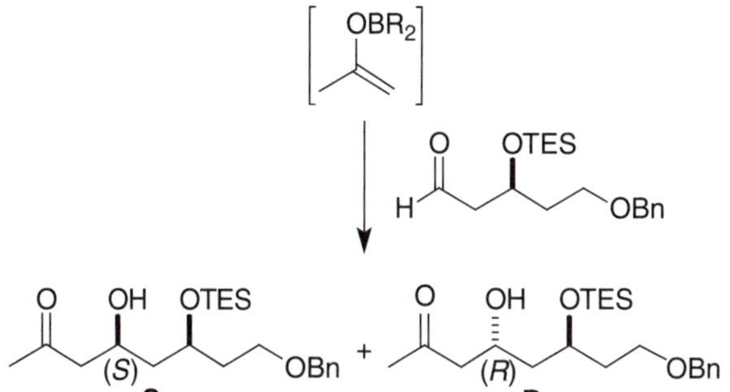

	−BR$_2$	Diastereoselectivity (C : D)	
achiral ---▶	−BChx$_2$	3 : 1	aldehyde DS
	−B[(+)-(Ipc)]$_2$	1 : 1.3	mismatched case
	−B[(−)-(Ipc)]$_2$	13.3 : 1	matched case

Ideally, double (and triple)[128] asymmetric synthesis is most effective when the DS of the chiral reagent is significantly greater than the DS of all other reaction components, thus overwhelming any mismatch with a substrate and thereby dictating the stereochemical outcome based solely on the chirality of the reagent. One of the most powerful examples involving a reagent with an overwhelming DS is in Sharpless asymmetric epoxidation (see Section 5.1).

6.8 CONDENSATION REACTIONS OF ENOLS AND ENOLATES

The Mannich Reaction[129] α,β-Unsaturated carbonyl compounds are important substrates for C–C bond formation in Michael additions, Robinson annulations, and Diels-Alder reactions. There are a number of efficient methods available for the preparation of conjugated carbonyl compounds with a double bond positioned inside the chain or the ring. However, this is not the case when the double bond is located at the terminal position of a chain or is *exo* to a ring. These conjugated carbonyl compounds are vulnerable to attacks by nucleophiles or by radicals resulting in their polymerization or decomposition. For example, acrolein (propenal), the simplest conjugated carbonyl compound ($H_2C=CHCHO$), has a tendency to polymerize when used in conjugate addition reactions.

The ideal approach for the preparation of conjugated carbonyl compounds having an *exo*-methylene group would be via a mixed aldol condensation of an enolizable carbonyl substrate with formaldehyde as the electrophilic partner. However, formaldehyde is a very powerful electrophile and tends to react more than once with enols and enolates. This shortcoming can be circumvented by converting the formaldehyde to an iminium ion (Mannich reagent) by reaction with a secondary amine, usually dimethylamine, and a catalytic amount of HCl.

$$CH_2O + (CH_3)_2\overset{+}{N}H_2 \; I^- \longrightarrow H_2C=\overset{+}{N}(CH_3)_2 \; I^- + H_2O$$
$$\text{iminium salt}$$

Aminomethylation

In the presence of an enolizable aldehyde or a ketone, the resultant electrophilic iminium ion reacts in situ with the enol to produce, after neutralization, the corresponding β-amino ketone, also called a *Mannich base*. Unsymmetrically substituted ketones are aminomethylated preferentially at the more highly substituted carbon of the enol.

enol

Na$_2$CO$_3$ | H$_2$O workup

Mannich base

Utilization of preprepared *N,N*-dimethylmethyleneammonium iodide (Eschenmoser salt)[130] or chloride[131] gives higher yields of β-amino ketones than does the classical Mannich reaction. Silyl enol ethers also react with the Eschenmoser salt to give Mannich bases, as exemplified below.[132]

Enone Synthesis

In addition to the Mannich reaction being a valuable method for preparing amino ketones,[133] which are encountered in many drugs, the reaction is also important in organic synthesis in providing a stable equivalent of a conjugated *exo*-methylene moiety. Thus, addition of methyl iodide to the Mannich base converts it to the quaternary ammonium salt. Subsequent treatment with a base results in a β-elimination of trimethylamine to generate the α-methylene ketone.

Aminomethylation of lactone enolates with the Eschenmoser salt followed by neutralization (workup) yields the corresponding Mannich bases. Their conversion to quaternary ammonium iodides followed by treatment with DBU (1,5-diazabicyclo[5.4.0]undec-5-ene) leads to α-methylene lactones.

Asymmetric Mannich Reaction

A chiral promoter molecule such as L-proline can be used to catalytically mediate a stereoselective Mannich reaction.[134]

Michael Addition[135]

1,5-dicarbonyl

R′ = alkyl, H, phenyl, OR

Base-Catalyzed Michael Addition

The Michael-type addition, a nucleophilic addition of an anion to the carbon–carbon double bond of an α,β-unsaturated ketone, aldehyde, nitrile, nitro, sulphonyl, or carboxylic acid derivative, provides a powerful tool for carbon–carbon bond formation. The reaction is most successful with relatively nonbasic ("soft") nucleophiles such as thiols, cyanide, primary and secondary amines, and β-dicarbonyl compounds. There is often a competition between direct attack on the carbonyl carbon (1,2-addition) and conjugate addition (1,4-addition) when the substrate is an α,β-unsaturated carbonyl compound.

The reaction is facilitated by protonation of the enolate produced in the initial conjugate-addition step. The resultant enol rapidly rearranges to the more stable ketone form. Since the base is regenerated, a catalytic amount may be used. The Michael reaction is reversible. Hence, efficient protonation (EtOH, H_2O) of the adduct is necessary.

$$CH_2(CO_2Et)_2 + EtO^- \rightleftharpoons \bar{C}H(CO_2Et)_2 + EtOH$$

The reaction of an enolate with an α,β-unsaturated carbonyl compound yields a *1,5-dicarbonyl compound* as a product. Two different reaction paths can be envisioned for synthesis of these compounds. The superior pathway is always that which employs the most acidic carbonyl partner (least basic enolate) as the nucleophile.

Michael acceptors	Michael donors
$H_2C=CHCHO$	$RC(O)CH_2C(O)R′$
$H_2C=CHC(O)CH_3$	$RC(O)CH_2CO_2R′$
$H_2C=CHCO_2R$	$RO_2CCH_2CO_2R$
$H_2C=CHC\equiv N$	$RC(O)CH_2C\equiv N$
$H_2C=CHC(O)NR_2$	RCH_2NO_2
$H_2C=CHNO_2$	R_2CuLi

Enamines are excellent addends in many Michael-type reactions; no base is required.

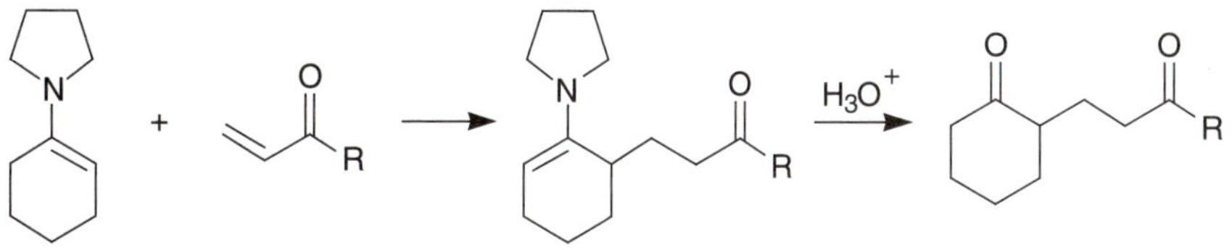

Although the Michael addition is most successful when the carbon acid is relatively acidic, the reaction also occurs with simple ketones. The Michael acceptor is mainly introduced at the more highly substituted position of unsymmetrical ketones via the thermodynamic enolate formed under equilibrating conditions (in contrast, *acylation* occurs at the less hindered position).

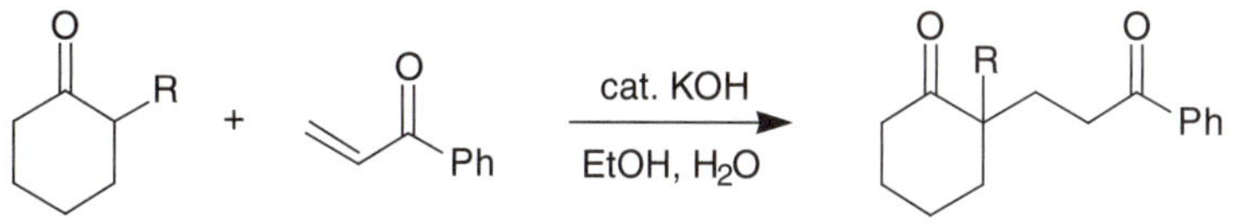

R = CH₃   64%
R = CO₂Et 68%

Acid-Catalyzed Michael Addition

The use of acidic conditions avoids base-catalyzed self-condensation and retro-Michael reactions and tolerates base-sensitive groups, as exemplified by the triflic-acid (CF_3SO_3H)-catalyzed Michael addition shown below.[136]

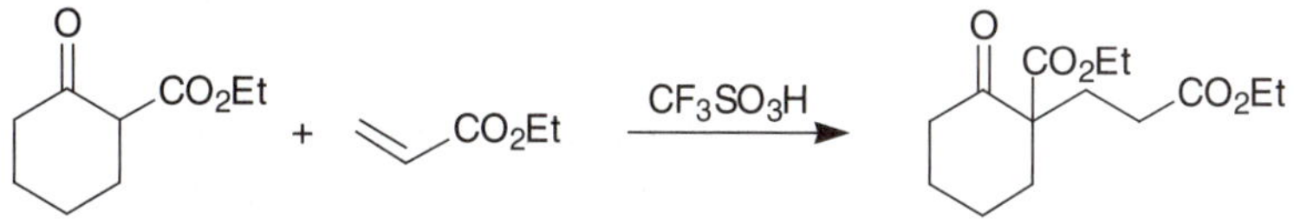

92%

Intramolecular Michael Addition

The intramolecular Michael reaction is a convenient means for the synthesis of small- to medium-size carbocycles.[137]

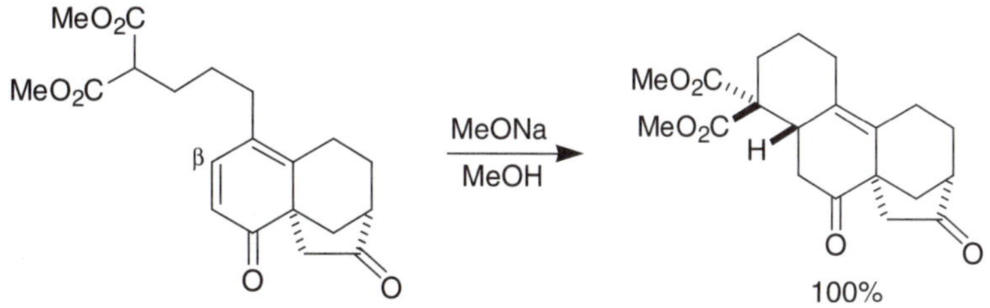

100%

After the conjugate addition, activating groups such as an ester or a sulfone often are removed by decarbalkoxylation or desulfonation methods, respectively, to afford the cyclic products.[138]

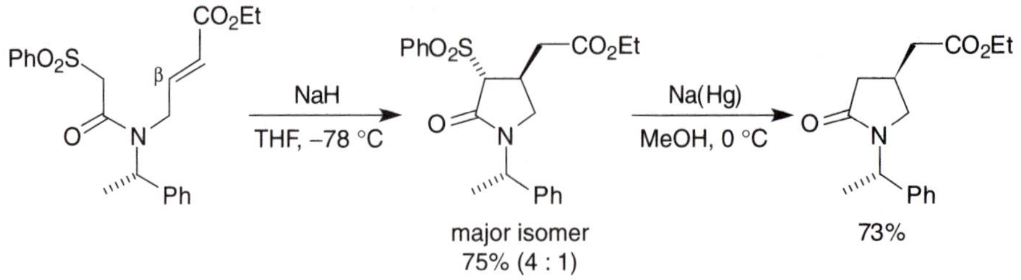

major isomer
75% (4 : 1)

73%

Stereochemistry

The alkylation of a cyclic enolate by a Michael acceptor proceeds *via* antiparallel addition.

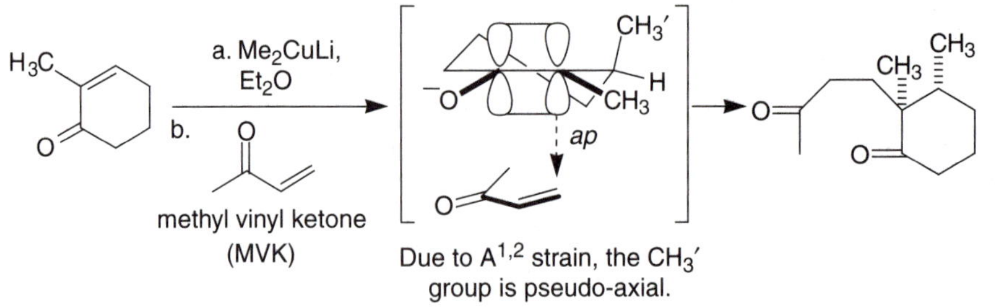

Due to A^{1,2} strain, the CH$_3'$ group is pseudo-axial.

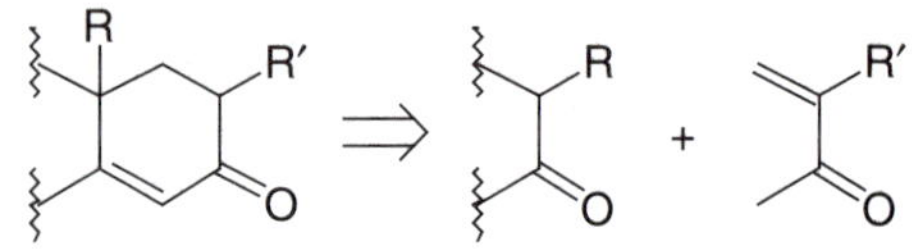

One of the most important reactions for the construction of six-member rings (the Diels-Alder reaction is another) is based on a tandem reaction sequence: a Michael addition reaction followed by an intramolecular aldol-dehydration reaction. This sequence is called the *Robinson annulation* (Sir Robert Robinson, Nobel Prize, 1947).[140]

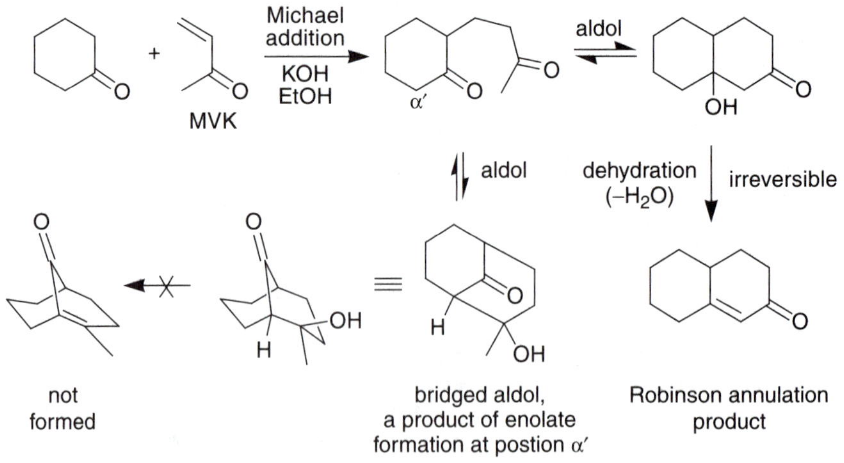

not formed

bridged aldol, a product of enolate formation at postion α'

Robinson annulation product

Note that under basic conditions, dehydration of the bridged aldol product would place a double bond at the bridgehead, forming a highly strained enone.

**Regiochemical Course of the
Robinson Annulation**

In the classical Robinson annulation, the Michael addition occurs at the more substituted carbon via the thermodynamic enolate, except when the "normal product" experiences severe nonbonded interactions.

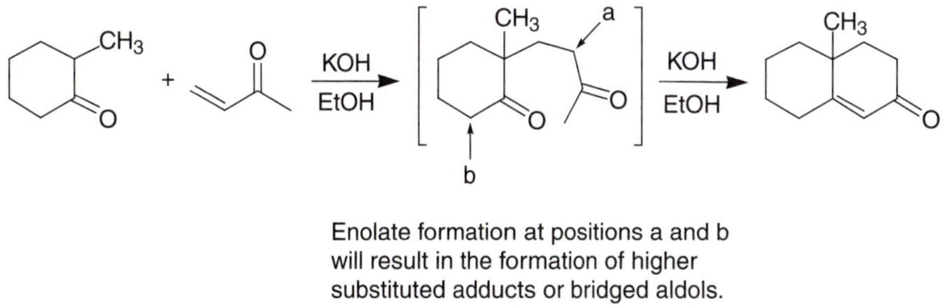

Enolate formation at positions a and b
will result in the formation of higher
substituted adducts or bridged aldols.

To repress polymerization of methyl vinyl ketone (MVK) under basic conditions, the following procedures may be followed:

- NaOEt in EtOH–Et$_2$O at –10 °C; at this temperature the polymerization of MVK is slow and the reaction stops at the β-hydroxy ketone, which is then dehydrated with oxalic acid,[141] or

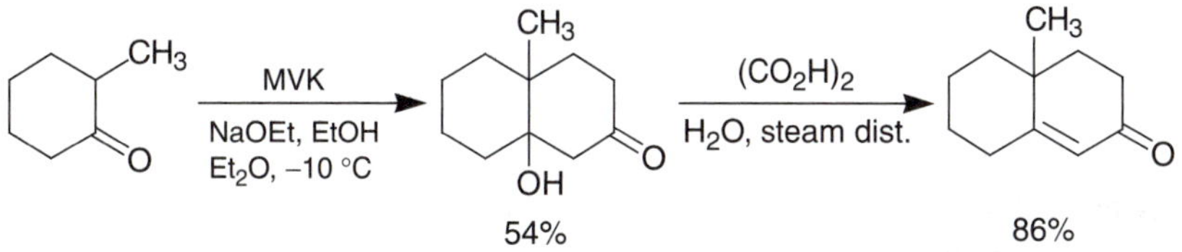

54%   86%

- conduct the reaction in the presence of a catalytic amount of H$_2$SO$_4$,[142] or

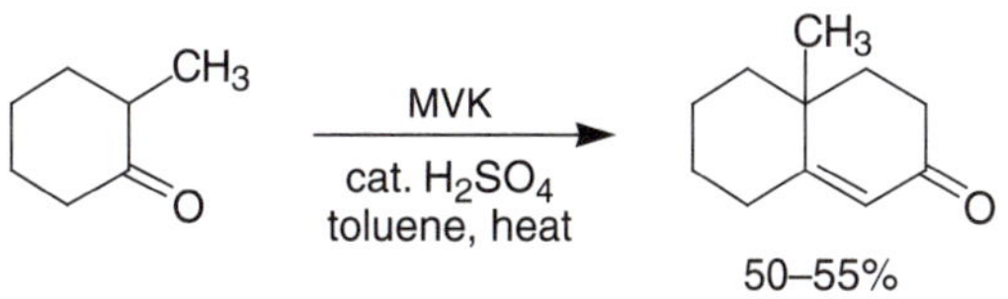

50–55%

- use an α-silylated vinyl ketone under kinetic, non-equilibrating conditions instead of MVK.[143] The α-silyl products are readily desilylated under the basic conditions.

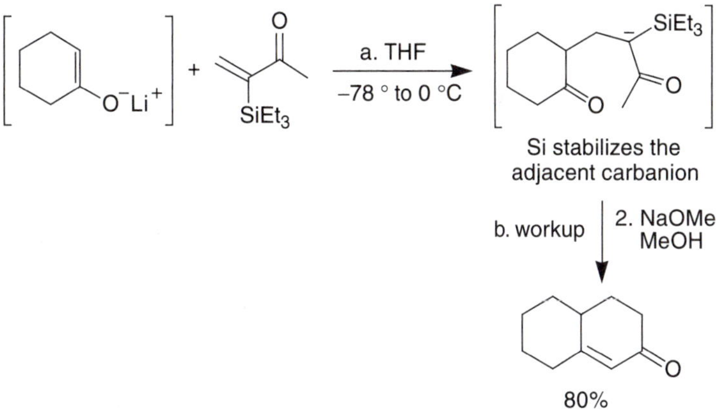

Si stabilizes the
adjacent carbanion

80%

Annulation at the less substituted carbon is accomplished either using enamines or an activating group, as shown below.[144]

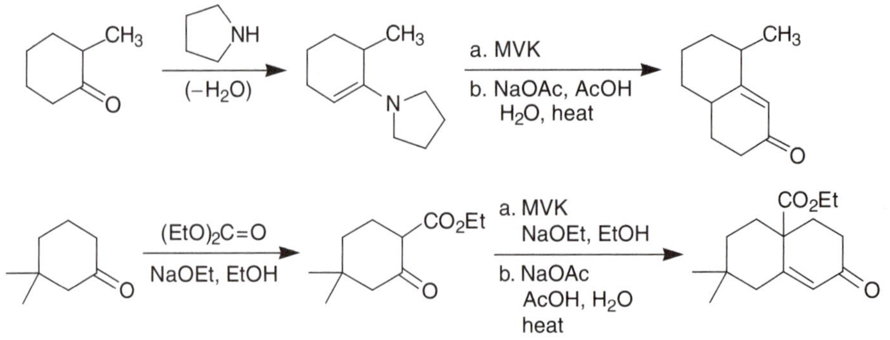

Another approach for annulation at the less substituted carbon is to first form the kinetic enolate using one of several methods (e.g., LDA, Birch reduction of an enone, conjugate addition of an organocuprate to an enone) followed by reaction with MVK and cyclization.

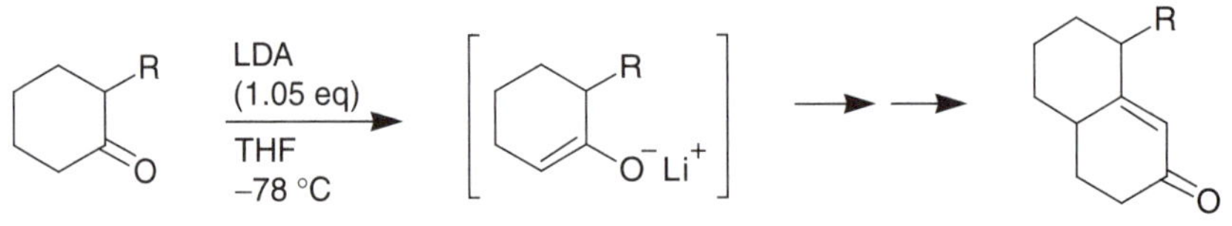

Stereochemical Course of the Robinson Annulation[145]

In the absence of strong steric interactions (synaxial), the Michael addition proceeds via an antiparallel approach of the electrophilic vinyl ketone to the enolate anion.

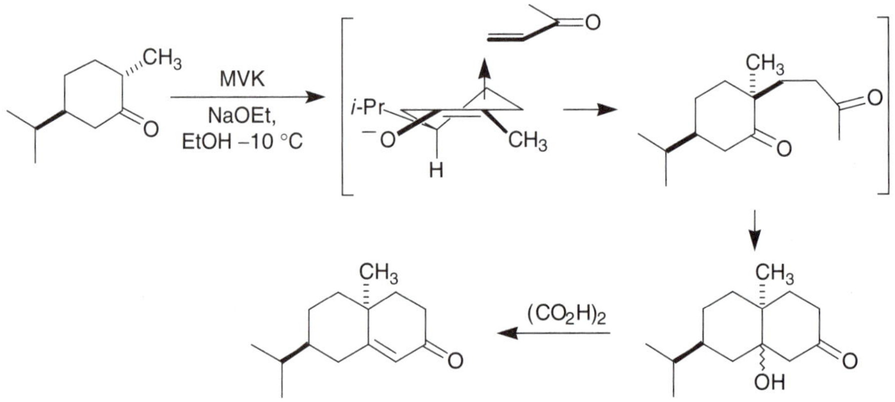

The cyclohexenone system generated by the Robinson annulation can be elaborated into a wide variety of synthetically useful structures via alkylation, conjugate addition, reduction, deconjugation, reductive deoxygenation, dissolving metal reduction, hydrogenation, etc.

An *asymmetric* Robinson annulation is available in which the ketone is reacted with a chiral amine to provide a chiral imine, which is then added to MVK to furnish, after annulation, the chiral product.[146]

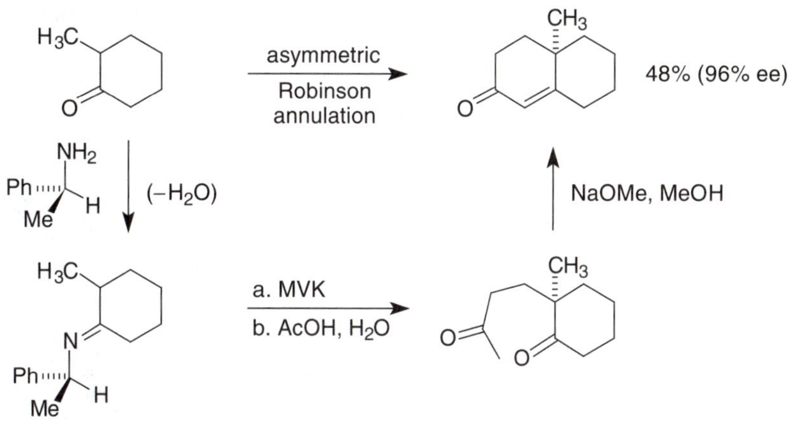
H3C
O
NH2
Ph
H
Me
(–H2O)
asymmetric
Robinson
annulation
CH3
O
48% (96% ee)
NaOMe, MeOH
H3C
N
Ph
H
Me
a. MVK
b. AcOH, H2O
CH3
O
O

PROBLEMS

1. **Reagents**. Give the structures of the major product(s) expected after each step of the following reactions. Be sure to indicate product stereochemistry where applicable.

a.

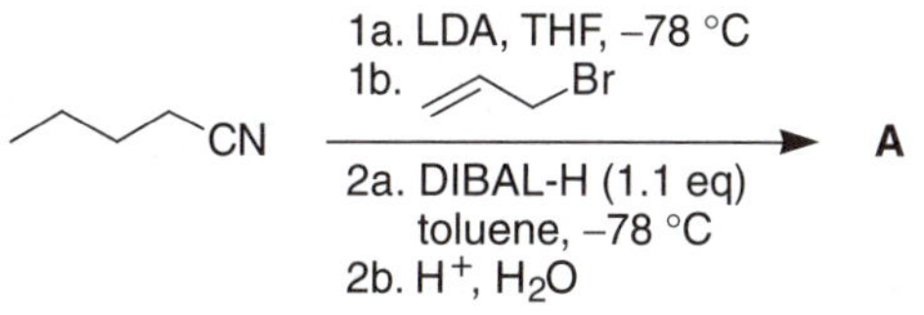
CN
1a. LDA, THF, –78 °C
1b. Br
2a. DIBAL-H (1.1 eq)
toluene, –78 °C
2b. H+, H2O
A

b.

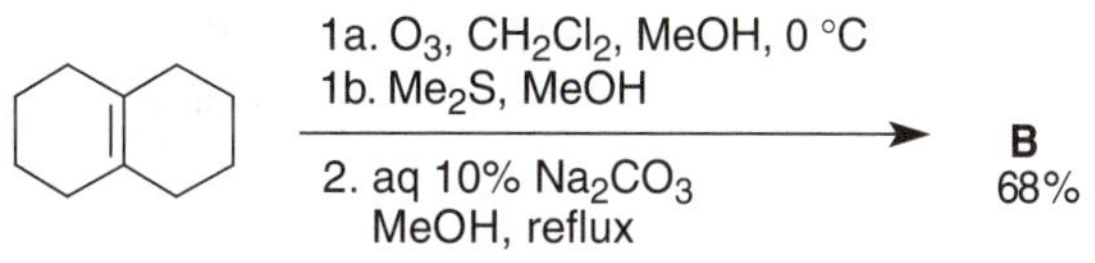
1a. O3, CH2Cl2, MeOH, 0 °C
1b. Me2S, MeOH
2. aq 10% Na2CO3
MeOH, reflux
B
68%

c.

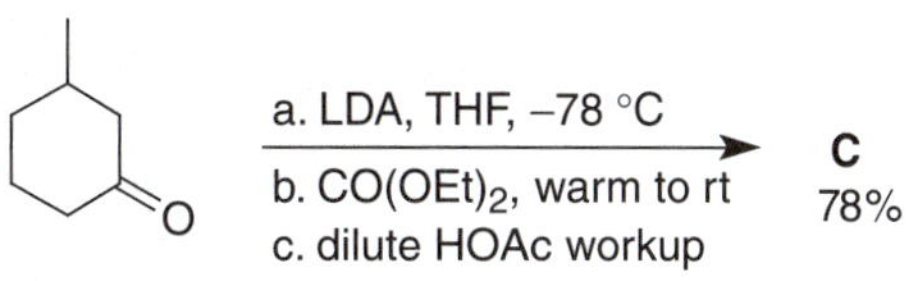
O
a. LDA, THF, –78 °C
b. CO(OEt)2, warm to rt
c. dilute HOAc workup
C
78%

d.

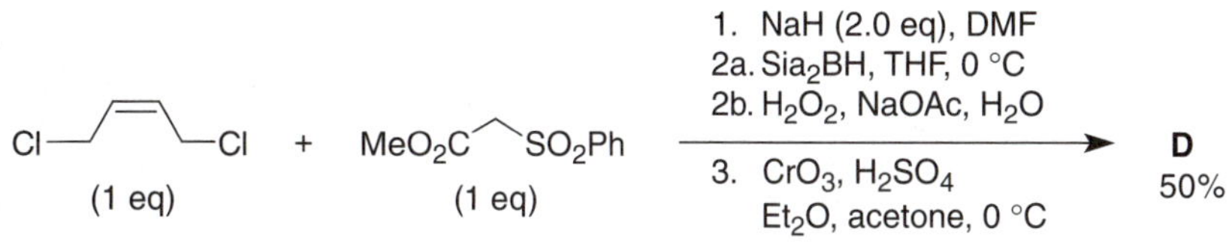
Cl
Cl
(1 eq)
+
MeO2C
SO2Ph
(1 eq)
1. NaH (2.0 eq), DMF
2a. Sia2BH, THF, 0 °C
2b. H2O2, NaOAc, H2O
3. CrO3, H2SO4
Et2O, acetone, 0 °C
D
50%

e.

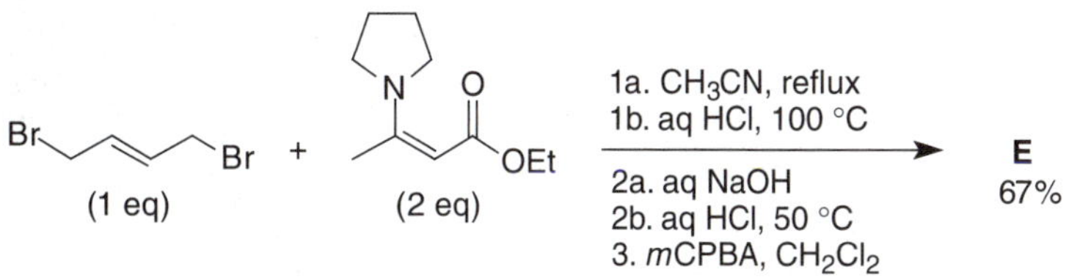
Br
Br
(1 eq)
+
N
O
OEt
(2 eq)
1a. CH3CN, reflux
1b. aq HCl, 100 °C
2a. aq NaOH
2b. aq HCl, 50 °C
3. mCPBA, CH2Cl2
E
67%

f.

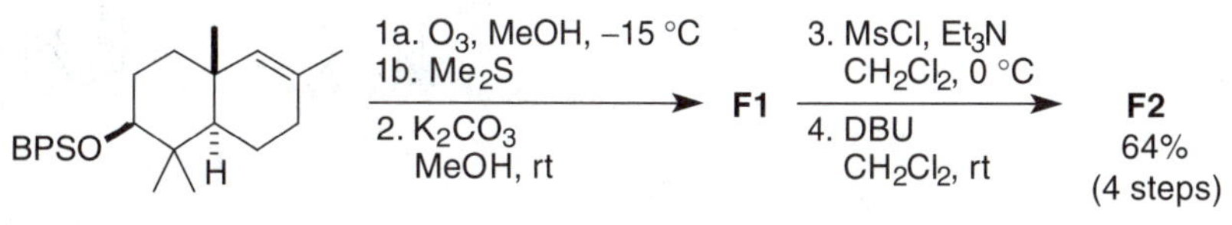
BPSO
H
1a. O3, MeOH, –15 °C
1b. Me2S
2. K2CO3
MeOH, rt
F1
3. MsCl, Et3N
CH2Cl2, 0 °C
4. DBU
CH2Cl2, rt
F2
64%
(4 steps)

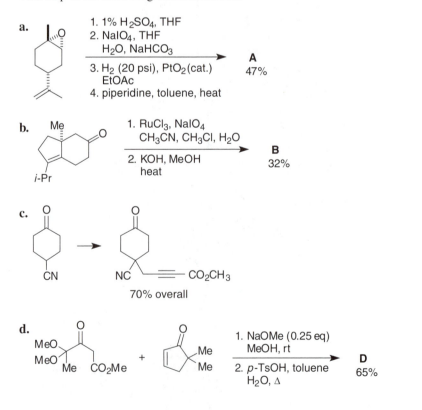

g. ∿∿CHO

1. *t*-BuNH$_2$, toluene
2a. LDA (1.05 eq), THF, −78 °C
2b. *n*-BuBr
2c. aq HCl workup

→ **G**

h. CbzHN—(CH)$_5$—C(O)—OH

1a. (imid)$_2$C=O, THF, 0 °C
1b. (MeO$_2$CCH$_2$CO$_2$)$_2$Mg, warm to rt
1c. aq HCl workup
2a. NaBH$_4$, MeOH
2b. aq NaHCO$_3$ workup

→ **H**
62%

i.

1. 30% aq H$_2$C=O
   Me$_2$NH·HCl, dioxane
2. MeI (excess), MeOH, rt
3. DMF, 80 °C

→ **I**
67%

*j.

1. Na (>2 eq), NH$_3$ (l)
   t-BuOH, Et$_2$O
2. (Ph$_3$P)$_3$RhCl (cat.)
   H$_2$, toluene
3a. (CO)$_2$Cl$_2$
3b. EtOH

→ **J1**

4. OsO$_4$ (cat.), NaIO$_4$ (excess)
   H$_2$O, dioxane
5. Et$_3$N, heat

→ **J2**

*k.

+ TBSO ... CHO

1. MgCl$_2$, Et$_3$N, TMSCl
   EtOAc, rt
2. TESCl, imidazole
   CH$_2$Cl$_2$, rt
3. LiBH$_4$, MeOH, THF

→ **K**

2. **Selectivity.** Show the product(s) obtained or appropriate reagent(s) to be used for each step of the following transformations.

a.

1. 1% H$_2$SO$_4$, THF
2. NaIO$_4$, THF
   H$_2$O, NaHCO$_3$
3. H$_2$ (20 psi), PtO$_2$(cat.)
   EtOAc
4. piperidine, toluene, heat

→ **A**
47%

b.

1. RuCl$_3$, NaIO$_4$
   CH$_3$CN, CH$_3$Cl, H$_2$O
2. KOH, MeOH
   heat

→ **B**
32%

c.

→

70% overall

d.

+

1. NaOMe (0.25 eq)
   MeOH, rt
2. *p*-TsOH, toluene
   H$_2$O, Δ

→ **D**
65%

*e.

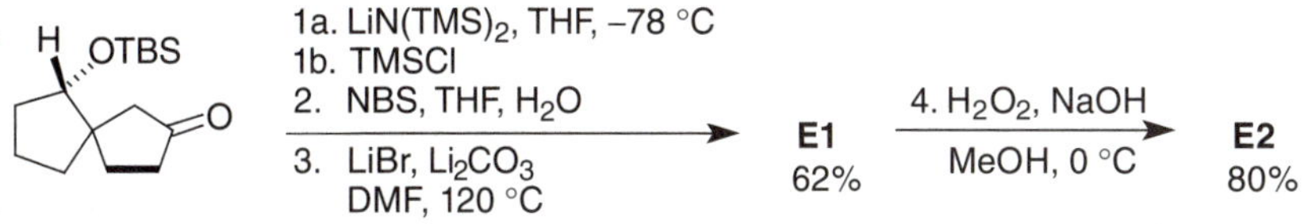
H
OTBS
O
1a. LiN(TMS)2, THF, –78 °C
1b. TMSCl
2. NBS, THF, H2O
3. LiBr, Li2CO3
DMF, 120 °C
E1
62%
4. H2O2, NaOH
MeOH, 0 °C
E2
80%

3. **Stereochemistry.** Give the structure and predict the stereochemistry of the major product formed in each step for each of the following reactions. Give an explanation for your choice.

a.

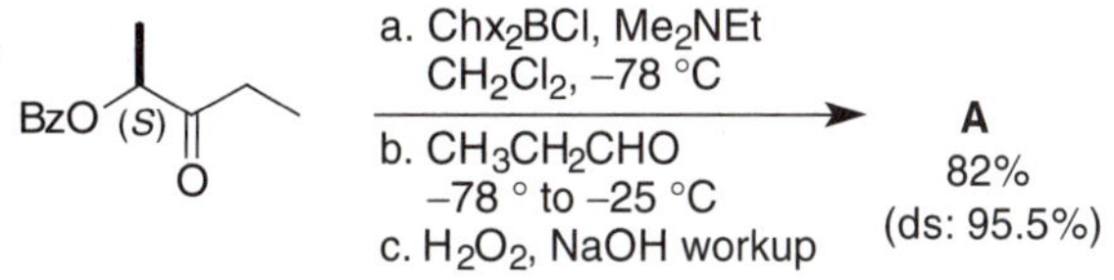
BzO
(S)
O
a. Chx2BCl, Me2NEt
CH2Cl2, –78 °C
b. CH3CH2CHO
–78 ° to –25 °C
c. H2O2, NaOH workup
A
82%
(ds: 95.5%)

b.

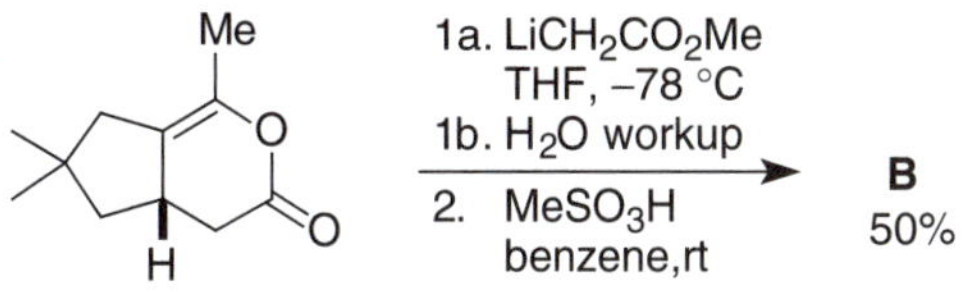
Me
O
O
H
1a. LiCH2CO2Me
THF, –78 °C
1b. H2O workup
2. MeSO3H
benzene,rt
B
50%

c.

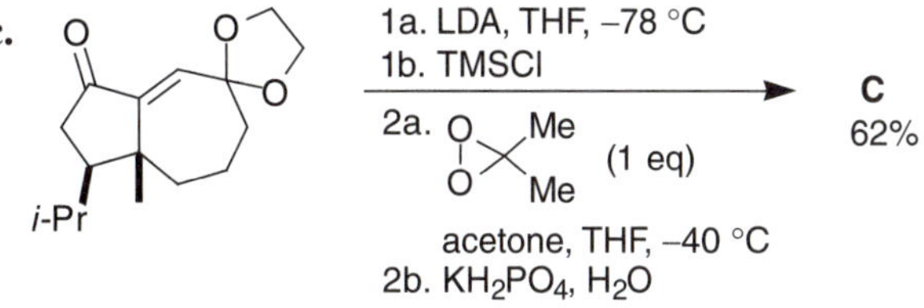
O
O
O
i-Pr
1a. LDA, THF, –78 °C
1b. TMSCl
2a.
Me
Me
(1 eq)
acetone, THF, –40 °C
2b. KH2PO4, H2O
C
62%

d.

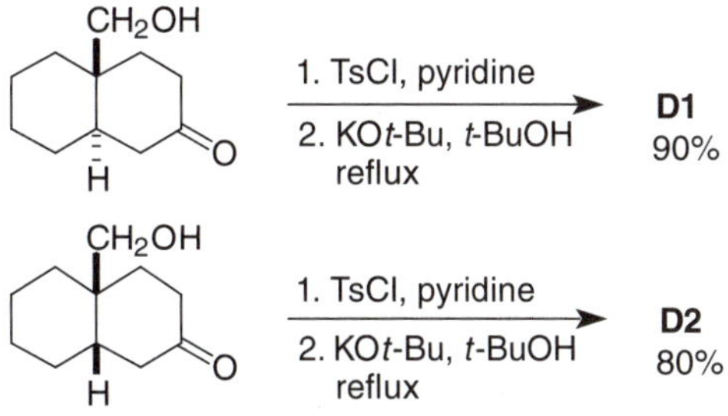
CH2OH
H
O
1. TsCl, pyridine
2. KOt-Bu, t-BuOH
reflux
D1
90%
CH2OH
H
O
1. TsCl, pyridine
2. KOt-Bu, t-BuOH
reflux
D2
80%

*e.

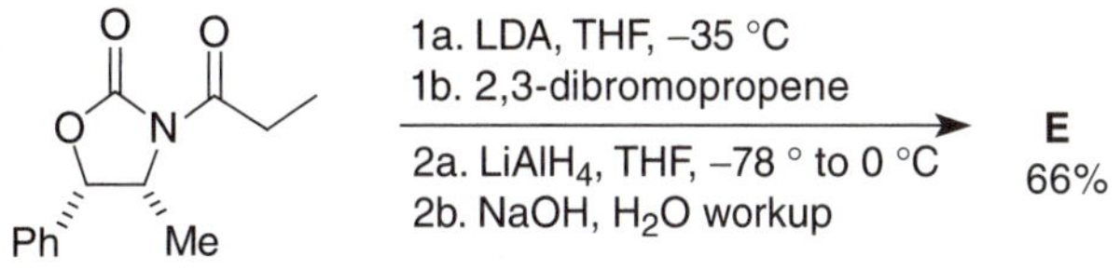
O
O
O
N
Ph
Me
1a. LDA, THF, –35 °C
1b. 2,3-dibromopropene
2a. LiAlH4, THF, –78 ° to 0 °C
2b. NaOH, H2O workup
E
66%

*f.

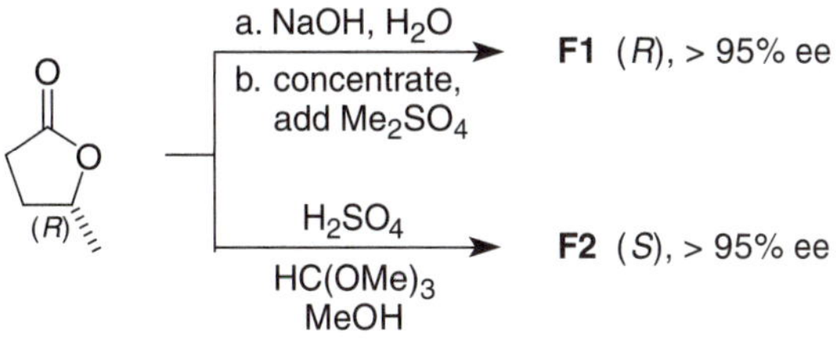
O
O
(R)
a. NaOH, H2O
b. concentrate,
add Me2SO4
F1 (R), > 95% ee
H2SO4
HC(OMe)3
MeOH
F2 (S), > 95% ee

4. **Reactivity**. Propose mechanisms for each of the following transformations.

a.

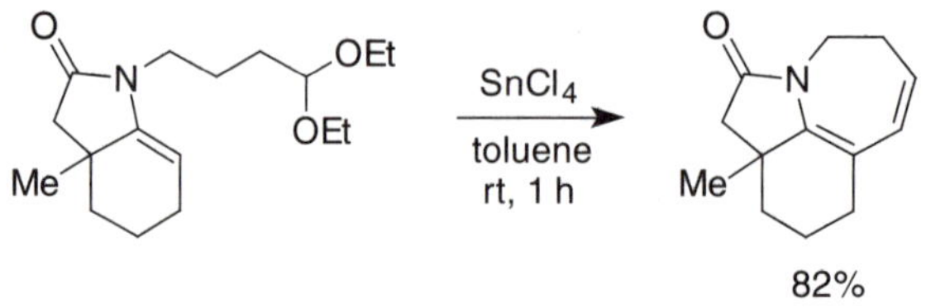

82%

b.

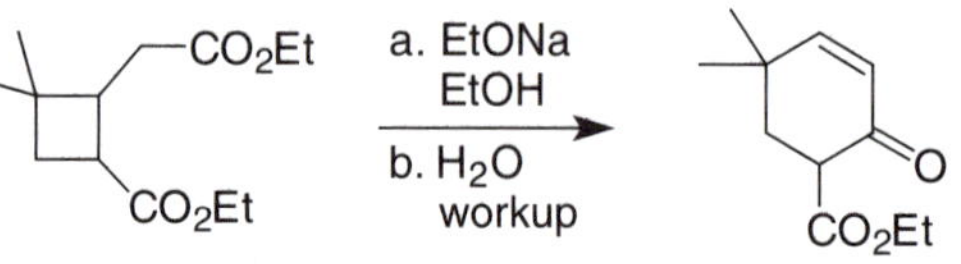

c.

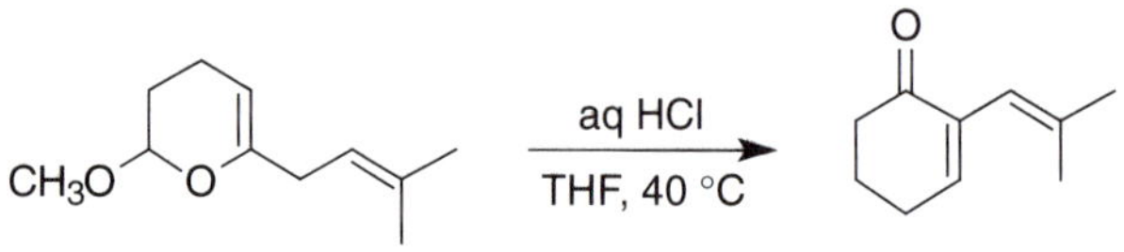

d.

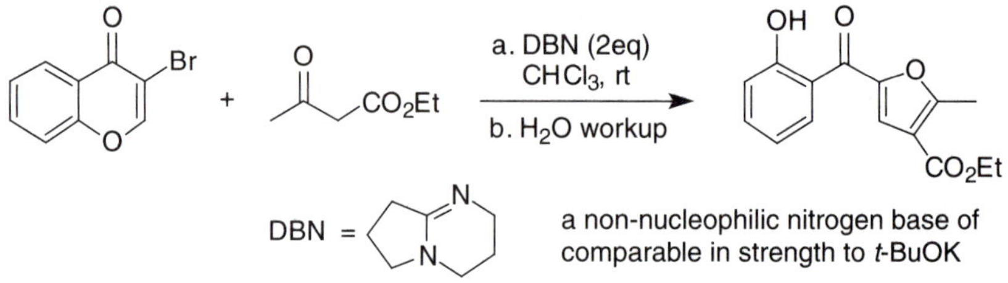

DBN =

a non-nucleophilic nitrogen base of comparable in strength to *t*-BuOK

*e.

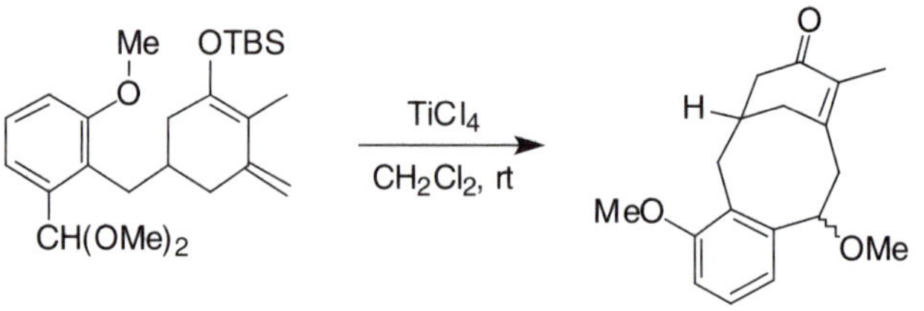

*f.

1. MeI, acetone, rt

2. *t*-BuNH₂, aq NaOH
   H₂C=CHCO₂H, 60 °C
   (acrylic acid)

80%

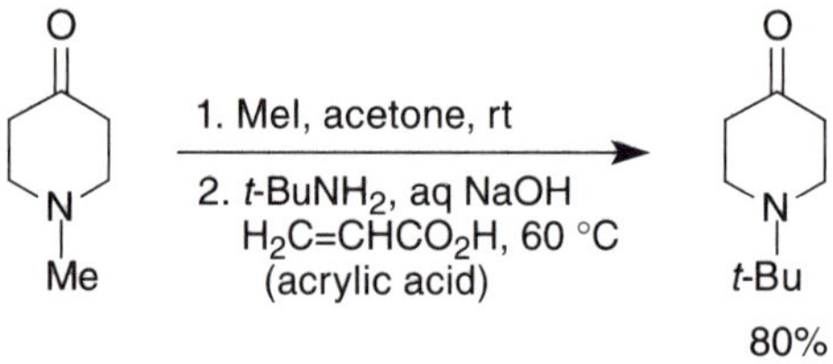

5. **Synthesis.** Supply the reagents required to accomplish each of the following syntheses. Give the structures of the intermediates obtained after each step and show their relative stereochemistries where applicable.

a.

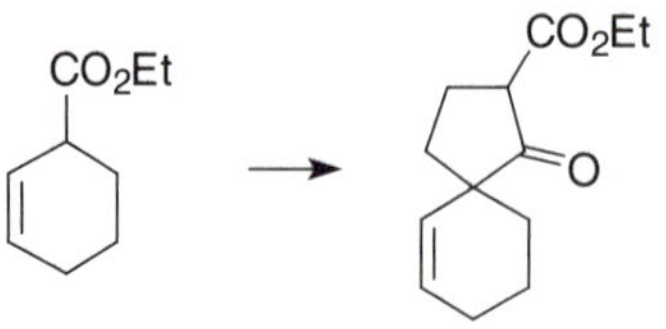

b.

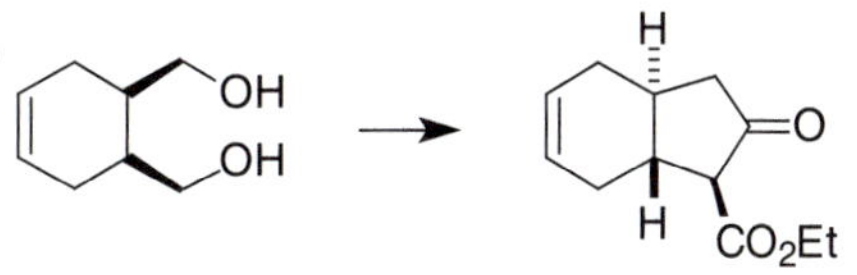
OH
OH
H
O
H
CO_2Et

c.

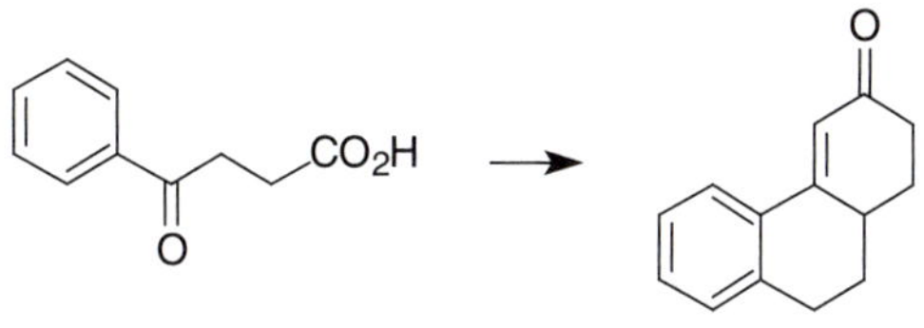
CO_2H
O
O

d.

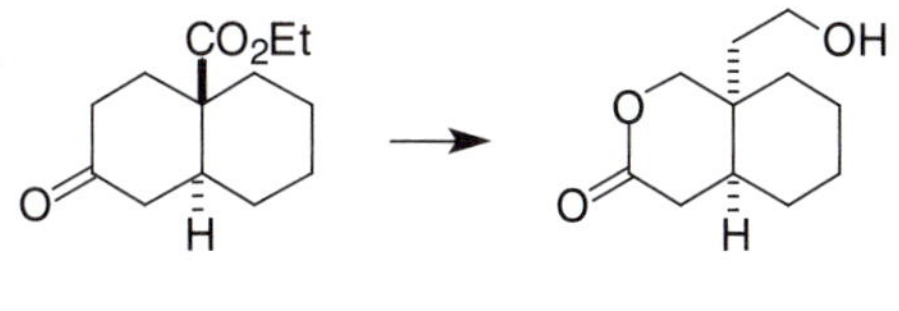
CO_2Et
O
H
OH
O
O
H

e.

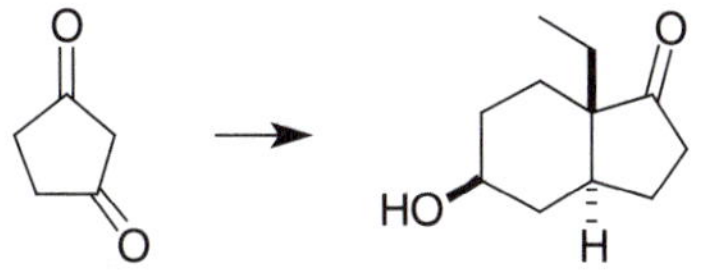
O
O
O
HO
H

f.

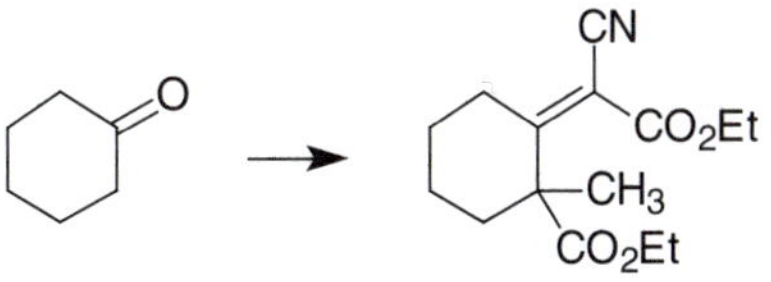
O
CN
CO_2Et
CH_3
CO_2Et

g.

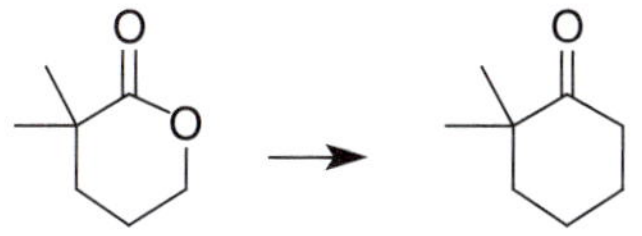
O
O
O

h.

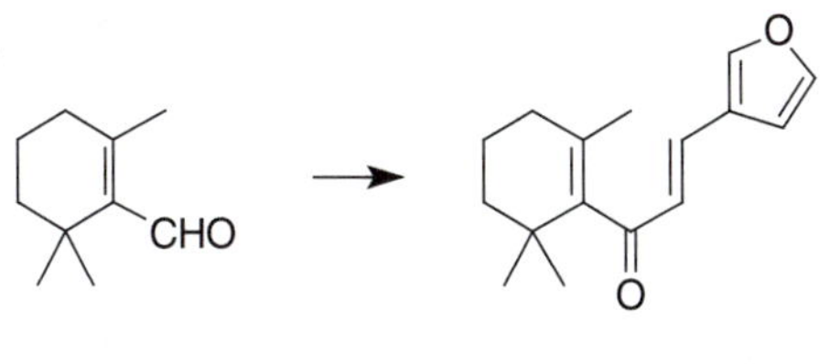
CHO
O
O

i.

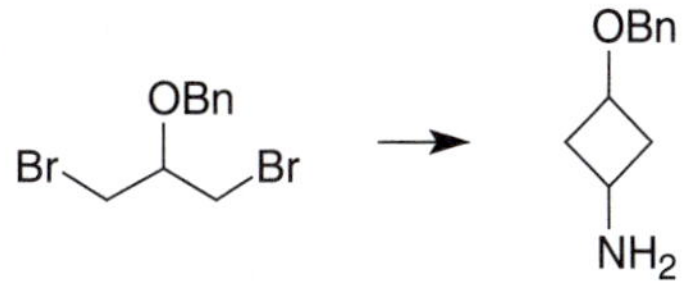
OBn
Br
Br
OBn
NH_2

***j.**

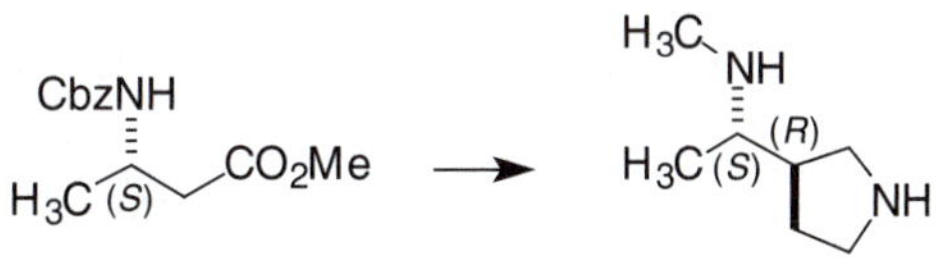
CbzNH
H_3C
(S)
CO_2Me
H_3C
NH
(R)
H_3C
(S)
NH

6. **Retrosynthetic Analysis.** Outline a synthetic scheme for preparing each of the following target molecules. Show (i) your retrosynthetic analysis, and (ii) all reagents and reaction conditions required to transform a commercially available starting material into the target molecule.

a.

b.

REFERENCES

1. (a) House, H. O. *Modern Synthetic Reactions*, 2nd ed., Benjamin: Menlo Park, CA, 1972. (b) Stowell, J. C. *Carbanions in Organic Synthesis*, Wiley-Interscience: New York, 1979. (c) Heathcock, C. H. In *Modern Synthetic Methods*, Scheffold, R., Ed., VCH: Weinheim, Germany, 1992, p. 1. (d) Perlmutter, P. *Conjugate Addition Reactions in Organic Synthesis*, Pergamon Press: Oxford, UK, 1992. (e) *Modern Carbonyl Chemistry*, Otera, J., Ed., Wiley-VCH: Weinheim, Germany, 2000. (f) Smith, M. B. *Organic Synthesis*, 2nd ed., McGraw-Hill: Boston, 2002, p. 714. (g) Buncel, E., Dust, J. M. *Carbanion Chemistry: Structures and Mechanisms*, Oxford University Press: Oxford, UK, 2003.

2. For more extensive tabulations of pK_a values, see (a) Smith, M. B., March, J. *March's Advanced Organic Chemistry*, 5th ed., Wiley: New York, 2001, pp. 329–331. (b) Negishi, E. *Organometallics in Organic Synthesis*, Wiley: New York, 1980, pp. 506–510.

3. (a) Cope, A. C., Holmes, H. L., House, H. O. *Org. React.* **1957**, *9*, 107. (b) Freeman, F. *Chem. Rev.* **1969**, *69*, 591.

4. Marvel, C. S. *Org. Synth.*, *Col. Vol. III* **1955**, 495.

5. Mariella, R. P., Raube, R. *Org. Synth.*, *Col. Vol. 4* **1963**, 288.

6. Toussaint, O., Capdevielle, P., Maumy, M. *Synthesis* **1986**, 1029.

7. (a) Jones, G. *Org. React.* **1967**, *15*, 204. (b) Ragoussis, N. *Tetrahedron Lett.* **1987**, *28*, 93.

8. (a) Andersen, N. H., Golec, F. A., Jr. *Tetrahedron Lett.* **1977**, 3783. (b) Nicolaou, K. C., van Delft, F., Ohshima, T., Vourloumis, D., Xu, J., Hosokawa, S., Pfefferkorn, J., Kim, S., Li, T. *Angew. Chem., Int. Ed.* **1997**, *36*, 2520.

9. McNab, H. *Chem. Soc. Rev.* **1978**, *7*, 345.

10. (a) Junek, H., Ziegler, E., Herzog, U., Kroboth, H. *Synthesis* **1976**, 332. (b) Brooks, D. W., Castro de Lee, N., Peevey, R. *Tetrahedron Lett.* **1984**, *25*, 4623.

11. Hauser, C. R., Hudson, B. E., Jr. *Org. React.* **1942**, *1*, 266.

12. Schaefer, J. P., Bloomfield, J. J. *Org. React.* **1967**, *15*, 1.

13. Dieckmann, W. *Ber.* **1894**, *27*, 102.

14. Stevens, R. V., Gaeta, F. C. A. *J. Am Chem. Soc.* **1977**, *99*, 6105.

15. Davis, B. R., Garratt, P. J. In *Comprehensive Organic Synthesis*, Trost, B. M., Fleming, I., Eds., Pergamon Press: Oxford, UK. 1991, Vol. 2, p. 795.

16. (a) McElvain, S. M. *Org. React.* **1948**, *4*, 256. (b) Bloomfield, J. J., Owsley, D. C., Nelke, J. M. *Org. React.* **1976**, *23*, 259. (c) Finley, K. T. *Chem. Rev.* **1964**, *64*, 573.

17. Ronsheim, M. D., Zercher, C. K. *J. Org. Chem.* **2003**, *68*, 1878.

18. Krapcho, A. P., Diamanti, J., Cayen, C., Bingham, R. *Org. Synth., Col. Vol. V* **1973**, 198.

19. Mander, L. N., Sethi, S. P. *Tetrahedron Lett.* **1983**, *24*, 5425.

20. House, H. O. *Modern Synthetic Reactions*, 2nd ed., Benjamin: Menlo Park, CA, 1972, pp. 735–760.

21. Hutchinson, C. R., Nakane, M., Gollman, H., Knutson, P. L. *Org. Synth., Col. Vol. VII* **1990**, 323.

22. Brooks, D. W., Lu, L. D.-L., Masamune, S. *Angew. Chem, Int. Ed.* **1979**, *18*, 72.

23. Staab, H. A. *Angew. Chem., Int. Ed.* **1962**, *1*, 351.

24. (a) Durham, T. B., Miller, M. J. *J. Org. Chem.* **2003**, *68*, 35. (b) Coulon, E., Cristina, M., De Andrade, C., Ratovelomanana-Vidal, V., Genet, J.-P. *Tetrahedron Lett.* **1998**, *39*, 6467.

25. Duplantier, A. J., Masamune, S. *J. Am. Chem. Soc.* **1990**, *112*, 7079.

26. For a text on dianions, see Thompson, C. M., *Dianions in Organic Synthesis*, CRC Press: Boca Raton, FL, 1994.

27. Huckin, S. N., Weiler, L. *J. Am. Chem. Soc.* **1974**, *96*, 1082.

28. Schlessinger, R. H., Wood, J. L., Poss, A. J., Nugent, R. A., Parsons, W. H. *J. Org. Chem.* **1983**, *48*, 1146.

29. Krapcho, A. P., Weimaster, J. F., Eldridge, J. M., Jahngen, E. G. E., Jr., Lovey, A. J., Stephens, W. P. *J. Org. Chem.* **1978**, *43*, 138.

30. Magatti, C. V., Kaminski, J. J., Rothberg, I. *J. Org. Chem.* **1991**, *56*, 3102.

31. (a) Wehrli, P. A., Chu, V. *J. Org. Chem.* **1973**, *38*, 3436. (b) Ho, T.-L. *Synth. Commun.* **1981**, *11*, 7.

32. Taber, D. F., Amedio, J. C., Jr., Gulino, F. *J. Org. Chem.* **1989**, *54*, 3474.

33. Johnson, W. S., Lunn, W. H., Fitzi, K. *J. Am. Chem. Soc.* **1964**, *86*, 1972.

34. Piers, E., Grierson, J. R. *J. Org. Chem.* **1977**, *42*, 3755.

35. Choudhary, A., Baumstark, A. L. *Synthesis* **1989**, 688.

36. Petragnani, N., Yonashiro, M. *Synthesis* **1982**, 521.

37. (a) Cregge, R. J., Herrmann, J. L., Lee, C. S., Schlessinger, R. H. *Tetrahedron Lett.* **1973**, 2425. (b) Petraganami, N., Brocksom, T. J., Ferraz, H. M. C., Constantino, M. G. *Synthesis* **1977**, 112.

38. Rathke, M. W., Lindert, A. *J. Am. Chem. Soc.* **1971**, *93*, 2318.

39. Seebach, D., Beck, A. K., Studer, A. In *Modern Synthetic Methods*, Ernst, B., Leumann, C., Eds., VCH: Weinheim, Germany, 1995, p. 1.

40. Arseniyadis, S., Kyler, K. S., Watt, D. S. *Org. React.* **1984**, *31*, 1.

41. Pfeffer, P. E., Silbert, L. S., Chirinko, J. M., Jr. *J. Org. Chem.* **1972**, *37*, 451.

42. Lucht, B. L., Collum, D. B. *Acc. Chem. Res.* **1999**, *32*, 1035.

43. Sung, M. J., Lee, H. I., Lee, H. B., Cha, J. K. *J. Org. Chem.* **2003**, *68*, 2205.

44. Albarella, J. P. *J. Org. Chem.* **1977**, *42*, 2009.

45. (a) d'Angelo, J. *Tetrahedron* **1976**, *32*, 2979. (b) Jackman, L. M., Lange, B. C. *Tetrahedron* **1977**, *33*, 2737. (c) Pollack, R. M. *Tetrahedron* **1989**, *45*, 4913. (d) Arya, P., Qin, H. *Tetrahedron* **2000**, *56*, 917.

46. (a) Williard, P. G., Salvino, J. M. *J. Org. Chem.* **1993**, *58*, 1. (b) Pratt, L. M., Newman, A., Cyr, J. S., Johnson, H., Miles, B., Lattier, A., Austin, E., Henderson, S., Hershey, B., Lin, M., Balamraju, Y., Sammonds, L., Cheramie, J., Karnes, J., Hymel, E., Woodford, B., Carter, C. *J. Org. Chem.* **2003**, *68*, 6387.

47. (a) Negishi, E., Chatterjee, S. *Tetrahedron Lett.* **1983**, *24*, 1341. (b) Paquette, L. A., Roberts, R. A., Drtina, G. J. *J. Am. Chem. Soc.* **1984**, *106*, 6690.

48. Orban, J., Turner, J. V., Twitchin, B. *Tetrahedron Lett.* **1984**, *25*, 5099.

49. (a) Grieco, P. A., Ferrino, S., Oguri, T. *J. Org. Chem.* **1979**, *44*, 2593. (b) Stork, G., Danheiser, R. L. *J. Org. Chem.* **1973**, *38*, 1775.

50. Lee, R. A., McAndrews, C., Patel, K. M., Reusch, W. *Tetrahedron Lett.* **1973**, 965.

51. For the preparation of Ph$_3$CLi, see Greene, A. E., Muller, J.-C., Ourisson, G. *J. Org. Chem.* **1974**, *39*, 186.

52. Brownbridge, P. *Synthesis* **1983**, 1.

53. (a) Noyori, R., Nishida, I., Sakata, J. *Tetrahedron Lett.* **1980**, *21*, 2085. (b) Kigoshi, H., Imamura, Y., Mizuta, K., Niwa, H., Yamada, K. *J. Am. Chem. Soc.* **1993**, *115*, 3056.

54. (a) Kleschick, W. A., Buse, C. T., Heathcock, C. H. *J. Am. Chem. Soc.* **1977**, *99*, 247. (b) Harmon, K. M., Southworth, B. A., Wilson, K. E., Keefer, P. K. *J. Org. Chem.* **1993**, *58*, 7294.

55. Smith, H. A., Huff, B. J. L., Powers, W. J., III, Caine, D. *J. Org. Chem.* **1967**, *32*, 2851.

56. Danishefsky, S., Chackalamannil, S., Harrison, P., Silvestri, M., Cole, P. *J. Am. Chem. Soc.* **1985**, *107*, 2474.

57. Fortunato, J. M., Ganem, B. *J. Org. Chem.* **1976**, *41*, 2194.

58. (a) Whitesell, J. K., Whitesell, M. A. *Synthesis* **1983**, 517. (b) Seebach, D. *Angew. Chem., Int. Ed.* **1988**, *27*, 1624.

59. Rathke, M. W., Lindert, A. *Synth. Comm.* **1978**, *8*, 9.

60. (a) Negishi, E., Chatterjee, S. *Tetrahedron Lett.* **1983**, *24*, 1341. (b) Paquette, L. A., Roberts, R. A., Drtina, G. J. *J. Am. Chem. Soc.* **1984**, *106*, 6690.

61. Tenud, L., Farooq, S., Seibl, J., Eschenmoser, A. *Helv. Chim. Acta* **1970**, *53*, 2059.

62. (a) Baldwin, J. E. *J. Chem. Soc., Chem. Commun.* **1976**, *18*, 734. (b) Baldwin, J. E., Lusch, M. J. *Tetrahedron* **1982**, *38*, 2939. (c) Deslongchamps, P. *Stereoelectronic Effects in Organic Chemistry*, Pergamon Press: Oxford, UK, 1983. (d) For radical cyclizations, see Chatgilialoglu, C., Ferreri, C., Guerra, M., Timokhin, V., Froudakis, G., Gimisis, T. *J. Am. Chem. Soc.* **2002**, *124*, 10765.

63. (a) Piers, E., Zbozny, M., Wigfield, D. C. *Can. J. Chem.* **1979**, *57*, 1064. (b) Hutchinson, D. K., Fuchs, P. L. *J. Am. Chem. Soc.* **1987**, *109*, 4755.

64. Pollack, R. M. *Tetrahedron* **1989**, *45*, 4913.

65. (a) Zimmerman, H. E., Linder, L. W. *J. Org. Chem.* **1985**, *50*, 1637. (b) Paquette, L. A., Belmont, D. T., Hus, Y.-L. *J. Org. Chem.* **1985**, *50*, 4667.

66. Matthews, R. S., Hyer, P. K., Folkers, E. A. *J. Chem. Soc. D* **1970**, 38.

67. Seebach, D., Naef, R., Calderari, G. *Tetrahedron* **1984**, *40*, 1313.

68. (a) Stork, G., Dowd, S. R. *J. Am. Chem. Soc.* **1963**, *85*, 2178. (b) Kieczkowaski, G. R., Schlessinger, R. H., Sulsky, R. B.

Tetrahedron Lett. **1976**, 597. (c) Wender, P. A., Schaus, J. M. *J. Org. Chem.* **1978**, *43*, 782, and references cited therein.

69. LeBel, N. A., Balasubramanian, N. *J. Am. Chem. Soc.* **1989**, *111*, 3363.

70. Galiano-Roth, A. S., Collum, D. D. *J. Am. Chem. Soc.* **1989**, *111*, 6772.

71. (a) Enders, D., Eichenauer, H. *Chem. Ber.* **1979**, 112, 2933. (b) Enders, D., Eichenauer, H., Baus, U., Schubert, H., Kremer, K. A. H. *Tetrahedron* **1984**, *40*, 1345. (c) Enders, D., Fey, P., Kipphardt, H. *Org. Synth.* **1987**, *65,* 173. (d) Tomioka, K. *Synthesis* **1990**, 541. (e) Enders, D., Wortmann, L., Peters, R. *Acc. Chem. Res.* **2000**, *33*, 157.

72. (a) Enders, D., Eichenauer, H. *Angew. Chem., Int. Ed.* **1976**, *15*, 549. (b) Enders, D., Eichenauer, H. *Angew. Chem., Int. Ed.* **1979**, *18*, 397. (c) Enders, D., Eichenauer, H., Baus, U., Schubert, H., Kremer, K. A. M. *Tetrahedron* **1984**, *40*, 1345. (d) Enders, D., Kipphardt, H., Fey, P. *Org. Synth.* **1987**, *65*, 183.

73. Enders, D., Gatzweiler, W., Jegelka, U. *Synthesis* **1991**, 1137.

74. (a) Stork, G., Terrell, R., Szmuszkovicz, J. *J. Am. Chem. Soc.* **1954**, *76*, 2029. (b) Cook, A. G. *Enamines: Synthesis, Structure and Reactions*, Dekker: New York, 1988. (c) Hickmott, P. W. *Tetrahedron* **1982**, *38*, 1975.

75. Stork, G., Brizzolara, A., Landesman, H., Szmuszkovicz, J., Terrell, R. *J. Am. Chem. Soc.* **1963**, *85*, 207.

76. (a) Whitesell, J. K. *Acc. Chem. Res.* **1985**, *18*, 280. (b) Whitesell, J. K. *Chem. Rev.* **1989**, *89*, 1581.

77. (a) Schlessinger, R. H., Iwanowicz, E. J. *Tetrahedron Lett.* **1987**, *28*, 2083. (b) Short, R. P., Kennedy, R. M., Masamune, S. *J. Org. Chem.* **1989**, *54*, 1755.

78. (a) Whitesell, J. K., Felman, S. W. *J. Org. Chem.* **1980**, *45*, 755. (b) Schlessinger, R. H., Iwanowicz, E. J., Springer, J. P. *J. Org. Chem.* **1986**, *51*, 3070.

79. (a) For the first reported example of ketone self-condensation, see Kane, R. *J. Prakt. Chem.* **1838**, *15*, 129. (b) Nielsen, A. T., Houlihan, W. J. *Org. React.* **1968**, *16*, 1. (c) Mukaiyama, T. *Org. React.* **1982**, *28*, 203. (d) Braun, M. *Angew. Chem., Int. Ed.* **1987**, *26*, 24. (e) Heathcock, C. H. In *Comprehensive Organic Synthesis*, Trost, B. M., Fleming, I., Eds., Pergamon Press: Oxford, UK, 1991, Vol. 2, p. 181. (f) Kim, B. M., Williams, S. F., Masamune, S. In *Comprehensive Organic Synthesis*, Trost, B. M., Fleming, I., Eds., Pergamon Press: Oxford, UK, 1991, Vol. 2, p. 239. (g) Carreira, E. M. In *Modern Carbonyl Chemistry*, Otera, J., Ed., Wiley-VCH: Weinheim, Germany, 2000, p. 227.

80. Guthrie, J. P., Guo, J. *J. Am. Chem. Soc.* **1996**, *118*, 11472.

81. Danishefsky, S., Zimmer, A. *J. Org. Chem.* **1976**, *41*, 4059.

82. Bouillon, J.-P., Portella, C., Bouquant, J., Humbel, S. *J. Org. Chem.* **2000**, *65*, 5823.

83. (a) Schmidt, J. G. *Ber.* **1880**, *13*, 2342. (b) Claisen, L. *Ber.* **1890**, *23*, 976. (c) Marvel, C. S., King, W. O. *Org. Synth. Col.* Vol. I **1941**, 252.

84. Hatsuda, M., Kuroda, T., Seki, M. *Synth. Commun.* **2003**, *33*, 427.

85. (a) Mukaiyama, T., Banno, K., Narasaka, K. *J. Am. Chem. Soc.* **1974**, *96*, 7503. (b) Mukaiyama, T. *Org. React.* **1982**, *28*, 203. (c) Duthaler, R. O., Hafner, A. In *Transition Metals for Organic Sythesis*, Beller, M., Bolm, C., Eds., Wiley-VCH: Weinheim, Germany, 1998, Vol. 1, p. 447. (d) Mukaiyama, T. *Tetrahedron* **1999**, *55*, 8609.

86. Mukaiyama, T., Ohno, T., Han, J. S., Kobayashi, S. *Chem. Lett.* **1991**, 949.

87. Evans, D. A., Allison, B. D., Yang, M. G. *Tetrahedron Lett.* **1999**, *40*, 4457.

88. Oppolzer, W., Starkmann, C. *Tetrahedron Lett.* **1992**, *33*, 2439.

89. Wittig, G., Reiff, H. *Angew. Chem., Int. Ed.* **1968**, *7*, 7.

90. Birkofer, L., Kim, S. M., Engels, H. D. *Chem. Ber.* **1962**, *95*, 1495.

91. (a) Heathcock, C. H. In *Asymmetric Synthesis*, Morrison, J. D., Ed., Academic Press: New York, 1984, Vol. 3, pp. 111–212. (b) Gennari, C., Todeschini, R., Beretta, M. G., Favini, G., Scolastico, C. *J. Org. Chem.* **1986**, *51*, 612. (c) Masamune, S. *Pure Appl. Chem.* **1988**, *60*, 1587. (d) Li, Y., Paddon-Row, M. N., Houk, K. N. *J. Org. Chem.* **1990**, *55*, 481. (e) Nasipuri, D. *Stereochemistry of Organic Compounds*, Wiley: New York, 1991, p. 414. (f) Brown, M., In *Advances in Carbanion Chemistry*, Snieckus, V., Ed., Jai Press, Greenwich, CT 1992, Vol. 1, pp. 177–247. (g) Atkinson, R. S. *Stereoselective Synthesis*, Wiley: Chichester, UK, 1995, pp. 218–242. (h) Gawley, R. E., Aubé, J. *Principles of Asymmetric Synthesis*, Pergamon Press: Oxford, UK, 1996. (i) Lin, G.-Q., Li, Y.-M., Chan, A. S. *Principles and Applications of Asymmetric Synthesis*, Wiley Interscience: New York, 2001, pp. 135–193.

92. Heathcock, C. H., Buse, C. T., Kleschick, W. A., Pirrung, M. C., Sohn, J. E., Lampe, J. *J. Org. Chem.* **1980**, *45*, 1066.

93. (a) Fellmann, P., Dubois, J. E. *Tetrahedron* **1978**, *34*, 1347. (b) Heathcock, C. H. In *Modern Synthetic Methods*, Scheffold, R., Ed., VCH: Weinheim, Germany, 1992, pp. 51–70.

94. Ireland, R. E., Mueller, R. H., Willard, A. K. *J. Am. Chem. Soc.* **1976**, *98*, 2868.

95. (a) Mukaiyama, T., Inoue, T. *Chem. Lett.* **1976**, 559. (b) Inoue, T., Mukaiyama, T. *Bull. Chem. Soc. Jpn.* **1980**, 174. (c) Kim, B. M., Williams, S. F., Masamune, S. In *Comprehensive Organic Synthesis*, Trost, B. M., Ed., Pergamon Press: Oxford, UK, 1991, Vol. 2, p. 239.

96. (a) Brown, H. C., Ganesan, K., Dhar, R. K. *J. Org. Chem.* **1993**, *58*, 147. (b) Ganesan, K., Brown, H. C. *J. Org. Chem.* **1993**, *58*, 7162.

97. (a) Nerz-Stormes, M., Thornton, E. R. *Tetrahedron Lett.* **1986**, *27*, 897. (b) Harrison, C. R. *Tetrahedron Lett.* **1987**, *28*, 4135.

98. (a) Paterson, I., Tillyer, R. D. *Tetrahedron Lett.* **1992**, *33*, 4233. (b) Mukaiyama, T., Kobayashi, S. *Org. React.* **1994**, *46*, 1.

99. Evans, D. A., Clark, J. S., Metternich, R., Novack, V. J., Sheppard, G. S. *J. Am. Chem. Soc.* **1990**, *112*, 866.

100. Evans, D. A., Rieger, D. L., Bilodeau, M. T., Urpi, F. *J. Am. Chem. Soc.* **1991**, *113*, 1047.

101. Zimmerman, H. E., Traxler, M. D. *J. Am. Chem. Soc.* **1957**, *79*, 1920.

102. (a) Ishihara, K., Yamamoto, H., Heathcock, C. H. *Tetrahedron Lett.* **1989**, *30*, 1825. (b) Faunce, J. A., Friebe, T. L., Grisso, B. A., Losey, E. N., Sabat, M., Mackenzie, P. B. *J. Am. Chem. Soc.* **1989**, *111*, 4508.

103. Heathcock, C. H., Young, S. D., Hagen, J. P., Pilli, R., Badertscher, U. *J. Org. Chem.* **1985**, *50*, 2095.

104. Bal, B., Buse, C. T., Smith, K., Heathcock, C. H. *Org. Synth., Col. Vol. VII* **1990**, 185.

105. White, C. T., Heathcock, C. H. *J. Org. Chem.* **1981**, *46*, 191.

106. Brown, H. C., Dhar, R. K., Bakshi, R. K., Pandiarajan, P. K., Singaram, B. *J. Am. Chem. Soc.* **1989**, *111*, 3441.

107. Evans, D. A., Nelson, J. V., Vogel, E., Taber, T. R. *J. Am. Chem. Soc.* **1981**, *103*, 3099.

108. (a) Hirama, M., Masamune, S. *Tetrahedron Lett.* **1979**, 2225. (b) Gennari, C., Moresca, D., Vieth, S., Vulpetti, A. *Angew. Chem., Int. Ed.* **1993**, *32*, 1618. (c) Gennari, C., Moresca, D., Vulpetti, A., Pain, G. *Tetrahedron Lett.* **1994**, *35*, 4623. (d) Gennari, C., Vulpetti, A., Moresca, D. *Tetrahedron Lett.* **1994**, *35*, 4857.

109. McGarvey, G. J., Hiner, R. N., Williams, J. M., Matasubara, Y., Poarch, J. W. *J. Org. Chem.* **1986**, *51*, 3742.

110. (a) Anderson, R. J., Henrick, C. A., Rosenblum, L. D. *J. Am. Chem. Soc.* **1974**, *96*, 3654. (b) Blanchette, M. A., Malamas, M. S., Nantz, M. H., Roberts, J. C., Somfai, P., Whritenour, D. C., Masamune, S., Kageyama, M., Tamura, T. *J. Org. Chem.* **1989**, *54*, 2817.

111. (a) Heathcock, C. H. In *Modern Synthetic Methods*, Scheffold, R., Ed., VCH: Weinheim, Germany, 1992, p. 64. (b) Arya, P., Qin, H. *Tetrahedron* **2000**, *56*, 917.

112. Cowden, C. J., Paterson, I. *Org. React.* **1997**, *51*, 1.

113. (a) Heathcock, C. H., Pirrung, M. C., Buse, C. T., Hagen, J. P., Young, S. D., Sohn, J. E. *J. Am. Chem. Soc.* **1979**, *101*, 7077. (b) Heathcock, C. H., Pirrung, M. C., Lampe, J., Buse, C. T., Young, S. D. *J. Org. Chem.* **1981**, *46*, 2290.

114. (a) Masamune, S., Ali, S. A., Snitman, D. L., Garvey, D. S. *Angew. Chem., Int. Ed.* **1980**, *19*, 557. (b) Masamune, S., Choy, W., Kerdesky, F. A. J., Imperiali, B. *J. Am. Chem. Soc.* **1981**, *103*, 1566. (c) Masamune, S., Hirama, M., Mori, S., Ali, S. A., Garvey, D. S. *J. Am. Chem. Soc.* **1981**, *103*, 1568.

115. (a) Evans, D. A., Bartroli, J., Shih, T. L. *J. Am. Chem. Soc.* **1981**, *103*, 2127. (b) Ager, D. J., Prakash, I., Schaad, D. R. *Aldrichimica Acta* **1997**, 30, 3.

116. Gage, J. R., Evans, D. A. *Org. Synth.* **1990**, *68*, 77.

117. Gage, J. R., Evans, D. A. *Org. Synth.* **1990**, *68*, 83.

118. Evans, D. A., Tedrow, J. S., Shaw, J. T., Downey, C. W. *J. Am. Chem. Soc.* **2002**, *124*, 392.

119. Evans, D. A., Britton, T. C., Ellman, J. A. *Tetrahedron Lett.* **1987**, *28*, 6141.

120. (a) Basha, A., Lipton, M., Weinreb, S. W. *Tetrahedron Lett.* **1977**, *18*, 4171. (b) Levin, J. I., Turos, E., Weinreb, S. M. *Synth. Commun.* **1982**, *12*, 989.

121. (a) Nahm, S., Weinreb, S. M. *Tetrahedron Lett.* **1981**, *22*, 3815. (b) Sibi, M. P. *Org. Prep. Proced. Int.* **1993**, *25*, 15. (c) Mentzel, M., Hoffmann, H. M. R. *J. Prakt. Chem.* **1997**, *339*, 517. (d) Singh, J., Satyamurthi, N., Aidhen, I. S. *J. Prakt. Chem.* **2000**, *342*, 340.

122. Dias, L. C., de Oliveira, L. G., de Sousa, M. A. *Org. Lett.* **2003**, *5*, 265.

123. Rizzacasa, M., Perkins, M. In *Stoichiometric Asymmetric Synthesis*, Sheffield Academic Press: Malden, UK, 2000, pp. 72–122.

124. (a) Masamune, S., Sato, T., Kim, B.-M., Wollmann, T. A. *J. Am. Chem. Soc.* **1986**, *108*, 8279. (b) Reetz, M. T., Rivadeneira, E., Niemeyer, C. *Tetrahedron Lett.* **1990**, *31*, 3863. (c) Paterson, I., Goodman, J. M., Lister, M. A., Schumann, R. C., McClure, C. K., Norcross, R. D. *Tetrahedron* **1990**, *46*, 4663. (d) Gennari, C., Hewkin, C. T., Molinari, F., Bernardi, A., Comotti, A., Goodman, J. M., Paterson, I. *J. Org. Chem.* **1992**, *57*, 5173.

125. Masamune, S., Choy, W., Peterson, J. S., Sita, L. R. *Angew. Chem., Int. Ed.* **1985**, *24*, 1.

126. (a) Paterson, I., McClure, C. K. *Tetrahedron Lett.* **1987**, *28*, 1229. (b) Paterson, I., Lister, M. A. *Tetrahedron Lett.* **1988**, *29*, 585.

127. Paterson, I., Oballa, R. M., Norcross, R. D. *Tetrahedron Lett.* **1996**, *37*, 8581.

128. (a) Duplantier, A. J., Nantz, M. H., Roberts, J. C., Short, R. P., Somfai, P., Masamune, S. *Tetrahedron Lett.* **1989**, *30*, 7357. (b) Paterson, I., Florenze, G. A., Gerlach, K., Scott, J. P. *Angew. Chem., Int. Ed.* **2000**, *39*, 377.

129. (a) Kleinman, E. F. In *Comprehensive Organic Synthesis*, Trost, B. M., Fleming, I., Eds., Pergamon Press: Oxford, UK, 1991, Vol 2., p. 893. (b) Tramontini, M., Angiolini, L. *Mannich Bases, Chemistry and Uses*, CRC Press: Boca Raton, FL, 1994.

130. (a) Schreiber, J., Maag, H., Hashimoto, N., Eschenmoser, A. *Angew. Chem., Int. Ed.* **1971**, *10*, 330. (b) Bryson, T. A., Bonitz, G. H., Reichel, C. J., Dardis, R. E. *J. Org. Chem.* **1980**, *45*, 524.

131. Kinast, G., Tietze, L.-F. *Angew. Chem., Int. Ed.* **1976**, *15*, 239.

132. Danishefsky, S., Kitahara, T., McKee, R., Schuda, P. F. *J. Am. Chem. Soc.* **1976**, *98*, 6715.

133. Roberts, J. L., Borromeo, P. S., Poulter, C. D. *Tetrahedron Lett.* **1977**, *18*, 1621.

134. (a) List, B. *J. Am. Chem. Soc.* **2000**, *122*, 9336. (b) Juhl, K., Gathergood, N., Jørgensen, K. A. *Angew. Chem., Int. Ed.* **2001**, *40*, 2995.

135. (a) Michael, A. *J. Prakt. Chem.* **1887**, *35*, 349. (b) Bergmann, E. D., Ginsburg, D., Pappo, R. *Org. React.* **1959**, *10*, 179. (c) Oare, D. A., Heathcock, C. H. *J. Org. Chem.* **1990**, *55*, 157. (d) Perlmutter, P. *Conjugate Addition Reactions in Organic Synthesis*, Pergamon Press: Oxford, UK, 1992. (e) Sibi, M. P., Manyem, S. *Tetrahedron* **2000**, *56*, 8033.

136. Kotsuki, H., Arimura, K., Ohishi, T., Maruzasa, R. *J. Org. Chem.* **1999**, *64*, 3770.

137. Little, R. D., Masjedizadeh, M. R., Wallquist, O., McLoughlin J. I. *Org. React.* **1995**, *47*, 315.

138. Galeazzi, R., Geremia, S., Mobbili, G., Orena, M. *Tetrahedron: Asymmetry.* **1996**, *7*, 79.

139. (a) Jung, M. E. *Tetrahedron* **1976**, *32*, 3. (b) Gawley, R. E. *Synthesis* **1976**, 777.

140. Rapson, W. S., Robinson, R. *J. Chem. Soc.* **1935**, 53, 1285.

141. Marshall, J. A., Fanta, W. I. *J. Org. Chem.* **1964**, *29*, 2501.

142. Heathcock, C. H., Ellis, J. E., McMurry, J. E., Coppolino, A. *Tetrahedron Lett.* **1971**, *12*, 4995.

143. (a) Stork, G., Ganem, B. *J. Am. Chem. Soc.* **1973**, *95*, 6152. (b) Boeckman, R. K., Jr. *J. Am. Chem. Soc.* **1973**, *95*, 6867. (c) Stork, G., Singh, J. *J. Am. Chem. Soc.* **1974**, *96*, 6181. (d) Boeckman, R. K., Jr. *Tetrahedron* **1983**, *39*, 925.

144. Metzger, J. D., Baker, M. W., Morris, R. J. *J. Org. Chem.* **1972**, *37*, 789.

145. Varner, M. A., Grossman, R. B. *Tetrahedron* **1999**, 55, 13867.

146. Revial, G., Pfau, M. *Org. Synth.* **1991**, *70*, 35.

CHAPTER 7

Formation of Carbon-Carbon Bonds via Organometallic Reagents

Research on transition metal templates in catalytic reactions has suggested that these easily tailored templates may become the "chemists' enzymes."

Barry M. Trost

Organometallic reagents play a key role in carbon-carbon-bond-forming reactions, which are the backbone of organic synthesis. The negatively polarized carbon-metal bond ($C^{\delta-}M^{\delta+}$) is especially well suited for this purpose, offering a convenient site for reactions with organic molecules. The reactivity of an organometallic reagent generally *increases* with the ionic character of the carbon-metal bond and is related to the electronegativity value EN (electrostatic force exerted by a nucleus on the valence electrons) difference between the carbon atom and the metal center, EN_C–EN_{Met} (Table 7.1).[1]

The percent ionicity (ionic character) is related to the difference between the EN values of the atoms of the C–Met bond (EN_C–EN_{Met}).[1] These are estimated values, which are affected by the nature of the substituents on carbon. Nevertheless, they indicate that the C–Li, C–Mg, C–Ti, and C–Al bonds are more ionic than C–Zn, C–Cu, C–Sn, and C–B, which form mainly covalent bonds with carbon. Manipulation of certain organometallic reagents requires special techniques.[2]

7.1 ORGANOLITHIUM REAGENTS[3]

Organolithium reagents react with a wide variety of organic substrates to form carbon-carbon bonds and serve as precursors for the preparation of other organometallic reagents. The following sections describe methods for the preparation of various organolithiums and their utilization in organic syntheses.

Organolithiums from Alkyl Halides and Lithium Metal

The scope of this method is broad and is especially suited for the preparation of *alkyl-* and *aryllithiums*. It is, however, less general than the corresponding method for preparing Grignard reagents in that *allylic, benzylic,* and *propargylic halides* cannot be successfully converted into the corresponding organolithiums because they tend to

Table 7.1	Electronegativity Values and Ionic Character								
Element	Li	Mg	Ti	Al	Zn	Cu	Si	Sn	B	C
EN*	0.97	1.23	1.32	1.47	1.66	1.75	1.74	1.72	2.01	2.50
% Ionicity	43	35	30	22	15	12	12	11	6	

*Allred-Rochow electronegativity (EN).

undergo Wurtz coupling, in which the lithium reagents initially formed react competitively with the R–X to produce homocoupled products.

$$RCH{=}CHCH_2I \ + \ 2\ Li \ \longrightarrow \ RCH{=}CHCH_2Li \ + \ LiI$$

$$\downarrow RCH{=}CHCH_2I \ + \ 2\ Li$$

$$RCH{=}CHCH_2CH_2CH{=}CHR \ + \ LiI$$

Important points to consider when preparing and using organolithiums are

- *Atmosphere*. Reactions with organolithium compounds must be carried out in an inert atmosphere (Ar and He are best; N_2 tarnishes lithium metal by forming lithium nitride).

- *Nature of the halide*. Bromides generally are best; iodides have a tendency to undergo the Wurtz reaction. With chlorides, use Li containing 1–2% Na.

- *Purity and physical state of the metal*. The metal surface should be clean and have a large surface area. Li wire typically is flattened with a hammer and then cut into small pieces. Li dispersions in mineral oil may be employed in place of Li wire. The oil is removed by washing with hydrocarbon solvents such as *n*-hexane.

- *Solvent*. Most R–Li reagents are prepared in hydrocarbon solvents. However, phenyllithium, methyllithium, and vinyllithium, which are almost insoluble in hydrocarbon solvents, are quite soluble in Et_2O. *n*-BuLi, *sec*-BuLi, and *t*-BuLi react at room temperature with Et_2O and THF, so they must be used at low temperature in these solvents.

$$n\text{-BuCl} \ + \ \text{Li wire} \ \xrightarrow{\text{pentane}} \ n\text{-BuLi } (> 90\%) \ + \ \text{LiCl}$$

$$t\text{-BuCl} \ + \ \text{Li–2\% Na dispersion} \ \xrightarrow{\text{pentane}} \ t\text{-BuLi } (> 80\%)$$

$$\text{PhBr} \ + \ \text{Li dispersion} \ \xrightarrow{Et_2O} \ \text{PhLi}$$

- *Analysis of organolithium reagents*. Many analytical procedures for the assay of alkyllithiums in solution have been reported.[4]

- *Organolithium aggregation*. Organolithiums associate in solution to form oligomeric species in which the monomeric units are held together via multicenter bonding. Coordinating solvents such as Et_2O and THF influence their aggregation[5] and reactivity.[6]

MeLi in Et_2O — tetrameric   *n*-BuLi in hexane — hexameric
      in THF — tetrameric   in THF — tetrameric + dimeric

- *Reactivity*. The basicity of organolithium reagents decreases with increasing stability of the carbanion moiety (e.g., *t*-BuLi > *s*-BuLi > *n*-BuLi).

Organolithium reagents exhibit reactivities similar to those of Grignard reagents, with the notable exception that they react with CO_2 to produce ketones[7] on workup, whereas Grignard reagents furnish carboxylic acids (Table 7.2).

$$RLi \ + \ CO_2 \ \longrightarrow \ RCO_2^- Li^+ \ \xrightarrow{RLi} \ R_2C \overset{O^- Li^+}{\underset{O^- Li^+}{\Big\langle}} \ \xrightarrow{H^+, H_2O} \ R_2C{=}O$$

$$CH_3CH_2\underset{MgBr}{\overset{|}{C}}HCH_3 \ + \ CO_2 \ \longrightarrow \ CH_3CH_2\underset{CO_2^- \ MgBr^+}{\overset{|}{C}}HCH_3 \ \xrightarrow{H^+, H_2O} \ CH_3CH_2\underset{CO_2H}{\overset{|}{C}}HCH_3$$

76–86%

Table 7.2	Retrosyntheses Using Organolithium, RLi (A), or Grignard, RMgX (B), Reagents

1. Preparation of alcohols

$RCH_2OH \implies CH_2=O$ + A or B

$RCH_2CH_2OH \implies$ + A or B

$\begin{array}{c} R' \\ R \end{array}\!\!C\!\!\begin{array}{c} H \\ OH \end{array} \implies R'CHO$ + A or B

$\begin{array}{c} R' \\ R \end{array}\!\!C\!\!\begin{array}{c} R'' \\ OH \end{array} \implies \begin{array}{c} R' \\ R'' \end{array}\!\!C\!\!=\!\!O$ + A or B

$\begin{array}{c} R \\ R \end{array}\!\!C\!\!\begin{array}{c} H \\ OH \end{array} \implies HCOOEt$ + 2A or 2B

$\begin{array}{c} R' \\ R \end{array}\!\!C\!\!\begin{array}{c} R \\ OH \end{array} \implies \begin{array}{c} R' \\ X \end{array}\!\!C\!\!=\!\!O$ + 2A or 2B

2. Preparation of aldehydes

$RCHO \implies \underset{DMF}{(CH_3)_2N\!-\!CHO}$ + A or B

3. Preparation of ketones

$\begin{array}{c} R' \\ R \end{array}\!\!C\!\!=\!\!O \implies R'\!-\!CO_2Li$ + A

$\begin{array}{c} R' \\ R \end{array}\!\!C\!\!=\!\!O \implies R'\!-\!CN$ + A or B

4. Preparation of carboxylic acids

$RCO_2H \implies CO_2$ + B

The high reactivity of organolithium reagents toward most functional groups requires special reaction conditions for the preparation of functionally substituted organolithiums, such as working at low temperature or generating the organolithium in the presence of the electrophile (Barbier-type reaction).[8]

Organolithiums via Lithium–Halogen Exchange

$$R\!-\!X + R'\!-\!Li \rightleftharpoons R\!-\!Li + R'\!-\!X$$

This reaction proceeds in the forward direction when the new lithium reagent RLi formed is a weaker base (more stable carbanion) than the starting organolithium R'Li. The method is best suited for exchanges between C_{sp^3}–Li (stronger base) and C_{sp^2}–X to give alkenyllithiums, C_{sp^2}–Li (weaker base).[9]

Alkenyllithium Reagents

A problem encountered in the preparation of alkenyllithiums via lithium-halogen exchange may be the coupling of the newly formed alkyl halide (e.g., *n*-BuBr) with the alkenyllithium.

The alkylation problem can be circumvented by using two equivalents of *tert*-butyllithium. The second equivalent of *t*-BuLi is involved in the dehydrohalogenation (E2 reaction) of the *t*-BuBr formed in situ.[10]

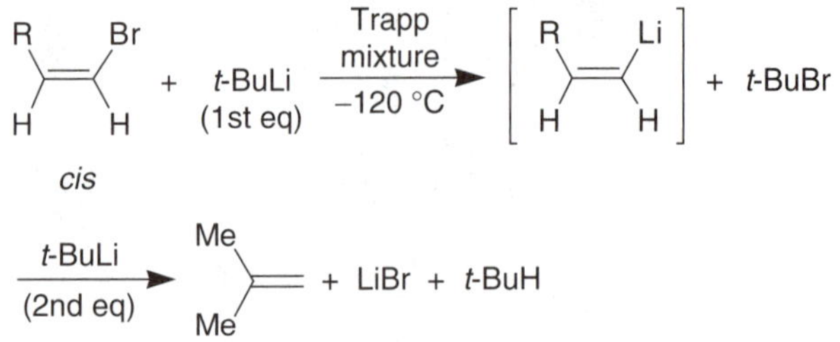

(E)- and (Z)-alkenyllithiums are configurationally stable at low temperatures. The preparation of certain (Z)-alkenyllithiums should be carried out in Et_2O rather than in THF.[11] When working at $-100\ °C$ or below, a 4 : 1 : 1 mixture of THF : Et_2O : n-pentane, known as the Trapp mixture,[12] is required as reaction solvent.

The alkenyllithium reagents are used for stereospecific syntheses of alkenes and functionally substituted alkenes.[13]

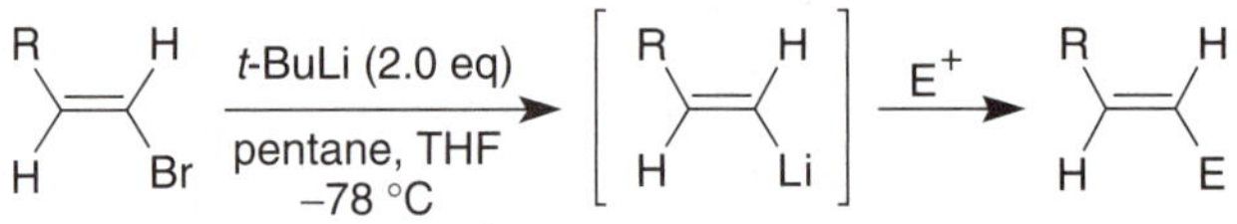

E^+ = primary RBr or RI, CO_2, DMF, RCHO, R_2CO, or epoxides

Aryllithium Reagents

Metal-halogen exchange provides an efficient route to aryllithiums and heteroaromatic lithium reagents that are inaccessible by metal-hydrogen exchange.[14] The lithium-halogen exchange reaction is very fast, even at low temperatures, particularly in electron-donating solvents. Therefore, competitive alkylation and metal-hydrogen exchange (metalation) reactions are usually not a problem. Caution should be used when employing TMEDA (tetramethylethylenediamine) as a promoter for metal-halogen exchange reactions, since it accelerates metalations more than it does metal-halogen exchange.

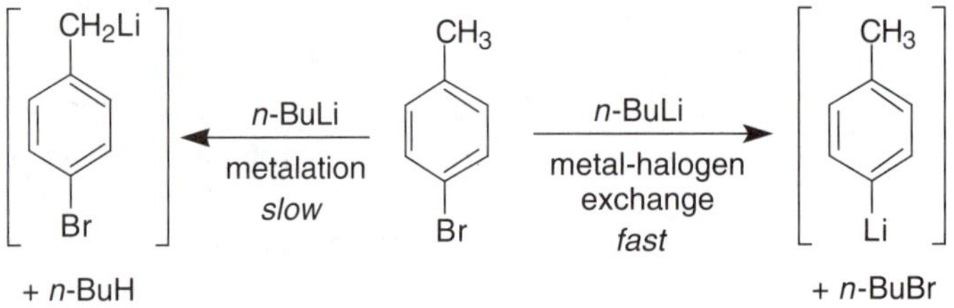

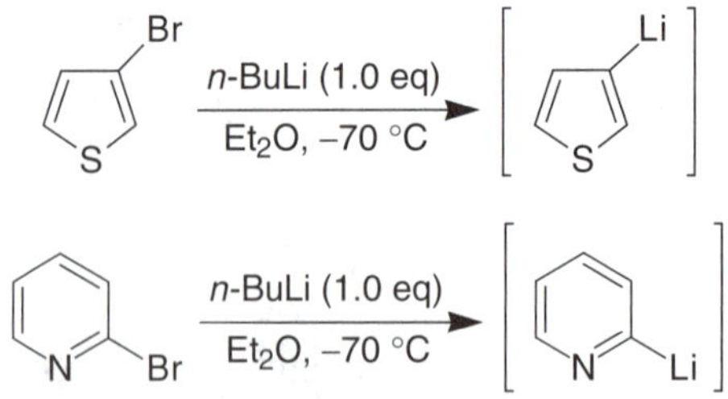

Functionally substituted aryllithiums, such as the lithiobenzonitrile and lithionitrobenzene shown below, are only stable at low temperature and thus require trapping with a reactive electrophile.[15]

Organolithiums via Lithium-Metal Exchange (Transmetalation)

$$R–M \ + \ R'–Li \ \rightleftharpoons \ R–Li \ + \ R'–M$$

Transmetalation is used to prepare *allylic*, *benzylic*, and *propargylic* lithium reagents, which are difficult to obtain by other routes.

The above conversion of the readily available allylic Grignard reagent into the corresponding allylic lithium reagent involves two metal-metal exchanges. These reactions proceed in the forward direction because (1) in the Mg-Sn exchange, the more electropositive Mg preferentially exists as the more ionic salt MgBrCl, and (2) in the Sn-Li exchange, the more electropositive Li is associated with the more electronegative allylic ligand.

Organolithiums via Lithium-Hydrogen Exchange (Metalation)

Metal-hydrogen exchange provides a general route to organolithium compounds. The tendency to form the C–Li bond (and thus the reactivity of the C–Li bond) depends on the stability of the R group as a negative ion. The most important measure of stability is the acidity of the corresponding carbon acid. A difference of 2–3 pK_a units is sufficient to drive the reaction to completion (98%), although a greater pK_a difference is desirable (for a list of pK_a values, see Chapter 6[16]).

The following factors influence the acidity of C–H bonds:

- Hybridization (s character of the C–H bond)—higher % s character, lower pK_a

$$pK_a\!: \ C–H \sim 50, \ C=C–H \sim 44, \ C\equiv C–H \sim 25$$

- Effect of substitution—lower carbanion stability, higher pK_a

$$\text{Carbanion stability: } RCH_2\!:^- > R_2CH\!:^- > R_3C\!:^-$$

- Resonance—an adjacent electron withdrawing group, lower pK_a

Acidity of $-\overset{|}{\underset{\mathbf{H}}{C}}-R$ decreases in the following order:

$$R = CHO > C(O)R' > CO_2R' > C(O)NR'_2 \sim CO_2^- > SO_2R' > Ph \sim C=C$$

Alkyllithium and Arylithium Reagents for Metalation

An important development in organolithium chemistry was the discovery that certain solvents such as THF (tetrahydrofuran), DME (dimethoxyethane), diglyme (diethylene glycol dimethyl ether), and various additives can greatly alter their reactivity.[17] The addition of chelating agents such as TMEDA (tetramethylethylenediamine), HMPA (hexamethylphosphoramide, potential carcinogen), *tertiary* amines, crown ethers, and t-BuO$^-$K$^+$ increases the basicity and/or the nucleophilicity of organolithiums. For example, TMEDA or HMPA function to deoligomerize the hexameric n-BuLi in hexane to the kinetically more reactive monomer by coordination of the Li$^+$ atom. These strong complexing agents generally are used in stoichiometric amounts or in slight excess. An excellent replacement solvent for the carcinogenic HMPA in a variety of reactions is DMPU [N,N'-dimethylpropyleneurea; 1,3-dimethyl-3,4,5,6-tetrahydro-2(1H)pyrimidinone].[18]

TMEDA	HMPA	DMPU	16-crown-4 ether (16 atoms, 4 oxygens)

The commonly used lithium dialkylamides are LDA (lithium diisopropylamide), LTMP (lithium 2,2,6,6-tetramethylpiperidide), and LHMDS (lithium hexamethyldisilazide). They are available by reacting the appropriate amine with an organolithium reagent in Et$_2$O or in THF solvent, as shown for the preparation of LDA.[19]

	i-Pr$_2$NLi		(Me$_3$Si)$_2$NLi
	LDA	LTMP	LHMDS
pK_a of amine	~38	~37	~30

$$i\text{-Pr}_2\text{NH} + n\text{-BuLi} \xrightarrow[-78\,°C]{\text{THF}} i\text{-Pr}_2\text{NLi} + n\text{-Bu-H}$$

pK_a ~ 38
stronger acid

pK_a ~ 50
weaker acid

Chemoselectivity

The choice of the metalating agent is especially crucial when the substrate molecule contains functional groups that can be attacked by bases and nucleophiles, as is usually the case.

R₂NLi (e.g., LDA) are non-nucleophilic, strong bases.

RLi are powerful nucleophiles as well as strong bases.

Interestingly, R₂NLi reagents are generally more effective metalating agents than the thermodynamically more basic RLi reagents. The increased kinetic basicity of heteroatom bases may be rationalized by the availability of the free electron pair, which permits formation of a four-centered transition state, thus avoiding the free carbanion.[20] A similar transition state has been proposed for the deprotonation of ketones by R₂NLi.[21]

Benzylic Metalation

The preparation of benzyllithium from benzyl halides and alkyllithiums is not feasible because the benzyllithium initially formed reacts with the starting benzyl halides, producing 1,2-diphenylethane. Metalation of toluene with n-BuLi in the presence of TMEDA at 30 °C results in a 92 : 8 ratio of benzyllithium and ring metalated products.[22] Metalation of toluene with n-BuLi in the presence of potassium tert-butoxide, and treatment of the resultant organopotassium compound with lithium bromide, affords pure benzyllithium in 89% yield.[23] Alternatively, benzyllithiums are accessible by cleavage of alkyl benzyl ethers with lithium metal.[24]

Allylic Metalation

The reaction of allylic organometallics with electrophilic reagents is a very important tool for the formation of carbon-carbon bonds in acyclic systems[25] and for controlling their stereochemistry.[26] Crotyl organometallic (2-butenylmetal) species undergo a 1,3-shift of the metal at room temperature. For the stereocontrolled use of allylmetals in synthesis, it is important to avoid their equilibration.

M = Li, MgX, ZnX, BR$_2$, AlR$_2$, TiL$_3$, ZrL$_3$
L = ligand

Treatment of propene or isobutylene with *n*-BuLi in Et$_2$O in the presence of TMEDA provides a convenient route to allyllithium and methallyllithium, respectively.[27]

The rate of deprotonation of weakly acidic compounds by alkyllithiums may be changed by several orders of magnitude simply by altering the cation. Potassium *tert*-butoxide activates *n*-butyllithium (Schlosser's "super base"), allowing metalation of allylic C–H bonds of olefins in the low acidity range (pK_a ~ 40).[25,28] Although the true nature of the Super Base is not known, it is probably an organopotassium/lithium alcoholate aggregate.

Crotyllithium and crotylpotassium compounds can assume either the *endo*- or *exo*-configuration. Due to their planarity, both forms are stabilized by electron delocalization.

M is above the plane

endo   exo

While equilibration of the *endo*- and *exo*-forms of crotyllithium is very fast, the corresponding potassium reagents are remarkably stable and may be intercepted with electrophiles (E$^+$).[29] However, after several hours, the crotylpotassium compounds also equilibrate, surprisingly favoring the *endo*-form over the sterically less hindered *exo*-form .

endo

exo

Crotyllithium reagents are ambident nucleophiles and can react with electrophiles either at the α- or γ-carbon. The regiochemistry of attack depends on many factors, such as structure, the electrophile, and the solvent. Generally, unhindered carbonyl compounds preferentially add to crotyllithiums at the γ-position.

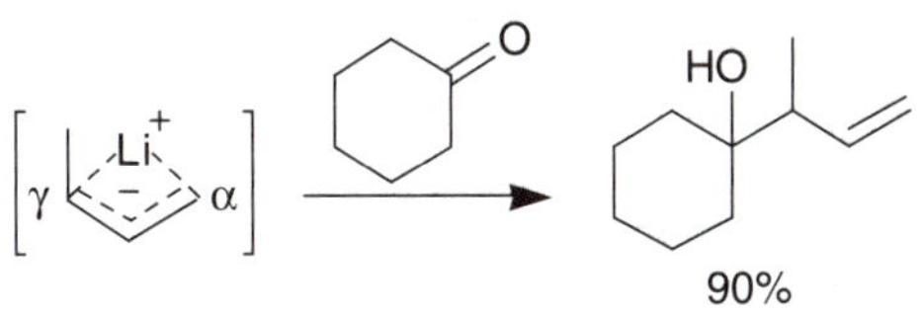

90%

Allylic potassium organometallics derived from BuLi–*t*-BuOK react with electrophiles predominantly at the α-position.[30]

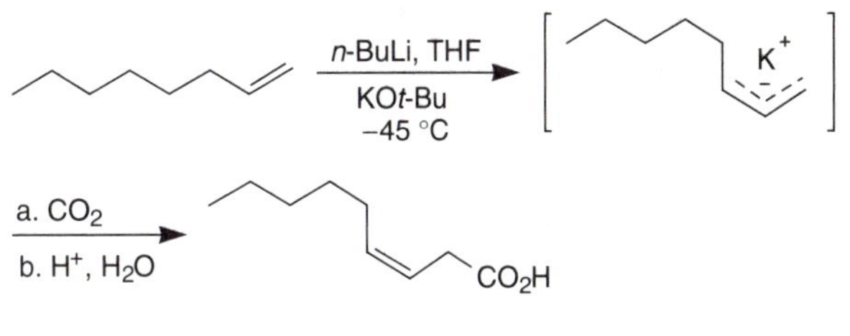

Metalations of α-Heteroatom Substituted Alkenes

Protons attached to sp^2 carbons are more acidic than protons attached to nonallylic sp^3 carbons. Also, the inductive effect of a heteroatom further increases the acidity of an adjacent sp^2 C–H bond, facilitating α-lithiation. The relative activating effect of heteroatoms is sulfur > oxygen > nitrogen. Thus, treatment of 2-ethoxy-1-(phenylthio)ethylene with *t*-BuLi results in exclusive lithiation at the phenylthio substituted carbon.[31]

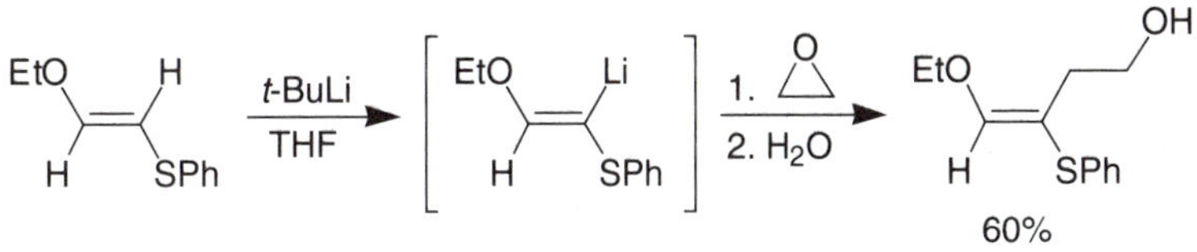

60%

Metalation of dihydropyran with *n*-BuLi in the presence of TMEDA occurs at the α-vinylic position rather than at the allylic position. Abstraction of an allylic proton proceeds at a slower rate than abstraction of the vinylic proton of the sp^2-carbon bonded to the inductively electron-withdrawing oxygen.

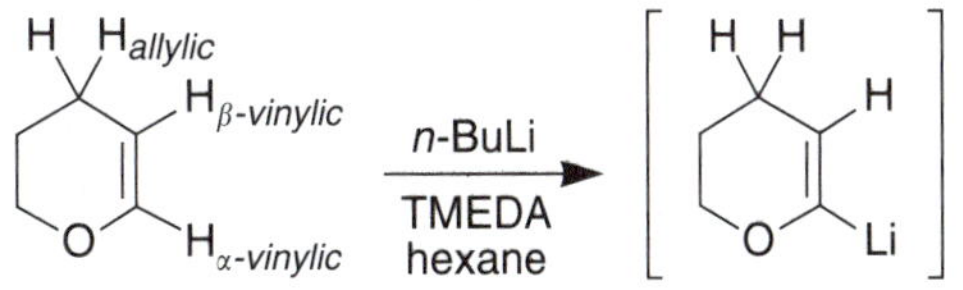

Metalation of methyl vinyl ether or phenyl vinyl sulfides furnishes an α-metalated vinyl ether or vinyl sulfide, respectively. These carbanions represent *acyl anion equivalents*.[32]

Ortho-Metalation of Substituted Benzenes and Heteroaromatic Compounds[33]

Direct metalation of certain aromatic substrates permits regioselective preparation of substituted benzene derivatives and heterocycles. Thus, replacement of a C_{sp^2}–H by organolithium reagents is facilitated at the *ortho*-position to a functional group with nonbonding electrons, such as nitrogen or oxygen. Coordination of the lithium reagent with the nitrogen or oxygen holds the organolithium in proximity to the ortho-hydrogens.[34]

$$X = -NR_2, -OR, -CH_2OR, -CH_2NR_2, -CH(OR)_2, -CONR_2$$

$$E^+ = CO_2, DMF, RCHO, R_2CO, \text{epoxides, primary alkyl halides}$$

Because of the greater coordinating ability of nitrogen as compared to oxygen, treatment of *p*-methoxy-*N,N*-dimethylbenzylamine with *n*-BuLi results in metalation *ortho* to the –CH$_2$NMe$_2$. However, in the presence of the strongly complexing TMEDA, coordination of lithium with the nitrogen of –CH$_2$NMe$_2$ is suppressed. In this case, the most acidic proton *ortho* to the –OMe group is removed preferentially.[35]

Metalation of the heteroaromatic compounds furan and thiophene with alkyllithium reagents furnishes the corresponding 2-lithio derivatives.[33] For example, sequential treatment of 2-methylfuran with *t*-BuLi in THF, followed by alkylation of the organolithium intermediate and hydrolysis of the resultant bis-vinyl ether, produces an unsaturated 1,4-diketone.

Sulfur is more effective than oxygen in stabilizing an adjacent carbanion. Thus, using an equimolar mixture of furan and thiophene, the thiophene is selectively metalated when using one equivalent of *n*-BuLi.

Metalation of 1-Alkynes (Preparation of Lithium Alkynylides)[36]

A filled sp orbital is lower in energy than filled sp^2 or sp^3 orbitals since it is closer to the positively charged nucleus. This imparts sufficiently greater acidity to acetylene and 1-alkynes (pK_a 24–26) so that bases such as alkyllithiums, lithium dialkylamides, sodium amide in liquid ammonia, and ethylmagnesium bromide may be used to generate the alkynyl anions (see Section 8.2).

Conjugate Addition Reactions of Lithium Reagents

Although alkyl- and aryllithium reagents usually attack the carbonyl group of α, β-unsaturated carbonyl compounds (1, 2-addition), conjugate addition (1, 4-addition) is observed with very hindered esters where approach to the carbonyl group is impeded, as in 2,6-di-*tert*-butyl-4-methylphenyl esters (butylated hydroxytoluene, BHT esters) and 2,6-di-*tert*-butyl-4-methoxyphenyl esters (butylated hydroxyanisole; BHA esters).[37]

7.2 ORGANOMAGNESIUM REAGENTS[38]

The Grignard reaction, reported in 1900 by Victor Grignard (Nobel Prize, 1912),[39] provides the synthetic chemist with one of the most powerful tools for connecting carbon moieties.

Preparation of Grignard Reagents[40]

Alkyl Grignard reagents are prepared by the reaction of an alkyl chloride, bromide, or iodide[41] with (1) "activated" magnesium turnings in Et_2O or THF solvent or (2) with *Rieke magnesium.*[42] Although it is also in the metallic state, Rieke magnesium differs from the bulk metal by being in the form of highly reactive small particles with a large surface area.

Alkenyl and phenyl Grignard reagents are usually prepared from the corresponding bromides or iodides in THF.[43] In Et_2O, Grignard reagents derived from (E)- and (Z)-alkenyl halides are configurationally unstable, producing mixtures of isomers.

Allylic Grignard reagents, prepared from allylic halides and magnesium, are often accompanied by allylic halide coupling products. This problem can be obviated by using the highly reactive Rieke-Mg or by mixing the allylic halide, the aldehyde or ketone, and magnesium together in what is called a Barbier-type reaction. As the Grignard reagent forms, it reacts immediately with the electrophile before it has a chance to couple with unreacted allylic halide.

Alkynyl Grignard reagents are obtained by deprotonation of 1-alkynes with ethylmagnesium bromide in THF.[44] For the preparation of ethynylmagnesium bromide ($HC\equiv CMgBr$), a solution of ethylmagnesium bromide in THF is slowly added to a cooled solution of THF containing the acetylene.[45]

Although 100 years have passed since Grignard published the preparation of ethereal solutions of organomagnesium halides, the actual mechanism(s) for the formation of the reagents and their structures are still not completely understood. The overall reaction for the formation of Grignard reagents involves an insertion of magnesium into the carbon-halogen bond via an oxidative addition, thereby changing its oxidation state from Mg(0) to Mg(II).[46]

$$R-X \ + \ Mg(0) \ \xrightarrow{\text{Et}_2\text{O or THF}} \ R-Mg(II)X$$

It is generally accepted that the structure of RMgX can be represented by the Schlenk equilibrium shown below.[38b,47]

$$2 \ RMgX \ \rightleftharpoons \ R_2Mg \ + \ MgX_2 \ \rightleftharpoons \ R_2Mg \cdot MgX_2$$

Reactions of Grignard Reagents with Carbonyl Compounds

Grignard reagents are capable of nucleophilic additions to hetero double bonds such as those in carbonyl compounds. The carbonyl reactivity toward Grignard reagents decreases in the order aldehyde > ketone > ester > amide. The high reactivity of Grignard reagents toward carbonyl groups is due primarily to the polarization of the C=O π-bond and the weak C–Mg bond.

Additions of Grignard reagents to carbonyl groups may proceed either via a polar-concerted or a stepwise electron-transfer mechanism.[48] A possible mechanistic scheme for the polar-concerted reaction of a Grignard reagent with an aldehyde or a ketone is depicted below. Coordination of the Lewis acidic magnesium to the Lewis basic carbonyl oxygen further polarizes the carbonyl group while enhancing the nucleophilicity of the R group.

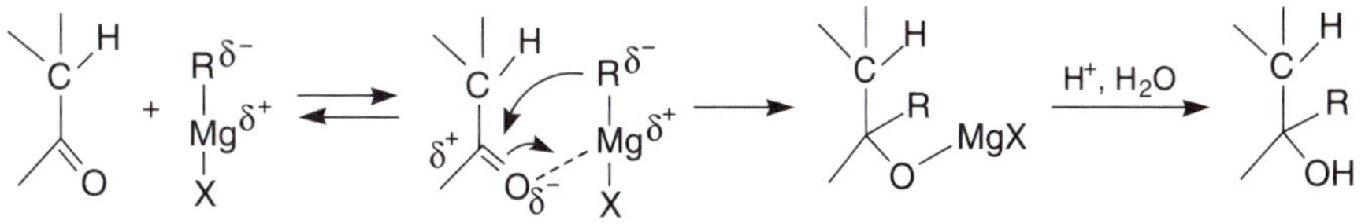

A summary of the utility of Grignard reagents in synthesis is shown in Table 7.2.

In spite of the versatility and broad synthetic utility of the Grignard reaction for coupling two carbon moieties, it is often accompanied by competing side reactions such as enolization, reduction, or aldol condensation of the carbonyl substrate. Organomagnesium compounds can act not only as nucleophiles, but also as bases, thereby converting ketones with enolizable hydrogens to the corresponding magnesium enolates. Loss of the carbanion moiety as R–H and hydrolytic workup leads to the starting ketone.

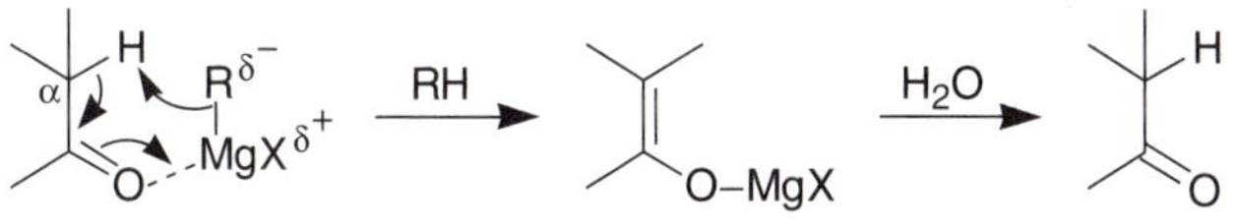

enolization

If the Grignard reagent has a hydrogen in the β-position, reduction of the carbonyl group by hydride transfer may compete with the addition reaction (Table 7.3).

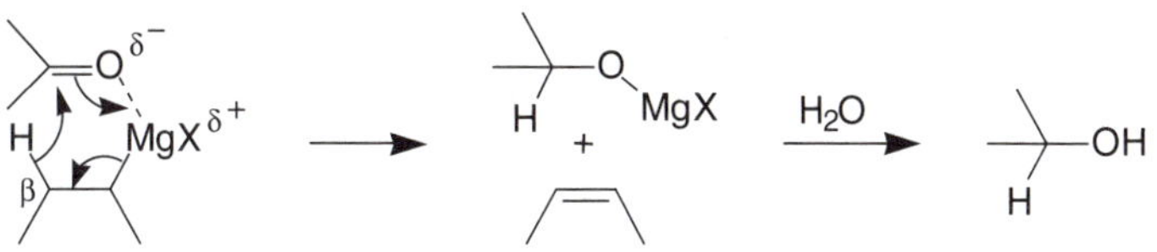

reduction

To suppress these side reactions, use the smallest-possible group for the Grignard reagent, or use the corresponding lithium reagents, which give less reduction and enolization products.

Table 7.3 Influence of Structure on Grignard Reactivity

| | Product distribution (%) | | |
Grignard reagent	Addition	Enolization	Reduction
CH_3MgX	95	—	—
$(CH_3)_3CMgX^a$	0	35	65
$(CH_3)_3CCH_2MgX^b$	0	90	0

$^a\beta$-hydrogens; ^{b}no β-hydrogens.

Limitations

Certain functional groups present in a molecule interfere with the preparation of Grignard reagents. Thus, –NH, –OH, and –SH groups will protonate the Grignard reagent once it is formed. Carbonyl and nitrile groups attached to the molecule containing the halogen substituent will undergo addition reactions. However, iodine-magnesium exchange reactions of functionalized aryl iodides and heteroaryl iodides recently are reported to produce functionalized Grignard reagents at low temperature.[49]

FG = Br, C(O)NR$_2$, CN, CO$_2$Et, CO$_2$t-Bu

E$^+$ = RCHO, allyl-Br, etc.

7.3   ORGANOTITANIUM REAGENTS[50]

Many chemoselectivity problems are associated with organolithium and Grignard C–C-bond-forming reactions. In some cases, competing nucleophilic additions to aldehydes, ketones, and esters are not well differentiated by these organometallics, and other electrophilic functional groups such as cyano, nitro, or halo may interfere. With α,β-unsaturated carbonyl substrates, regioselectivity (i.e., 1,2- vs. 1,4-addition) is often an issue. Also, as noted above, the basicity of organolithium and Grignard reagents may cause unwanted side reactions when they act as bases instead of as nucleophiles with carbonyl substrates, causing proton abstraction α to the carbonyl group. Fortunately, with *organotitanium* compounds, derived from organolithium and organomagnesium compounds, the reactivity and basicity of their respective carbanions are tempered considerably.[50,51]

The highly chemo- and regioselective titanium reagents may be generated in situ by adding a Ti-alkoxide or amide to organolithium or Grignard reagents.[51a] Transmetalation with titanium is essentially instantaneous, even at low temperatures. A variety of groups may be attached to titanium. These include primary, secondary, and even tertiary alkyl groups, as well as allylic, propargylic, vinylic, and aryl groups.

$$CH_3Li + ClTi(O\text{-}iPr)_3 \xrightarrow[-58\ °C]{Et_2O} H_3C\text{—}Ti(Oi\text{-}Pr)_3 + LiCl\ (ppt)$$

Table 7.4   Selectivity in the Addition Reactions of CH$_3$Ti(Oi-Pr)$_3$[51a]

More reactive substrate	Less reactive substrate	Selectivity (%)
A	**B**	**C : B** (unreacted)
		> 99.9
		92 : 8
		> 97 : <3
		> 94 : 6

The most characteristic difference between organotitanates, RTi(OR)$_3$, and the RLi and RMgX reagents is the selectivity of titanium reagents. Branched and unbranched aldehydes, cyclic and acyclic ketones, and saturated and α,β-unsaturated ketones can be distinguished with excellent selectivity, as exemplified by the competition experiments shown in Table 7.4.[51a] Functional groups such as COOR, CONR$_2$, C-halogen, and epoxides do not interfere with the additions of RTi(OR)$_3$ to aldehyde groups.

7.4   ORGANOCERIUM REAGENTS[52,53]

Especially useful for introducing alkyl groups onto hindered and easily enolizable ketones such as β,γ-enones are the weakly basic and strongly nucleophilic organocerium reagents. These are readily accessible from the corresponding organolithium or organomagnesium compounds via transmetalation with anhydrous CeCl$_3$. Again, the metal-metal exchange proceeds so that the more electropositive metal exists as the more ionic inorganic salt. The organocerium reagents are used directly without isolation.

$$\text{RLi} + \text{CeCl}_3 \xrightarrow[-78\ °C]{\text{THF}} \text{``RCeCl}_2\text{''} + \text{LiCl}$$

$$\text{RMgX} + \text{CeCl}_3 \xrightarrow[0\ °C]{\text{THF}} \text{``RCeCl}_2\text{''} + \text{MgClX}$$

Table 7.5 Selectivity in the 1,2-Addition Reactions of RCeCl$_2$ Reagents[54]

| M = Li: | n-BuLi | 0%[a] |
| M = Ce: | n-BuCeCl$_2$ (from n-BuLi) | 99% |

| M = Mg: | i-PrMgBr | 3% | 88% |
| M = Ce: | i-PrCeCl$_2$ (from i-PrMgBr) | 80% | trace |

| M = Mg: | i-PrMgBr | 12% | 53% |
| M = Ce: | i-PrCeCl$_2$ | 91% | 5% |

[a] Main reaction is likely Li-I exchange.

A distinguishing feature of organocerium reagents is their high regioselectivity toward α,β-unsaturated ketones. While Grignard reagents often afford mixtures of 1,2- and 1,4-addition products, the corresponding organocerium regents furnish predominately 1,2-adducts. The high selectivity of organocerium reagents for nucleophilic 1,2-additions is contrasted in Table 7.5 with those of organolithium and Grignard reagents.[54]

7.5 ORGANOCOPPER REAGENTS[55]

Use of organocopper reagents offers a very efficient method for coupling of two different carbon moieties. Since copper is less electropositive than lithium and magnesium, the C–Cu bond is less polarized than the C–Li and C–Mg bonds. This difference produces three useful changes in reactivity:

- The organocopper reagents react with alkyl-, alkenyl-, and aryl halides to give alkylated products.

- The organocopper reagents are more selective and can be acylated with acid chlorides without concomitant attack on ketones, alkyl halides, and esters. Relative reactivity: RCOCl > RCHO > tosylates, iodides > epoxides > bromides >> ketones > esters > nitriles.

- In reactions with α,β-unsaturated carbonyl compounds, the organocopper reagents prefer 1,4-over 1,2-addition.

Preparation of Organocuprates[56]

Homocuprate Reagents (Gilman Reagents: R_2CuLi, R_2CuMgX)

Homocuprates are widely used organocopper reagents. They are prepared by the reaction of copper(I) bromide or preferably copper(I) iodide with 2 equivalents of the appropriate lithium or Grignard reagents in ether or THF solvent.[57] The initially formed organocopper species $(RCu)_n$ are polymeric and insoluble in Et_2O and THF but dissolve on addition of a second equivalent of RLi or RMgX. The resultant organocuprates are thermally labile and thus are prepared at low temperatures.

$$RM \ + \ Cu(I)Br, I \ \xrightarrow[- MBr, I]{Et_2O \ or \ THF} \ (RCu)_n \ \underset{\xleftarrow{\hspace{1cm}}}{\xrightarrow{RM}} \ R_2CuM$$
$$M = Li, MgX$$

Heterocuprate Reagents

Since only one of the organic groups of homocuprates is usually utilized, a non-transferable group bonded to copper, such as $RC\equiv C$, 2-thienyl, PhS, t-BuO, R_2N, Ph_2P, or Me_3SiCH_2, is employed for the preparation of heterocuprate reagents. These cuprates are usually thermally more stable (less prone toward β-elimination of Cu–H), and a smaller excess of the reagent may be used.

$$R_2NLi + CuBr \cdot SMe_2 \ \xrightarrow{THF} \ R_2NCu \cdot SMe_2 \cdot LiBr \ \xrightarrow{R'Li} \ [R_2NCuR'] \ Li$$

$$RLi \ + \ \left[\underset{S}{\overset{}{\bigcirc}}\!\!-Li \right] \ + \ CuI \ \longrightarrow \ [(2\text{-thienyl})CuR] \ Li$$
$$2\text{-thienyl}$$

$$RLi \ + \ Me_3SiCH_2Li \ + \ CuI \ \longrightarrow \ [(Me_3SiCH_2)CuR] \ Li$$

Higher-Order Cyanocuprates (Lipshutz Reagents)

Cyanocuprates exhibit the reactivity of homocuprates and the thermal stability of heterocuprates. They are readily available by the reaction of $CuC\equiv N$ with 2 equivalents of RLi.[58,59] The cyanocuprates are especially useful for substitution reactions of secondary halides and epoxides.

$$CuCN \ + \ 2 \ RLi \ \xrightarrow[-78° \ to \ 0 \ °C]{THF \ or \ Et_2O} \ R_2Cu(CN)Li_2$$

Grignard-Copper(I) Reagents

Copper-catalyzed reactions of RMgX reagents are attractive when compatible with the functionality present in the starting material. The use of Grignard reagents is often the method of choice since they are readily available and only catalytic amounts of Cu(I) halides are required.[55d,60]

$$\underset{MgCl}{\diagup\!\!\diagup} \ \xrightarrow[\substack{b. \ n\text{-}C_7H_{15}I \ (Br, \ OTs) \\ THF, \ 0 \ °C}]{a. \ CuI \ (0.2 \ eq)} \ \underset{80\text{–}90\%}{\diagup\!\!\diagup n\text{-}C_7H_{15}}$$

The Cu-catalyzed alkylation of organomagnesium reagents by alkyl bromides and iodides in the presence of NMP (N-methylpyrrolidinone, a nontoxic, polar, aprotic solvent) represents an attractive alternative to the classical cuprate alkylation reaction.

Only a slight excess of the Grignard reagent is required, and the reaction tolerates keto, ester, amide and nitrile groups. This method is especially suited for large-scale preparations.[61]

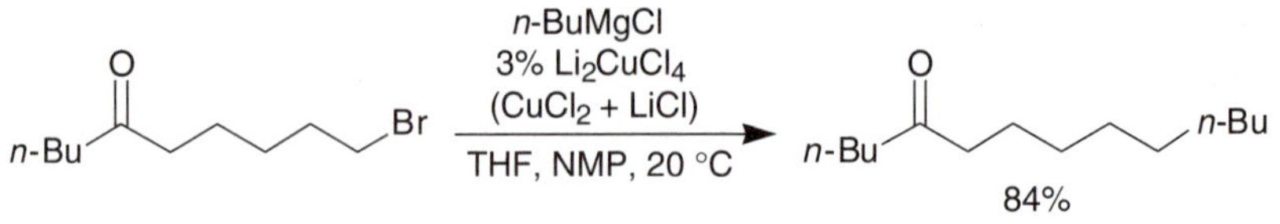

Reactions of Organocuprates

Substitution of Alkyl Halides

As depicted below, the coupling of a primary alkyl iodide with an organocuprate is more economical when using a heterocuprate than a homocuprate.[62]

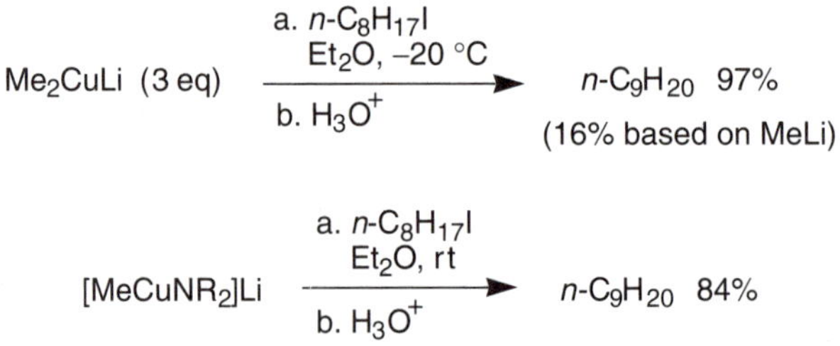

While homocuprates readily undergo substitution reactions at primary positions, they do not couple well with unactivated secondary halides. However, cyanocuprates undergo substitution reactions even at unactivated secondary carbon centers.[63]

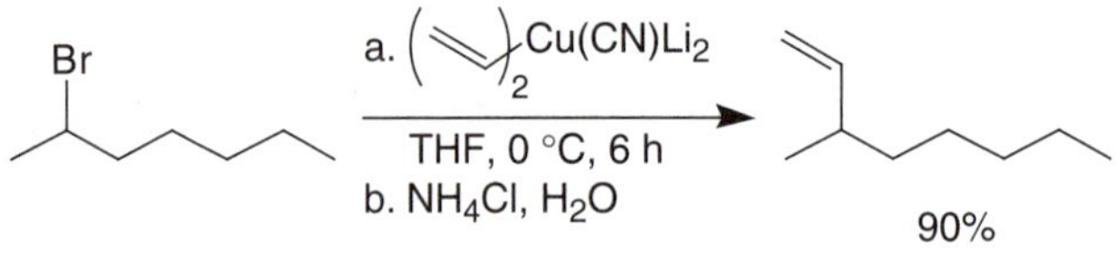

The mechanism for the substitution reaction is complex, depending on the nature of the cuprate reagent, the substrate, and the solvent used. The reaction may proceed via a S_N2 displacement or via an oxidative addition followed by reductive elimination.

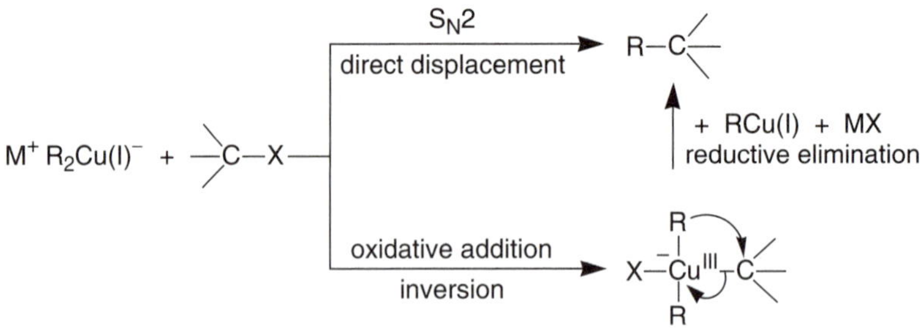

Substitution of Allylic Halides

Alkylation of allylic halides with organocuprates usually produces mixtures of products due to competing S_N2 and S_N2' reactions. Substitution with complete allylic rearrangement (S_N2' reaction) is observed with $RCu \cdot BF_3$ as the alkylating agent.[64]

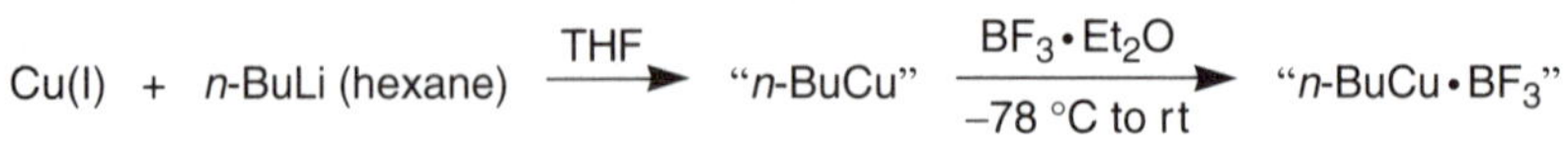

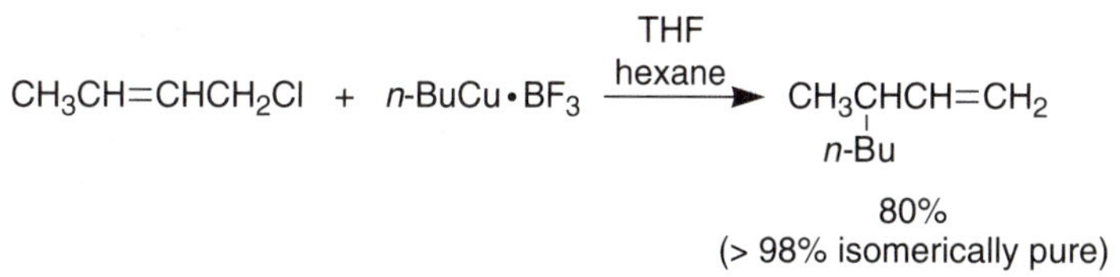

$$CH_3CH=CHCH_2Cl \ + \ n\text{-}BuCu \cdot BF_3 \ \xrightarrow[\text{hexane}]{\text{THF}} \ CH_3\underset{n\text{-}Bu}{CH}CH=CH_2$$

80%
(> 98% isomerically pure)

Reaction with Vinyl Halides

Coupling of alkenyl bromides or iodides with organocuprates proceeds with high stereoselectivity.[65]

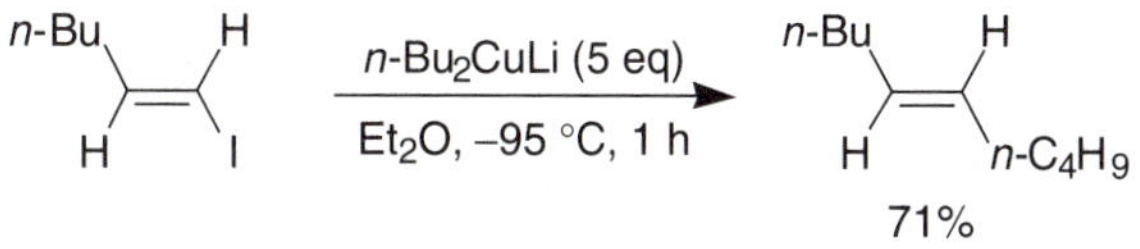

71%

Alternatively, iron-catalyzed alkenylation of organomagnesium compounds provides a highly stereo- and chemoselective synthesis of substituted alkenes.[66]

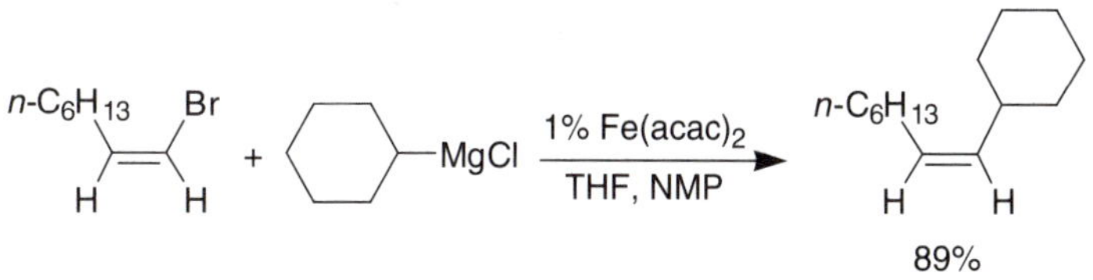

89%

Acylation

The reaction of organocopper reagents with acid chlorides affords the corresponding ketones in high yields.[55c,d] Retrosynthetically, the reaction amounts to an alkylation of a carboxylic acid.

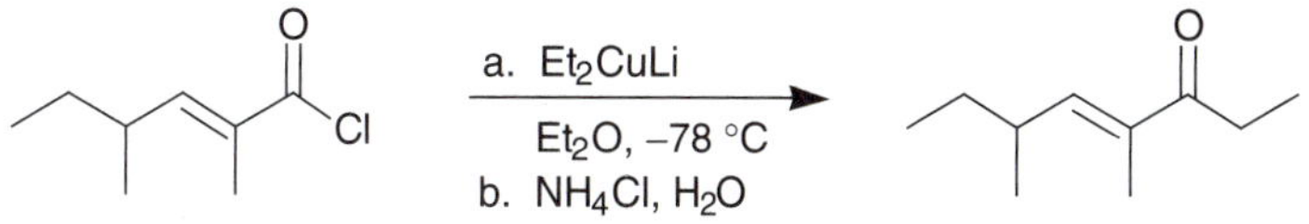

a. Et$_2$CuLi
Et$_2$O, –78 °C
b. NH$_4$Cl, H$_2$O

In the presence of a catalytic amount of CuI, Grignard reagents convert acid chlorides chemoselectively to the corresponding ketones via a transiently formed cuprate reagent, which reacts competitively with the initial Grignard reagent.[67]

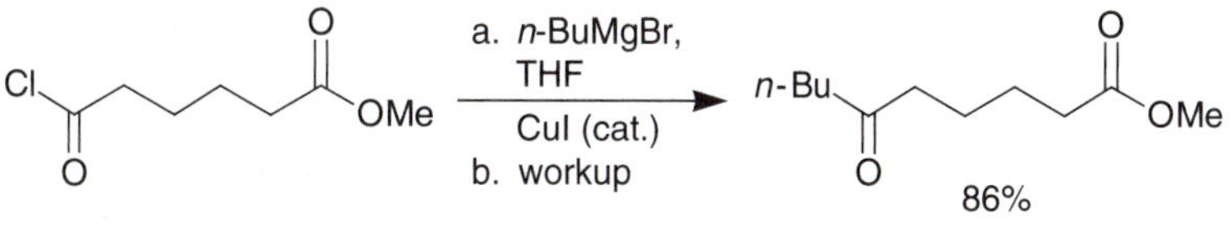

a. n-BuMgBr,
THF
CuI (cat.)
b. workup

86%

1,2-Additions to Aldehydes and Ketones

Organocuprates undergo 1,2-additions to aldehydes, ketones, and imines. These reactions are often highly diastereoselective.[68]

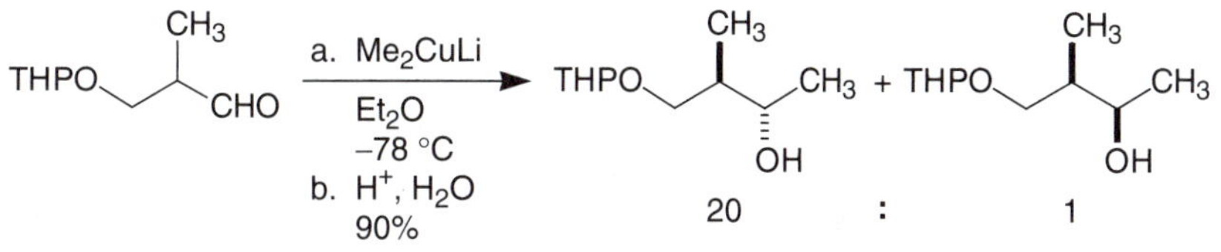

a. Me$_2$CuLi
Et$_2$O
–78 °C
b. H$^+$, H$_2$O
90%

20   :   1

Epoxide Cleavage Reactions

$R_2Cu(CN)Li_2$ reagents are among the mildest and most efficient reagents available for generating carbon-carbon bonds by way of epoxide cleavage using organocopper chemistry. The nucleophilic addition occurs at the less sterically hindered carbon of the oxirane ring.[59,63,69]

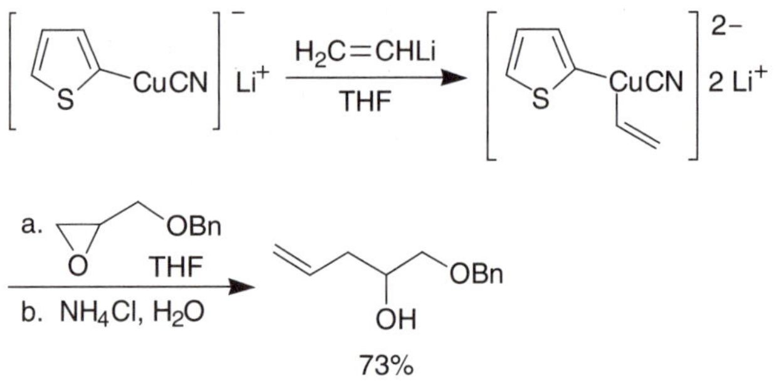

Stereospecific S_N2 opening of cyclic epoxides with cyanocuprates furnishes, after workup, the *trans*-2-hydroxy-alkylated products.

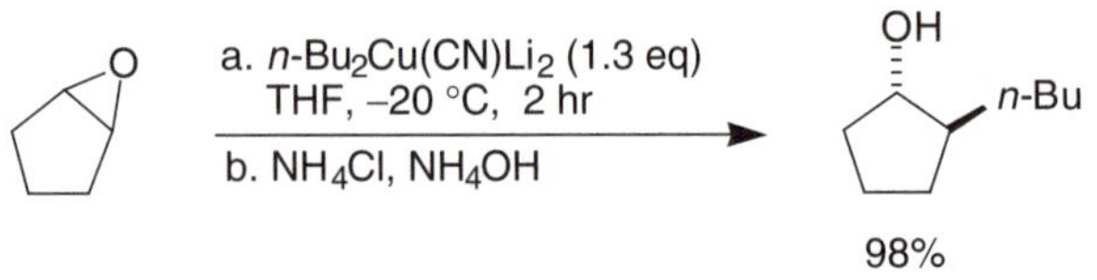

However, the unsaturated epoxide shown below reacts with cyanocuprates via an *anti*-S_N2'-type mechanism. Directed epoxidation of the resultant allylic alcoholate produces a hydroxy epoxide containing four stereodefined carbon centers.

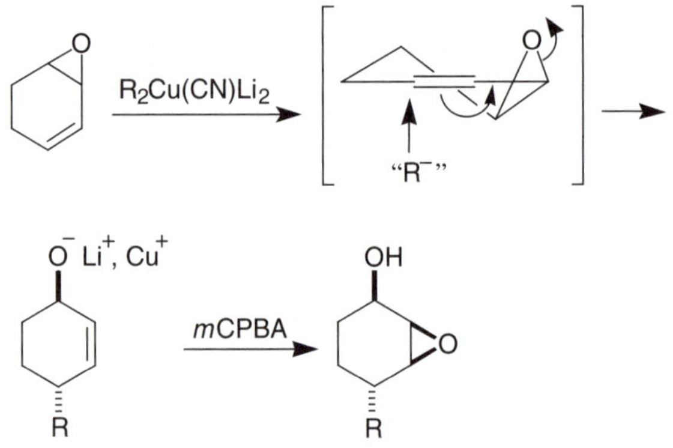

Conjugate Addition[55b,55c,70]

Conjugate addition is an important C–C bond formation strategy available to the organic chemist. Organometallic reagents may add in a 1,2- or 1,4-manner to α,β-unsaturated carbonyl compounds (Table 7.6). *1,4-Addition* (conjugate addition) is most successful

Table 7.6	Regioselectivity in Addition of RLi, RMgX, and Organocopper Reagents to α, β-Unsaturated Carbonyl Compounds	
Nucleophile	**1,2-Addition**	**1,4-Addition**
RLi	+	—
RMgX	+	—
R_2CuLi	—	+
RMgX • CuX	—	+

with "soft" (relatively nonbasic) nucleophiles such as ⁻C≡N, RNH_2, R_2NH, RSH, enolates derived from β-dicarbonyl compounds, and organocuprates. *1,2-Addition* is most successful with "hard" (relatively basic) nucleophiles such as hydride, organolithiums, and Grignard reagents. The classification of hard and soft bases (nucleophiles) and hard and soft acids (electrophiles) has recently been reviewed.[71]

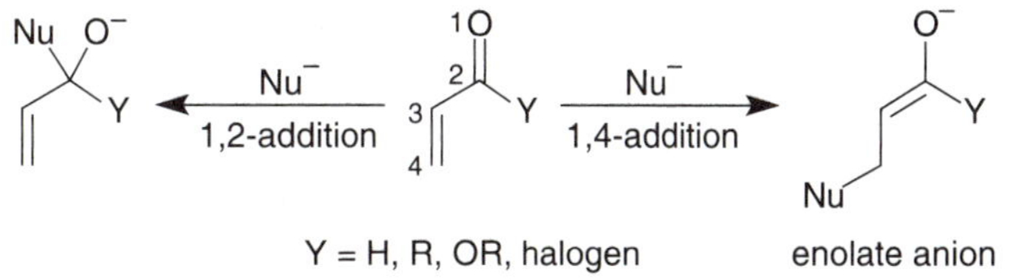

Y = H, R, OR, halogen   enolate anion

The organocopper reagents used for conjugate additions to enones are homocuprates, heterocuprates, higher-order cuprates, and Grignard reagents in the presence of catalytic amounts of copper salts (CuX).

Addition of organocopper reagents to α,β-unsaturated carbonyl compounds (enones and conjugated esters) generates enolates with concomitant introduction of an organic group at the β-position. In the bicyclic system shown below, the addition is chemoselective, involving the less hindered double bond of the dienone. The reaction is also stereoselective in that introduction of the "Me" group occurs preferentially from the less hindered side of the molecule.

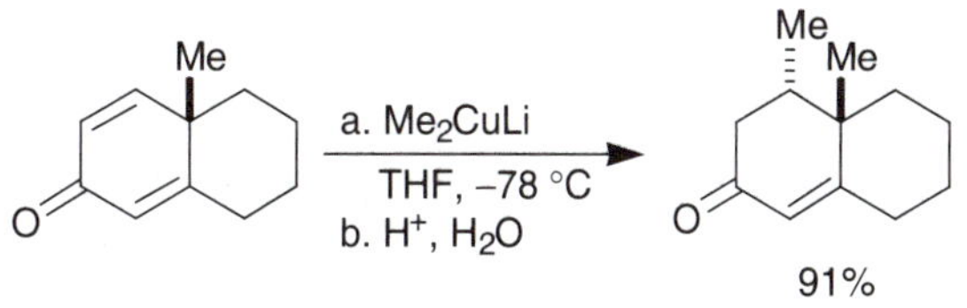

The mechanistic picture for addition of organocuprates to α,β-unsaturated carbonyl compounds is no less complex than that for substitution reactions. On the basis of current information, conjugate addition of lithiocuprates to α, β-unsaturated ketones and esters may proceed via a initial reversible copper(I)-olefin-lithium association, which then undergoes oxidative addition followed by reductive elimination.[72]

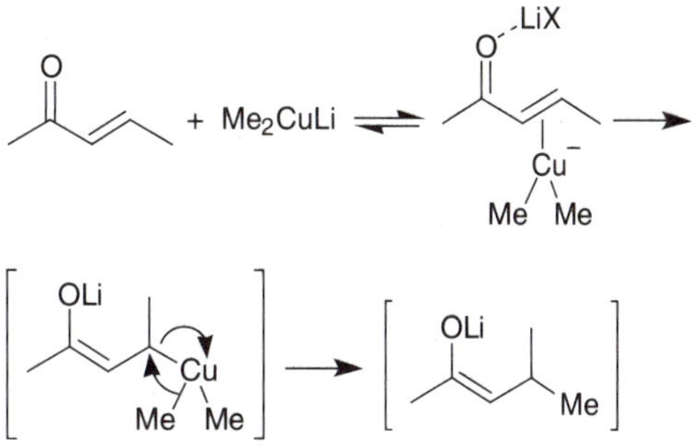

Conjugate additions of organocopper reagents with large steric requirements and/or when there is steric hindrance at the reaction center of the enone may be difficult. Addition of Me_3SiCl accelerates the conjugate additions of copper reagents to such enones, probably by activating the carbonyl group.[73] For example, 3-methylcyclohexenone is essentially inert to n-Bu_2CuLi at –70 °C in THF. However, in the presence of Me_3SiCl the enolate initially formed is trapped to give the β-disubstituted silyl enol ether in 99% yield. Hydrolysis of the silyl enol ether regenerates the carbonyl group.

Reactions of β,β-disubstituted enones with organocuprates are often not very successful because of steric crowding of the double bond. In these cases, use of $R_2CuLi–BF_3 \cdot OEt_2$ often obviates the problem.[74,75] Possibly, the Lewis acid BF_3 further polarizes and activates the ketone by coordination.

Grignard reagents in the presence of CuX or in the presence of a mixture of $MnCl_2$ and CuI undergo 1,4-addition to hindered enones.[76]

The reaction of dialkylcuprates with α,β-unsaturated aldehydes results in the preferential 1,2-addition to the carbonyl group. However, in the presence of Me_3SiCl, conjugate addition prevails to furnish, after hydrolysis of the resultant silyl enol ether, the saturated aldehyde.

Conjugate additions of dialkylcuprates to β-substituted-α,β-unsaturated acids and esters give low yields. Addition of boron trifluoride etherate, $BF_3 \cdot OEt_2$, to certain dialkylcuprates and higher-order cuprates enhances their reactivity in Michael additions to conjugated acids and esters.[77]

Tandem 1,4-Addition—Enolate Trapping

One of the fundamental contributions of organocopper chemistry to organic synthesis is the ability to transfer a variety of ligands in a 1,4-manner to α,β-unsaturated carbonyl compounds to produce enolate anions in a *regioselective* manner. These may be trapped by a variety of electrophiles in *tandem*-type reactions.[78] The enolates produced from conjugate additions of organocuprates to α, β-unsaturated carbonyl compounds possess two nucleophilic sites, reacting either at the oxygen or the carbon terminus. Electrophiles having a high affinity toward oxygen such as chlorosilanes and phosphorochloridates tend to give *O*-trapping products, whereas alkyl halides, aldehydes, α-halocarbonyl compounds, and halogens furnish *C*-trapping products.

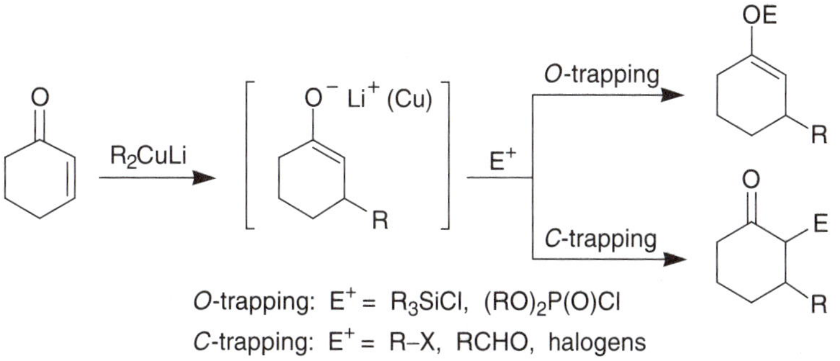

$$O\text{-trapping: } E^+ = R_3SiCl, (RO)_2P(O)Cl$$
$$C\text{-trapping: } E^+ = R\text{–}X, RCHO, \text{halogens}$$

O-Trapping. The enolate generated from the enone shown below reacts at oxygen with chlorotrimethylsilane in the presence of triethylamine to produce the trimethylsilyl enol ether.[78] Silyl enol ethers are valuable intermediates for the preparation of regiodefined enolates (see Chapter 6).

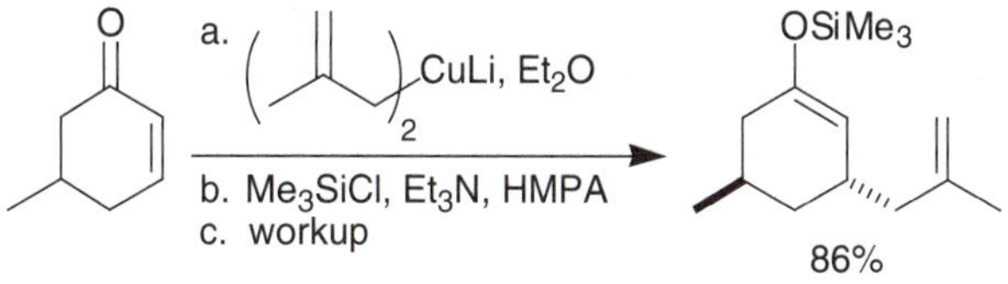

Treatment of enolates with $(RO)_2P(O)Cl$ also results in *O*-trapping to yield the corresponding enol phosphates. Dissolving metal reduction of enol phosphates is a useful procedure for the deoxygenation of ketones with concomitant, regiospecific formation of the alkene.[79]

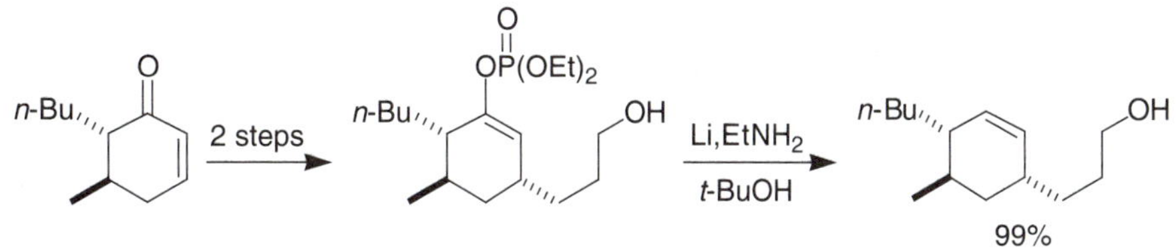

C-Trapping. Alkylation or hydroxyalkylation (i.e., reaction with RCHO) of enolates derived from conjugate addition of organocuprates affords vicinal dialkylated products. However, the reaction is confined to highly reactive alkylating agents such as methyl, allyl, propargyl, benzyl, and α-halocarbonyl compounds or aldehydes.

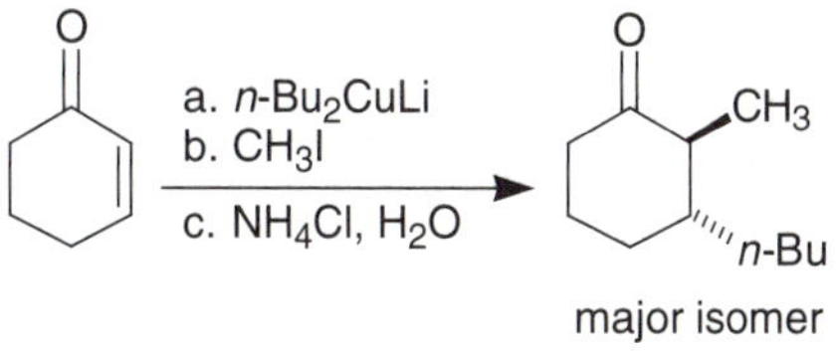

major isomer

Enolates obtained from conjugate addition of either homocuprates or copper-catalyzed Grignard reagents undergo aldol condensation with aldehydes in the presence of $ZnCl_2$ to give stereoisomeric mixtures of aldol products.[78]

Stereochemistry of 1,4-Addition Reactions

Factors controlling the stereochemistry of conjugate additions are not well understood. Mixtures of isomers are often produced, but generally one isomer predominates. Both steric and electronic factors play a role.[80] Generally, Michael-type additions have late and hence productlike—and chairlike—transition states. In the example shown below, for stereoelectronic reasons antiparallel attack by the nucleophilic "CH_3" is favored over parallel attack.

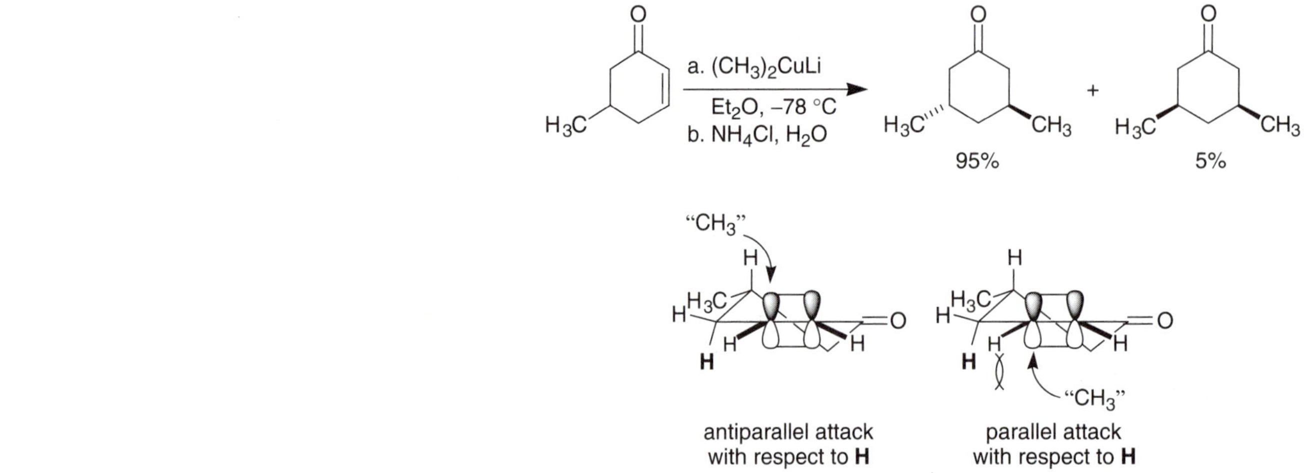

antiparallel attack
with respect to **H**

parallel attack
with respect to **H**

Note that the 4-methylcyclohexenone gives preferentially the *trans*-product. In this case, the reaction proceeds via a conformation with the Me group being pseudoaxial due to $A^{1,2}$ strain.

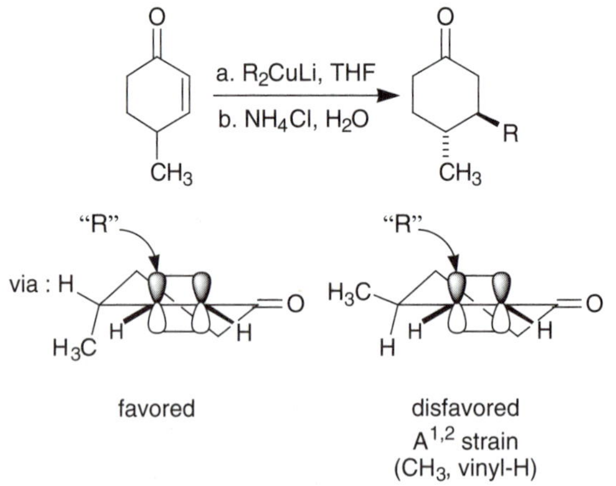

favored

disfavored
$A^{1,2}$ strain
(CH_3, vinyl-H)

As pointed out earlier, addition of organocuprates to enones followed by alkylation of the resultant enolates generates two carbon-carbon bonds in a single reaction. *Alkylation of an enolate* proceeds via an early, hence a reactantlike, transition state. Thus, steric factors in the ground state play an important role.

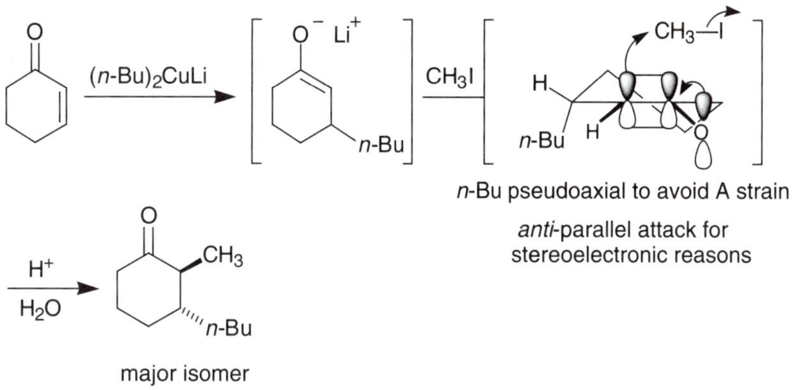

n-Bu pseudoaxial to avoid A strain

anti-parallel attack for
stereoelectronic reasons

major isomer

Preparation of Enones: Substrates for Conjugate Additions

α,β-Unsaturated ketones may be prepared using one of several methods described below:

- From allylic alcohols via oxidation (see Section 4.3)
- From ketones via bromination-dehydrobromination[81]

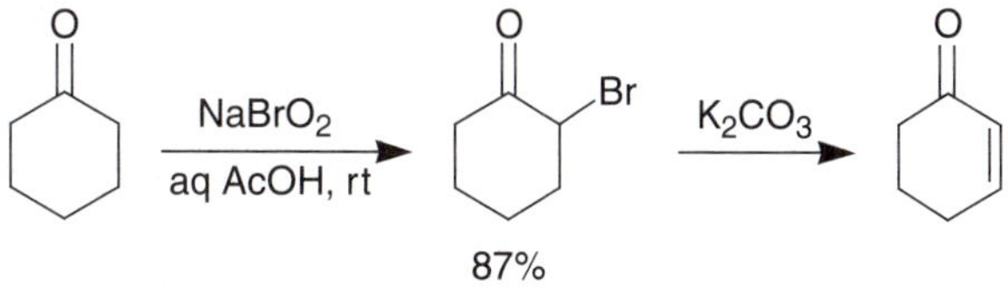

87%

- From silyl enol ethers by iodination (*N*-iodosuccinimide) followed by dehydroiodination[82]

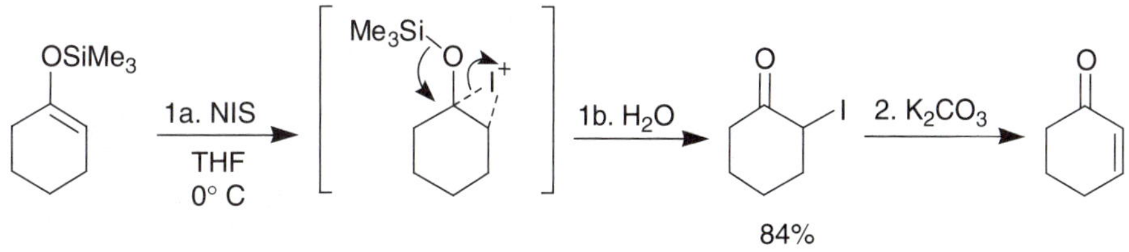

84%

- From alcohols or ketones via oxidation with *o*-iodoxybenzoic acid (IBX)[83]

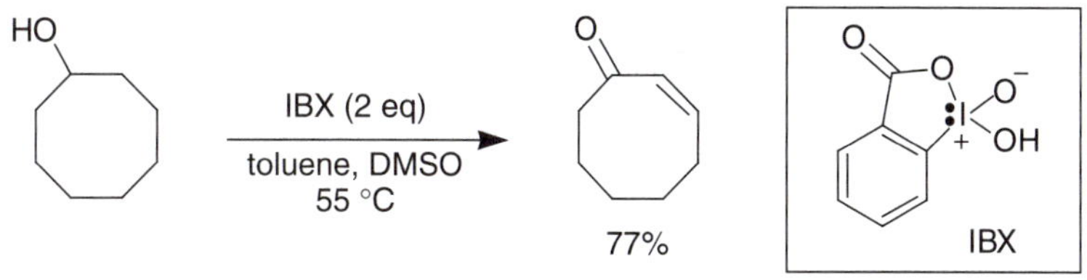

77%   IBX

- By *Saegusa-Ito oxidation*.[84] Dehydrosilylation of silyl enol ethers with palladium(II) acetate and *p*-benzoquinone in acetonitrile occurs regioselectively to furnish the corresponding α,β-unsaturated compound. The Saegusa-Ito oxidation is particularly powerful when used in tandem with cuprate chemistry to restore the α,β-unsaturation.

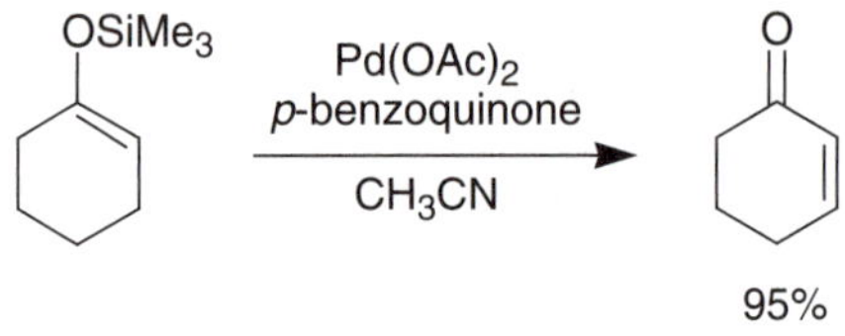

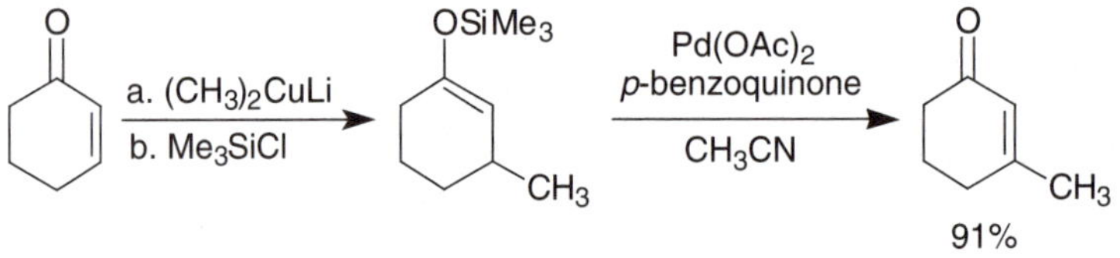

95%

91%

- Aldehydes can be converted to α,β-unsaturated aldehydes in a one-pot transformation by in situ silyl enol ether formation followed by the Pd(II)-catalyzed dehydrosilylation.[85]

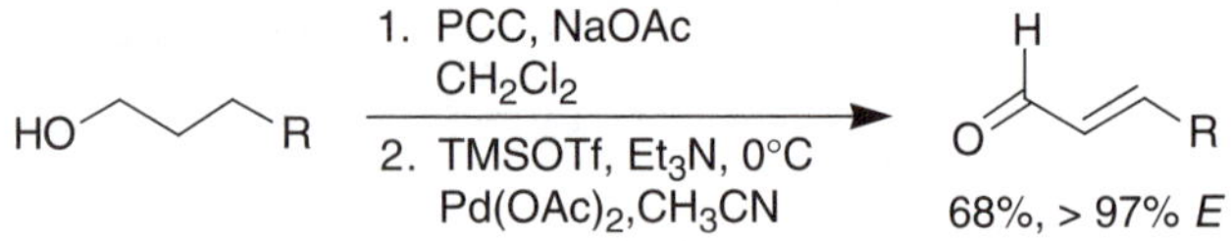

1. PCC, NaOAc
   CH₂Cl₂
2. TMSOTf, Et₃N, 0°C
   Pd(OAc)₂,CH₃CN

68%, > 97% E

- From enolates via selenoxides.[86] Lithium enolates derived from ketones, lactones, and esters react with PhSe-SePh or with phenylselenyl bromide or chloride (PhSeX) to form α-(phenylseleno)carbonyl compounds. These can be oxidized to the corresponding selenoxides with subsequent *syn*-elimination of benzeneselenic acid to form enones.

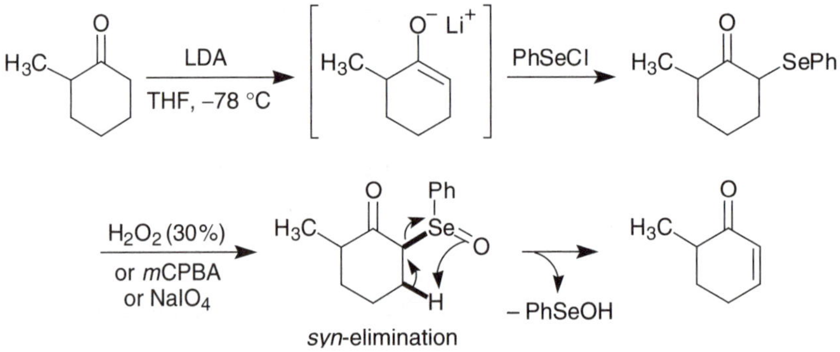

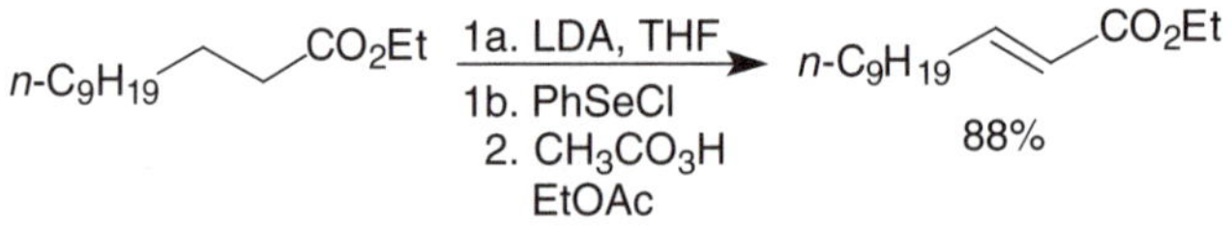

syn-elimination

88%

7.6   ORGANOCHROMIUM REAGENTS[87]

The Nozaki-Hiyama Reaction[88]   Although not derived from organolithium or Grignard reagents, organochromium compounds are especially useful for coupling an allylic moiety to an aldehyde or to a ketone. The organochromium reagents are obtained by oxidative addition of Cr(II) chloride to allylic halides or tosylates. Their reaction with aldehydes and ketones produces homoallylic alcohols in which the more substituted γ-carbon of the allyl group becomes attached to the carbonyl carbon.

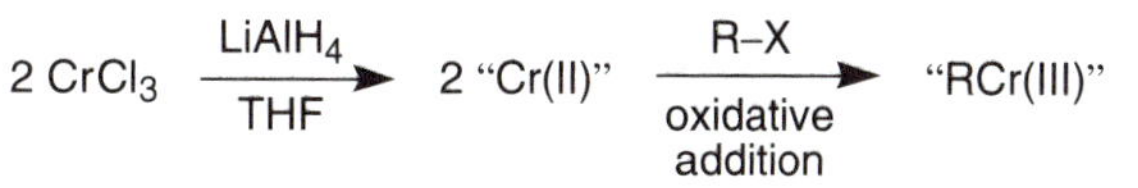

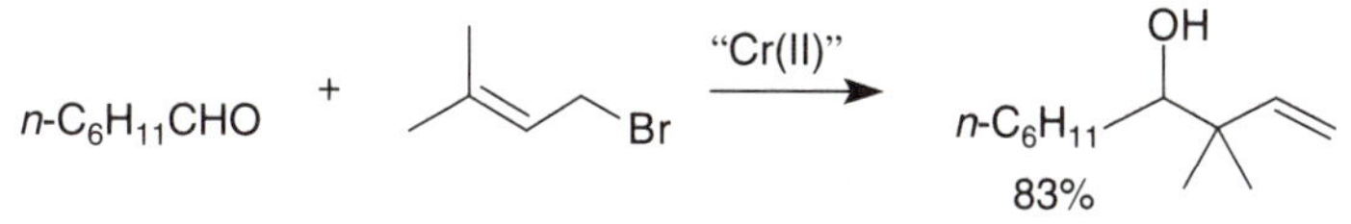

$$2 \; CrCl_3 \xrightarrow[\text{THF}]{\text{LiAlH}_4} 2 \; \text{"Cr(II)"} \xrightarrow[\substack{\text{oxidative} \\ \text{addition}}]{\text{R–X}} \text{"RCr(III)"}$$

Reaction of vinyl or aryl iodides or bromides with chromium(II) chloride produces chromium(III) species that react chemoselectively with aldehydes to give allylic or benzylic alcohols, respectively, in high yields.[89a]

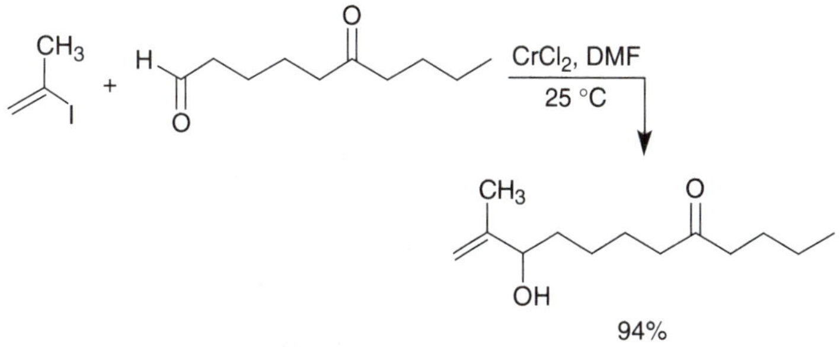

The reduction of *gem*-diiodo- or *gem*-triiodoalkanes, such as iodoform, with $CrCl_2$ in the presence of aldehydes furnishes (*E*)-alkenes,[89b] as exemplified below.[89c]

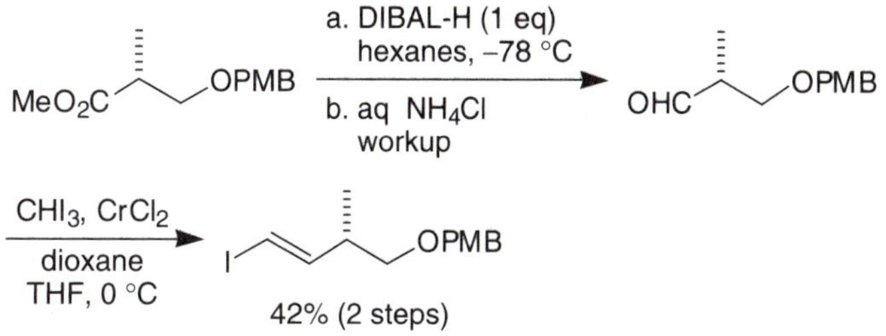

The Nozaki-Takai-Hiyama-Kishi Coupling[90]

Traces of nickel salts exert a catalytic effect on the formation of the C–Cr bond, allowing $CrCl_2$–$NiCl_2$-mediated coupling of vinyl bromides, iodides,[90c] and triflates (shown below[90d]) with aldehydes. The stereochemistry of the vinyl halide is retained in the allylic alcohol product.

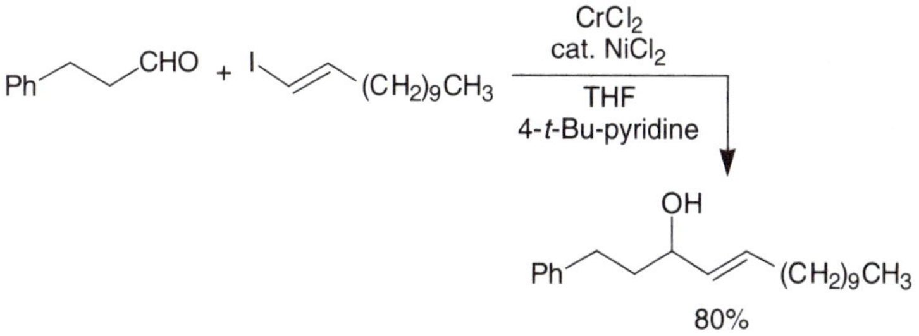

7.7 ORGANOZINC REAGENTS[91]

In contrast to the polar nature of C–Li and C–MgX bonds, the C–Zn bond is highly covalent and hence less reactive, allowing the preparation of functionalized derivatives. Utilization of organozinc reagents in organic synthesis has mainly centered around the preparation and utilization of functional organozinc compounds in organic syntheses (Reformatsky reaction), cyclopropanation (Simmons-Smith reaction), and transmetalations with transition metals.

Preparation of Organozinc Compounds[92]

Alkylzinc Iodides

Primary and secondary alkylzinc iodides (RZnI) are best prepared by direct insertion of zinc metal (zinc dust activated by 1,2-dibromoethane or chlorotrimethylsilane) into alkyl iodides or by treating alkyl iodides with *Rieke* zinc.[93] The zinc insertion shows a remarkable functional group tolerance, permitting the preparation of polyfunctional organozinc reagents.[94]

$$ZnCl_2 \xrightarrow{\text{Li-naphthalenide}} \underset{\text{(Rieke zinc)}}{Zn} \xrightarrow[\text{THF, 25–60 °C}]{\textbf{FG}\text{—RX}} \left[\textbf{FG}\text{—RZnX}\right]$$

R: alkyl, aryl, benzyl, allyl
X: Br, I
FG: CO$_2$R, enolate, CN, halide, etc.

Dialkylzincs

Unfunctionalized dialkylzincs (R$_2$Zn) are obtained by transmetalation of zinc halides, such as ZnCl$_2$, with organolithium or Grignard reagents. Iodide-zinc exchange reactions catalyzed by CuI provide a practical way for preparing functionalized dialkylzincs.

$$2\ i\text{-BuMgBr} + ZnCl_2 \longrightarrow i\text{-Bu}_2Zn + 2\ MgClBr$$

$$\textbf{FG}\text{—RCH}_2\text{I} \xrightarrow[\text{neat}]{\text{Et}_2\text{Zn, CuI (cat.)}} (\textbf{FG}\text{—RCH}_2)_2\text{Zn}$$

Zinc Carbenoids (Carbene-like Species)

The oxidative addition of zinc metal to diiodomethane affords an iodomethylzinc iodide, tentatively assigned as ICH_2ZnI (Simmons-Smith reagent), which is used for cyclopropanation of alkenes. Alkyl group exchange between diethylzinc and diiodomethane produces the iodomethyl zinc carbenoid species, tentatively assigned as $EtZnCH_2I$ (Furukawa's reagent).

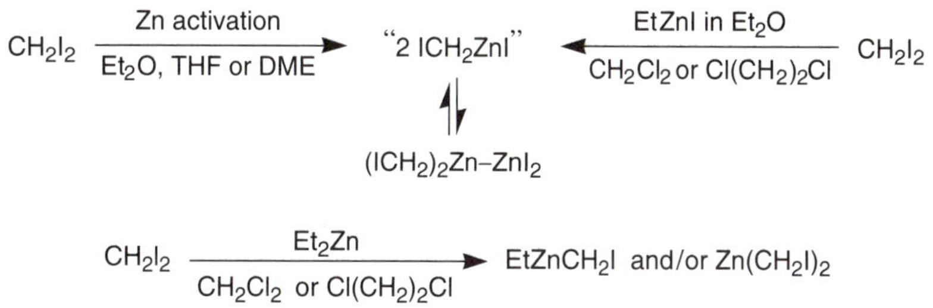

Reactions of Organozinc Compounds

The Reformatsky Reaction[95]

The Reformatsky reaction involves condensation of ester-derived zinc enolates with aldehydes or ketones to furnish the corresponding β-hydroxy esters. The zinc enolates are generated by addition of an α-haloester in THF, DME, Et_2O, benzene, or toluene to an activated zinc, such as a Zn-Cu couple or zinc obtained by reduction of zinc halides with potassium (Rieke zinc). An example of a Reformatsky condensation using Rieke zinc is shown below.[93a]

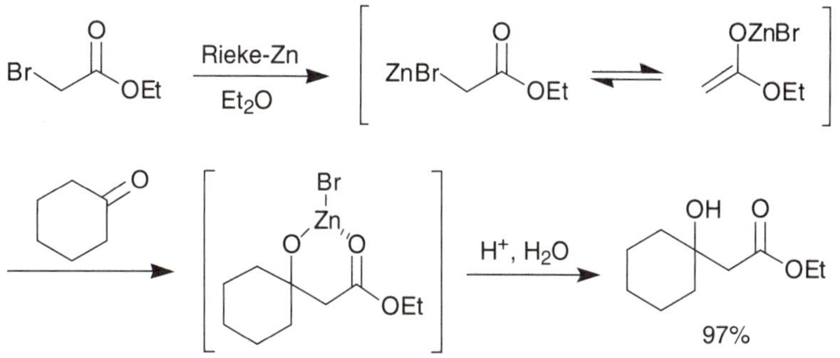

An important application of the Reformatsky reaction is the conversion of β-hydroxy esters to α, β-unsaturated esters. Acid-catalyzed dehydration usually leads to a mixture of α, β- and β, γ-unsaturated esters. However, conversion of the initially formed β-hydroxy esters to their corresponding acetates by treatment with acetyl chloride, followed by base-catalyzed dehydration with NaOEt, produces conjugated esters in high purity.[96] This sequence of reactions provides an alternative route to the Horner-Wadsworth-Emmons olefination of ketones (see Chapter 8).

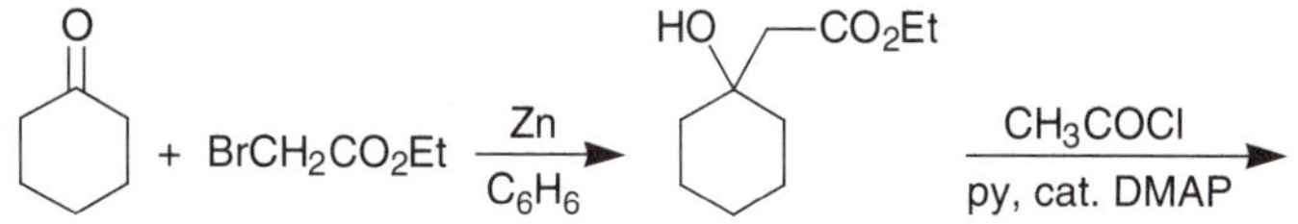

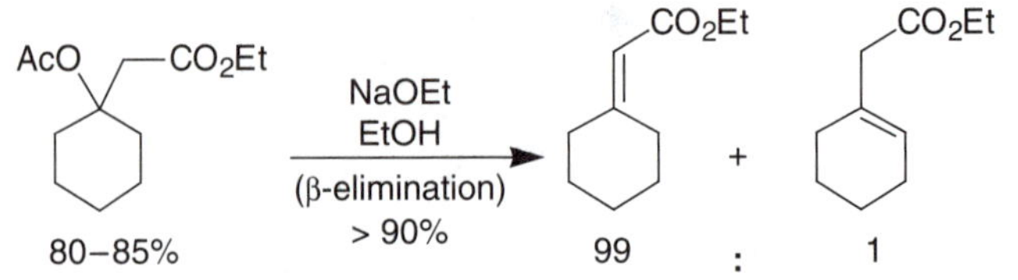

80–85%   99 : 1

β-Hydroxy esters and α,β–unsaturated esters may also be prepared either via a single-step *Barbier-type condensation* of a *t*-Bu-bromoester with carbonyl compounds in the presence of Mg metal or via a two-step condensation with Li-enolates, as depicted below.

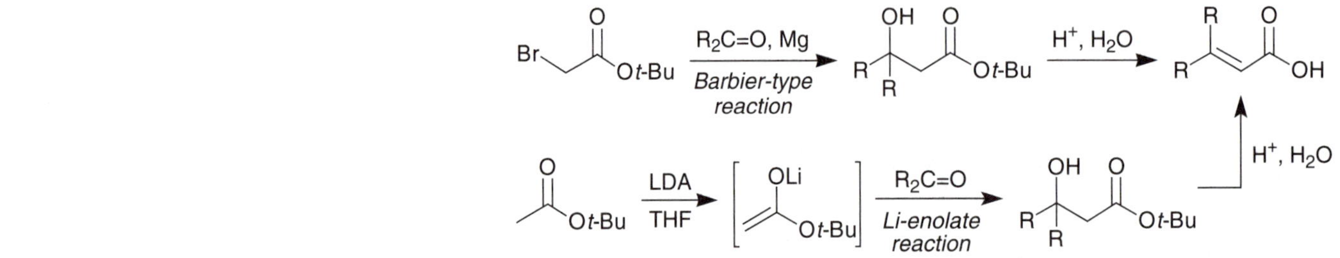

A feature distinguishing Reformatsky enolates from base-generated enolates is that zinc enolates add to highly hindered as well as to easily enolizable ketones, such as cyclopentanones, thus avoiding formation of condensation products. Moreover, there is no danger of a Claisen-type self-condensation since zinc-enolates do not react with esters but react readily with aldehydes and ketones to furnish aldol-type products.[97]

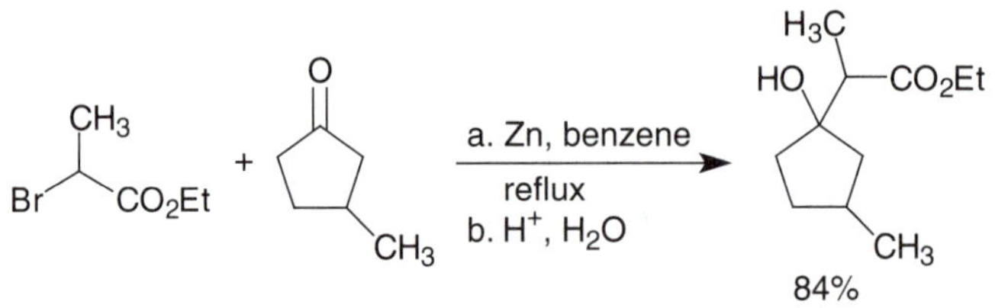

84%

Enantioselective addition of organozinc reagents to carbonyl compounds furnishes chiral alcohols.[98]

Reactions of Functionally Substituted RZnI[99]

The application of functionally substituted organozincs allows for the construction of carbon-carbon bonds while circumventing tedious protection-deprotection strategies, as exemplified below.[100, 101] Note that tosyl cyanide reacts with alkenyl or arylzinc reagents to provide α,β-unsaturated alkenyl or aromatic nitriles, respectively.[102]

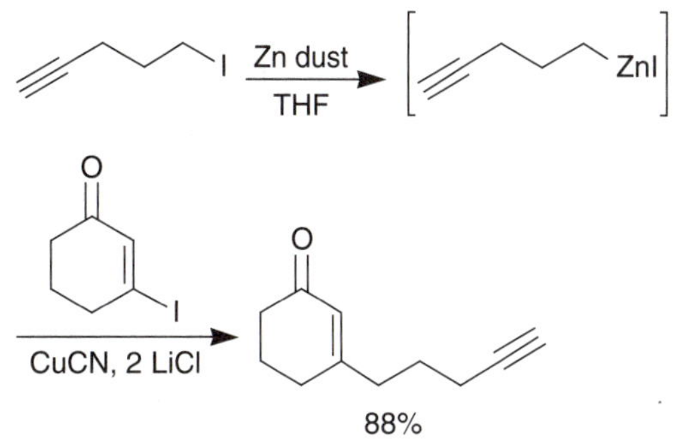

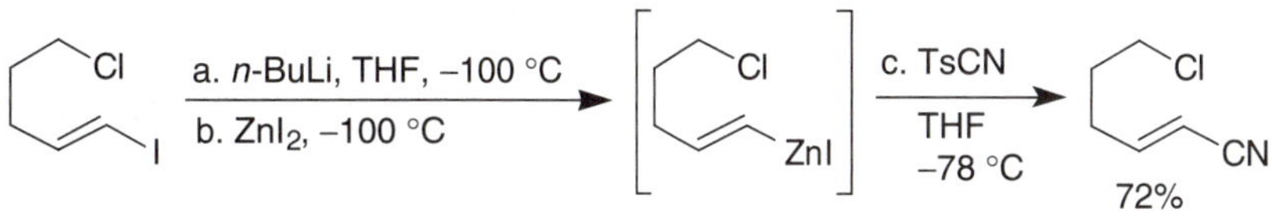

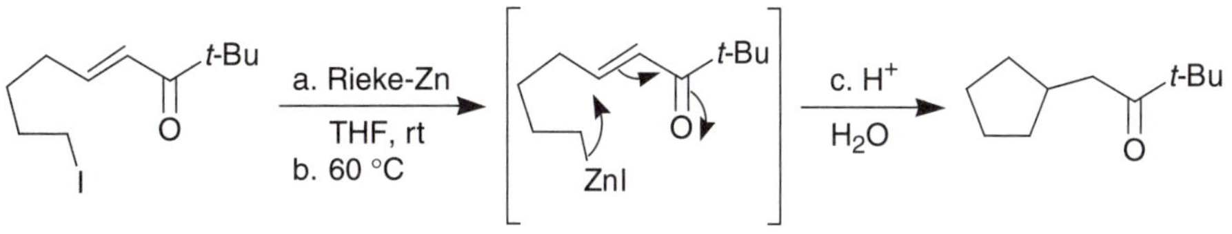

Cyclopropanation[103]

Cyclopropane rings are encountered in many natural products possessing interesting biological activities, such as pyrethrin (shown below), a natural insecticide from the pyrethrum daisy. Moreover, the cyclopropane moiety represents a useful synthon for further synthetic transformations.[104]

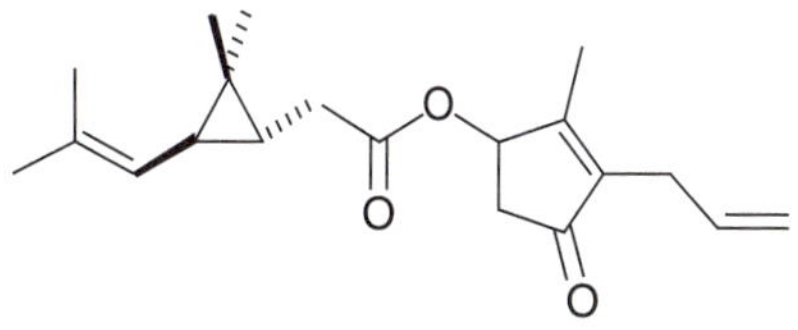

In 1958, Simmons and Smith reported that treatment of a zinc-copper couple with diiodomethane in ether produces a reagent that adds to alkenes to form cyclo-propanes.[105]

$$CH_2I_2 \ + \ Zn(Cu) \xrightarrow[35\ °C]{Et_2O} ICH_2ZnI$$

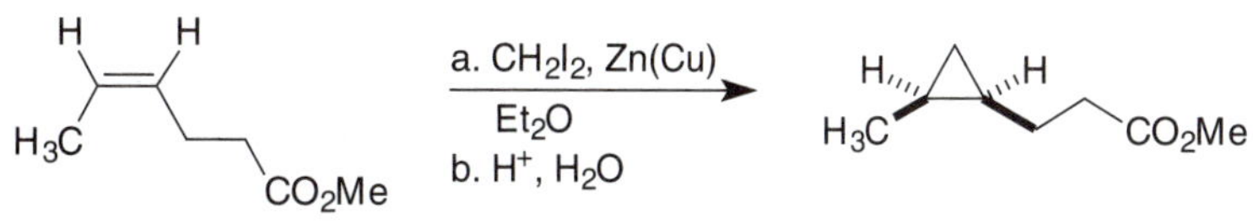

The cyclopropanation reaction of simple alkenes appears to proceed via *stereo-specific syn*-addition of a Zn-carbenoid (carbene-like species) to the double bond without the involvement of a free carbene.

The synthetic utility of the reaction stems from the following characteristics: (1) stereospecificity (retention of olefin geometry); (2) tolerance of a variety of functional groups, such as Cl, Br, OH, OR, CO_2R, C=O, and CN; (3) the *syn*-directing effect of hydroxyl and ether functions; and (4) chemoselectivity—zinc carbenoids are electrophilic and react chemoselectively with the more nucleophilic double bond in dienes and polyenes.[106]

A number of modifications of the original Simmons-Smith cyclopropanation procedure have been reported.[107] Furukawa's reagent, (iodomethyl)zinc derived from diethylzinc and diiodomethane,[108] or its modification using chloroiodomethane instead of diiodomethane,[109] allows more flexibility in the choice of solvent. The reagent is homogeneous and the cyclopropanation of olefins can be carried out in non-complexing solvents, such as dichloromethane or 1,2-dichloroethane, which greatly increase the reactivity of the zinc carbenoids.[110]

An alternative to the Simmons-Smith and Furukawa reagents is iodomethylzinc phenoxide, readily accessible by deprotonation of phenol with Et_2Zn and subsequent metal-halogen exchange with CH_2I_2.[111] An economically attractive method for cyclopropanation of alkenes is to use CH_2Br_2, which is considerably less expensive and easier to purify and store than CH_2I_2.[112]

Directed Simmons-Smith Cyclopropanation

A particularly interesting aspect of the Simmons-Smith reaction is the *stereoelectronic* control exhibited by proximal OH, OR groups, which favor cyclopropanation to occur from the same face of the double bond as the oxy substituents.[113,] The following order of decreasing directive effects has been observed: OH > OR > C=O.

Allylic alcohols undergo cyclopropanation faster than unfunctionalized alkenes,[114] and excellent diastereoselectivities have been observed in reactions of the Furukawa reagent with acyclic chiral allylic alcohols.[115]

81%

syn : anti > 100 : 1

87%

Transmetalation Reactions

Although the C–Zn bond is rather unreactive toward electrophiles such as aldehydes, ketones, esters, and nitriles, it undergoes transmetalations with many transition metal complexes to furnish new organometallics capable of reacting with a variety of electrophilic substrates (see Section 7.10).

7.8   ORGANOBORON REAGENTS[116]

There is a vast array of carbon-carbon-bond-forming reactions involving organoboron compounds. This section covers transfer reactions of carbon moieties from boron to an adjacent carbon. Examples include the transfer of CO, CN, carbanions derived from dichloromethyl methyl ether and dichloromethane, and allyl groups in reactions of allylic boranes with aldehydes.

Tricoordinate boron compounds are Lewis acids and form complexes when treated with nucleophiles. If the resultant tetracoordinate organoborate (*ate*-complex) possesses an α-leaving group at the boron atom, a 1,2-migration of an alkyl group from boron to the migrating terminus can occur as in the carbonylation, cyanidation, dichloromethyl methyl ether (DCME), and boronic ester homologation reactions.[117] The key features to all of these reactions are formation of the borate anion as the transient species and intramolecular anionotropic 1,2-migration with *retention of configuration of the migrating group*. The products of 1,2-migration reactions are new organoboranes that can be converted into a variety of organic compounds via oxidation, protonolysis, etc.

Carbonylation[116f,118]

The reaction of organoboranes with carbon monoxide [(−):C≡O:(+)], an ylide, gives rise to three possible rearrangements. Depending on the reaction conditions, single, double, and triple migrations of groups may occur leading, after the appropriate workup, to a variety of aldehydes and ketones as well as primary, secondary, and tertiary alcohols.

Synthesis of Aldehydes and Primary Alcohols

Addition of CO to trialkylboranes leads to the formation of an *ate*-complex (structure A below). Subsequent migration of an R group to the CO ligand yields intermediate B. In the presence of a mild reducing agent, such as lithium trimethoxyaluminum hydride or potassium triisopropoxyborohydride, B is converted to the mono-migration product C. Oxidation (H_2O_2) of C furnishes the corresponding aldehyde. On the other hand, treatment of C with $LiAlH_4$ followed by oxidative workup produces the primary alcohol.

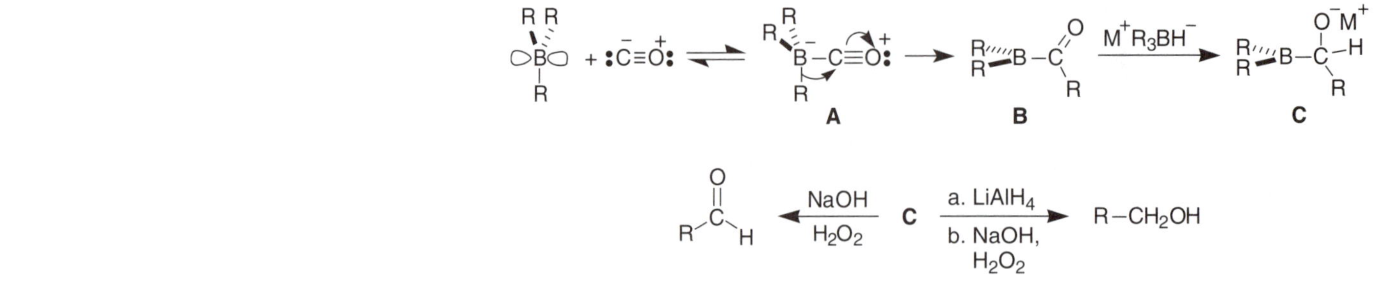

To maximize utilization of valuable alkyl groups, alkenes used for the transfer reaction are hydroborated with 9-BBN. The resultant alkyl-9-BBN derivatives undergo selective migration of the alkyl group when treated with CO in the presence of a reducing agent. As shown below, applying this sequence of reactions to 2-methylcyclopentene produces *trans*-2-methylcyclopentane carboxaldehyde, indicating retention of configuration of the migrating group.[119]

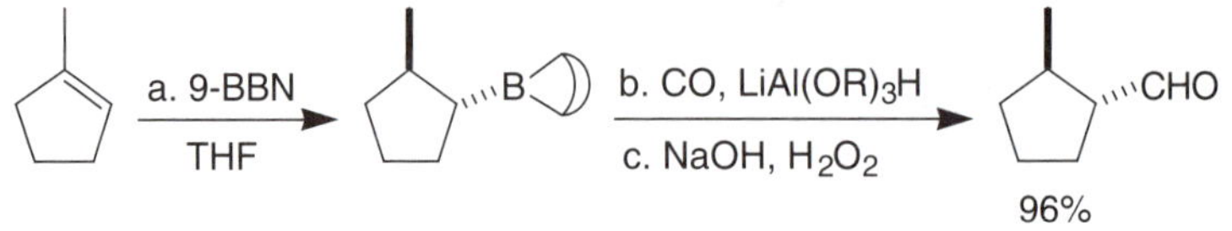

Synthesis of Ketones

If the carbonylation reaction is done in the presence of a small amount of water at 100 °C, a second alkyl group migrates from B to the adjacent carbon to furnish, after oxidative workup, the corresponding ketone.

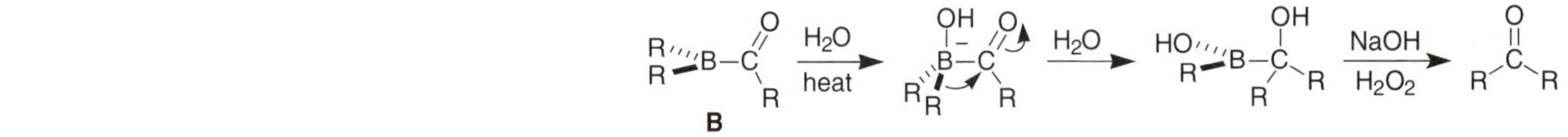

The use of thexylborane (1,1,2-trimethylpropylborane) as the hydroborating agent permits (a) the synthesis of mixed trialkyboranes, and (b) cyclic hydroboration of dienes. When followed by carbonylation, these hydroboration-carbonylation sequences generate a variety of unsymmetrically substituted ketones[120] and cyclic ketones,[121] respectively. Since the thexyl moiety exhibits a low migratory aptitude in carbonylation reactions, it serves as an anchor group.

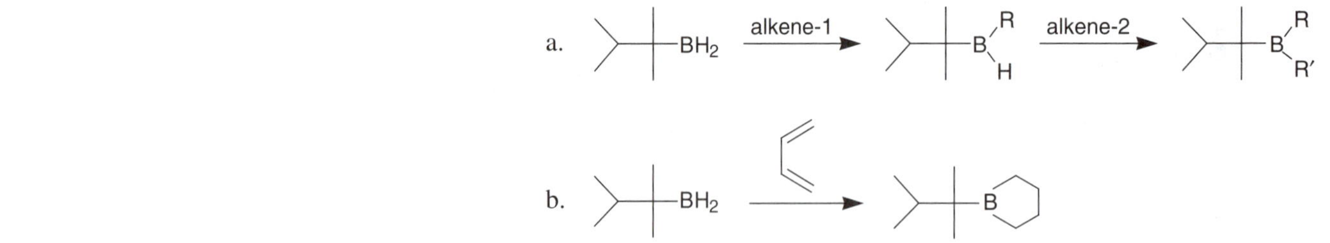

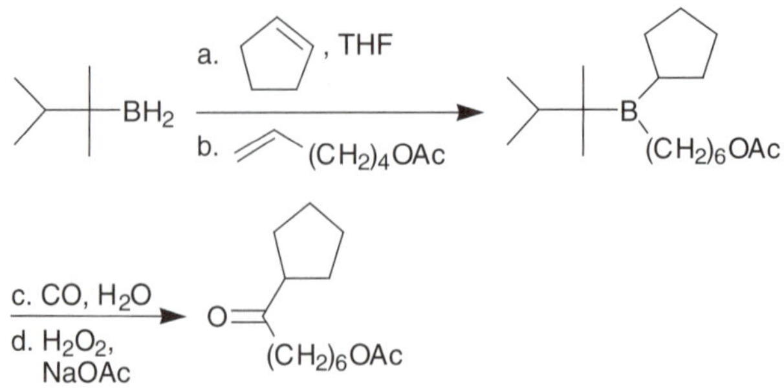

Synthesis of *tert*-Alcohols

Carbonylation of trialkylboranes in the presence of ethylene glycol results in migration of a second and a third alkyl group to give, after oxidation, the corresponding *tert*-alcohols.[122]

Hydroboration of polyenes followed by carbonylation and oxidation provides access to carbocyclic systems.[123]

62%

Cyanidation[124] A useful alternative to the carbonylation route to ketones and trialkylmethanols from alkylboranes is the cyanidation reaction. The nitrile anion [(−):C≡N:] is isoelectronic with CO and also reacts with R_3B. However, the cyanoborate salts are thermally stable and therefore require an electrophile such as benzoyl chloride or trifluoroacetic anhydride (TFAA) to induce 1,2-migration.

The formation of ketones and trialkylmethanols occurs under milder conditions than when using CO. Again, the thexyl group may serve as an anchor group in the preparation of ketones.[125] It is important that the NaCN be dry.

Using an excess of TFAA results in a third migration to furnish, after oxidation, trialkylmethanols as exemplified by the conversion of cyclohexene to tricyclohexyl-methanol.[126]

86%

Dichloromethyl Methyl Ether Reaction[127]

The reaction of organoboranes with nucleophiles containing more than one leaving group results in multiple migrations. Thus, on treatment of R_3B with α,α-dichloromethyl methyl ether (DCME) in the presence of a sterically hindered base, such as Li-triethylmethoxide ($LiOCEt_3$), all three groups are transferred, and oxidation of the product affords the corresponding tertiary alcohol. The $LiCCl_2OMe$ is generated in situ from dichloromethyl methyl ether with lithium triethylmethoxide.

95%

Matteson's Boronic Ester Homologation[128]

Homologation of chiral alkylboronic esters with dichloromethyllithium provides a highly effective method for introduction of a chiral center while forming a carbon-carbon bond. The required boronic esters are readily accessible from Grignard reagents and trimethylborate or from lithium reagents and triisopropylborate. Hydrolysis of the resultant alkylboronic esters gives the corresponding alkylboronic acids, which on treatment with either (r)- or (s)-pinanediols furnish the stable (r)- or (s)-pinanediol alkylboronic esters.

(r)-pinanediol
alkylboronic ester

The chiral directing groups are pinanediols derived from osmium tetroxide-catalyzed oxidation of either (+)-α-pinene or (–)-α-pinene with trimethylamine oxide or with NMO (N-methyl-morpholine-N-oxide). The (s) and (r) notations shown in the abbreviations refer to the configuration of the chiral center in the α-chloroboronic ester using the appropriate pinanediol.

The reaction of an (s)-pinanediol boronic ester A with dichloromethyllithium is rapid at –100 °C and forms the borate complex B, which on warming to 0–25 °C rearranges to the (1S)-1-chloroalkylboronic ester C. Zinc chloride catalyzes the rearrangement and results in improved diastereoselection. Reaction of C with a Grignard or lithium reagent leads to borate complex D which rearranges with inversion to a *sec*-alkylboronic ester E. Oxidation with alkaline hydrogen peroxide furnishes the secondary alcohol in greater than 97% enantiomeric excess.

An important feature of the boron ester homologation reaction is that the boronic ester product E can itself be used as a starting boronic ester so that the cycle can be repeated to introduce a second chiral center. Matteson has used this methodology to synthesize insect pheromones and sugars.[129]

Brown's Asymmetric Crotylboration[130]

The β-methyl homoallylic alcohol moiety of both *anti*- and *syn*-configurations is a characteristic structural element of a number of macrolides and polyether antibiotics. Reactions of crotylmetal (2-butenylmetal) reagents with carbonyl substrates provide access to acyclic stereo- and enantioselective syntheses of β-methyl homoallylic alcohols. The alkene moiety of these alcohols can be further elaborated into aldehydes by oxidative cleavage of the double bond, leading to aldol-type products.

Crotyl organometallic species undergo 1,3-shifts of the metal at room temperature. For the stereocontrolled use of allylic organometallic reagents in synthesis, it is important that the stereoisomeric reagents not equilibrate under the reaction conditions and add to carbonyl compounds regioselectively and irreversibly. Of the various allylic organometallic reagents, allylboronic esters and allyldialkylboranes are especially suited for acyclic stereoselective syntheses of homoallylic alcohols.[129,131]

The rate of interconversion of crotyl boron reagents varies with the nature of the R groups on boron: crotyldialkylborane (crotyl-BR$_2$) > crotylalkylborinate (crotyl-BR(OR) > crotylboronate (crotyl-B(OR)$_2$).

Crotylboronic esters (2-butenylboronates) are thermally stable and isolable compounds at room temperature and undergo nearly quantitative additions to aldehydes. The required crotylboronic esters may be prepared by reaction of the crotyl potassium reagents derived from *cis-* or *trans-*2-butene with *n*-butyllithium and potassium *tert*-butoxide followed by addition of the appropriate trialkyl borates.[132]

The reactions of crotylboronic esters with aldehydes is regioselective, generating two new stereochemical relationships and potentially *four* possible stereoisomeric products. Thus, there are two stereochemical aspects: *enantioselection* (*Re-* vs. *Si-*face addition) and *diastereoselection* (*syn* vs. *anti*).

(Z)-crotylboronate

a. RCHO
b. NaOH, H_2O_2

enantiomeric *syn*-homoallylic alcohols

(E)-crotylboronate

a. RCHO
b. NaOH, H_2O_2

enantiomeric *anti*-homoallylic alcohols

The transfer of the allylic moieties from boron to the electrophilic carbonyl carbon proceeds via rearrangement to form intermediate boronic esters C and D (see below). The reaction is highly *diastereoselective*. The (E)-crotylboronate reacts to give the *anti*-homoallylic alcohol and the (Z)-crotylboronate reacts to afford the *syn*-homoallylic alcohol.[133] This behavior has been interpreted in terms of the Zimmerman-Traxler chair-type transition state model.[134] Because of the double bond geometry, coordination of the (E)-crotylboronic ester places the Me preferentially equatorial, whereas coordination of the (Z)-crotylboronic ester places the Me axial, as illustrated in the cyclohexane chair-form transition state conformations A and B, respectively. In both cases, the R moiety of the aldehyde must occupy a pseudoequatorial position to avoid steric repulsion by one of the OR substituents on boron.

anti-alcohol

syn-alcohol

The use of enantiopure allylic boranes in reactions with achiral aldehydes results not only in high *diastereoselection*, but also in high *enantioselection*.[135] Pure (Z)- and (E)-crotyldiisopinocampheylboranes can be prepared at low temperature from (Z)- or (E)-crotylpotassium and B-methoxydiisopinocampheylborane, respectively, after treatment of the resultant *ate*-complexes with $BF_3 \cdot OEt_2$. The B-methoxydiisopinocampheylboranes are prepared by reacting (−)-diisopinocampheylborane, derived from (+)-α-pinene, or (+)-diisopinocampheylborane, derived from (−)-α-pinene, with methanol.

The reaction of the (Z)-crotyldiisopinocampheylborane derived from (+)-α-pinene with aldehydes at −78 °C, followed by oxidative workup, furnishes the corresponding *syn*-β-methylhomoallyl alcohols with 99% diastereoselectivity and 95% enantioselectivity. Use of (Z)-crotyldiisopinocampheylborane derived from (−)-α-pinene also produces *syn*-alcohols with 99% diastereoselectivity but with opposite enantioselectivity, an example of reagent control.

enantioselectivity (using (+)-α-pinene): 95% 5%
diastereoselectivity: 99%

enantioselectivity (using (−)-α-pinene): 4% 96%
diastereoselectivity: 99%

Treatment of (*E*)-crotyldiisopinocampheylborane, derived from (+)-α-pinene, with aldehydes furnishes, after oxidation, the *anti*-alcohols with 95% enantioselectivity. The reaction of (*E*)-crotyldiisopinocampheylborane, derived from (−)-α-pinene, with aldehydes yields the enantiomeric *anti*-alcohols, also with 95% enantioselectivity. By this approach, using *cis*-and *trans*-2-butene and (+)-α-pinene and (−)-α-pinene, all four stereoisomeric homoallyl alcohols may be obtained.[136]

enantioselectivity (using (+)-α-pinene): 95% 5%
diastereoselectivity: 99%

enantioselectivity (using (−)-α-pinene): 4% 96%
diastereoselectivity: 99%

7.9 ORGANOSILICON REAGENTS[137]

The utility of organosilicon compounds in organic synthesis is based on the stability of the C–Si bond and on the regio- and stereoselective reactions of organosilanes with both nucleophilic and electrophilic reagents. The following discussion will present an overview of the properties of organosilicon compounds and on the preparation and utilization of alkynyl-, alkenyl-, and allylic silanes in organic synthesis.

Properties of Bonds to Silicon[138]

Electronegativity
Since silicon is more electropositive than carbon (see Table 7.1), the carbon-silicon bond is polarized in the direction $Si^{\delta+}–C^{\delta-}$. This polarization has two important

effects: (1) it makes the silicon susceptible to nucleophilic attack, and (2) it imparts stabilization to β-carbocations and α-carbanions.

Bond Energy

Organosilicon compounds are much more stable than other typical organometallic compounds. Single bonds from Si to electronegative elements such as O, Cl, and F are very strong (Table 7.7). There is little doubt that p_π–d_π bonding plays a role in these cases. Since such strong bonds are formed, substitution at Si is especially easy when the nucleophile is O^-, Cl^-, or F^-. The very strong affinity of silicon for F^- is evidenced by the facile desilylation of silyl ethers by n-$Bu_4N^+F^-$ (see Chapter 3).

Bond Length

The Si–C bond (1.89 Å) is significantly longer than a typical C–C bond (1.54 Å), suggesting that the $SiMe_3$ group is sterically less demanding than a *tert*-butyl group. In fact, the A value for $SiMe_3$ is approximately that of the isopropyl group (~ 2.1 kcal/mol).

Stabilization of β-Carbocations and α-Carbanions

The $Si^{\delta+}$–$C^{\delta-}$ polarization supports the formation or development of carbocations in the β-position, a stereoelectronic phenomenon known as the *β-effect*.[140] This is commonly attributed to the stabilizing interaction between the Si–C bond and the empty p_π orbital of the β-carbocation (σ–π conjugation). The β-effect is conformationally dependent in that the carbon-silicon σ-bond and the vacant p-orbital must be in a common plane, as depicted in structure A below. As a corollary, cations α to silicon are destabilized relative to those with alkyl groups.[141]

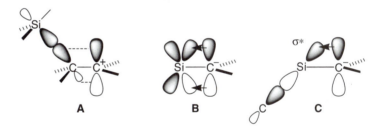

A   B   C

Second-row elements such as silicon are known to stabilize α-carbanions with considerably greater effectiveness than their first-row counterparts. Early rationalization of this phenomenon favored the p_π–d_π model B as a source of the stabilizing influence. More recently, stabilization by means of the high polarizability of Si and the presence of an empty σ^* orbital on the silicon have been proposed. Overlap with the filled orbital of a flanking carbanionic center as in C can account for the observed stabilization.[142]

	kcal/mol	kJ/mol
H_3C–C	83–88	347–368
H_3C–O	80	335
H_3C–Cl	84	351
H_3C–F	108	452
Si–C	76	318
Si–O	127	531
Si–Cl	113	471
Si–F	193	807

Table 7.7 Approximate Bond Dissociation Energies.[137b,139]

Preparation and Reactions of Alkynylsilanes, Alkenylsilanes, and Allylsilanes

Alkynylsilanes

Alkynylsilanes are readily accessible by metalation of terminal alkynes, followed by addition of the appropriate trialkylchlorosilane.

$$R'\!\!-\!\!\equiv\!\!-H \xrightarrow[\text{b. } R_3SiCl]{\substack{\text{a. MeLi or } n\text{-BuLi} \\ \text{or LDA, THF}}} R'\!\!-\!\!\equiv\!\!-SiR_3$$

Reactions of alkynylsilanes with electrophiles in the presence of a Lewis acid catalyst such as BF_3, $AlCl_3$, or $TiCl_4$ generally occur regioselectively at the silicon-bearing carbon, due to the β-effect, and under mild conditions.

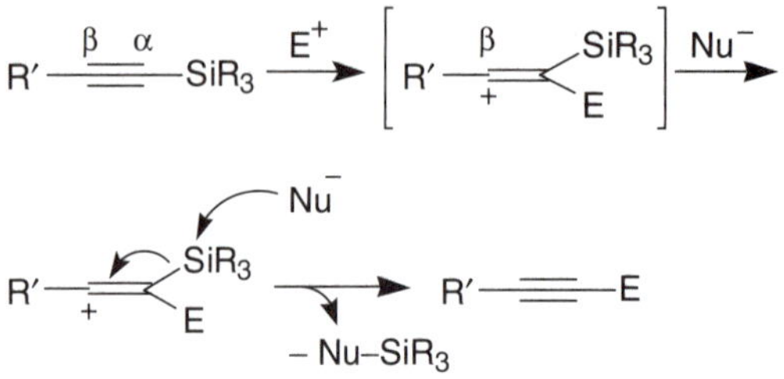

The $AlCl_3$-mediated electrophilic substitution of alkynylsilanes with acid chlorides provides a convenient route to alkynyl ketones. Transfer of the acetylenic moiety from silicon to the acyl chloride proceeds via the acyl cation (acylium ion) formed by reaction of the acyl chloride with $AlCl_3$.[143]

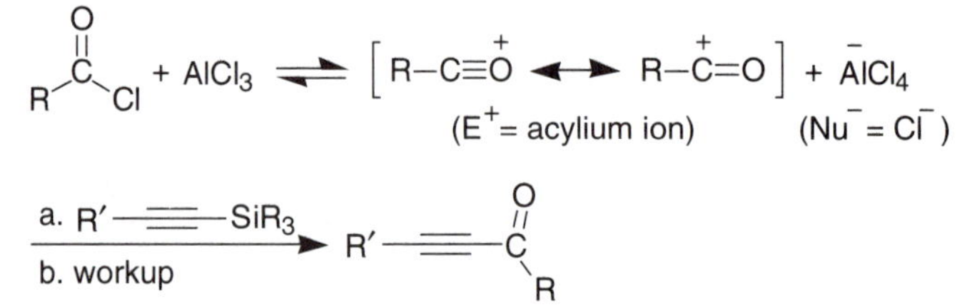

The remarkable silicon-directing reaction of electrophiles at the α-carbon of alkynylsilanes is further evident when comparing the intramolecular cyclizations of methyl- vs. silicon-substituted triple bonds, shown below.[144]

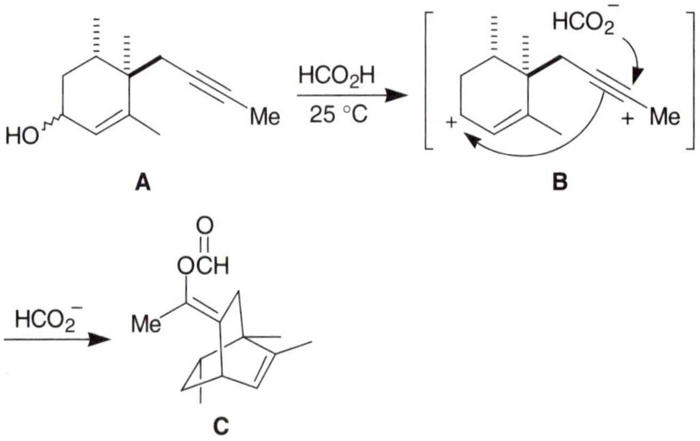

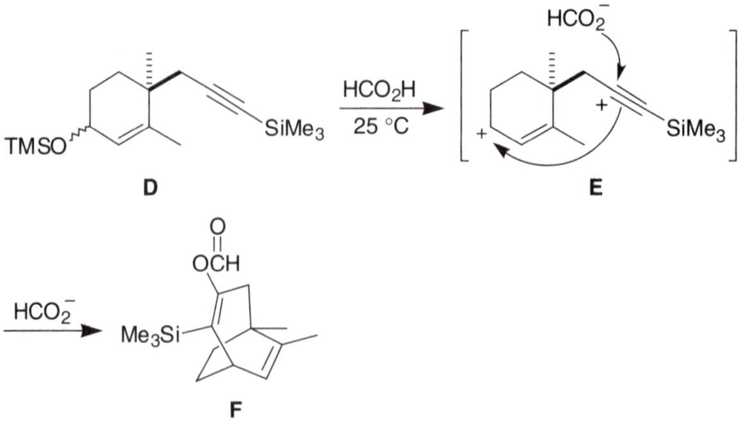

Protonation of the hydroxyl and trimethylsilyloxy groups of A and D, respectively, generates carbocations B and E. In B, attack of the methyl-substituted triple bond leads to formation of the preferred linear vinyl cation. Its trapping by the formate ion furnishes the substituted bicyclo[2.2.2]octene C. On the other hand, attack of the silicon-substituted triple bond in E occurs to form the silicon-stabilized β-carbocation, which, on trapping by formate ion, yields the substituted bicyclo[3.2.2]nonene F.

Alkenylsilanes

Among the many available procedures for preparing alkenylsilanes are hydrosilylation of alkynes and partial reduction of alkynylsilanes. Hydrosilylation of 1-alkynes with triethylsilane in the presence of catalytic chloroplatinic acid results in regioselective *syn-addition* of H–SiEt$_3$ to the triple bond to produce (E)-alkenylsilanes.[145]

Reduction of 1-(trialkylsilyl)alkynes with dialkylboranes[146] or with diisobutylaluminum hydride[147] proceeds in a regio- and stereoselective manner to yield, after protonolysis or hydrolytic workup, (Z)-alkenylsilanes.

Vinylsilanes react with a variety of electrophiles to give substitution or addition products. In substitution reactions, the silyl group is replaced by the electrophile because formation of the carbon-carbon π-bond is faster than capture of the incipient β-carbocation by the nucleophile.

In addition reactions, the nucleophile reacts with the carbocation before collapse of the Si–C bond. The resultant adduct can be induced to undergo *syn-* or *anti-*elimination depending on the reaction conditions. With few exceptions, electrophilic attack on the vinylsilyl moiety is highly regioselective, placing the developing carbocation at the β-position (β-effect).

Acylation of vinylsilanes with acid chlorides in the presence of a Lewis such as AlCl$_3$ provides a route to α, β-unsaturated ketones.[148]

77%

The reaction is not only regioselective but also stereospecific. Stabilization of the incipient carbocation occurs only if the C–Si bond is correctly aligned with the developing vacant p-orbital. Thus, parallel alignment of the C–Si σ-bond dictates the stereochemistry of the substitution product.

Isomerically pure vinyl halides are important starting materials for preparation of the corresponding vinyl lithium species (see Section 7.1) and for use in Pd-catalyzed coupling reactions (see Section 7.10). Halogenation of vinylsilanes offers a valuable route for the synthesis of both (E)- and (Z)-vinyl halides. These reactions involve addition of the halogen to the double bond, followed by elimination of the silicon and halide moieties. The stereochemical outcome of the reaction will depend on both the mode of addition and the mode of elimination. For example, bromination of (Z)-vinylsilanes with bromine in CH$_2$Cl$_2$ proceeds via an *anti*-addition. Treatment of the bromine adduct with sodium methoxide results in *anti*-desilicobromination. The conversion of the vinylsilane into the vinylbromide occurs with overall inversion of the double bond stereochemistry.[149]

(E)-vinylbromide
75–93% (E : Z, 99 : 1)

Allylsilanes

Allylsilanes play an important role in organic synthesis.[150] Whereas Li-, Mg-, Zn-, and B-allylic species undergo 1,3-shift of the metal, allylsilanes are stable.

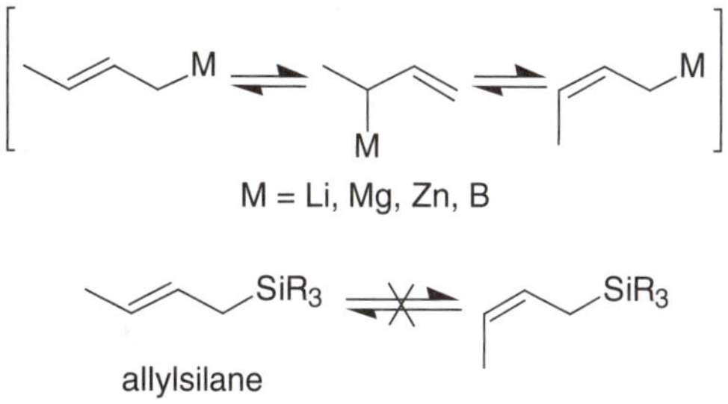

M = Li, Mg, Zn, B

allylsilane

Allylsilanes may be prepared by silylation of allyl-metal species, provided that the carbanions can be trapped regioselectively by R_3SiCl.

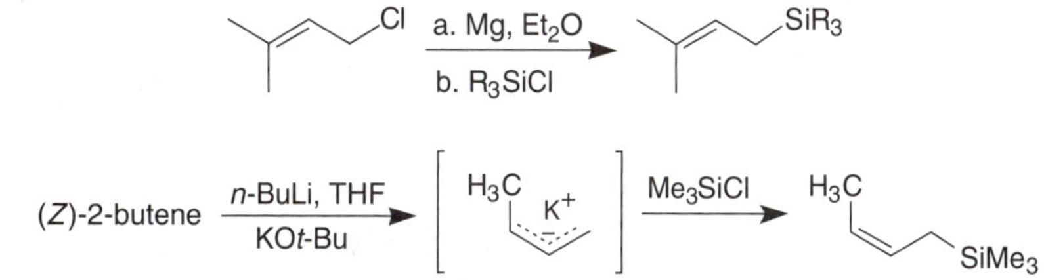

The nickel- or palladium-catalyzed reaction of stereodefined alkenyl halides with trimethylsilylmethylmagnesium chloride (TMSCH$_2$MgCl) delivers allylsilanes with excellent regio- and stereocontrol.[151]

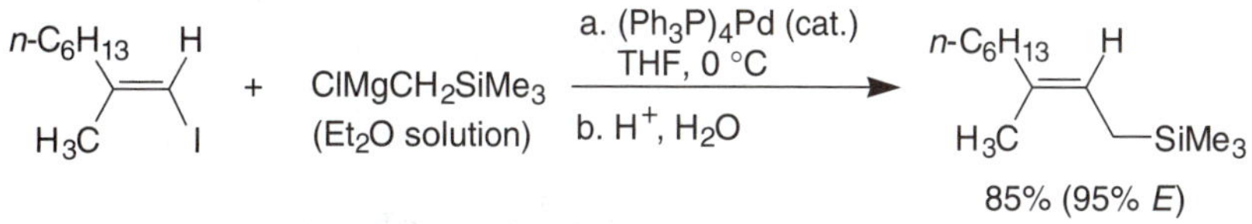

85% (95% *E*)

The hydroalumination-protonation of 1-trimethylsilyl-2-alkynes[152] and the Wittig reaction using silylated ylides[153] are other good approaches for the synthesis of allyl-silanes.

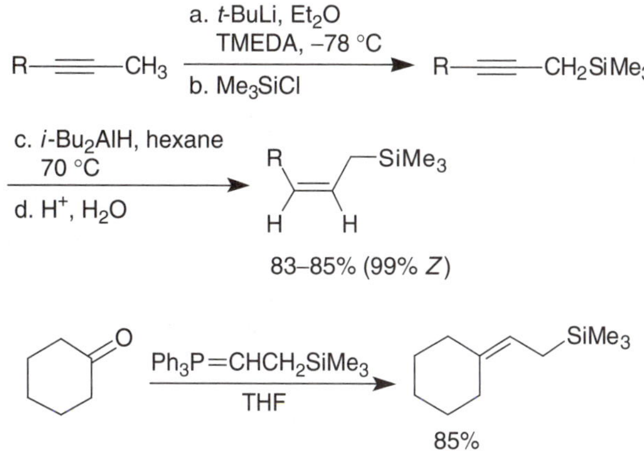

83–85% (99% *Z*)

85%

Attractive from a synthetic point of view is the readily available bifunctional reagent 2-bromo-3-(trimethylsilyl)propene. It represents a synthon for a 1,2-dianion, and offers an opportunity to store chemical reactivity which can be selectively unleashed as illustrated below.[154]

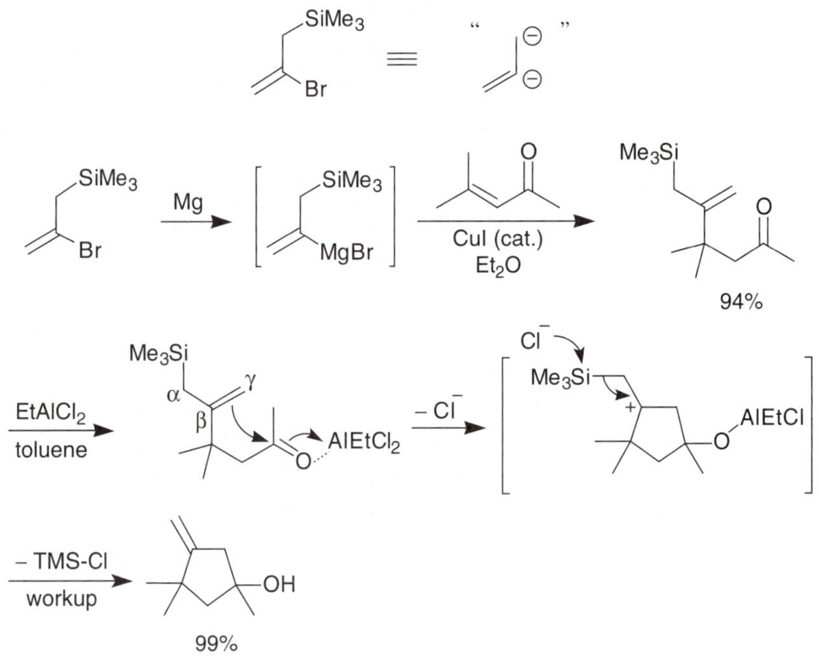

In the presence of a Lewis acid (such as Et$_2$AlCl), allylsilanes react with electrophiles in a regiospecific manner. The intermediate β-carbocation is stabilized by (σ–π)-conjugation with the C–Si bond. The most important feature of this reaction is that the electrophile reacts with the terminus (γ-carbon) of the allyl system, and the π-system is relocated adjacent to its original position. Even substituted allylic silanes can be acylated at the more hindered site. Because of this predictability and their high nucleophilicity, allylsilanes are valuable in many synthetic transformations.[155]

There are two types of reactions of allylsilanes with electrophiles: (1) Lewis acid-mediated reactions, and (2) reactions involving allyl anions.

Lewis Acid-Mediated Reactions. The highly nucleophilic double bond of allylsilanes reacts with a variety of electrophiles provided that the electrophiles are activated by a Lewis acid, such as TiCl$_4$, BF$_3$, or AlCl$_3$. The γ-position of the allylsilane attacks the electrophile to generate a stabilized carbocation β to silicon. Subsequent loss of the silyl group results in the transposition of the double bond.[156]

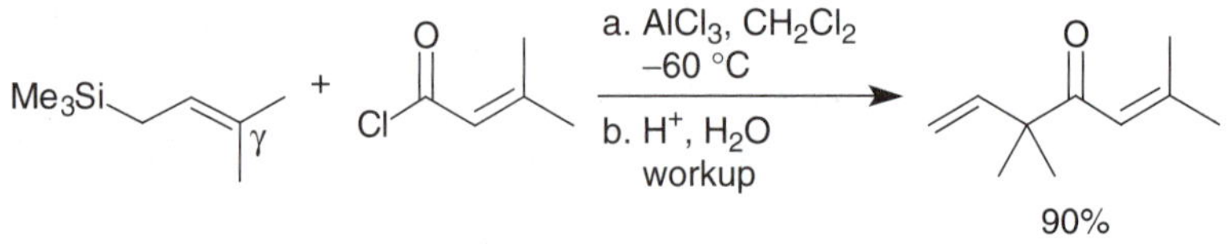

Both inter- and intramolecular reactions of allylsilanes with enones have been reported.[157] 1,4-Addition of allylsilane to the conjugated enone shown below permits direct introduction of an angular allyl group to a bicyclic enone.

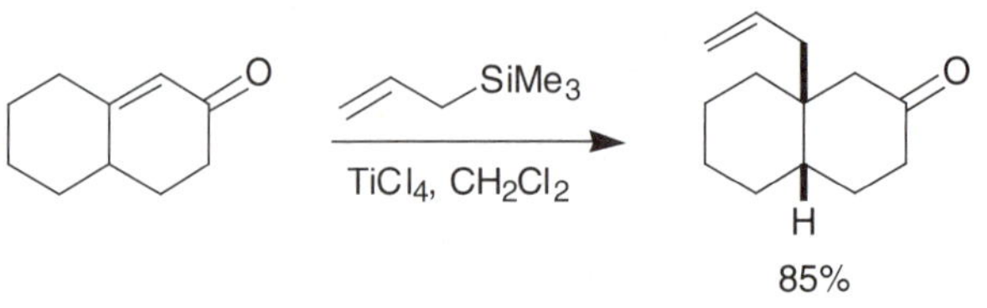

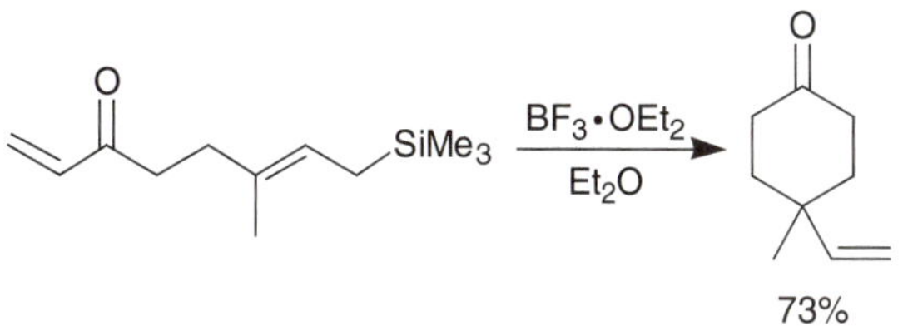

73%

The reaction of (E)-crotylsilane with RCHO in the presence of $TiCl_4$ proceeds with high *syn*-selectivity. On the other hand, the corresponding (Z)-silane furnishes *syn-anti* mixtures of homoallylic alcohols. It has been shown that the reaction proceeds via an acyclic linear and not via a cyclic six-member transition state. As shown below for the (E)-crotylsilane, transition state A leading to the *syn*-alcohol is sterically favored over the diastereomeric transition state B, which has a gauche Me-Et interaction.[158]

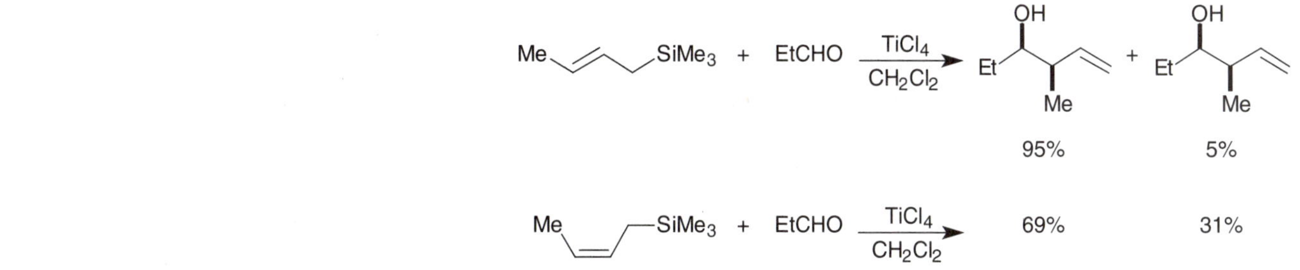

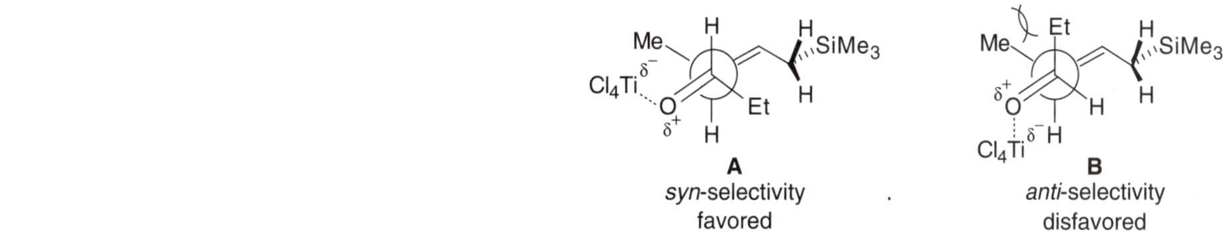

A	B
syn-selectivity	*anti*-selectivity
favored	disfavored

Reactions of Allylsilane Anions.[159] Generally, deprotonated silanes react in a γ-regio-selective manner with electrophiles to give vinylsilanes.

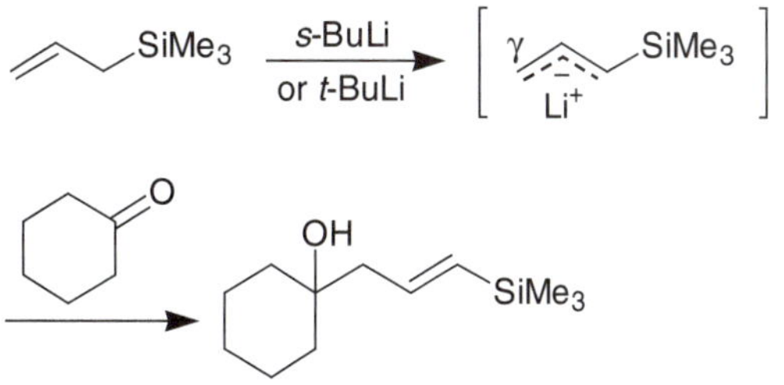

However, if the allylsilane anion is first complexed with certain metals or metalloids, such as boron or titanium, α-regioselectivity then predominates, and high *anti*-diastereo-selectivity is attained with aldehydes as electrophiles.[160] The resultant β-hydroxysilanes may be converted into (E)- and (Z)-terminal dienes using the Hudrlik-Peterson method.[161]

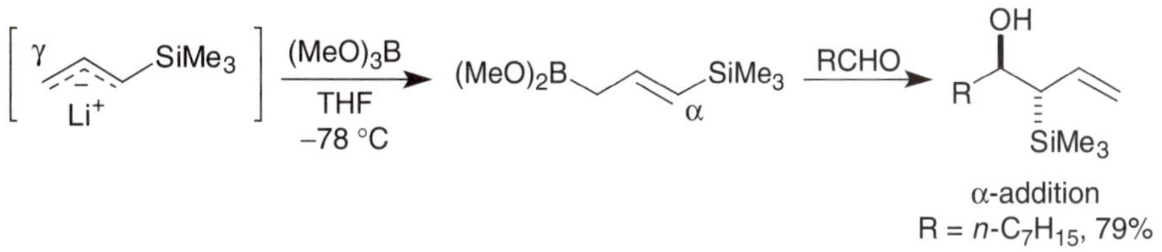

α-addition
R = *n*-C_7H_{15}, 79%

Acyl Silanes[162]   Acylsilanes are versatile intermediates for carbon-carbon bond formation reactions, and may serve as precursors for the synthesis of silyl enol ethers, aldehydes, or carboxylic acids. In the presence of a base or certain nucleophiles, they undergo the Brook rearrangement, where the silyl moiety migrates from carbon to oxygen (see below in this section).

Preparation

There are several procedures available for the preparation of acylsilanes.[163] A general synthesis of acyl silanes is via the dithiane route.[164] Hydrolysis of 2-silyl-1,3-dithianes which leads to acyl silanes can be accomplished with mercury(II) salts. To repress formation of aldehyde by-products, various modifications for the deprotection of silyl dithianes have been reported, including the use of HgO–F$_3$B•OEt$_2$[165] or chloramine-T.[166]

Hydroboration of 1-trimethylsilyl-1-alkynes with BH$_3$•SMe$_2$ in a 3:1 ratio, followed by oxidation of the resultant trivinylborane with anhydrous trimethylamine oxide, provides an operationally simple, one-pot synthesis of alkyl-substituted acylsilanes in good yield from readily available starting materials.[167]

Metalation-silylation-protonation of methyl vinyl ether furnishes acetyltrimethylsilane, the simplest acyl silane.[163]

Monohydroboration of bis(trimethylsilyl)acetylene and oxidation of the resultant trivinylborane with anhydrous trimethylamine oxide followed by hydrolytic workup affords [(trimethylsilyl)acetyl]trimethylsilane, which contains both α- and β-ketosilane structural features.[167b] This reagent is a versatile synthon for stereoselective syntheses of functionalized trisubstituted olefins.[168]

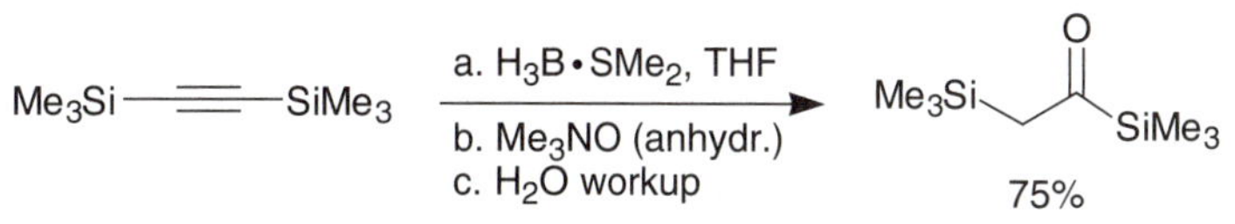

Reactions

Acylsilanes are sensitive to light and basic reaction conditions. They can be regarded as sterically hindered aldehyde synthons displaying a range of reactivity patterns with nucleophiles. For example, treatment of an acylsilane with carbon nucleophiles leads to α-silyl alkoxides **A,** which may be stable under the reaction conditions or may undergo a reversible rearrangement (Brook rearrangement)[169] of the silyl group from carbon to oxygen, providing a novel route to carbanions **B**. The carbanion formed may be protonated or trapped with an electrophile. The driving force for the rearrangement is likely formation of the strong Si–O bond. The silyl alkoxide rearrangement **A → B** is rapid only when R or R′ is a carbanion-stabilizing group, such as vinyl, aryl, or ethynyl. The reaction of acylsilanes with organolithium and Grignard reagents frequently gives complex product mixtures.[170]

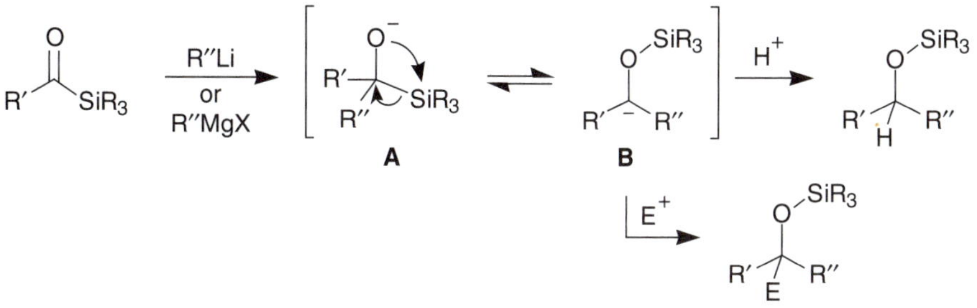

An instructive example showing the potential of acylsilanes as hindered aldehyde synthons is the preparation of the secondary alcohol shown below using (3-methylpentadienyl)lithium as the nucleophilic reagent. Since the pentadienyl anion possesses two nucleophilic sites, its reaction with an aldehyde could in principle furnish the conjugated and/or deconjugated addition product(s). The pentadienyl anion possesses the highest electron density at the internal γ-position, according to MO calculations.[171] Its reaction with acetaldehyde gave a mixture of isomers, favoring bond formation at the more hindered position of the lithium reagent. To enhance coupling of the carbonyl carbon with the less hindered site, the sterically more hindered synthon of acetaldehyde—the readily accessible acetyltrimethylsilane—was used. In this case, only the conjugated diene product was observed. The trimethylsilyl group then was removed via a Brook rearrangement.[172]

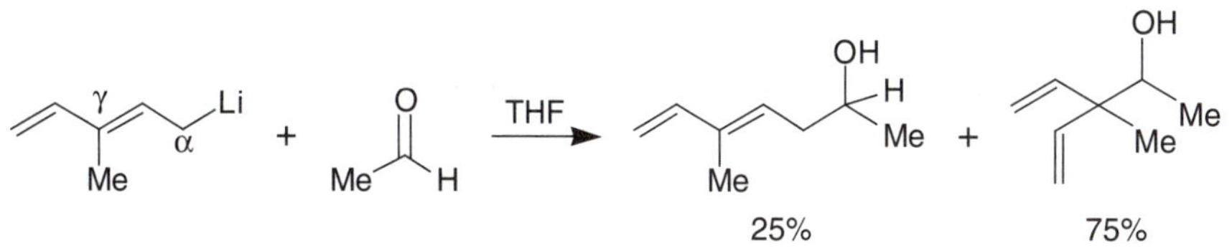

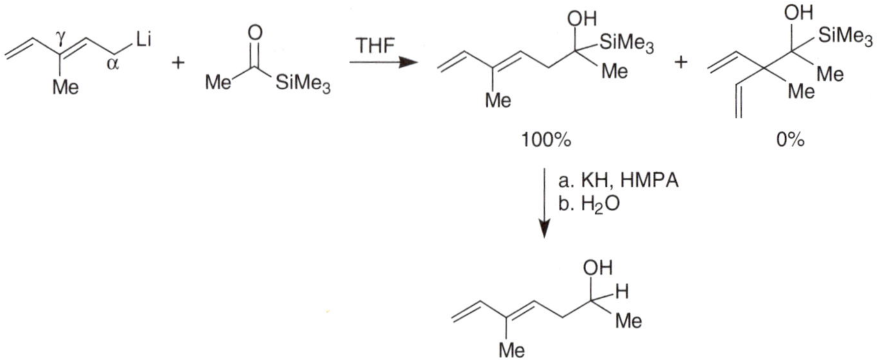

The reaction of acylsilanes with appropriate carbon nucleophiles, followed by the Brook rearrangement, provides a novel source of carbanions which can be further elaborated on treatment with a reactive electrophile, allowing the formation of two carbon-carbon bonds in a one-pot reaction.[173]

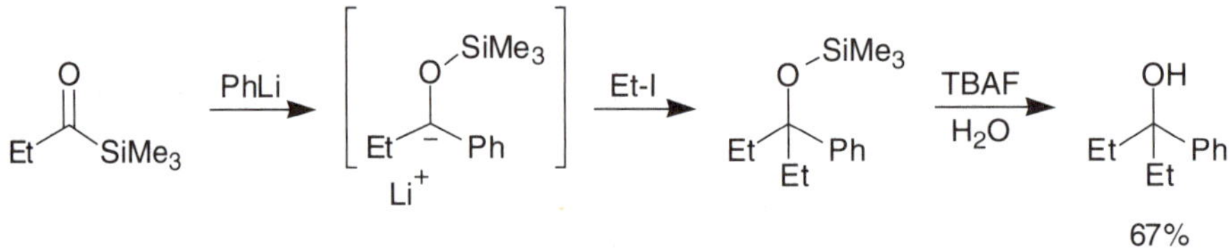

Addition of nucleophiles containing a leaving group, such as an α-sulfonyl anion,[174] to acylsilanes provides a route to silyl enol ethers, which have broad application as enolate equivalents.[175]

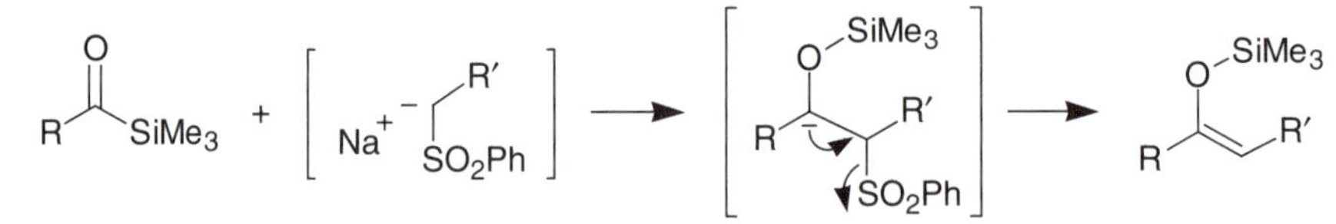

Acylsilanes may serve as synthetic equivalents of carboxylic acids when treated with alkaline hydrogen peroxide. Especially attractive is the one-pot conversion of the ethynyl group of 1-alkynes into monosubstituted acetic acids via sequential silylation, hydroboration, and oxidation, as depicted below.[176]

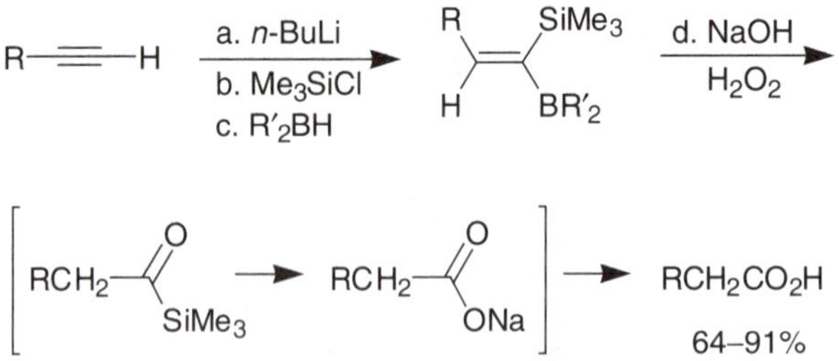

7.10   PALLADIUM-CATALYZED COUPLING REACTIONS[177]

During the past decades, palladium-catalyzed coupling reactions have emerged as versatile tools for the formation of carbon-carbon bonds in industrial processes and in the laboratory. The usually high chemo-, regio-, and stereoselectivities observed in these reactions, coupled with their atom economy (i.e., the lack of by-products; see Section 1.4), provide powerful strategies for the construction of complex carbon-carbon frameworks.

Transition metal complexes possess reaction paths not available to main group metals, allowing selective transformations that otherwise would be difficult or impossible to carry out, such as coupling of C_{sp^2}–C_{sp^2} centers. To understand the reactions of organotransition metal complexes and the reasons they react in a certain manner, it is important to grasp concepts such as the oxidation state of the metal, whether the metal is coordinatively saturated or unsaturated, and what effect the ligands will have on the reaction. The metal acts as a template to orient and activate coordinated reactants for further reaction. Transition metals have the ability to donate additional electrons or to accept electrons from organic substrates, and, in doing so, they change their oxidation state or coordination number. This facile, sometimes reversible oxidation-reduction process, plays an important role in catalytic reactions.

Oxidation state. There are various ways to determine the oxidation state of the metal. One approach is to assign the metal a formal oxidation state of (0), unless it is σ-bonded to ligands such as halogens, alkyl, aryl, or vinyl. In these cases, the metal and the ligands form bonds with shared electrons, with the ligands having a formal charge of -1. Neutral ligands such as Ph_3P and CO are not counted since they provide two of their own electrons. For example, palladium in $Pd(PPh_3)_4$ is in the zero oxidation state, whereas in $Pd(PPh_3)_2Cl_2$, the palladium is in the +2 oxidation state. It is assumed that the outer electrons of the metal are all d electrons. Thus, Pd(0) is a d^{10} rather than $4d^8 5s^2$ metal, and belongs in the transition metals Group 10, which includes Ni(0) and Pt(0). Pd(II) is a d^8 rather than a $4d^6 5s^2$ metal. Complexes derived from these metals are widely used in synthetic transformations.

Coordination number (CN). For monodentate and chelating ligands (i.e., ligands that form a metallocycle when bonded to a metal; see Table 7.8), the coordination number is defined as the number of atoms or ligands directly bonded to the metal atom. This definition does not encompass the metal bonding of alkenes or alkenyl fragments, such as ethylene or cyclopentadienyl (Cp), which generally can be counted according to the pairs of π-electrons these ligands donate to the metal (e.g., in ferrocene, three pairs of π-electrons from each Cp ring participate in bonding to the Fe; thus the CN is 6). Understanding the relationship between CN and the "18 electron rule"[178] allows us to make predictions about reactivity. By filling the one s-, three p-, and five d-orbitals with two electrons each, the metal attains the noble gas configuration of 18 electrons in its valence shell. For example, in tetrakis(triphenylphosphine)-palladium(0) A, the palladium has the d^{10} configuration, and each triphenylphosphine ligand donates two electrons, a total of eight electrons. Thus, the palladium complex has a total of 18 electrons, is coordinatively saturated, and has the maximum coordination number of four. In bis(triphenylphosphine)dichloropalladium(II) B, the Pd(II) possesses 16 electrons, four from the two triphenylphosphine ligands, four from the two chlorine ligands, and eight electrons from palladium(II). The 16-electron Pd(II)-complex B is coordinatively unsaturated and has the maximum coordination number of five.

oxidation state: Pd(0)
d^n count: d^{10}
max. coordination
number: 4

A

18 electrons

oxidation state: Pd(II)
d^n count: d^8
max. coordination
number: 5

B

16 electrons

Organic Synthesis with Palladium(0) Complexes

Palladium(0) complexes are readily accessible, easily prepared, and easily handled. The nature of the ligand plays an important role in palladium-catalyzed coupling reactions. For example, phosphine ligands (Table 7.8) provide a high electron density on the metal, making it a good nucleophile, thereby favoring oxidative additions with suitable substrates. Also, high electron density on the metal favors dissociation of the ligands to a coordinatively unsaturated complex. On the other hand, carbon monoxide is a strong π-acid ligand, lowering the electron density in the filled d-orbitals of palladium, thereby reducing its reactivity towards oxidative addition.

The chemistry of palladium is dominated by two stable oxidation states: the zerovalent state [Pd(0), d^{10}] and the +2 state [(Pd(II), d^8]. Each oxidation state has its own characteristic reaction pattern. Thus, Pd(0) complexes are electron-rich nucleophilic species, and are prone to oxidation, ligand dissociation, insertion, and oxidative-coupling reactions. Pd(II) complexes are electrophilic and undergo ligand association and reductive-coupling reactions. A large amount of literature deals with these reactions. However, a few fundamental principles, such as *oxidative addition, transmetalation,* and *reductive elimination,* provide a basis for applying the chemistry of palladium in research.

An important property of Pd(0) complexes is their reaction with a broad spectrum of halides having proximal π-bonds to form σ-organopalladium(II) complexes. This process is known as *oxidative addition* because the metal becomes formally *oxidized* from Pd(0) to Pd(II) while the substrate R–X (R = alkenyl, aryl; X = Br, I, OTf) adds to the metal. Oxidative addition is believed to occur on the 14-electron Pd(0) species resulting from ligand (L) dissociation in solution. This coordinatively unsaturated Pd(0) is then the active species undergoing oxidative addition with the substrate to form the 16-electron Pd(II)complex.

The most commonly used Pd(0) complex is tetrakis(triphenylphosphine)palladium, Pd(PPh$_3$)$_4$, a yellow crystalline material when freshly prepared. Since the catalyst

Table 7.8	Phosphine Ligands

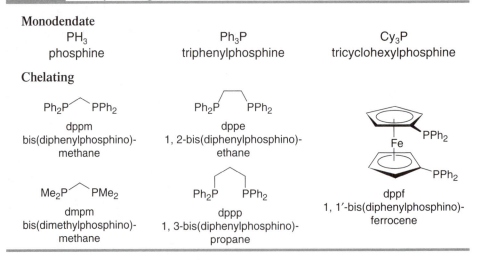

Monodendate

| PH$_3$
 phosphine | Ph$_3$P <br> triphenylphosphine | Cy$_3$P <br> tricyclohexylphosphine |

Chelating

Ph$_2$P ⌃ PPh$_2$
dppm
bis(diphenylphosphino)-
methane

Ph$_2$P ⌃ PPh$_2$
dppe
1, 2-bis(diphenylphosphino)-
ethane

Me$_2$P ⌃ PMe$_2$
dmpm
bis(dimethylphosphino)-
methane

Ph$_2$P ⌃ PPh$_2$
dppp
1, 3-bis(diphenylphosphino)-
propane

dppf
1, 1'-bis(diphenylphosphino)-
ferrocene

may lose its activity on storage, it is advisable to prepare it in small batches by reducing the more stable, commercially available Pd(II) complexes such as $PdCl_2$, $Pd(OAc)_2$, or $Pd(PPh_3)_2Cl_2$ in the presence of triphenylphosphine.[179] An active Pd-catalyst may be prepared in situ by reduction of a Pd(II) complex, e.g., $Pd(OAc)_2$ with PPh_3.[180] In this case, only 2 to 3 equivalents of phosphine may be needed for producing a coordinatively unsaturated Pd(0) catalyst. Another frequently used Pd(0) complex is bis(dibenzylideneacetone)palladium, $Pd_2(dba)_3$, prepared by boiling palladium(II) chloride with dibenzylideneacetone ($PhCH{=}CHCOCH{=}CHPh$) in methanol. The reduction of Pd(II) species to catalytically active Pd(0) species may also be achieved with amines, alkenes, or organometallics ($i\text{-}Bu_2AlH$, alkyllithiums, organostannanes).

The oxidative addition to palladium reagents proceeds readily with 1-haloalkenes at room temperature. However, the oxidative addition reactions of halides other than vinyl or aryl usually are very sluggish. Moreover, alkyl-Pd(II)-X complexes in which the alkyl moiety contains an sp^3-bonded hydrogen at the β-position may undergo rapid dehydropalladation by *syn*-β-hydrogen elimination, generating the hydridopalladium complex and a double bond. Thus, the substrates used for the oxidative addition reaction are usually restricted to vinyl and aryl halides and triflates.

The oxidative addition of alkenyl halides (see syntheses of alkenyl halides in Table 7.9) to Pd(0) complexes proceeds with retention of the double bond stereochemistry.

Table 7.9   Preparation of (*E*)- and (*Z*)-Alkenyl Halides[181]

Reagents	Alkenyl halide	Reference
a. R'_2BH; b. Br_2; c. NaOH		182
1a. R′Li; 1b. Br_2 or I_2; 2a. R'_2BH; 2b. AcOH		183
a. $i\text{-}Bu_2AlH$; b. Br_2 or I_2		184
a. $Cp_2Zr(H)Cl$; b. NBS or I_2		185
a. [R′MgX + CuX]; b. I_2		186
a. Me_3Al, Cp_2ZrCl_2; b. I_2		187
HI (from Me_3SiCl + NaI)		188

The resultant, coordinatively unsaturated, 16-electron alkenyl- (or aryl-) palladium(II) complexes can now acquire a second substrate by *insertion* of an alkene or by *transmetalation* with an organometallic reagent. The final step—coupling of the organopalladium intermediate to form a carbon-carbon bond—requires a *β-hydride elimination* for the insertion process and a *reductive elimination* for the transmetalation process. It should be noted that the Pd-carbon σ-bond is stable to many functional groups such as –NO$_2$, –Cl, –CO$_2$R, –CN, –NR$_2$, and –NHCOR. These coupling reactions, which bear the name of their discoverers, play a key role in organic synthesis and are described in the sections below.

The Heck Reaction—Palladium(0)-Catalyzed Olefin Insertion Reactions[177h,189]

The Heck reaction involves coupling of alkenyl or aryl halides with alkenes in the presence of a catalytic amount of a Pd(0) complex and a base to furnish alkenyl- and aryl-substituted alkenes.[190]

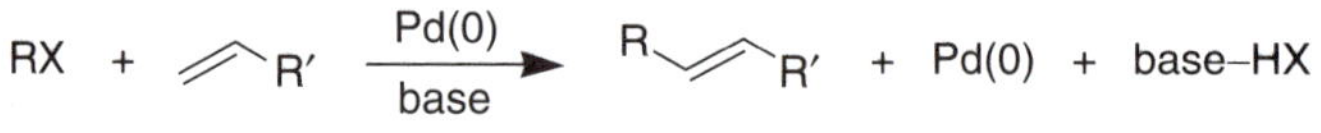

$$RX \ + \ \diagup\!\!\!\!\diagdown R' \ \xrightarrow[\text{base}]{\text{Pd(0)}} \ R\diagup\!\!\!\diagdown\!\!\!\diagup R' \ + \ Pd(0) \ + \ \text{base–HX}$$

R: allyl, alkenyl, aryl, alkynyl, benzyl
R′: alkyl, alkenyl, aryl, CO$_2$R, OR
base: Et$_3$N, NaOAc, aqueous Na$_2$CO$_3$

The catalytic cycle for the Heck reaction shown below presumably is initiated by oxidative addition of an alkenyl halide or an aryl halide to a coordinatively unsaturated 14-electron Pd(0) complex, generating a 16-electron σ-alkenyl- or σ-arylpalladium(II)-complex A, respectively. The presence of an alkene in the reaction mixture leads to its coordination with A. Subsequent *syn* insertion of the alkene double bond into the σ-alkenyl- (or the σ-aryl) Pd–C bond via a four-centered transition state produces intermediate B. The unsaturated ligand R of the palladium complex A becomes attached to the less hindered carbon of the alkene. Rotation of the newly formed carbon-carbon bond in B aligns a β-hydrogen with the palladium atom for a *syn* β-hydride elimination which leads to the coupling product C and the hydridopalladium complex D. Reductive elimination of HX from D by the added base regenerates the Pd(0) catalyst. For steric reasons (*anti*-periplanar arrangement of R and R′), the Heck reaction with terminal alkenes generally yields (*E*)-alkenes.

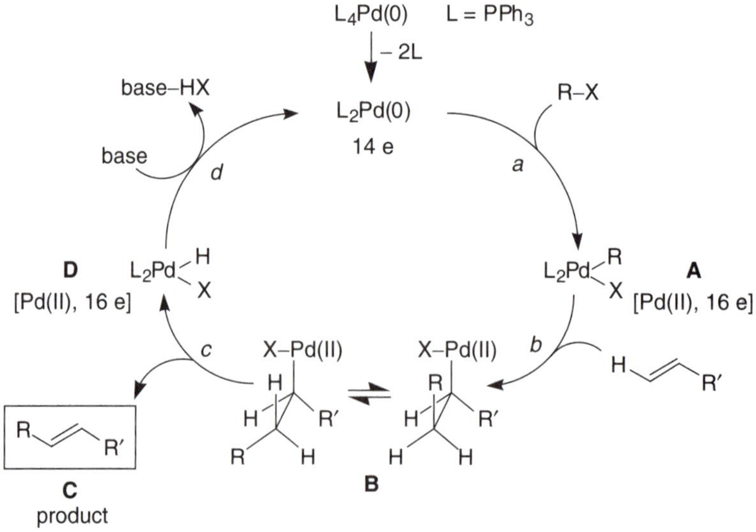

a = oxidative addition; b = carbopalladation; *syn*-insertion;
c = *syn* β-hydride elimination; d = reductive elimination

The Heck palladium-catalyzed reaction is widely used in organic synthesis with the distinguishing features that it can be applied to a variety of alkenes and is compatible with many functional groups.[191] A disadvantage of Heck-type reactions with organic halides is the relatively high temperature (~100 °C) required for the coupling reaction. Many modifications and improvements of the palladium catalyst have been reported since the original work of Heck. For example, addition of tetrabutylammonium chloride (TBAC) as a phase transfer agent markedly improves the reactivity of the catalyst.[192] Note that while most Heck reactions employ Pd(II) salts as the added regents, the Pd(II) is reduced in situ by amines or alkenes to the catalytically active Pd(0) species.

The regiochemistry of addition of the organopalladium intermediate to the olefin is dominated by steric effects, in that the organic group prefers the less substituted carbon of the double bond regardless of the substituents on either reactant. However, electronic effects play a significant role in the case of some olefins. Thus, electron-releasing groups on the double bond lead to increased addition to the most electron-deficient carbon of the double bond.[193]

The stereochemistry of the ultimate product alkene is the result of a *syn*-addition of the organopalladium followed by a *syn*-elimination of palladium hydride.

If the organopalladium intermediate contains two β-hydrogens, *syn*-elimination can yield a mixture of (*E*)- and (*Z*)-alkenes. However, the thermodynamically more stable alkene is formed preferentially. Aryl substituents such as –Cl, –C≡N, –CO₂R, –CHO, and –NMe₂ do not interfere with the coupling reaction.

The reactivity order of aryl halides in oxidative additions is as follows:

$$\text{Ar–I} > \text{Ar–OTf} > \text{Ar–Br} \gg \text{Ar–Cl}$$

Vinylation of alkenes with vinyl halides[194] or vinyl triflates furnishes conjugated dienes. Vinyl triflates are especially attractive in coupling reactions since they can be prepared regio- and stereoselectively from enolizable ketones by treatment with a hindered base and trapping the enolate oxygen with triflating agents such as triflic anhydride[195] or N-phenyltrifluoromethanesulfonimide (PhNTf$_2$).[196] For example, the palladium-catalyzed reaction of vinyl triflates with acrolein, which generally cannot be used in the Heck reaction because the acrolein polymerizes, produces α,β-unsaturated aldehydes.

Under the usual reaction conditions, coupling of (E)- or (Z)-1-halo-1-alkenes with alkenes leads to mixtures of stereoisomeric dienes. However, stereoselective coupling reactions have been achieved by using Pd(OAc)$_2$ in the presence of TBAC and K$_2$CO$_3$.[197]

	E, E	E, Z	Yield
(E)-isomer	99	1	96%
(Z)-isomer	5	95	90%

Intramolecular Heck reactions provide a powerful tool for the construction of carbocyclic and heterocyclic ring systems.[198] Note in the example below that unlike the other Heck reactions shown, the double bond does not end up in its original position due to the requirement of a syn-β-hydride elimination after the syn-alkene insertion.

In the presence of a chiral auxiliary, the intramolecular Heck reaction provides for enantioselective syntheses of a variety of compounds.[199]

Related to the Heck reaction is the *Larock annulation* of internal alkynes, which is a general route to heterocyclic and carbocyclic systems. Especially attractive is the construction of the pharmaceutically important indole ring system via palladium-catalyzed coupling of 2-iodo-aniline and the corresponding N-methyl, acetyl, and tosyl derivatives with a wide variety of internal alkynes. The catalytic process appears to involve arylpalladium formation, regioselective addition to the carbon-carbon triple bond, and subsequent intramolecular palladium displacement.[200]

Palladium-Catalyzed Cross-Coupling with Organometallic Reagents

Palladium-catalyzed cross-coupling reactions have common features with the Heck reaction but differ in that alkenyl and aryl organometallics instead of alkenes are involved in the coupling process. The role of the organometallic reagents is to transfer an alkenyl or an aryl group onto the species $R–PdL_2X$ in exchange for X (halide *or* triflate). Although a variety of organometallics for transmetalation have been reported, reagents containing zinc, aluminum or zirconium (Negishi reactions), boron (Suzuki reaction), and tin (Stille reaction) are the most widely used. These organometallics are more compatible with functional groups such as esters, amides, nitriles, and nitro compounds than organolithium and organomagnesium reagents.

The palladium(0)-catalyzed cross-coupling of alkenyl or aryl halides and triflates with organometallics proceeds via sequential oxidative addition (to species A below), transmetalation (usually rate determining), isomerization, and reductive elimination processes. The catalysts commonly used are the Pd(0)-complexes $Pd(PPh_3)_4$ or $Pd_2(dba)_3$.

R–X = alkenyl and aryl halides and triflates

R′ M = alkenyl and aryl zinc, aluminum, zirconium, or boron and tin

a = oxidative addition; *b* = transmetalation; *c* = *trans-cis*-isomerization; *d* = reductive elimination

The Negishi Reaction: Coupling Reactions with Organozinc, Organoaluminum, and Organozirconium Compounds[177j,201]

The Negishi palladium-catalyzed cross-coupling reaction of alkenyl, aryl, and alkynyl halides with unsaturated organozinc reagents provides a versatile method for preparing stereodefined arylalkenes, arylalkynes, conjugated dienes, and conjugated enynes.[202]

The advantages of using organozinc reagents for cross-coupling reactions are efficient transmetalation to palladium and ready availability, even directly from functionalized halides.[203]

The reaction of 2-methylphenylzinc chloride with 4-bromonitrobenzene using Pd(PPh3)4 (cat.), THF, then 3N HCl workup gives 2-methyl-4'-nitrobiphenyl in 78%.

The reaction of 4-cyanophenylzinc bromide with ethyl 4-iodobenzoate using Pd(PPh3)4 (cat.), THF, then NH4Cl gives ethyl 4'-cyanobiphenyl-4-carboxylate in 82%.

Utilization of stereodefined alkenylalanes or alkenylzirconium reagents in palladium-catalyzed cross-coupling reactions greatly enhances the versatility of Negishi-type coupling reactions. These organometallics are readily available by hydroalumination, carboalumination, and hydrozirconation of alkynes, respectively.

Hydroalumination. The treatment of alkynes with diisobutylaluminum hydride in hydrocarbon solvents results in a *cis*-addition of the Al–H bond to the triple bond to produce stereodefined alkenylalanes.[204] The hydroalumination of alkynes is more limited in scope than the corresponding hydroboration reaction of alkynes (see Chapter 5) with regard to accommodation of functional groups and regioselectivity. Whereas hydroalumination of 1-alkynes is highly regioselective, placing the aluminum at the terminal position of the triple bond, unsymmetrically substituted alkynes produce mixtures of isomeric alkenylalanes.

Palladium-catalyzed cross-coupling of alkenylalanes with (E) or (Z) alkenyl halides leads to efficient, stereoselective syntheses of 1,3-dienes, as exemplified below.[205]

Carboalumination. The treatment of terminal alkynes and symmetrically disubstituted alkynes with trimethylaluminum in the presence of a catalytic amount of Cp_2ZrCl_2 [bis(cyclopentadienyl)zirconium dichloride] results in a regioselective *cis*-addition of the Me–Al moiety to the triple bond to furnish β-methyl-substsituted alkenylalanes.[206]

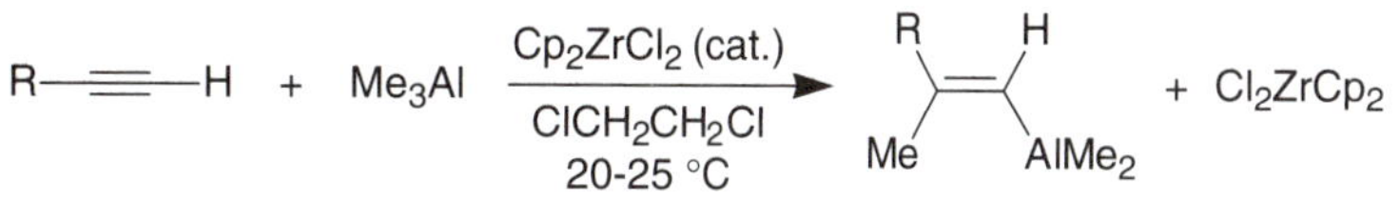

An example of a palladium-catalyzed stereospecific and regioselective coupling of an allylic chloride with an alkenylalane derived via Zr-catalyzed methylalumination in the synthesis of α-farnesene is shown below.[207]

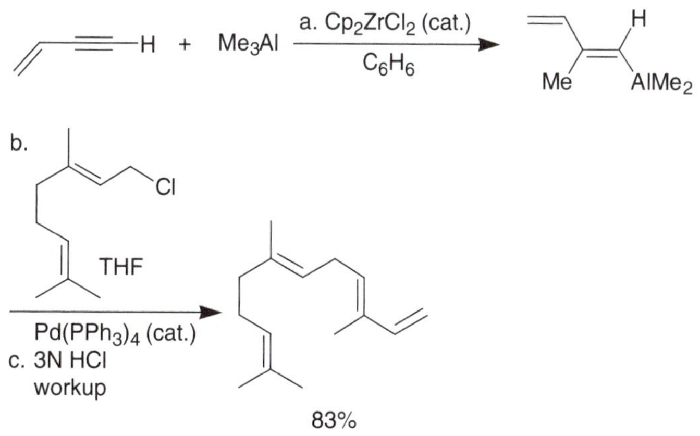

83%

Attempts to introduce a primary alkyl group other than methyl onto the triple bond produced a mixture of regioisomers.

Hydrozirconation. The treatment of 1-alkynes with the Schwartz reagent, $Cp_2Zr(H)Cl$ [bis-(cyclopentadienyl)zirconium chloride hydride] results in highly regio- and stereospecific *cis*-addition of the Zr–H to the triple bond, with the zirconium occupying the less substituted carbon in the resultant alkenylzirconium compound.[208] Procedures for the preparation of $Cp_2Zr(H)Cl$ and its in situ generation are available.[209] Hydrozirconation of unsymmetrically substituted alkynes leads to mixtures of regioisomers. Interestingly, in the presence of an excess of $Cp_2Zr(H)Cl$, the adduct initially formed isomerizes to yield preferentially the alkenylzirconium having the metal at the sterically less hindered position of the double bond.[210]

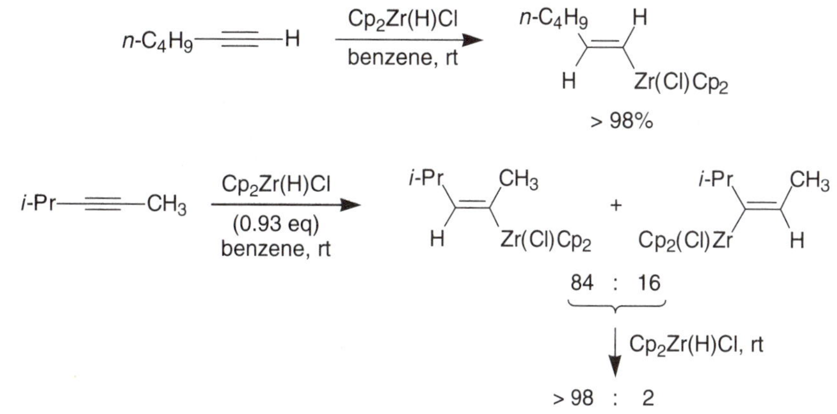

Alkenylzirconium compounds couple efficiently with alkenyl and aryl halides in the presence of Pd(0) catalysts, maintaining the stereochemical integrity of both partners in the product. Since the hydrozirconation and oxidative addition steps tolerate the presence of some functional groups, this cross-coupling methodology permits the synthesis of functionally substituted dienes.[211]

Transmetalations of trisubstituted alkenylalanes derived from hydroalumination or carboalumination of symmetrically disubstituted alkynes, as well as alkenylzirconium compounds obtained from hydrozirconation of disubstituted alkynes to Pd(0), are sluggish, resulting in low yields of cross-coupling products. However, addition of $ZnCl_2$ or $ZnBr_2$ as a co-catalyst has a remarkable effect on both the rate of the coupling and yield of product. Although it is not clear what causes the accelerating and yield effects, it is conceivable that the alkenyl groups may be transferred from Al or Zr to Pd via Zn, resulting in a more energetically favorable process.[212]

1 week, 25 °C, no $ZnCl_2$:   < 1%
1 hr, 25 °C, $ZnCl_2$:   88%

The Suzuki Reaction: Coupling Reactions with Organoboron Compounds[213,214]

The Suzuki reaction provides a versatile, general method for the stereo- and regiospecific synthesis of conjugated dienes, enynes, aryl substituted alkenes, and biaryl compounds via Pd-catalyzed cross-coupling of vinyl halides or aryl halides with vinyl-, aryl-, or alkynylboron reagents.[215] The wide use of this carbon-carbon bond forming reaction in organic synthesis stems from the ability to preserve the alkene geometry of both the haloalkene and alkenylboron in the product, the tolerance of functional groups, and the ready availability of the starting materials.

The catalytic cycle for the Suzuki cross-coupling reaction involves an oxidative addition (to form RPd(II)X)-transmetalation-reductive elimination sequence. The transmetalation between the RPd(II)X intermediate and the organoboron reagent does not occur readily until a base, such as sodium or potassium carbonate, hydroxide or alkoxide, is present in the reaction mixture. The role of the base can be rationalized by its coordination with the boron to form the corresponding *ate*-complex A, thereby enhancing the nucleophilicity of the organic group, which facilitates its transfer to palladium. Also, the base R'O$^-$ may activate the palladium by formation of R–Pd–OR' from R–Pd–X.

a = oxidative addition; b = hydroxypalladation;
c = transmetalation; d = reductive elimination

The organoboron reagents in Suzuki cross-coupling reactions are either organoboranes, 1-alkenyl- and arylboronic acids (RB(OH)$_2$), or boronate esters (RB(OR')$_2$). The traditional synthesis of 1-alkenyl- and arylboronic acids and boronate esters involves treatment of Grignard or lithium reagents with trialkyl borates to produce the corresponding boronate esters. Hydrolysis of these furnishes the corresponding boronic acids.[213] However, several problems are associated with the preparation of these boron reagents, such as purification, polymerization, and lack of atom economy.[216] These shortcomings can be obviated by converting the boronate esters to potassium organotrifluoroborates with potassium hydrogen fluoride (KHF$_2$). Organotrifluoroborates are crystalline solids, readily isolated and stable to air and moisture, and are excellent partners in Suzuki coupling reactions.[216]

(*E*)-and (*Z*)-alkenylboranes, or the corresponding boronic esters or potassium trifluoroborates, are conveniently available via hydroboration of terminal alkynes or 1-halo-1-alkynes using either disiamylborane, dicyclohexylborane, dibromoborane dimethyl sulfide, or catecholborane as hydroborating agents.[213,214b,216a]

Preparation of (E)-Alkenylboron Reagents

Preparation of (Z)-Alkenylboron Reagents

A variety of conjugated dienes, whose structural features are frequently encountered in natural products and insect pheromones, as well as dienophiles in Diels-Alder reactions, are accessible via the Suzuki reaction.[217]

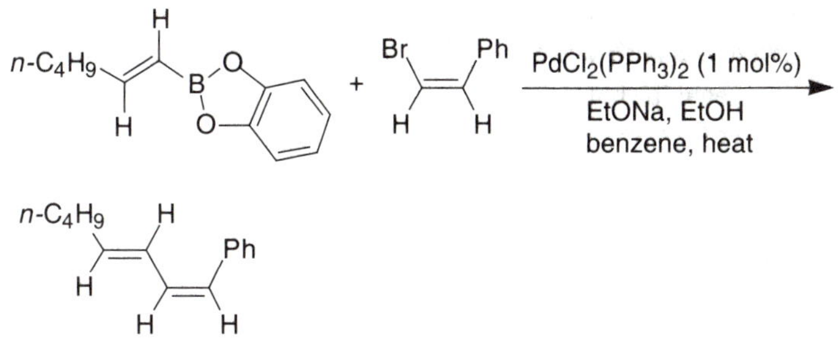

82% (> 98% *E, Z* isomer)

Cross-coupling of 1-alkenylboron compounds with 1-halo-1-alkynes leads to conjugated enynes, which can serve as precursors for conjugated dienes.[217]

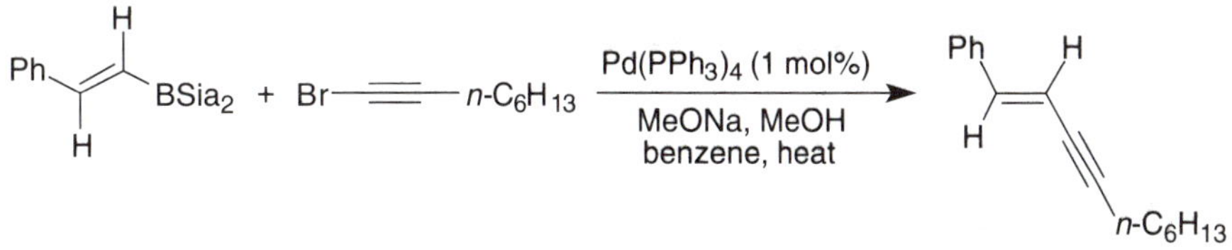

95% (99% *E*-isomer)

Pd(0)-mediated arylation of 1-alkenylboronates or 1-alkenyltrifluoroboronates with haloarenes is a stereoselective method for the synthesis of aryl-substituted alkenes.[218,219]

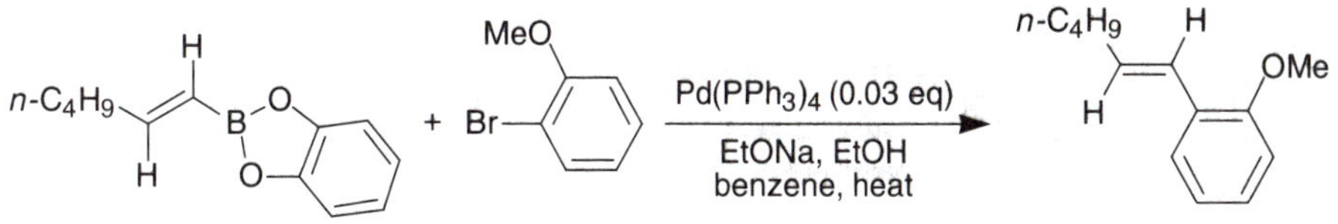

81%

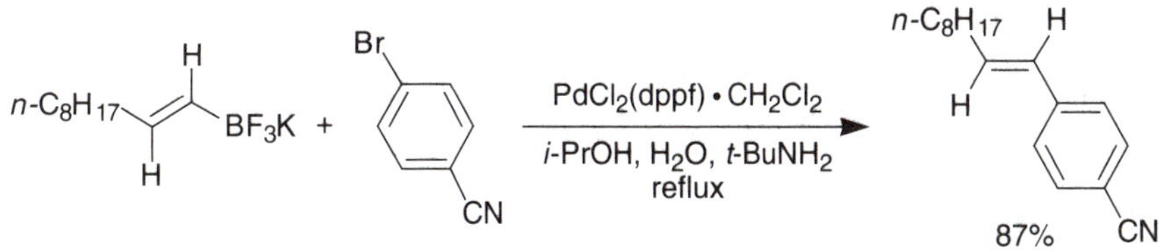

87%

The Suzuki coupling reaction is a powerful tool for the construction of biaryl compounds and their homologues, which are key structural elements of various natural products, polymers, and compounds of medicinal interest. Aryl boronic acids and their esters are the usual substrates in reactions with aryl or heteroaromatic halides and aryl triflates.[220]

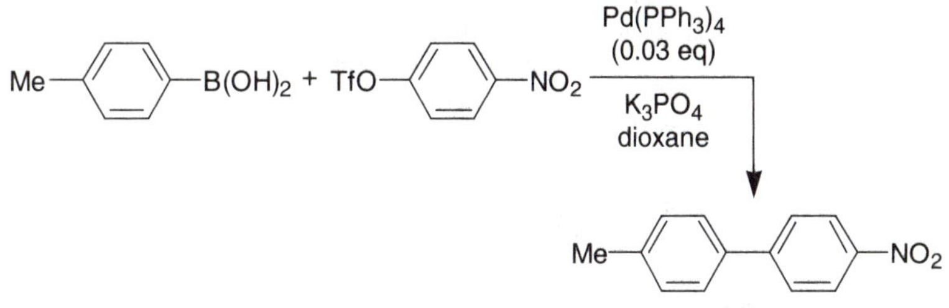

82%

73%

An important feature of the Suzuki reaction is that it allows *alkyl*-alkenyl or *alkyl*-aryl coupling of *B-alkyl*-9-borabicyclo[3.3.3]nonane derivatives with haloalkenes and haloarenes without concomitant β-hydride elimination. The required organoboranes are obtained by hydroboration of the appropriate alkenes or functionally substituted alkenes with 9-BBN. The coupling reaction tolerates the presence of functional groups in both reaction partners, thus circumventing a requirement of their prior protection.[221]

PdCl$_2$(dppf): dichloro[1,1'-bis(diphenylphosphine)
ferrocene]palladium(II)

The palladium-catalyzed carbonylative coupling reaction of alkenyl halides with organoboron compounds in the presence of carbon monoxide is a valuable procedure for the synthesis of unsymmetrical ketones.[222]

The probable catalytic cycle for the carbonylative coupling reaction involves oxidative addition, carbon monoxide insertion into the R–Pd complex, transmetalation, *trans-cis*-isomerization, and reductive elimination.

a = oxidative addition; b = carbon monoxide insertion;
c = transmetalation followed by *trans-cis*-isomerization; d = reductive elimination

The Stille Reaction: Coupling Reactions with Organotin Compounds[223]

The Stille reaction involves the palladium-catalyzed cross-coupling of organostannanes with electrophiles such as organic halides, triflates, or acid chlorides. The coupling of the two carbon moieties is stereospecific and regioselective, occurs under mild conditions, and tolerates a variety of functional groups (CHO, CO_2R, CN, OH) on either coupling partner. These properties make the Stille reaction frequently the method of choice in syntheses of complex molecules.[224]

$$R-X \xrightarrow[\text{R'Sn}(n\text{-Bu})_3]{\text{Pd}(L_n)} R-R' \quad + \quad X-Sn(n\text{-Bu})_3$$

R = acyl, allyl, aryl, vinyl, benzyl
R' = aryl, vinyl
$Pd(L_n)$ = $Pd(PPh_3)_4$, $(MeCN)_2PdCl_2$, $(PhCN)_2Pd(PPh_3)_2$

A problem of the Stille reaction is the toxicity of organotin reagents, especially the lower-molecular weight alkyl derivatives. Thus, the preparation of organostannanes and their reactions must be carried out in a well-ventilated hood.[225]

Electrophiles used in the Stille reaction are allylic chlorides, aryl or vinylbromides, iodides and triflates. They are employed in a polar solvent such as THF or DMF. Because of the greater reactivity of alkenyl iodides compared to that of the corresponding bromides and triflates, their reaction can be carried out at lower temperature, thereby increasing the stereoselectivity of the alkenyl moiety being transferred to palladium. The catalyst precursors for coupling aryl and vinyl halides are 1–2 mol % $Pd(PPh_3)_4$, $(MeCN)_2PdCl_2$, or $Pd(OAc)_2$, which are rapidly reduced to catalytically active Pd(0) species by the stannanes. The catalytic cycle proposed for the Stille reaction follows the general principles of transition metal-mediated cross-coupling reactions: (*a*) oxidative addition, (*b*) transmetalation, (*c*) *trans-cis*-isomerization, and (*d*) reductive elimination.

The most widely used groups in transmetalations from tin to carbon are those with proximal π-bonds such as alkenyl-, alkynyl-, and arylstannanes.[226] The transfer of an R′ group from R′Sn(n-Bu)$_3$ follows the order:

$$RC{\equiv}C > RCH{=}CH > Ar > RCH{=}CHCH_2 \approx ArCH_2 \gg alkyl$$

Thus, alkyl groups such as methyl or n-butyl in mixed stannanes [e.g., R′SnMe$_3$, R′Sn(n-Bu)$_3$] serve as nontransferable ligands, ensuring that acetylenic, vinyl, and aryl groups (R′) are selectively transferred to palladium. The requisite mixed organostannanes are available by various routes such as treatment of trialkyltin halides with unsaturated organolithium or organomagnesium reagents and radical-induced or palladium-promoted addition of tin hydrides to alkynes.[226,227] Cross-coupling reactions of alkenyl[228] and aryl[229] halides with organostannanes proceed in high yields. Using allylic chlorides as electrophiles poses a regioselectivity problem. However, coupling generally occurs by attack of the organostannane at the less hindered terminus of the allylic moiety.[230]

Vinyl triflates are especially attractive in Stille reactions since they can be prepared regioselectively from enolizable ketones by trapping the enolate oxygen with triflating agents (e.g., Tf$_2$O,[231] PhNTf$_2$[232]). For reactions of vinyl triflates with organostannanes, $Pd(PPh_3)_4$ is the catalyst of choice. An excess of lithium chloride is required.[233]

regioselectivity = 95 : 5

regioselectivity = 98 : 2

When the Stille reaction is carried out under a CO atmosphere, the carbonylative coupling proceeds in a manner similar to that described for the Suzuki reaction; namely, carbonyl insertion into the Pd–C bond of the oxidative addition complex. Transmetalation, followed by *cis-trans*-isomerization and reductive elimination, generates the ketone product.[234]

65%

78%

Cross-Coupling Reactions Involving sp-Carbons

This section deals with specific procedures for conjugation of -ene and -yne fragments via arylation and alkenylation of 1-alkynes. These unsaturated structural features are often found in natural products, especially pheromones, in pharmaceuticals and in organic materials such as molecular wires.

The Castro-Stephens Reaction: Preparation of Disubstituted Alkynes and Enynes

The Castro-Stephens reaction entails coupling of alkynylcopper(I) reagents with aryl or heteroaromatic halides in refluxing pyridine to furnish arylacetylenes. The cuprous alkynylides are prepared by adding the 1-alkynes to an aqueous ammonia solution of cuprous iodide.[235] Since dry copper(I) alkynylides have the tendency to explode, they must be handled with great care and should only be used in a slightly damp state.[236]

A milder route to arylacetylenes is the *Sonogashira reaction*, the direct cross-coupling of terminal alkynes with aryl halides or aryl triflates in the presence of catalytic amounts of $Pd(PPh_3)_4$ or $(PPh_3)_2PdCl_2$ and using cuprous iodide as a cocatalyst in amine solvents.[237] The reaction can also be applied to the vinylation of 1-alkynes using vinyl halides or triflates to yield conjugated enynes.[238]

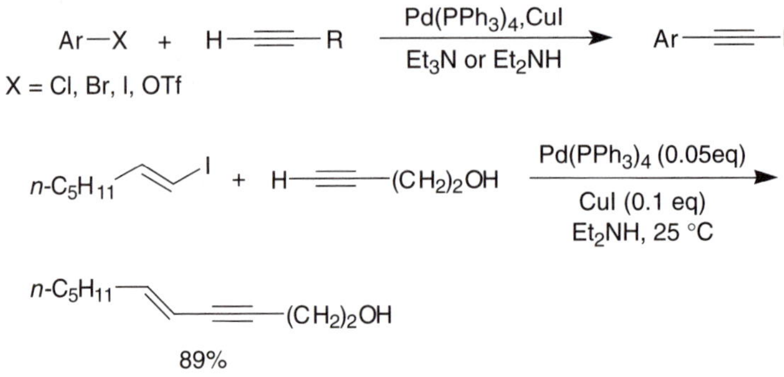
Ar—X + H—≡—R
Pd(PPh3)4,CuI
Et3N or Et2NH
Ar—≡—R
X = Cl, Br, I, OTf
n-C5H11
I
H—≡—(CH2)2OH
Pd(PPh3)4 (0.05eq)
CuI (0.1 eq)
Et2NH, 25 °C
n-C5H11
(CH2)2OH
89%

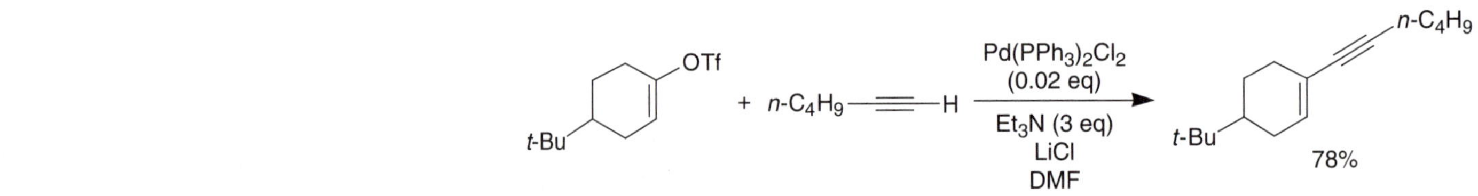
OTf
t-Bu
+ n-C4H9—≡—H
Pd(PPh3)2Cl2
(0.02 eq)
Et3N (3 eq)
LiCl
DMF
n-C4H9
t-Bu
78%

Instead of using palladium(0) complexes, the more stable bis(triphenylphosphine)Pd(II) chloride, which is rapidly reduced in situ to a coordinatively unsaturated 14-electron, catalytically active Pd(0) species, is often the catalyst of choice.[239]

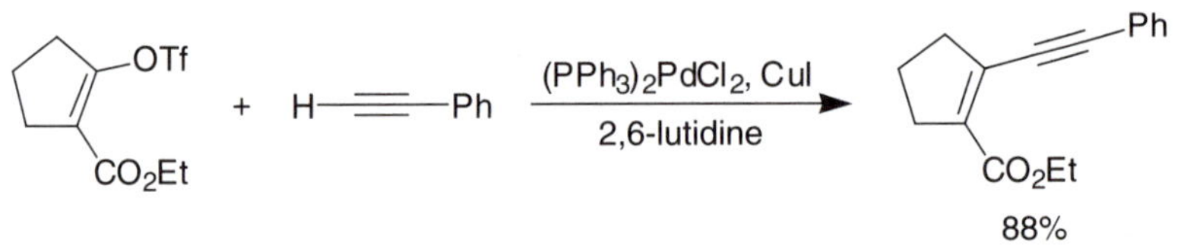
OTf
CO2Et
+ H—≡—Ph
(PPh3)2PdCl2, CuI
2,6-lutidine
Ph
CO2Et
88%

The cross-coupling reaction of the sp carbon with the sp^2 carbon entails oxidative addition, Cu(I) catalyzed transmetalation, and reductive elimination.

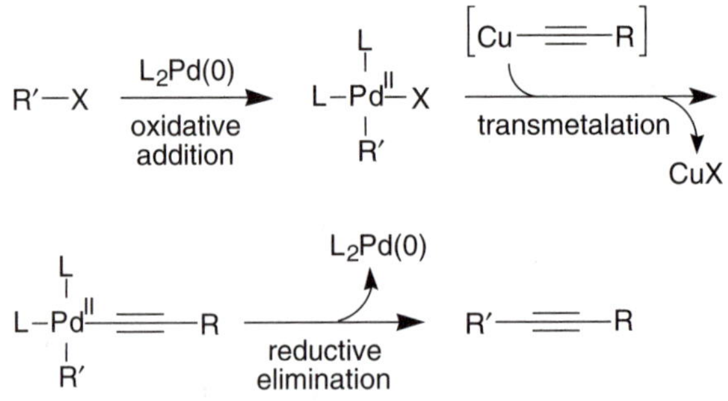
R′—X
L2Pd(0)
oxidative
addition
L
L–PdII–X
R′
[Cu—≡—R]
transmetalation
CuX
L
L–PdII—≡—R
R′
L2Pd(0)
reductive
elimination
R′—≡—R

A more general method for arylation of terminal alkynes as well as electron-deficient alkynes is the *Negishi Pd-catalyzed cross-coupling* of aryl halides with alkynylzinc reagents.[240] When using functionally substituted alkynylzincs, the deprotonation of 1-alkynes must be done with LDA instead of alkylithiums.

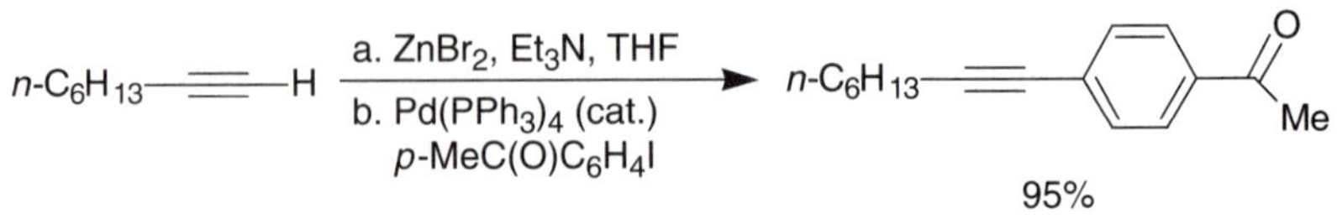
n-C6H13—≡—H
a. ZnBr2, Et3N, THF
b. Pd(PPh3)4 (cat.)
p-MeC(O)C6H4I
n-C6H13
O
Me
95%

Preparation of Conjugated Enediynes

The Pd-catalyzed homocoupling of terminal alkynes with (*Z*)-1,2-dichloroethylene furnishes the corresponding (*Z*)-conjugated enediynes, whose structural feature is encountered in a class of antitumor antibiotics and is the object of current intense synthetic and mechanistic studies.[241] Unsymmetrical (*E*)- and (*Z*)-enediynes are prepared in a one-pot procedure by two sequential Pd-catalyzed cross-coupling reactions of (*E*)- and (*Z*)-1,2-dichloroethylene with the appropriate 1-alkynes.[242] Reduction of enediynes A with a zinc copper-silver couple in aqueous methanol provides access to (*Z*)-trienes.[243]

A one-pot tandem process for the synthesis of (*E*)-enediynes that allows incorporation of ethynyl groups is via Pd-catalyzed coupling of (*E*)-bromoiodoethylene with zinc alkynylides.[244]

Preparation of 1,3-Diynes

The synthesis of conjugated diynes via the *Glaser* coupling reaction[245] is the classical method for homocoupling of terminal alkynes. The coupling reaction is catalyzed by CuCl or Cu(OAc)$_2$ in the presence of an oxidant and ammonium chloride or pyridine to yield symmetrically substituted diynes.[246] The oxidative dimerization appears to proceed via removal of the acetylenic proton, formation of an alkynyl radical, and its dimerization.

Homocoupling of 1-trimethylsilylalkynes with CuCl in DMF under aerobic conditions also produces the corresponding 1,3-butadiynes.[247]

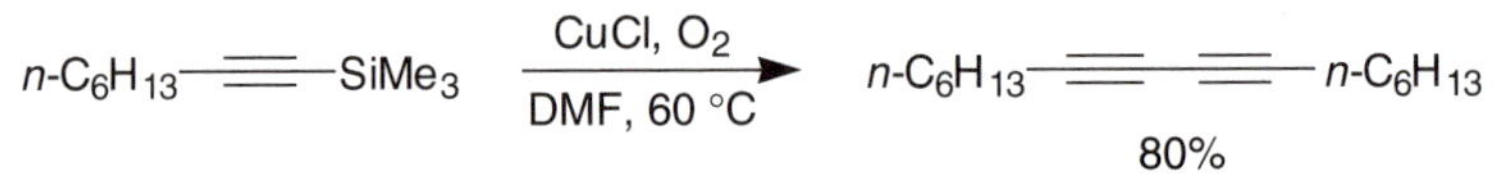

n-C6H13—≡—SiMe3
CuCl, O2
DMF, 60 °C
n-C6H13—≡—≡—n-C6H13
80%

Unsymmetrically substituted 1,3-diynes can be obtained via the *Cadiot-Chodkiewicz* copper(I)-catalyzed coupling of 1-alkynes with 1-halo-1-alkynes in the presence of hydroxylamine hydrochloride and an amine.[248]

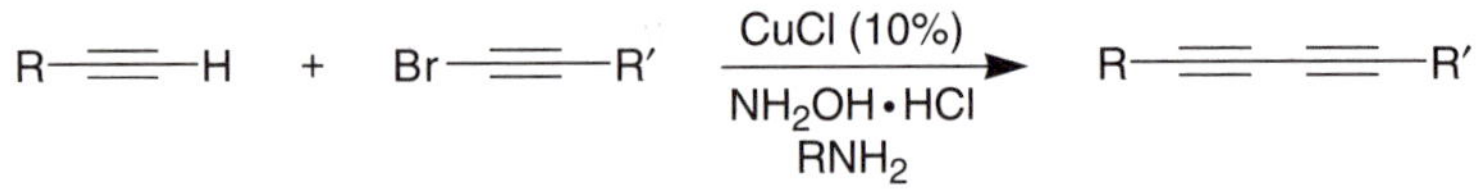

R—≡—H + Br—≡—R′
CuCl (10%)
NH2OH•HCl
RNH2
R—≡—≡—R′

A problem that may be encountered when using the original procedure is that some functionally substituted acetylenes do not survive the acidic reaction conditions. Furthermore, one or both undesired homocoupled diynes may be formed, and these are often difficult to separate from the desired product. A number of methods are available that minimize or circumvent these shortcomings. For example, coupling of terminal alkynes with in situ preformed copper alkynylides accommodates the cleavage-prone Me_3Si moiety.[249]

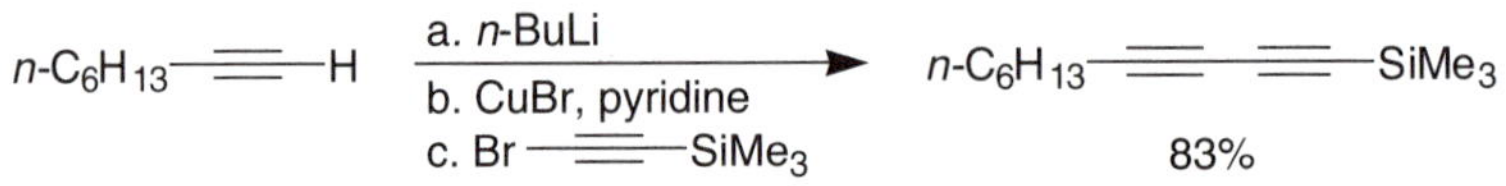

n-C6H13—≡—H
a. n-BuLi
b. CuBr, pyridine
c. Br—≡—SiMe3
n-C6H13—≡—≡—SiMe3
83%

Especially attractive are syntheses of 1,3-diynes via Pd(0)-catalyzed coupling reactions using 1,2-dihaloethylenes. For example, treatment of an acetylene with (*Z*)-1,2-dichloroethylene followed by dehydrohalogenation of the resultant halo-enyne produces monosubstituted 1,3-diynes.[250]

(EtO)2CH—≡—H +
CuI (cat.)
Pd(PPh3)4
n-BuNH2
(EtO)2CH
n-Bu4NF
THF
(EtO)2CH—≡—≡—H
70%

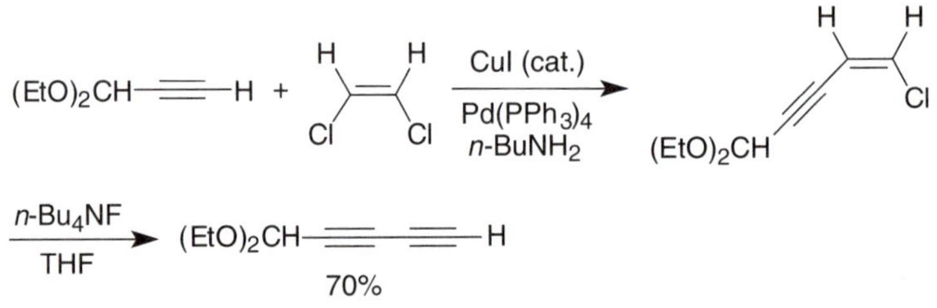

The Pd-catalyzed cross-coupling of diynylzincs, formed as shown below, with aryl halides leads to disubstituted conjugated diynes. This procedure allows great flexibility in the choice of reaction partners.[251]

Me
1a. n-BuLi, THF
1b. ZnBr2
1c. (E)-ClCH=CHI
2% Pd(PPh3)4
1d. aq NH4Cl
Me
Cl
90%

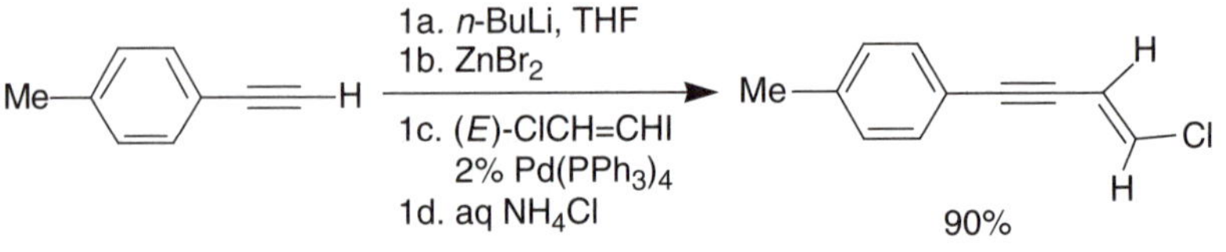

74%

The Trost-Tsuji Reaction: Palladium-Catalyzed Allylic Substitution[252]

Allylic substrates with good leaving groups are excellent reagents for joining an allyl moiety with a nucleophile. However, these reactions suffer from loss of regioselectivity because of competition between S_N2 and S_N2' substitution reactions (see Section 1.10). Palladium-catalyzed nucleophilic substitution of allylic substrates allows the formation of new carbon-carbon or carbon-hetero bonds with control of both regio- and stereochemistry. Moreover, use of chiral ligands increases the potential for asymmetric syntheses, which has been the subject of intensive investigations.

The palladium-mediated allylation proceeds via an initial oxidative addition of an allylic substrate to Pd(0). The resultant π-allylpalladium(II) complex A is electrophilic and reacts with carbon nucleophiles generating the Pd(0) complex B, which undergoes ligand exchange to release the product and restart the cycle for palladium. With substituted allylic compounds, the palladium-catalyzed nucleophilic addition usually occurs at the less substituted side.

The palladium catalyst generally used is Pd(PPh₃)₄, which can be formed in situ from Pd(OAc)₂ and PPh₃.[253] The most often used allylic substrates are those having an ester or a carbonate as a leaving group, although –OPO(OR)₂, –OPh, –Cl, or –Br will also work. Soft nucleophiles of the malonate-type generally give the best results for carbon-carbon bond formation. The reaction is usually irreversible and thus proceeds under kinetic control. Other soft carbon nucleophiles are anions from nitromethane, enolates, and enamines.

With the introduction of the highly reactive allylic carbonates, coupling reactions can be carried out under neutral conditions. This is especially important when dealing with compounds that are sensitive to bases. The oxidative addition of allylic carbonates is followed by decarboxylation as the irreversible step to produce the π-allylpalladium alkoxides. Since the alkoxides produced are rather poor nucleophiles, they do not compete with the carbon nucleophile for the π-allylpalladium complex but do deprotonate the active methylene compound. Attack of the Nu⁻ on the π-allylpalladium complex leads to the substitution product.

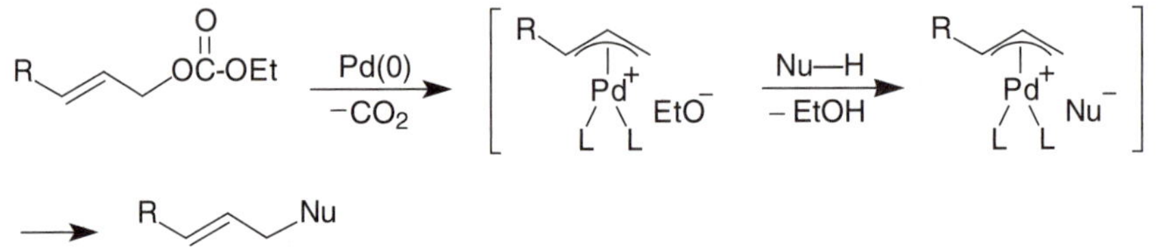

An example of a neutral, chemoselective palladium-catalyzed carbon-allylation is the reaction of allyl ethyl carbonate with one equivalent of the nitroacetic ester shown below.[254]

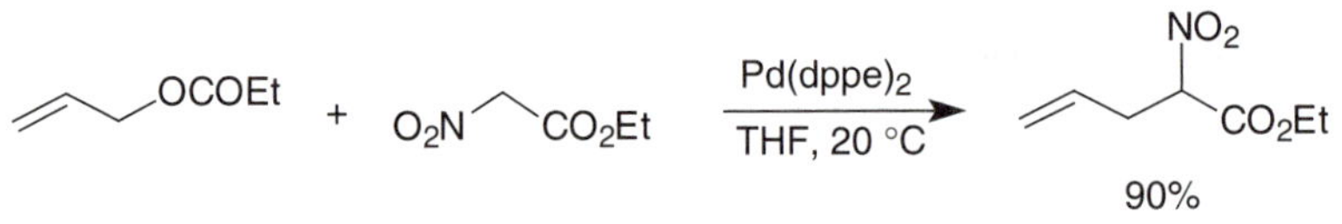

Secondary amines are especially good nucleophiles for allylic substitutions. Alkylation-amination of the (Z)-bis-allylic compound shown below in the presence of $Pd(PPh_3)_4$ with one equivalent of malonate furnishes the bis-substitution product with (E)-selectivity.[255] Alkene isomerization due to *syn-anti* interconversion in the π-allyl intermediate is often observed.

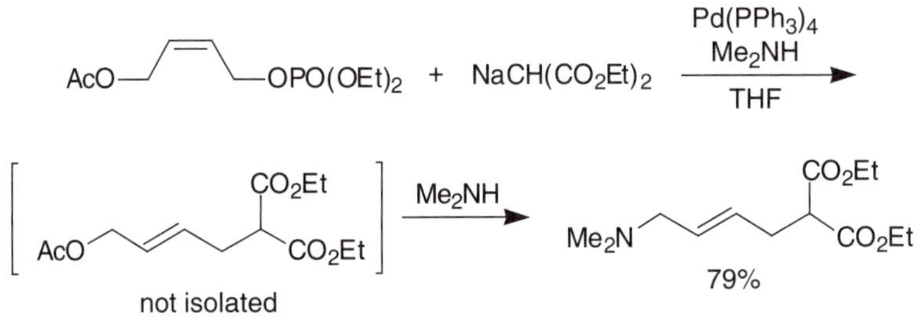

Palladium-catalyzed displacement reactions with carbon nucleophiles are not only regioselective but also highly stereoselective. In the first step, displacement of the leaving group by palladium to form the π-allylpalladium complex occurs from the less hindered face with inversion. Subsequent nucleophilic substitution of the intermediate π-allylpalladium complex with soft nucleophiles such as amines, phenols, or malonate-type anions also proceeds with inversion of the stereochemistry. The overall process is a retention of configuration as a result of the double inversion.

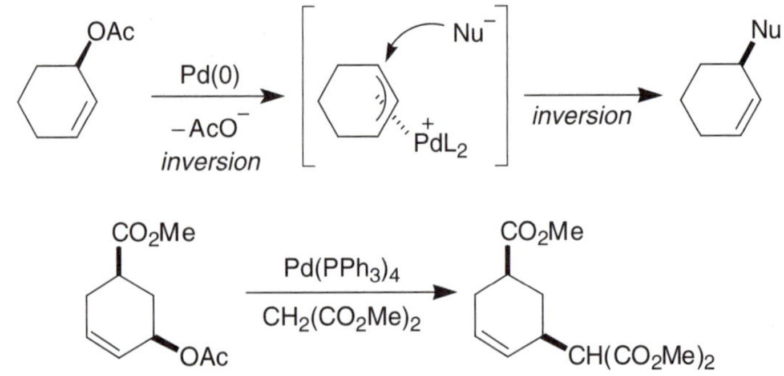

Intramolecular allylic substitution reactions provide a valuable tool for the construction of carbo- and heterocyclic systems.[256]

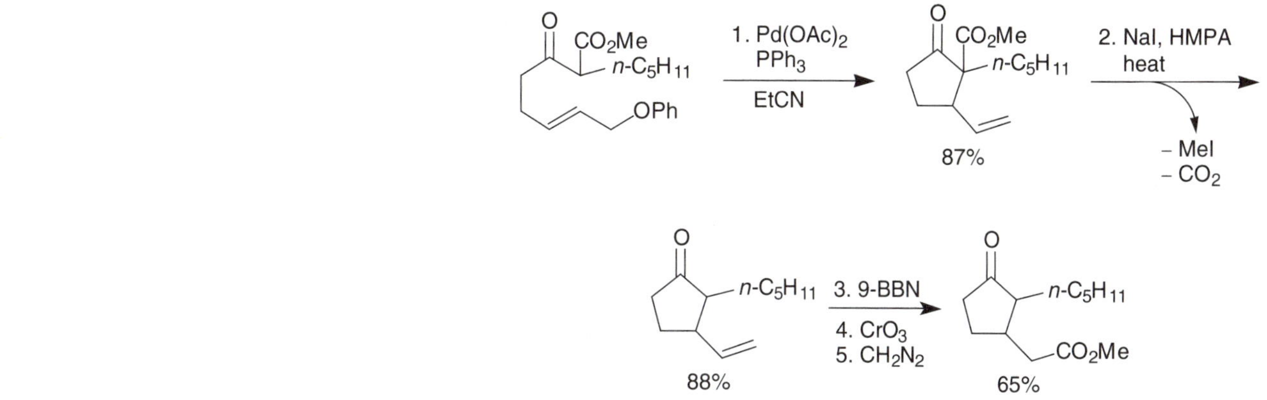

In the example shown below, the regiochemistry of attack of the nucleophilic amine at the six-member ring favors, for steric reasons, formation of a five-member heterocyclic ring.[257]

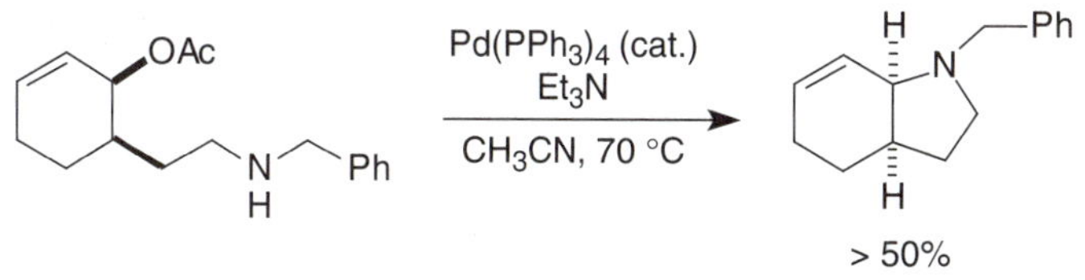

Ammonium formate hydrogenolysis of π-allyl-palladium(II) complexes formed via oxidative addition of Pd(0) to propargylic carbonate esters is a useful method for preparing disubstituted alkynes,[258] as depicted below.[259] This method of deoxygenation avoids the use of expensive and toxic reagents often associated with the free-radical Barton-McCombie deoxygenation.[260]

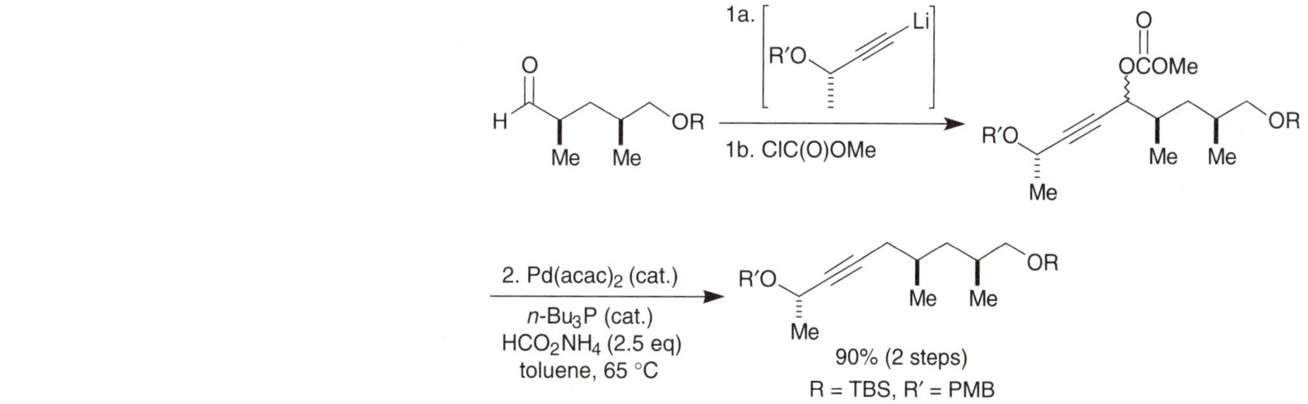

PROBLEMS

1. **Reagents**. Give the structures of the intermediates obtained in each step and the final major product expected for each of the following reaction sequences. Be sure to indicate product stereochemistry where applicable.

   a.

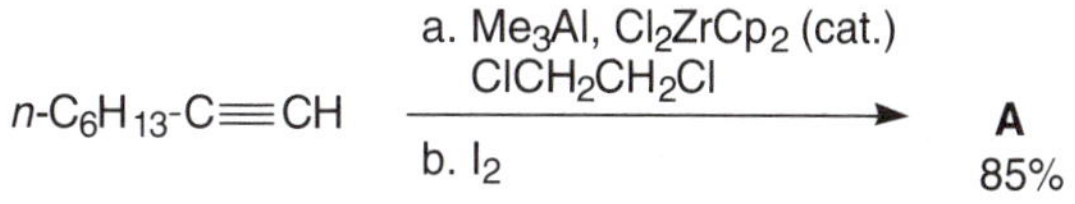

b.
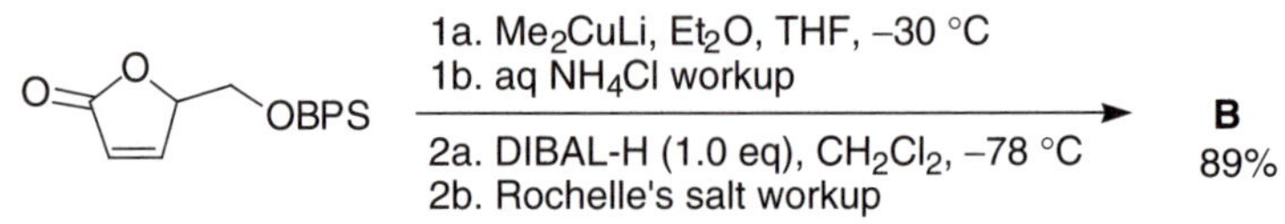
OBPS
1a. Me2CuLi, Et2O, THF, –30 °C
1b. aq NH4Cl workup
2a. DIBAL-H (1.0 eq), CH2Cl2, –78 °C
2b. Rochelle's salt workup
B
89%

c.
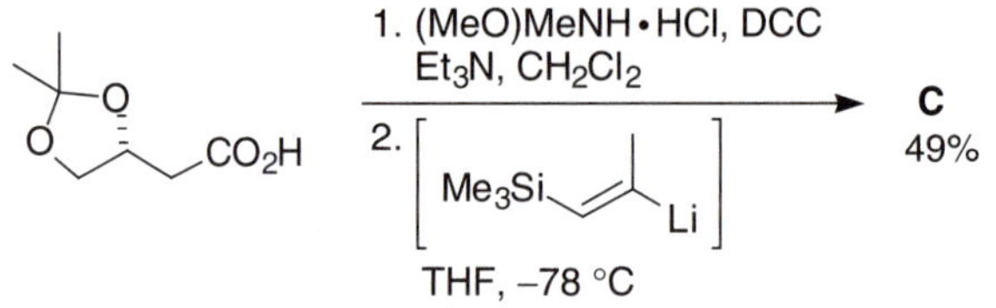
CO2H
1. (MeO)MeNH•HCl, DCC
Et3N, CH2Cl2
2.
Me3Si
Li
THF, –78 °C
C
49%

d.
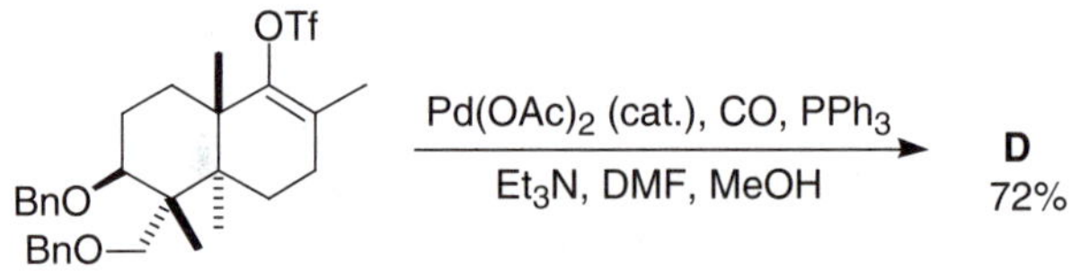
OTf
BnO
BnO
Pd(OAc)2 (cat.), CO, PPh3
Et3N, DMF, MeOH
D
72%

e.
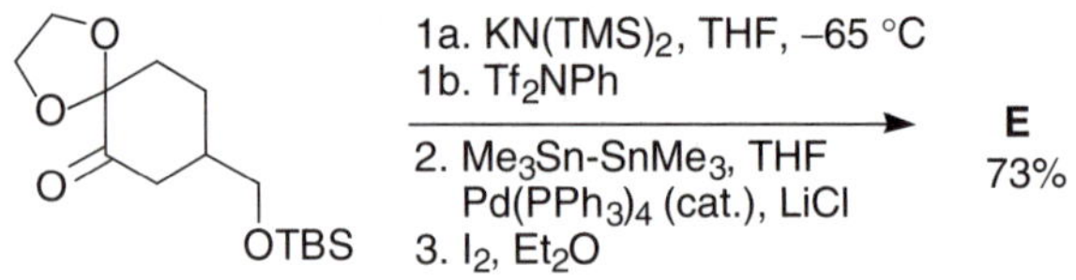
OTBS
1a. KN(TMS)2, THF, –65 °C
1b. Tf2NPh
2. Me3Sn-SnMe3, THF
Pd(PPh3)4 (cat.), LiCl
3. I2, Et2O
E
73%

f.
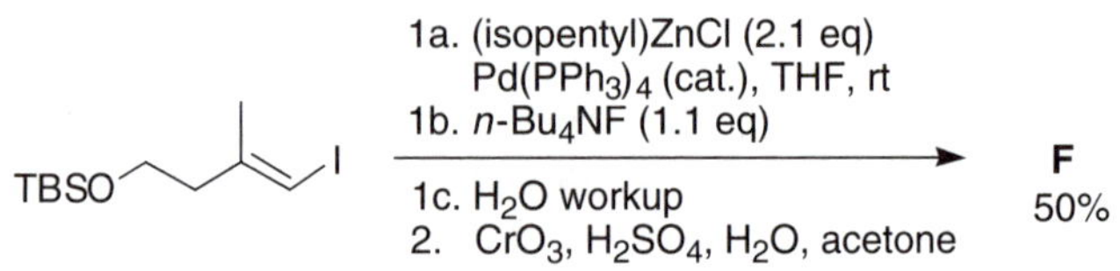
TBSO
I
1a. (isopentyl)ZnCl (2.1 eq)
Pd(PPh3)4 (cat.), THF, rt
1b. n-Bu4NF (1.1 eq)
1c. H2O workup
2. CrO3, H2SO4, H2O, acetone
F
50%

g.
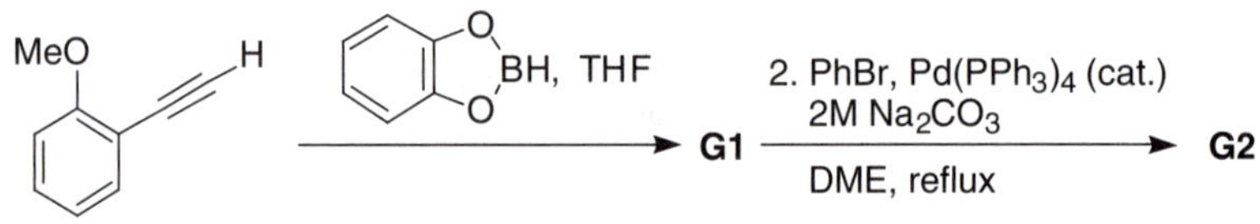
MeO
H
BH, THF
G1
2. PhBr, Pd(PPh3)4 (cat.)
2M Na2CO3
DME, reflux
G2

h.
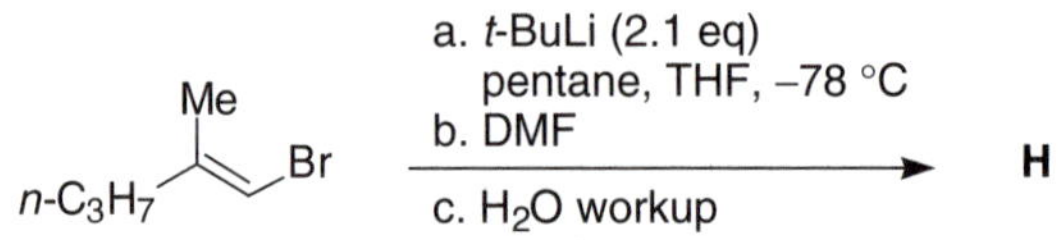
Me
Br
n-C3H7
a. t-BuLi (2.1 eq)
pentane, THF, –78 °C
b. DMF
c. H2O workup
H

i.
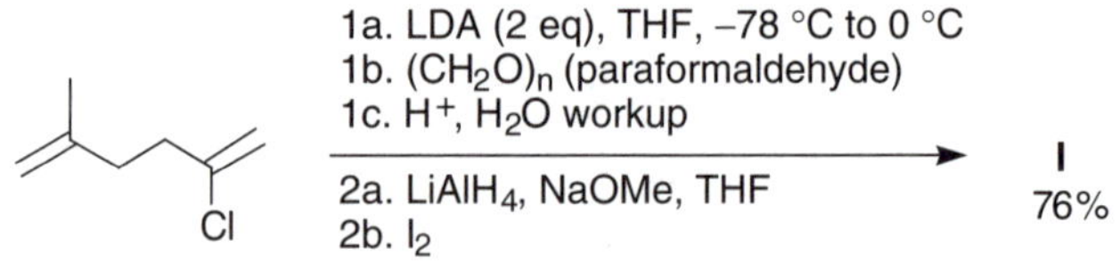
Cl
1a. LDA (2 eq), THF, –78 °C to 0 °C
1b. (CH2O)n (paraformaldehyde)
1c. H+, H2O workup
2a. LiAlH4, NaOMe, THF
2b. I2
I
76%

j.
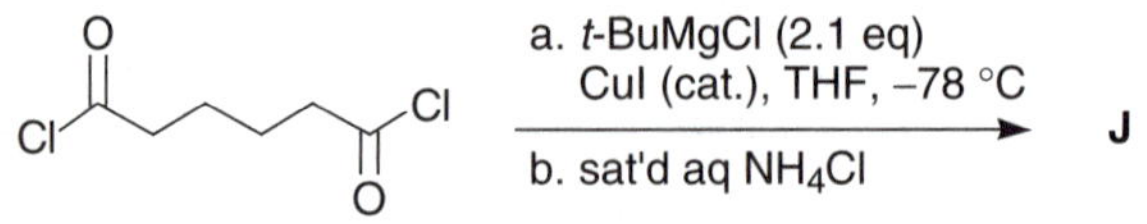
Cl
Cl
a. t-BuMgCl (2.1 eq)
CuI (cat.), THF, –78 °C
b. sat'd aq NH4Cl
J

k.
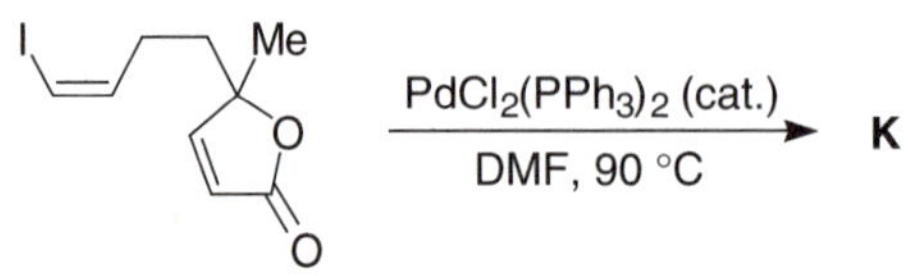
I
Me
PdCl2(PPh3)2 (cat.)
DMF, 90 °C
K

l.

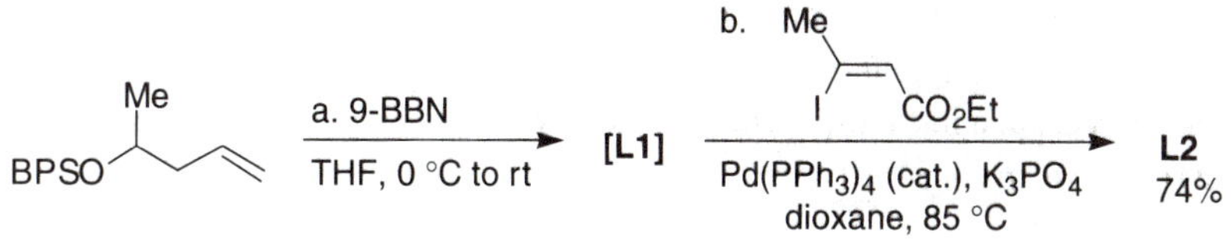

Me

BPSO ⟍⟍ → **[L1]**

a. 9-BBN
THF, 0 °C to rt

b. Me
        I   CO₂Et

Pd(PPh₃)₄ (cat.), K₃PO₄
dioxane, 85 °C

L2
74%

m.

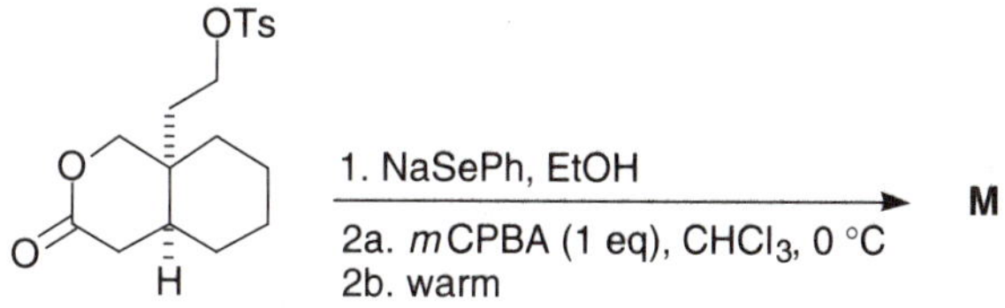

OTs

1. NaSePh, EtOH

2a. *m*CPBA (1 eq), CHCl₃, 0 °C
2b. warm

M

n.

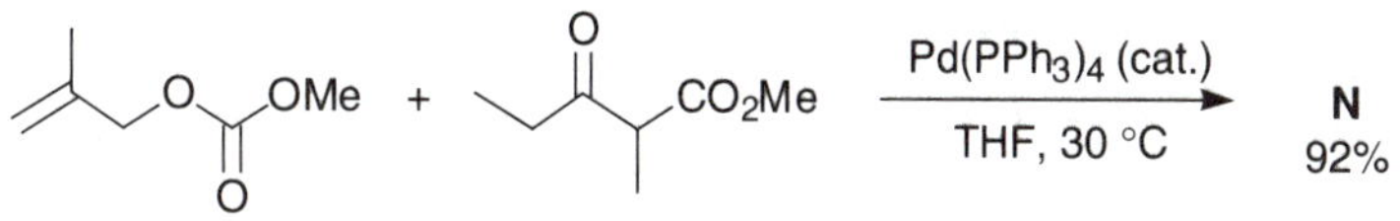

        O   OMe   +   O   CO₂Me

        O

Pd(PPh₃)₄ (cat.)
THF, 30 °C

N
92%

*o.

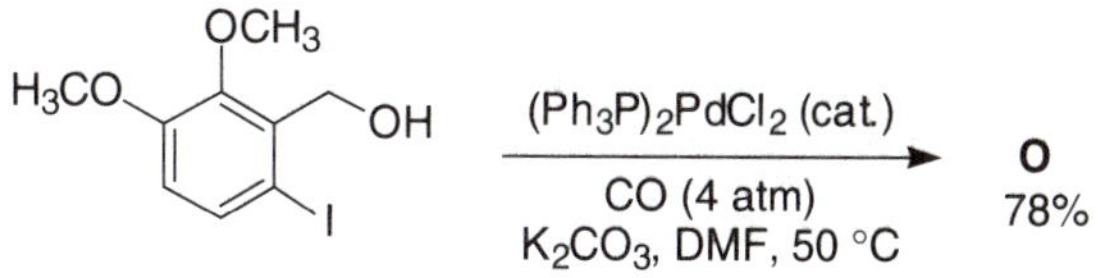

OCH₃
H₃CO   OH

I

(Ph₃P)₂PdCl₂ (cat.)
CO (4 atm)
K₂CO₃, DMF, 50 °C

O
78%

*p.

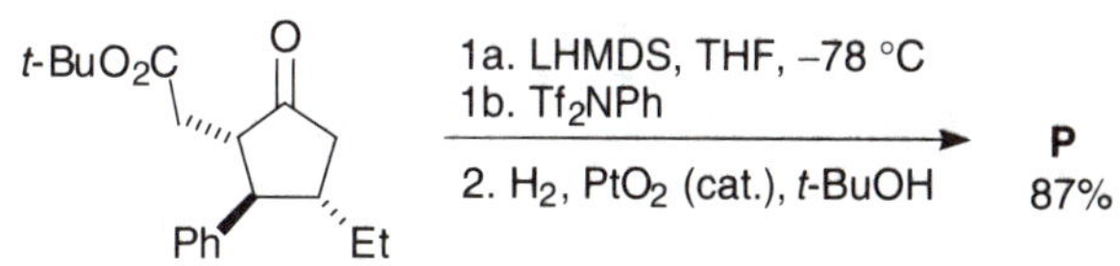

t-BuO₂C   O

Ph   Et

1a. LHMDS, THF, –78 °C
1b. Tf₂NPh

2. H₂, PtO₂ (cat.), *t*-BuOH

P
87%

2. Selectivity. Give the structures of the intermediates obtained after each step, and show the major product obtained for each of the following transformations.

a.

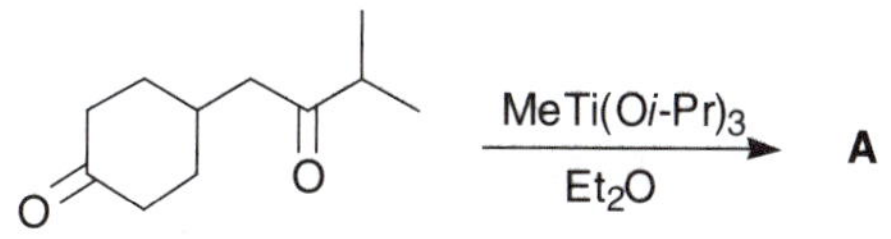

O   O

MeTi(O*i*-Pr)₃
Et₂O

A

b.

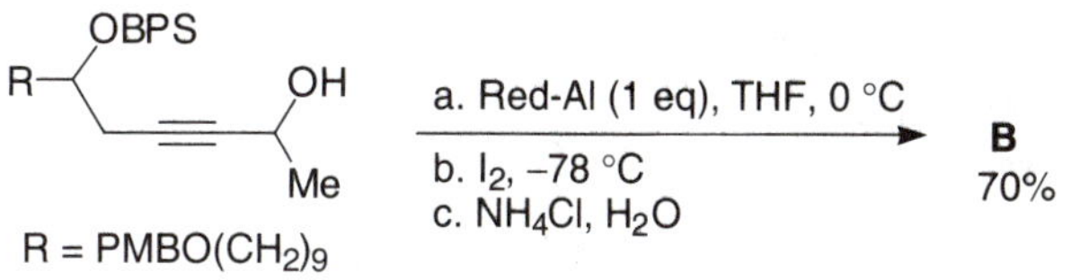

OBPS
R   OH

Me

R = PMBO(CH₂)₉

a. Red-Al (1 eq), THF, 0 °C
b. I₂, –78 °C
c. NH₄Cl, H₂O

B
70%

c.

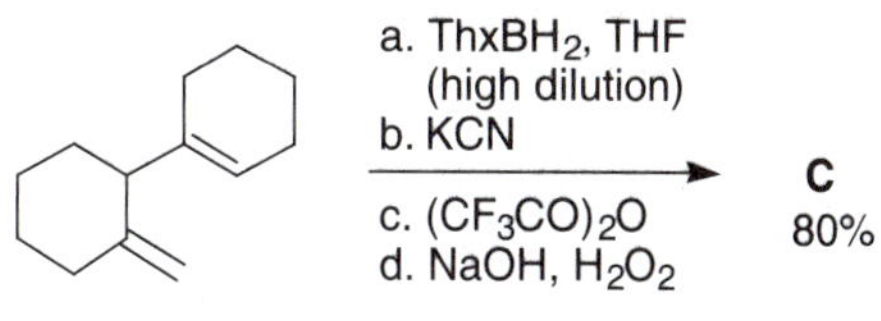

a. ThxBH₂, THF
   (high dilution)
b. KCN

c. (CF₃CO)₂O
d. NaOH, H₂O₂

C
80%

3. Stereochemistry. Predict the stereochemistry of the major product formed for each of the following reactions. Give an explanation for your choice.

a.

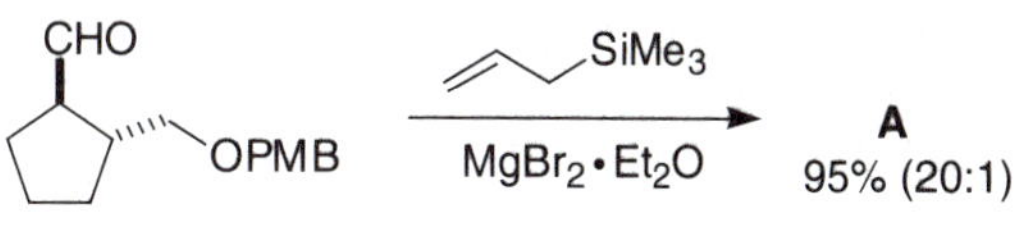

CHO

OPMB

⟍⟍SiMe₃

MgBr₂•Et₂O

A
95% (20:1)

b.

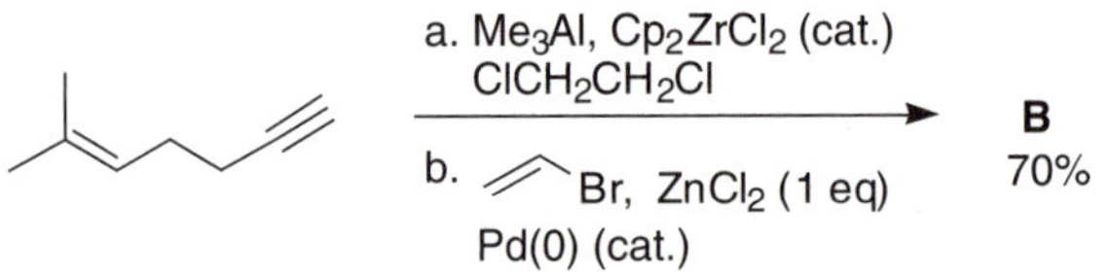
a. Me3Al, Cp2ZrCl2 (cat.)
ClCH2CH2Cl
b. Br, ZnCl2 (1 eq)
Pd(0) (cat.)
B
70%

c.

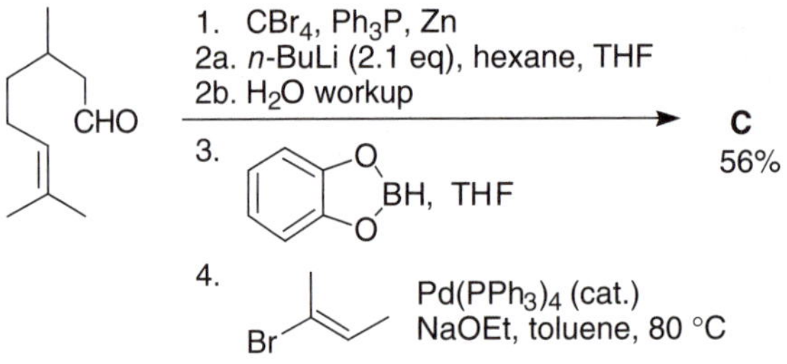
CHO
1. CBr4, Ph3P, Zn
2a. n-BuLi (2.1 eq), hexane, THF
2b. H2O workup
3.
BH, THF
4.
Br
Pd(PPh3)4 (cat.)
NaOEt, toluene, 80 °C
C
56%

d.

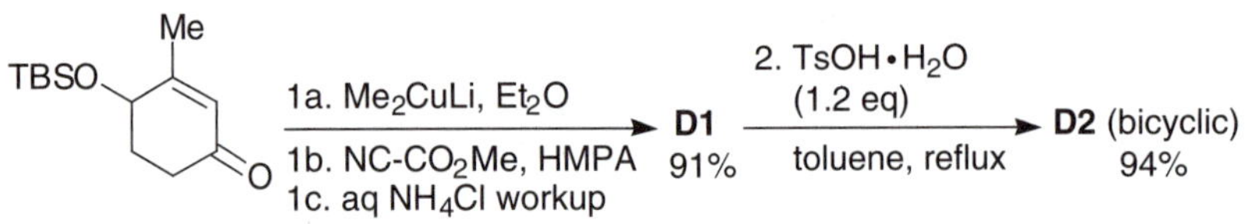
Me
TBSO
O
1a. Me2CuLi, Et2O
1b. NC-CO2Me, HMPA
1c. aq NH4Cl workup
D1
91%
2. TsOH•H2O
(1.2 eq)
toluene, reflux
D2 (bicyclic)
94%

***e.**

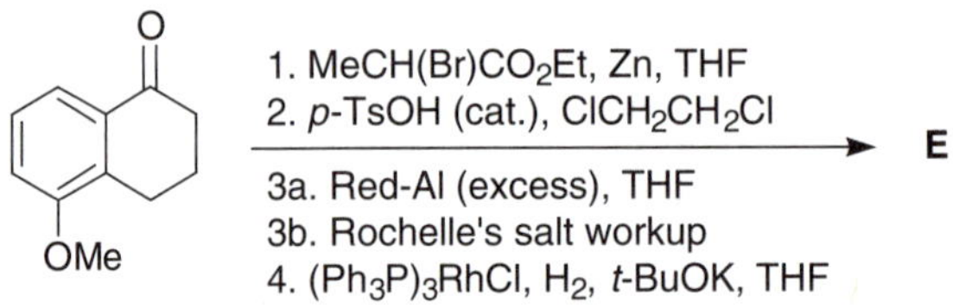
O
OMe
1. MeCH(Br)CO2Et, Zn, THF
2. p-TsOH (cat.), ClCH2CH2Cl
3a. Red-Al (excess), THF
3b. Rochelle's salt workup
4. (Ph3P)3RhCl, H2, t-BuOK, THF
E

***f.** Explain the observed stereoselectivities.

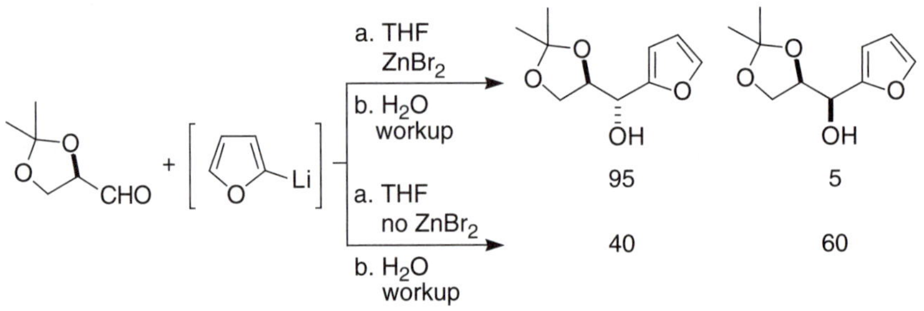
CHO
Li
a. THF
ZnBr2
b. H2O
workup
a. THF
no ZnBr2
b. H2O
workup
OH
OH
95
5
40
60

***g.** Supply the missing reagents and structures.

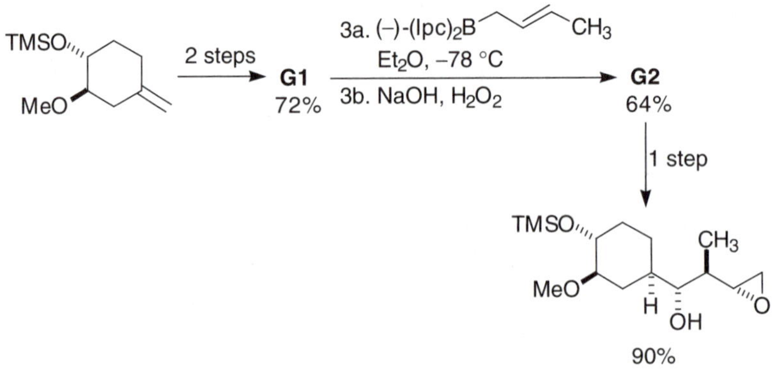
TMSO
MeO
2 steps
G1
72%
3a. (−)-(Ipc)2B CH3
Et2O, −78 °C
3b. NaOH, H2O2
G2
64%
1 step
TMSO
MeO
H
OH
CH3
90%

4. Reactivity. Propose a mechanism for each of the following transformations to explain the observed regioselectivity and stereochemistry.

a.

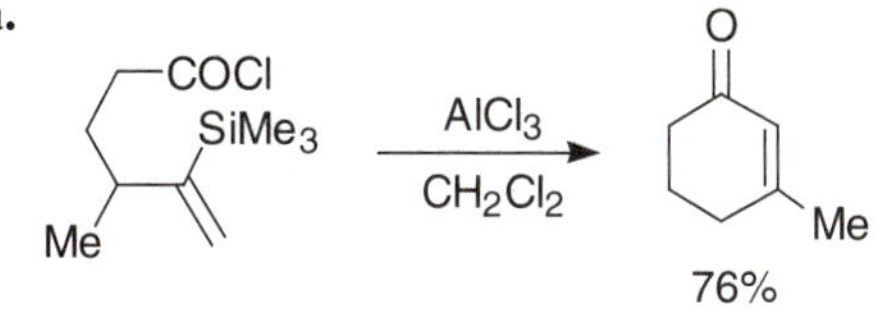

76%

b.

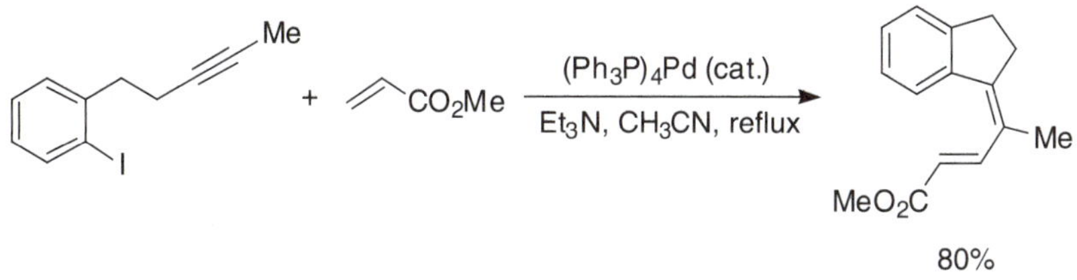

80%

c.

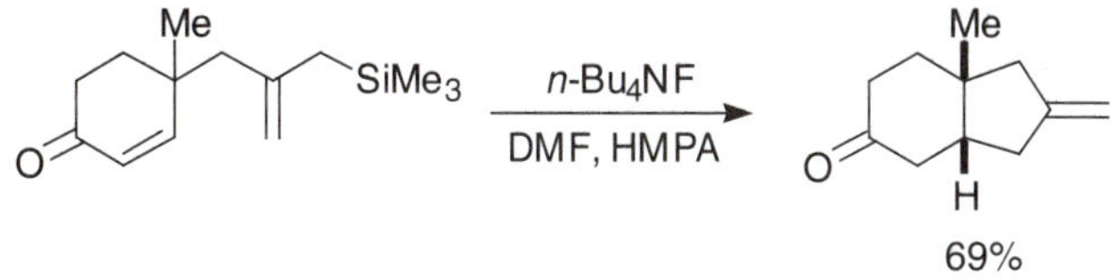

69%

d.

85%

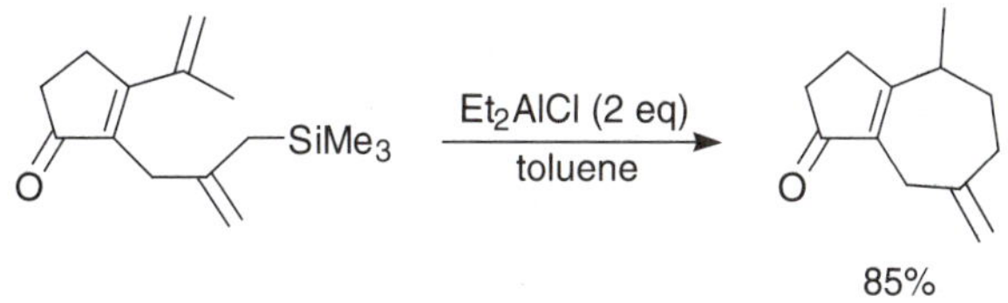

5. Synthesis. Supply the reagents required to accomplish each of the following syntheses. Show the structures of the intermediates obtained after each step and their relative stereochemistry where applicable.

a.

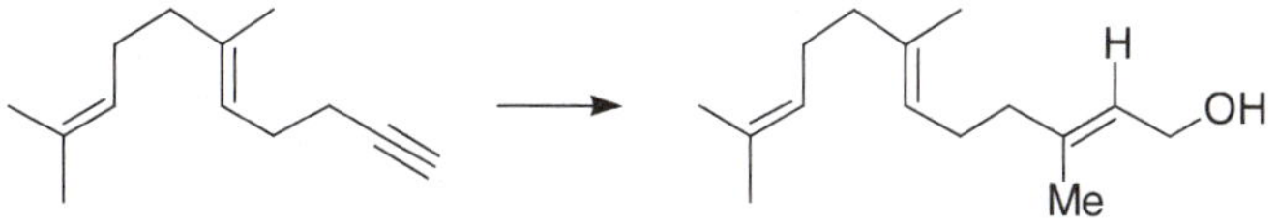

b.

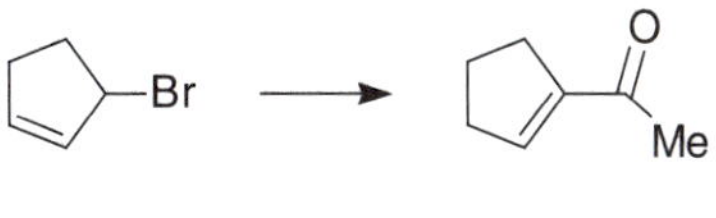

c.

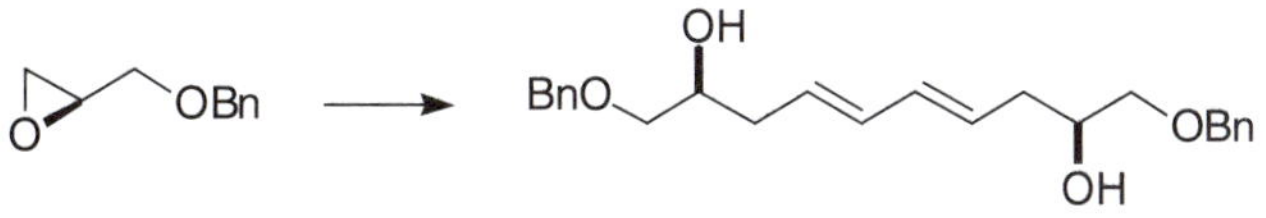

d.

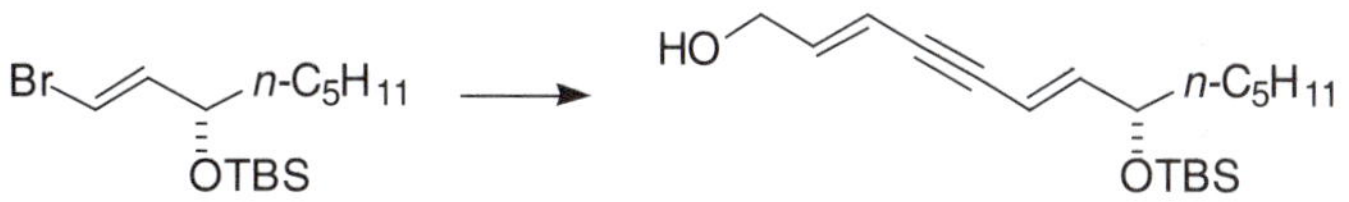

e.

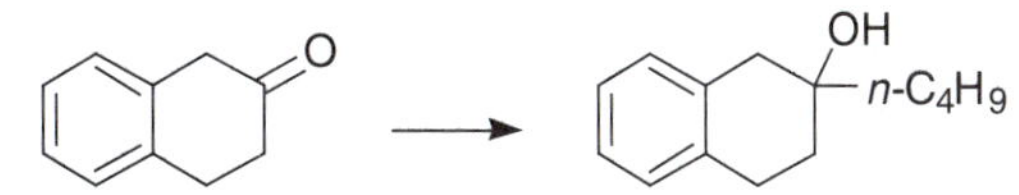

consider the pK_a

f.

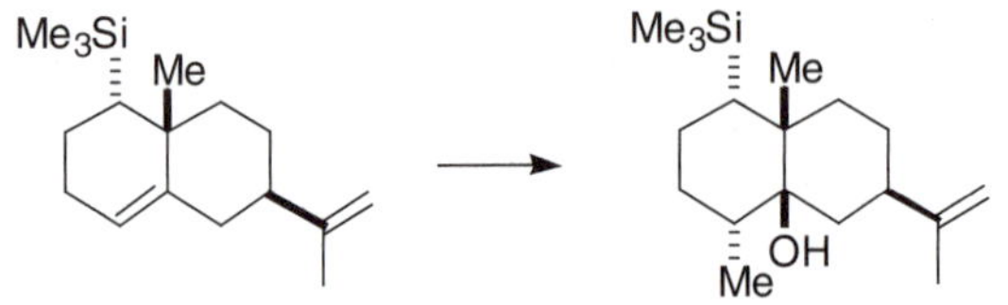

g.

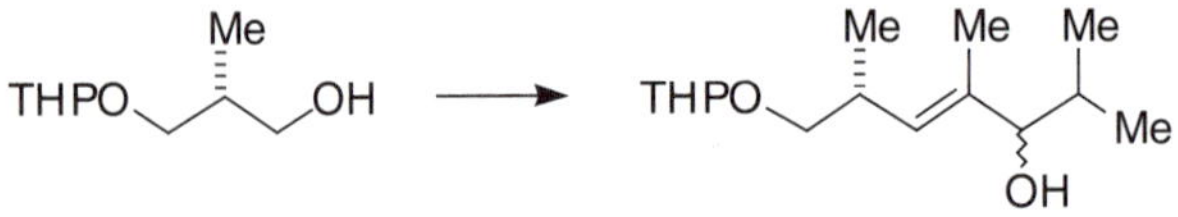

h.

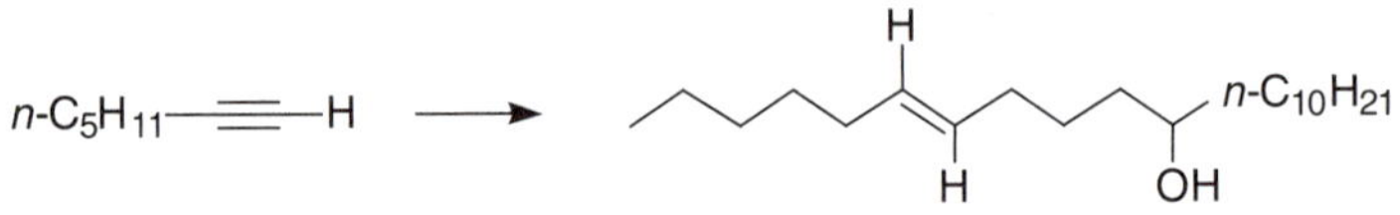

$n\text{-}C_5H_{11}\!=\!=\!\!H \longrightarrow$

i.

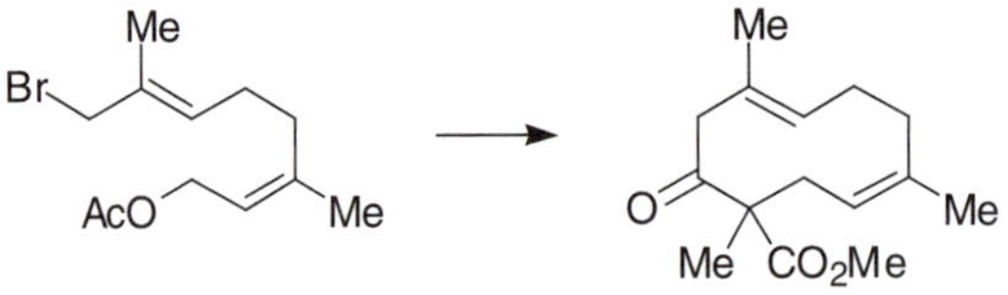

*j.

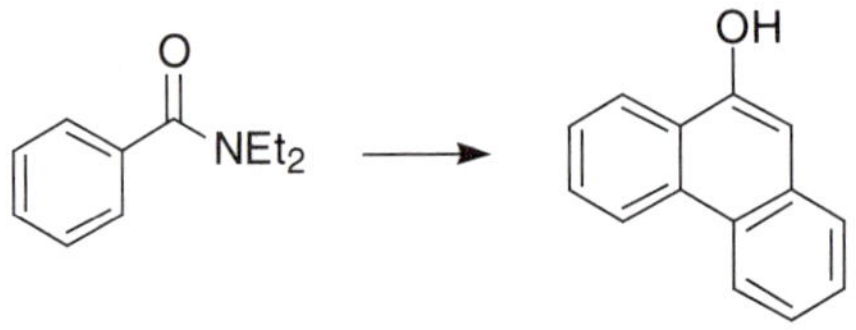

6. **Retrosynthetic Analysis.** Propose syntheses of the following compounds using a Suzuki coupling procedure as a key step. Show (1) your retrosynthetic analysis and (2) all reagents and reaction conditions required to transform commercially available starting materials into the target molecules.

a.

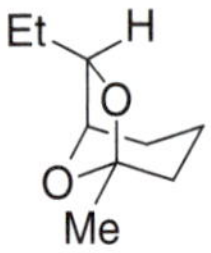

b.

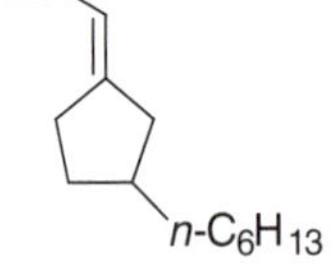

REFERENCES

1. Negishi, E. *Organometallics in Organic Synthesis*, Wiley: New York, 1980.

2. (a) Brown, H. C., Kramer, G. W., Levy, A. B., Midland, M. M. In *Laboratory Operations with Air-Sensitive Substances: Survey in Organic Syntheses via Boranes*, Wiley: New York, 1975, p. 191. (b) *The Manipulation of Air-Sensitive Compounds*, 2nd ed., Shriver, D. F., Drezdozon, M. A., Eds., Wiley: New York, 1986.

3. (a) Wakefield, B. J. *The Chemistry of Organolithium Compounds*, Pergamon Press: Oxford, UK, 1974. (b) Brandsma, L., Verkruijsse, H. *Preparative Polar Organometallic Chemistry 1*, Springer-Verlag: Berlin, 1987. (c) Wakefield, B. J. *Organolithium Methods*, Academic Press: London, 1988. (d) Brandsma, L. *Preparative Polar Organometallic Chemistry 2*, Springer-Verlag: Berlin, 1990. (e) Willard, P. G. In *Comprehensive Organic Synthesis*, Trost,

B. M., Fleming, I., Eds., Pergamon Press: Oxford, UK, 1991, Vol 1., p. 1. (f) *Main-Group Metal Organometallics in Organic Synthesis*. In *Comprehensive Organometallic Chemistry II*, Abel, E. W., Stone, F. G. A., Wilkinson, G., Eds., Pergamon Press: Oxford, UK, 1995, Vol 11. (g) Schlosser, M. In *Organometallics in Synthesis: A Manual*, 2nd ed., Schlosser, M., Ed., Wiley: Chichester, UK, 2002, pp. 1–352.

4. Watson, S. C., Eastham, J. F. *J. Organomet. Chem.* **1967**, *9*, 165.

5. (a) Seebach, D., Hässig, R., Gabriel, J. *Helv. Chim. Acta* **1983**, *66*, 308. (b) McGarrity, J. F., Ogle, C. A. *J. Am. Chem. Soc.* **1985**, *107*, 1805.

6. McGarrity, J. F., Ogle, C. A., Birch, Z., Loosli, H.-R. *J. Am. Chem. Soc.* **1985**, *107*, 1810.

7. Jorgenson, M. *Org. React.* **1970**, *18*, 1.

8. (a) Blomberg, C. *The Barbier Reaction and Related One-Step Processes*, Springer-Verlag: Berlin, 1993. (b) Boudier, A., Bromm, L. O., Lotz, M., Knochel, P. *Angew. Chem., Int. Ed.* **2000**, *39*, 4415.

9. (a) Millon, J., Lorne, R., Linstrumelle, G. *Synthesis* **1975**, 434. (b) Cahiez, G., Bernard, D., Normant, J. F. *Synthesis* **1976**, 245.

10. Neumann, H., Seebach, D. *Chem. Ber.* **1978**, *111*, 2785.

11. Evans, D. A., Crawford, T. C., Thomas, R. C., Walker, J. A. *J. Org. Chem.* **1976**, *41*, 3947.

12. Köbrich, G., Trapp, H. *Chem. Ber.* **1966**, *99*, 680.

13. (a) Miller, R. B., McGarvey, G. *Synth. Comm.* **1979**, *9*, 831. (b) Meyers, A. I., Spohn, R. F. *J. Org. Chem.* **1985**, *50*, 4872.

14. (a) Parham, W. E., Bradsher, C. K. *Acc. Chem. Res.* **1982**, *15*, 300. (b) Narasimhan, N. S., Mali, R. S. *Synthesis* **1983**, 957.

15. Parham, W. E., Jones, L. D. *J. Org. Chem.* **1976**, *41*, 1187.

16. For more extensive tabulations of pK_a values, see (a) Smith, M. B., March, J. *March's Advanced Organic Chemistry*, 5th ed., Wiley: New York, 2001, pp. 329–331; (b) Negishi, E. *Organometallics in Organic Synthesis*, Wiley: New York, 1980, pp. 506–510.

17. Fraser, R. R., Mansour, T. S. *Tetrahedron Lett.* **1986**, *27*, 331.

18. Seebach, D., Beck, A. K., Studer, A. In *Modern Synthetic Methods*, Ernst, B., Leumann, C., Eds., VCH: Weinheim, Germany, 1995, p. 1.

19. (a) Kopka, I. E., Fataftah, Z. A., Rathke, M. W. *J. Org. Chem.* **1987**, *52*, 448. (b) Podraza, K. F., Bassfield, R. L. *J. Org. Chem.* **1988**, *53*, 2643.

20. Huisgen, R., Sauer, J. *Chem. Ber.* **1959**, *92*, 192.

21. (a) Baigrie, L. M., Seiklay, H. R., Tidwell, T. T. *J. Am. Chem. Soc.* **1985**, *107*, 5391. (b) Laube, T., Dunitz, J. D., Seebach, D. *Helv. Chim. Acta* **1985**, *68*, 1373.

22. Broaddus, O. D. *J. Org. Chem.* **1970**, *35*, 10.

23. Schlosser, M., Hartmann, J. *Angew. Chem., Int. Ed.* **1973**, *12*, 508.

24. Gilmam, H., Schwebke, G. L. *J. Org. Chem.* **1962**, *27*, 4259.

25. Schlosser, M. *Pure & Applied Chemistry* **1988**, *60*, 1627.

26. (a) Yamamoto, Y., Maruyama, K. *Heterocycles* **1982**, *18*, 357. (b) Hoffmann, R. W. *Angew. Chem., Int. Ed.* **1982**, *21*, 555.

27. Akiyama, S., Hooz, J. *Tetrahedron Lett.* **1973**, 4115.

28. Schlosser, M. In *Modern Synthetic Methods*, Scheffold, R., Ed., VCH: Weinheim, Germany, 1992, p. 227.

29. (a) Schlosser, M., Hartmann, J., David, V. *Helv. Chim. Acta* **1974**, *57*, 1567. (b) Rauchschwalbe, G., Schlosser, M. *Helv. Chim. Acta* **1975**, *58*, 1094.

30. Schlosser, M., Desponds, O., Lehmann, R., Moret, E., Rauchschwalbe, G. *Tetrahedron* **1993**, *49*, 10175.

31. Vlattas, I., Vecchia, L. D., Lee, A. O. *J. Am. Chem. Soc.* **1976**, *98*, 2008.

32. Baldwin, J. E., Höfle, G. A., Lever, O. W., Jr., *J. Am. Chem. Soc.* **1974**, *96*, 7125.

33. Gschwend, H. W., Rodriguez, H. R. *Org. React.* **1979**, *26*, 1.

34. Ellison, R. A., Kotsonis, F. N. *Tetrahedron* **1973**, *29*, 805.

35. Slocum, D. W., Jennings, C. A. *J. Org. Chem.* **1976**, *41*, 3653.

36. Brandsma, L. *Preparative Acetylenic Chemistry*, Elsevier: Amsterdam, 1988.

37. Cooke, M. P., Jr., *J. Org. Chem.* **1986**, *51*, 1637.

38. (a) Karasch, M. S., Reinmuth, M. S. O. *Grignard Reactions of Nonmetallic Substances*, Prentice Hall: New York, 1954. (b) *Handbook of Grignard Reagents*, Silverman, G. S., Rakita, P. E., Eds., Marcel Dekker: New York, 1996. (c) *Grignard Reagents: Novel Developments*, Richey, H. G., Jr., Ed., Wiley-VCH: Chichester, UK, 2000.

39. Grignard, V. *Comptes Rendus, France,* **1900**, *130*, 1322.

40. Wakefield, B. J. *Organomagnesium Methods in Organic Synthesis*, Academic Press: London, 1995.

41. For a discussion on mechanism, see Hoffmann, R. W., Brönstrup, M., Müller, M. *Org. Lett.* **2003**, *5*, 313.

42. (a) Rieke, R. D., Bales, S. E., Hudnall, P. M., Burns, T. P., Poindexter, G. S. *Org. Synth.*, *Col. Vol. VI*, **1988**, 845, endnote 19. (b) Sell, M. S., Klein, W. R., Rieke, R. D. *J. Org. Chem.* **1995**, *60*, 1077.

43. Matteson, D. S., Liedtke, J. D. *J. Am. Chem. Soc.* **1965**, *87*, 1526.

44. Umeno, M., Suzuki, A. In *Handbook of Grignard Reagents*, Silverman, G. S., Rakita, P. E., Eds., Marcel Dekker: New York, 1996, p. 645.

45. (a) Skattebøl, L., Jones, E. H., Whiting, M. *Org. Synth.*, *Col. Vol. IV*, **1963**, 792. (b) Amos, R. A., Katzenellenbogen, J. A. *J. Org. Chem.* **1978**, *43*, 555.

46. Garst, J. F., Ungváry, F. In *Grignard Reagents—New Developments*, Richey, H. G., Jr., Ed., Wiley: Chichester, UK, 2000, p. 185.

47. (a) Wakefield, B. J. *Organomagnesium Methods in Organic Synthesis*, Academic Press: London, 1995. (b) Sakamoto, S., Imamoto, T., Yamaguchi, K. *Org. Lett.* **2001**, *3,* 1793.

48. (a) Ashby, E. C. *Acc. Chem. Res.* **1988,** *21*, 414. (b) Holm, T., Crossland, I. In *Grignard Reagents—New Developments*, Richey, H. G., Ed., Wiley: Chichester, UK, 2000, p. 1.

49. Boymond, L., Rottländer, M., Cahiez, G., Knochel, P. *Angew. Chem., Int. Ed.* **1998**, *37*, 1701.

50. Reetz, M. T. *Organotitanium Reagents in Organic Synthesis*, Springer-Verlag: Berlin, 1992.

51. (a) Seebach, D. In *Modern Synthetic Methods*, Wiley: Chichester, UK, 1983, Vol. 3, p. 217. (b) Reetz, M. T. *Top. Curr. Chem.* **1982**, *106*, 1. (c) Saito, S., Yamamoto, H. In *Modern Carbonyl Chemistry,* Otera, J., Ed., Wiley-VCH: Weinheim, Germany, 2000, p. 42. (d) Reetz, M. T. In *Organometallics in Synthesis: A Manual*, 2nd ed., Schlosser, M., Ed., Wiley: New York, 2002, p. 817.

52. Imamoto, T. In *Comprehensive Organic Synthesis*, Trost, B. M., Fleming, I., Eds., Pergamon Press: Oxford, 1991, Vol. 1, p. 231.

53. (a) Imamoto, T., Takiyama, N., Nakamura, K., Hatajiama, T., Kamija, Y. *J. Am. Chem. Soc.* **1989**, *111,* 4392. (b) Liu, H.-J., Shia, K.-S., Shang, X., Zhu, B.-Y. *Tetrahedron* **1999**, *55*, 3803.

54. Liu, H.-J., Shia, K.-S., Shang, X., Zhu, B.-Y. *Tetrahedron* **1999**, *55*, 3803.

55. (a) Lipshutz, B. H. In *Comprehensive Organometallic Chemistry II*, Abel, E. W., Stone, F. G. A., Wilkinson, G., Eds., Pergamon Press: Oxford, UK, 1995, Vol. 12, pp. 59–130. (b) Posner, G. H. *Org. React.* **1972**, *19*, 1. (c) Posner, G. H. *Org. React.* **1975**, *22*, 253. (d) Lipshutz, B. H., Sengupta, S. *Org. React.* **1992**, *41*, 135. (e) *Organocopper Reagents: A Practical Approach*, Taylor, R. J. K., Ed., Oxford University Press: Oxford, UK, 1994. (f) Lipshutz, B. H. In *Organometallics in Synthesis: A Manual*, 2nd ed., Schlosser, M., Ed., Wiley: New York, 2002, p. 665. (g) Krause, N. *Modern Organocopper Chemistry*; Wiley-VCH: Weinheim, Germany, 2002.

56. (a) Taylor, R. J. K., Casey, G. In *Organocopper Reagents: A Practical Approach*, Taylor, R. J. K., Ed., Oxford University Press: Oxford, UK, 1994, p. 27. (b) Rieke, R. D., Klieg, W. R. In *Organocopper Reagents: A Practical Approach*, Taylor,

R. J. K., Ed., Oxford University Press: Oxford, UK, 1994, p. 73.

57. Bertz, S. H., Gibson, C. P., Dabbagh, G. *Tetrahedron Lett.* **1987**, *28*, 4251.

58. (a) Lipshutz. B. H. *Synthesis* **1987**, 325. (b) Lipshutz, B. H., Moretti, R., Crow, R. *Org. Synth.* **1990**, *69*, 80.

59. Lipshutz, B. H. In *Organocopper Reagents: A Practical Approach*, Taylor, R. J. K., Ed., Oxford University Press: Oxford, UK, 1994, p. 105.

60. (a) Derguini-Boumechal, F., Lorne, R., Linstrumelle, G. *Tetrahedron Lett.* **1977**, 1181. (b) Derguini-Boumechal, F., Bessière, Y., Linstrumelle, G. *Synth. Commun.* **1981**, *11*, 859.

61. Cahiez, G., Chaboche, C., Jézéquel, M. *Tetrahedron* **2000**, *56*, 2733.

62. Bertz, S. H., Dabbagh, G., Villacorta, G. M. *J. Am. Chem. Soc.* **1982**, *104*, 5824.

63. Lipshutz, B. H., Wilhelm, R. S., Kozlowski, J. A., Parker, D. *J. Org. Chem.* **1984**, *49*, 3928.

64. Maruyama, K., Yamamoto, Y. *J. Am. Chem. Soc.* **1977**, *99*, 8069.

65. Corey, E. J., Posner, G. H. *J. Am. Chem. Soc.* **1968**, *90*, 5615.

66. Cahiez, G., Avedissian, H. *Synthesis* **1998**, 1199.

67. Fujisawa, T., Sato, T. *Org. Synth.* **1987**, *66*, 116.

68. Lipshutz, B. H. In *Comprehensive Organic Synthesis,* Trost, B. M., Fleming, I., Eds., Pergamon Press: Oxford, UK, 1991, Vol. 1, p. 107.

69. Lipshutz, B. H., Moretti, R., Crow, R. *Org. Synth.* **1990**, *69*, 80.

70. Perlmutter, P. *Conjugate Addition Reactions in Organic Synthesis*, Tetrahedron Organic Chemistry Series No. 9, Pergamon Press: Oxford, UK, 1992.

71. Woodward, S. *Tetrahedron* **2002**, *58*, 1017.

72. Smith, R. A. J. In *Organocopper Reagents: A Practical Approach*, Taylor, R. J. K., Ed., Oxford University Press: Oxford, UK, 1994, p. 293.

73. Nakamura, E. In *Organocopper Reagents: A Practical Approach*, Taylor, R. J. K., Ed., Oxford University Press: Oxford, UK, 1994, p. 129.

74. Yamamoto, Y. *Angew. Chem., Int. Ed.* **1986**, *25*, 947.

75. Lipshutz, B. H., Ellsworth, E. L., Siahaan, T. *J. Am. Chem. Soc.* **1989**, *111*, 1351.

76. Alami, M., Marquais, S., Cahiez, G. *Org. Synth.* **1993**, *72*, 135.

77. Ibuka, T., Yamamoto, Y. In *Organocopper Reagents: A Practical Approach*, Taylor, R. J. K., Ed., Oxford University Press: Oxford, UK, 1994, p. 143.

78. Suzuki, M., Noyori, R. In *Organocopper Reagents: A Practical Approach*, Taylor, R. J. K., Ed., Oxford University Press: Oxford, UK, 1994, p.185.

79. Walts, A. E., Roush, W. R. *Tetrahedron* **1985**, *41*, 3463.

80. Yamamoto, Y., Chounan, Y., Nishii, S., Ibuka, T., Kitahava, H. *J. Am. Chem. Soc.* **1992**, *114*, 7652.

81. (a) Bloch, R. *Synthesis* **1978**, 140. (b) Kageyama, T., Tabito, Y., Katoh, A., Ueno, Y., Okawara, M. *Chem. Lett.* **1983**, 1481.

82. Vankar, Y. D., Kumaravel, G. *Tetrahedron Lett.* **1984**, *25*, 233.

83. Nicolaou, K. C., Zhong, Y.-L., Baran, P. C. *J. Am. Chem. Soc.* **2000**, *122*, 7596.

84. Ito, Y., Hirao, T., Saegusa, T. *J. Org. Chem.* **1978**, *43*, 1011.

85. Overman, L. E., Thompson, A. S. *J. Am. Chem. Soc.* **1988**, *110*, 2248.

86. (a) Clive, D. L. J. *Tetrahedron* **1978**, *34*, 1049. (b) Reich, H. J., Wollowitz, S. *Org. React.* **1993**, *44*, 1.

87. (a) Saccomano, N. A. In *Comprehensive Organic Synthesis*, Trost, B. M., Fleming, I., Eds., Pergamon Press: Oxford, UK, 1991, Vol. 1, p. 173. (b) Semmelhack, M. F. In *Organometallics in Synthesis: A Manual*, 2nd ed., Schlosser, M., Ed., Wiley: London, 2002, p. 1003. (c) Takai, K. *Org. React.* **2004**, *64*, 253.

88. (a) Okude, Y., Hirano, S., Hiyama, T., Nozaki, H. *J. Am. Chem. Soc.* **1977**, *99*, 3179. (b) Cintas, P. *Synthesis* **1992**, 248. (c) Wessjohann, L. A., Scheid, G. *Synthesis* **1999**, 1. (d) Fürstner, A. *Chem. Rev.* **1999**, *99*, 991.

89. (a) Takai, K., Kimura, K., Kuroda, T., Hiyama, T., Nozaki, H. *Tetrahedron Lett.* **1983**, *24*, 5281. (b) Okazoe, T., Takai, K., Utimoto, K. *J. Am. Chem. Soc.* **1987**, *109*, 951. (c) Paquette, L. A., Guevel, R., Sakamoto, S., Kim, I. H., Crawford, J. *J. Org. Chem.* **2003**, *68*, 6096.

90. (a) Jin, H., Uenishi, J.-I., Christ, W. J., Kishi, Y. *J. Am. Chem. Soc.* **1986**, *108*, 5644. (b) Takai, K., Tagashira, M., Kuroda, T., Oshima, K. Utimoto, K., Nozaki, H. *J. Am. Chem. Soc.* **1986**, *108*, 6048. (c) Stamos, D. P., Sheng, X. C., Chen, S. S., Kishi, Y. *Tetrahedron Lett.* **1997**, *38*, 6355. (d) Hirai, K., Ooi, H., Esumi, T., Iwabuchi, Y., Hatakeyama, S. *Org. Lett.* **2003**, *5*, 857.

91. (a) Erdik, E. *Organozinc Reagents in Organic Synthesis*, CRC Press: Boca Raton, FL, 1996. (b) Knochel, P., Millot, N., Rodriguez, A. L., Tucker, C. E. *Org. React.* **2001**, *58*, 417. (c) Nakamura, E. In *Organometallics in Synthesis: A Manual*, 2nd Ed., Schlosser, M., ed., Wiley: London, 2002, p. 579.

92. (a) Knochel, P., Singer, R. D. *Chem. Rev.* **1993**, *93*, 2117. (b) *Organozinc Reagents: A Practical Approach,* Knochel, P., Jones, P., Eds., Oxford University Press: Oxford, UK, 1999.

93. (a) Rieke, R. D., Hanson, M. V. In *Organozinc Reagents: A Practical Approach*, Knochel, P., Jones, P., Eds., Oxford University Press: Oxford, UK, 1999, p. 23. (b) Rieke, R. D. *Aldrichimica Acta* **2000**, *33*, 52.

94. Boudier, A., Bromm, L. O., Lotz, M., Knochel, P. *Angew. Chem., Int. Ed.* **2000**, *39*, 4414.

95. (a) Reformatsky, S. *Chem. Ber.* **1887**, *20*, 1210. (b) Rathke, M. W. *Org. React.* **1975**, *22*, 423. (c) Rathke, M. W., Weipert, P. In *Comprehensive Organic Synthesis*, Trost, B. M., Fleming, I., Eds., Pergamon Press: Oxford, UK, 1991, Vol. 2, p. 277. (d) Fürstner, A. In *Organozinc Reagents: A Practical Approach*, Knochel, P., Jones, P., Eds., Oxford University Press: Oxford, UK, 1999, p. 289.

96. Fung, K. H., Schmalzl, K. J., Mirrington, R. N. *Tetrahedron Lett.* **1969**, 5017.

97. Korte, F., Falbe, J., Zschocke, A. *Tetrahedron* **1959**, *6*, 201.

98. (a) Soai, N., Niwa, S. *Chem. Rev.* **1992**, *92*, 833. (b) Pu, L., Yu, H.-B. *Chem. Rev.* **2001**, *101*, 757.

99. Boudier, A., Bromm, L. O., Lotz, M., Knochel, P. *Angew. Chem., Int. Ed.* **2000**, *39*, 4414.

100. Knoess, H. P., Furlong, M. T., Rozeman, M. J., Knochel, P. *J. Org. Chem.* **1991**, *56*, 5974.

101. Bronk, B. S., Lippard, S. J., Danheiser, R. L. *Organometallics* **1993**, *12*, 3340.

102. Klement, I., Lennick, K., Tucker, C. E., Knochel, P. *Tetrahedron Lett.* **1993**, *34*, 4623.

103. (a) Simmons, H. E., Cairns, T. L., Vladuchick, S. A., Hoiness, C. M. *Org. React.* **1972**, *20*, 1. (b) Charette, A. B., Beauchemin, A. *Org. React.* **2001**, *58*, 1.

104. Salaün, J. *Top. Curr. Chem.* **2000**, *207*, 1.

105. Simmons, H. E., Smith, R. D. *J. Am. Chem. So*c. **1958**, *80*, 5323.

106. Wender, P. A., Eck, S. L. *Tetrahedron Lett.* **1982**, *23*, 1871.

107. Charette, A. B., Beauchemin, A. *Org. React.* **2001**, *58*, 1.

108. Furukawa, J., Kawabata, N., Nishimura, J. *Tetrahedron* **1968**, *24*, 53.

109. Denmark, S. E., Edwards, J. P. *J. Org. Chem.* **1991**, *56*, 6974.

110. Charette, A. B., Marcoux, J.-F. *Synlett* **1995**, 1197.

111. Charette, A. B., Francoeur, S., Martel, J., Wilb, N. *Angew. Chem., Int. Ed.* **2000**, *39*, 4539.

112. (a) Friedrich, E. C., Domek, J. M., Pong, R. Y. *J. Org. Chem.* **1985**, *50*, 4640. (b) Friedrich, E. C., Lewis, E. J. *J. Org. Chem.* **1990**, *55*, 2491.

113. (a) Poulter, C. D., Friedrich, E. C., Winstein, S. *J. Am. Chem. Soc.* **1969**, *91*, 6892. (b) Hoveyda, A. H., Evans, D. A., Fu, G. C. *Chem. Rev.* **1993**, *93*, 1307.

114. Moss, R. A., Chen, E. Y., Banger, J., Matsuo, M. *Tetrahedron Lett.* **1978**, 4365.

115. Charette, A. B., Lebel, H. *J. Org. Chem.* **1995**, *60*, 2966.

116. (a) Brown, H. C., Kramer, G. W., Levy, A. B., Midland, M. M. *Organic Syntheses via Boranes*, Wiley: New York, 1975. (b) Brown, H. C., Zaidlewicz, M., Negishi, E. In *Comprehensive Organometallic Chemistry,* Wilkinson, G.,

Stone, F. G. A., Abel, E. W., Eds., Pergamon Press: Oxford, UK, 1982, Vol. 7, pp. 111–363. (c) Suzuki, A., Dhillon, R. S. *Top. Curr. Chem.* **1985**, p. 130. (d) Negishi, E., Idacavage, M. J. *Org. React.* **1985**, *33*, 1. (e) Pelter, A., Smith, K., Brown, H. C. *Borane Reagents*, Academic Press: London, 1988. (f) Thomas, S. E. *Organic Synthesis: The Roles of Boron and Silicon*, Oxford University Press: Oxford, UK, 1991. (g) Matteson, D. S. *Stereodirected Synthesis with Organoboranes*, Springer-Verlag: Berlin, 1995. (h) Vaultier, M., Carboni, B. In *Comprehensive Organometallic Chemistry II*, Abel, E. W., Stone, F.G.A., Wilkinson, G., Eds., Pergamon Press: Oxford, UK, 1995, Vol. 11, p. 191. (i) Smith, K. In *Organometallics in Synthesis: A Manual*, 2nd. ed., Schlosser, M., Ed., Wiley: Chichester, UK, 2002, p. 465.

117. Negishi, E. *Organometallics in Organic Synthesis*, Wiley: New York, 1980, p. 302.

118. Pelter, A., Smith, K., Brown, H. C. *Borane Reagents*, Academic Press: New York, 1988.

119. Brown, H. C., Hubbard, J. L., Smith, K. *Synthesis* **1979**, 701.

120. Negishi, E., Brown, H. C. *Synthesis* **1972**, 196.

121. (a) Brown, H. C., Negishi, E. *J. Am. Chem. Soc.* **1967**, *89*, 5477. (b) Brown, H. C., Negishi, E. *J. Chem. Soc., Chem. Commun.* **1968**, 594.

122. (a) Brown, H. C., Rathke, M. W. *J. Am. Chem. Soc.* **1967**, *89*, 4528. (b) Yamamoto, Y., Kondo, K., Moritani, I. *J. Org. Chem.* **1975**, *40*, 3644.

123. Brown, H. C., Negishi, E. *J. Am. Chem. Soc.* **1969**, *91*, 1224.

124. Pelter, A., Smith, K., Brown, H. C. *Borane Reagents*, Academic Press: New York, 1988, p. 280.

125. Pelter, A., Hutchings, M. G., Smith, K. *J. Chem. Soc., Chem. Commun.* **1971**, 1048.

126. Pelter, A., Hutchings, M. G., Rowe, K. Smith, K. *J. Chem. Soc., Perkin Trans. I* **1975**, 138.

127. (a) Brown, H. C., Carlson, B. A. *J. Org. Chem.* **1973**, *38*, 2422. (b) Brown, H. C., Katz, J.-J., Carlson, B. A. *J. Org. Chem.* **1973**, *38*, 3968. (c) Pelter, A., Smith, K. In *Comprehensive Organic Chemistry*, Barton, D. H. R., Ollis, W. D. Eds., Pergamon Press: Oxford, UK, 1979, Vol. 3, p. 689.

128. Matteson, D. S. *Tetrahedron*, **1989**, *45*, 1859.

129. Matteson, D. S. *Stereodirected Synthesis with Organoboranes*, Springer-Verlag: Berlin, 1995.

130. Brown, H. C., Bhat, K. S. *J. Am. Chem. Soc.* **1986**, *108*, 5919.

131. (a) Hoffmann, R. W. *Angew. Chem.,Int. Ed.* **1982**, *21*, 555. (b) Pelter, A., Smith, K., Brown, H. C. *Borane Reagents*, Academic Press: New York, 1988. (c) Negishi, E. In

Comprehensive Organometallic Chemistry, Wilkinson, G., Stone, F. G. A., Abel, E. W., Eds., Pergamon Press: Oxford, UK, 1982, Vol. 7, p. 349.

132. (a) Fujita, K., Schlosser, M. *Helv. Chim. Acta* **1982**, *65*, 1258. (b) Hoffmann, R. W., Zeiss, H.-J. *Angew. Chem., Int. Ed.* **1979**, *18*, 306. (c) Roush, W. R., Ano, K., Powers, D. B., Palkowitz, A. D., Haltermann, R. L. *J. Am. Chem. Soc.* **1990**, *112*, 6339.

133. (a) Hoffmann, R. W. *Angew. Chem., Int. Ed.* **1987**, *26*, 489. (b) Yamamoto, Y., Maruyama, K. *Heterocycles* **1982**, *18*, 357. (c) Yamamoto, Y. *Acc. Chem. Res.* **1987**, *20*, 243. (d) Roush, W. R. In *Comprehensive Organic Synthesis*, Trost, B. M., Fleming, I., Eds., Pergamon Press: Oxford, UK, 1990, Vol. 2, p. 1.

134. (a) Zimmerman, H. E., Traxler, M. D. *J. Am. Chem. Soc.* **1957**, *79*, 1920. (b) Li, Y., Houk, K. N. *J. Am. Chem. Soc.* **1989**, *111*, 1236 and references cited therein.

135. (a) Brown, H. C., Bhat, K. S. *J. Am. Chem. Soc.* **1986**, *108*, 5919. (b) Brown, H. C., Bhat, K. S., Randad, R. S. *J. Org. Chem.* **1989**, *54*, 1570. (c) Brown, H. C., Racherla, U. S., Liao, Y., Khanna, V. V. *J. Org. Chem.* **1992**, *57*, 6608.

136. Brown, H. C., Bhat, K. S. *J. Am. Chem. Soc.* **1986**, *108*, 5919.

137. (a) Chan, T. H., Fleming, I. *Synthesis* **1979**, 761. (b) Colvin, E. W. *Silicon in Organic Synthesis,* Butterworths: London, 1981. (c) Magnus, P. D., Sarkar, T., Djuric, S. In *Comprehensive Organometallic Chemistry*, Wilkinson, G., Stone, F. G. A., Abel, E. W., Eds., Pergamon Press: Oxford, UK, 1982, Vol. 7, p. 515. (d) Fleming, I. In *Comprehensive Organic Chemistry*, Barton, D. H. R., Ollis, W. D., Eds., Pergamon Press: Oxford, UK, 1979, Vol. 3, p. 541. (e) Weber, W. P. *Silicon Reagents in Organic Synthesis*, Springer-Verlag: Berlin, 1983. (f) Carruthers, W. *Some Modern Methods of Organic Synthesis*; 3rd ed., Cambridge University Press: Cambridge, MA, 1986, pp. 317–343. (g) Colvin, E. W. *Silicon Reagents in Organic Synthesis*, Academic Press: London, 1988. (h) Thomas, S. E. *Organic Synthesis: The Roles of Boron and Silicon*, Oxford Press: Oxford, UK, 1991.

138. Negishi, E. *Organometallics in Organic Synthesis*, Wiley: New York, 1980.

139. Kerr, J. A. *Chem. Rev.* **1966**, *66*, 465.

140. (a) Wierschke, S. G., Chandrasekhar, J., Jorgensen, W. L. *J. Am. Chem. Soc.* **1985**, *107*, 1496. (b) Lambert, J. B., Wang, G., Finzel, R. B., Teramura, D. H. *J. Am. Chem. Soc.* **1987**, *109*, 7838.

141. Mayr, H., Pock, R. *Tetrahedron* **1986**, *42*, 4111.

142. Gilday, J. P., Gallucci, J. C., Paquette, L. A. *J. Org. Chem.* **1989**, *54*, 1399, and references 6, 7, and 20 therein.

143. Colvin, E. W. *Silicon Reagents in Organic Synthesis*, Academic Press: London, 1988, p. 47.

144. (a) Kozar, L. G., Clark, R. D., Heathcock, C. H. *J. Org. Chem.* **1977**, *42*, 1386. (b) Blumenkopf, T. A., Overman, L. E. *Chem. Rev.* **1986**, *86*, 857.

145. Colvin, E. W. *Silicon Reagents in Organic Synthesis*, Academic Press: London, 1988, p. 9.

146. Miller, R. B., Reichenbach, T. *Tetrahedron Lett.* **1974**, 543.

147. Zweifel, G., Miller, J. A. *Org. React.* **1984**, *32*, 375.

148. Fleming, I., Pearce, A. *J. Chem. Soc., Perkin Trans. 1* **1980**, 2485.

149. Miller, R. B., McGarvey, G. *J. Org. Chem.* **1979**, *44*, 4623.

150. (a) Parnes, Z. N., Bolestova, G. I. *Synthesis* **1984**, 991. (b) Hosomi, A. *Acc. Chem. Res.* **1988**, *21*, 200. (c) Schinzer, D. *Synthesis* 1988, 263.

151. Negishi, E., Luo, F., Rand, C. L. *Tetrahedron Lett.* **1982**, *23*, 27.

152. (a) Rajagopalan, S., Zweifel, G. *Synthesis* **1984**, 111. (b) Rajagopalan, S., Zweifel, G. *Synthesis* **1984**, 113.

153. Seyferth, D., Wursthorn, K. R., Mammarella, R. *J. Org. Chem.* **1977**, *42*, 3104.

154. Trost, B. M., Coppala, B. P. *J. Am. Chem. Soc.* **1982**, *104*, 6879.

155. (a) Masse, C. E., Panek, J. S. *Chem. Rev.* **1995**, *95*, 1293. (b) Langkopf, E., Schinzer, D. *Chem. Rev.* **1995**, *95*, 1375.

156. Pillot, J.-P., Dunogues. J., Calas, R. *Tetrahedron Lett.* **1976**, 1871.

157. (a) Hosomi, A., Sakurai, H. *J. Am. Chem. Soc.* **1977**, *99*, 1673. (b) Wilson, S. R., Price, M. F. *J. Am. Chem. Soc.* **1982**, *104*, 1124.

158. Hayashi, T., Kabeta, K., Hamachi, I., Kumada, M. *Tetrahedron Lett.* **1983**, *24*, 2865.

159. Chan, T. M., Wang, D. *Chem. Rev.* **1995**, *95*, 1279.

160. (a) Tsai, D. J. S., Matteson, D. S. *Tetrahedron Lett.* **1981**, *22*, 2751. (b) Sato, F., Suzuki, Y., Sato, M. *Tetrahedron Lett.* **1982**, *23*, 4589.

161. Hudrlik, P. F., Peterson, D. *J. Am. Chem. Soc.* **1975**, *97*, 1464.

162. (a) Ager, D. J. *Chem. Soc. Rev.* **1982**, *11*, 493. (b) Bulman Page, P. C., Klair, S. S., Rosenthal, S. *Chem. Soc. Rev.* **1990**, *19*, 147. (c) Reich, H. J., Kelly, M., Olson, R. E., Holton, R. *Tetrahedron* **1983**, *39*, 949. (d) Ricci, A., Degl Innocenti, A. *Synthesis* **1989**, 647. (e) Cirillo, P. F., Panek, J. S. *Org. Prep. Proced. Int.* **1992**, *24*, 553.

163. Soderquist, J. A. *Org. Synth., Col. Vol. VIII*, **1993**, 19.

164. Brook, A. G., Duff, J. M., Jones, P. F., Davis, N. R. *J. Am. Chem. Soc.* **1967**, *89*, 431. (b) Corey, E. J., Seebach, D., Freedman, R. *J. Am. Chem. Soc.* **1967**, *89*, 434.

165. (a) Vedejs, E., Fuchs, P. L. *J. Org. Chem.* **1971**, *36*, 366. (b) Chuang, T.-H., Fang, J.-M., Jiaang, W.-T., Tsai, Y.-M. *J. Org. Chem.* **1996**, *61*, 1794.

166. Reich, H. J., Eisenhart, E. K., Olson, R. E., Kelly, M. J. *J. Am. Chem. Soc.* **1986**, *108*, 7791.

167. (a) Hassner, A., Soderquist, J. A. *J. Organomet. Chem.* **1977**, *131*, C1. (b) Miller, J. A., Zweifel, G. *Synthesis* **1981**, 288.

168. Miller, J. A., Zweifel, G. *J. Am. Chem. Soc.* **1981**, *103*, 6217.

169. Brook, A. G. *Acc. Chem. Res.* **1974**, *7*, 77.

170. Brook, A. G., Pascoe, J. D. *J. Am. Chem. Soc.* **1971**, *93*, 6224.

171. Hine, J. *Adv. Phys. Org. Chem.* **1977**, *15*, 1.

172. Wilson, S. R., Hague, M. S., Misra, R. N. *J. Org. Chem.* **1982**, *47*, 747.

173. Reich, H. J., Olson, R. E., Clark, M. C. *J. Am. Chem. Soc.* **1980**, *102*, 1423.

174. Reich, H. J., Eisenhart, E. K. *J. Org. Chem.* **1984**, *49*, 5282.

175. Rasmussen, J. K. *Synthesis* **1977**, 91.

176. Zweifel, G., Backlund, S. J. *J. Am. Chem. Soc.* **1977**, *99*, 3184.

177. (a) Tsuji, J. *Palladium Reagents and Catalysts: Innovations in Organic Synthesis*, Wiley: Chichester, UK, 1995. (b) Nicolaou, K. C., Sorensen, E. J. *Classics in Total Synthesis*, VCH: Weinheim, Germany, 1996, pp. 566, 712. (c) Malleron, J.-L., Fiaud, J.-C; Legros, J.-Y. *Handbook of Palladium Catalyzed Reactions*, Academic Press: San Diego, 1997. (d) Browning, A. F., Greeves, N. In *Transition Metals in Organic Synthesis: A Practical Approach*, Gibson, S. E., Ed., Oxford University Press: Oxford, UK, 1997. (e) Brandsma, L., Vasilevsky, S. F., Verkruijsse, H. D. *Application of Transition Metal Catalysts in Organic Synthesis*, Springer-Verlag: Berlin, 1998. (f) *Metal-Catalyzed Cross-Coupling Reactions*, Diederich, F., Stang, P. J., Eds., Wiley-VCH: Weinheim, Germany, 1998. (g) Trost, B. M. In *Transition Metals for Organic Synthesis*, Beller, M., Bolm, C., Eds., Wiley-VCH: Weinheim, Germany, 1998, Vol. 1, p. 3. (h) Hegedus, L. S. *Transition Metals in the Synthesis of Complex Organic Molecules*: University Science Books: Sausalito, CA, 1999. (i) Hegedus, L. S. In *Organometallics in Synthesis: A Manual,* 2nd ed., Schlosser, M., Ed., Wiley: London, 2002, p. 1123. (j) Negishi, E. *Handbook of Organopalladium Chemistry for Organic Synthesis*, Wiley: New York, 2002.

178. Tolman, C. A. *Chem. Soc. Rev.* **1972**, *1*, 337.

179. Coulson, D. R. *Inorg. Synth.* **1972**, *13*, 121.

180. Amatore, C., Jutand, A., M., Barki, M. A. *Organaometallics* **1992**, *11*, 3009.

181. For a comprehensive list of these reagents, see Larock, R. C., *Comprehensive Organic Transformations*, Wiley-VCH: New York, 1999, pp. 426–435.

182. Brown, H. C., Hamaoka, T., Ravindran, N. *J. Am. Chem. Soc.* **1973**, *95*, 6456.

183. Zweifel, G., Arzoumanian, H. *J. Am. Chem. Soc.* **1967**, *89*, 5086.

184. Zweifel, G., Whitney, C. C. *J. Am. Chem. Soc.* **1967**, *89*, 2753.

185. Hart, D. W., Blackburn, T. F., Schwartz, J. *J. Am. Chem. Soc.* **1975**, *97*, 679.

186. Normant, J. F., Alexakis, A. *Synthesis* **1981**, 841.

187. Negishi, E., Van Horn, D. E., King, A. O., Okukado, N. *Synthesis* **1979**, 501.

188. Kamiya, N., Chikami, Y., Ishii, Y. *Synlett* **1990**, 675.

189. (a) Heck, R. F. In *Palladium Reagents in Organic Syntheses*, Katritzky, A. R., Ed., Academic Press: London, 1985, p. 179. (b) Nicolaou, K. C., Sorensen, E. J. *Classics in Total Synthesis,* VCH: Weinheim, Germany, 1996, p. 566. (c) Bräse, S., de Meijere, A. In *Metal-Catalyzed Cross-Coupling Reactions*, Diederich, F., Stang, P. J., Eds., Wiley-VCH: Weinheim, Germany, 1998, p. 99.

190. (a) Heck, R. F. *J. Am. Chem. Soc.* **1968**, *90*, 5518. (b) Heck, R. F. *Org. React.* **1982**, *27*, 345.

191. Heck, R. F., Nolley, J. P., Jr. *J. Org. Chem.* **1972**, *37*, 2320.

192. Jeffery, T. *J. Chem. Soc., Chem. Commun.* **1984**, 1287.

193. Heck, R. F. *Acc. Chem. Res.* **1979**, *12*, 146.

194. Dieck, H. A., Heck, R. F. *J. Am. Chem. Soc.* **1974**, *96*, 1133.

195. Stang, P. J., Treptow, W. *Synthesis* **1980**, 283.

196. (a) McMurry, J. E., Scott, W. J. *Tetrahedron Lett.* **1983**, *24*, 979. (b) Ritter, K. *Synthesis* **1993**, 735.

197. Jeffery, T. *Tetrahedron Lett.* **1985**, *26*, 2667.

198. (a) Abelman, M. M., Oh, T., Overman, L. E. *J. Org. Chem.* **1987**, *52*, 4133. (b) Link, J. T. *Org. React.* **2002**, *60*, 157. (c) Link, J. T., Overman, L. E. In *Metal-Catalyzed Cross-Coupling Reactions*, Diederich, F., Stang, P. J. , Eds., Wiley-VCH: Weinheim, Germany, 1998, p. 231.

199. Beller, M., Riermeier, T. H., Stark, G. In *Transition Metals for Organic Synthesis*, Beller, M., Bolm, C., Eds., Wiley-VCH: Weinheim, Germany, 1998, Vol. 1, p. 208.

200. (a) Larock, R. C., Yum, E. K., Refvik, M. D. *J. Org. Chem.* **1998**, *63*, 7652. (b) Larock, R. C. *J. Organomet. Chem.* **1999**, *576*, 111.

201. Negishi, E., Liu, F. In *Metal-Catalyzed Cross-Coupling Reactions*, Diederich, F., Stang, P. J., Eds., Wiley-VCH: Weinheim, Germany, 1998, p. 1.

202. (a) Negishi, E. *Pure & Applied. Chemistry* **1981**, *53*, 2333. (b) Negishi, E. *Acc. Chem. Res.* **1982**, *15*, 340.

203. (a) Negishi, E., Takahashi, T., King, A. O. *Org. Synth.* **1988**, *66*, 67. (b) Zhu, L., Wehmeyer, R. M., Rieke, R. D. *J. Org. Chem.* **1991**, *56*, 1445.

204. Zweifel, G., Miller, J. A. *Org. React.* **1984**, *32*, 375.

205. Baba, S., Negishi, E. *J. Am. Chem. Soc.* **1976**, *98*, 6729.

206. (a) Van Horn, D. E., Negishi, E. *J. Am. Chem. Soc.* **1978**, *100*, 2252. (b) Negishi, E., Kondakov, D. Y., Choueiry, D., Kasai, K., Takahashi, T. *J. Am. Chem. Soc.* **1996**, *118*, 9577. (c) Negishi, E. In *Catalytic Asymmetric Synthesis*, 2nd ed., Ojima, T., Ed., Wiley-VCH: Weinheim, Germany, 2000, p. 165.

207. Negishi, E., Matsushita, H. *Org. Synth.* **1984**, *62*, 31.

208. (a) Hart, D. W., Blackburn, T. F., Schwartz, J. *J. Am. Chem. Soc.* **1975**, *97*, 679. (b) Negishi, E., Takahashi, T. *Synthesis* **1988**, 1. (c) Wipf, T. *Pure & Applied Chemistry* **1998**, *70*, 1077. (d) Negishi, E. In *Organometallic Synthesis: A Manual*, 2nd ed., Schlosser, M., Ed., Wiley: Chichester, UK, 2002, p. 925.

209. (a) Buchwald, S. L., LaMaire, S. L., Nielsen, R. B., Watson, B. T., King, S. M. *Tetrahedron Lett.* **1987**, *28*, 3895. (b) Negishi, E., Takahashi, T. *Synthesis* **1988**, 1.

210. Hart, D. W., Blackburn, T. F., Schwartz, J. *J. Am. Chem. Soc.* **1975**, *97*, 679.

211. Okukado, N., Van Horn, D. E., Dlima, W., Negishi, E. *Tetrahedron Lett.* **1978**, 1027.

212. Negishi, E., Okukado, N., King, A. O., Van Horn, D. E., Spiegel, B. I. *J. Am. Chem. Soc.* **1978**, *100*, 2254.

213. (a) Suzuki, A., Brown, H. C. *Suzuki Coupling: Organic Syntheses via Boranes*, Aldrich Chemical Co.: Milwaukee, 2002, Vol. 3. (b) Suzuki, A. *Organoboranes in Organic Syntheses*, Hokkaido University: Sapporo, Japan, 2004.

214. (a) Nicolaou, K. C., Sorensen, E. J. *Classics in Total Synthesis*; VCH: Weinheim, Germany, 1996, p. 586. (b) Suzuki, A. In *Metal-Catalyzed Cross-Coupling Reactions*, Diederich, F., Stang, P. J., Eds., Wiley-VCH: Weinheim, Germany, 1998, p. 49. (c) Miyaura, N. *Topics in Current Chemistry*, Springer-Verlag: Berlin, 2002, Vol. 219, p. 11.

215. (a) Miyaura, N., Suzuki, A. *Chem. Commun.* **1979**, 866. (b) Suzuki, A. *Pure & Applied. Chemistry.* **1985**, *57*, 1749. (c) Miyaura, N., Suzuki, A. *Chem. Rev.* **1995**, *95*, 2457. (d) Chemler, S. R., Trauner, D., Danishefsky, S. J. *Angew. Chem., Int. Ed.* **2001**, *40*, 4545.

216. (a) Campbell, J. B., Molander, G. A. *J. Organomet. Chem.* **1978**, *156*, 71. (b) Molander, G. A., Figueroa, R. *Aldrichimica Acta* **2005**, *38*, 49. (c) Molander, G. A., Biolatto, B. *J. Org. Chem.* **2003**, *68*, 4302.

217. Miyaura, N., Yamada, K., Suginome, H., Suzuki, A. *J. Am. Chem. Soc.* **1985**, *107*, 972.

218. (a) Miyaura, N., Suzuki, A. *J. Chem. Soc., Chem. Commun.* **1979**, 866. (b) Suzuki, A. *Pure & Applied Chemistry* **1985**, *57*, 1749.

219. Molander, G. A., Bernardi, C. R. *J. Org. Chem.* **2002**, *67*, 8424.

220. (a) Miyaura, N., Maruoka, K. In *Synthesis of Organometallic Compounds,* Komiya, S., Ed., Wiley: Chichester, UK, 1997, p. 345. (b) Stanforth, S. *Tetrahedron* **1998**, *54*, 285. (c) Murata, M., Oyama, T., Watanabe, S., Masuda, Y.

J. Org. Chem. **2000**, *65*, 164. (d) Baudoin, O., Guénard, D., Guéritte, F. *J. Org. Chem.* **2000**, *65,* 9268.

221. Miyaura, N., Ishiyama, T., Sasaki, H., Ishikawa, M., Satoh, M., Suzuki, A. *J. Am. Chem. Soc.* **1989**, *111*,314.

222. Ishiyama, N., Miyaura, N., Suzuki, A. *Bull. Chem. Soc. Jpn.* **1991**, *64*, 1999.

223. (a) Farina, V., Krishamurthy, V., Scott, W. J. *The Stille Reaction*, Wiley: New York, 1998. (b) Mitchell, T. N. In *Metal-Catalyzed Cross-Coupling Reactions*, Diederich, F., Stang, P. J., Eds., Wiley-VCH: Weinheim, Germany, 1998, p. 167. (c) Marshall, J. A. In *Organometallics in Synthesis: A Manual*, 2nd. ed., Schlosser, M., Ed., Wiley: Chichester, UK, 2002, p. 353. (d) See p. 1158 in the article by Hegedus, L. S., in *Organometallics in Synthesis: A Manual*, 2nd ed., Schlosser, M., Ed., Wiley: Chichester, UK, 2002, p. 1123.

224. Nicolaou, K. C., Sorensen, E. J. *Classics in Total Synthesis*, VCH: Weinheim, Germany,1996, p. 591.

225. For the disposal of organotin residues from reaction mixtures, see Salomon, C. J., Danelon, G. O., Mascaretti, O. A. *J. Org. Chem.* **2000**, *65*, 9220.

226. (a) Stille, J. K. *Angew. Chem., Int. Ed.* **1986**, *25,* 508. (b) Farina, V., Krishnamurthy, V., Scott, W. J. *Org. React.* **1997**, *50*, 1.

227. (a) Pereyre, M., Quintard, J.-P., Rahm, A. *Tin in Organic Synthesis*, Butterworths: London, 1987. (b) Zhang, H. X, Guibé, F., Balavoine, G. *J. Org. Chem.* **1990**, *55*, 1857.

228. Stille, J. K., Groh, B. L. *J. Am. Chem. Soc.* **1987**, *109*, 813.

229. Bailey, T. R. *Tetrahedron Lett.* **1986**, *27*, 4407.

230. Sheffy, F. K., Godschalx, J. P., Stille, J. K. *J. Am. Chem. Soc.* **1984**, *106*, 4833.

231. Stang, P. J., Treptow, W. *Synthesis* **1980**, 283.

232. (a) McMurry, J. E., Scott, W. J. *Tetrahedron Lett.* **1983**, *24*, 979. (b) Ritter, K. *Synthesis* **1993**, 735.

233. Crisp, G. T., Scott, W. J., Stille, J. K. *J. Am. Chem. Soc.* **1984**, *106*, 7500.

234. (a) Goure, W. F., Wright, M. E., Davis, P. D., Labadie, S. S., Stille, J. K. *J. Am. Chem. Soc.* **1984**, *106*, 6417. (b) Baillargeon, V. P., Stille, J. K. *J. Am. Chem. Soc.* **1986**, *108*, 452.

235. (a) Castro, C. E., Stephens, D. D. *J. Org. Chem.* **1963**, *23*, 2163. (b) Stephens, R. D., Castro, C. E. *J. Org. Chem.* **1963**, *28*, 3313. (c) Cao, D., Kolshorn, H., Meier, H. *Tetrahedron Lett.* **1995**, *36*, 4487.

236. Curtis, R. F., Taylor, J. A. *J. Chem. Soc. C* **1971**, 186.

237. (a) Sonogashira, K., Tohda, Y., Hagihara, N. *Tetrahedron Lett.* **1975**, 4467. (b) Sonogashira, K. In *Comprehensive Organic Synthesis*, Trost, B. M., Fleming, I., Eds., Pergamon Press: New York, 1991, Vol. 3, p. 521. (c) Sonogashira, K. In *Metal-Catalyzed Cross-Coupling Reactions*, Diederich, F., Stang, P. J., Eds., Wiley-VCH: Weinheim, Germany,1998, p. 203.

238. Scott, W. J., Pena, M. R., Swörd, K., Stoessel, S. J., Stille, J. K. *J. Org. Chem.* **1985**, *50*, 2302.

239. Nakatani, K., Isoe, S., Maekawa, S., Saito, I. *Tetrahedron Lett.* **1994**, *35*, 605.

240. Anastasia, L., Negishi, E. *Org. Lett.* **2001**, *3*, 3111, and references cited therein.

241. (a) Grissom, J. W., Gunawardena, G. U., Klingberg, D., Huang, D. *Tetrahedron* **1996**, *52*, 6453. (b) Maier, M. E. *Synlett* **1995**, 13. (c) Nicolaou, K. C., Dai, W.-M. *Angew. Chem., Int. Ed.* **1991**, *30*, 1387.

242. Alami, M., Crousse, B., Linstrumelle, G. *Tetrahedron Lett.* **1994**, *35,* 3543.

243. Boland, W., Schrorer, N., Sieler, C., Feigel, M. *Helv. Chim. Acta* **1987**, *70*, 1025.

244. Negishi, E., Alimardanov, A., Xu, C. *Org. Lett.* **2000**, *2,* 65, and references cited therein.

245. (a) Glaser, C. *Chem. Ber.* **1869**, *2*, 422. (b) Brandsma, L., Vasilevski, S., Verkruijsse, H. D. *Application of Transition Metal Catalysts in Organic Synthesis*, Springer-Verlag: Berlin, 1999, p. 67.

246. (a) Heilbron, I., Jones, E. R. H., Sondheimer, F. *J. Chem. Soc.* **1947**, 1586. (b) Hébert, N., Beck, A., Lennox, R. B., Just, G. *J. Org. Chem.* **1992**, *57*, 1777.

247. Nishihara, Y., Ikegashira, K., Hirabayashi, K., Ando, J., Mori, A., Hiyama, T. *J. Org. Chem.* **2000**, *65*, 1780.

248. (a) Cadiot, P., Chodkiewicz, W. *Compte Rend.* **1955**, *241*, 1055. (b) P. Cadiot, P., Chodkiewicz, W. In *Chemistry of Acetylenes*, Viehe, H. G., Ed., Marcel Dekker: New York, 1969, p. 609. (c) Negishi, E., Okukado, N., Lovich, S. F., Luo, F. T. *J. Org. Chem.* **1984**, *49*, 2629. (d) Brandsma, L., Vasilevski, S., Verkruijsse, H. D. *Application of Transition Metal Catalysis in Organic Synthesis*, Springer-Verlag: Berlin, 1999, p. 49.

249. (a) Miller, J. A., Zweifel, G. *Synthesis* **1983**, 128.

250. Kende, A. S., Smith, C. A. *J. Org. Chem.* **1988**, *53*, 2655.

251. Negishi, E., Hata, M., Xu, C. *Org. Lett.* **2000**, 2, 3687.

252. (a) Tsuji, J. *Tetrahedron* **1986**, *42*, 4361. (b) Tsuji, J. *Palladium Reagents and Catalysts, Innovations in Organic Synthesis*, Wiley: Chichester, UK, 2000, p. 116. (c) Trost, B. M., Verhoeven, T. R. In *Comprehensive Organometallic Chemistry*, Wilkinson, G., Stone, F. G. A., Abel, E. W., Eds., Pergamon Press: Oxford, UK, 1982, Vol. 8, p. 799. (d) Godleski, S. A. In *Comprehensive Organic Synthesis,* Trost, B. M., Fleming, I., Eds, Pergamon Press: Oxford, UK, 1991, Vol. 4, p. 585. (e) Heumann, A. In *Transition Metals for Organic Synthesis*, Beller, M., Bolm, C. Eds., Wiley-VCH: Weinheim, Germany, 1998, Vol. 1, p. 251.

253. Amadore, C., Jutand, A., M'Barki, M. A. *Organometallics* **1992**, *11*, 3009.

254. Genet, J. P., Ferroud, D. *Tetrahedron Lett.* **1984**, *25*, 3579.

255. Tanigawa, Y., Nishimura, K., Kawasaki, A., Murahashi, S. *Tetrahedron Lett.* **1982**, *23*, 5549.

256. (a) Tsuji, J., Kobayashi, Y., Kataoka, H., Takahashi, T. *Tetrahedron Lett.* **1980**, *21*, 1475. (b) Yamamoto, K., Tsuji, J. *Tetrahedron Lett.* **1982**, *23*, 3089.

257. Trost, B. M., Genet, J. P. *J. Am. Chem. Soc.* **1976**, *98*, 8516.

258. Radinov, R., Hutchings, S. D. *Tetrahedron Lett.* **1999**, *40*, 8955.

259. Nagamitsu, T., Takano, D., Fukuda, T., Otoguro, K., Kuwajima, I., Harigaya, Y., Omura, S. *Org. Lett.* **2004**, *6*, 1865.

260. (a) Barton, D. H. R., McCombie, S. W. *J. Chem. Soc., Perkin Trans. I*, **1975**, 1574. (b) Crich, D., Quintero, L. *Chem. Rev.* **1989**, *89*, 1413. (c) McCombie, S. W. In *Comprehensive Organic Synthesis*, Trost, B. M., Ed., Pergamon Press: New York, 1991; Vol. 8, Chap. 4.2.

CHAPTER 8

Formation of Carbon-Carbon π-Bonds

In the field of observation, chance only favors those minds which have been prepared.

Louis Pasteur

The carbon-carbon double and triple bond structural features are frequently encountered in nature, and their π-systems are rich repositories for a variety of functional group interconversions.

8.1   FORMATION OF CARBON-CARBON DOUBLE BONDS[1]

Stereoselective synthesis of substituted olefins is a major challenge to organic chemists since these are fundamental structural units found in numerous natural products and compounds having biological activities. Several methods may be used for the preparation of alkenes, and computer-assisted analysis is available for guidance in stereoselective olefin synthesis.[2] The following sections deal with syntheses of olefins via commonly used methods such as (1) introduction of a double bond by functional group interconversion (e.g., 1,2-elimination reactions), (2) conversion of alkynes to stereodefined olefins (e.g., catalytic hydrogenation, hydrometalation, carbometalation), and (3) construction of double bonds where the olefin C_{sp2} carbons are joined in a convergent reaction (e.g., Wittig, Peterson, and Julia olefination reactions).

β-Elimination Reactions[3]   The synthesis of stereodefined *acyclic* alkenes via β-elimination reactions—such as (1) dehydration of alcohols, (2) base-induced eliminations of alkyl halides or sulfonates (tosyl or mesyl esters), and (3) Hofmann eliminations of quaternary ammonium salts—often suffers from a lack of regio- and stereoselectivity, producing mixtures of isomeric alkenes.

Dehydration of alcohols in the presence of an acid such as H_3PO_4 or H_2SO_4 proceeds most readily with substrates that ionize in an E1-type elimination to give a relatively stable tertiary, allylic, or benzylic carbocation. Secondary alcohols require more vigorous reaction conditions, and the dehydration of primary alcohols occurs only at high temperatures. E1 eliminations are subject to side reactions such S_N1 substitution, hydrogen or alkyl group shift, and isomerization of the initially formed double bond. In cases where the E1 elimination could lead to two regioisomeric alkenes, the more substituted (more stable) double bond is generally favored. Also, the *trans*-isomer usually predominates.

major product   minor product

E2-type elimination reactions proceed under milder reaction conditions and allow more control over the stereoselectivity of double bond formation. Moreover, since the reactions do not involve a carbocation, typical side reactions observed in acid-catalyzed eliminations are not a problem. Although the –OH group is not a leaving group in E2 elimination reactions, it can be converted into a halide (usually bromide or iodide), a sulfonate ester (tosylate or mesylate ester), or a dichlorophosphate intermediate. These derivatives are good leaving groups for *anti*- or *syn*-elimination in the presence of an appropriate base. The commonly used procedure for the preparation of tosylates involves stirring the alcohol with *p*-toluenesulfonyl chloride in pyridine at 0–25 °C.[4] The function of the pyridine is to form a low concentration of alkoxide species as well as to neutralize the HCl formed in the reaction.

$$ROH + pyridine \rightleftharpoons RO^- [pyridine-H]^+ \xrightarrow{TsCl} ROTs + Cl^- [pyridine-H]^+$$

The tosylation reaction is chemoselective in that a less hindered hydroxyl group reacts preferentially with tosyl chloride.

Tosylate esters may be prepared in the absence of pyridine as solvent by converting the hydroxyl group to a lithium salt by addition of methyl or butyllithium. The resultant lithium alkoxide is then treated with tosyl chloride.[5] This approach is recommended for the preparation of tosylates from very sensitive alcohols, provided that they do not contain other functional groups that react with the alkyl lithium reagents.

The treatment of alcohols with methanesulfonyl chloride and triethylamine provides a rapid, convenient method for the preparation of mesylates,[6] and these can be eliminated to furnish alkenes.

Mesylation of diols containing both a secondary- and a tertiary-OH group with slightly more than one equivalent of methanesulfonyl chloride results in a chemoselective esterification of the sterically less hindered secondary-OH group.[7]

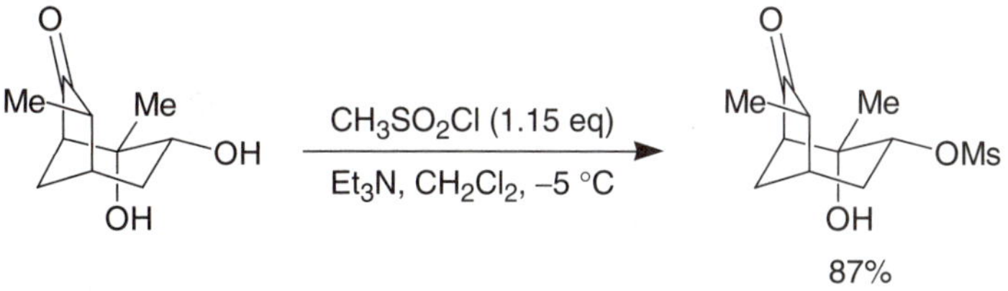

87%

Dehydration of secondary and tertiary alcohols with phosphorus oxychloride ($POCl_3$) in pyridine leads directly to alkenes without isolating the dichlorophosphate intermediate.[8]

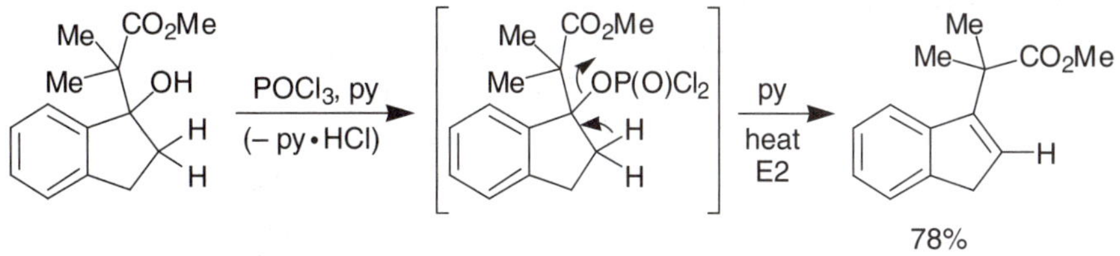

78%

Secondary and tertiary alcohols on treatment with dialkoxydiarylsulfuranes $[Ph_2S(OR)_2]$ react under very mild conditions without isolation of intermediates to give good yields of alkenes.[9]

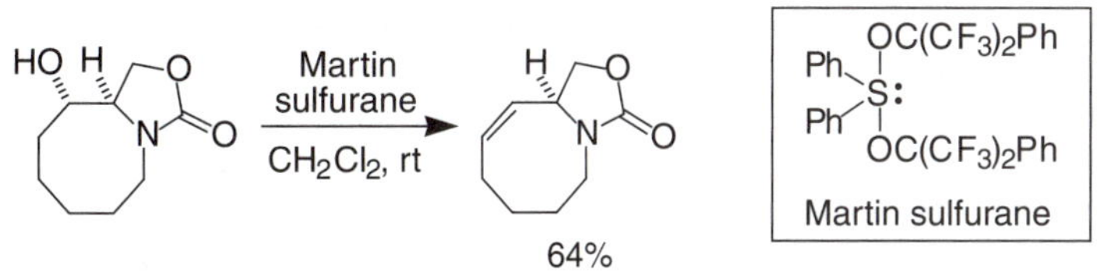

64%

Regarding the stereochemistry of E2 elimination reactions, *anti*-elimination requires an *antiperiplanar* arrangement of the hydrogen atom and the leaving group, whereas *syn*-eliminations are observed when the antiperiplanar arrangement is precluded. To suppress competing S_N2 reactions and to permit some control over the regiochemistry of elimination, strong bases with large steric requirements such as *t*-BuOK in *t*-BuOH or in DMSO, or the amidine bases 1,5-diazabicyclo[4.3.0]-non-5-ene (DBN) and 1,8-diazabicyclo[5.4.0]undec-7-ene (DBU) should be utilized. Both DBN and DBU are strong bases with large steric requirements and are frequently the reagents of choice for dehydrohalogenation reactions.[10]

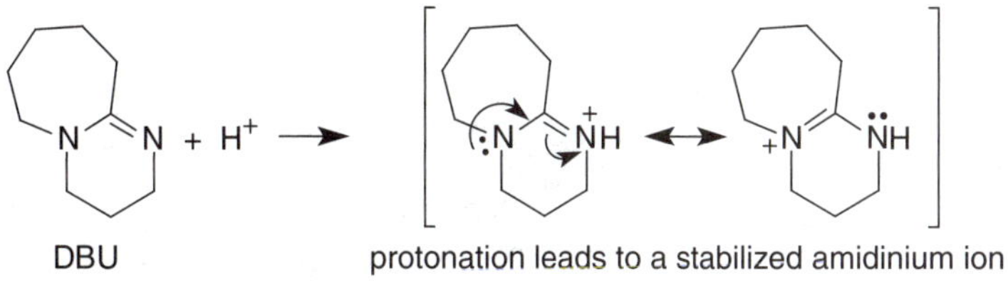

DBU

protonation leads to a stabilized amidinium ion

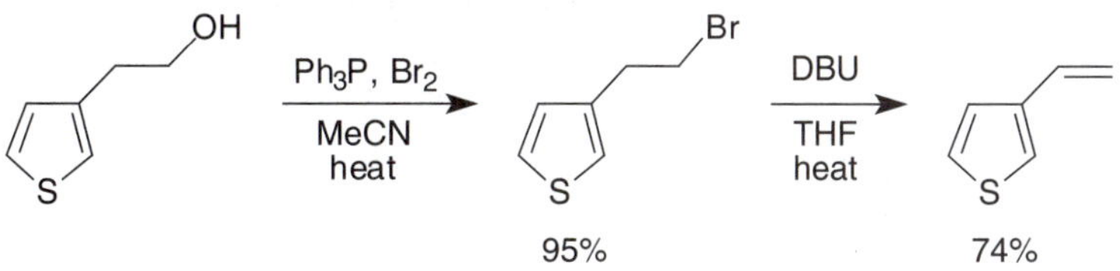

95%

74%

Only with diastereomeric acyclic alkyl halides or sulfonates having only one β-hydrogen is formation of a single geometrical isomer observed.

Pyrolytic *syn*-Elimination Reactions

Alkenes are formed by the thermal decomposition of esters, xanthates, amine oxides, sulfoxides, and selenoxides that contain at least one β-hydrogen atom. These elimination reactions require a *cis*-configuration of the eliminated group and hydrogen and proceed by a concerted process. If more than one β-hydrogen is present, mixtures of alkenes are generally formed. Since these reactions proceed via cyclic transition states, conformational effects play an important role in determining the composition of the alkene product.

Pyrolysis of Esters

High temperatures are required for these reactions, and some alkenes formed may not be stable under these conditions.[11]

~65%

Pyrolysis of Xanthates: The Chugaev Reaction[12]

The loss of C(O)S and MeSH occur at lower temperatures than the carboxylic ester pyrolysis and thus is more suitable for the preparation of sensitive alkenes. For example, heating the xanthate derived by sequential treatment of *cis*-2-benzylcyclopentanol with sodium in toluene followed by addition of excess carbon disulfide and methyl iodide furnishes, via a *syn*-elimination, 3-benzylcyclopentene in 90% yield with 98% purity. On the other hand, dehydration of the alcohol in the presence of H_3PO_4 produces a mixture of isomeric olefins containing only 20% of 3-benzyl-cyclopentene.[13]

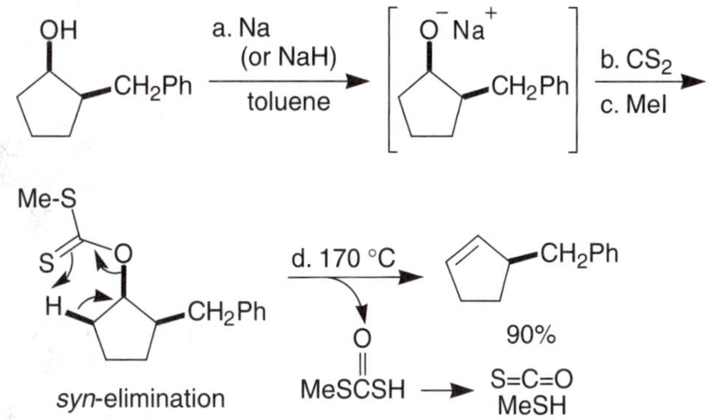

Xanthates and also thiocarbonates are useful for deoxygenation of an alcohol moiety.[14] The reaction, called the Barton-McCombie deoxygenation, involves treatment of the derived xanthate or thiocarbonate with n-Bu$_3$SnH in the presence of an initiator such as AIBN, as shown below.[15]

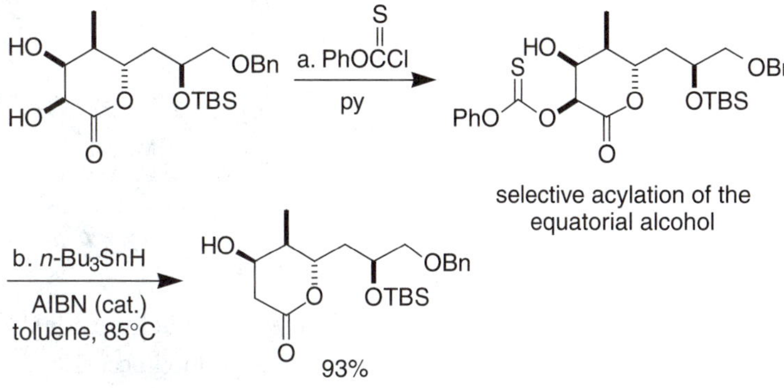

Pyrolysis of Amine Oxides: The Cope Elimination Reaction[16]

Amine oxides are derived from oxidation of *tert*-amines with peroxyacids or hydrogen peroxide. In the Cope elimination, the negatively charged oxygen acts as a base in accepting a β-hydrogen atom via a *syn* transition state.

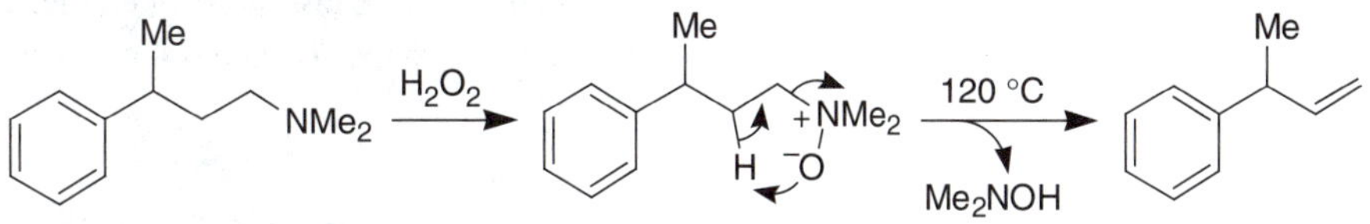

Pyrolysis of Sulfoxides

Sulfoxides are readily available via oxidation of sulfides with peroxycarboxylic acids (e.g., *m*CPBA, 1.0 eq) or sodium or potassium periodate. The sulfides themselves can be prepared by nucleophilic displacement of tosylate or mesylate esters with sodium alkyl- or phenylsulfides. In the example shown below, the sulfoxide approach worked better than direct E2 elimination of the mesylate precursor.[17]

The reaction of enolates with dimethyl disulfides (MeSSMe) or diphenyl disulfides (PhSSPh) leads to the corresponding sulfides. Their oxidation to sulfoxides followed by heating provides a route to functionally substituted alkenes, such as α,β-unsaturated carbonyl compounds.[18] In the example below, the exocyclic alkene is formed since that is the only possible *syn*-elimination product.

Syn-Elimination of Selenoxides[19]

Selenoxide elimination occurs under relatively mild conditions in comparison to the elimination reactions described above. Selenoxides undergo spontaneous *syn*-elimination at room temperature or below and thus have been used for the preparation of a variety of unsaturated compounds. The selenide precursors can be obtained by displacement of halides or sulfonate esters with PhSeNa. Oxidation of the selenides with hydrogen peroxide or *tert*-butyl hydroperoxide, sodium periodate, or peroxycarboxylic acids furnishes the corresponding selenoxides. Their eliminations usually favor formation of the less substituted olefin in the absence of heteroatom substituents or delocalizing groups.[20] Since selenium compounds are toxic, they should be handled with care.

An important strategy for the regio- and stereoselective preparation of α,β-unsaturated carbonyl compounds is via selenoxide elimination.[21] Treatment of enolates derived from ketones, esters, amides, and lactones with diphenyl diselenide, or with benzeneselenenyl chloride or bromide, leads to α-selenenylation. Oxidation of the selenium intermediates produces the selenoxides, which undergo spontaneous elimination to furnish the corresponding α,β-unsaturated carbonyl compounds, with the *trans*-isomer usually being formed in better than 98% purity.[22]

Direct selenenylation of aldehydes with LDA followed by addition of PhSeCl gives low yields of selenides contaminated with complex mixtures of products. However, conversion of aldehydes to enamines followed by selenenylation, oxidation, and elimination leads to an efficient synthesis of *trans*-α,β-unsaturated aldehydes.[23]

Conjugate addition of organocuprates to enones followed by selenenylation of the resultant enolates and oxidation provides a valuable method for β-alkylation with enone restoration, although with double bond rearrangement in the example below.[24]

Syntheses of Stereodefined Alkenes from Alkynes

An alternate strategy for the stereoselective synthesis of alkenes is to start with an alkyne and then transform the carbon-carbon triple bond to the double bond of the target molecule. This approach plays an important role in syntheses of alkenes because a number of reliable methods exist for the preparation of alkynes and for their stereo-selective elaboration into substituted alkenes via reduction, hydrometalation, and car-bometalation reactions.

Via Reduction

Semireduction of internal alkynes in the presence of a transition metal catalyst (e.g., Ni_2B, Pd/C) provides disubstituted *cis*-alkenes. On the other hand, dissolving metal reduction of alkynes or reduction of propargylic alcohols with $LiAlH_4$ or with Red-Al [sodium bis(2-methoxyethoxy)aluminum hydride] furnishes *trans*-disubstituted alkenes.[25]

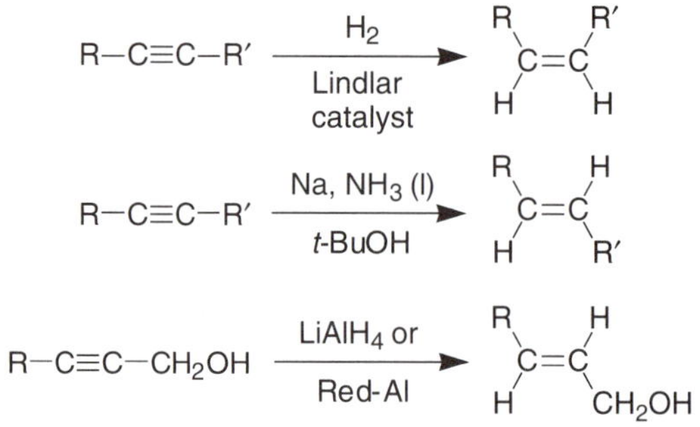

Chemoselective semireduction of a propargylic alcohol triple bond in the presence of an isolated triple bond is feasible.[25]

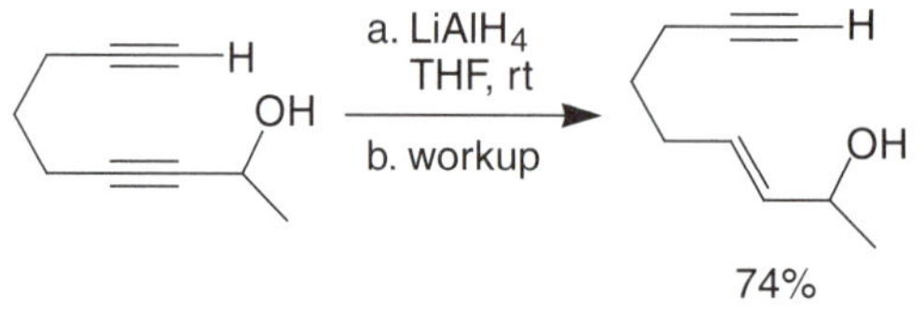

74%

Via Hydrometalation

The regio- and stereoselective monohydroboration of 1-alkynes or functionally substituted alkynes with dialkylboranes followed by protonolysis of the resultant alkenylboranes with a carboxylic acid is a powerful tool for the synthesis of 1-alkenes, and *cis*-disubstituted and functionally substituted alkenes.[26]

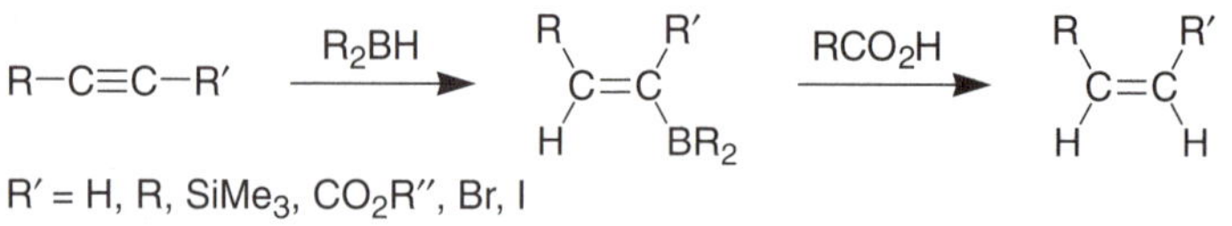

R′ = H, R, $SiMe_3$, $CO_2R″$, Br, I

Monohydroboration of unsymmetrically dialkylsubstituted alkynes with dialkylboranes produces mixtures of regioisomeric alkenylboranes. However, their protonolysis leads to *cis*-disubstituted alkenes.

R ≠ R′

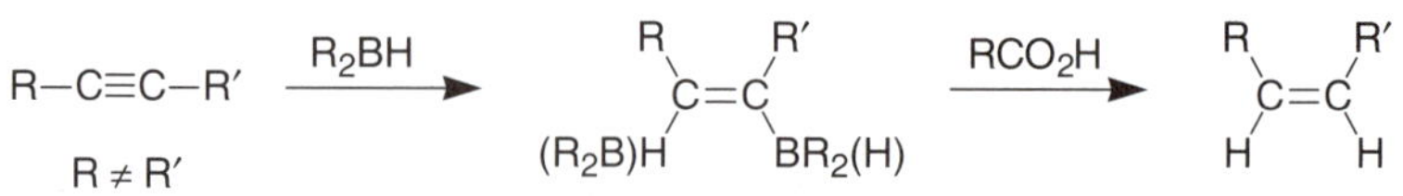

Reduction of mono- or disubstituted alkynes with diisobutylaluminum hydride proceeds, like the corresponding hydroboration reaction, via stereoselective *cis*-addition of the Al–H bond to the triple bond.[27] Hydrolysis of the resultant *trans*-alkenylalanes produces 1-alkenes and *cis*-alkenes, respectively.

Diisobutylaluminum hydride reacts chemoselectively with a triple bond in competition with a double bond in *enynes*. However, a carbonyl moiety contained in an alkyne is reduced preferentially.

Although alkenylaluminum and alkenylboron compounds are structurally related, the C–Al and C–B bonds exhibit different reactivities with inorganic and organic electrophiles. Thus, reactions of vinylaluminum compounds with electrophilic reagents generally proceed with stereospecific intermolecular transfer of the alkenyl moiety *with retention of configuration of the double bond* to afford functionally substituted olefins. For example, *trans*-1-alkenyldiisobutylalanes derived from 1-alkynes on treatment with N-bromosuccinimide or with I_2 provides *trans*-1-bromo and *trans*-1-iodo-1-alkenes, respectively.[27]

In contrast to the reactivity behavior observed with alkenylalanes, halogenation of alkenyldialkylboranes takes a different course. Thus, bromination of alkenyl-dialkylboranes proceeds *via* an initial *trans*-addition of the bromine to the double bond. *Trans*-elimination of dialkylboron bromide during hydrolytic workup leads to *cis*-1-alkenylbromides.[28]

Hydroboration of 1-hexyne with bis(*trans*-2-methylcyclohexyl)borane followed by treatment with iodine and aqueous NaOH results in the transfer of one alkyl group from boron to the adjacent carbon. The hydroxide-mediated deboronoiodination gives isomerically pure *cis*-alkenes.[29]

The formation of the *cis*-alkene can be rationalized by migration of the *trans*-2-methylcyclohexyl group proceeding with inversion at the migration terminus, and the deboronoiodination occurring in a *trans*-manner.

The transfer of a *trans*-alkenyl group from alkenylalanes derived from 1-alkynes onto carbon electrophiles is sluggish except with methyl and ethyl chloroformate.

However, conversion of alkenylalanes into the corresponding *ate*-complexes (tetracoordinated aluminum) by addition of methyllithium or *n*-butyllithium enhances their reactivity toward carbon electrophiles.[30]

Via Carbometalation[31]

Carbometalation of alkynes is a widely used method for the synthesis of stereodefined di- and trisubstituted alkenes. These structural features are often encountered in acyclic natural products, such as insect pheromones, where small amounts of the wrong stereoisomer inhibits the bioactivity. Carbometalation reactions exhibit high regio- and stereoselectivity in the addition of a carbon and metal to triple bonds. The resultant alkenyl organometallic intermediates can be elaborated into highly functionalized olefins in a one-pot procedure. The following discussion focuses on (1) the zirconium-catalyzed carboalumination of alkynes and (2) the addition of organocopper reagents to alkynes.

Carboalumination.[32] The zirconium-catalyzed carboalumination of terminal alkynes was briefly discussed in connection with palladium-catalyzed cross-coupling reactions. An important facet of the carboalumination reaction is that it allows the preparation of trisubstituted olefins containing a methyl substituent, a structural feature encountered in a wide variety of natural products, especially in terpenoids. The *Negishi methylalumination* of 1-alkynes is a highly regio- and stereoselective reaction that proceeds via *syn*-addition of an Me-Al bond to a triple bond with the aluminum being attached to the terminal carbon. The alkenylaluminum intermediate formed (or the derived *ate*-complex thereof) reacts with a variety of electrophiles with retention of the double bond configuration, as exemplified below.[33] Since trimethylalane is highly pyrophoric, it must be kept and used under an inert atmosphere.

The efficiency of the above one-pot methylalumination procedure applied to the synthesis of *trans*-farnesol compares favorably[32] with the reported three-step procedure involving preparation of the propargylic alcohol, LiAlH$_4$ reduction of the triple bond, iodination of the vinyl carbon-aluminum bond, and final coupling of the alkenyl iodide with lithium dimethylcuprate.[34]

Carbocupration.[31a,35] The carbocupration of alkynes, introduced by J. F. Normant in 1971,[36] shares some common features with the methylalumination reaction in regard to its regio- and *syn*-stereoselectivity, but is wider in scope in that it allows the introduction of a variety of alkyl groups onto the triple bond of acetylene, 1-alkynes, and functionally substituted alkynes. The organocopper reagents used in carbocupration reactions are derived either from Grignard reagents[37] (RCu, MgX_2 and RCu•SMe_2, MgX_2) or from organolithium reagents (homocuprates R_2CuLi, and heterocuprates RCuXLi, where X = SPh, O*t*-Bu, or C≡CR′).

A possible mechanistic scheme for carbocupration may involve initial complexation of the RCu(I)$MgBr_2$ species with the triple bond, oxidative addition (insertion) of RCu to the activated triple bond, transfer of the R to the vinylic carbon, reductive elimination of Cu(I), and finally metal-metal exchange to furnish the 2,2-disubstituted vinylcopper intermediate.

The formation of vinylcopper reagents is most successful when R and R′ are primary alkyl groups. Utilization of the dimethylsulfide-cuprous bromide complex, CuBr•SMe_2, is recommended since SMe_2 stabilizes the vinylcopper species toward side reactions such as metal-hydrogen exchange, dimerization, and vinyl-alkyl coupling. Carbocupration of acetylene with homocuprates usually results in the utilization of both ligands to produce divinylcuprates.[31a, 38]

Vinylcopper reagents react with a wide variety of electrophilic reagents such as halogens, alkyl halides, allylic halides, acid chlorides, epoxides, α,β-unsaturated ketones, and α,β-acetylenic esters with complete retention of the double bond stereochemistry.[31a] To enhance the reactivity of vinylcopper intermediates toward carbon electrophiles, the coupling is often carried out in the presence of activators such as HMPT, DMPU, and/or P(OEt)₃ (triethylphosphite). Some representative examples of stereospecific syntheses of di- and trisubstituted alkenes via the carbocupration route are shown below.[37b,39]

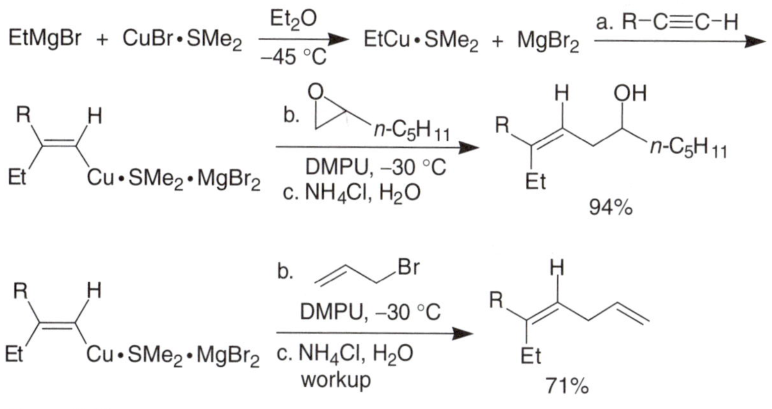

R = n-C$_6$H$_{13}$

Under controlled conditions, divinylcuprates derived from acetylene carbocupration can be converted to the (1Z, 3Z)-bis(dienyl)cuprates using an excess of acetylene. Reacting the divinylcopper intermediates with electrophiles affords conjugated (Z, Z)-dienes in a highly stereoselective manner.[40]

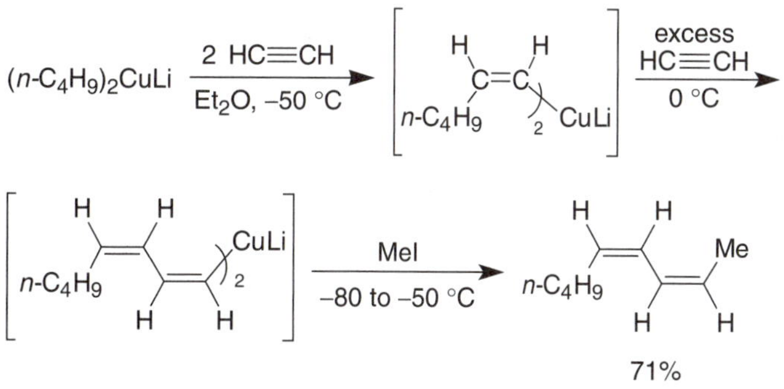

Addition to α,β-Unsaturated Ketones. 2,2-Disubstituted alkenylcopper-dimethylsulfide complexes react by conjugate addition with a number of α, β-unsaturated carbonyl compounds to furnish trisubstituted olefins containing a keto group.[37b]

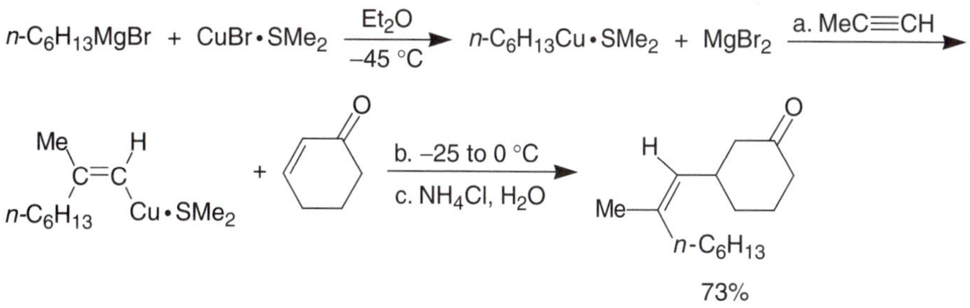

Addition to α,β-Acetylenic Esters. Addition of alkylcuprates to acetylenic esters proceeds in a stereoselective manner to yield, after hydrolytic work, trisubstituted α,β-unsaturated esters.[41]

Conjugate additions of divinylcuprates to ethyl propiolate joins two acetylenic units to furnish isomerically pure (2E, 4Z)-conjugated dienoates.[42]

The Wittig, Horner-Wadsworth-Emmons, Peterson, and Julia Olefination Reactions

In each of the reactions described below, the double bond is formed by connecting two separate molecules. The steps leading to the alkene involve addition of a carbanion to the carbonyl carbon of an aldehyde or a ketone followed by an elimination process.

Wittig Olefinations[43]

In the 1950s, Georg Wittig (Nobel Prize, 1979) reported that the reaction of phosphorus ylides with aldehydes or ketones produces alkenes.[44] The Wittig olefination is one of the most powerful tools for the construction of double bonds and has found wide application in natural product synthesis as well in the pharmaceutical industry for the manufacture of drugs. The prime utility of the Wittig reaction lies in the ease with which the reaction occurs under mild conditions and that no ambiguity exists concerning the location of the double bond in the product. A drawback of the Wittig reaction is its susceptibility to steric hindrance. Whereas aldehydes usually give high yield of alkenes, ketones often react less satisfactorily. The synthesis of tetrasubstituted alkenes via the Wittig reaction is problematic. The following discussion will be confined to reactions of ylides with aldehydes and ketones. A literature review is available for phosphorous ylide reactions with ester- and amide-type substrates.[43d]

The Wittig reagent is prepared by alkylation of a phosphine followed by treatment of the resulting phosphonium salt with a base (RLi, NaH, etc.). The ylide formed has a Lewis contributing structure with a positive charge on phosphorus and a negative charge residing on carbon. The ylide is stabilized by electron delocalization into the phosphorus 3d orbitals, as represented by the ylide-alkylidene resonance form. Ylide addition to a carbonyl group results in the formation of an alkene and a phosphine oxide.

The reactivity and stability of phosphoranes depend on the nature of the substituents at the ylide carbanion carbon. Accordingly, the ylides used in Wittig reactions are divided into *nonstabilized* and *stabilized* ylides.

Nonstabilized Ylides. These ylides are devoid of electron-withdrawing substituents at the anionic center; they carry hydrogen or electron-donating alkyl groups on the ylide carbon and phenyl groups on the phosphorus. Nonstabilized ylides are very nucleophilic and react with CO_2, O_2, and H_2O; hence they must be handled in an inert atmosphere. Primary as well as secondary alkyl halides may be used for the preparation of phosphonium salts and subsequently ylides.

Stereochemistry. Wittig reactions of aldehydes with nonstabilized ylides generally produce (Z)-alkenes. However, the presence of soluble lithium salts may affect the stereochemical outcome by decreasing the (Z)-alkene selectivity. Therefore, sodium or potassium hexamethyldisilazide (NaHMDS or KHMDS) or *t*-BuOK in THF, which produce the less soluble sodium or potassium salts, should be used for deprotonation of the phosphonium salt instead of organolithium reagents. The solvent plays an important role for achieving high (Z)-stereoselectivity. Generally, ethers such as Et_2O, THF, or DME are the solvents of choice. For carbonyl compounds that show a high tendency for enolate formation, toluene is often the most suitable solvent. DMSO and protic solvents such as alcohols must be avoided because they adversely affect both the yields and the stereoselectivities.[45] Also, the aldehyde or ketone must be added dropwise to the ylide at low temperature (-78 °C). As exemplified below, the Wittig reaction tolerates the presence of functional groups such as ester, ether, halogen, or double and triple bonds.

base, solvent	(Z) : (E) ratio
t-BuOK, THF	94 : 6
NaH, DMF	94 : 6
n-BuLi, Et_2O	78 : 22

The (Z)-stereoselectivity of salt-free Wittig olefinations leading to the thermodynamically less stable *cis*-alkenes has long been the subject of intense investigations.[43e] At one time, the olefination was thought to proceed via an ionic stepwise process involving a zwitterionic betaine and a 1,2-oxaphosphetane intermediate. However, ^{31}P NMR spectroscopy studies revealed the oxaphosphetane as the *only* observed intermediate.[46]

As mentioned above, the ratio of (Z)- to (E)-alkenes in Wittig reactions is influenced by a variety of factors such as the structure of the ylide and carbonyl components, the presence of lithium salts, the solvent, and reaction temperature. To incorporate these reaction characteristics into a mechanistic scheme that explains the high (Z)-alkene selectivity observed in reactions of salt-free nonstabilized ylides with aldehydes is no easy task, and there may not be a single unifying mechanism for *all* Wittig-type reactions.[47] Current views to rationalize the (Z)-alkene stereoselectivity of Wittig olefinations of nonstabilized ylides and aldehydes favor a one-step, nonsynchronous cycloaddition mechanism. Early carbon-carbon bond formation by attack of the ylide at the carbonyl group in a four-centered, puckered transition state, keeping the substituent R′ of the ylide, the Ph moieties of the phosphorus, and the R group of the aldehyde positioned to minimize steric interactions, leads directly to the kinetic (*cis*)-oxaphosphetane. For detailed discussions on mechanistic schemes for the Wittig reaction, the original literature should be consulted.[43e,48]

Generally, oxaphosphetanes are thermodynamically unstable and fragment into alkenes and triphenylphosphine oxide. *This elimination step is stereospecific* with oxygen and phosphorus departing in a *syn*-periplanar mode to produce (Z)-alkenes, the driving force being formation of the very stable P = O bond (130–140 kcal/mol, 544–586 kJ/mol bond dissociation energy).

Interestingly, methylenetriphenylphosphorane, $Ph_3P=CH_2$, the simplest Wittig reagent, reacts sluggishly with hindered ketones. Although camphor is converted to 2-methylenebornane on treatment with $Ph_3P=CH_2$ in DMSO at 50 °C, the sterically more hindered fenchone is unaffected when reacted with $Ph_3P=CH_2$ at 50° C in THF, THF-HMPA, or in DMSO.[49] A more reactive methenylating agent for hindered ketones is the lithiated ylide $Ph_3P=CHLi$ derived from deprotonation of $Ph_3P=CH_2$ with *t*-BuLi. Thus, treatment of fenchone with $Ph_3P=CHLi$ in the presence of HMPA followed by decomposition of the adduct with *t*-BuOH furnishes the *exo*-methylene derivative.[50]

Schlosser Modification.[51] Almost pure *trans*-olefins are obtained from nonstabilized ylides by the Schlosser modification of the Wittig reaction (*Wittig-Schlosser reaction*). For example, treatment of the (*cis*)-oxaphosphetane intermediate A with *n*-BuLi or PhLi at –78 °C results in lithiation of the acidic proton adjacent to phosphorus to produce the β-oxido phosphonium ylide B. Protonation of B with *t*-BuOH leads to the *trans*-1,2-disubstituted alkene C.

Treatment of β-oxido ylides with electrophiles other than proton donors provides a route to stereospecific trisubstituted alkenes. For example, trapping the β-oxido phosphonium ylide B with formaldehyde (generated from paraformaldehyde) leads to dioxido phosphonium derivative D to yield, after elimination of triphenylphosphine oxide, the trisubstituted allylic alcohol E.[52]

γ-Oxido ylides are accessible by the reaction of epoxides with either $Ph_3P=CH_2$ or with $Ph_3P=CHLi$.[50,53] They react with aldehydes to stereoselectively form homoallylic alcohols, as exemplified by the conversion of cyclopentene oxide to the *trans, trans*-homoallylic alcohol, thereby generating two carbon-carbon bonds and three stereocenters.

Intramolecular Wittig reactions provide a methodology for the preparation of five-, six-, and seven-member rings.[54]

The phosphorane derived from triphenylphosphine and chloromethyl methyl ether (MOM-Cl, carcinogenic) reacts with aldehydes or ketones to furnish the corresponding enol ethers. These on acid hydrolysis produce one-carbon homologated aldehydes or lead to the conversion of a keto into an aldehyde group, respectively.[55]

Stabilized Ylides.[56] The carbanion center of phosphorus ylides is further stabilized by conjugation, usually with a carbonyl group (–CHO, –COR, –CO₂R). Hence, these ylides are less reactive and weaker bases, but are more stable toward oxygen and protic solvents than nonstabilized ylides. This type of stabilized ylide is mainly used in olefinations with aldehydes.

Synthesis of the synthetically important carbethoxymethylenetriphenylphosphoranes ($Ph_3P=CHCO_2Et$ and $Ph_3P=CRCO_2Et$) involves treatment of α-bromoesters with Ph$_3$P and deprotonation of the phosphonium salts formed with NaH, metal alkoxides, or aqueous NaOH.[57] Organolithium reagents are not suitable for deprotonation since they would react with the ester carbonyl group.

The corresponding formyl analogue, formylmethylenetriphenylphosphorane, is available by treatment of methylenetriphenylphosphorane with ethyl formate.[58]

While stabilized Wittig reagents in nonpolar solvents produce preferentially (*E*)-olefinic products, in polar solvents the (*Z*)-isomer may predominate. Depending on the substitution pattern of the ylide, its reaction with aldehydes can lead to di- or trisubstituted olefins of predictable stereochemistry. Olefinations in nonpolar solvents furnish olefins with the electron-withdrawing group of the ylide and the R group of the aldehyde in a *trans*-relationship. For example, the reactions of Ph$_3$P = CHCO$_2$Et[59] and Ph$_3$P = CHCHO[60] with aldehydes leads to an (*E*)-α,β-unsaturated ester or an (*E*)-α,β-unsaturated aldehyde, respectively. The (*Z*)-α,β-unsaturated aldehyde A can be prepared by condensing Ph$_3$P=CHCH(OEt)$_2$ (*Bestmann reagent*)[61] with an aldehyde followed by mild, selective hydrolysis of the resultant unsaturated acetal.

The mechanism proposed for the (E)-alkene selectivity involves a nonsynchronous cycloaddition with a relatively advanced, productlike transition state leading to the kinetic *trans*-oxaphosphetane intermediate. Extrusion of triphenylphosphine oxide produces the (E)-alkene.[62]

The Horner-Wadsworth-Emmons Reaction[43e,56,63]

An important addition to the Wittig *trans*-olefination procedure is the introduction of phosphonate-stabilized carbanions as olefin-forming reagents, referred to as the *Horner-Wadsworth-Emmons* or HWE reaction. The HWE olefination offers several advantages over the Wittig reaction using stabilized ylides:

- Phosphonate carbanions are more nucleophilic than phosphonium ylides. Thus, they can be used in condensations with ketones as well as with aldehydes under mild conditions.

- Separation of the olefin product from the water-soluble phosphate ester by-product formed from phosphonates circumvents the problem often encountered in removing $Ph_3P = O$.

- Reaction conditions are available for the preparation of alkenes enriched in either the (E)- or the (Z)-isomer.

The HWE reaction works best when the phosphonate bears a strong electron-withdrawing group, such as –COR, –COOR, –CN, that can stabilize a carbanion. Owing to their increased nucleophilicity, phosphonate carbanions often give higher yields of olefins than the corresponding stabilized phosphoranes. In the case of nonstabilized phosphonate reagents, decomposition of the condensation products to alkenes is slow.

A number of methods are available for the preparation of phosphonate reagents.[63] For example, treatment of triethylphosphite with ethyl bromoacetate (*Arbuzov reaction*) produces the phosphonoacetate A. Its reaction with a suitable base such as NaH gives the carbanion B, which, on treatment with cyclohexanone, furnishes cyclohexylideneacetate C in 70% yield. This compares favorably to the 25% yield obtained when using the triphenylphosphorane $Ph_3P=CHCO_2Et$.

The Arbuzov reaction is not suitable for the preparation of α-ketophosphonates. However, these synthetically useful reagents[64] can be prepared from the lithium salt of dimethyl methylphosphonate and an ester.[65] Treatment of α-ketophosphonates with sodium hydride or metal alkoxides furnishes the corresponding carbanions, which on further reaction with aldehydes provides the synthetically valuable conjugated enones, as exemplified below.[66]

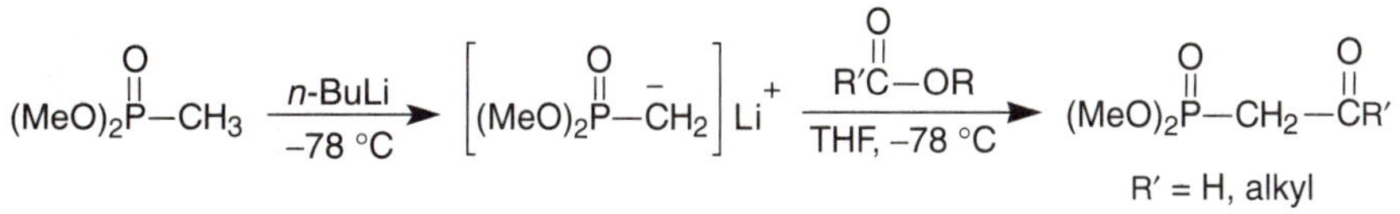

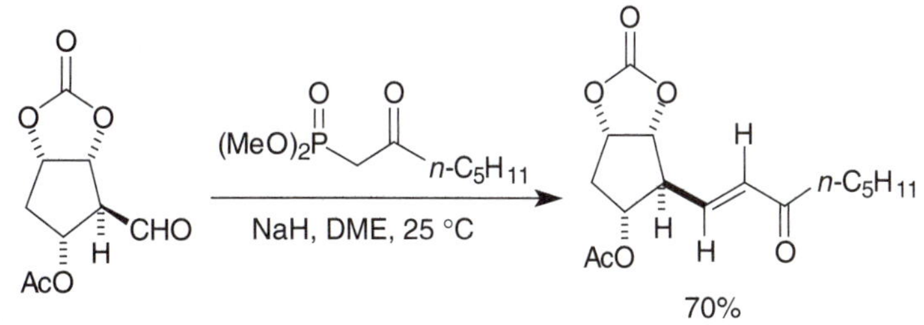

The stereoselectivity of the HWE olefination reaction depends on the nature of the RO groups on phosphorus, structural features of the ylide, the solvent, and the reaction temperature. Generally, the HWE reactions give preferentially the more stable *trans*-disubstituted olefins.

The preparation of trisubstituted olefins via the HWE reaction can be achieved either by using α-branched phosphonate reagents with aldehydes or by using ketones as the carbonyl reagents. The stereoselectivity of these reactions depends on the structure of the phosphonate anion and on the solvent used. For example, treatment of the α-methyl-substituted phosphonate anion with the sterically encumbered aldehyde depicted below gives the (Z)-unsaturated ester.[67] The corresponding (E)-unsaturated ester is accessible by condensation of the aldehyde with carbomethoxyethylidenetriphenylphosphorane. Note that it is difficult to obtain high (E)-selectivity when the HWE reagent has an α-substituent that is larger than a methyl group.[63]

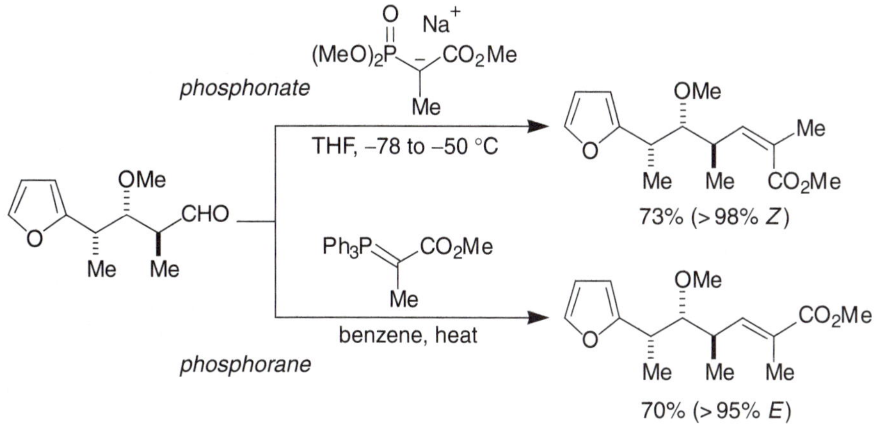

Relatively strong bases are used for the deprotonation of phosphonate reagents, and the phosphonate-stabilized carbanions formed are more basic than the corresponding phosphorane reagents. Such conditions may be incompatible with base-sensitive aldehydes and ketones, causing epimerization of chiral compounds or condensation reactions. Mild HWE olefination procedures circumventing these problems involve condensation with a *tertiary*-amine as the base in the presence of LiCl[68] or MgBr$_2$.[69] The function of the Lewis acid cation (Li$^+$ or Mg^{+2}) is to lower the pK_a of the methylene α-hydrogens of A by coordination with both the ester carbonyl group and the phosphonate oxygen (as in B) so that a weaker base is sufficient to form metal enolate C. Also, complexation of the carbonyl oxygen of the aldehyde by the metal cation increases the electrophilic character of the carbonyl carbon, thereby facilitating nucleophilic addition by the phosphono enolate. LiCl-mediated olefinations are now widely used in organic syntheses.

By proper choice of the phosphonate reagent, the HWE reaction shows great flexibility in controlling the stereochemistry of the double bond of vinylic esters. Whereas condensation of dialkylphosphonoacetates with aldehydes gives preferentially the more stable (*E*)-unsaturated esters, the reaction of methyl bis(trifluoroethyl)phosphonoacetate with aliphatic and aromatic aldehydes in the presence of a mixture of KHMDS (potassium hexamethyldisilazide) and 18-crown-6 produces (*Z*)-α, β-unsaturated esters stereoselectively.[70]

An alternative, economical procedure for generating (Z)-unsaturated esters from phosphonates and aldehydes avoiding the use of the expensive 18-crown-6 is the olefination with ethyl diarylphosphonoacetates in the presence of NaH or Triton B (benzyltrimethylammonium hydroxide).[71] Condensations of (diarylphosphono)acetates with base-labile or functionalized aldehydes are best carried out in the presence of NaI and DBU in THF solvent.[72]

The (E)- and (Z)-selectivity in HWE reactions is determined by a combination of the stereoselectivity in the initial carbon-carbon bond formation and the reversibility of the intermediate adducts. The (E)-selectivity has been explained by the formation of the thermodynamically more stable *threo*-adduct, which then decomposes via the oxaphosphetane intermediate to the (E)-olefin. The (Z)-selectivity has been attributed to the predominant formation of the *erythro*-adduct which collapses irreversibly via a transient oxaphosphetane intermediate to the (Z)-olefin.[71]

Peterson Olefination[73]

The *Peterson olefination* is a connective alkene synthesis and represents a useful alternative to the Wittig reaction. The precursors for the Peterson olefination are β-hydroxy-alkyltrimethylsilanes which undergo β-elimination of trimethylsilanol under basic or acidic conditions to furnish stereodefined alkenes. This olefination method is especially valuable for the preparation of terminal and *exo*-cyclic double bonds and for the methylenation of hindered ketones where the Wittig reaction is problematic. Also, the by-product of the reaction, hexamethyldisiloxane (Me₃Si–O–SiMe₃, bp 101°C), is readily removed by distillation.

Several procedures are available for the preparation of the requisite β-hydroxysilanes such as addition of α-silyl carbanions to aldehydes and ketones, reaction of organocuprates with α,β-epoxysilanes, reduction of β-ketosilanes, and addition of organometallic regents to β-ketosilanes. The selection of a particular procedure is dictated by the structure and stereochemistry of the desired alkene.

The synthesis of terminal alkenes entails addition of either (trimethylsilyl)methylmagnesium chloride (Me₃SiCH₂MgCl) or (trimethylsilyl)methyllithium (Me₃SiCH₂Li) to an aldehyde or a ketone to yield the corresponding β–hydroxysilane, which undergoes base- or acid-mediated elimination to furnish the alkene, as exemplified below.[74]

Instead of using the highly basic lithium reagent Me₃SiCH₂Li for the methylenation of readily enolizable carbonyl compounds, the less basic and more nucleophilic cerium reagent (presumably "RCeCl₂") derived by combining (trimethylsilyl)methyllithium with anhydrous cerium(III) chloride should be employed. To facilitate isolation of the β-hydroxy silane, one equivalent of TMEDA per equivalent of CeCl₃ is added before hydrolytic workup. The conversion of 2-indanone to 2-methyleneindane is representative.[75]

The Peterson olefination may also be applied to the preparation of α,β-unsaturated esters, thus providing an attractive alternative to the HWE phosphonate method. For example, deprotonation of trimethylsilylacetate with R₂NLi followed by addition of cyclohexanone yields the intermediate β-hydroxysilane, which, on elimination of the silicon-oxygen moieties, furnishes ethyl cyclohexylideneacetate.[76]

$$\text{Me}_3\text{SiCH}_2\text{CO}_2\text{R} \xrightarrow[\text{THF, } -78\ ^\circ\text{C}]{\text{LDA or } (\text{Chx})_2\text{NLi}} \left[\text{Me}_3\text{SiCHLiCO}_2\text{R}\right]$$

R = Et, *t*-Bu

a.

b. H⁺, H₂O

−78 °C

=CHCO₂R

95–98%

An attractive feature of the Peterson reaction is that a single diastereomer of a β-hydoxyalkylsilane can serve as precursor for the stereospecific preparation of both (*Z*)- and (*E*)-alkenes, depending on whether the elimination is carried out under acidic (H_2SO_4 or $BF_3 \cdot Et_2O$ in THF) or basic (KH in THF) conditions. For example, treatment of the *syn* diastereomer of 5-trimethylsilyl-4-octanol with H_2SO_4 in THF followed by workup gave the (*Z*)-4-octene. The same *syn*-β-hydroxysilane, but using potassium hydride in THF to induce elimination, produced the (*E*)-octene. Subjecting the *anti*-β-hydroxysilane under similar reaction conditions to acid gives the (*E*)-4-octene and to base the (*Z*)-isomer.[77] The acid-mediated elimination is stereospecific, proceeding via an antiperiplanar transition state. In the presence of base, the β-hydroxysilane is deprotonated and the oxyanion attacks the silicon atom to form, presumably, the oxasiletane anion A, which undergoes a *syn*-periplanar elimination to furnish the alkene.[78]

In the *β-ketosilane approach,* the condensation of α-silyl carbanions with carbonyl compounds produces mixtures of diastereomeric β-hydroxysilanes. Therefore, the preparation of stereodefined alkenes via the Peterson reaction hinges on the availability of just one diastereomer. To overcome this shortcoming, procedures have been developed to prepare stereoselectively β-hydroxyalkylsilanes via the β-ketosilane or the epoxysilane routes, as exemplified below.

The following example illustrates the preparation of (*Z*)-4-octene via the β-ketosilane approach. Addition of ethyllithium to vinyltrimethylsilane in THF generated the α-trimethylsilylbutyl carbanion, which reacted with *n*-butanal to give a diastereomeric mixture of β-hydroxysilanes. Oxidation with CrO_3-pyridine produced the β-ketosilane A, which on reduction with *i*-Bu₂AlH yielded stereoselectively the *syn*-β-hydroxysilane B. The observed *syn*-stereochemistry is in agreement with that predicted based on the preferred approach of the hydride to the carbonyl group of the β-ketosilane using the Felkin-Anh model. Treatment of the *syn* β-hydroxysilane with conc. H_2SO_4 in THF afforded (*Z*)-4-octene.[77]

The reaction of β-ketosilanes with alkyllithium reagents produces predominately one diastereomer of two possible β-hydroxysilanes. For example, sequential treatment of 5-trimethylsilyl-4-decanone with methyllithium followed by basic workup yielded preferentially the trisubstituted (*E*)-alkene.[79]

Diastereomerically enriched β-hydroxysilanes are also accessible from α,β-epoxysilanes and reaction of these with organocuprate reagents. The epoxysilanes are synthesized by epoxidation of (*E*)- or (*Z*)-vinylsilanes with *m*-chloroperbenzoic acid. The required (*Z*)- and (*E*)-vinylsilanes can be obtained by hydroboration-protonolysis of 1-trimethylsilyl-1-alkynes or by hydrosilylation of 1-alkynes, respectively.

The reaction of α,β-epoxysilanes with organocuprate reagents proceeds with the regioselective opening of the epoxide ring to form β-hydroxyalkylsilanes, which are then converted to stereodefined alkenes, as exemplified below.[80]

Julia Olefination[81]

The Julia olefination reaction is highly regioselective and (*E*)-stereoselective, providing a valuable alternative to the Schlosser reaction for making *trans*-disubstituted olefins. The reaction involves condensation of a metalated alkyl phenyl sulfone with an aldehyde to yield a β-hydroxysulfone, which is then subjected to a reductive elimination to produce the alkene. There are limitations to the preparation of tri- and tetra-substituted alkenes via the sulfone route because the β-alkoxy sulfones derived from addition of the sulfone anion to ketones may be difficult to trap and isolate or they may revert back to their ketone and sulfone precursors.

The alkyl phenyl sulfone precursors for the olefination may be prepared by the reaction of alkyl halides with sodium benzenethiolate, followed by oxidation of the alkyl phenyl sulfides formed with *m*CPBA. Alternatively, displacement of an alkyl bromide or iodide with the less nucleophilic sodium benzenesulfinate $PhSO_2Na$ leads directly to the alkyl phenyl sulfone.

Alkyl phenyl sulfones (pK_a 27) are nearly as acidic as esters; hence they are readily deprotonated by *n*-BuLi, LDA in THF, or EtMgBr in THF to give α-metalated sulfones. Their reaction with aldehydes gives a mixture of diastereomeric β-phenylsulfone alkoxide adducts. Reductive elimination of the benzenesulfinate moiety from the adduct to produce the alkene is usually slow. To minimize side reactions, the hydroxyl group is first converted to an acetate, benzoate, mesylate, or *p*-toluenesulfonate and then treated with an excess of sodium amalgam [Na(Hg), prepared by adding small pieces of sodium to mercury][82] in methanol to furnish the *trans*-alkene.[83]

Reduction of the mixture of *erythro*- and *threo*-sulfones with Na(Hg) leads stereoselectively to the *trans*-alkene, indicating that the alkene-forming steps for both diastereomeric sulfones must involve a common intermediate. The (*E*)-selectivity of the elimination step may involve two successive electron transfers from the sodium to the sulfone, resulting in the loss of benzenesulfinate. The anion that is formed must have a lifetime sufficient to equilibrate and assume the low-energy conformation, which places (1) the p-orbital of the carbanion antiperiplanar to the acetoxy leaving group and (2) the bulky substituents as far apart as possible. Therefore, the stereochemical course of the elimination is governed by the steric interactions of the substituents.

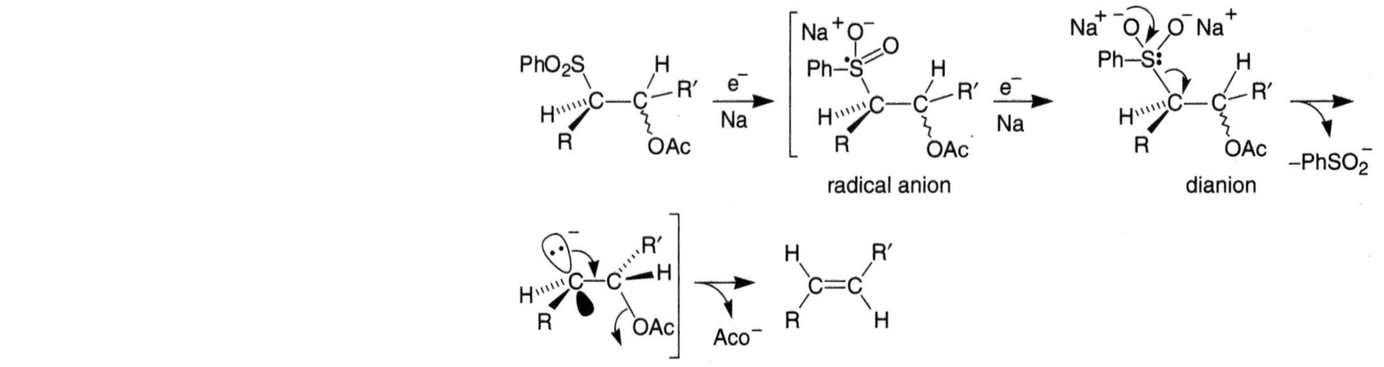

Conjugate addition of sodium benzenesulfinate to α,β-unsaturated carbonyl compounds produces β-ketosulfones, which may be elaborated into *trans*-α,β-unsaturated aldehydes or ketones, respectively, by the sequence of reactions shown below.[84]

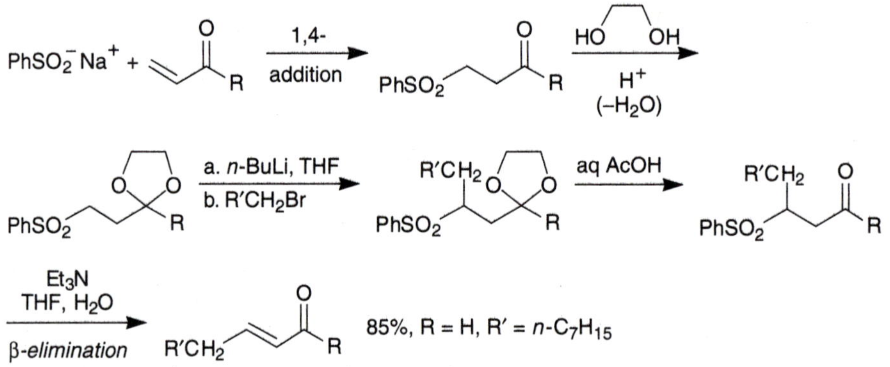

The utility of alkyl arylsulfones in organic synthesis extends beyond the olefination reaction by adding to the arsenal of carbon-carbon-bond-forming reactions. The sulfone may serve as temporary activating group for alkylation, acylation, and addition reactions by providing a source of sulfur-stabilized carbanions. After having served its purpose, the benzenesulfinyl moiety is removed by reductive elimination.[85]

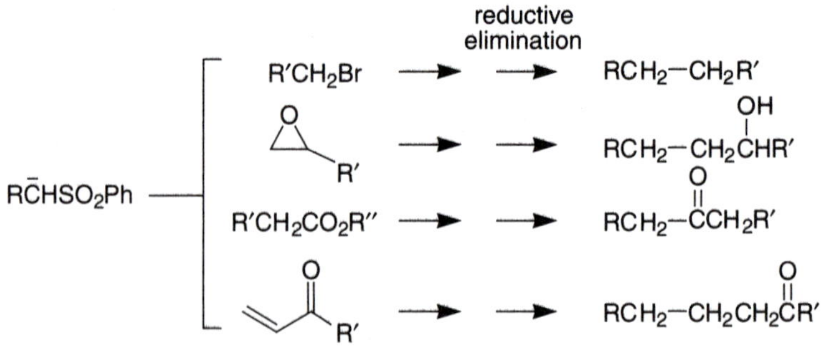

In the *Kocienski-modified Julia olefination*,[86] replacement of the phenylsulfone moiety with a heteroarylsulfone such as the 1-phenyl-1*H*-tetrazol-5-yl group (PT), shown below, profoundly alters the course of the Julia coupling. The olefin product is obtained in a one-pot procedure as opposed to the three-step protocol of the classical Julia olefination.[87]

The PTSO$_2$ moiety is introduced either via an S$_N$2 displacement-oxidation sequence or by using the Mitsunobu reaction followed by oxidation, as shown below.[88]

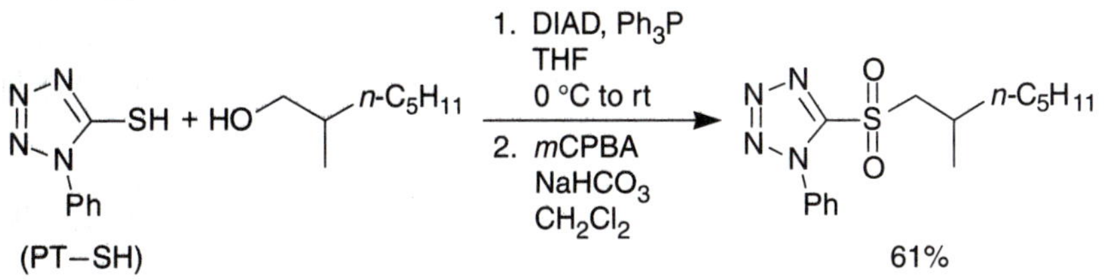

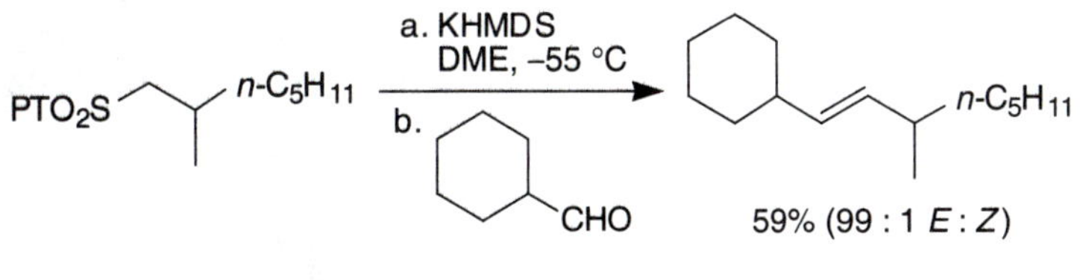

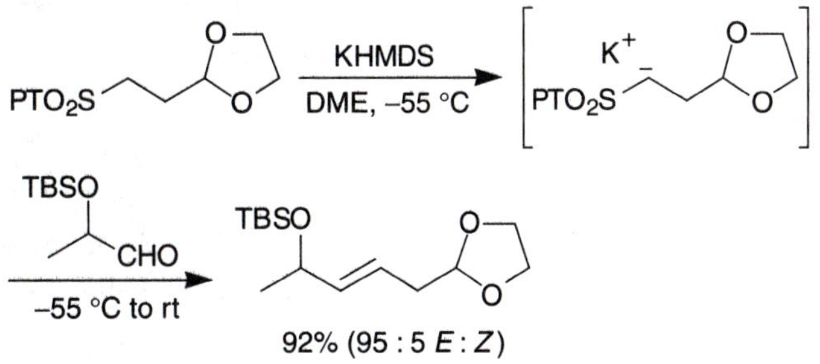

The Shapiro Reaction[89]

The *Shapiro reaction* is a nonconnective procedure for interconversion of a keto group and its α-methylene moiety into an alkene, providing a powerful method for the preparation of alkenyllithiums that are not readily available by current methodologies. Further elaboration of the intermediate alkenyllithiums leads to a variety of structurally diverse alkenes.

Condensation of ketones with an arylsulfonylhydrazine gives the corresponding arylhydrazones. These on treatment with a strong base such as MeLi, *n*-BuLi, *s*-BuLi, or an LDA fragment to vinyl carbanions, with loss of arylsulfinate and nitrogen. Depending on the reaction conditions, the intermediate vinyl carbanions may be protonated by the solvent to produce alkenes or they may be intercepted with various electrophiles to give alkene products.[90]

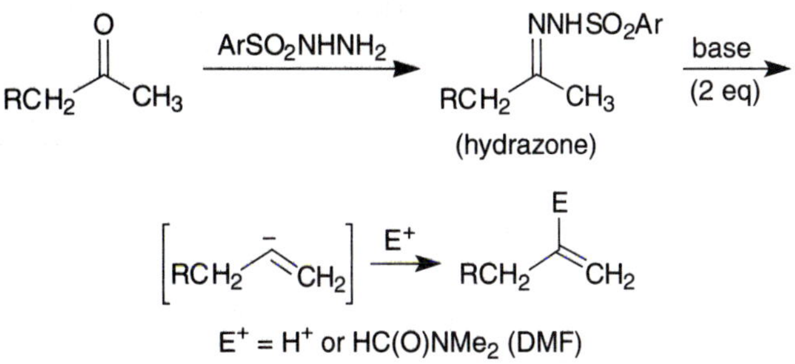

$E^+ = H^+$ or $HC(O)NMe_2$ (DMF)

The proposed mechanism for the formation of the vinyl carbanion intermediate involves removal of the acidic N–H proton from the hydrazone by the strong base to form the mono-anion A. In the presence of a second equivalent of base, the α-proton is removed to give the dianion B. Elimination of the arylsulfinate anion in the rate-limiting step leads to the alkenyl diazenide C, which loses nitrogen to produce the alkenyllithium D. Depending on the nature of the aryl moiety of the ketone hydrazone and the reaction conditions, the intermediate vinyl anion D may undergo in situ protonation by the solvent to give the alkene or it may be trapped by an added electrophile.

(E,Z) mixture
of hydrazone isomers

Although the original procedure using p-toluenesulfonylhydrazones (tosylhydrazones) and methyllithium in ether furnished alkenes in good yields, as exemplified below,[91] it was not suited for the generation of alkenyllithiums and their trapping with electrophiles because of protonation by the solvent (usually Et$_2$O or THF).

> 80% overall

Subsequently, conditions were reported that not only gave high yields of alkenes, but also allowed the incorporation of electrophiles.[92] However, for efficient trapping of the alkenyllithium intermediates a large excess of base (≥ 3 eq of n-BuLi) was needed, instead of the two equivalents required by the mechanism, because of competing metalation of the *ortho*-and/or *benzylic*-hydrogens of the tosyl group.

The problem necessitating the use of excess n-BuLi, and consequently an excess of the electrophile, was obviated by using ketone-derived 2,4,6-triisopropylbenzene-sulfonylhydrazones (trisylhydrazones).[93] Elimination to give the carbanion intermediate is much faster with the trisylhydrazone-derived dianion than with that obtained from tosylhydrazones, presumably because of relief of strain. This shortens the time of exposure of the carbanion to the solvent, thereby reducing the chance of its protonation before it can be trapped by an electrophilic reagent.

The trisylhydrazide can be prepared in nearly quantitative yield by treatment of a solution of 2,4,6-triisopropylbenzenesulfonylchloride in THF with hydrazine hydrate at 0 °C. The trisylhydrazide formed reacts rapidly with a wide variety of ketones (and aldehydes) in the absence of an acid catalyst to give the corresponding trisylhydrazones as mixtures of (E,Z) isomers, with the (E)-hydrazone usually being present in excess.[94]

no *ortho*-hydrogens

benzylic-H is hindered

trisylhydrazone

Alkenyllithiums generated from trisylhydrazones react with electrophiles such as primary alkyl bromides, aldehydes, ketones, dimethylformamide, CO_2, chloro-trimethylsilane, or 1,2-dibromoethane to furnish substituted alkenes, allylic alcohols, α,β-unsaturated aldehydes, α,β-unsaturated acids, alkenylsilanes, or alkenyl bromides, respectively, as exemplified below for the preparation of an allylic alcohol and an α,β-unsaturated aldehyde.[95]

90%; R = *i*-Pr, R'= *n*-C_6H_{13}

95%   60%

An important feature of tosyl- and trisylhydrazones as a source of alkenyllithium reagents is that generally deprotonation of the mono-anion is regioselective and occurs predominately at the less substituted α-position, following the order of acidity: $RCH_3 > R_2CH_2 > R_3CH$. Fragmentation of the dianion formed produces the less substituted vinyllithium. For the removal of a secondary α-hydrogen, two equivalents of the stronger base *sec*-butyllithium should be used.[96]

Alkenylithium reagents derived from symmetrical acyclic ketone hydrazones possess the *trans*-configuration, which is consistent with a *syn*-deprotonation of the hydrazone anion in a conformation that places the alkyl group R *anti* to the hydrazone moiety during dianion formation. The presence of an α-branch such as an isopropyl group, however, diminishes the stereoselectivity.

favored   major isomer

disfavored   minor isomer

Transposition of Double Bonds—The Claisen Rearrangement[97]

The *Claisen rearrangement*[98] and its variants are powerful tools for the homologation of allylic alcohols into stereodefined γ,δ-unsaturated aldehydes, ketones, esters, acids, and amides and play an important role as methods for acyclic stereocontrol. Because of the predictable stereo-, regio-, and enantioselectivity, and diversity of functionality, the Claisen rearrangement is often a crucial step in the synthesis of natural products. The prototype of the Claisen rearrangement involves a pericyclic, thermally induced [3,3]-sigmatropic rearrangement of an allyl vinyl ether or an allyl aryl ether to produce a γ,δ-unsaturated carbonyl compound or an *o*-allylphenol, respectively. Pericyclic reactions take place in a single step without intermediates and involve bond breaking and bond forming through a cyclic array of interacting orbitals.

With few exceptions, the rearrangement of acyclic systems occurs via a chair-like transition state where the two groups connected by the σ-bond, C(1)–O(1′), migrate to position 3 of the allyl group and to position 3′ of the vinyl ether (hence the term [3,3] shift) while minimizing 1,3-diaxial or pseudodiaxal interactions.[99] The conformation of the transition state determines the stereochemistry of the newly formed double bond. The strong preference for the formation of products with the (*E*)-configuration is a characteristic feature of Claisen rearrangements. In cyclic systems, because of conformational constraints of the chairlike transition state, the rearrangement may proceed via a boatlike transition state.

The Claisen rearrangement not only controls the stereochemistry of the newly formed double bond, but also the stereochemistry of any substituents at the newly formed single bond. The new stereocenters are formed in direct relation to the geometry of the starting allyl and vinyl ether double bonds, as shown below.[100]

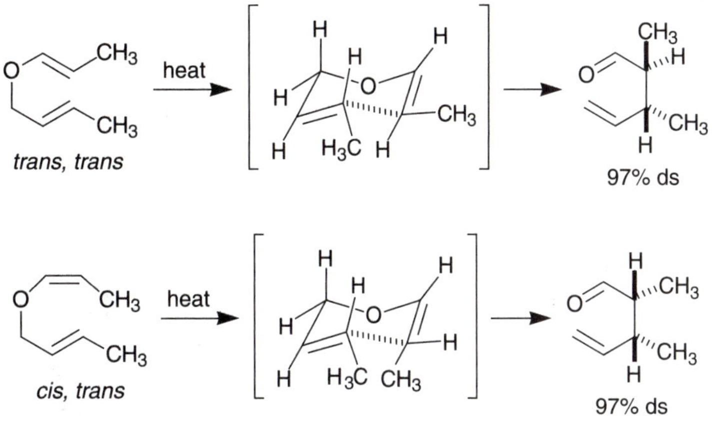

The stereochemical course of the rearrangement is controlled by suprafacial in-phase interactions of the participating molecular [$\pi^2s+\sigma^2s+\pi^2s$] orbitals and is consistent with that predicted by frontier molecular orbital theory[101] and orbital symmetry rules.[102]

The highly ordered transition state of [3,3]-sigmatropic Claisen rearrangements allows the chirality transfer of a preexisting chiral center along the allylic system to a new position with high enantioselectivity. In the course of the rearrangement, the original chiral center is destroyed (self-immolative process) and a new chiral center is created whose configurations is related to that of the starting material. An example of the simultaneous stereo- and enantioselective transposition of a double bond and a chiral center is shown below.[103] Either enantiomer of the unsaturated carbonyl compound may be obtained by changing the configuration of the double bond of the starting material.

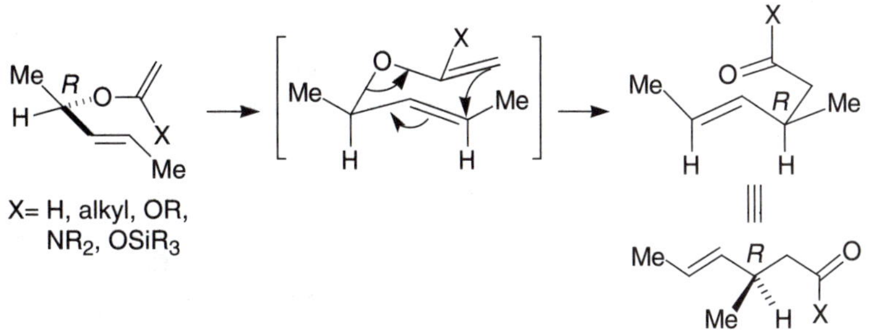

The versatility of the Claisen rearrangement for the synthesis of functionally substituted γ, δ-unsaturated carbonyl compounds has been greatly enhanced by the introduction of various vinyl ether appendages. These not only participate in the stereochemical control of the rearrangement, but also determine the nature of the functional group in the product (–CHO, –COR, –COOH, –COOR, –CONR$_2$).

The pyrolysis of allyl vinyl ethers derived by acid- or mercuric ion-catalyzed ether exchange of ethyl vinyl ether or substituted vinyl ethers with allylic alcohols produces the corresponding γ, δ-unsaturated aldehydes or ketones, respectively.[104]

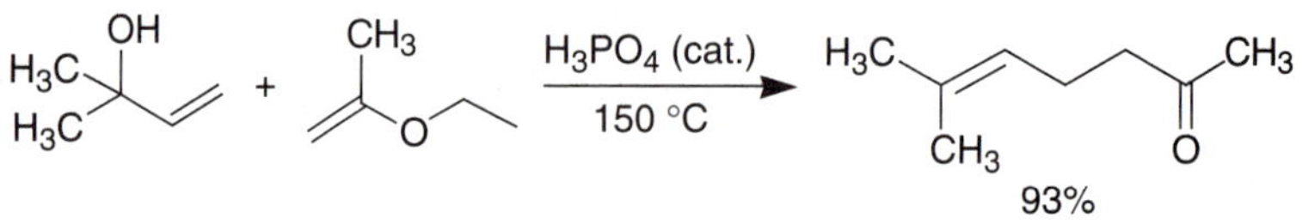

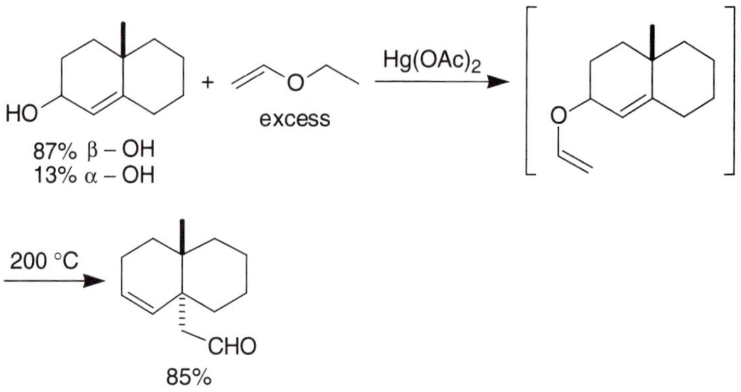

An alternative route to γ,δ-unsaturated ketones is via the *Carroll-Claisen rearrangement*, which uses allylic esters of β-keto acids (which exist mainly in the enol form) as substrates. These are readily prepared by condensation of allylic alcohols with acetoacetic esters or diketene. Following rearrangement, the intermediate β-keto acid undergoes in situ decarboxylation on heating.[105]

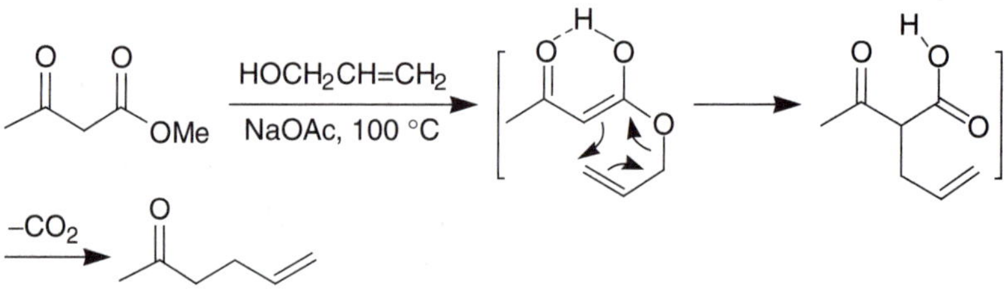

A modification of the Carroll reaction that enables the preparation of γ,δ-unsaturated ketones under milder conditions uses the dianion of allylic acetoacetates in a sequence of reactions depicted below.[106]

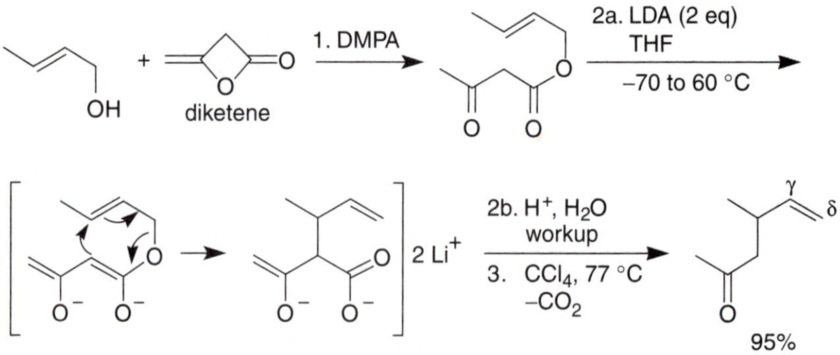

The *Johnson ortho-ester* variant of the Claisen rearrangement provides access to γ,δ-unsaturated esters.[107] The reaction entails heating the allylic alcohol with an *ortho*-ester in the presence of a carboxylic acid to form a ketene acetal, which then rearranges to the *trans*-unsaturated ester.[108] An elevated reaction temperature is necessary for the in situ formation of the ketene acetal but not for the rearrangement.

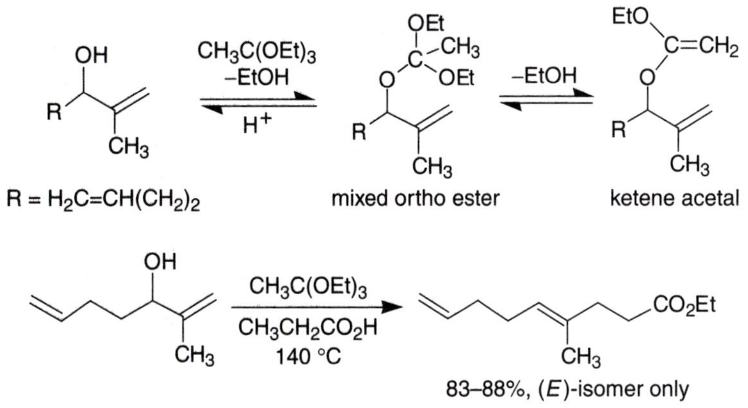

$R = H_2C=CH(CH_2)_2$ mixed ortho ester   ketene acetal

83–88%, (*E*)-isomer only

The *ortho*-ester Claisen rearrangement is also an effective tool for the transfer of chirality. As shown below, simply changing the stereochemistry of the double bond of the allyl vinyl ether while maintaining the configuration of the chiral center leads to the other enantiomer of the unsaturated ester.[109]

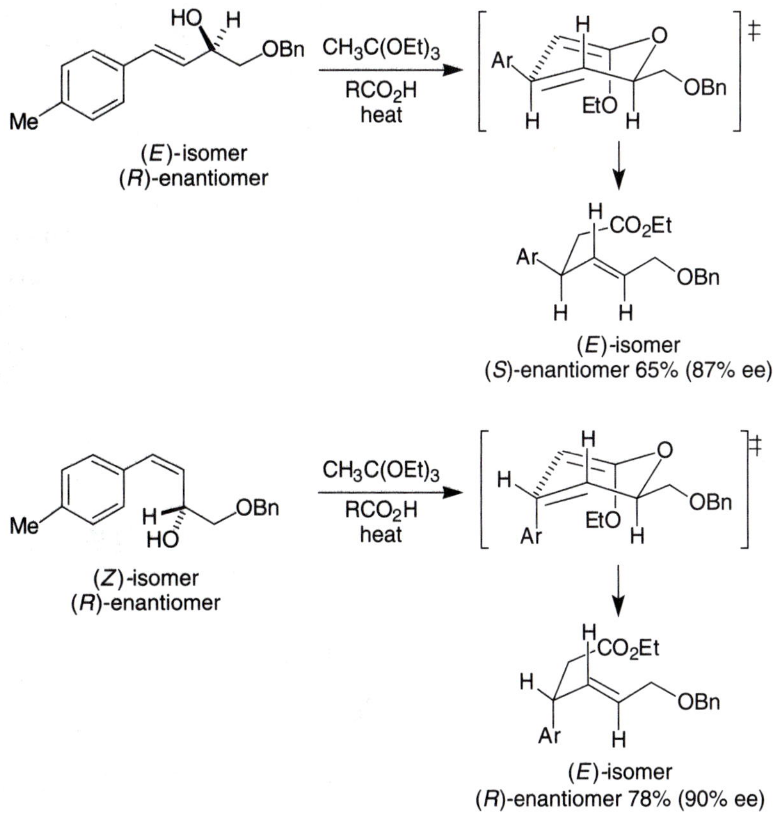

(*E*)-isomer
(*R*)-enantiomer

(*E*)-isomer
(*S*)-enantiomer 65% (87% ee)

(*Z*)-isomer
(*R*)-enantiomer

(*E*)-isomer
(*R*)-enantiomer 78% (90% ee)

Chirality may also be transferred from a side chain onto a ring system via a chair-like transition state.[110]

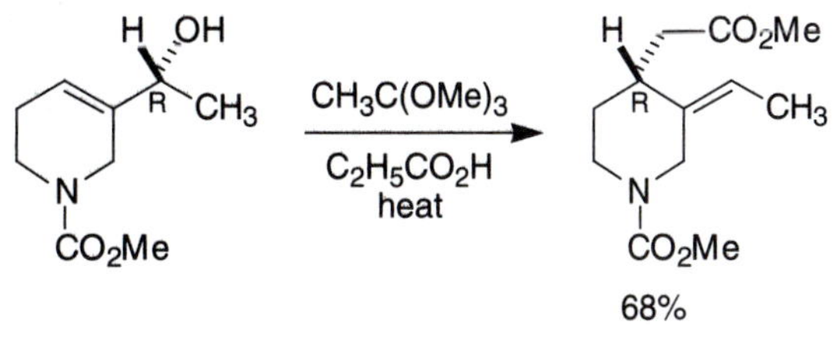

68%

Related to the *ortho*-ester rearrangement is the *Eschenmoser variation*, which involves exchange of *N,N*-dimethylacetamide dimethyl acetal with an allylic alcohol and in situ rearrangement of the intermediate ketene *N,O*-acetal to furnish the corresponding unsaturated amides. Again, transfer of stereochemistry from the allylic alcohol to the product is observed.[111]

The *Ireland modification* utilizes silyl ketene acetals derived from allyl ester enolates and provides a general method to effect stereocontrolled Claisen rearrangements under mild conditions, making it possible to apply the reaction to acid-sensitive and thermally labile substrates. Moreover, by proper choice of the reaction conditions, one can control the geometry of the enol ethers and hence the stereochemistry of the new C–C bond that is produced in the rearrangement step.[112]

Treatment of allylic esters with a strong, non-nucleophilic base such as LDA in THF produces preferentially the (*E*)-enolate, while the use of a mixture of THF and HMPA favors the (*Z*)-enolate. To avoid formation of unwanted aldol products and to preserve the integrity of the enolate double bond stereochemistry, the lithium enolates are converted to the corresponding stable silyl enol ethers by treatment with a trialkylchlorosilane (usually *tert*-butyldimethylchlorosilane). The sigmatropic rearrangement occurs at ambient or slightly higher temperature to furnish, after hydrolytic workup of the intermediate silyl esters, the *erythro*- or *threo*-γ,δ-unsaturated carboxylic acids, respectively (the reversal of configuration of the enolate and the ketene acetal is due to the higher priority of Si over Li). In acyclic systems, the Ireland-Claisen rearrangement exhibits the well-established preference for chairlike transition states.[113]

A remarkably facile tandem 1,4-conjugate addition Claisen rearrangement is observed by utilizing the copper enolate derived by addition of lithium dimethylcopper to 2-(allyloxy)-2-cyclohexenone. The resultant hydroxy ketone is obtained as a single stereoisomer in nearly quantitative yield.[114]

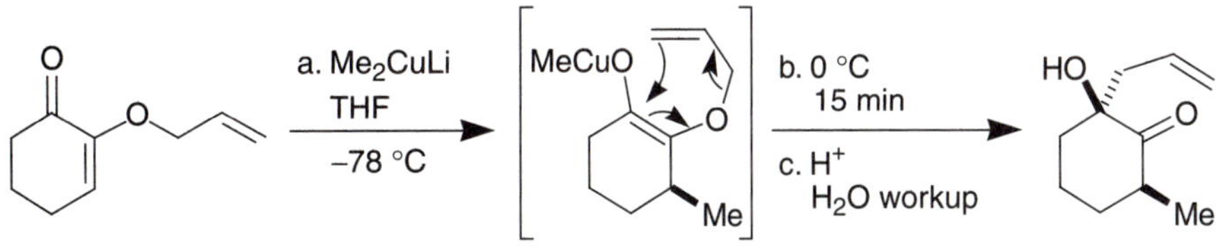

Yamamoto Asymmetric Claisen Rearrangement[115]

Several chiral organoaluminum Lewis acids catalyze the Claisen rearrangement of achiral allyl vinyl ethers to furnish chiral β, γ-unsaturated aldehydes with good enantioselectivity. Among the most effective catalysts is ATBN-F, a chiral aluminum tris(β-naphthoxide) species prepared from enantiomerically pure binapthol.[116]

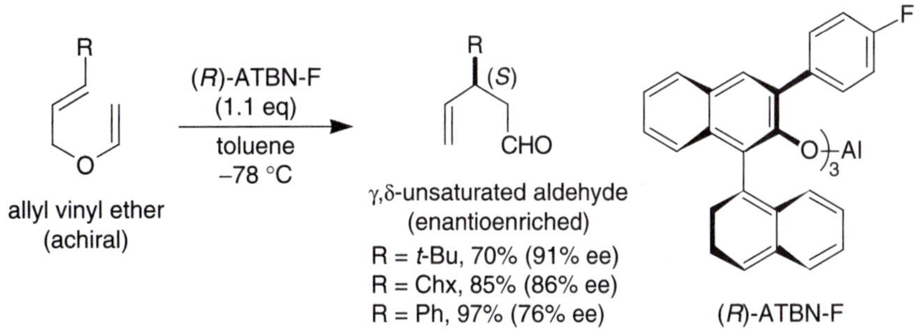

ATBN-F may selectively activate (bind) one of the two possible six-member chair-like transition states in which the (R)-substituent is equatorial.

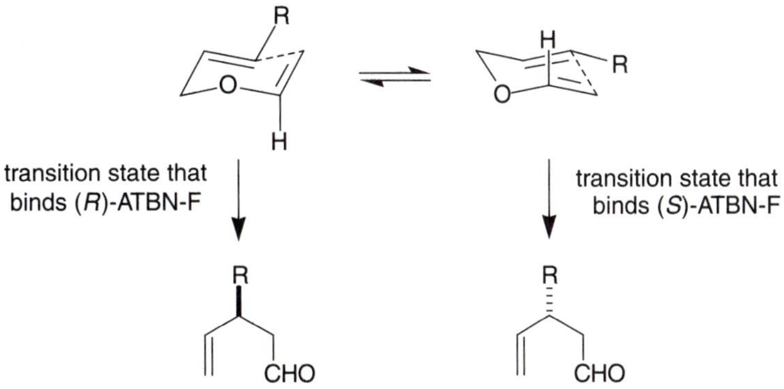

Allyl Vinyl Ether Synthesis

Allyl vinyl ethers may be prepared via stereospecific copper-catalyzed coupling of allyl alcohols and vinyl halides.[117] For example, the copper catalyst derived from the tetramethyl 1,10-phenanthroline ligand shown below facilitates C–O bond formation between (E)-vinyl iodides and allyl alcohols when the reaction mixture is heated in the presence of air.[118]

8.2 FORMATION OF CARBON-CARBON TRIPLE BONDS[119]

Alkynes play a pivotal role in organic chemistry. Although they are not as wide-spread in nature as alkenes, triple bonds are embedded in many natural products, drugs, and antitumor agents. From a synthetic viewpoint, the ethynyl moiety is an exceedingly versatile group because it is readily introduced onto a variety of organic substrates and provides a site for further modifications, as already demonstrated in the preceding chapters. A plethora of procedures exist for the preparation of acetylenes[120]; the following discussion will focus on some selected methods for their preparation.

Alkynes and functionally substituted alkynes may be synthesized either from starting materials that do not already contain the triple bond by elimination reactions, or the ethynyl and alkynyl groups may be introduced onto substrates by nucleophilic or electrophilic substitution reactions.

Elimination Reactions[121]

Triple bonds are generated when 1,1- or 1,2-dihaloalkanes or vinyl halides are subjected to dehydrohalogenation with an appropriate base. The requisite 1,1-dihaloalkanes are generally obtained by chlorination of aldehydes or ketones with PCl_5, by treatment of an alkyl or cycloalkyl chloride with vinyl chloride in the presence of anhydrous aluminum chloride ($RCH_2Cl + H_2C=CHCl + AlCl_3 \rightarrow RCH_2CH_2CHCl_2$) or by treatment of 1,2-dichloroalkanes by halogenation of alkenes with chlorine or bromine.

Base-induced dehydrohalogention of vicinal and geminal dihaloalkanes proceeds via the intermediacy of the vinyl halides and requires three equivalents of the base to generate the triple bond.

Elimination of the first HX from dihaloalkanes can be effected with weaker bases such as *tert*-amines or *t*-BuOK, whereas elimination of the second HX usually requires stronger bases, such as $NaNH_2$ or lithium dialkylamides.[119g,122]

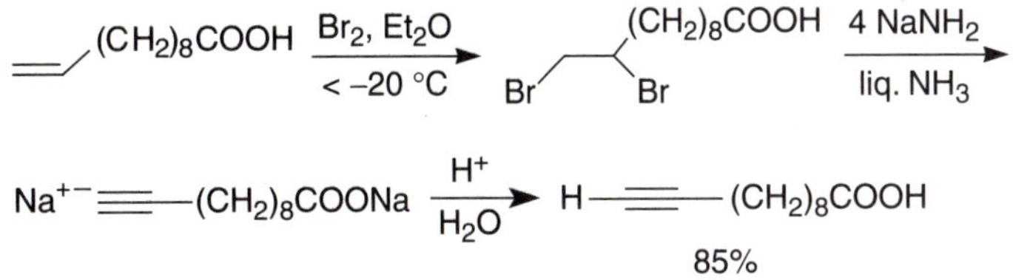

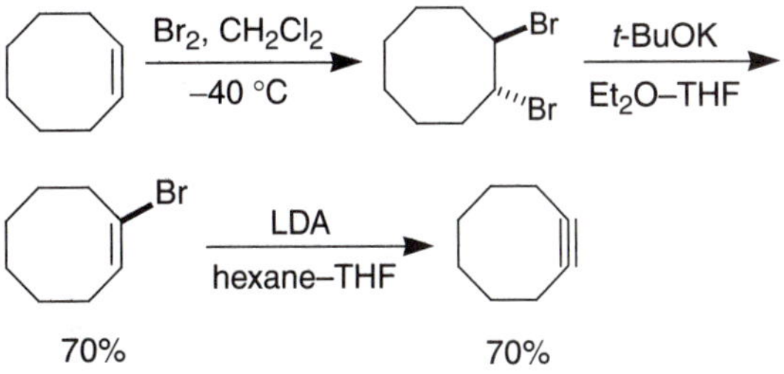

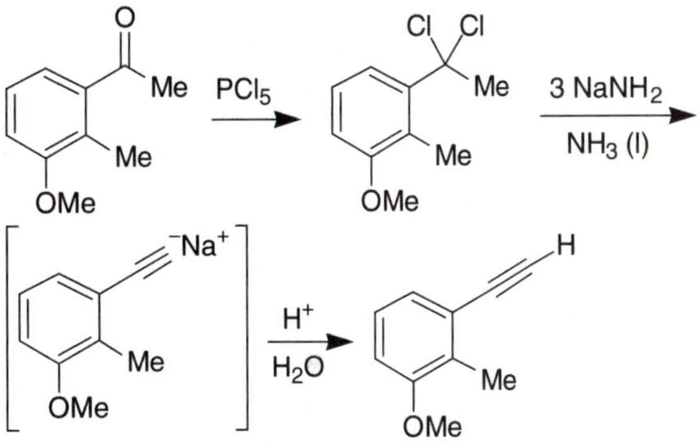

The choice of base for the elimination reaction depends on the structure of the substrate and the acidity of the proton being removed. For example, dehydrohalogenation of the vicinal dibromo carboxylic acid shown below occurs readily with alcoholic potassium hydroxide.[123]

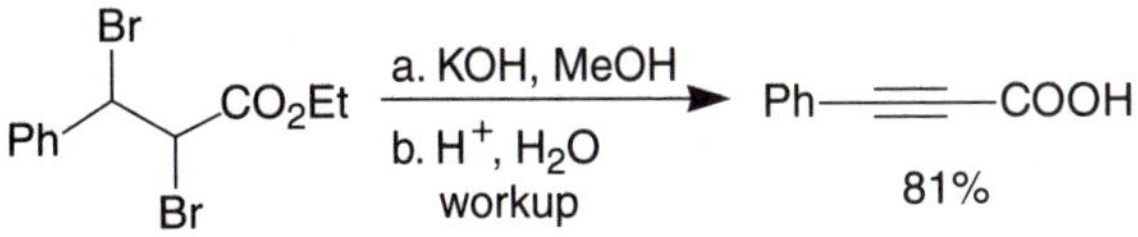

Bromomethylenation of aldehydes with dibromomethylenphosphorane, a Wittig-type reaction, followed by dehydrobromination of the resultant *geminal* dibromoalkene is an expeditious method for preparing mono- and disubstituted acetylenes. The overall reaction represents a chain extension of an aldehyde to an acetylene with one additional carbon.[124] The phosphorane is generated by addition of Ph_3P (4 eq) to CBr_4 (2 eq) or by adding zinc dust (2 eq) and Ph_3P (2 eq) to CBr_4 (2 eq). The latter procedure is preferable since less Ph_3P is required, which facilitates isolation of the product.

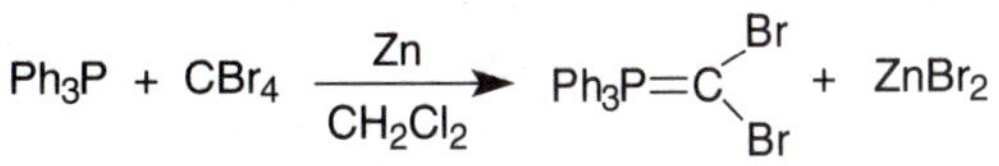

Addition of the appropriate aldehyde to the dibromomethylenphosphorane prepared in situ furnishes the dibromoalkene, which, on treatment with two equivalents of *n*-BuLi, undergoes dehydrobromination to produce the corresponding lithium alkynylide. Its protonation furnishes the 1-alkyne. An attractive feature of this reaction sequence is that the intermediate lithium alkynylide formed is itself an important precursor for the formation of carbon-carbon bonds (e.g., by reaction with electrophiles).

Another route to terminal alkynes having one carbon more than the aldehyde precursor is via alkylation of primary alkyl bromides with dichloromethyllithium in the presence of HMPA to yield the 1,1-dichloroalkane. Subsequent dehydrochlorination with three equivalents of *n*-BuLi followed by hydrolytic workup affords the corresponding terminal alkyne.[125]

A mild and efficient synthesis of terminal alkynes starts with readily accessible methyl ketones and converts them to the corresponding enolates with LDA. The enolates produced are trapped with diethyl chlorophosphate to give enol phosphates, which possess a good leaving group for elimination. Subsequent treatment of the enol phosphates with LDA furnishes the corresponding lithium alkynylides and on protonation of these, the corresponding terminal acetylenes.[126]

85% overall

Alkylation[122,127]   The most utilized method for alkylation of alkynes is via alkynylide anions. Their reaction with electrophilic reagents provides access to both terminal and internal alkynes as well as to functionally substituted alkynes. The higher electronegativity of carbon in the sp-hybridization state imparts relatively greater acidity to acetylene and 1-alkynes (pK_a 24–26), so that bases such as alkyllithiums, lithium dialkylamides, sodium amide in liquid ammonia, and ethyl magnesium bromide[128,129] can be used to generate the carbanions.

Lithium acetylide is obtained commercially as a complex with ethylenediamine [HC≡CLi • EDA].[130] However, the EDA-complexed lithium acetylide is less reactive than uncomplexed lithium acetylide, which is readily prepared by Midland's procedure.[131]

To avoid addition of alkyllithiums to functionally substituted alkynes, their metalation is best carried out using LDA.[132]

Displacement Reactions

Alkylation of lithium acetylide[133] or the lithium acetylide·EDA complex with primary alkyl bromides affords monosubstituted alkynes. The reaction of the lithium acetylide·EDA complex with α,ω-dibromoalkanes produces terminal diynes.[134]

Lithium or sodium alkynylides derived from 1-alkynes react with primary alkyl halides in the presence of HMPA to give disubstituted alkynes.[135] This method is limited to primary alkyl halides that are not branched at the β-position. Secondary and tertiary alkyl halides tend to undergo dehydrohalogenation (E2-elimination).

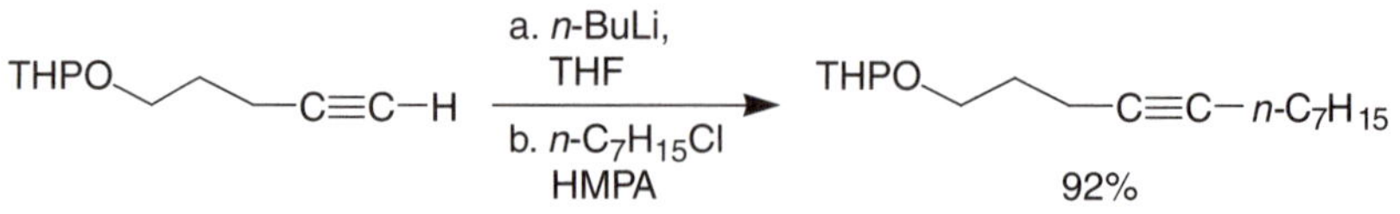

THPO$\diagdown$$\diagup$$\diagdown$$-$C≡C$-$H $\xrightarrow[\text{b. } n\text{-C}_7\text{H}_{15}\text{Cl} \atop \text{HMPA}]{\text{a. } n\text{-BuLi,} \atop \text{THF}}$ THPO$\diagdown$$\diagup$$\diagdown$$-$C≡C$-n$-C$_7H_{15}$

92%

Disubstituted acetylenes having a secondary and a primary alkyl substituent at the triple bond carbons are accessible by proper choice of the starting materials, as exemplified below.

Favorable synthesis:

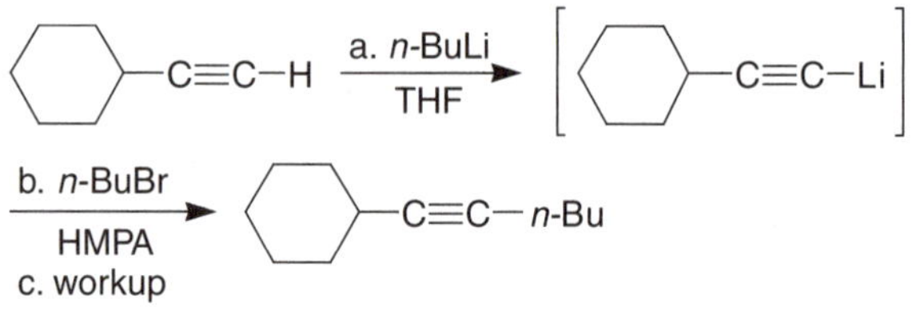

⬡$-$C≡C$-$H $\xrightarrow[\text{THF}]{\text{a. } n\text{-BuLi}}$ $\left[⬡-\text{C≡C}-\text{Li} \right]$

$\xrightarrow[\text{c. workup}]{\text{b. } n\text{-BuBr} \atop \text{HMPA}}$ ⬡$-$C≡C$-n$-Bu

Unfavorable synthesis:

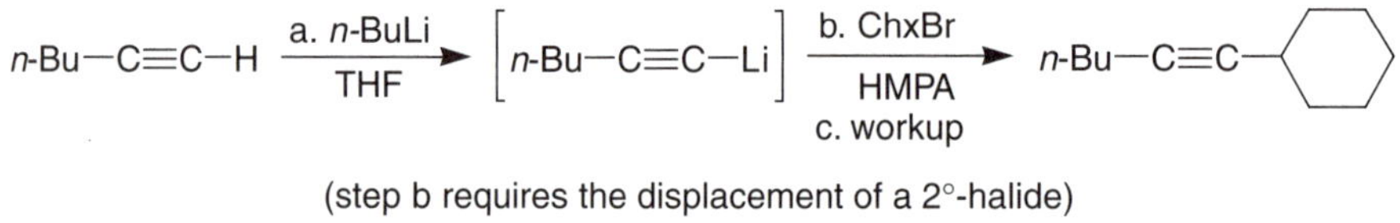

n-Bu$-$C≡C$-$H $\xrightarrow[\text{THF}]{\text{a. } n\text{-BuLi}}$ $\left[n\text{-Bu}-\text{C≡C}-\text{Li} \right]$ $\xrightarrow[\text{c. workup}]{\text{b. ChxBr} \atop \text{HMPA}}$ n-Bu$-$C≡C$-$⬡

(step b requires the displacement of a 2°-halide)

No general method is available for *tert-alkyl-alkynyl*-coupling based on sodium or lithium alkynylide reagents because of prevalent elimination. However, trialkynylalanes obtained from the corresponding alkynyllithiums and anhydrous aluminum trichloride undergo clean reaction with *tert*-halides to produce *tert*-alkyl substituted acetylenes.[136]

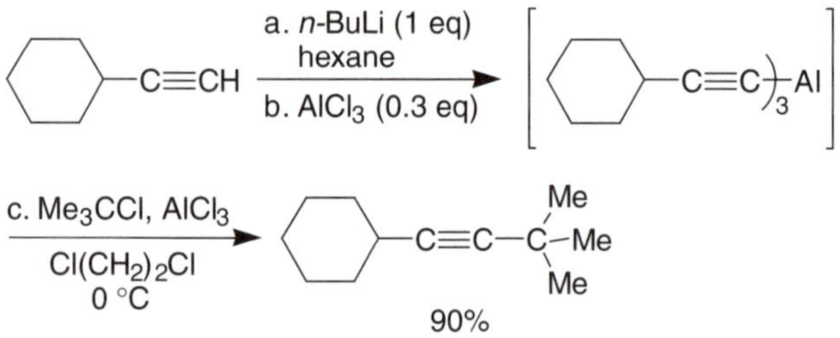

⬡$-$C≡CH $\xrightarrow[\text{b. AlCl}_3 \text{ (0.3 eq)}]{\text{a. } n\text{-BuLi (1 eq)} \atop \text{hexane}}$ $\left[⬡-\text{C≡C} \right]_3 \text{Al}$

$\xrightarrow[\substack{\text{Cl(CH}_2)_2\text{Cl} \\ 0\ °\text{C}}]{\text{c. Me}_3\text{CCl, AlCl}_3}$ ⬡$-$C≡C$-$C$\begin{smallmatrix}\text{Me}\\|\\\text{C}\\|\\\text{Me}\end{smallmatrix}$Me

90%

Carbonyl Addition

α-Hydroxysubstituted alkynes are important intermediates for synthetic transformations and are often used as starting materials for the synthesis of medicinally valuable drugs. They are generally prepared by treatment of 1-lithio-1-alkynes with the appropriate aldehydes or ketones.[122g] Since the carbonyl carbon is more electrophilic than the carbon-halogen bond in alkyl halides, alkynyl-carbonyl coupling proceeds with greater ease. Addition of lithium alkynylides to ketones with enolizable hydrogens results in competition between deprotonation to form the enolates and addition to the carbonyl carbon, leading to low yields of propargylic alcohols. However, addition of LiBr markedly increases the yield of propargylic alcohols.[137]

Acylation of metallated 1-alkynes with dimethylformamide or ethyl chloroformate leads to conjugated acetylenic aldehydes or esters, respectively.[119g] These are important substrates for conjugate addition reactions.

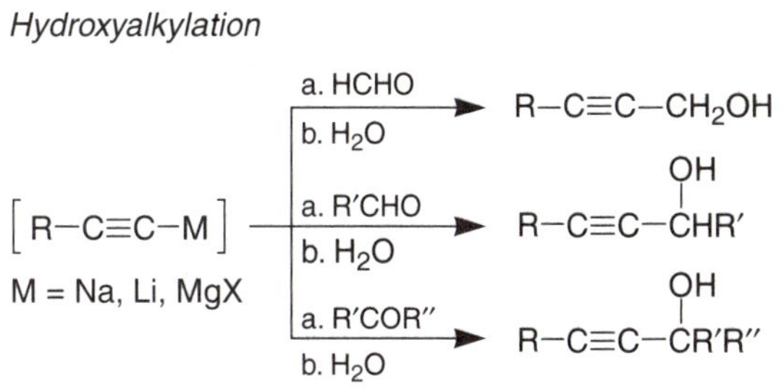

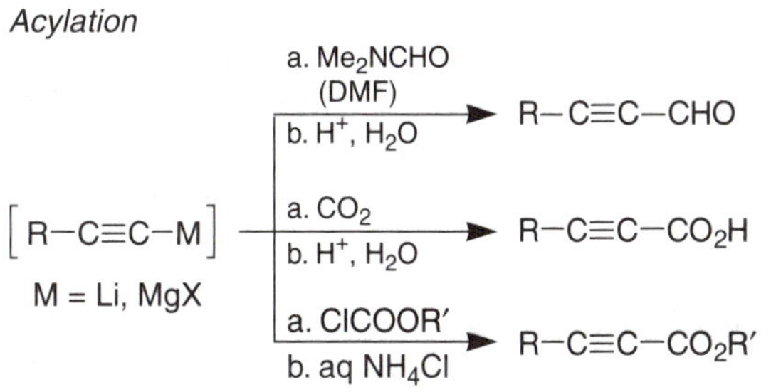

Hydroxyalkylation

$$\left[R-C\equiv C-M \right]$$

M = Na, Li, MgX

a. HCHO
b. H_2O → $R-C\equiv C-CH_2OH$

a. R′CHO
b. H_2O → $R-C\equiv C-\overset{\overset{\displaystyle OH}{|}}{C}HR'$

a. R′COR″
b. H_2O → $R-C\equiv C-\overset{\overset{\displaystyle OH}{|}}{C}R'R''$

Acylation

$$\left[R-C\equiv C-M \right]$$

M = Li, MgX

a. Me_2NCHO
(DMF)
b. H^+, H_2O → $R-C\equiv C-CHO$

a. CO_2
b. H^+, H_2O → $R-C\equiv C-CO_2H$

a. ClCOOR′
b. aq NH_4Cl → $R-C\equiv C-CO_2R'$

Propargylic Alkylation

Tandem chain extension at both the 1-alkyne and propargylic positions of 1-alkynes by sequential (1) alkylation and (2) hydroxy-alkylation (1,2-addition) of 1,3-dilithiated alkynes provides an attractive, one-pot preparation of functionally disubstituted alkynes. The required 1,3-dilithiated species are obtained by lithiation of 1-alkynes at both the 1- and 3- positions with two equivalents of *n*-BuLi in the presence of TMEDA (tetramethylethylenediamine). The two-stage chain extension depicted below for propyne involves initial alkylation of the dilithiated propyne with the less reactive electrophile, *n*-BuBr, at the more nucleophilic propargylic position. Subsequent hydroxyalkylation at the less nucleophilic alkynylide position with the more electrophilic formaldehyde furnishes the α-hydroxyalkyne.[138]

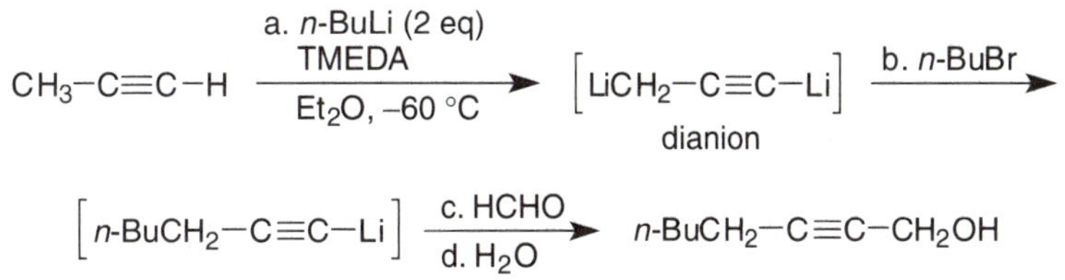

$$CH_3-C\equiv C-H \xrightarrow[\text{Et}_2\text{O, }-60\text{ °C}]{\substack{\text{a. }n\text{-BuLi (2 eq)}\\ \text{TMEDA}}} \left[LiCH_2-C\equiv C-Li \right] \xrightarrow{\text{b. }n\text{-BuBr}}$$

dianion

$$\left[n\text{-BuCH}_2-C\equiv C-Li \right] \xrightarrow[\text{d. }H_2O]{\text{c. HCHO}} n\text{-BuCH}_2-C\equiv C-CH_2OH$$

In view of the differential reactivities of the 1,3-dilithiated positions of 1-alkynes, it is possible to selectively alkylate the more reactive propargylic position.[139]

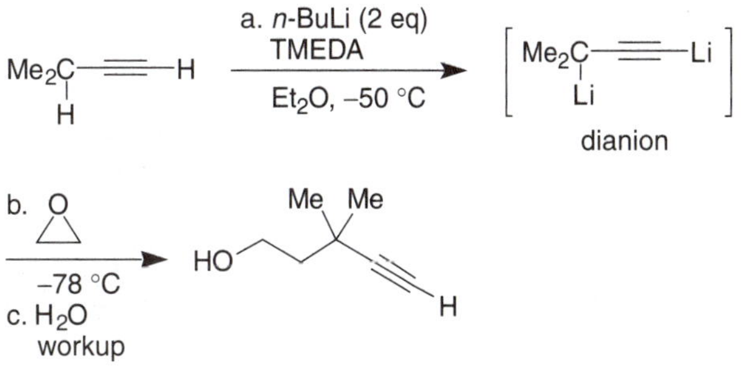

$$Me_2\overset{\overset{\displaystyle }{|}}{\underset{\underset{\displaystyle H}{|}}{C}}{=}{=}H \xrightarrow[\text{Et}_2\text{O, }-50\text{ °C}]{\substack{\text{a. }n\text{-BuLi (2 eq)}\\ \text{TMEDA}}} \left[Me_2\overset{}{\underset{\underset{\displaystyle Li}{|}}{C}}{=}{=}Li \right]$$

dianion

b. O
△
$\xrightarrow[\substack{\text{c. }H_2O\\ \text{workup}}]{-78\text{ °C}}$ HO—\\//—H (Me, Me)

Fuchs Alkynylation

Even unactivated hydrocarbons can be alkynylated to furnish disubstituted alkynes.[140] For example, treatment of cyclohexane with the acetylenic triflone ($RC\equiv C-SO_2CF_3$)

shown below in the presence of a catalytic amount of AIBN in THF leads, via a radical intermediate, to the 1-phenyl-2-cyclohexyl acetylene.

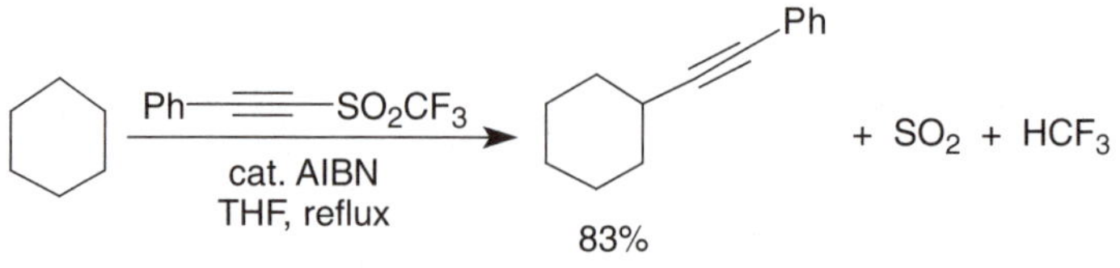

$$ \text{cyclohexane} \xrightarrow[\substack{\text{cat. AIBN} \\ \text{THF, reflux}}]{Ph-\!\!\!\equiv\!\!\!-SO_2CF_3} \text{cyclohexyl-C≡C-Ph} \quad + \text{SO}_2 + \text{HCF}_3 $$

83%

Isomerization

In the presence of a base, the triple bond of certain terminal and internal alkynes can be shifted to a new position along the chain, providing a valuable tool for interconverting acetylenic compounds. Thus, straight chain 1-alkynes, on treatment with t-BuOK in DMSO at room temperature or above, isomerize cleanly to the thermodynamically more stable 2-alkynes.[141]

$$ n\text{-C}_5\text{H}_{11}\text{CH}_2\text{—C}\!\!\equiv\!\!\text{C—H} \xrightarrow[\text{25 to 80 °C}]{t\text{-BuOK, DMSO}} n\text{-C}_5\text{H}_{11}\text{—C}\!\!\equiv\!\!\text{C—CH}_3 $$

90%

The *acetylene zipper reaction* is the shifting of a triple bond in any position of a straight chain internal alkyne to the terminus of the chain in the presence of a very strong base. This multipositional isomerization of the triple bond is especially interesting when applied to acetylenic alcohols as a route to long chain structures with chemically differentiated functionality.

Potassium 3-aminopropylamide (KAPA), prepared by adding potassium hydride[142] to an excess of 3-aminopropylamine (APA), causes the rapid migration of triple bonds from internal positions to the terminus of the carbon chain at 0 °C within minutes to from acetylide anions. The migration of the triple bond to the terminal position is blocked by an alkyl branch. The corresponding terminal acetylenes are obtained on hydrolytic workup.[143]

$$ \text{KH} + \text{H}_2\text{N(CH}_2)_3\text{NH}_2 \longrightarrow \left[\text{KHN(CH}_2)_3\text{NH}_2\right] \text{ in APA} $$

APA (excess)   KAPA

$$ \text{CH}_3\text{(CH}_2)_5\text{—C}\!\!\equiv\!\!\text{C—(CH}_2)_5\text{CH}_3 \xrightarrow[\text{b. H}_2\text{O workup}]{\text{a. KAPA, APA}} n\text{-C}_{12}\text{H}_{25}\text{—C}\!\!\equiv\!\!\text{C—H} $$

89%

$$ \text{CH}_3\text{(CH}_2)_7\text{—C}\!\!\equiv\!\!\text{C—(CH}_2)_6\text{—OH} \xrightarrow[\text{b. H}_2\text{O workup}]{\text{a. KAPA, APA}} \text{H—C}\!\!\equiv\!\!\text{C—(CH}_2)_{14}\text{—OH} $$

90%
(< 1% internal alkyne isomers)

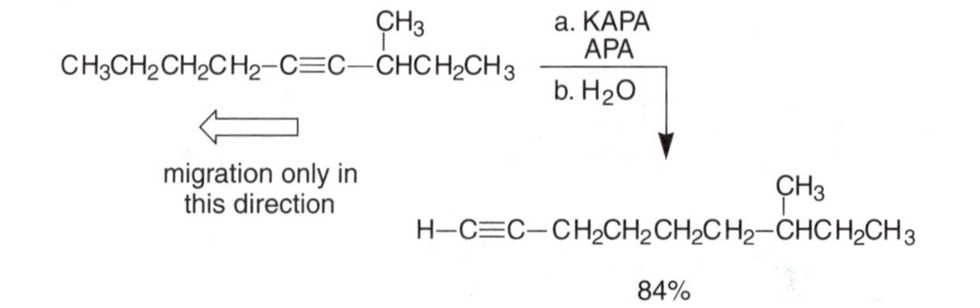

$$ \underset{\text{migration only in this direction}}{\underset{\Longleftarrow}{\text{CH}_3\text{CH}_2\text{CH}_2\text{CH}_2\text{—C}\!\!\equiv\!\!\text{C—}\overset{\overset{\text{CH}_3}{|}}{\text{CH}}\text{CH}_2\text{CH}_3}} \xrightarrow[\text{b. H}_2\text{O}]{\substack{\text{a. KAPA} \\ \text{APA}}} $$

$$ \text{H—C}\!\!\equiv\!\!\text{C—CH}_2\text{CH}_2\text{CH}_2\text{CH}_2\text{—}\overset{\overset{\text{CH}_3}{|}}{\text{CH}}\text{CH}_2\text{CH}_3 $$

84%

The following example depicts a chain elongation of 2-propyne-1-ol. The resultant internal triple bond is then shifted to the terminal position using the KAPA reagent.[144]

$$H-C\equiv C-CH_2OH \xrightarrow[\text{liq. NH}_3]{\text{LiNH}_2 \text{ (2 eq)}} \left[Li-C\equiv C-CH_2OLi \right] \xrightarrow{n\text{-C}_5\text{H}_{11}\text{Br}}$$

$$\underset{85\%}{n\text{-C}_5\text{H}_{11}-C\equiv C-CH_2OH} \xrightarrow[\text{liq. NH}_3]{\text{KAPA}} \underset{90\%}{H-C\equiv C-(CH_2)_6OH}$$

A modification of the "acetylene zipper" method employs lithium 3-amino-propylamide and circumvents the use of potassium hydride. The reagent is prepared by adding lithium metal to APA followed by addition of t-BuOK.[145]

$$\underset{}{HOCH_2-C\equiv C-(CH_2)_6CH_3} \xrightarrow[\text{b. H}_2\text{O}]{\substack{\text{a. LiHN(CH}_2)_3\text{NH}_2 \\ t\text{-BuOK, APA}}} \underset{83-88\%}{HO(CH_2)_8-C\equiv C-H}$$

The isomerization of a thermodynamically more stable internal triple bond to a thermodynamically less stable terminal triple bond may proceed via a sequence of deprotonation and proton donation steps involving allenic anions and/or allene intermediates, eventually leading to the more stable acetylide anion.[146]

Base-induced isomerization of terminal allenes provides convenient access to branched terminal alkynes that are not readily accessible by displacement of 2°-halides.[147]

One-Step, One-Carbon Homologation—Gilbert's Reagent

The elaboration of aldehydes to terminal alkynes in one step may be accomplished by one-carbon homologation using methyl (diazomethyl)phosphonate (Gilbert's reagent).[148] The mechanism of alkyne formation may proceed via a rapid Wittig-type reaction (e.g., a Horner-Wadsworth-Emmons olefination) to form a diazoethene intermediate, which undergoes a 1,2-shift of hydrogen and concomitant loss of nitrogen to furnish 1-alkynes.

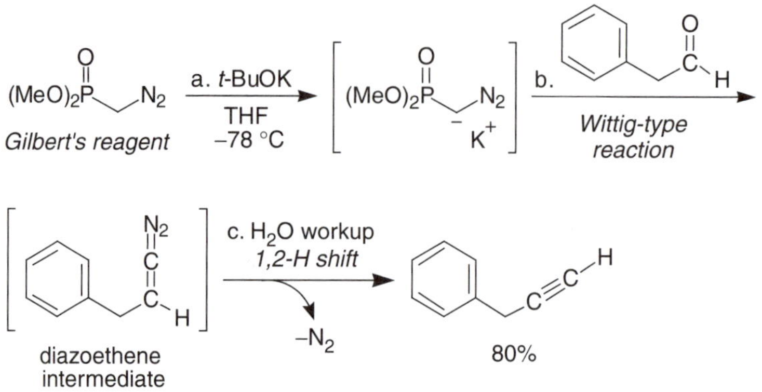

The mild (low temperature) conditions used makes this approach the method of choice for the conversion of enolizable aldehydes to chain-extended 1-alkynes. It is especially useful in cases where the aldehyde α-carbon is an enolisable stereocenter, as exemplified below.[149]

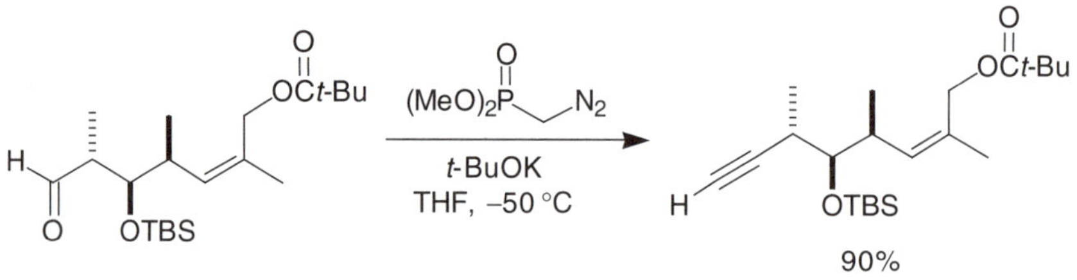

PROBLEMS

1. **Reagents**. Give the structure of the major product expected from each of the following reactions. Be sure to indicate product stereochemistry.

a.

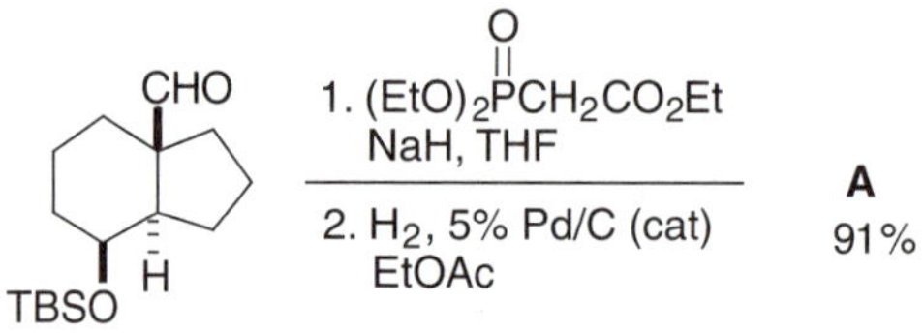

b.

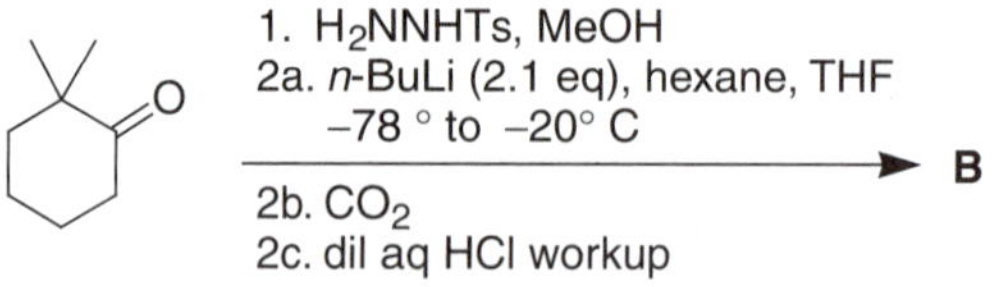

c.

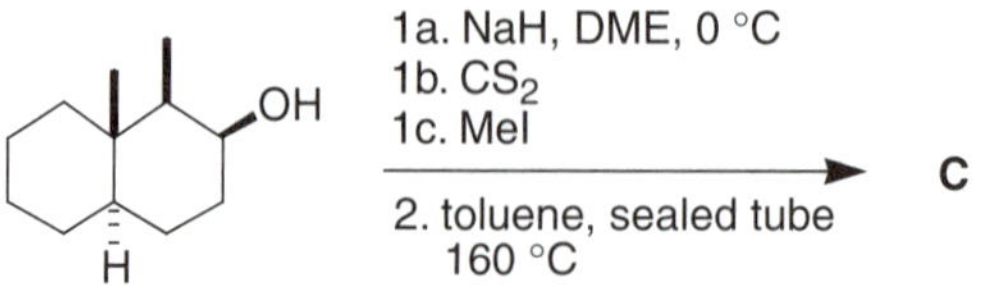

d.

1. TBSCl, imidazole, DMF
2a. TMSCH$_2$MgBr, Et$_2$O
2b. HO$_2$C-CO$_2$H, H$_2$O

3. *n*-Bu$_4$NF, THF

→ **D**

e.

1. PPh$_3$, CBr$_4$, Et$_3$N, CH$_2$Cl$_2$, –78 °C to rt
2a. DIBAL-H (3 eq), toluene, –78 °C
2b. Rochelle's salt workup

3. *n*-Bu$_4$NF, THF, rt
4a. *n*-BuLi (5 eq), hexane, –78 °C to rt
4b. aq HCl

→ **E**
71%

f.

1a. NaH (1.1 eq), DMF, 0 °C
1b. PMB-Cl
2. PCC, NaOAc, Celite, CH$_2$Cl$_2$, rt

3. Ph$_3$P=CHCO$_2$Et, benzene, reflux
4a. DIBAL-H (2.1 eq), CH$_2$Cl$_2$
4b. Rochelle's salt workup

→ **F**
52%

g.

1a. LDA (1.05 eq), THF, –78 °C
1b.

1c. 1N HCl, acetone, rt
2. NaH, DME

→ **G**
74%

h.

MeC(OMe)$_3$, CH$_3$CH$_2$CO$_2$H

140 °C

→ **H**
65%

i.

1. KCN, DMSO

2a. DIBAL-H (1.2 eq), CH$_2$Cl$_2$ –78 °C to rt
2b. Rochelle's salt workup

→ **I1**
87%

3. (MeO)$_2$PCH$_2$CO$_2$Me
THF, –78 °C to rt

→ **I2**
83%

j.

1a. [PhSCH$_2$Li], THF, 0 °C
1b. dil aq HCl workup

2. *m*CPBA (2.2 eq), CH$_2$Cl$_2$, 0 °C
3. POCl$_3$, pyridine, rt

→ **J1**

4a. *n*-BuLi (1.05 eq), THF, –78 °C
4b. PhCHO
4c. Ac$_2$O
5. Na(Hg), MeOH, EtOAc

→ **J2**
67%

k.

1.

K$_2$CO$_3$, acetone
2. N,N-dimethylaniline heat

→ **K1**
86%

3. BnBr, K$_2$CO$_3$, acetone
4a. 2N NaOH, THF
4b. aq HCl
5. aq NaHCO$_3$, Br$_2$, CHCl$_3$

→ **K2**
78%

l.

1a. PhMgCl (excess), THF, –78 °C
1b. sat'd. NH$_4$Cl quench at –78 °C

2. Ph$_2$S[OC(CF$_3$)$_2$Ph]$_2$ (5 eq) CH$_2$Cl$_2$

→ **L**
86%

2. **Selectivity.** Show the reaction intermediates obtained after each step and the major product obtained as well as provide the reagents and conditions when not given to accomplish the following transformations.

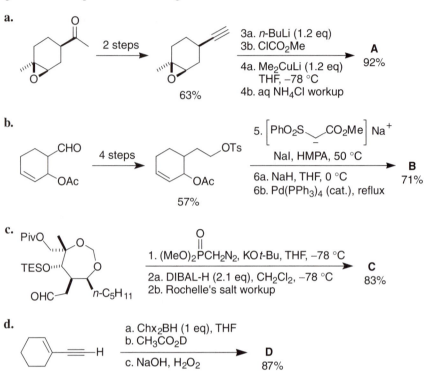

a.

2 steps

63%

3a. *n*-BuLi (1.2 eq)
3b. ClCO₂Me

4a. Me₂CuLi (1.2 eq)
THF, −78 °C
4b. aq NH₄Cl workup

A
92%

b.

4 steps

57%

5. [PhO₂S⌇CO₂Me] Na⁺

NaI, HMPA, 50 °C

6a. NaH, THF, 0 °C
6b. Pd(PPh₃)₄ (cat.), reflux

B
71%

c.

1. (MeO)₂PCH₂N₂, KO*t*-Bu, THF, −78 °C

2a. DIBAL-H (2.1 eq), CH₂Cl₂, −78 °C
2b. Rochelle's salt workup

C
83%

d.

a. Chx₂BH (1 eq), THF
b. CH₃CO₂D
c. NaOH, H₂O₂

D
87%

3. **Stereochemistry.** Predict the stereochemistry of the major product formed for each of the following reactions. Give an explanation for your choice.

a.

n-C₅H₁₁ ≡ CH(OEt)₂

1. NaBH₄ (excess), NiCl₂, MeOH
2. TsOH, acetone, H₂O
3. (EtO)₂P(O)CH₂CO₂Et, NaH, THF

A
46%
(88:12 mixture)

*b.

R = (CH₂)₃OBn

1.

DMAP (cat.)
THF, rt

2. Al₂O₃, 60 °C

B1 + **B2** (94 : 4 mixture of isomers)
72%

c.

1a. NaH (1.0 eq), THF, 0 °C
1b. TBSCl

2a. Ph₃P, DEAD, PhCO₂H, THF
2b. LiOH, MeOH, H₂O

C1
93%

3. ⌇OEt

Hg(OAc)₂
4. 200 °C

C2
89%

d.

1. PhSeCl, CH₂Cl₂, −78 °C

2. H₂O₂, THF

D

(Hint: phenylselenolactonization)

4. **Reactivity**. Explain the regioselectivity and stereochemistry observed in each of the following transformations.

a.

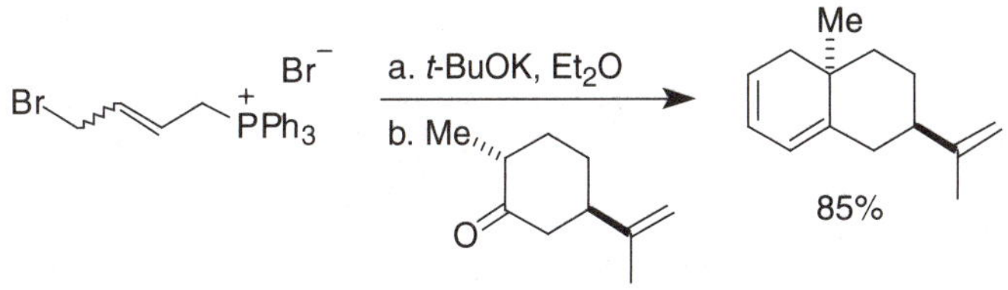
Br
Br⁻
+
PPh3
a. t-BuOK, Et2O
b. Me,,,
O
Me
85%

b.

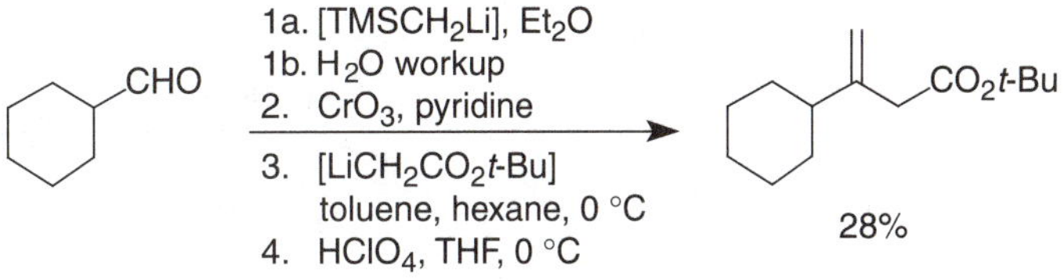
CHO
1a. [TMSCH2Li], Et2O
1b. H2O workup
2. CrO3, pyridine
3. [LiCH2CO2t-Bu]
toluene, hexane, 0 °C
4. HClO4, THF, 0 °C
CO2t-Bu
28%

***c.**

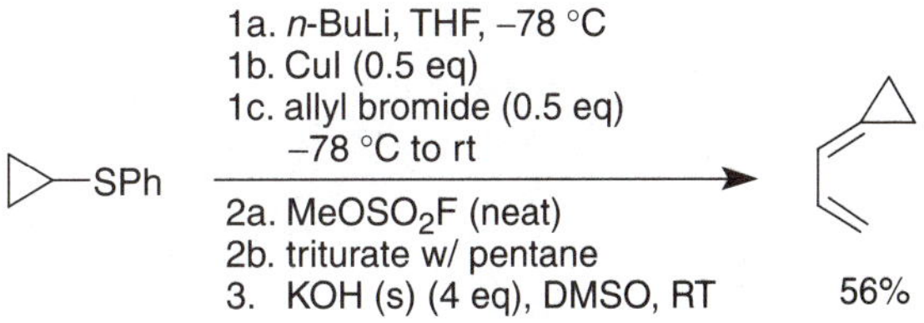
SPh
1a. n-BuLi, THF, –78 °C
1b. CuI (0.5 eq)
1c. allyl bromide (0.5 eq)
–78 °C to rt
2a. MeOSO2F (neat)
2b. triturate w/ pentane
3. KOH (s) (4 eq), DMSO, RT
56%

5. **Synthesis.** Supply the missing reagents required to accomplish each of the following syntheses. Be sure to control the relative stereochemistry.

a.

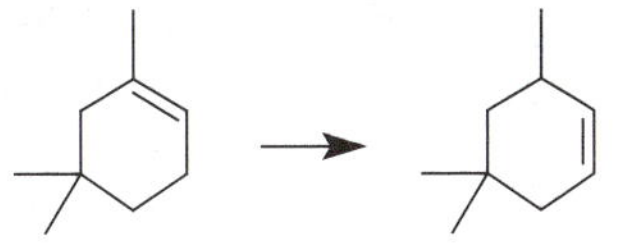

b.

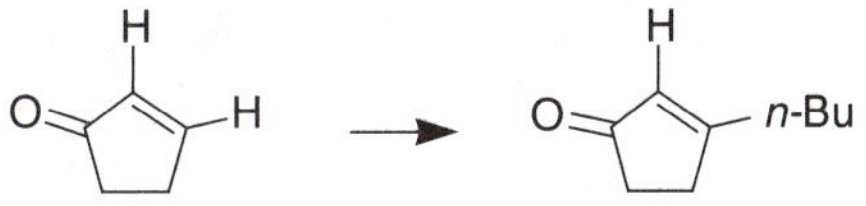
H
O
H
H
O
n-Bu

c.

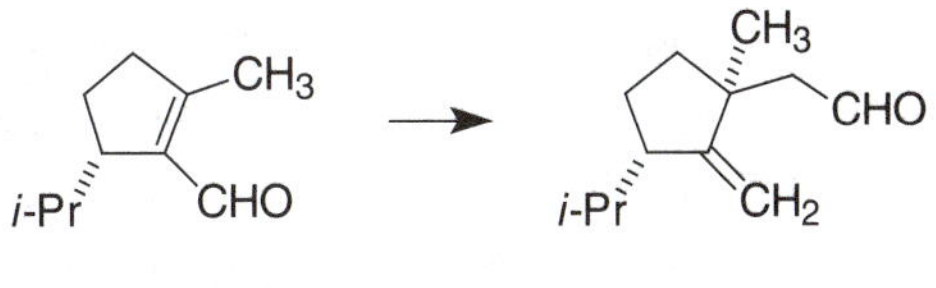
CH3
i-Pr
CHO
CH3
CHO
i-Pr
CH2

d.

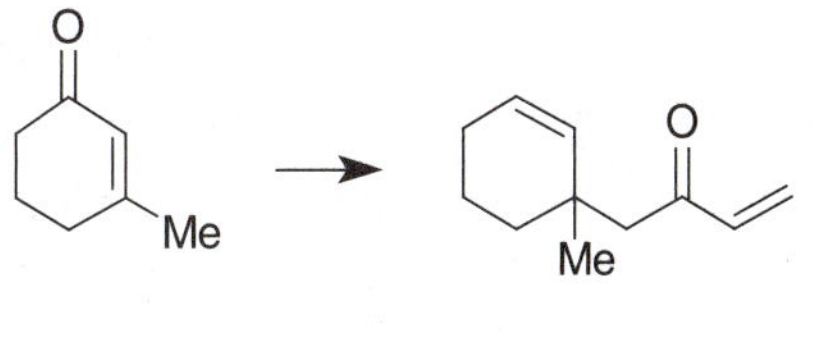
O
Me
O
Me

e.

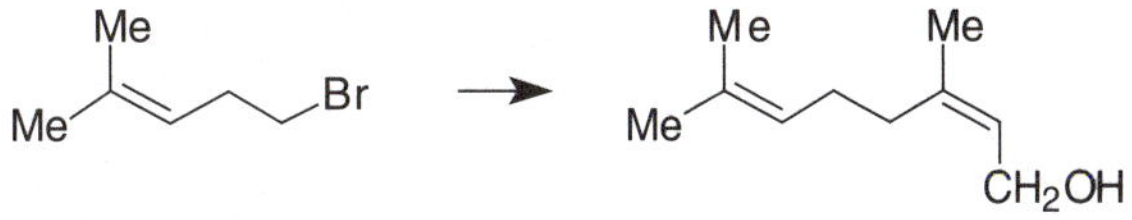
Me
Me
Br
Me
Me
Me
CH2OH

f.

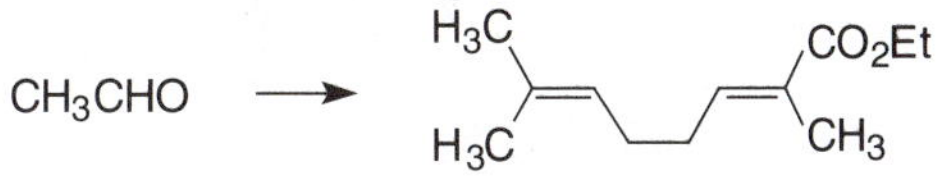
CH3CHO
H3C
H3C
CO2Et
CH3

g.

h.

i.

REFERENCES

1. (a) Whitham, G. H. In *Comprehensive Organic Chemistry*, Barton, D., Ollis, W. D., Eds., Pergamon Press: Oxford, UK, 1979, Vol. 1, p. 121. (b) Kelly, S. E. In *Comprehensive Organic Synthesis*, Trost, B. M., Fleming, I., Eds., Pergamon Press: Oxford, UK, 1991, Vol. 1, p. 729. (c) Larock, R. C. *Comprehensive Organic Transformations*, 2nd Ed., Wiley-VCH: New York, 1999, pp. 251–315. (d) Organ, M. G., Cooper, J. T., Rogers, L. R., Soleymanzadeh, F., Paul, T. *J. Org. Chem.* **2000**, *65*, 7959, and references cited therein.

2. Corey, E. J., Long, A. K. *J. Org. Chem.* **1978**, *43*, 2208.

3. (a) Carey, F. A., Sundberg, R. J. *Advanced Organic Chemistry: Reactions and Synthesis*, Part B, 4th ed., Plenum Press: New York, 2001, pp. 408–414. (b) Carruthers, W., Coldham, I., *Modern Methods of Organic Synthesis*, 4th ed., Cambridge University Press: Cambridge, MA, 2004, pp. 105–110.

4. Tipson, R. S. *J. Org. Chem.* **1944**, *9*, 235.

5. Brown, H. C., Bernheimer, R., Kim, C. J., Scheppele, S. E. *J. Am. Chem. Soc.* **1967**, *89*, 370.

6. Crossland, R. K., Servis, K. L. *J. Org. Chem.* **1970**, *35*, 3195.

7. Larsen, S. D., Monti, S. A. *J. Am. Chem. Soc.* **1977**, *99*, 8015.

8. Bunce, R. A., Schilling. C. L. III *J. Org. Chem.* **1995**, *60*, 2748.

9. (a) Arhart, R. J., Martin, J. C. *J. Am. Chem. Soc.* **1972**, *94*, 5003. (b) Winkler, J. D., Stelmach, J. E., Axten, J. *Tetrahedron Lett.* **1996**, *37*, 4317. (c) Yokokawa, F., Shioiri, T. *Tetrahedron Lett.* **2002**, *43*, 8679.

10. (a) Vogel, E., Klärner, F. G. *Angew. Chem., Int. Ed.* **1968**, *7*, 374. (b) Nicolas, M., Fabré, B., Pilard, J.-F., Simonet, J. *J. Heterocycl. Chem.* **1999**, *36*, 1105.

11. Hückel, W., Rücker, D. *Liebigs Ann.* **1963**, *666*, 30.

12. Nace, H. R. *Org. React.* **1962**, *12*, 57.

13. McNamara, L. S., Price, C. C. *J. Org. Chem.* **1962**, *27*, 1230.

14. (a) Barton, D. H. R., McCombie, S. W. *J. Chem. Soc., Perkin Trans. 1* **1975**, 1574. (b) Crich, D., Quintero, L. *Chem. Rev.* **1989**, *89*, 1413.

15. Ramachandran, P. V., Prabhudas, B., Chandra, J. S., Reddy, M. V. R. *J. Org. Chem.* **2004**, *69*, 6294.

16. Cope, A. C., Trumbull, E. R. *Org. React.* **1960**, *11*, 317.

17. Fukuyama, T., Wang, C.-L. J., Kishi, Y. *J. Am. Chem. Soc.* **1979**, *101*, 260.

18. Grieco, P. A., Reap, J. J. *Tetrahedron Lett.* **1974**, 1097.

19. (a) Clive, D. L. J. *Tetrahedron* **1978**, *34*, 1049. (b) Nicolaou, K. C., Petasis, N. A. In *Selenium in Natural Product Synthesis*, CIS Inc.: Philadelphia, 1984. (c) Paulmier, C. In *Selenium Reagents and Intermediates in Organic Synthesis*, Pergamon Press: Oxford, UK, 1986. (d) Liotta, D. *Organoselenium Chemistry,* Wiley: New York, 1987. (e) Back, T. G. *Organoselenium Chemistry: A Practical Approach*, Oxford University Press: Oxford, UK, 1999. (f) *Organoselenium Chemistry: Modern Developments in Organic Synthesis*, Wirth, T., Ed., Springer-Verlag: Berlin, 2000.

20. Sharpless, K. B., Young, M. W., Lauer, R. F. *Tetrahedron Lett.* **1973**, 1979.

21. Reich, H. J., Wollowitz, S. *Org. React.* **1993**, *44*, 1.

22. Sharpless, K. B., Lauer, R. F., Teranishi, A. Y. *J. Am. Chem. Soc.* **1973**, *95*, 6137.

23. Mancini, I., Guella, G., Pietra, F. *Helv. Chim. Acta* **1991**, *74*, 941.

24. Oppolzer, W., Bättig, K. *Helv. Chim. Acta* **1981**, *64*, 2489.

25. Trost, B. M., Lee, D. C. *J. Org. Chem.* **1989**, *54*, 2271.

26. Brown, H. C., Kramer, G. W., Levy, A. B., Midland, M. M. *Organic Synthesis via Boranes*, Wiley: New York, 1975.

27. Zweifel, G., Miller, J. A. *Org. React.* **1984**, *32,* 375.

28. Brown, H. C., Bowman, D. H., Misumi, S., Unni, M. K. *J. Am. Chem. Soc.* **1967**, *89*, 4531.

29. (a) Zweifel, G., Arzoumanian, H., Whitney, C. C. *J. Am. Chem. Soc.* **1967**, *89*, 3652. (b) Zweifel, G., Fisher, R. P., Snow, J. T., Whitney, C. C. *J. Am. Chem. Soc.* **1971**, *93*, 6309.

30. Zweifel, G., Steele, R. B. *J. Am. Chem. Soc.* **1967**, *89*, 2754.

31. (a) Normant, J. F., Alexakis, A. *Synthesis* **1981**, 841. (b) Marek, I., Normant, J. F. In *Metal-Catalyzed Cross-Coupling Reactions*, Diederich, F., Stang, P. J., Eds., Wiley-VCH: Weinheim, Germany, 1988, p. 271. (c) Marek, I., Normant, J. F. In *Transition Metals for Organic Synthesis*, Beller, M., Bolm, C., Eds., Wiley-VCH: Weinheim, Germany, 1998, Vol. 1, p. 514. (d) Negishi, E. In *Catalytic Asymmetric Synthesis*, Ojima, I., Ed., 2nd ed., VCH: New York, 2000, p. 165.

32. (a) Negishi, E., Valente, L. F., Kobayashi, M. *J. Am. Chem. Soc.* **1980**, *102*, 3298. (b) Negishi, E. *Pure & Applied Chemistry* **1981**, *53*, 2333, and references cited therein.

33. Okukado, N., Negishi, E. *Tetradedron Lett.* **1978**, 2357.

34. Corey, E. J., Katzenellenbogen, J. A., Posner, G. H. *J. Am. Chem. Soc.* **1967**, *89*, 4245.

35. (a) Lipshutz, B. H., Sengupta, S. *Org. React.* **1992**, *41*, 171. (b) Normant, J. F. In *Organocopper Reagents: A Practical Approach*, Taylor, R. J. K., Ed., Oxford University Press: Oxford, UK, 1994, p. 237.

36. Normant. J. F., Bourgain, M. *Tetrahedron Lett.* **1971**, 2583.

37. (a) Marfat, A., McGuirk, P. R., Helquist, P. *J. Org. Chem.* **1979**, *44*, 1345. (b) Marfat, A., McGuirk, P. R., Helquist, P. *J. Org. Chem.* **1979**, *44*, 3888, and references cited therein.

38. (a) Levy, A. B., Talley, P., Dunford, J. A. *Tetrahedron Lett.* **1977**, 3545. (b) Alexakis, A., Cahiez, G., Normant, J. F. *Org. Synth. Col. Vol. VII*, **1990**, 290.

39. Iyer, R. S., Helquist, P. *Org. Synth. Col. Vol. VII* **1990**, 236.

40. Furber, M., Taylor, J. K., Burford, S. C. *J. Chem. Soc. Perkin Trans. 1* **1986**, 1809.

41. (a) Corey, E. J., Katzenellenbogen, J. A. *J. Am. Chem. Soc.* **1969**, *91*, 1851. (b) Siddall, J. B., Biskup, M., Fried, J. H. *J. Am. Chem. Soc.* **1969**, *91*, 1853.

42. (a) Alexakis, A., Commercon, A., Villiéras, J., Normant, J. F. *Tetrahedron Lett.* **1976**, 2313. (b) Alexakis, A., Normant, J. F., Villiéras, J. *Tetrahedron Lett.* **1976**, 3461.

43. (a) Maercker, A. *Org. React.* **1965**, *14*, 270. (b) Schlosser, M. *Top. Stereochem.* **1970**, *5*, 1. (c) Bestmann, H. J., Vostrowsky, O. *Top. Curr. Chem.* **1983**, *109*, 85. (d) Murphy, P. J., Brennan, J. *Chem. Soc. Rev.* **1988**, *17*, 1. (e) Maryanoff, B. E., Reitz, A. B. *Chem. Rev.* **1989**, *89*, 863. (f) Kelly, S. E. In *Comprehensive Organic Synthesis*, Trost, B. M., Fleming, I., Eds., Pergamon Press: Oxford, UK, 1991, Vol. 1, p. 755. (g) Smith, M. B. *Organic Synthesis*, 2nd d., McGraw-Hill: Boston, 2002, pp. 656–670.

44. (a) Wittig, G., Geissler, G. *Liebigs Ann. Chem.* **1953**, *580*, 44. (b) Wittig, G., Schöllkopf, U. *Chem. Ber.* **1954**, *87*, 1318.

45. Schlosser, M., Schaub, B., de Oliveira-Neto, J., Jeganathan, S. *Chimia*, **1986**, *40*, 244.

46. Vedejs, E., Snoble, K. A. *J. Am. Chem.*, *Soc.* **1973**, *95*, 5778.

47. Vedejs, E., Meier, G. P., Snoble, K. A. J. *J. Am. Chem. Soc.* **1981**, *103*, 2823.

48. (a) Vedejs, E., Marth, C. F. *J. Am. Chem. Soc.* **1988**, *110*, 3948. (b) Vedejs, E., Marth, C. F. *J. Am. Chem. Soc.* **1990**, *112*, 3905. (c) Mari, F.: Lahti, P. M., McEwan, W. E. *J. Am. Chem. Soc.* **1992**, *114*, 813.

49. Greenwald, R., Chaykovsky, M., Corey, E. J. *J. Org. Chem.* **1963**, *28*, 1128.

50. Corey, E. J., Kang, J. *J. Am. Chem. Soc.* **1982**, *104*, 4724.

51. (a) Schlosser, M., Christmann, K.-F. *Angew. Chem.*, *Int. Ed.* **1966**, *5*, 126. (b) Schlosser, M., Christmann, F.-K. *Synthesis* **1969**, 38.

52. Corey, E. J., Yamamoto, H. *J. Am. Chem. Soc.* **1970**, *92*, 226.

53. Salmond, W. G., Barta, M. A., Havens, J. L. *J. Org. Chem.* **1978**, *43*, 790.

54. (a) Becker, K. B. *Helv. Chim. Acta* **1977**, *60*, 69. (b) Becker, K. B. *Tetrahedron* **1980**, *36*, 1717, and references cited therein.

55. Novak, J., Salemink, C. A. *Tetrahedron Lett.* **1981**, *22*, 1063.

56. Wadsworth, W. S. Jr., *Org. React.* **1977**, *25*, 73.

57. Denney, V. B., Ross, S. T. *J. Org. Chem.* **1962**, *27*, 998.

58. Trippett, S., Walker, D. M. *J. Chem. Soc.* **1961**, 1266.

59. Lui, Y. Y., Thom, E., Liebman, A. A. *J. Heterocycl. Chem.* **1979**, *16*, 799.

60. Katsuki, T., Lee, A. W. M., Ma, P., Martin, V. S., Masamune, S., Sharpless, K. P., Tuddenham, D., Walker, F. J. *J. Org. Chem.* **1982**, *47*, 1373.

61. Bestmann, H. J., Roth, K., Ettlinger, M. *Angew. Chem.*, *Int. Ed.* **1979**, *18*, 687.

62. (a) Vedejs, E., Fleck, T., Hara, S. *J. Org. Chem.*, **1987**, *52*, 4637. (b) Vedejs, E., Marth, C. F. *J. Am. Chem. Soc.* **1988**, *110*, 3948.

63. Boutagy, J., Thomas, R. *Chem. Rev.* **1974**, *74*, 87.

64. Takahashi, H., Fujiwara, K., Ohta, M. *Bull. Chem. Soc. Jpn.* **1962**, *35*, 1498.

65. Corey, E. J., Kwiatkowski, G. T. *J. Am. Chem. Soc.* **1966**, *88*, 5655.

66. Corey, E. J., Weinshenker, N. M., Schaaf, T. K., Huber, W. *J. Am. Chem. Soc.* **1969**, *91*, 5675.

67. Schmid, G., Fukuyama, T., Akasaka, K., Kishi, Y. *J. Am. Chem. Soc.* **1979**, *101*, 259.

68. Blanchette, M. A., Choy, W., Davis, J. T., Essenfeld, A. P., Masamune, S., Roush, W. R., Sakai, T. *Tetrahedron Lett.* **1984**, *25*, 2183.

69. Rathke, M. W., Nowak, M. *J. Org. Chem.* **1985**, *50*, 2624.

70. Still, W. C., Gennari, C. *Tetrahedron Lett.* **1983**, *24*, 4405.

71. (a) Ando, K. *J. Org. Chem.* **1997**, *62*, 1934. (b) Ando, K. *J. Org. Chem.* **1998**, *63*, 8411.

72. Ando, K., Oishi, T., Hirama, M., Ohno, H., Ibuka, T. *J. Org. Chem.* **2000**, *65*, 4745.

73. (a) Peterson, D. J. *Organomet. Chem. Rev. A* **1972**, *7*, 295. (b) Hudrlik, P. F. *J. Organomet. Chem. Lib.* **1976**, *1*, 127. (c) Chan, T-H. *Acc. Chem. Res.* **1977**, *10*, 442. (d) Ager, D. J. *Synthesis* **1984**, 384. (e) Colvin, E. W. *Silicon Reagents in Organic Synthesis*, Academic Press: London, 1988, p. 63. (f) Hudrlik, P. F., Agwaramgbo, E. L. O. In *Silicon Chemistry*, Corey, E. R., Corey, J. Y., Gaspar, P. P., Eds., Ellis Horwood: Chichester, UK, 1988, p. 95. (g) Ager, D. J. *Org. React.* **1990**, *38*, 1.

74. Peterson, D. J. *J. Org. Chem.* **1968**, *33*, 780.

75. Johnson, C. R., Tait, B. D. *J. Org. Chem.* **1987**, *52*, 281.

76. (a) Shimoji, K., Taguchi, H., Oshima, K., Yamamoto, H., Nozaki, H., *J. Am. Chem. Soc.* **1974**, *96*, 1620. (b) Hartzell, S. L., Sullivan, D. F., Rathke, M. W. *Tetrahedron Lett.* **1974**, 1403.

77. Hudrlik, P. F., Peterson, D. J. *J. Am. Chem. Soc.* **1975**, *97*, 1464.

78. Hudrlik, P. F., Agwaramgbo, E. L. O., Hudrlik, A. M. *J. Org. Chem.* **1989**, *54*, 5613.

79. Utimoto, K., Obayashi, M., Nozaki, H. *J. Org. Chem.* **1976**, *41*, 2940.

80. Hudrlik, P. F., Peterson, D., Rona, R. *J. Org. Chem.* **1975**, *40*, 2263.

81. (a) Julia, M., Paris, J-M. *Tetrahedron Lett.* **1973**, 4833. (b) Kocienski, P., Lythgoe, B., Ruston, S. *J. Chem. Soc. Perkin 1* **1978**, 829. (c) Kocienski, P. *Phosphorus Sulfur* **1985**, *24*, 97. (d) Julia, M. *Pure Appl. Chem.* **1985**, *57*, 763. (e) Kelly, S. K. In *Comprehensive Organic Synthesis*, Trost, B. M., Fleming, I., Eds., Pergamon Press: Oxford, UK, 1991, Vol. 1, p. 792.

82. Chang, T. C. T., Rosenblum, M., Simms, N. *Org. Synth. Col. Vol. VIII,* **1993**, 479.

83. Kocienski. P. J, Lythgoe, B., Ruston, S. *J. Chem. Soc., Perkin Trans. 1* **1978**, 829.

84. Kondo, K., Tunemoto, D. *Tetrahedron Lett.* **1975**, 1007.

85. Magnus, P. D. *Tetrahedron* **1977**, *40*, 150.

86. Blakemore, P. R. *J. Chem. Soc., Perkin Trans. 1* **2002**, 2563.

87. Lear, M. J., Hirama, M. *Tetrahedron Lett.* **1999**, *40*, 4897.

88. Blakemore, P. R., Cole, W. J., Kocienski, P. J., Morley, A. *Synlett* **1998**, 26.

89. (a) Shapiro, R. H. *Org. React.* **1976**, *23*, 405. (b) Chamberlin, A. R., Bloom, S. H. *Org. React.* **1990**, *39*, 1.

90. Shapiro, R. H., Heath, M. J. *J. Am. Chem. Soc.* **1967**, *89*, 5734.

91. Scott, W. L., Evans, D. A. *J. Am. Chem. Soc.* **1972**, *94*, 4779.

92. (a) Shapiro, R. H., Lipton, M. F., Kolonko, K. J., Buswell, R. L., Capuano, L. A. *Tetrahedron Lett.* **1975**, 1811. (b) Stemke, J. E., Bond, F. T. *Tetrahedron Lett.* **1975**, 1815.

93. (a) Chamberlin, A. R., Stemke, J. E., Bond, F. T. *J. Org. Chem.* **1978**, *43*, 147. (b) Adlington, R. M., Barrett, A. G. M. *Acc. Chem. Res.* **1983**, *16*, 55.

94. Cusack, N. J., Reese, C. B., Risius, A. C., Roozpeikar, B. *Tetrahedron* **1976**, *32*, 2157.

95. (a) Martin, S. F., Daniel, D., Cherney, R. J., Liras, S. *J. Org. Chem.* **1992**, *57*, 2523. (b) α,β-unsaturated aldehyde: Wasserman, H. H., Fukuyama, J. M. *Tetrahedron Lett.* **1984**, *25*, 1387.

96. Chamberlin, A. R., Bond, F. T. *Synthesis* **1979**, 44.

97. (a) Rhoads, S. J., Raulins, N. R. *Org. React.* 1975, *22*, 1. (b) Bennett, G. B. *Synthesis* 1977, 589. (c) Carruthers, W., Coldham, I. *Modern Methods of Organic Synthesis*, 4th ed., Cambridge University Press: Cambridge, MA, 2004, pp. 244–252. (d) Ziegler, F. E. *Chem. Rev.* **1988**, *88*, 1423. (e) Blechert, S. *Synthesis* **1989**, 71. (f) Wipf, P. In *Comprehensive Organic Synthesis*, Trost, B. M., Fleming, I., Eds., Pergamon Press: Oxford, UK, 1991, Vol. 5, p. 827. (g) Smith, M. B. *Organic Synthesis*, 2nd Ed., McGraw-Hill: Boston, 2002, p. 1021.

98. Claisen, L. *Chem. Ber.* **1912**, *45*, 3157.

99. (a) Vance, R. L., Rondan, N. G., Houk, K. N., Jensen, F., Borden, W. T., Komornicki, A., Wimmer, E. *J. Am. Chem. Soc.* **1988**, *110*, 2314. (b) Ganem, B. *Angew. Chem., Int. Ed.* **1996**, *35*, 936.

100. Hansen, J. J., Schmid, H. *Tetrahedron* **1974**, *30*, 1959.

101. Fleming, I. *Pericyclic Reactions*, Oxford University Press: Oxford, UK, 1999.

102. (a) Woodward, R. B., Hoffmann, R. *Angew. Chem., Int. Ed.* **1969**, *8*, 781. (b) Woodward, R. B., Hoffmann, R. *The Conservation of Orbital Symmetry*, VCH: Weinheim, Germany, 1970.

103. Kametani, T., Suzuki, T., Sato, E., Nishimura, M., Unno, K. *J. Chem. Soc., Chem. Commun.* **1982**, 123.

104. (a) Saucy, G., Marbet, R. *Helv. Chim. Acta* **1967**, *50*, 2091. (b) Dauben, W. G., Dietsche, T. J. *J. Org. Chem.* **1972**, *37*, 1212.

105. (a) Carroll, M. F. *J. Chem. Soc.* **1940**, 704. (b) Carroll, M. F. *J. Chem. Soc.* **1940**, 1266. (c) Carroll, M. F. *J. Chem. Soc.* **1941**, 507. (d) Kimel, W., Cope, A. C. *J. Am. Chem. Soc.* **1943**, *65*, 1992.

106. Wilson, S. R., Price, M. F. *J. Org. Chem.* **1984**, *49*, 722.

107. Johnson, W. S., Werthemann, L., Bartlett, W. R., Brocksom, T. J., Li, T., Faulkner, D. J., Petersen, M. R. *J. Am. Chem. Soc.* **1970**, *92*, 741.

108. Trust, R. I., Ireland, R. E. *Org. Synth. Col. Vol. VI* **1988**, 606.

109. Takano, S., Sugihara, T., Samizu, K., Akiyama, M., Ogasawara, K. *Chem. Lett.* **1989**, 1781.

110. Uskokovic, M. R., Lewis, R. L., Partridge, J. J., Despreaux, C. W., Pruess, D. L. *J. Am. Chem. Soc.* **1979**, *101*, 6742.

111. Wick, A. E., Felix, D., Steen, K., Eschenmoser, A. *Helv. Chim. Acta* **1964**, *47*, 2425.

112. (a) Ireland, R. E., Mueller, R. H. *J. Am. Chem. Soc.* **1972**, *94*, 5897. (b) Ireland, R. E., Mueller, R. H., Willard, A. K. *J. Am. Chem. Soc.* **1976**, *98*, 2868. (c) Ireland, R. E., Wilcox, C. S. *Tetrahedron Lett.* **1977**, 2839. (d) Mulzer, J., Altenbach, H.-J., Braun, H., Krohn, K., Reissig, H.-U. *Organic Synthesis Highlights*, VCH: Weinheim, Germany, 1991, p. 111. (e) Pereira, S., Srebnik, M. *Aldrichimica Acta* **1993**, *26*, 17.

113. Ireland, R. E., Wipf, P., Xiang, J.-N. *J. Org. Chem.* **1991**, *56*, 3572.

114. Koreeda, M., Luengo, J. I. *J. Am. Chem. Soc.* **1985**, *107*, 5572.

115. Maruoka, K., Banno, H., Yamamoto, H. *J. Am. Chem. Soc.* **1990**, *112*, 7791.

116. Maruoka, K., Saito, S., Yamamoto, H. *J. Am. Chem. Soc.* **1995**, *117*, 1165.

117. Keegstra, M. A. *Tetrahedron* **1992**, *48*, 2681.

118. Nordmann, G., Buchwald, S. L. *J. Am. Chem. Soc.* **2003**, *125*, 4978.

119. (a) Raphael, R. A. *Acetylenic Compounds in Organic Synthesis*, Butterworth: London, 1955. (b) Rutledge, T. F. *Acetylenic Compounds*, Reinhold Book Corp.: New York, 1968. (c) Viehe, H. G., Ed. *Chemistry of Acetylenes*, Marcel Dekker: New York, 1969. (d) Brandsma, L. *Preparative Acetylene Chemistry*, 2nd ed., Elsevier: Amsterdam, 1988. (e) Brandsma, L, Verkruijsse, H. D. *Synthesis of Acetylenes, Allenes and Cumulenes*, Elsevier: Amsterdam, 1981. (f) Stang, P. J., Diederich, F., Eds. *Modern Acetylene Chemistry*, VCH: Weinheim, Germany, 1995. (g) Brandsma, L. *Synthesis of Acetylenes, Allenes and Cumulenes*, Elsevier: Amsterdam, 2004.

120. Larock, R. C. *Comprehensive Organic Transformations*, 2nd. ed., Wiley-VCH: New York, 1999, pp. 565–605.

121. Krebs, A., Swienty-Busch, J. In *Comprehensive Organic Synthesis*, Trost, B. M., Fleming, I., Eds., Pergamon Press: Oxford, 1991, Vol. 6, p. 949.

122. Fringuelli, F., Mancini, V., Taticchi, A. *Tetrahedron* **1969**, *25*, 4249.

123. Abbott, T. W. *Org. Synth. Col. Vol II* **1943**, 515.

124. Corey, E. J., Fuchs, P. L. *Tetrahedron Lett.* **1972**, 3769.

125. Villiéras, J., Periot, P., Normant, J. F. *Synthesis* **1979**, 502.

126. (a) Negishi, E., King, A. O., Klima, W. L., Patterson, W., Silveira, A., Jr., *J. Org. Chem.* **1980**, *45*, 2526. (b) Negishi, E., King, A. O., Tour, J. M. *Org. Synth.* **1986**, *64*, 44.

127. Garratt, P. J. In *Comprehensive Organic Synthesis*, Trost, B. M., Fleming, I., Eds., Pergamon Press: Oxford, UK, 1991, Vol. 3, p. 271.

128. Umeno, M., Suzuki, A. In *Handbook of Grignard Reagents*, Silverman, G. S., Rakita, P. E., Eds., Marcel Dekker: New York, 1996.

129. (a) Skattebøl, L., Jones, E. H., Whiting, M. *Org. Synth.Col. Vol. IV* **1963**, 792. (b) Amos, R. A., Katzenellenbogen, J. A. *J. Org. Chem.* **1978**, *43*, 555.

130. Beumel, O. F., Jr., Harris, R. F. *J. Org. Chem.* **1977**, *42*, 566.

131. Midland, M. M., McLoughlin, J. I., Werley, R. T. Jr. *Org. Synth.* **1990**, *68*, 14.

132. Yamada, K., Miyaura, N., Itoh, M., Suzuki, A. *Synthesis* **1977**, 679.

133. Brandsma, L. *Synthesis of Acetylenes, Allenes and Cumulenes*, Elsevier: Amsterdam, 2004, pp. 85–118.

134. Smith, W. N., Beumel, O. F. Jr., *Synthesis* **1974**, 441.

135. (a) Schwarz, M., Waters, M. *Synthesis* **1972**, 567. (b) Brattesani, D. N., Heathcock, C. H. *Synth. Commun.* **1973**, *3*, 245. (c) Sagar, A. J. G., Scheinmann, F. *Synthesis* **1976**, 321.

136. Negishi, E., Baba, S. *J. Am. Chem. Soc.* **1975**, *97*, 7385.

137. (a) van Rijn, P. E., Mommers, S., Visser, R. G., Verkruijsse, H. D., Brandsma, L. *Synthesis* **1981**, 459. (b) Seebach, D., Beck, A. K., Studer, A. In *Modern Synthetic Methods*, Ernst, B., Leumann, C., Eds., VCH: Weinheim, Germany, 1995, p. 1.

138. Bhanu, S., Scheinmann, F. *J. Chem. Soc., Chem. Commun.* **1975**, 817.

139. McMurry, J., Matz, J. R., Kees, K. L., Bock, P. A. *Tetrahedron Lett.* **1982**, *23*, 1777.

140. Gong, J., Fuchs, P. L. *J. Am. Chem. Soc.* **1996**, *118*, 4486.

141. Brandsma, L. *Synthesis of Acetylenes, Allenes and Cumulenes*, Elsevier: Amsterdam, 2004, pp. 319–340.

142. For handling KH, see Brown, C. A. *J. Org. Chem.* **1974**, *39*, 3913.

143. (a) Brown, C. A., Yamashita, A. *J. Am. Chem. Soc.* **1975**, *97*, 891. (b) Brown, C. A., Yamashita, A. *Chem. Commun.* **1976**, 959.

144. Waanders, P. P., Thijs, L., Zwanenburg, B. *Tetrahedron Lett.* **1987**, *28*, 2409.

145. Abrams, S. R., Shaw, A. C. *Org. Synth. Col. Vol. VIII* **1993**, 146.

146. For mechanistic interpretations, see (a) Wotiz, J. H., Barelski, P. M., Koster, D. R. *J. Org. Chem.* **1973**, *38*, 489. (b) Abrams, S. R., Shaw, A. C. *J. Org. Chem.* **1987**, *52*, 1835. (c) Brandsma, L. *Synthesis of Acetylenes, Allenes and Cumulenes*, Elsevier: Amsterdam, 2004, p. 35.

147. Spence, J. D., Wyatt, J. K., Bender, D. M., Moss, D. K., Nantz, M. H. *J. Org. Chem.* **1996**, *61*, 4014.

148. Gilbert, J. C., Weerasooriya, U. *J. Org. Chem.* **1979**, *44*, 4997.

149. Hung, D. T., Nerenberg, J. B., Schreiber, S. L. *J. Am. Chem. Soc.* **1996**, *118*, 11054.

CHAPTER 9

Syntheses of Carbocyclic Systems

The Diels-Alder reaction has both enabled and shaped the art and science of total synthesis over the last few decades to an extent which, arguably, has yet to be eclipsed by any other transformation in the current synthetic repertoire.

K. C. Nicolaou

In the preceding chapters, various methods for the preparation of *small* carbocyclic systems were presented. Among these were the Birch reduction of aromatic compounds, cyclopropanation, Dieckmann-type reactions, intramolecular alkylations, intramolecular aldol reactions, the intramolecular Michael reaction, and the Robinson annulation. The following sections describe methods for the synthesis of carbocyclic systems either by connecting two atoms in a single molecule (*intra*molecular reaction) or by joining together two molecules (*inter*molecular reaction).[1]

A set of rules governing intramolecular cyclizations (Baldwin's rules) based on entropy (reflected in the probability of the two ends of a chain coming together) and enthalpy (ring strain) effects was presented in Chapter 6. When planning an intramolecular cyclization, both of these factors must be considered. For example, although the entropy factor favors formation of small-member rings, the enthalpy factor with these is less favorable. The situation is reversed for formation of large-member carbocyclic systems. Depending on the number of carbons, carbocyclic systems are described as *small-* (3 to 6 carbons), *medium-* (7 to 11 carbons) or *large-* (≥12 carbons) member rings. The large-member ring compounds also are often referred to as "macrocycles."

9.1 INTRAMOLECULAR FREE RADICAL CYCLIZATIONS

Intramolecular free radical cyclization of dicarboxylic esters leads to α-hydroxy ketones (*acyloins*). Reductive coupling of dicarbonyl compounds provides 1,2-diols (*pinacols*) and further reaction of these yields cycloalkenes (McMurry reaction). These cyclization reactions are especially valuable for the preparation of medium and large rings that are not readily accessible by other methods.

Acyloin Condensation[2]

Treatment of α,ω-diesters with dispersions of sodium or a sodium-potassium alloy in refluxing toluene or xylene results in joining of the two ester moieties to furnish cyclic α-hydroxy ketones (acyloins).

$n = 2$ to 13 and higher

an acyloin

The diester is usually added to the suspension of molten, dispersed sodium in the appropriate solvent over a prolonged period, especially for the preparation of medium and large carbocyclic systems. This is to limit intermolecular coupling from competing with the desired cyclization. Since the intermediates formed during the reaction are prone to oxidation in the basic reaction medium, the cyclization should be carried out under a nitrogen atmosphere.

In the initial step of the reductive coupling, electron transfer from Na or K to the diester leads to a di(radical anion) species (two ketyl radical anions), which then combine. Elimination of alkoxide produces the 1,2-diketone, which is further reduced to an enediolate by transfer of two electrons. Protonation of the dianion with dilute aqueous acid leads to the enediol, which tautomerizes to the α-hydroxy ketone.[3]

Undesirable side-reactions, such as Dieckmann ring closure and Claisen condensation reactions, often compete with the acyloin condensation.

53%
acyloin condensation
(**6**-member ring)

75%
Dieckmann condensation
(**5**-member ring)

These side reactions can be minimized by adding trimethylsilyl chloride to the reaction mixture as an alkoxide scavenger. This traps the enediolate dianion as a *bis*-silyl enol ether, and traps the sodium or potassium alkoxides, which are catalysts for the Dieckmann ring closure, as neutral silyl ethers.[4] The resultant bis-siloxy cycloalkenes are either isolated or converted in situ to α-hydroxy ketones by alcoholysis[5] or by acid hydrolysis.[6]

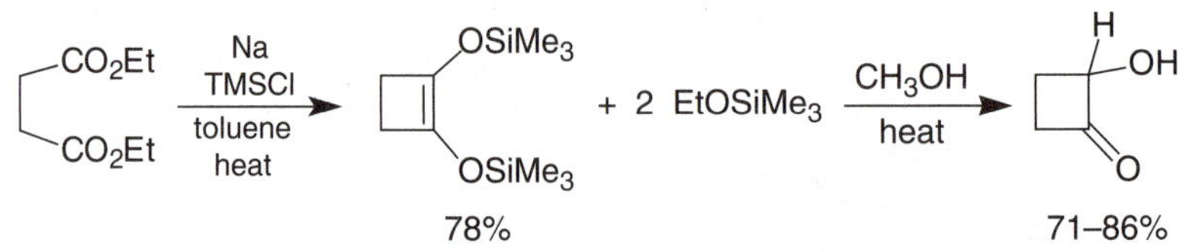

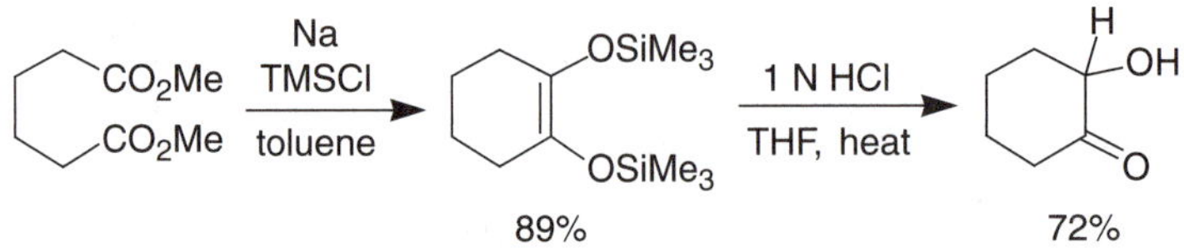

78% 71–86%

89% 72%

The acyloin coupling reaction is suitable for the synthesis of medium and large (≥ 12 membered) carbocyclic systems, using high-dilution techniques.[3,7]

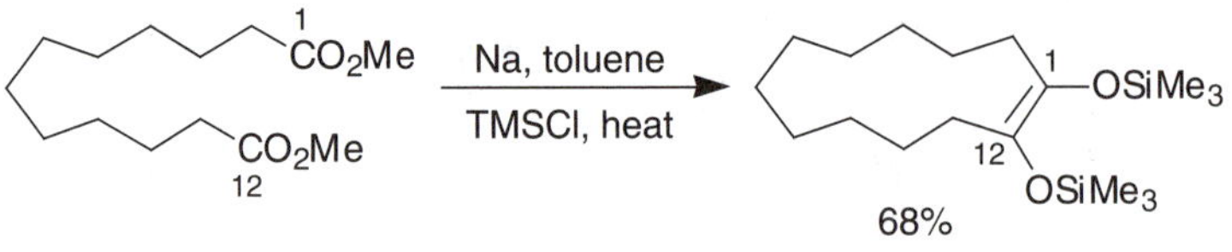

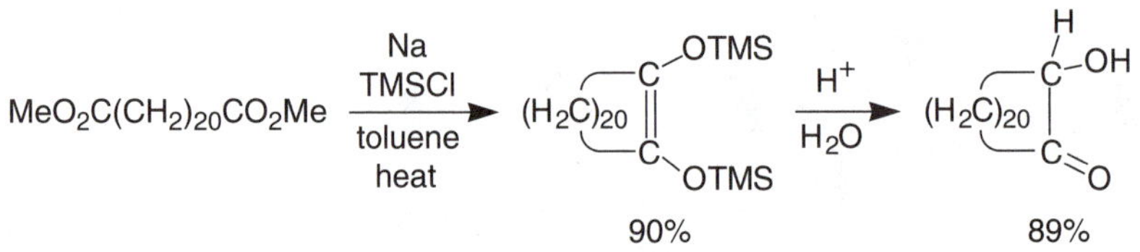

68%

90% 89%

Acyloins are versatile intermediates in organic synthesis since they offer the possibility for a variety of functional group interconversions. For example, treatment of an acyloin with zinc dust and acetic anhydride in glacial acetic acid furnishes the corresponding ketone.[8]

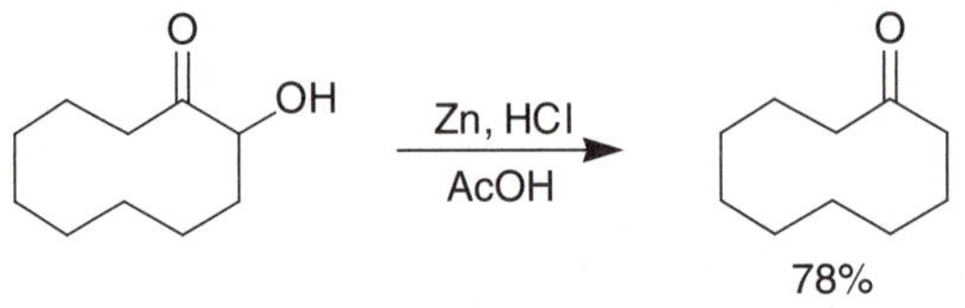

78%

Intermolecular dimerization of mono-esters via acyloin condensation is a viable tool for the preparation of acyclic, symmetrically substituted α-hydroxy ketones.[9] However, attempted coupling of two different mono-esters usually produces mixtures of products.

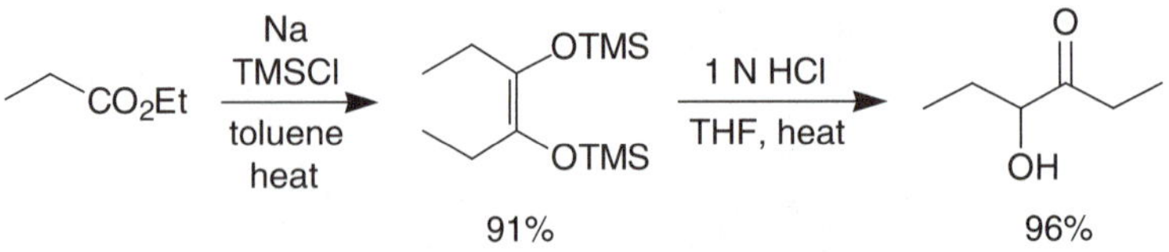

91% 96%

Pinacol Coupling[10]

The inter- and intramolecular coupling of two carbonyl groups of aldehydes or ketones in the presence of a low-valent titanium species produces a C–C bond with two adjacent stereocenters, a 1,2-diol (a pinacol). These may be further elaborated into ketones by the pinacol rearrangement or be deoxygenated to alkenes (McMurry reaction).

Low-valent titanium metal species [Ti(II), Ti(I), Ti(0)] are generated by reduction of $TiCl_4$ with magnesium amalgam[11] or by reduction of $TiCl_3$ with either Li, K, Zn-Cu couple, or $LiAlH_4$.[10a] Although the following discussion focuses on the use of low-valent titanium species, other metals have been used for pinacol-type coupling reactions.[10b]

A mechanistic scheme for the titanium-mediated coupling of two carbonyl groups involves an initial one-electron transfer from titanium metal to the aldehyde or ketone carbonyl group to produce radical anions (ketyl species), which then dimerize to a Ti-pinacolate. Hydrolysis of the pinacolate generates the vicinal diol.

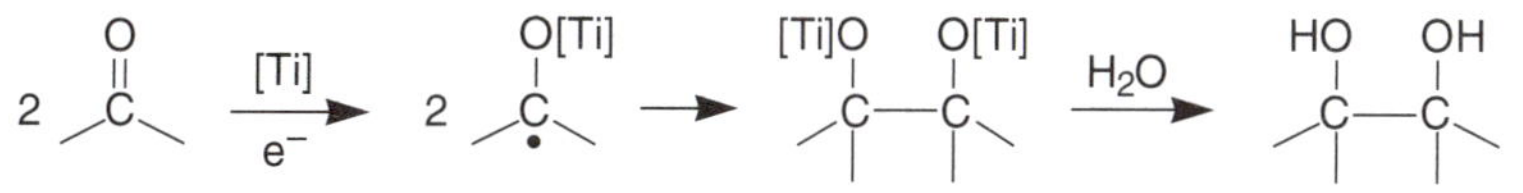

Intermolecular pinacol reactions provide access to either symmetrical or unsymmetrical 1,2-diols. In the latter case, however, an excess of the smaller carbonyl compound has to be used. The diols produced can be converted to ketones by treatment with acids (*pinacol rearrangement*). The following reaction scheme illustrates the intermolecular pinacol coupling of cyclopentanone, and conversion of the pinacol to the spirodecanone.[11] When employing low-valent Ti-species, it is important to carry out the coupling at low temperature to avoid deoxygenation of the initially formed 1,2-diol to the corresponding alkene.

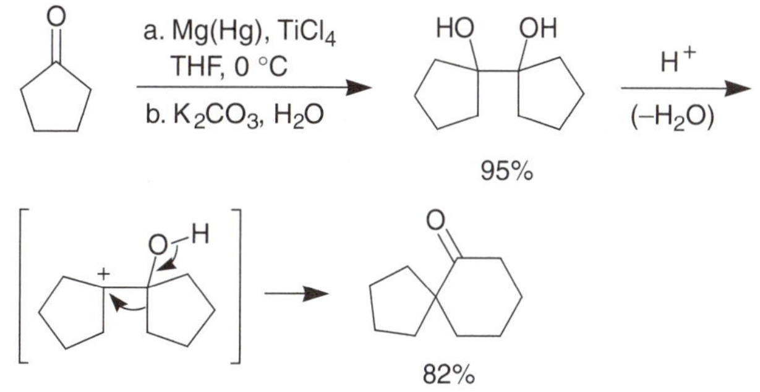

The titanium-mediated *intramolecular* pinacol coupling is suitable for the construction of both small-ring and macrocyclic 1,2-diols. Interestingly, the yields of cyclic diols obtained are nearly independent of the ring size. The success of ring formation over a such a wide range may be due to a template effect, where the two carbonyl moieties are brought together on the active titanium surface to form a C–C bond. In intramolecular pinacol coupling reactions, this template effect may overcome the strain involved in the formation of small rings as well as the entropy factors associated with formation of large rings.[12]

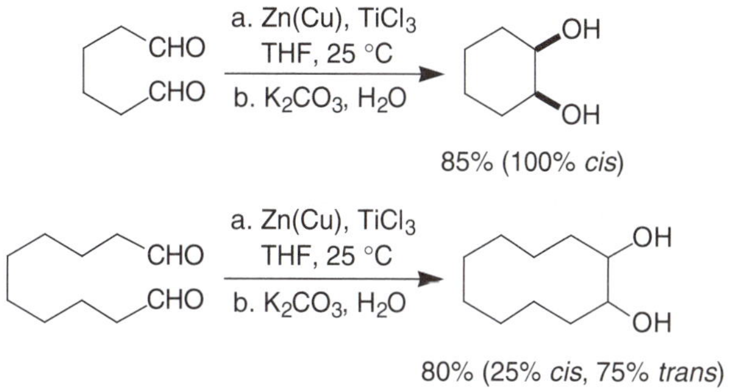

The McMurry Reaction— Synthesis of Alkenes[10a,13]

Finely divided Ti(0) metal particles produced by reduction of $TiCl_3$ with potassium or lithium metal or with a zinc-copper couple[14] reductively couple aldehydes and ketones at elevated temperature to produce alkenes.

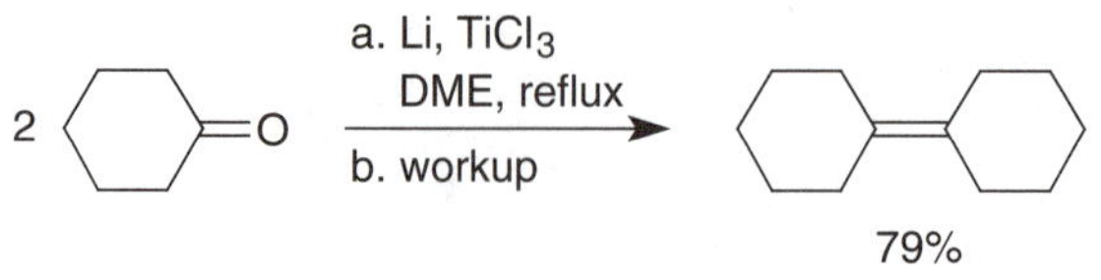

79%

Formation of the alkene is thought to proceed via initial formation of a Ti-pinacolate followed by its deoxygenation to produce the double bond. In support of the proposed mechanistic scheme are the facts that (1) pinacol intermediates can be isolated when the coupling reaction is carried out at 0 °C rather than at solvent reflux temperature and (2) treatment of isolated pinacols with low-valent titanium at 60 °C deoxygenates them to the corresponding alkenes.

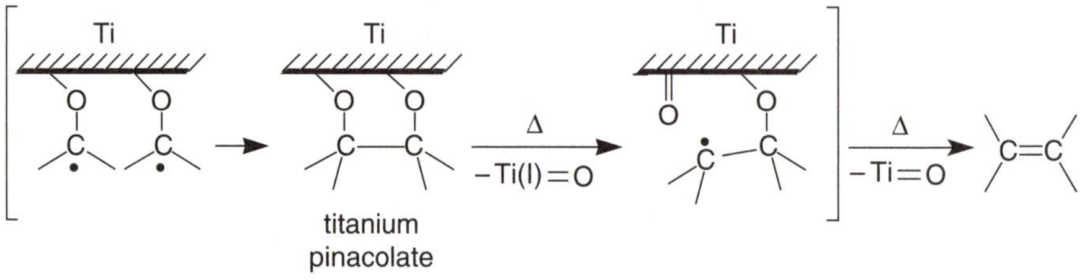

titanium
pinacolate

The stereochemistry of the alkene formed by coupling of aliphatic carbonyl compounds is related to the energy difference between the (E) and (Z)-alkene. If ΔE is larger than 4–5 kcal/mol, the (E)-isomer is formed selectively.[10a]

The *intermolecular* McMurry carbonyl coupling reaction has been applied to the synthesis of a variety of symmetrically and, in certain cases, unsymmetrically substituted alkenes.[15]

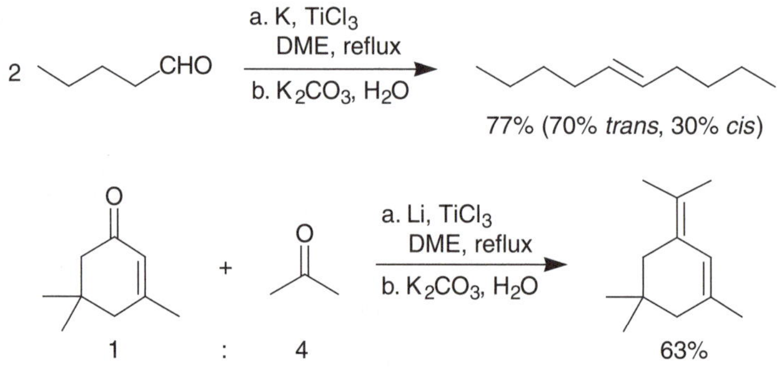

77% (70% *trans*, 30% *cis*)

63%

An especially active Ti(0) metal is obtained by converting $TiCl_3$ into its crystalline $TiCl_3$-dimethoxyethane solvate [$TiCl_3$(DME)], which is then reduced with a zinc-copper couple to Ti(0) metal.[15] The Ti-mediated coupling may also be carried out in the presence of carbon-carbon double bonds, as exemplified below.[16]

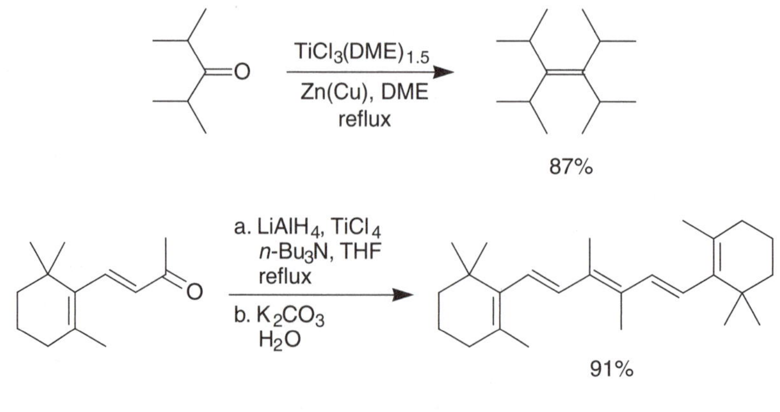

87%

91%

From a synthetic point of view, the titanium-induced *intra*molecular cyclization of dicarbonyl compounds is especially attractive in that carbocyclic rings of all sizes may be prepared in good yields, competitive with those obtained using the Dieckmann and acyloin cyclization methods. The synthesis of medium- and large-ring cycloalkenes by the McMurry reaction requires high-dilution conditions (< 0.01M) to repress intermolecular polymerization.[15,16]

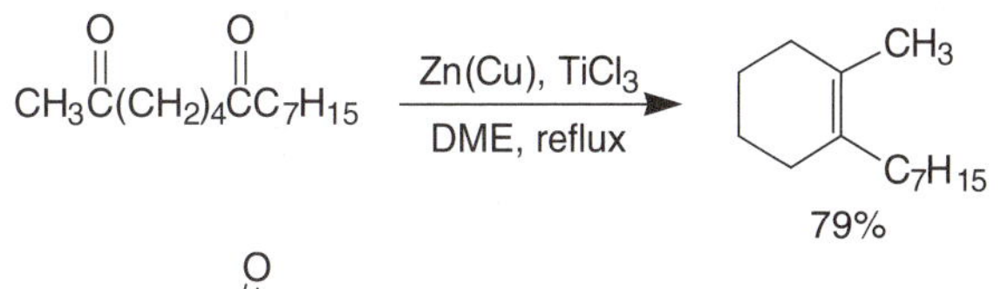

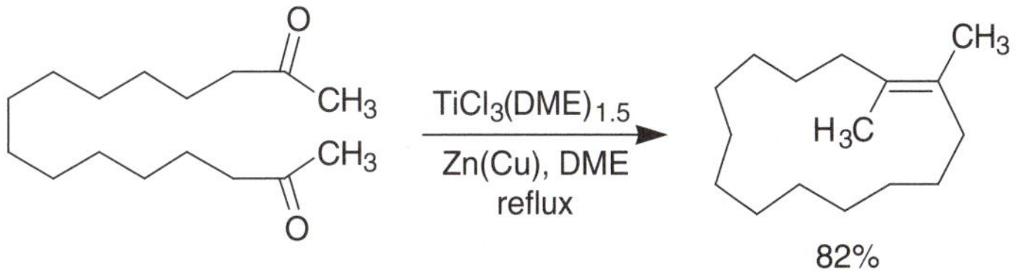

$$CH_3C(CH_2)_4CC_7H_{15} \xrightarrow[\text{DME, reflux}]{\text{Zn(Cu), TiCl}_3} \quad 79\%$$

$$\xrightarrow[\substack{\text{Zn(Cu), DME} \\ \text{reflux}}]{\text{TiCl}_3(\text{DME})_{1.5}} \quad 82\%$$

Extension of the low-valent titanium dicarbonyl coupling reaction to ketoesters leads to cyclic enol ethers, and hydrolysis of these affords the corresponding cycloalkanones.[17] This methodology has been applied to the synthesis of various natural products, as exemplified by the preparation of the highly unsaturated $C_{(14)}$ macrocyclic ketone shown below.[18]

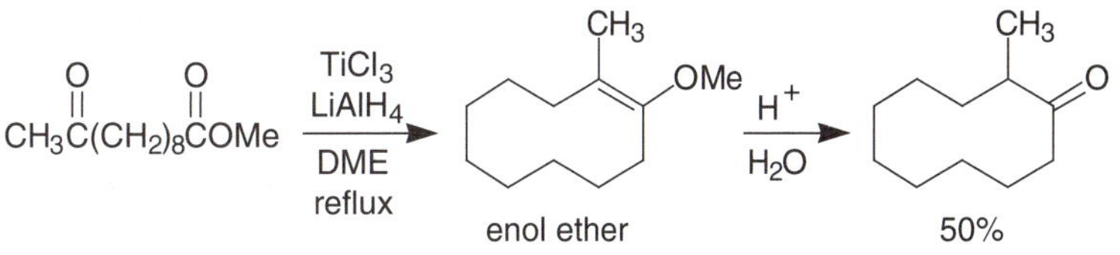

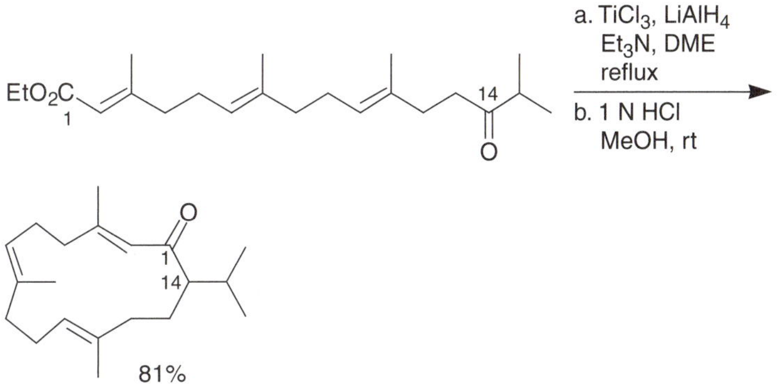

$$CH_3C(CH_2)_8COMe \xrightarrow[\substack{\text{DME} \\ \text{reflux}}]{\substack{\text{TiCl}_3 \\ \text{LiAlH}_4}} \text{enol ether} \xrightarrow[\text{H}_2\text{O}]{\text{H}^+} 50\%$$

a. TiCl$_3$, LiAlH$_4$
   Et$_3$N, DME
   reflux

b. 1 N HCl
   MeOH, rt

81%

9.2 CATION-π CYCLIZATIONS[19]

Cation-π cyclization is an effective method for constructing five- and six-member ring systems. A carbocation of an acyclic molecule that is in proximity to a π-bond may react intramolecularly with the nucleophilic double bond to form a C–C bond with concomitant ring closure and generation of a new cation in the product (Table 9.1). This type of cyclization works best if a relatively stable carbocation can be generated under mild conditions using protonic acids (H_2SO_4, HCO_2H, CF_3CO_2H) or Lewis acids ($BF_3 \cdot OEt_2$, $TiCl_4$, $SnCl_4$). Alkenes, tertiary alcohols, allylic alcohols, epoxides, and acetals are common precursors of the cations.

Table 9.1 Initiation and Propagation Steps in Cation-π Cyclizations of Unsaturated Substrates

Starting substrate	Mode of cyclization
diene; polyene	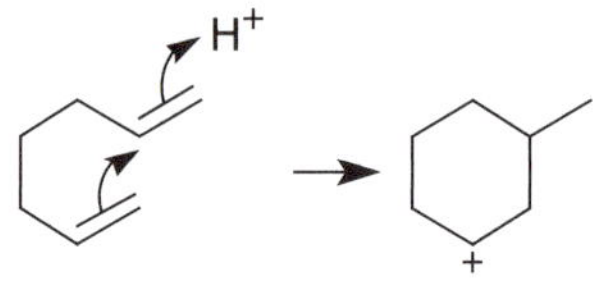
alcohol	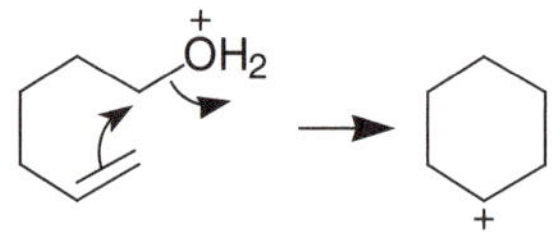
aldehyde or ketone (*Prins reaction*)	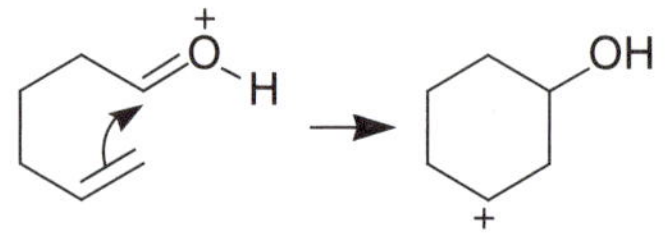
epoxide	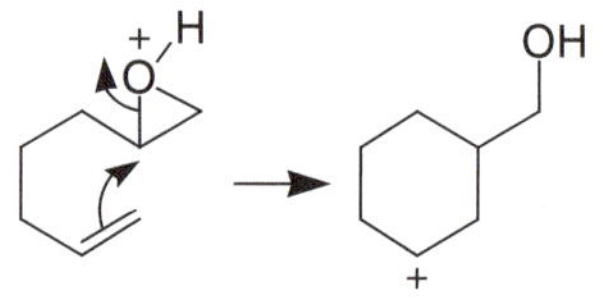
acetal	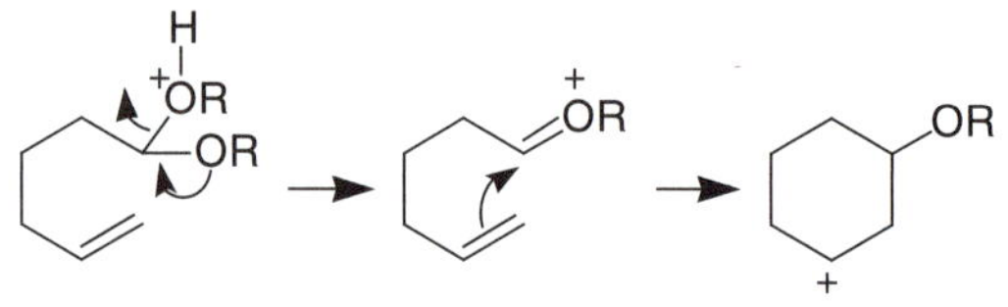

The success of a given cyclization depends on an interplay between the method of generation of the carbocation (*initiation*), the nucleophilicity of the double bond undergoing bond formation with the cationic center (*propagation*), and the method for intercepting the resulting carbocation by an internal or external nucleophile or by elimination of a proton to form an alkene (*termination*). All of these steps can proceed in a concerted or in a stepwise manner depending on the nature of the substrate and the reaction conditions. It is important to terminate the cyclization process by trapping the carbocation in the final product in an efficient manner to minimize side reactions. Proton elimination is usually the observed termination step with tertiary cations and also takes place with secondary cations when the cyclizing reagent is a Lewis acid. Termination by nucleophilic attack is often observed when a protonic acid such as formic acid is used as a cyclizing reagent and is usually stereoselective. Without careful control of the reaction conditions, complex mixtures of products may be obtained.

The monocyclization reaction shown below is initiated by protonation of the trisubstituted double bond to give a transient tertiary carbenium ion. Markovnikov addition of this electrophilic species to the proximal double bond forms a σ-bond and generates a new tertiary carbocation. Loss of a proton produces the methylenecyclohexane derivative.[20]

Bicyclization of polyenes provides access to fused ring systems. When the cyclization leads to a cyclohexane ring, the reaction usually proceeds via a chairlike transition state. Thus, the allylic cation-promoted cyclization of the butenylcyclo-hexenol below presumably proceeds via formation of the allylic carbocation followed by cyclization. Equatorial interception of the resultant cation by the solvent furnishes the bicyclic formate ester.[21]

Formation of ring junctions with substrates containing an (E)-alkene in a 5,6-relationship to the carbocation initiator leads to the thermodynamically more stable *trans*-fused rings, as exemplified by the cyclization of diene A to the *trans*-bicyclic product B. Interestingly, the same product B was obtained when the (Z)-isomer C was subjected to the same cyclization conditions.[22] It appears that both dienes A and C cyclize initially to a common monocyclic intermediate E. Cyclization of E and trapping of the incipient cation by formic acid proceeds in a *trans*-manner to give the bicyclic formate ester F, which is hydrolyzed under the reaction conditions to the alcohol B.[23]

Five-member rings are widely encountered in the structures of natural products such as in the D ring of steroids. A methyl[24] or trialkylsilyl[25] substituted triple bond located in the 5,6-position to a developing cationic center is an effective cyclization terminator, leading to a *trans*-fused five-membered ring. Where ring junctions are formed in bi- and tricyclizations, the stereochemistry of the alkene that becomes the bridgehead is maintained.[19c] In the example below, the (*E*)-alkene cyclizes to the *trans*-fused product in 90% yield.

In summary, the stereochemical course of cation-π cyclizations is determined by stereoelectronic and conformational effects. Concerted cation-π cyclizations usually involve a stereospecific *trans*-addition (axial attack) of the carbocation to the double bond. This is exemplified by Johnson's biomimetic synthesis (±)-progesterone, which possesses a *trans, anti, trans*-fused ring system.[26]

Pericyclic reactions take place in a single step without intermediates and involve a cyclic redistribution of bonding electrons through a *concerted* process in a cyclic transition state. An example is the *Diels-Alder* reaction, a pericyclic reaction between a 1,3-diene and a dienophile that forms a six-member ring adduct.

| diene | dienophile | | adduct |

A related intramolecular pericyclic reaction is the *electrocyclic* ring closure of 1,3,5-polyenes.

The Diels-Alder Reaction[1,28]

The discovery of the [$4\pi + 2\pi$] cycloaddition reaction by Otto Diels (Nobel Prize, 1950) and Kurt Alder (Nobel Prize, 1950), a landmark in synthetic organic chemistry,[29] permits the regio- and stereoselective preparation of both carbocyclic and heterocyclic ring systems. Its application can result simultaneously in an increase of (1) the number of rings, (2) the number of asymmetric centers, and (3) the number of functional groups. *The reaction controls the relative stereochemistry at four contiguous centers.* The Diels-Alder reaction can be depicted as a concerted $\pi^{4s} + \pi^{2s}$ (suprafacial) cycloaddition. While depicted as a concerted process, the reaction has been proposed to proceed in a nonsynchronous manner via an unsymmetrical transition state.[30]

Cycloadditions are controlled by *orbital symmetry* (Woodward-Hoffman rules) and can take place only if the symmetry of all reactant molecular orbitals is the same as the symmetry of the product molecular orbitals.[31] Thus, an analysis of all reactant and product orbitals is required. A useful simplification is to consider only the *frontier molecular orbitals.*[32] These orbitals are the *highest occupied molecular orbitals* (HOMO) and the *lowest unoccupied molecular orbitals* (LUMO). The orbital symmetry must be such that bonding overlap of the terminal lobes can occur with suprafacial geometry; that is, both new bonds are formed using the same face of the diene.

The Diels-Alder reaction is reversible, and many adducts, particularly those formed from cyclic dienes, dissociate into their components at higher temperatures.[33] Indeed, a *retro*-Diels-Alder reaction is the principal method for preparing cyclopentadiene prior to its use in cycloaddition reactions.[34]

Dienes

The diene must be able to adopt the s-*cis* geometry (s refers to the single bond) in order for cycloaddition to occur. Thus, mono-substituted (*E*)-dienes are more reactive than (*Z*)-dienes since (*E*)-dienes more readily adopt the reactive s-*cis* conformation.[35]

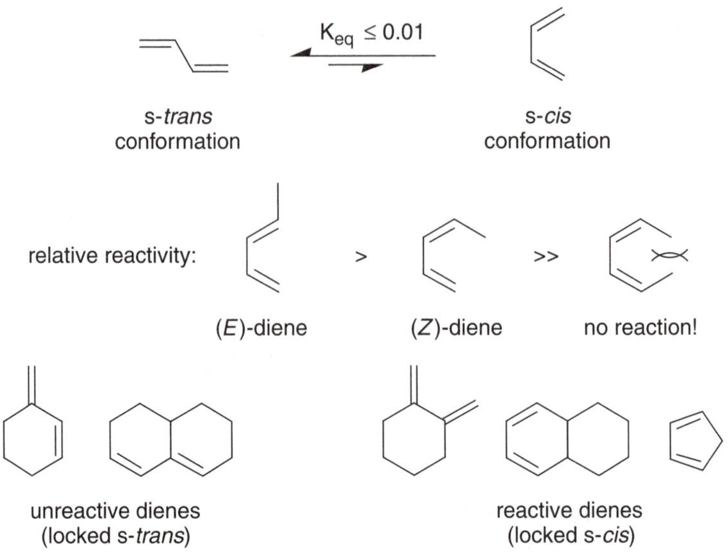

s-*trans*
conformation

s-*cis*
conformation

$K_{eq} \leq 0.01$

relative reactivity:

(*E*)-diene > (*Z*)-diene >> no reaction!

unreactive dienes
(locked s-*trans*)

reactive dienes
(locked s-*cis*)

In *electron-demand* Diels-Alder reactions, dienes are activated by electron-donating substituents, such as alkyl, $-NR_2$, and $-OR$. Electron-rich dienes accelerate the reaction with electron-deficient dienophiles, as illustrated by the relative reactivity trend shown below.

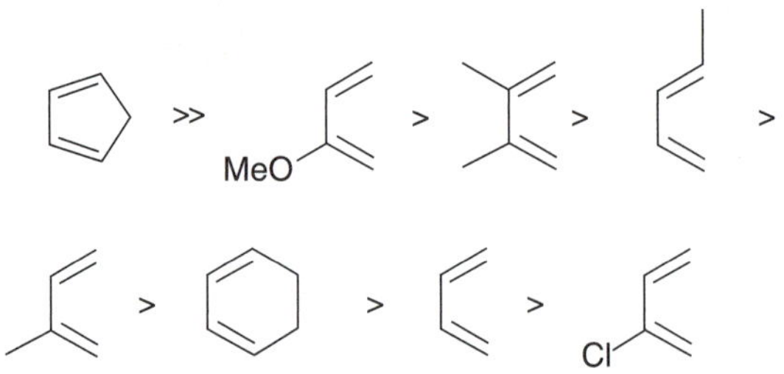

Among the more reactive and synthetically useful dienes are doubly and triply activated alkoxy- and amino-substituted dienes, such as (*E*)-1-methoxy-3-(trimethylsilyloxy)-1,3-butadiene (Danishefsky's diene),[36] (*E*)-1-(dimethylamino)-3-(*tert*-butyldimethylsilyloxy)-1,3-butadiene (Rawal's diene)[37], and 1,3-dimethoxy-1-(trimethylsilyloxy)-1,3-butadiene (Brassard's diene).[38] As illustrated below, the cyclo-addition products arising from these dienes can either be hydrolyzed or treated with fluoride ion to remove the silyl group, which is followed by β-elimination to provide conjugated cyclohexenones.

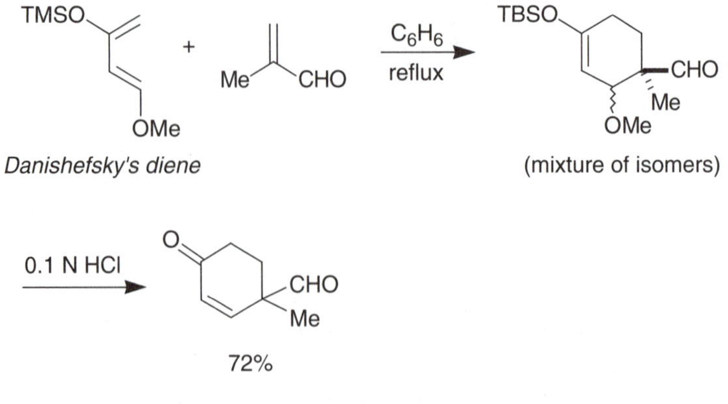

Danishefsky's diene

C_6H_6
reflux

(mixture of isomers)

0.1 N HCl

72%

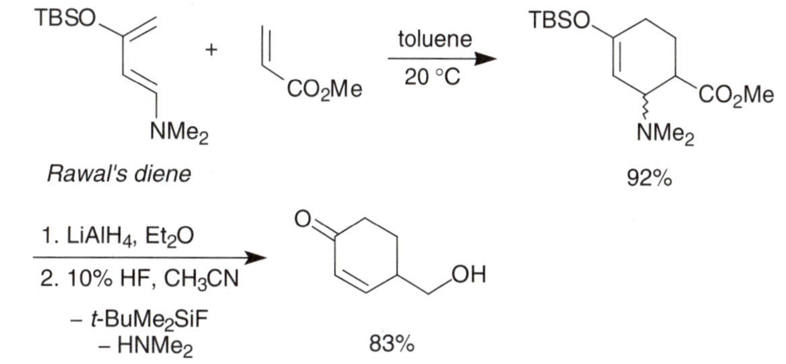

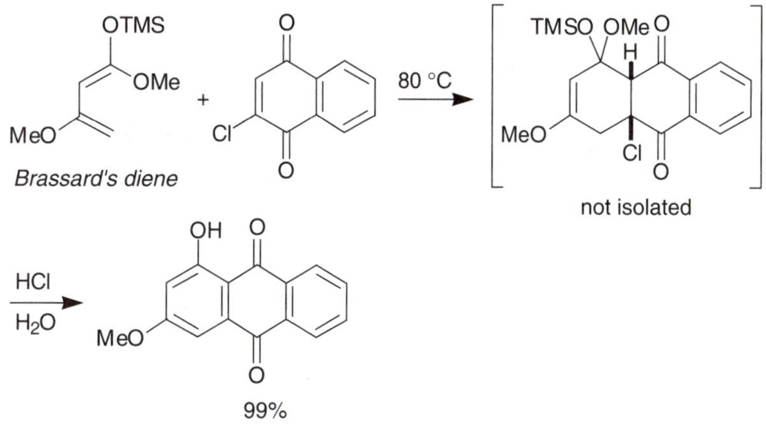

Rawal's diene

92%

83%

Brassard's diene

not isolated

99%

Dienophiles

Dienophiles are activated by electron-withdrawing substituents.[39] Alkyl groups, by means of inductive electron donation and steric effects, tend to reduce the rate of cycloaddition.

Relative dienophile reactivity

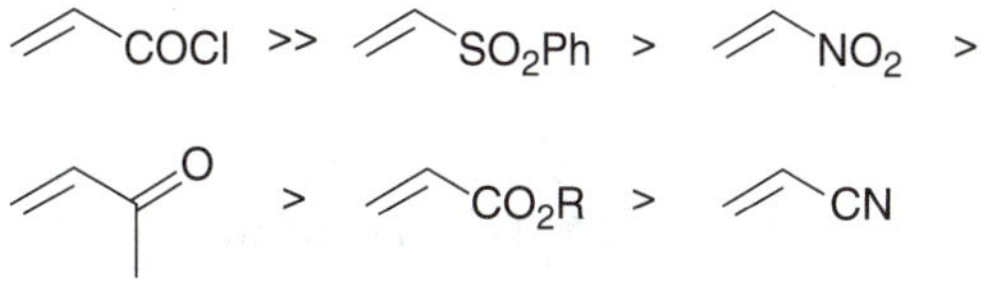

The reactivity of dienophiles may be increased by conjugation with additional electron-withdrawing groups. Doubly activated alkynes and 1,4-benzoquinones are particularly good participants in the Diels-Alder reaction.[40]

Relative dienophile reactivity

EWG = electron-withdrawing group

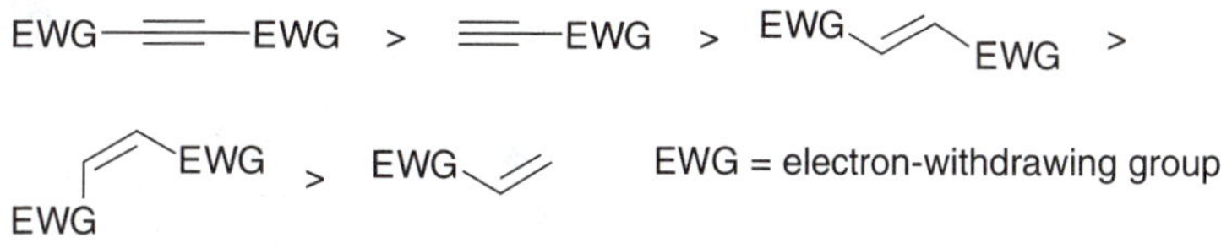

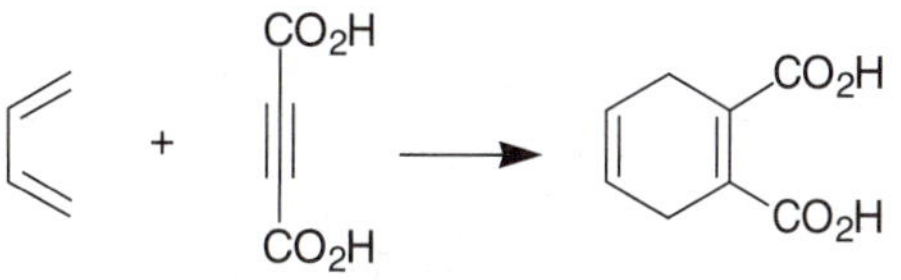

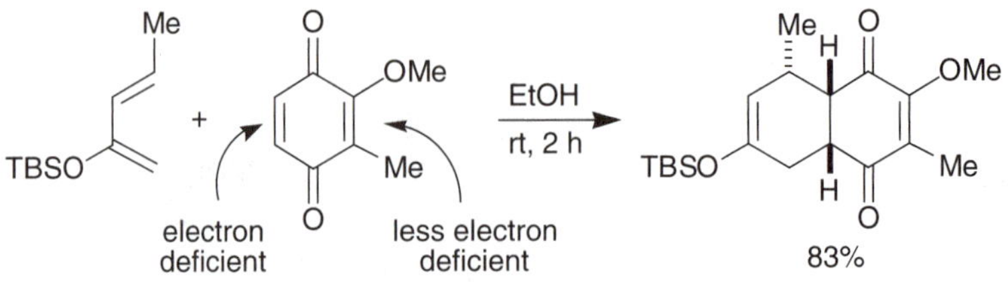

In situ-generated cyclopentadienones are among the most reactive dienophiles.[41] The cyclopentadienone moiety is generated by eliminative processes in the presence of a diene, as shown below.[42] The doubly activated alkene reacts in preference to the singly activated one.

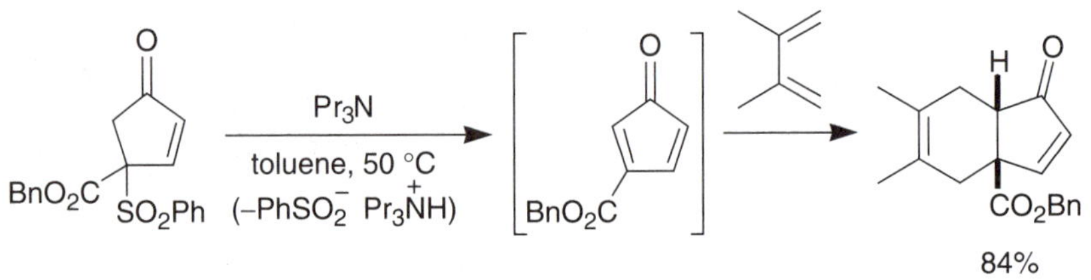

Cyclopentadienone equivalents, such as 4-acetoxy-2-cyclopenten-1-one, below, may serve as conjunctive reagents for tandem Diels-Alder reactions providing polycyclic adducts.[43]

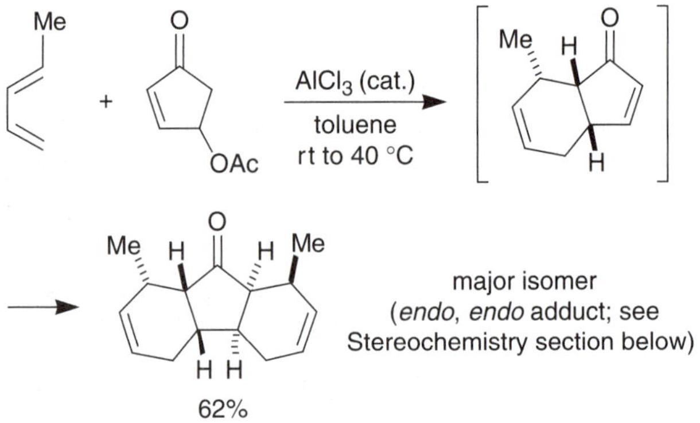

major isomer
(*endo, endo* adduct; see
Stereochemistry section below)

62%

Lewis Acid Activation

Many Diels-Alder reactions are accelerated by Lewis acid catalysts[44] such as $BF_3 \cdot OEt_2$, $AlCl_3$, Et_2AlCl, $SnCl_4$, $TiCl_4$, and $InCl_3$.[45] These increase the rate of reaction by complexation with conjugated C=O and C≡N groups in the dienophile.[46,47]

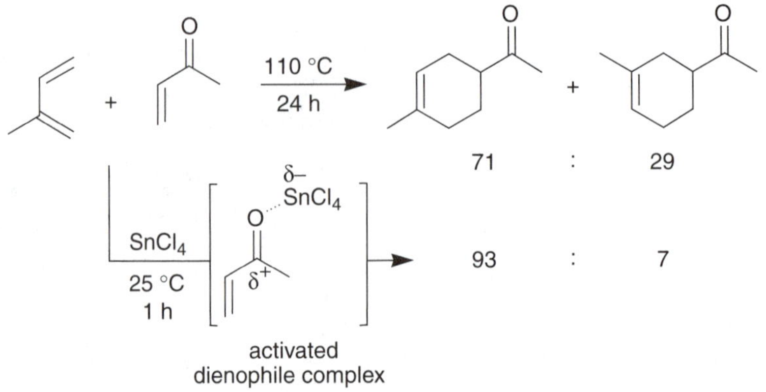

The reaction selectivity is often improved when using Lewis acids.[45b,48,49]

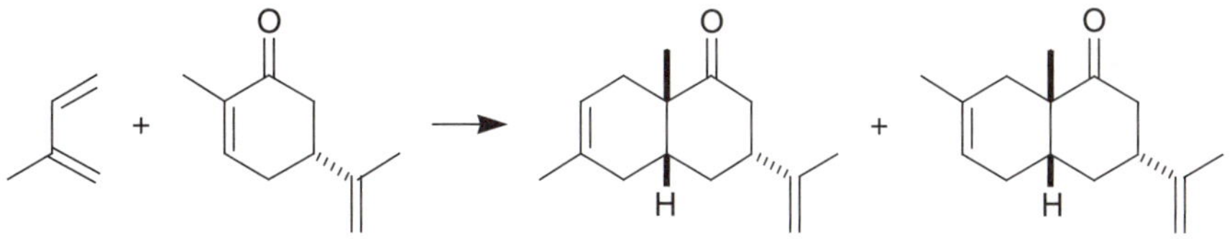

conditions	yield	isomer ratio
150 °C, 142 h	20%	65 : 35
AlCl$_3$, 25 °C, 17 h	52%	90 : 10 (80% de)

Lewis acid activation is particularly important for the catalysis of *hetero*-Diels-Alder reactions involving aldehydes.[50] Lanthanide complexes of Yb and Eu[51] as well as samarium diiodide[52] are mild catalysts for this reaction.

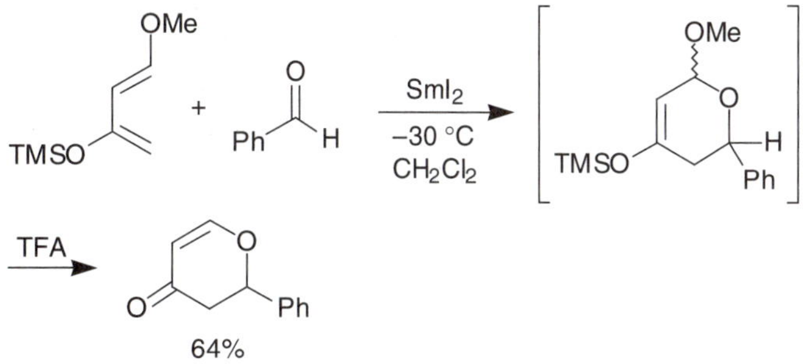

Regiochemistry

The regioselectivity of Diels-Alder reactions ranges from moderate to very high.[53] As shown below in Tables 9.2 and 9.3, the cycloaddition reactions of monosubstituted dienes proceed with good selectivity.[54] Generally, the more powerful the electronic effect of the diene substituent is, the more regioselective is the reaction.

Table 9.2   1-Substituted Butadienes React to Give Mainly the "ortho" Product

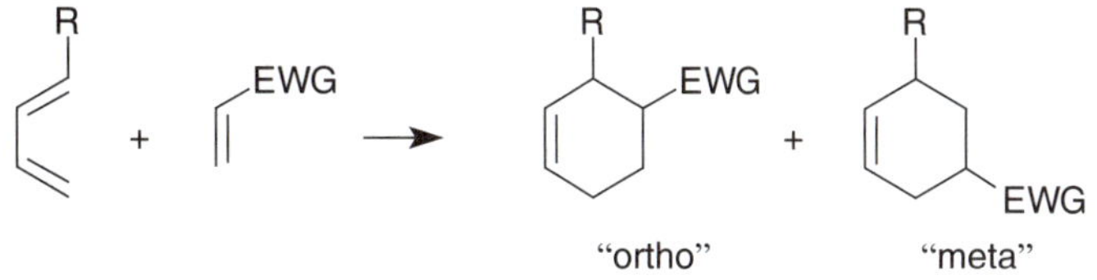

R	EWG	Temp	Yield	ortho : meta
NEt$_2$	CO$_2$H	20 °C	94%	100 : 0
CH$_3$	CO$_2$H	20 °C	64%	95 : 5
CO$_2$H	CO$_2$H	70 °C	86%	91 : 9

Table 9.3   2-Substituted Butadienes React to Give Mainly the "para" Product

R	EWG	Temp	Yield	meta : para
OEt	CO_2CH_3	160 °C	50%	0 : 100
CH_3	CO_2H	20 °C	54%	16 : 84
CN	CO_2CH_3	95 °C	86%	0 : 100

A "simplistic" approach to predicting the regiochemical course of a Diels-Alder reaction is to consider the polarization of the diene and of the dienophile by examining the resonance forms, and then join the atoms with unlike charges to form a six-member ring, as exemplified below. However, this approach fails to account for some reactions that occur with good regiochemistry.[55]

D = electron-donating group
EWG = electron-withdrawing group

To rigorously predict the regiochemistry of Diels-Alder reactions, one has to apply frontier molecular orbital theory.[56]

- Estimate the energies of the HOMO and the LUMO of both components.
- Identify which HOMO-LUMO pair is closer in energy.
- Using this HOMO-LUMO pair, estimate the relative sizes of the coefficients of the atomic orbitals on the atoms at which bonding is to occur.
- Match up the larger coefficient of one component with the larger on the other.

Note: In the example above, the HOMO of the diene possessing an electron-donating group (OMe) has the largest coefficient at the terminus of that group, and the LUMO has the larger coefficient at the end of the electron-withdrawing CO_2Me group.

Stereochemistry

An important aspect of the Diels-Alder reaction is its stereospecificity, wherein the relative stereochemical relationships present in the starting materials are preserved throughout the course of the reaction.[57]

***cis*-Principle.** The D–A reaction is *concerted* and *suprafacial* with respect to diene and dienophile. Hence, the stereochemistries of both the diene and dienophile are retained in the adduct. Note that the initial suprafacial cycloadduct formed may in some cases be prone to isomerization.

Suprafacial with respect to diene:

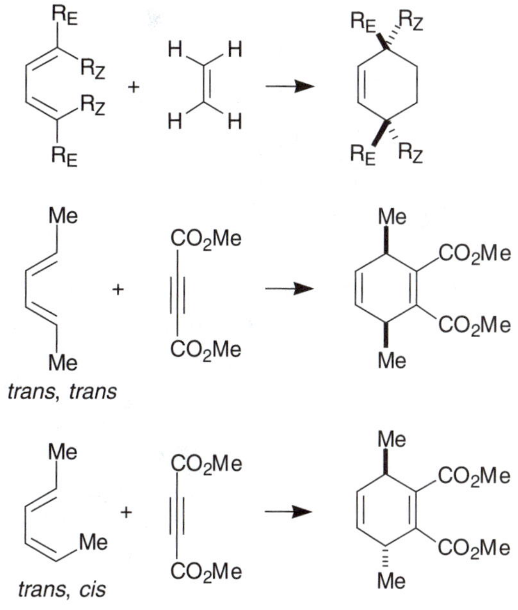

trans, trans

trans, cis

Suprafacial with respect to dienophile:

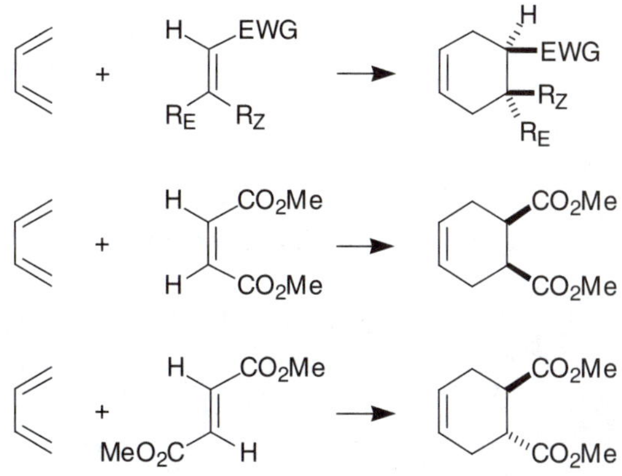

***endo*-Principle.** The dienophile can undergo reaction via two different orientations with respect to the plane of the diene (*endo* or *exo*). Maximum overlap of the π-orbitals of the diene and dienophile favors formation of the *endo*-adduct (*secondary orbital overlap* between the dienophile's activating substituent and the diene).[58]

exo-Transition state (−CO$_2$Me away from π-system of diene)[59]

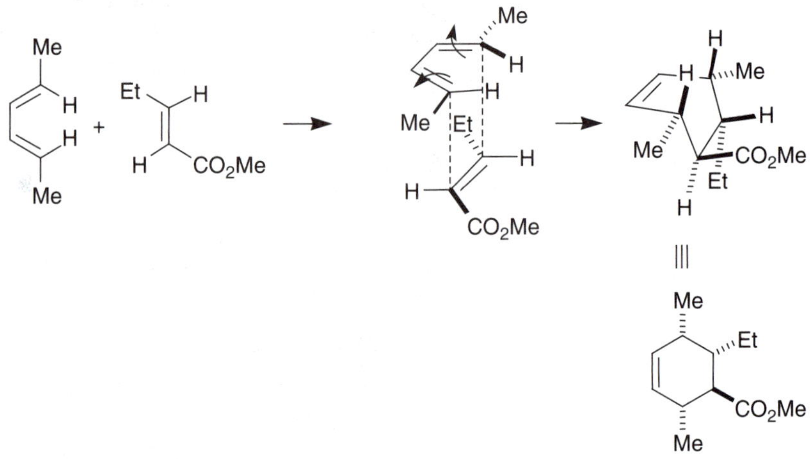

endo-Transition state (−CO$_2$Me toward π-system of diene)[59]

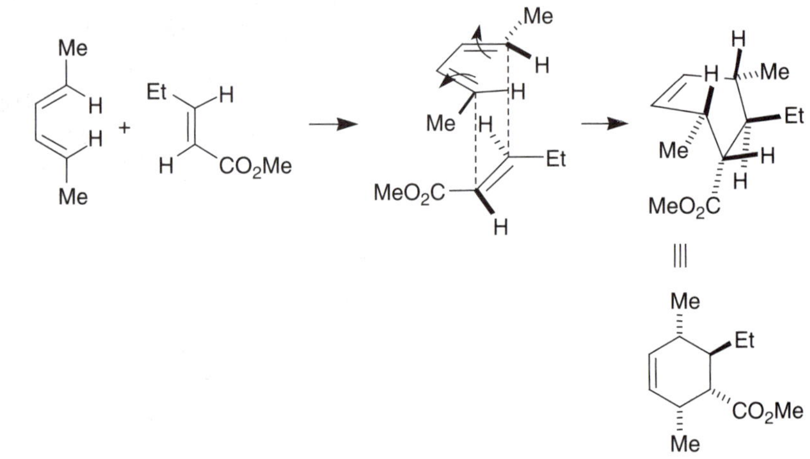

Note that the *endo*-product has the electron-withdrawing group *cis* to the (*E, E*)-substituents of the diene. Only one mode of stacking of the reagents is shown, with the diene *above* the dienophile. The other possibility is with the diene *below* the dienophile, leading to the enantiomer.

In general, *endo-selectivity* is determined by a balance of electronic and steric effects. The *endo*-selectivity may be improved by using Lewis acid catalysis to lower the temperature required for cycloaddition, as shown below.

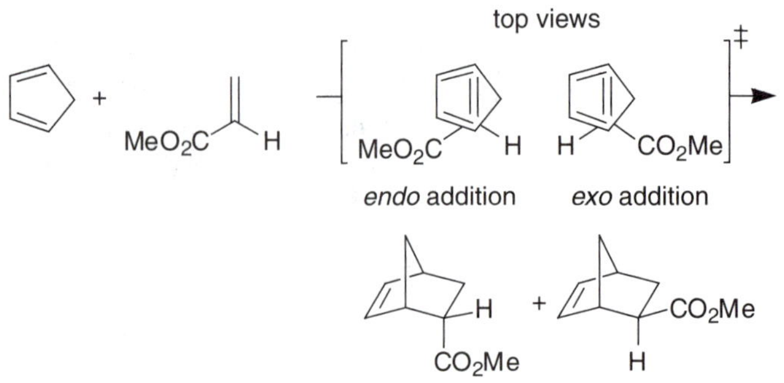

	endo		*exo*
in CH$_2$Cl$_2$ at 0 °C	80	:	20
in CH$_2$Cl$_2$ at −70 °C + BF$_3$•OEt$_2$	100	:	0

The diene will approach the dienophile from the face opposite a bulky substituent, and a dienophile will approach the less hindered face of the diene.

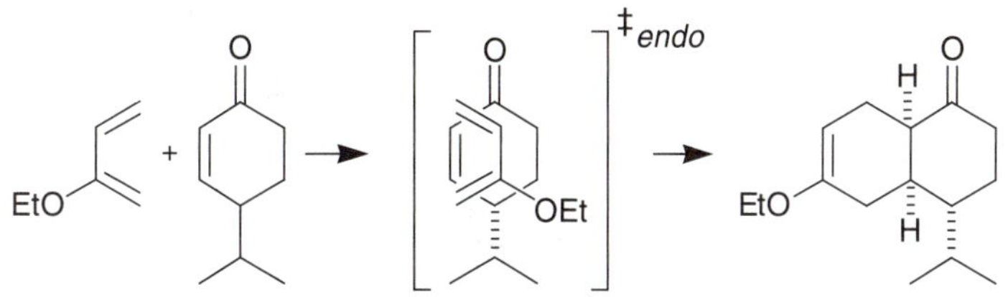
EtO
O
OEt
‡ endo
H
H

diene assumes *endo*-TS on face *opposite* the *i*-Pr group

Intramolecular Diels-Alder Reactions[28i,60]

In the intramolecular Diels-Alder reaction, two rings are formed in one step. Notable advantages of intramolecular cycloadditions include (1) entropy assistance (2) well-defined regiochemistry, and (3) greater control of *endo*- versus *exo*- selectivity.

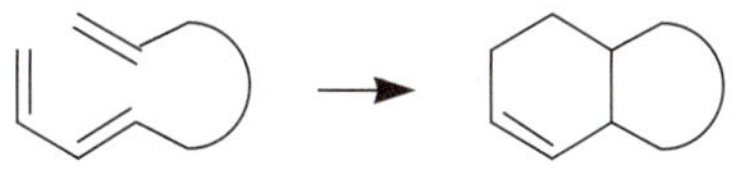

fused

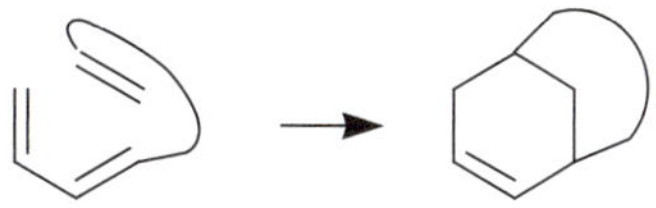

bridged

Of the two possible regiochemical modes of addition, one that leads to a fused product usually predominates over one that leads to a bridged product, especially for (*E*)-dienes. Although products formed via an *endo*-transition state may predominate, steric constraints in transition states may favor the *exo*- products. As a retrosynthetic strategy, the intramolecular Diels-Alder reaction should be considered for any synthesis of a molecule containing a six-member ring fused to a six- or five-member ring.

(Z)-Dienes can attain only a single transition state (on either face of the dienophile) as a consequence of their geometry. Hence, intramolecular cycloadditions of (Z)-dienes are highly diastereoselective.[61]

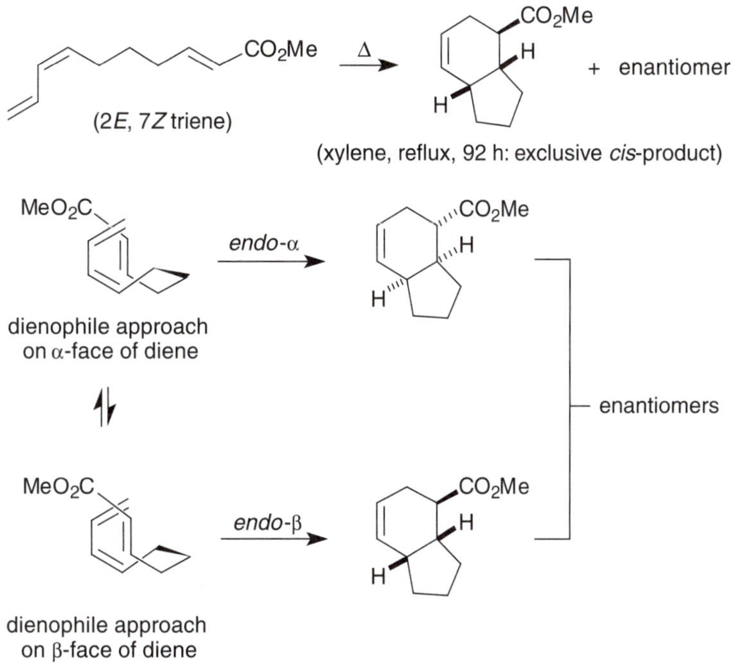
CO2Me
Δ
CO2Me
H
H
+ enantiomer
(2E, 7Z triene)
(xylene, reflux, 92 h: exclusive cis-product)
MeO2C
endo-α
CO2Me
H
H
dienophile approach
on α-face of diene
enantiomers
MeO2C
endo-β
CO2Me
H
H
dienophile approach
on β-face of diene

In contrast, (*E*)-dienes are more likely to afford *cis*- and *trans*-fused cycloadducts since both the *endo*- and *exo*-transition states are conformationally possible.

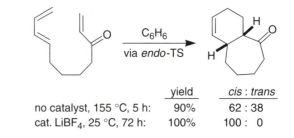

As with intermolecular D-A reactions, the diastereoselectivity of intramolecular cycloadditions can be improved by the addition of Lewis acids.[62,63]

	yield	cis : trans
no catalyst, 155 °C, 5 h:	90%	62 : 38
cat. LiBF$_4$, 25 °C, 72 h:	100%	100 : 0

	yield			
C$_6$H$_6$, 150 °C, 72 h:	65%	60	:	40
EtAlCl$_2$, rt, 36 h:	60%	100	:	0

Asymmetric Diels-Alder Reactions

Catalysis of D-A reactions by Lewis acids makes it possible to conduct the cycloadditions under mild conditions, which promotes higher levels of diastereoselection and enantioselection in comparison to the thermally induced reactions. Control over the formation of single diastereomers or enantiomers in D-A reactions may be achieved using chiral promoters functioning either as chiral auxilliaries[64] (*substrate control*) or chiral catalysts[65] (*reagent control*).

Substrate Control

Camphor-derived scaffolds can function as chiral auxilliaries in Lewis acid–promoted D-A reactions to direct the cycloaddition in a predictable manner with nearly complete asymmetric induction.[66]

chiral auxilliary
derived from (–)-camphor

98% (> 99% de)

Reagent Control

Chiral oxazaborolidine salts are useful catalysts not only for enantioselective reductions (i.e., CBS reduction), but also for promoting enantioselective D-A reactions.[67] The cationic Lewis acids formed by reaction of chiral oxazaborolidines with triflic acid [TfOH] or trifluoromethanesulfonimide [(Tf)$_2$NH] coordinate to dienophiles to direct subsequent cycloadditions in a highly controlled fashion. The D-A reactions using these chiral salts proceed in CH$_2$Cl$_2$ under very mild conditions at temperatures as low as –95 °C (Table 9.4).[67]

Ar = Ph or 3,5-dimethylphenyl
X$^-$ = CF$_3$SO$_3^-$ or (CF$_3$O$_2$S)$_2$N$^-$

CBS salts for enantioselective
Diels-Alder additions

The dienophile α-face is blocked;
only the β-face is available for cycloaddition

Table 9.4 Enantioselective Diels-Alder Reactions Using Oxazaborolidine Catalysis

R	Cat. mol %	Temp, Time	Yield (*endo : exo*)	% ee
H	6	–95 °C, 2 h	90 (92 : 8)	69
Et	20	–20 °C, 2 h	99 (94 : 6)	97
OEt	20	–20 °C, 16 h	96 (97 : 3)	> 99

R	Cat. mol %	Temp, Time	Yield (*endo : exo*)	% ee
H	20	–95 °C, 1.5 h	36 (84 : 16)	63
Et	20	–20 °C, 2 h	97 (69 : 31)	65
OEt	13	+4 °C, 72 h	46 (91 : 9)	> 98

1,3-Dipolar Cycloaddition Reactions: [3+2] Cycloadditions[68]

Rolf Huisgen developed the concept of 1,3-dipolar cycloadditions in the early 1960s.[69] These cycloadditions are closely related to the Diels-Alder reaction in that they also are 6 π-electron pericyclic reactions. However, unlike with the Diels-Alder reaction, the 4 π-electron partner is a 3-atom moiety containing at least one heteroatom. These charged molecules are called *1,3-dipoles*, having charge-separated resonance structures with opposite charges in a 1,3-relationship. The 2 π-electron partner is the *dipolarophile*, usually an alkene or an alkyne. The following example illustrates the basic principles involved in a 1,3-dipolar cycloaddition reaction.

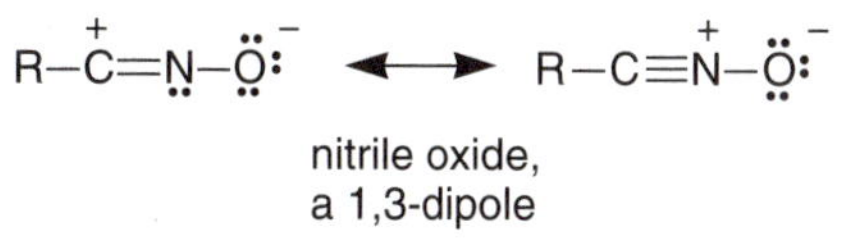

$$R-\overset{+}{C}=\overset{\cdot\cdot}{N}-\overset{\cdot\cdot}{\underset{\cdot\cdot}{O}}:^{-} \longleftrightarrow R-C\equiv\overset{+}{N}-\overset{\cdot\cdot}{\underset{\cdot\cdot}{O}}:^{-}$$

nitrile oxide,
a 1,3-dipole

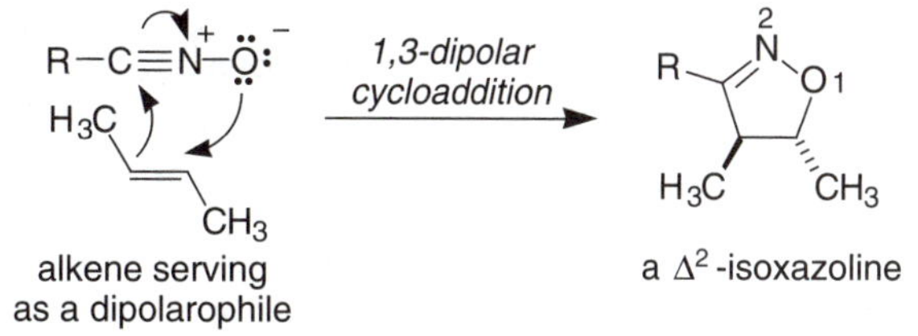

alkene serving
as a dipolarophile

a Δ^2-isoxazoline

The products of [3 + 2] cycloadditions are *five-member heterocyclic systems*. A large body of information is available concerning the scope of 1,3-dipolar addition reactions and their mechanistic interpretation; a detailed discussion is outside the realm of this book.

9.4 RING-CLOSING OLEFIN METATHESIS (RCM)[70]

Olefin metathesis has emerged as a powerful tool for the preparation of cyclic organic compounds. Metathesis involves the redistribution of carbon-carbon double bonds in the presence of metal carbene complexes ($[M]=CR_2$). The reaction of these metal carbenes with α,ω-dienes leads to well-defined carbocyclic systems in what is termed *ring-closing olefin metathesis* (RCM).[71]

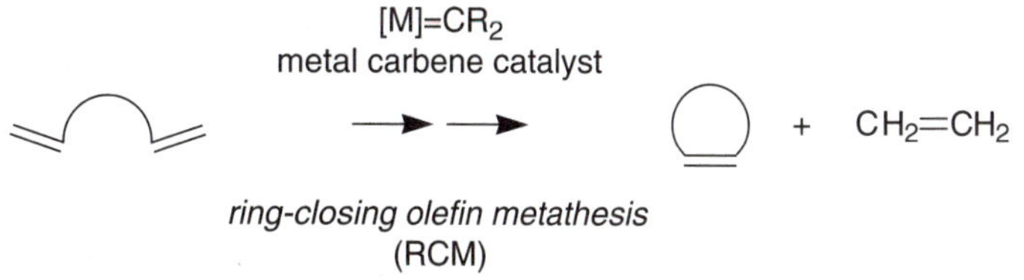

[M]=CR_2
metal carbene catalyst

ring-closing olefin metathesis
(RCM)

The introduction of molybdenum[72] by R. Schrock (Nobel Prize, 2005) and ruthenium catalyst systems by R. Grubbs (Nobel Prize, 2005), especially the commercially available *Grubbs catalyst*[73] [bis(tricyclohexylphosphine)benzylidene ruthenium(IV) dichloride], A, and its second-generation variant, B, shown below,[74,75] has led to many RCM applications in organic synthesis.[70]

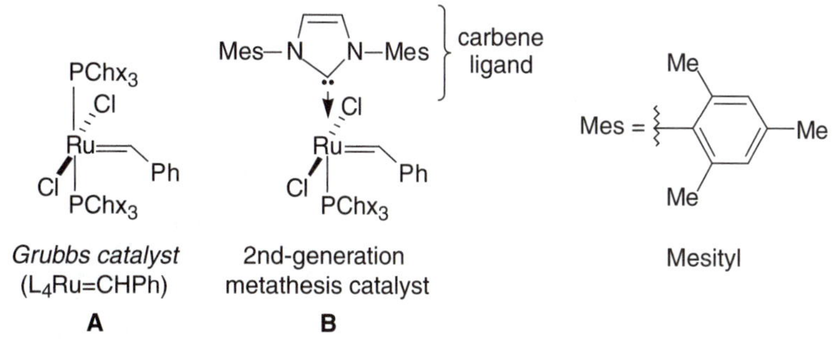

Grubbs catalyst
($L_4Ru=CHPh$)
A

2nd-generation
metathesis catalyst
B

Mesityl

RCM may proceed by the formation of a series of metallocyclobutane intermediates that break apart to form new alkenes and new metal carbenes, which propagate the reaction.[70f,76] The cycloalkene product accumulates as the reaction proceeds since the reverse reaction, ring-opening metathesis, is kinetically disfavored.

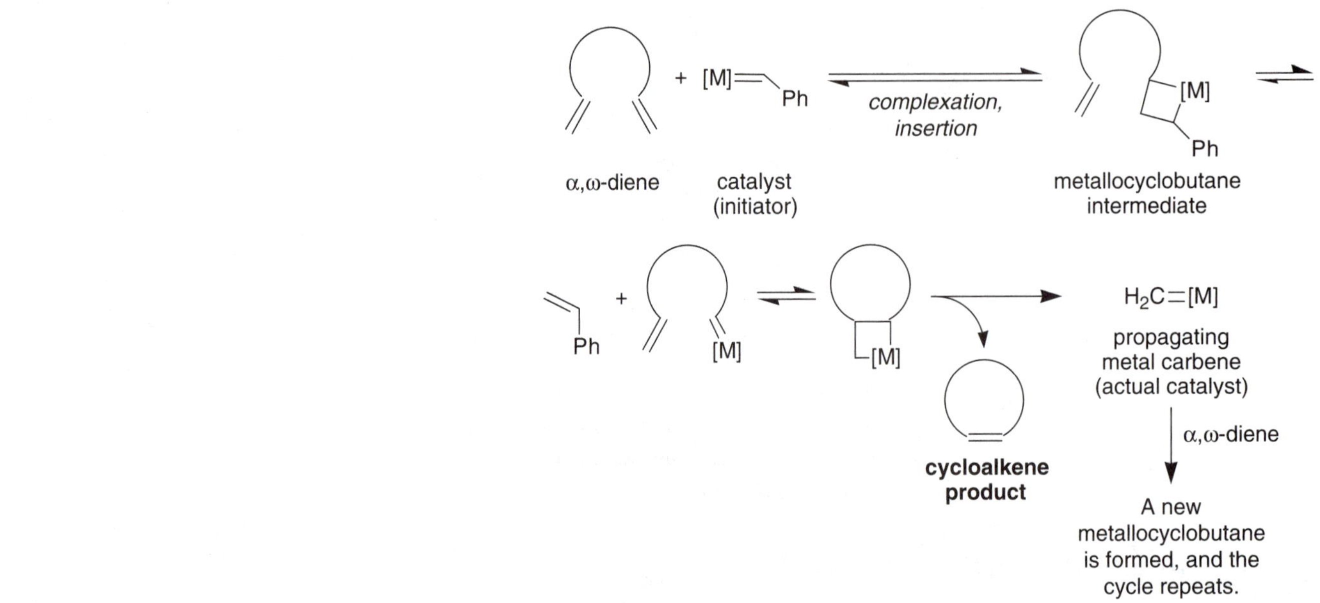

The reaction tolerates a wide range of functional groups, including base-sensitive (ester) and acid-sensitive (acetonide) functionality, as shown in the following examples.[77]

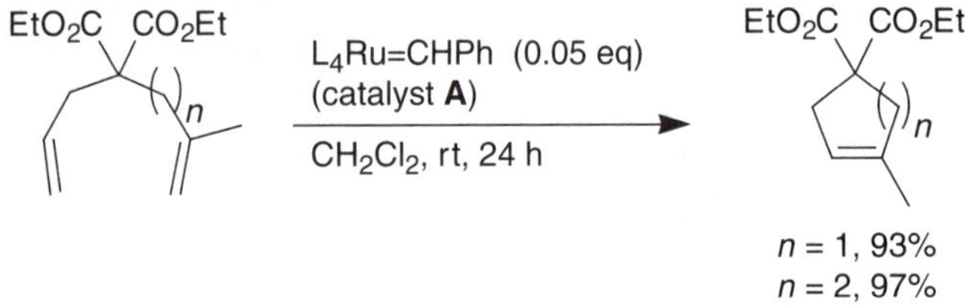

n = 1, 93%
n = 2, 97%

Note that the ruthenium catalyst also can be used in the presence of hydroxyl groups.[78]

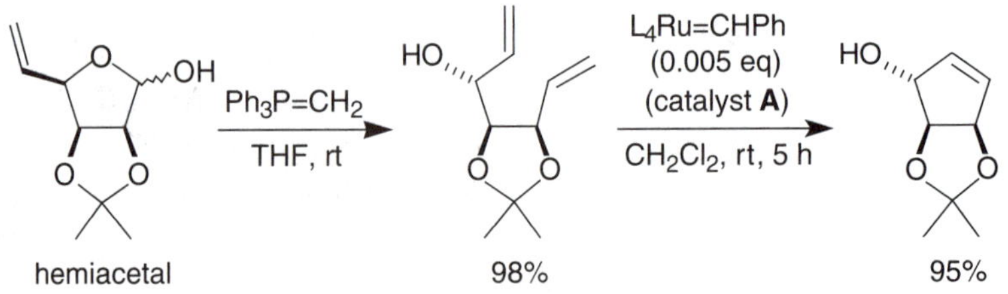

hemiacetal   98%   95%

RCM is an excellent strategy for the synthesis of spiro-fused compounds.[79]

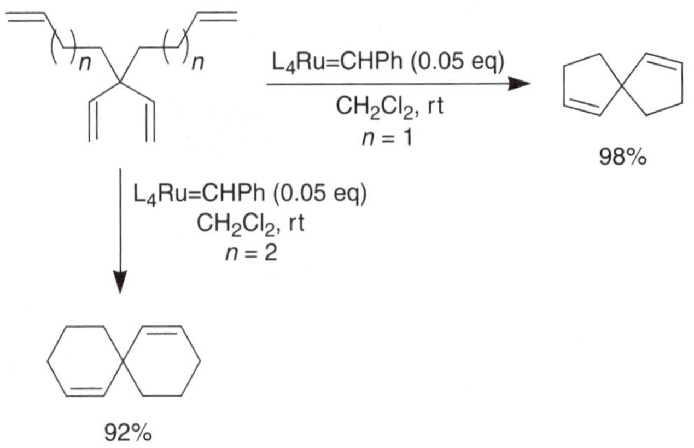

98%

92%

Synthesis of Heterocycles

The olefin metathesis reaction is often used to prepare heterocyclic rings.[80] The mild conditions tolerate a wide variety of oxygen and nitrogen functionalities, such as the epoxide[81] and cyclic amines, as exemplified below.[82,83]

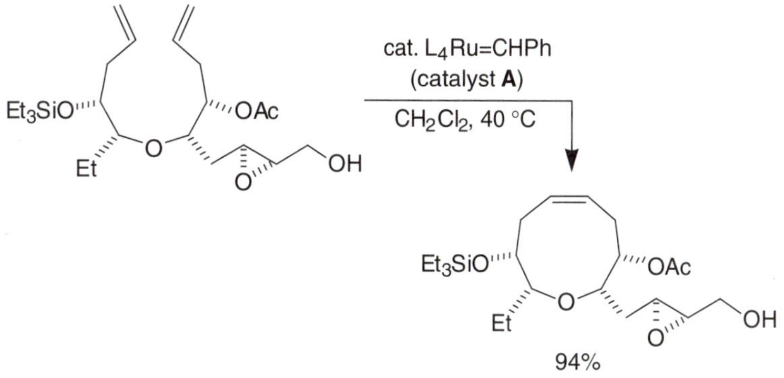

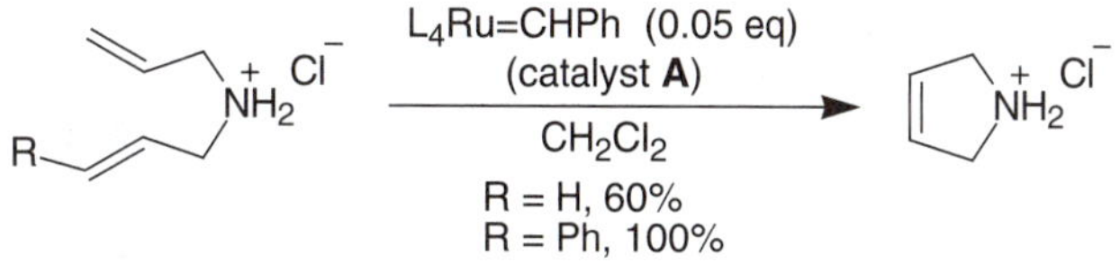

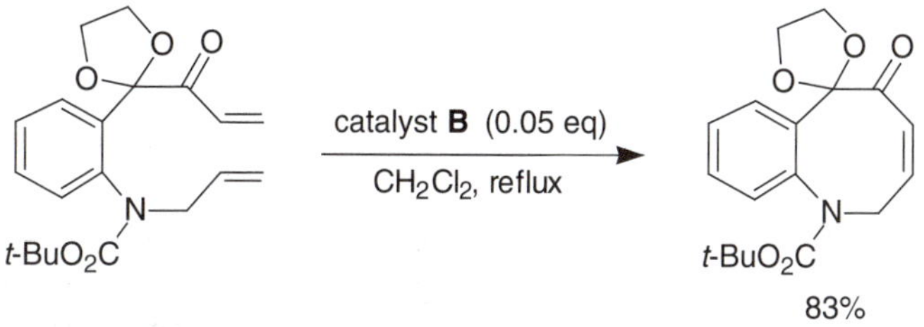

When larger amounts of ruthenium catalyst are used (0.3–0.5 eq), the removal of colored, toxic ruthenium metal by-products is problematic and requires multiple purifications by silica gel column chromatography interspersed with activated carbon treatments.[84,85]

Large-Ring Synthesis

The RCM reaction is one of the most reliable methods for the formation of large-ring systems. The preparation of macrocyclic systems via RCM reactions may lead to E/Z mixtures in which the (E)-isomer generally predominates because of further metathesis reactions that isomerize the product to a thermodynamically favorable E/Z ratio.[86]

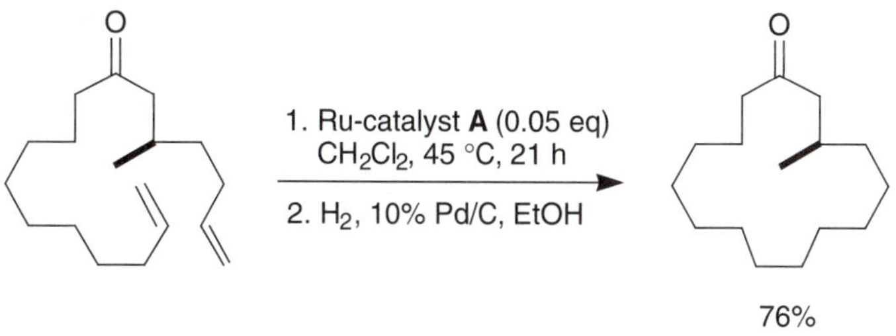

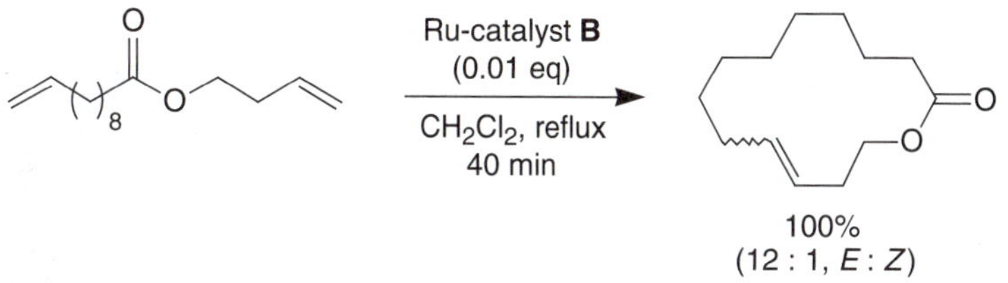

PROBLEMS

1. **Reagents**. Give the structure of the major product expected for each of the following reactions. Be sure to indicate product stereochemistry where applicable.

a.

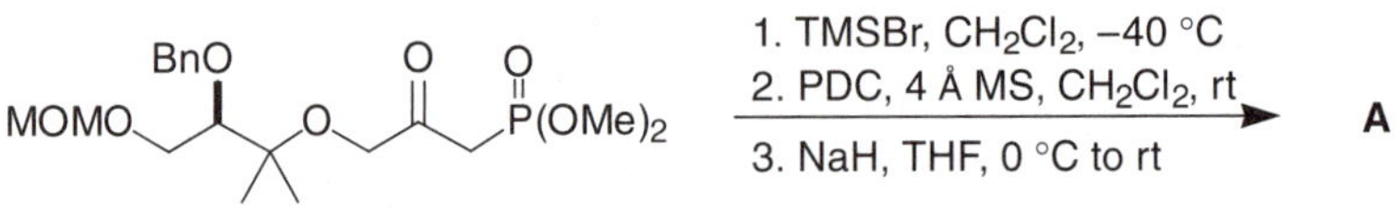

MOMO, BnO, P(OMe)$_2$

1. TMSBr, CH$_2$Cl$_2$, –40 °C
2. PDC, 4 Å MS, CH$_2$Cl$_2$, rt
3. NaH, THF, 0 °C to rt

A

b.

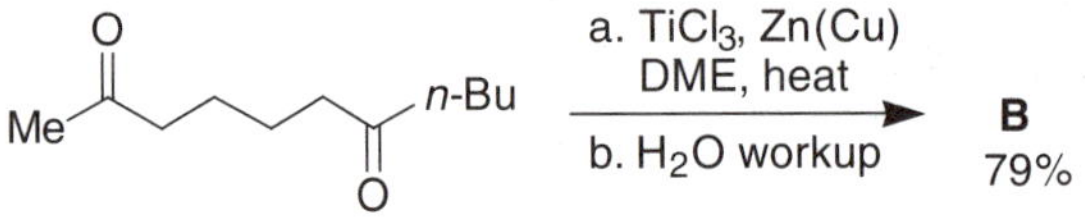

Me, n-Bu

a. TiCl$_3$, Zn(Cu) DME, heat
b. H$_2$O workup

B 79%

c.

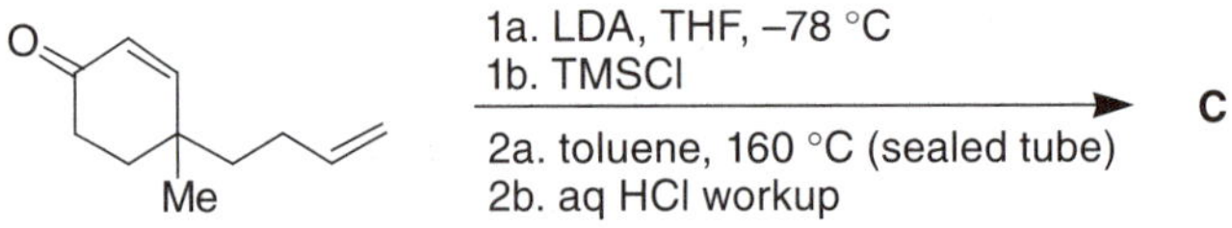

Me

1a. LDA, THF, –78 °C
1b. TMSCl
2a. toluene, 160 °C (sealed tube)
2b. aq HCl workup

C

d.

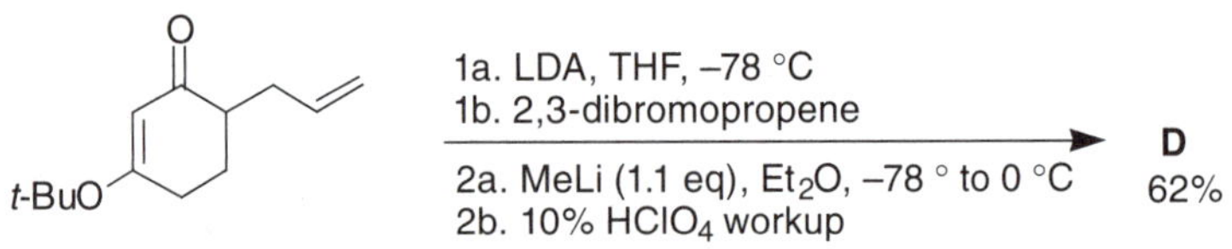

t-BuO

1a. LDA, THF, –78 °C
1b. 2,3-dibromopropene
2a. MeLi (1.1 eq), Et$_2$O, –78 ° to 0 °C
2b. 10% HClO$_4$ workup

D 62%

e.

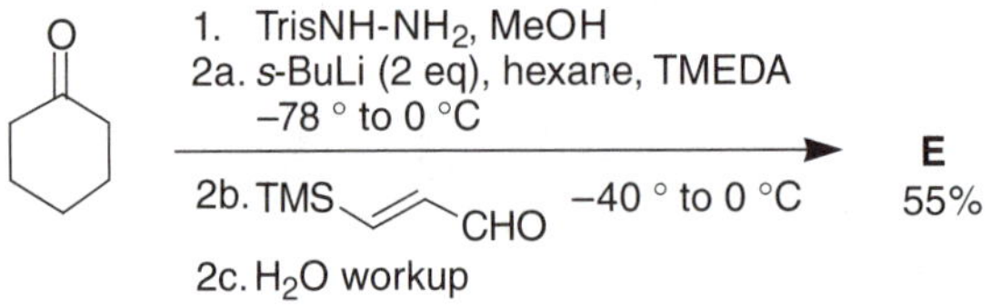

1. TrisNH-NH$_2$, MeOH
2a. s-BuLi (2 eq), hexane, TMEDA –78 ° to 0 °C
2b. TMS   CHO   –40 ° to 0 °C
2c. H$_2$O workup

E 55%

f.

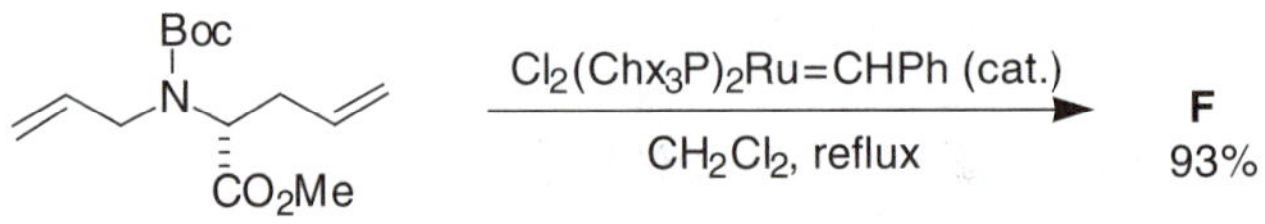

Boc, N, CO$_2$Me

Cl$_2$(Chx$_3$P)$_2$Ru=CHPh (cat.)
CH$_2$Cl$_2$, reflux

F 93%

g.

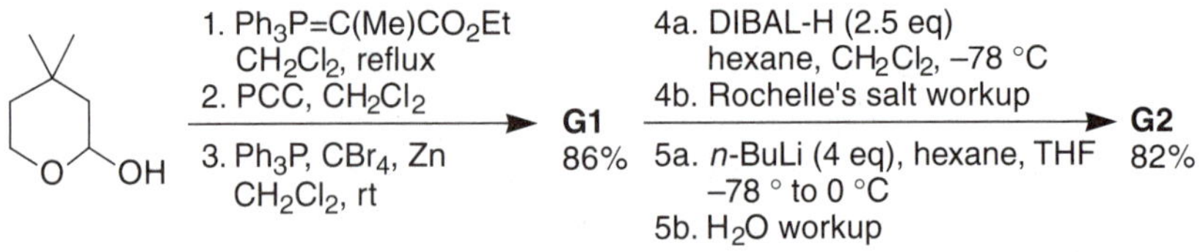

OH, O

1. Ph$_3$P=C(Me)CO$_2$Et CH$_2$Cl$_2$, reflux
2. PCC, CH$_2$Cl$_2$
3. Ph$_3$P, CBr$_4$, Zn CH$_2$Cl$_2$, rt

G1 86%

4a. DIBAL-H (2.5 eq) hexane, CH$_2$Cl$_2$, –78 °C
4b. Rochelle's salt workup
5a. n-BuLi (4 eq), hexane, THF –78 ° to 0 °C
5b. H$_2$O workup

G2 82%

h.

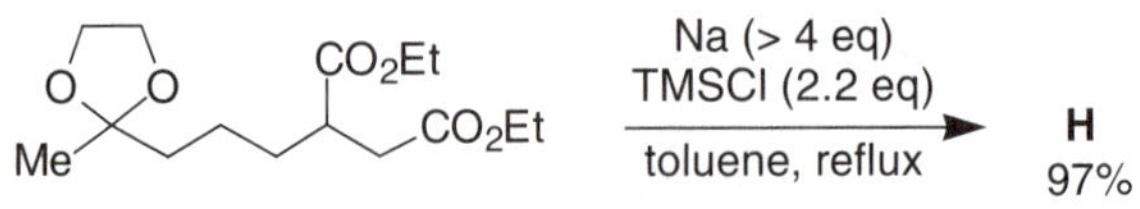

Me, CO$_2$Et, CO$_2$Et

Na (> 4 eq)
TMSCl (2.2 eq)
toluene, reflux

H 97%

i.

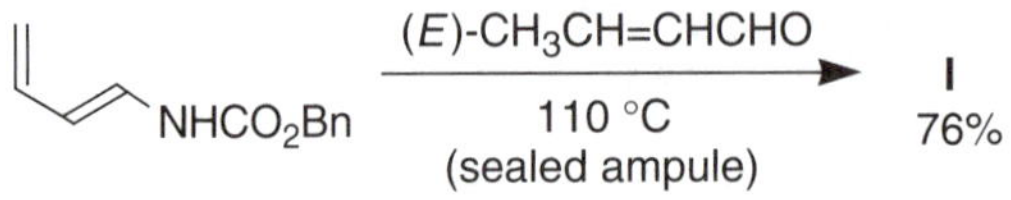

NHCO$_2$Bn

(E)-CH$_3$CH=CHCHO
110 °C
(sealed ampule)

I 76%

*j.

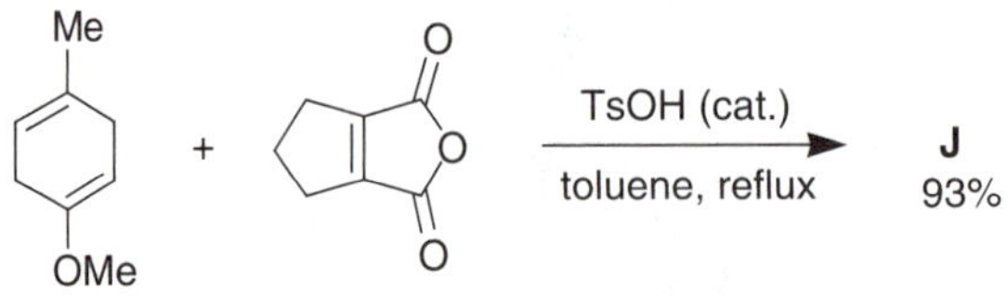

Me, OMe   +   O, O

TsOH (cat.)
toluene, reflux

J 93%

2. Selectivity. Show the product(s) obtained and the missing reagent(s) necessary for each of the following transformations.

a.

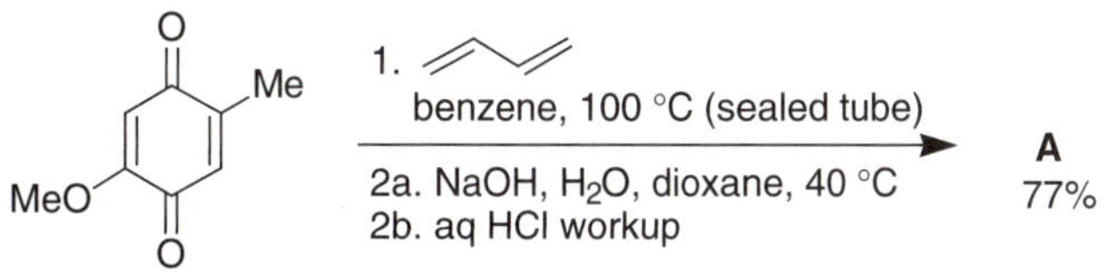
Me
MeO
1.
benzene, 100 °C (sealed tube)
2a. NaOH, H_2O, dioxane, 40 °C
2b. aq HCl workup
A
77%

b.

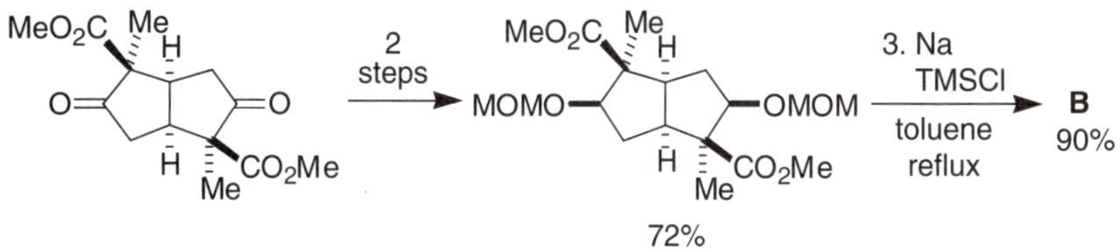
MeO_2C Me H
H CO_2Me Me
2 steps
MOMO OMOM
72%
3. Na
TMSCl
toluene
reflux
B
90%

c.

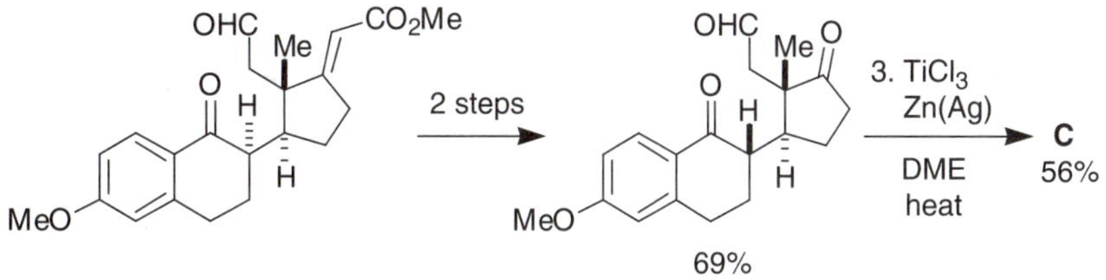
OHC CO_2Me Me
H
H
MeO
2 steps
69%
3. $TiCl_3$
Zn(Ag)
DME
heat
C
56%

***d.**

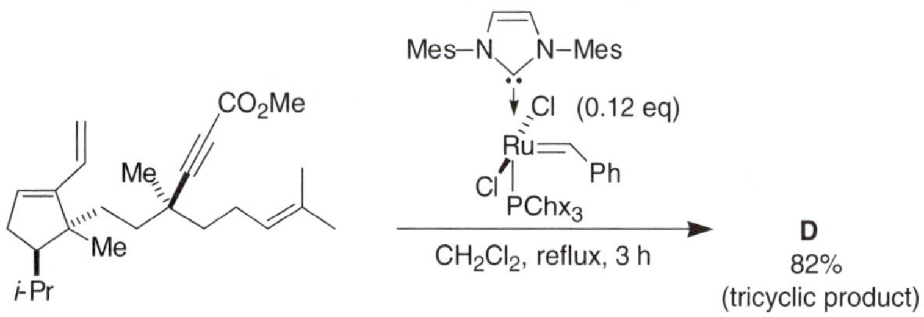
CO_2Me
Me
Me
i-Pr
Mes–N N–Mes
Cl (0.12 eq)
Ru
Ph
Cl
$PChx_3$
CH_2Cl_2, reflux, 3 h
D
82%
(tricyclic product)

***e.**

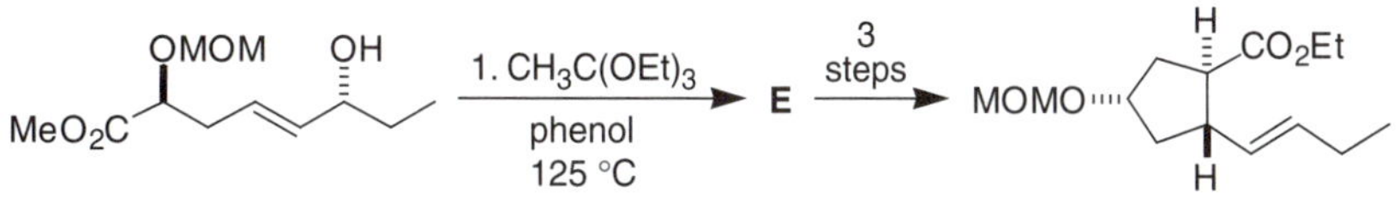
OMOM OH
MeO_2C
1. $CH_3C(OEt)_3$
phenol
125 °C
E
3 steps
MOMO
H CO_2Et
H

3. Stereochemistry. Give the structure of the major product formed for each of the following reactions. Provide an explanation for your choice of stereochemistry.

a.

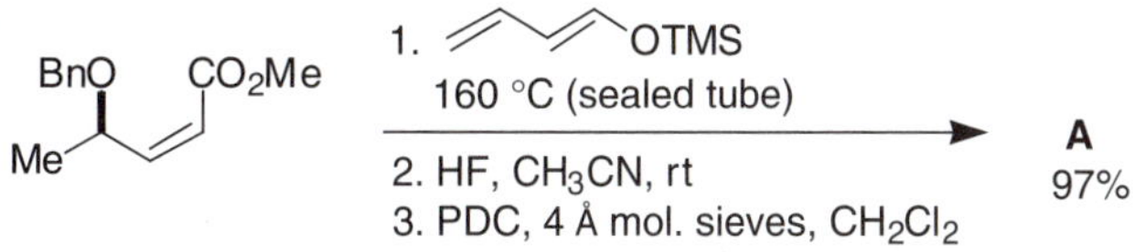
BnO CO_2Me
Me
1. OTMS
160 °C (sealed tube)
2. HF, CH_3CN, rt
3. PDC, 4 Å mol. sieves, CH_2Cl_2
A
97%

b.

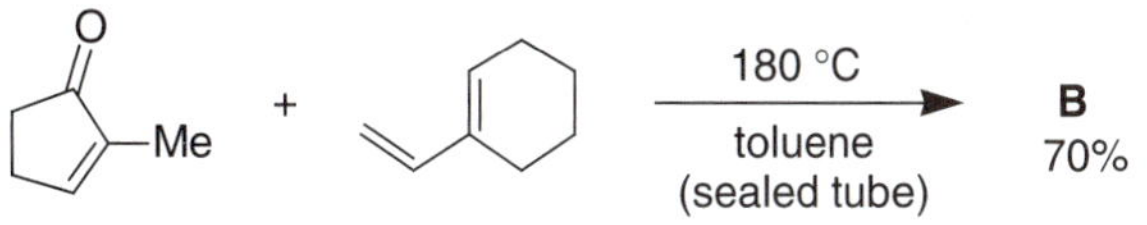
Me
+
180 °C
toluene
(sealed tube)
B
70%

c.

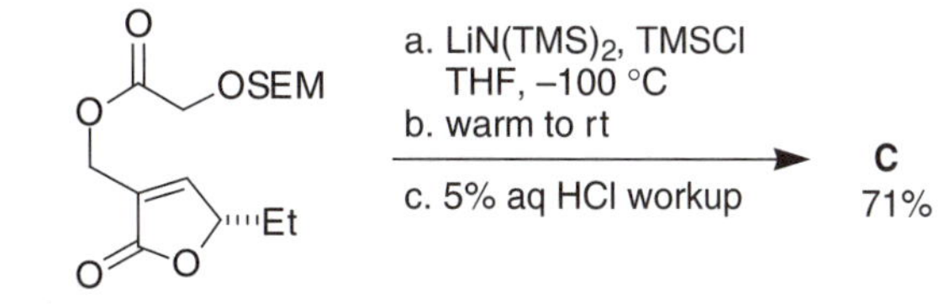
OSEM
Et
a. $LiN(TMS)_2$, TMSCl
THF, –100 °C
b. warm to rt
c. 5% aq HCl workup
C
71%

SEM = $CH_2CH_2SiMe_3$

d.

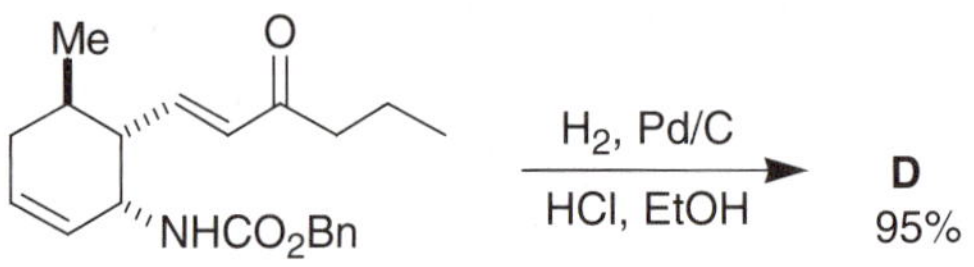

$$\xrightarrow[\text{HCl, EtOH}]{\text{H}_2, \text{Pd/C}} \quad \textbf{D}$$

95%

*e.

$$\xrightarrow[\text{b. aq NaHCO}_3 \text{ workup}]{\text{a. BF}_3 \cdot \text{OEt}_2, \text{CH}_2\text{Cl}_2, -78°\text{C}} \quad \textbf{E}$$

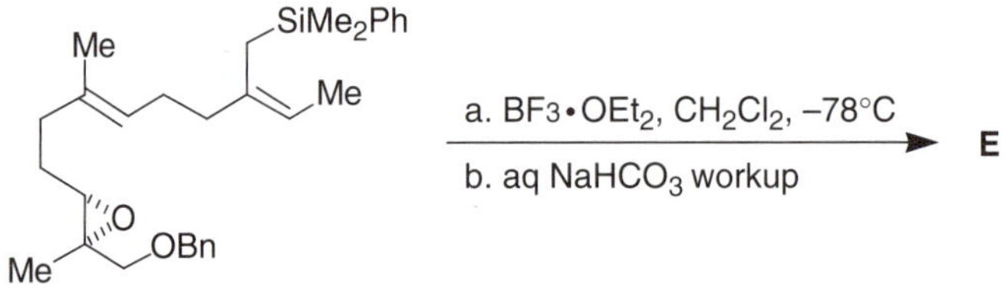

4. **Reactivity**. Propose a mechanism for each of the following transformations that explains the observed regioselectivity and stereochemistry.

a.

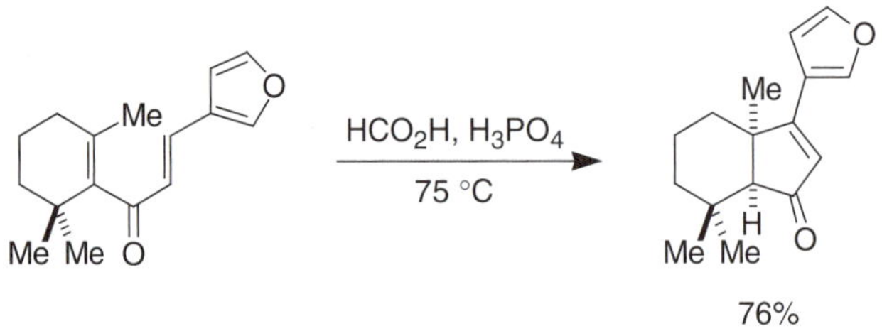

$$\xrightarrow[75 \text{ °C}]{\text{HCO}_2\text{H, H}_3\text{PO}_4}$$

76%

b.

$$\xrightarrow{\text{CF}_3\text{CO}_2\text{H}}$$

c.

$$\xrightarrow[\substack{\text{b. CF}_3\text{CO}_2\text{H} \\ \text{c. NaCN}}]{\substack{\text{a. CH}_3\text{CN} \\ \text{4 Å mol.} \\ \text{sieves}}}$$

79%

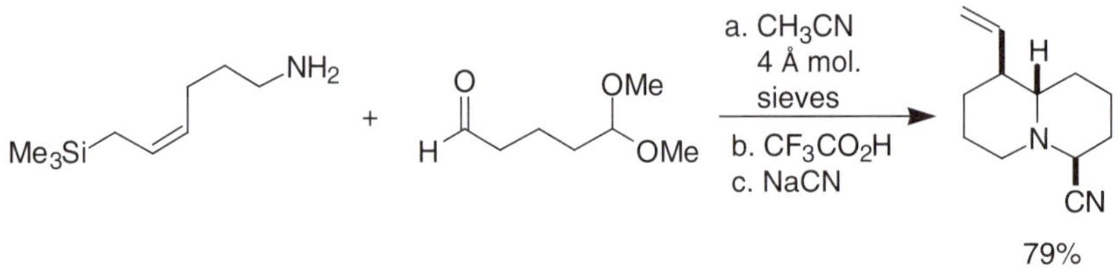

*d.

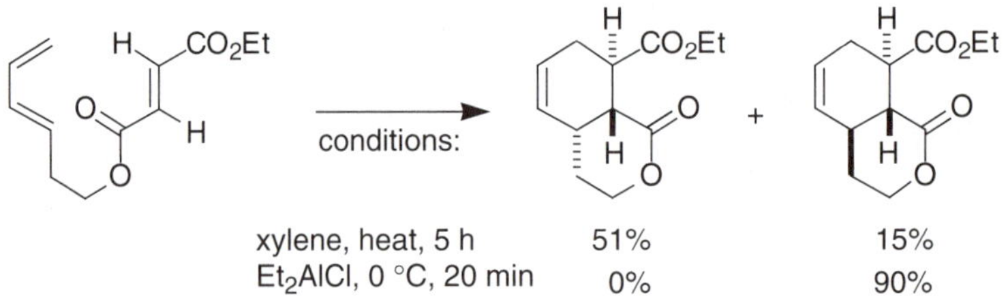

conditions:

xylene, heat, 5 h   51%   15%
Et$_2$AlCl, 0 °C, 20 min   0%   90%

*e.

$$\xrightarrow[\text{b. H}_2\text{O workup}]{\substack{\text{a. TiCl}_4 \text{ CH}_2\text{Cl}_2 \\ -78 \text{ to } 0°\text{C}}}$$

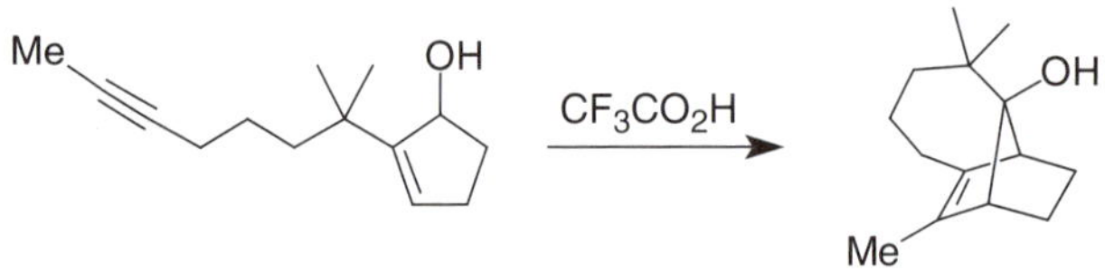

95%

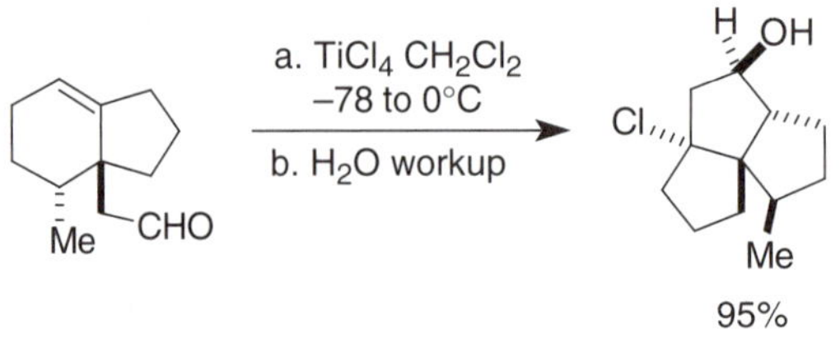

5. **Synthesis**. Supply the missing reagents necessary to accomplish each of the following syntheses. Be sure to control the relative stereochemistry.

a.

b.

c.

d.

e.

f.

g.

6. **Retrosynthetic Analysis**

a. Propose a synthesis of (−)-coniceine using a ring-closing metathesis (RCM) procedure as a key step. Show (1) your retrosynthetic analysis and (2) all reagents and reaction conditions required to transform a commercially available starting material into the target molecule.

(−)-coniceine

b. Propose syntheses of pumiliotoxin C using (1) an intramolecular Diels-Alder cycloaddition as a key step and (2) a conjugate addition of a side chain, either the C(5)-methyl or C(2)-n-propyl group.

pumiliotoxin C

REFERENCES

1. Mundy, B. P. *Concepts of Organic Synthesis: Carbocyclic Chemistry*, Marcel Decker: New York, 1979, pp. 50–64.

2. (a) Finely, K. T. *Chem. Rev.* **1964**, *64*, 573. (b) House, H. O. *Modern Synthetic Reactions,* 2nd ed., Benjamin: Menlo Park, CA, 1972, p. 169. (c) Bloomfield, J. J., Owsley, D. C., Nelke, J. M. *Org. React.* **1976**, *23*, 259. (e) Brettle, G. In *Comprehensive Organic Synthesis*, Trost, B. M., Fleming, I., Eds., Pergamon Press: Oxford, UK, 1991, Vol. 3, p. 613.

3. Bloomfield, J. J., Owsley, D. C., Ainsworth, C., Robertson, R. E. *J. Org. Chem.* **1975**, *40*, 393.

4. Rühlmann, H. *Synthesis* **1971**, 236.

5. Bloomfield, J. J., Nelke, J. M. *Org. Synth., Col. Vol. VI* **1988**, 167.

6. Schräpler, U., Rühlmann, K. *Chem. Ber.* **1964**, *97*, 1383.

7. Ashkenazi, P., Kettenring, J., Migdal, S., Gutman, A. L., Ginsburg, P. *Helv. Chim. Acta* **1985**, *68*, 2033.

8. Cope, A. C., Barthel, J. W., Smith, R. D. *Org. Synth., Coll. Vol IV* **1963**, 218.

9. Schräpler, U., Rühlmann, K. *Chem. Ber.* **1963**, *96*, 2780.

10. (a) McMurry, J. E. *Chem. Rev.* **1989**, *89*, 1513. (b) Robertson, G. M. In *Comprehensive Organic Synthesis*, Trost, B. M., Fleming, I., Eds., Pergamon Press: Oxford, UK, 1991, Vol. 3, p. 563. (c) Wirth, T. In *Organic Synthesis Highlights III*, J. Mulzer and H. Waldmann, Eds., Wiley-VCH: Weinheim, Germany, 1998, p. 113. (d) Fu, G. C. In *Modern Carbonyl Chemistry*, Otera, J., Ed., Wiley-VCH: Weinheim, Germany, 2000, p. 69.

11. Corey, E. J., Danheiser, R. L., Chandrasekaran, S. *J. Org. Chem.* **1976**, *41*, 260.

12. McMurry, J. E., Rico, J. G. *Tetrahedron Lett.* **1989**, *30*, 1169.

13. (a) McMurry, J. E. *Acc. Chem. Res.* **1983**, *16*, 405. (b) Fürstner, A. In *Transition Metals for Organic Synthesis*, Beller, M. Bolm, Eds., Wiley-VCH: Weinheim, Germany, 1998, Vol. 1, p. 381.

14. McMurry, J. E., Fleming, M. P., Kees, K. L., Krepski, L. R. *J. Org. Chem.* **1978**, *43*, 3255.

15. McMurry, J. E., Lectka, T., Rico, J. G. *J. Org. Chem.* **1989**, *54*, 3748.

16. Ishida, A., Mukaiyama, T. *Chem. Lett.* **1976**, 1127.

17. McMurry, J. E., Miller, D. *J. Am. Chem. Soc.* **1983**, *105*, 1660.

18. Li, W., Li, Y., Li, Y. *Synthesis* **1994**, 678.

19. (a) Johnson, W. S. *Acc. Chem. Res.* **1968**, *1*, 1. (b) Johnson, W. S. *Angew. Chem., Int. Ed.*, **1976**, *15*, 9. (c) Sutherland, J. K. In *Comprehensive Organic Synthesis,* Trost, B. M., Fleming, I., Eds., Pergamon Press: Oxford, UK, 1991, Vol. 3, p. 341.

20. (a) Grütter, H., Schinz, H. *Helv. Chim. Acta* **1952**, *35*, 1656. (b) Kappeler, H., Schinz, H., Eschenmoser, A., Stauffacher, D. *Helv. Chim. Acta* **1954**, *37*, 957. (c) Grütter, H., Vogt, H. R., Schinz, H. *Helv. Chim. Acta* **1954**, *37*, 1791.

21. Johnson, W. S., Lunn, W. H., Fitzi, K. *J. Am. Chem. Soc.* **1964**, *86*, 1972.

22. Stadler, P. A., Nechvatal, A., Frey, A. J., Eschenmoser, A. *Helv. Chim. Acta* **1957**, *40*, 1373.

23. Stork, G., Burgstahler, A. W. *J. Am. Chem. Soc.* **1955**, *77*, 5068.

24. Johnson, W. S., Gravestock, M. B., Parry, R. J., Bryson, T. A., Miles, D. H. *J. Am. Chem. Soc.* **1971**, *93*, 4330.

25. (a) Johnson, W. S., Yarnell, T. M., Myers, R. F., Morton, D. R. *Tetrahedron Lett.* **1978**, 2549. (b) Johnson, W. S., Yarnell, T. M., Myers, R. F., Morton, D. R., Boots, S. G. *J. Org. Chem.* **1980**, *45*, 1254.

26. (a) Johnson, W. S., Gravestock, M. B., McCarry, B. E. *J. Am. Chem. Soc.* **1971**, *93*, 4332. (b) Gravestock, M. B., Johnson, W. S., McCarry, B. E., Parry, R. J., Ratcliffe, B. E. *J. Am. Chem. Soc.* **1978**, *100*, 4274. (c) Nicolaou, K. C., Sorensen, E. J. *Classics in Total Synthesis*, VCH: Weinheim, Germany, 1996, p. 83.

27. (a) Houk, K. N. In *Pericyclic Reactions*, Marchand, A. P., Lehr, R. I., Eds., Academic Press: New York, 1977, Vol. 2, p. 203. (b) Fleming, I. *Pericyclic Reactions*, Oxford University Press: Oxford, UK, 1999.

28. (a) Kloetzel, M. C. *Org. React.* **1948**, *4*, 1. (b) Holmes, H. L. *Org. React.* **1948**, *4*, 60. (c) Sauer, J., Sustmann, R. *Angew. Chem., Int. Ed.* **1980**, *19*, 779. (d) Corey, E. J., Cheng, X.-M. *The Logic of Chemical Synthesis*, Wiley: New York, 1989. (e) Carruthers, W. *Cycloaddition Reactions in Organic Synthesis*, Pergamon Press: Oxford, UK, 1990. (f) Oppolzer, W. In *Comprehensive Organic Synthesis,* Trost, B.M., Fleming, I., Eds., Pergamon Press: Oxford, UK, 1991, Vol. 5, p. 315. (g) Fringuelli, F., Taticchi, A. *The Diels-Alder Reaction. Selected Practical Methods*, Wiley: New York, 2002. (h) Smith, M. B. *Organic Synthesis*, 2nd ed., McGraw-Hill: New York, 2002, pp. 917–977. (i) Nicolaou, K. C., Snyder, S. A., Montagnon, T., Vassilikogiannakis, G. *Angew. Chem., Int. Ed.* **2002**, *41*, 1669.

29. Diels, O., Alder, K. *Justus Liebigs Ann. Chem.* **1928**, *460*, 98.

30. Dewar, M. J. S., Pierini, A. B. *J. Am. Chem. Soc.* **1984**, *106*, 203.

31. (a) Woodward, R. B., Hoffmann, R. *Angew. Chem., Int. Ed.* **1969**, *8*, 781. (b) Woodward, R. B., Hoffmann, R. *The Conservation of Orbital Symmetry*, VCH: Weinheim, Germany, 1971.

32. (a) Fukui, K. In *Molecular Orbitals in Chemistry, Physics and Biology*, Löwdin, P.-O, Pullman, B., Eds., Academic Press: New York, 1964, p. 513. (b) Houk, K. *Acc. Chem. Res.* **1975**, *8*, 361. (c) Fleming, I. *Froutier Orbitals and Organic Chemical Reactions*, Wiley: New York, 1976.

33. (a) Rickborn, B. *Org. React.* **1998**, *52*, 1. (b) Rickborn, B. *Org. React.* **1998**, *53*, 223. (c) Sweger, R.W., Czarnik, A. W. In *Comprehensive Organic Synthesis*, Trost, B.M., Fleming, I., Eds., Pergamon Press: Oxford, UK, 1991, Vol. 5, p. 551.

34. Magnusson, G. *J. Org. Chem.* **1985**, *50*, 1998.

35. (a) Fringuelli, F., Taticchi, A. *Dienes in the Diels-Alder Reaction*, Wiley: New York, 1990. (b) Boger, D. In *Comprehensive Organic Synthesis,* Trost, B.M., Fleming, I., Eds., Pergamon Press: Oxford, UK, 1991, Vol. 5, p. 451. (c) Bradley, A. C., Kociolek, M. G., Johnson, R. P. *J. Org. Chem.* **2000**, *65*, 7134.

36. Danishefsky, S., Kitahara, T. *J. Am. Chem. Soc.* **1974**, *96*, 7807.

37. (a) Kozmin, S. A., Rawal, V. H. *J. Org. Chem.* **1997**, *62*, 5252. (b) Kozmin, S. A., Rawal, V. H. *J. Am. Chem. Soc.* **1997**, *119*, 7165. (c) Kozmin, S. A., Janey, J. M., Rawal, V. H. *J. Org. Chem.* **1999**, *64*, 3039.

38. (a) Savard, J., Brassard, P. *Tetrahedron Lett.* **1979**, 4911. (b) For a recent application, see Fan, Q., Lin, L., Liu, J., Huang, Y., Feng, X., Zhang, G. *Org. Lett.* **2004**, *6*, 2185.

39. Weinreb, S. M. In *Comprehensive Organic Synthesis,* Trost, B. M., Fleming, I., Eds., Pergamon Press: Oxford, UK, 1991, Vol. 5, p. 401.

40. Nicolaou, K. C., Vassilikogiannakis, G., Mägerlein, W., Kranich, R. *Angew. Chem., Int. Ed.* **2001**, *40*, 2482.

41. Maier, G. *Chem. Ber.* **1985**, *118*, 3196.

42. Nantz, M. H., Fuchs, P. L. *J. Org. Chem.* **1987**, *52*, 5298.

43. Minuti, L., Selvaggi, R., Taticchi, A., Sandor, P. *Tetrahedron* **1993**, *49*, 1071.

44. For the first reported example of a Lewis acid–catalyzed Diels-Alder reaction, see Yates, P., Eaton, P. *J. Am. Chem. Soc.* **1960**, *82*, 4436.

45. (a) Carruthers, W. *Cycloaddition Reactions in Organic Synthesis*, Pergamon Press: Oxford, UK, 1990. (b) Shing, T. K. M., Lo, H. Y., Mak, T. C. W. *Tetrahedron* **1999**, *55*, 4643. (c) Babu, G., Perumal, P. T. *Aldrichimica Acta* **2000**, *33*, 16.

46. For a discussion on the complexation of a Lewis acid to benzaldehyde, see Reetz, M. T., Hüllmann, M., Massa, W., Berger, S., Rademacher, P., Heymanns, P. *J. Am. Chem. Soc.* **1986**, *108*, 2405.

47. Yamabe, S., Minato, T. *J. Org. Chem.* **2000**, *65*, 1830.

48. For theoretical aspects, see (a) Epiotis, N. D., Shaik, S. *J. Am. Chem. Soc.* **1978**, *100*, 1. (b) Branchadell, V., Oliva, A., Bertran, J. *Chem. Phys. Lett.* **1983**, *97*, 378.

49. Angell, E. C., Fringuelli, F., Pizzo, F., Porter, B., Taticchi, A., Wenkert, E. *J. Org. Chem.* **1985**, *50*, 4696.

50. (a) Danishefsky, S., Kerwin, J. F., Jr., Kobayashi, S. *J. Am. Chem. Soc.* **1982**, *104*, 358. (b) Danishefsky, S., Kerwin, J. F. Jr., *J. Org. Chem.* **1982**, *47*, 3183.

51. Bednarski, M., Danishefsky, S. *J. Am. Chem. Soc.* **1983**, *105*, 3716.

52. Van de Weghe, P., Collin, J. *Tetrahedron Lett.* **1994**, *35*, 2545.

53. (a) Alston, P. V., Ottenbrite, R. M., Cohen, T. *J. Org. Chem.* **1978**, *43*, 1864. (b) Kahn, S. D., Pau, C. F., Overman, L. E., Hehre, W. *J. Am. Chem. Soc.* **1986**, *108*, 7381.

54. (a) Sauer, J. *Angew. Chem., Int. Ed.* 1966, *5*, 221. (b) Sauer, J. *Angew. Chem., Int. Ed.* **1967**, *6*, 16.

55. (a) Houk, K. N. *J. Am. Chem. Soc.* **1973**, *95*, 4092. (b) Houk, K. N., Strozier, R. W. *J. Am. Chem., Soc.* **1973**, *95*, 4094.

56. Fleming, I. *Frontier Orbitals and Organic Chemical Reactions*, Wiley: New York, 1976.

57. Mehta, G., Uma, R. *Acc. Chem. Res.* **2000**, *33*, 278.

58. Garcia, J. I., Mayoral, J. A., Salvatella, L. *Acc. Chem. Res.* **2000**, *33*, 658.

59. Fuchs, P. L. *Understanding Organic Chemistry at the Molecular Level*, Stipes Publishing: Champaign, IL, 1991.

60. (a) Ciganek, E. *Org. React.* **1984**, *32*, 1. (b) Craig, D. *Chem. Soc. Rev.* **1987**, *16*, 187. (c) Roush, W. R. In *Comprehensive Organic Synthesis,* Trost, B. M., Fleming, I., Eds., Pergamon Press: Oxford, UK, 1991, Vol. 5, p 513. (d) Fallis, A. G. *Acc. Chem. Res.* **1999**, *32*, 464.

61. See, for example, (a) House, H. O., Cronin, T. H. *J. Org. Chem.* **1965**, *30*, 1061. (b) Pyne, S. G., Hensel, M. J., Byrn, S. R., McKenzie, A. T., Fuchs, P. L. *J. Am. Chem. Soc.* **1980**, *102*, 5960. (c) Yoshioka, M., Nakai, H., Ohno, M. *J. Am. Chem. Soc.* **1984**, *106*, 1133.

62. Smith, D. A., Houk, K. N. *Tetrahedron Lett.* **1991**, *32*, 1549.

63. Wulff, W. D., Powers, T. S. *J. Org. Chem.* **1993**, *58*, 2381.

64. (a) Choy, W., Reed, L. A., III, Masamune, S. *J. Org. Chem.* **1983**, *48*, 1137. (b) Masamune, S., Reed, L. A., III, Davis, J. T., Choy, W. *J. Org. Chem.* **1983**, *48*, 4441.

65. (a) Kagan, H. B., Riant, O. *Chem. Rev.* **1992**, *92*, 1007. (b) Corey, E. J. *Angew. Chem., Int. Ed.* **2002**, *41*, 1650.

66. Oppolzer, W., Chapuis, C., Dao, G. M., Reichlin, D., Godel, T. *Tetrahedron Lett.* **1982**, *23*, 4781.

67. (a) Corey, E. J., Shibata, T., Lee, T. W. *J. Am. Chem. Soc.* **2002**, *124*, 3808. (b) Ryu, D. H., Lee, T. W., Corey, E. J. *J. Am. Chem. Soc.* **2002**, *124*, 9992. (c) Ryu, D. H., Corey, E. J. *J. Am. Chem. Soc.* **2003**, *125*, 6388.

68. (a) Carruthers, W., Coldham, I. *Modern Methods of Organic Synthesis*, 4th ed., Cambridge University Press: Cambridge, MA, 2004, p. 222. (b) Padwa, A. In *Comprehensive Organic Synthesis,* Trost, B.M., Fleming, I., Eds., Pergamon Press: Oxford, UK, 1991, Vol. 4, p. 1069. (c) Wade, P. A. In *Comprehensive Organic Synthesis,* Trost, B.M., Fleming, I., Eds., Pergamon Press: Oxford, UK, 1991, Vol. 4, p. 1111.

69. (a) Huisgen, R. A*ngew. Chem., Int. Ed.* **1963**, *2*, 565. (b) For a recent synopsis of Huisgen's contributions, see *Tetrahedron* **2000**, *56*, vii–viii.

70. (a) Grubbs, R. H., Miller, S. J., Fu, G. C. *Acc. Chem. Res.* **1995**, *28*, 446. (b) Schuster, M., Blechert, S. *Angew. Chem., Int. Ed.* **1997**, *36*, 2037. (c) Grubbs, R. H., Chang, S. *Tetrahedron* **1998**, *54*, 4413. (d) Fürstner, A. *Angew. Chem., Int. Ed.* **2000**, *39*, 3012. (e) Trnka, T. M., Grubbs, R. H. *Acc. Chem. Res.* **2001**, *34*, 18. (f) Grubbs, R. H. *Tetrahedron* **2004**, *60*, 7117.

71. Dias, E. L., Nguyen, S. T., Grubbs, R. H. *J. Am. Chem. Soc.* **1997**, *119*, 3887.

72. (a) Schrock, R. R., Murdzek, J. S., Bazan, G. C., Robbins, J., DiMare, M., O'Regan, M. B. *J. Am. Chem. Soc.* **1990**, *112*, 3875. (b) Bazan, G. C., Khosravi, E., Schrock, R. R., Feast, W. J., Gibson, V. C., O'Regan, M. B., Thomas, J. K., Davis, W. J. *J. Am. Chem. Soc.* **1990**, *112*, 8378. (c) Bazan, G. C., Oskam, J. H., Cho, H.-N., Park, L. Y., Schrock, R. R. *J. Am. Chem. Soc.* **1991**, *113*, 6899.

73. (a) Schwab, P., France, M. B., Ziller, J. W., Grubbs, R. H. *Angew. Chem., Int. Ed.* **1995**, *34*, 2039. (b) Schwab, P., Grubbs, R. H., Ziller, J. W. *J. Am. Chem. Soc.* **1996**, *118*, 100.

74. (a) Weskamp, T., Schattenmann, W. C., Spiegler, M., Herrmann, W. A. *Angew. Chem., Int. Ed.* **1998**, *37*, 2490. (b) Scholl, M., Trnka, T. M., Morgan, J. P., Grubbs, R. H. *Tetrahedron Lett.* **1999**, *40*, 2247.

75. For a comparison of Ru-catalyst relative activities, see Trnka, T. M., Morgan, J. P., Sanford, M. S., Wilhelm, T. E., Scholl, M., Choi, T.-L., Ding, S., Day, M. W., Grubbs, R. H. *J. Am. Chem. Soc.* **2003**, *125*, 2546.

76. (a) Hérisson, J.-L., Chauvin, Y. *Makromol. Chem.* **1970**, *141*, 161. (b) Kress, J., Osborn, J. A., Greene, R. M. E., Ivin, K. J., Rooney, J. J. *J. Am. Chem. Soc.* **1987**, *109*, 899.

77. Kirkland, T., Grubbs, R. H. *J. Org. Chem.* **1997**, *62*, 7310.

78. Ovaa, H., Codée, J. D. C., Lastdrager, B., Overkleeft, H. S., van der Marel, G. A., van Boom, J. H. *Tetrahedron Lett.* **1998**, *39*, 7987.

79. Bassindale, M. J., Hamley, P., Leitner, A., Harrity, J. P. A. *Tetrahedron Lett.* **1999**, *40*, 3247.

80. (a) Clark, J. S., Hamelin, O. *Angew. Chem., Int. Ed.* **2000**, *39*, 372. (b) Maier, M. C. *Angew. Chem., Int. Ed.* **2000**, *39*, 2073.

81. Crimmins, M. T., Emmitte, K. A. *J. Am. Chem. Soc.* **2001**, *123*, 1533.

82. Kirkland, T. A., Lynn, D. M., Grubbs, R. H. *J. Org. Chem.* **1998**, *63*, 9904.

83. Papaioannou, N., Blank, J. T., Miller, S. J. *J. Org. Chem.* **2003**, *68*, 2728.

84. Cho, J. H., Kim, B. M. *Org. Lett.* **2003**, *5*, 531.

85. For other reported methods to remove ruthenium by-products, see (a) Maynard, H. D., Grubbs, R. H. *Tetrahedron Lett.* **1999**, *40*, 4137; (b) Paquette, L. A., Schloss, J. D., Efremov, I., Fabris, F., Gallou, F., Mendez-Andino, J., Yang, J. *Org. Lett.* **2000**, *2*, 1259; (c) Ahn, Y. M., Yang, K. L., Georg, G. I. *Org. Lett.* **2001**, *3*, 1411.

86. (a) Kamat, V. P., Hagiwara, H., Suzuki, T., Ando, M. *J. Chem. Soc., Perkin Trans. 1* **1998**, 2253. (b) Lee, C. W., Grubbs, R. H. *Org. Lett.* **2000**, *2*, 2145.

The Art of Synthesis

I believe that chemistry, including chemical synthesis, will be a key driver of progress in medicine and human health during the rest of the twenty-first century.

Elias J. Corey

The ultimate challenge for a synthetic organic chemist is to emulate nature's ability to assemble complex molecular structures with high efficiency and stereochemical precision. The motive for the total synthesis of a particular natural product or designed molecule may be to probe its biological origin, to develop new methodologies, or for its commercial exploitation. Many natural products of medicinal interest are produced in such minute quantities in nature that only laboratory syntheses will make them available in sufficient amounts for use in medicine. In the planning and execution stages of such syntheses, the chemist has to rely on analogy; that is, searching for published reactions that are related to the project. Thus, it is important to be intimately acquainted with the literature in the field. For those who are interested in the synthesis of natural products, *Classics in Total Synthesis*, by K. C. Nicolaou and E. J. Sorensen, is highly recommended. It provides a collection of imaginative, well-presented syntheses.

The previous nine chapters have provided a diligent student the knowledge necessary to design syntheses of complex target molecules, to address problems of regio- and stereocontrol, and to devise schemes for effective functional group protection and deprotection. You should now have reached the stage where you can apply your skill and imagination as an investigator to design procedures for total synthesis of the four natural products shown below.

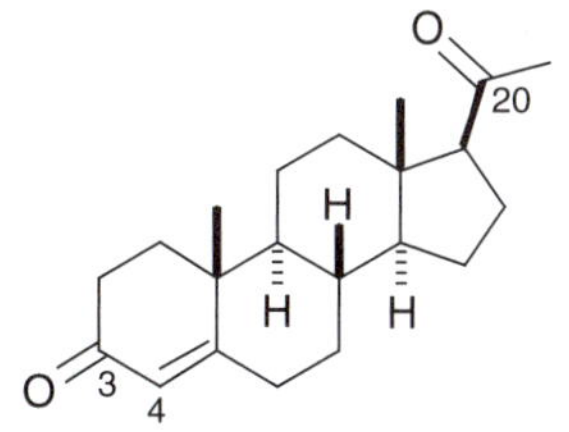

progesterone

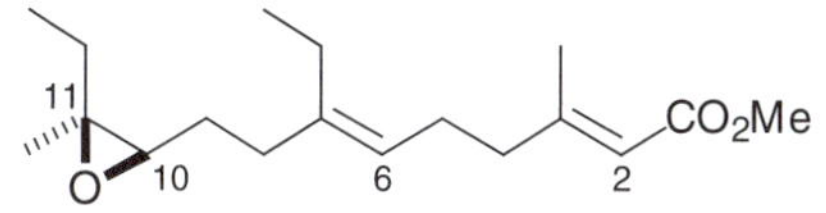

(+)-juvenile hormone I

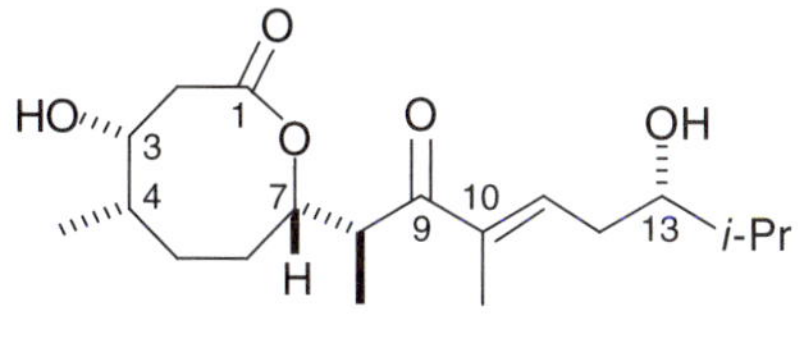

octalactin B

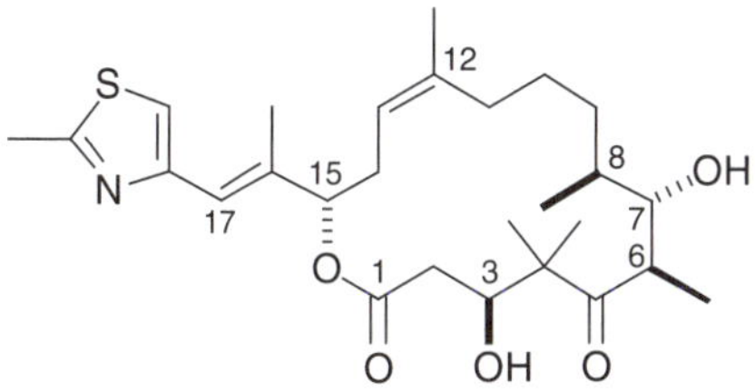

epothilone D

These natural products have each been synthesized via several routes and by different research groups. To provide an historical perspective, we have selected compounds that were synthesized in different decades, allowing for comparison of early synthetic approaches with modern ones. With your exposure to modern organic synthesis, we hope that your approaches to these target molecules will lead to new, ingenious routes for their synthesis.

Your protocol for proposing syntheses should include

- The strategic design: a detailed retrosynthetic analysis
- The synthetic execution: methodologies, reagents, and conditions to be used

Progesterone, a female sex hormone, has been prepared by total synthesis and by modifications of other steroids. Using established methodologies, such as Diels-Alder cycloadditions, Michael additions, Robinson annulations, aldol reactions, and ring contraction reactions, propose a total synthesis of progesterone. Discuss the merits of your approach in comparison with the total synthesis reported by Woodward, R. B., Sondheimer, F., Taub, D., Heusler, K., McLamore, W. M. *J. Am. Chem. Soc.* **1952**, *74*, 4223.

(+)-*Juvenile hormone I* is the vital hormone that regulates the metamorphosis of insects. Its unique structure stimulated a number of synthetic efforts to devise methods for controlling double-bond stereochemistry and the enantioselectivity of the epoxide. Propose a synthesis of juvenile hormone I as a single enantiomer and compare your approach with those reported by Imai, K., Marumo, S., Ohtaki, T. *Tetrahedron Lett.* **1976**, 1211 and Mori, K., Fujiwhara, M. *Tetrahedron* **1988**, *44*, 343.

(+)-*Octalactin B*, a marine-derived cytotoxic lactone, has been a target for total synthesis by several research groups. A common retrosynthetic strategy in many of the octalactin B syntheses has been the initial disconnection of the $C_{(9)}$–$C_{(10)}$ bond. Elaboration of the resultant two fragments must take into account the problem of how to control the introduction of the five stereocenters present in the target. This might be accomplished by employing asymmetric crotylboration and/or the Sharpless asymmetric epoxidation methodologies in the reaction protocol. Evaluate your strategy in terms of efficiency, selection of starting materials, and stereocontrol, and contrast it with three of the previously reported syntheses of octalactin B: Buszek, K. R., Sato, N., Jeong, Y. *J. Am. Chem. Soc.* **1994**, *116*, 5511; Andrus, M. B., Argade, A. B. *Tetrahedron Lett.* **1996**, *37*, 5049; and Inoue, S., Iwabuchi, Y., Irie, H., Hatakeyama, S. *Synlett* **1998**, 735.

Epothilone D, a microtubule-stabilizing 16-member macrolide that mimics the biological effects of Taxol (paclitaxel), features three contiguous stereocenters, $C_{(6)}$–$C_{(8)}$, as well as two trisubstituted alkene moieties. Its challenging molecular structure and important biological activity motivated intense research efforts culminating in several total syntheses. Creative applications of various olefination strategies, such as the Wittig reaction, the Horner-Wadsworth-Emmons modification, and ring-closing metathesis, have been used to assemble the side chain and the large ring of epothilone D. Contrast your proposed approach with those reported by Mulzer, J., Mantoulidis, A., Öhler, E. *J. Org. Chem.* **2000**, *65*, 7456; White, J. D., Carter, R. G., Sundermann, K. F., Wartmann, M. *J. Am. Chem. Soc.* **2001**, *123*, 5407 and Biswas, K., Lin, H., Njardarson, J. T., Chappell, M. D., Chou, T.-C., Guan, Y., Tong, W. P., He, L., Horwitz, S. B., Danishefsky, S. J. *J. Am. Chem. Soc.* **2002**, *124*, 9825.

ABBREVIATIONS

Ac ....acetyl, acetate
acac ....acetylacetonate
AIBN..............2,2′-azobisisobutyronitrile
anhydr ....anhydrous
APA................3-aminopropylamine
aq...................aqueous
atm.................atmosphere
9-BBN ....9-borabicyclo[3.3.1]nonane
Bn ....benzyl
Boc ....*tert*-butoxycarbonyl
bp...................boiling point
BPS ....*tert*-butyldiphenylsilyl
n-Bu...............*n*-butyl
s-Bu ....*sec*-butyl
t-Bu ....*tert*-butyl
Bz ....benzoyl
cat..................catalytic
CBS 2,5-oxazoborolidine
Cbz ....benzyloxycarbonyl
CDI................carbonyldiimidazole
Chx ....cyclohexyl
Cp ....cyclopentadienyl
CSA................10-camphorsulfonic acid
DABCO..........l,4-diazabicyclo[2.2.2]octane
DBN ....l,5-diazabicyclo[4.3.0]
                        non-5-ene
DBU ....l,8-diazabicyclo[5.4.0]
                        undec-7-ene
DCC................1,3-dicyclohexylcarbodiimide
DDQ...............2,3-dichloro-5,6-dicyano-
                        l,4-benzoquinone
de...................diastereomeric excess
DEAD ....diethyl azodicarboxylate
DET...............diethyl tartrate
DHP................3,4-dihydro-2*H*-pyran
DIAD ....diisopropyl azodicarboxylate
DIBAL-H ....diisobutylaluminum hydride
DIPT...............diisopropyl tartrate
DMAP...........4-dimethylaminopyridine
DMDO ....dimethyldioxirane
DME...............1,2-dimethoxyethane
DMF...............*N*,*N*-dimethylformamide
DMP...............Dess-Martin periodinane
DMPU ....*N*,*N*′-dimethylpropyleneurea

DMS..............dimethyl sulfide
DMSO............dimethyl sulfoxide
DS ....diastereofacial selectivity
ee ....enantiomeric excess
eq...................equivalents
Et ....ethyl
EWG ....electron-withdrawing group
FG ....functional group
h......................hour
HMDS ....hexamethyldisilazane
HMPA ....hexamethylphosphoramide
HQ.................hydroquinone
*h*ν ....light
IBX................*o*-iodoxybenzoic acid
imid ....imidazole
Ipc ....isopinocampheyl
KAPA ....potassium
                        3-aminopropylamide
KHMDS ....potassium
                        hexamethyldisilazide
L-Selectride....lithium
                        tri-*sec*-butylborohydride
LAH ....lithium aluminum hydride
LDA ....lithium diisopropylamide
LHMDS..........lithium
                        hexamethyldisilazide
*m*CPBA ....*m*-chloroperoxybenzoic acid
Me ....methyl
MEM 2-methoxyethoxymethyl
mol ....mole
MOM.............methoxymethyl
mp ....melting point
Ms ....mesyl (methanesulfonyl)
MS.................molecular sieves
MVK ....methyl vinyl ketone
NBS................*N*-bromosuccinimide
NCS................*N*-chlorosuccinimide
NMO ....*N*-methylmorpholine
                        N-oxide
NMP...............*N*-methyl-pyrrolidinone
PCC................pyridinium chlorochromate
PDC................pyridinium dichromate
PG ....protecting group
Ph ....phenyl

Piv ....pivaloyl
PMB 4-methoxybenzyl
PPTS ....pyridinium
                        p-toluenesulfonate
i-Pr.................*iso*-propyl
n-Pr................*n*-propyl
psi..................pounds per square inch
PT 1-phenyl-1*H*-tetrazol-5-yl
py...................pyridine
Ra-Ni.............Raney-nickel (usually
                        W-II type)
RCM..............ring-closing olefin
                        metathesis
Red-Al...........sodium bis
                        (2-methoxyethoxy)
                        aluminum hydride
rt ....room temperature
SAE...............Sharpless asymmetric
                        expoxidation
SE..................synthetic equivalent
SEM ....2-(trimethylsilyl)ethoxy-
                        methyl
Sia ....disiamyl
TBAF ....tetra-*n*-butylammonium
                        fluoride
TBS ....*tert*-butyldimethylsilyl
TES ....triethylsilyl
Tf....................trifluoromethanesulfonyl
TFA ....trifluoroacetic acid
TFAA ....trifluoroacetic anhydride
THF...............tetrahydrofuran
THP...............tetrahydropyranyl
Thx ....thexyl (Me$_2$CHMe$_2$C-)
TIPS ....triisopropylsilyl
TMEDA ....*N*,*N*,*N*′,*N*′-tetramethylethyl-
                        enediamine
TMS ....trimethylsilyl
TMSOTf.........trimethylsilyl
                        trifluoromethanesulfonate
TPAP ....tetra-*n*-propylammonium
                        perruthenate
Tr....................trityl (triphenylmethyl)
Ts...................tosyl (*p*-toluenesulfonyl)
Δ ....heat

ANSWERS TO SELECT END-OF-CHAPTER PROBLEMS

To give students the opportunity to put into practice what they have learned, relevant problems are included at the end of each chapter. Below is a selection of answers presented either as literature references, structures, or reaction schemes. By providing references we hope to encourage students to consult the original works to more fully appreciate how research is actually carried out.

Chapter 1

1b. 1. $SOCl_2$, 2. NH_3, 3a. $LiAlH_4$, THF, 3b. H^+, H_2O workup, 3c. NaOH, H_2O

1c. 1. $mCPBA$, CH_2Cl_2, 2a. [$CH_3CCMgBr$], THF, 2b. H_2O workup, 3. H_2, Pd·$BaSO_4$ (cat.), EtOH, quinoline, 4. $CH_3C(O)Cl$, pyridine

1e. McGuirk, P. R., Collum, D. B. *J. Org. Chem.* **1984**, *49*, 843.

1f. Takano, S., Tamura, N., Ogasawar, K. *J. Chem. Soc., Chem. Commun.* **1981**, 1155.

2a.

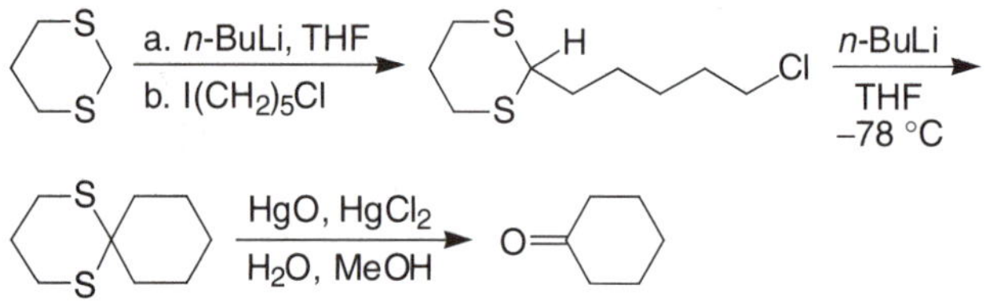

2c. Smith, R. G., Daves, Jr., G. D., Daterman, G. E. *J. Org. Chem.* **1975**, *40*, 1593.

2e. Kang, S. H., Lee, H. S. *Tetrahedron Lett.* **1995**, *36*, 6713.

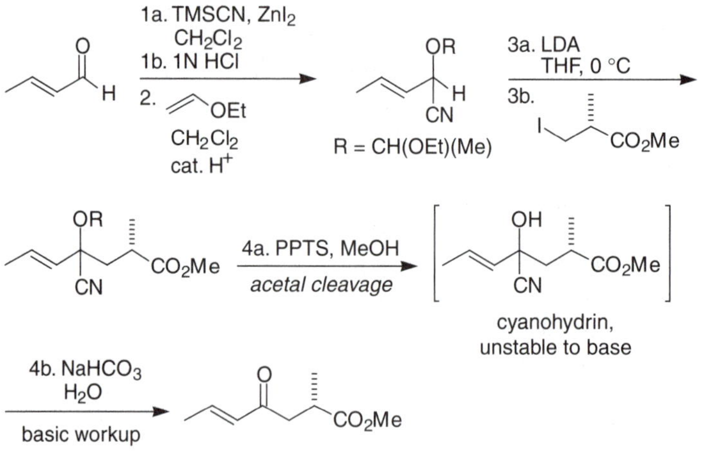

Step 1 furnishes the cyanohydrin.
Step 2 protects the OH as an acetal.

3a. Pearce, G. T., Gore, W. E., Silverstein, R. M. *J. Org. Chem.* **1976**, *41*, 2797.

3c.

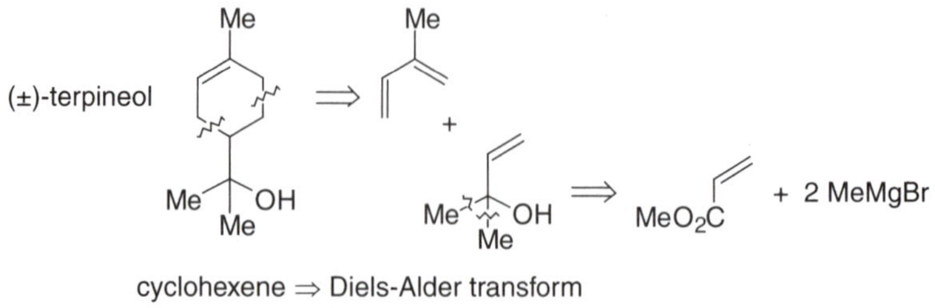

cyclohexene ⇒ Diels-Alder transform

The 3°-allylic alcohol is not well-suited as a dienophile (bulky, acid-sensitive). A better choice is methyl acrylate (electron deficient alkene). Addition of MeMgBr to the ester *after* cycloaddition furnishes the 3°-alcohol moiety.

3d. Denniff, P., Macleod, I., Whiting, D. A. *J. Chem. Soc., Perkin Trans. I* **1981**, 82.

Analysis:

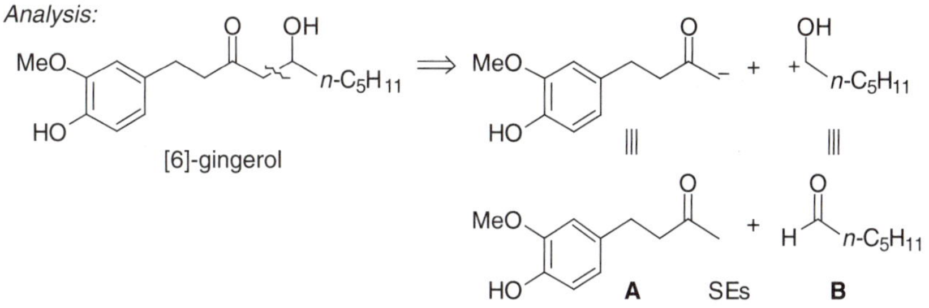

Synthesis:

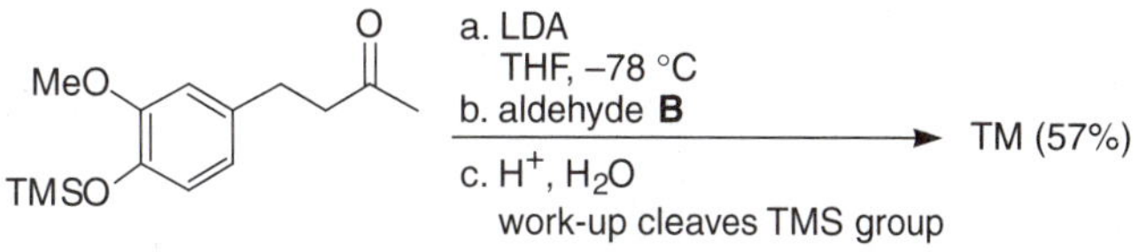

4a. Diana, G. D., Salvador, U. J., Zalay, E. S., Carabateas, P. M., Williams, G. L., Collins, J. C. *J. Med. Chem.* **1977**, *20*, 757.

4e. Gates, M. *J. Org. Chem.* **1980**, *45*, 1675.

4f. Sharma, A., Iyer, P., Gamre, S., Chattopadhyay, S. *Synthesis* **2004**, 1037.

Synthesis:

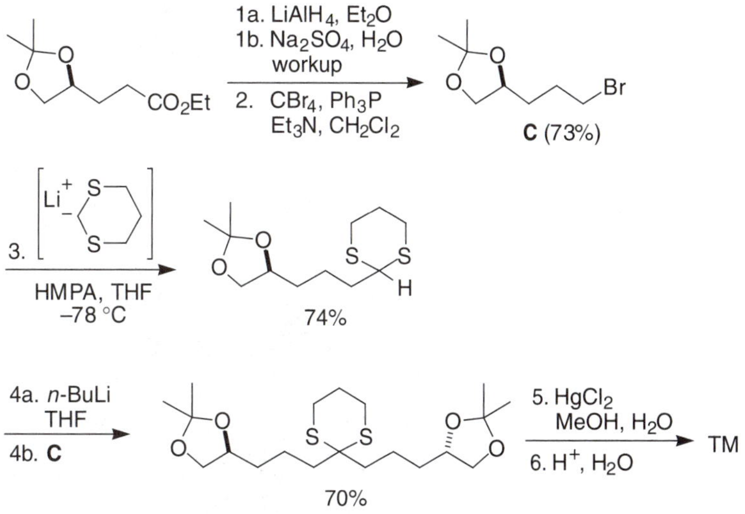

4h. Khanapure, S. P., Najafi, N., Manna, S., Yang, J.-J., Rokach, J. *J. Org. Chem.* **1995**, *60*, 7548.

Chapter 2

1a. $\Delta E_D = 6.15 - 1.0 = 5.15$ kcal/mol

1c. $\Delta E_D = 4.3 - 1.4 = 2.9$ kcal/mol

1e. $\Delta E_D = 4.5 - 2.4 = 2.1$ kcal/mol

1g. $E_D = 5.1 - 4.4 = 0.7$ kcal/mol

3a. 70% at 25 °C; 66% at 100 °C

3c. 99.9%

4. C > B > A

5. Marshall, J. A., Hochstetler, A. R. *J. Org. Chem.* **1966**, *31*, 1020.

6a.

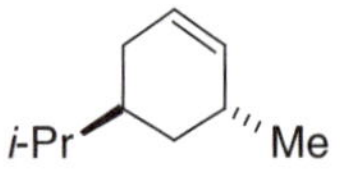
i-Pr
Me

6c.

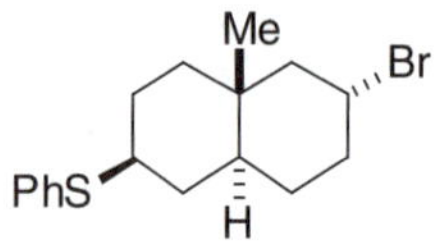
Me
Br
PhS
H

6e.

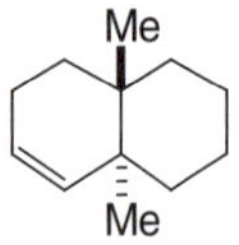
Me
Me

7. Suga, T., Shishibori, T. *Chemistry and Industry* **1971**, 733.

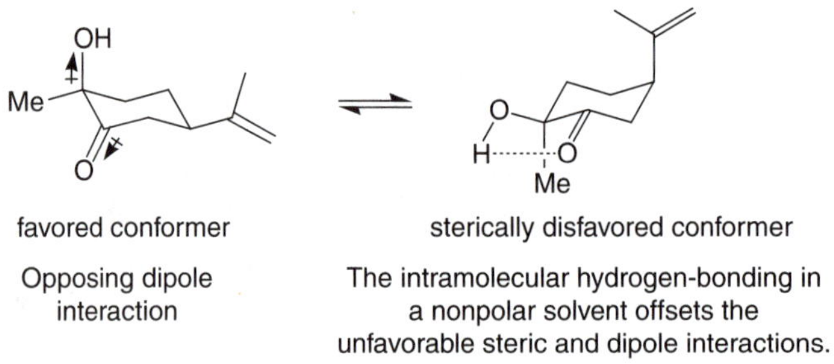
OH
Me
O
O
H
O
Me
favored conformer
Opposing dipole interaction
sterically disfavored conformer
The intramolecular hydrogen-bonding in a nonpolar solvent offsets the unfavorable steric and dipole interactions.

9. Quiñonero, D., Frontera, A., Capó, M., Ballester, P., Suñer, G. A., Garau, C., Deyà, P. M. *New J. Chem.* **2001**, *25*, 259.

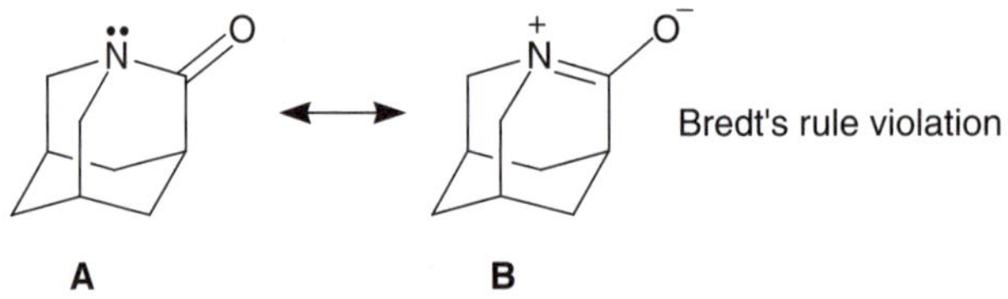
N
O
N
O
Bredt's rule violation
A
B

Resonance hybrid B violates Bredt's rule. Thus, there is little delocalization of the nitrogen lone pair to the carbonyl moiety.

Chapter 3

1a. Paterson, I., Delgado, O., Florence, G. J., Lyothier, I., O'Brien, M., Scott, J. P., Sereinig, N. *J. Org. Chem.* **2005**, *70*, 150.

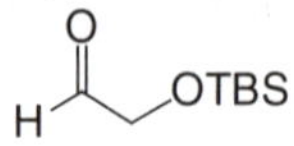
O
H
OTBS

1c. Gu, Y., Snider, B. B. *Org. Lett.* **2003**, *5*, 4385.

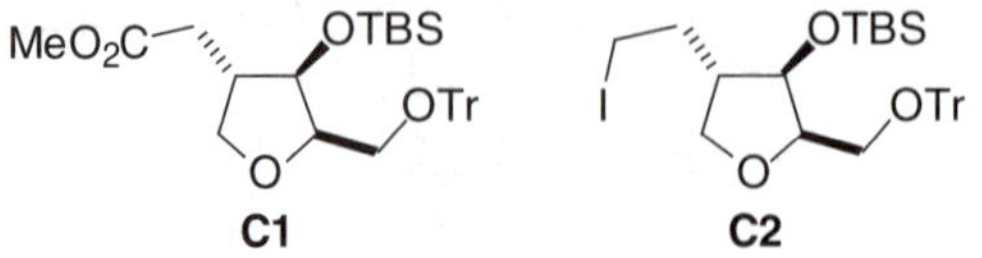
MeO2C
OTBS
OTr
O
C1
I
OTBS
OTr
O
C2

1e. Wang, Q., Sasaki, N. A. *J. Org. Chem.* **2004**, *69*, 4767.

1g. Nomura, I., Mukai, C. *J. Org. Chem.* **2004**, *69*, 1803.

2a. Gyergyói, K., Tóth, A., Bajza, I., Lipták, A. *Synlett* **1998**, 127.

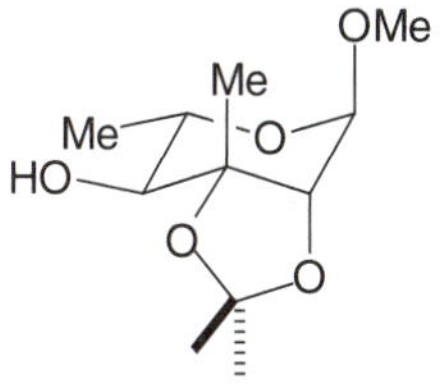

2c. Sefkow, M., Kelling, A., Schilde, U. *Eur. J. Org. Chem.* **2001**, 2735.

2e. Ella-Menye, J.-R., Sharma, V., Wang, G. *J. Org. Chem.* **2005**, *70*, 463.

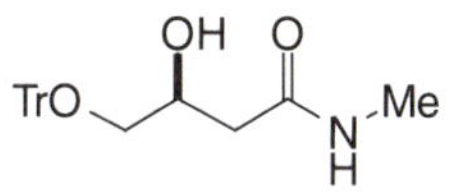

E1 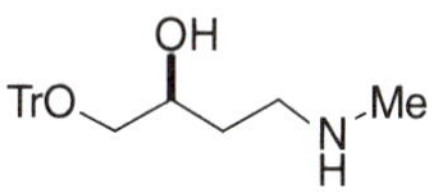   **E2**

2g.

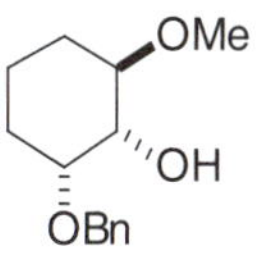

G1 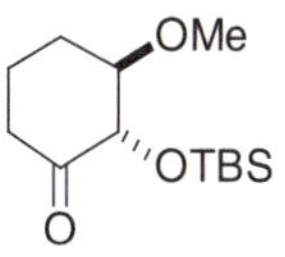   **G2**

3b. Magatti, C. V., Kaminski, J. J., Rothberg, I. *J. Org. Chem.* **1991**, *56*, 3102.

4a.

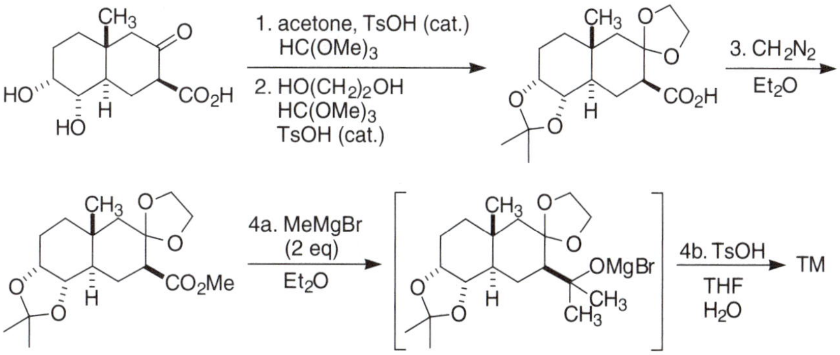

4b. Alibés, R., Ballbé, M., Busqué, F., de March, P., Elias, L., Figueredo, M., Font, J. *Org. Lett.* **2004**, *6*, 1813.

4e. Lanman, B. A., Myers, A. G. *Org. Lett.* **2004**, *6*, 1045.

4f.

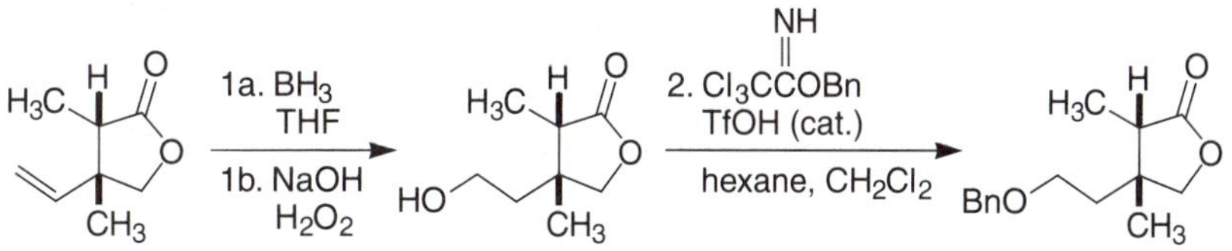

5. Cheol, E. L., Park, M., Yun, J. S. *J. Am. Chem. Soc.* **1995**, *117*, 8017.

Step 2. Reductive cleavage of the acetal and reduction of the ester groups
Step 3. Selective transacetalization of the 1,3-diol with benzaldehyde dimethyl acetal

Chapter 4

1b. Defosseux, M., Blanchard, N., Meyer, C., Cossy, J. *J. Org. Chem.* **2004**, *69*, 4626.

1d. Jiang, L., Martinelli, J. R., Burke, S. D. *J. Org. Chem.* **2003**, *68*, 1150.

1f.

1h. So, R. C., Ndonye, R., Izmirian, D. P., Richardson, S. K., Guerrara, R. L., Howell, A. R. *J. Org. Chem.* **2004**, *69*, 3233.

2a. Shi, B., Hawryluk, N. A., Snider, B. B. *J. Org. Chem.* **2003**, *68*, 1030.

2c. Ag_2CO_3 in acetone. MnO_2 or $Ba[MnO_4]_2$ can also be used for selective allylic oxidation.

2e. Momán, E., Nicoletti, D., Mouriño, A. *J. Org. Chem.* **2004**, *69*, 4615.

2g. Lee, H. K., Chun, J. S., Chwang, S. P. *J. Org. Chem.* **2003**, *68*, 2471.

G1   **G2**

3b. Clive, D. L. J., Magnuson, S. R., Manning, H. W., Mayhew, D. L. *J. Org. Chem.* **1996**, *61*, 2095.

3d. Mohr, P. J., Halcomb, R. L. *J. Am. Chem. Soc.* **2003**, *125*, 1712.

3f. Fettes, A., Carreira, E. M. *J. Org. Chem.* **2003**, *68*, 9274.

3h. Brown, H. C., Chandrasekharan, J., Ramachandran, P. V. *J. Am. Chem. Soc.* **1988**, *110*, 1539. The stereochemical outcome may be explained by considering the following transition state:

gem-dimethyl group points away from the axial-CH_3 of the pinane moiety

4a. Collum, D. B., McDonald, J. H., Still, W. C. *J. Am. Chem. Soc.* **1980**, *102*, 2118.

4c. Kang, S. H., Kang, S. Y., Kim, C. M., Choi, H., Jun, H.-Y., Lee, B. M., Park, C. M., Jeong, J. W. *Angew. Chem., Int. Ed.* **2003**, *42*, 4779.

5a. Su, Z., Paquette, L. A. *J. Org. Chem.* **1995**, *60*, 764.

5c. McDermott, T. S., Mortlock, A. A., Heathcock, C. H. *J. Org. Chem.* **1996**, *61*, 700.

5d. Clive, D. L. J., Magnuson, S. R., Manning, H. W., Mayhew, D. L. *J. Org. Chem.* **1996**, *61*, 2095.

Step 1. NaH treatment results in mono-alkoxide formation, which is critical to minimizing formation of a bis-silyl ether product. The authors report a 96% yield of the mono-silyl ether product.

5f. Pattenden, G., Plowright, A. T., Tornos, J. A., Ye, T. *Tetrahedron Lett.* **1998**, *39*, 6099.

5h. Dauben, W. G., Warshawsky, A. M. *J. Org. Chem.* **1990**, *55*, 3075.

Chapter 5

1c. Malkov, A. V., Pernazza, D., Bell, M., Bella, M., Massa, A., Teply, F., Meghani, P., Kocovsky, P. *J. Org. Chem.* **2003**, *68*, 4727.

1d. Lepage, O., Deslongchamps, P. *J. Org. Chem.* **2003**, *68*, 2183.

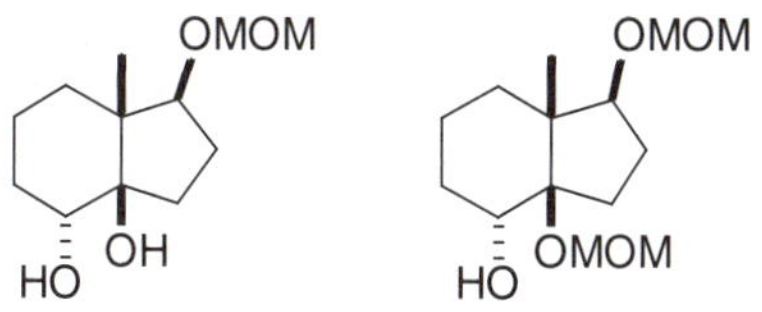

D1   **D2**

1e. Piers, E., Oballa, R. M. *J. Org. Chem.* **1996**, *61*, 8439.

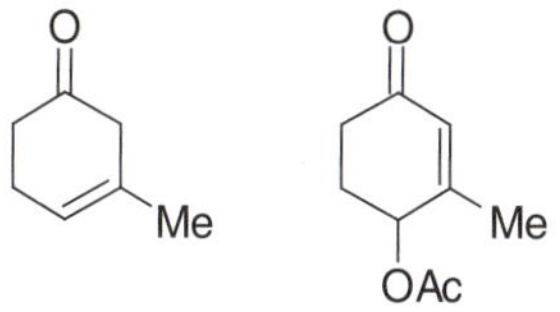

E1   **E2**

1f.

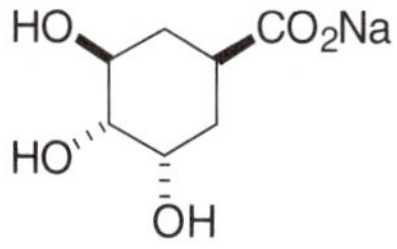

1i. Momán, E., Nicoletti, D., Mouriño, A. *J. Org. Chem.* **2004**, *69*, 4615.

1k. Chen, L., Wiemer, D. F. *J. Org. Chem.* **2002**, *67*, 7561.

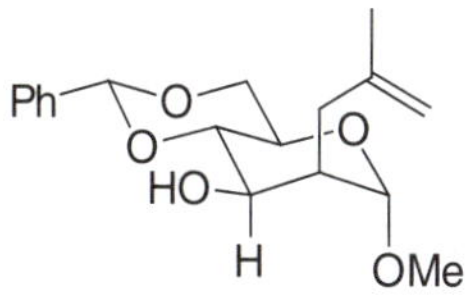

2b. Roush, W. R., Brown, R. J. *J. Org. Chem.* **1983**, *48*, 5093.

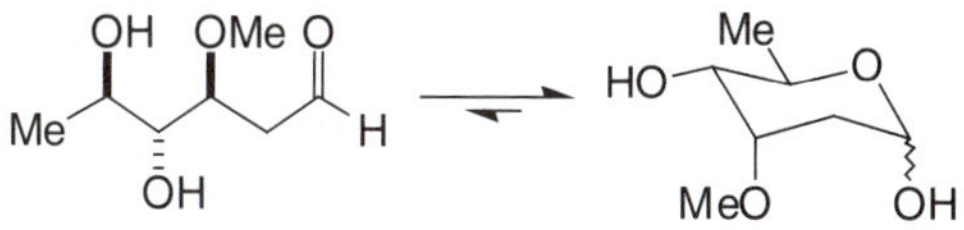

2d. Nozoe, S., Furukawa, J., Sankawa, U., Shibata, S. *Tetrahedron Lett.* **1976**, 195.

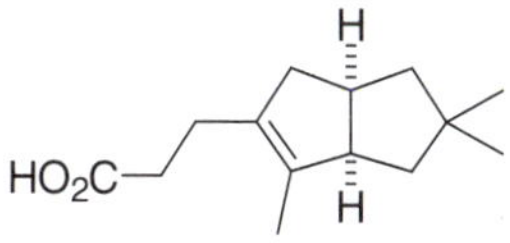

2f. Wovkulich, P. M., Shankaran, K., Kiegiel, J., Uskokovic, M. R. *J. Org. Chem.* **1993**, *58*, 832.

3a. Evans, D. A., Bender, S. L., Morris, J. *J. Am. Chem. Soc.* **1988**, *110*, 2506.

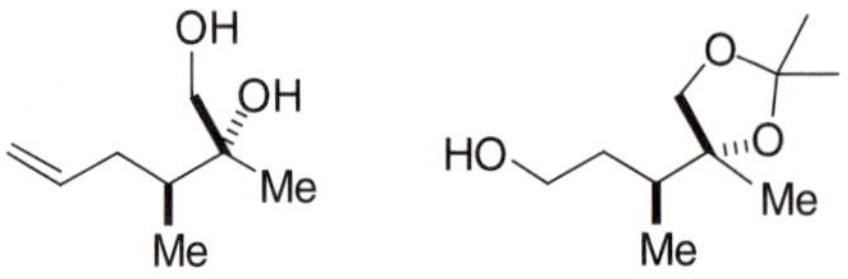

3b.

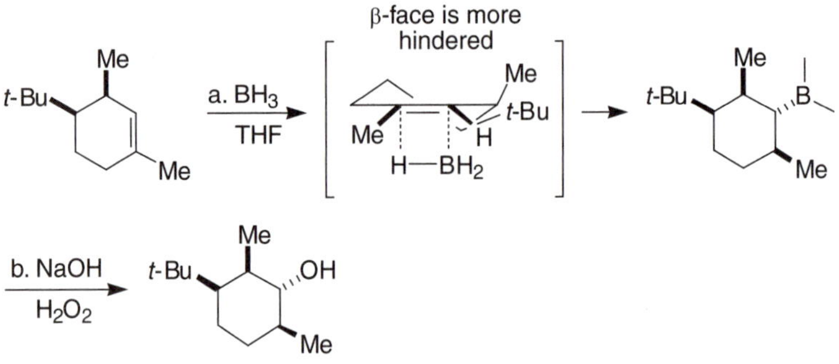

Step a. *syn*-Addition of the H–B bond at the sterically less hindered face. To minimize $A^{1,3}$ strain, the favored cyclohexene conformer has the *t*-Bu group equatorial and the allylic-Me group pseudoaxial.

3e. Martin, S. F., Zinke, P. W. *J. Org. Chem.* **1991**, *56*, 6600.

Reduction using $NaBH_4$ or DIBAL-H furnished a product mixture in which the equatorial allylic alcohol predominated. However, reduction using K-Selectride produced the axial alcohol as the major product (88% yield, 9.8:1 mixture of diastereomers).

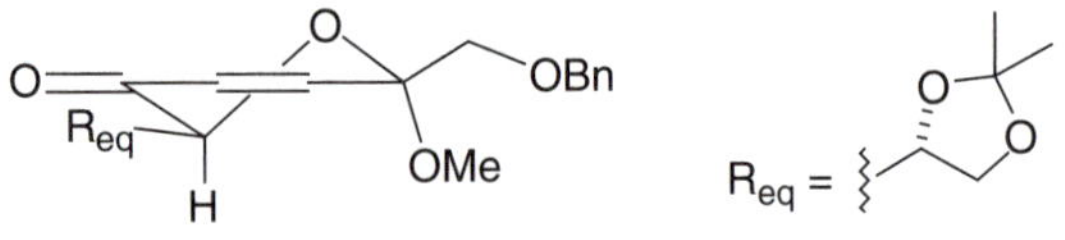

1. 1,4-Reduction is impeded by the adjacent disubsituted carbon center.
2. The sterically hindered reducing agent K-Selectride favors 1,2-addition from the face opposite of the R_{eq} substituent.

4c.

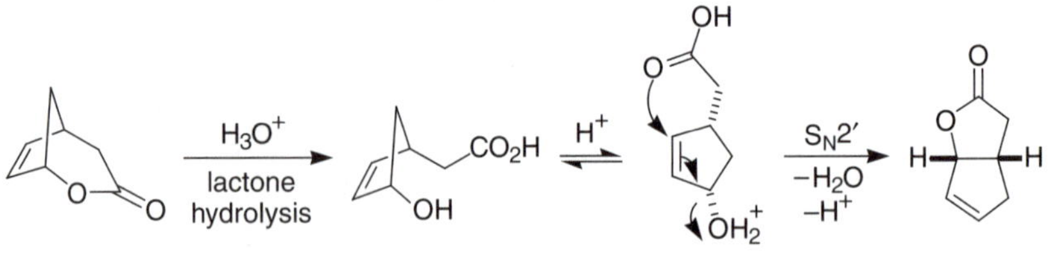

4d. Ekhato, I. V., Silverton, J. V., Robinson, C. H. *J. Org. Chem.* **1988**, *53*, 2180.
4e. Kende, A. S., Blacklock, T. J. *Tetrahedron Lett.* **1980**, *21*, 3119.
4f. Bergman, R., Magnusson, G. *J. Org. Chem.* **1986**, *51*, 212.

5c.

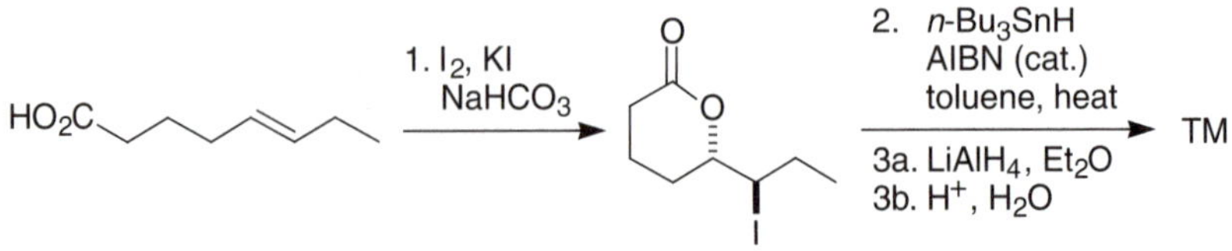

5d. Machinaga, N., Kibayashi, C. *Tetrahedron Lett.* **1993**, *34*, 5739.

5e. McWilliams, J. C., Clardy, J. *J. Am. Chem. Soc.* **1994**, *116*, 8378.

5f. Holoboski, M. A., Koft, E. *J. Org. Chem.* **1992**, *57*, 965.

5h. Altman, L. J., Han, C. Y., Bertolino, A., Handy, G., Laungani, D., Muller, W., Schwartz, S., Shanker, D., de Wolf, W. H., Yang, F. *J. Am. Chem. Soc.* **1978**, *100*, 3235.

6i. (a) **6ii.** (c)

7a. Kumar, P., Bodas, M. S. *J. Org. Chem.* **2005**, *70*, 360.

Retrosynthetic Analysis (PG = protecting group)

7b. Kang, J.-H., Siddiqui, M. A., Sigano, D. M., Krajewski, K., Lewin, N. E., Pu, Y., Blumberg, P. M., Lee, J., Marquez, V. E. *Org. Lett.* **2004**, *6*, 2413.

Retrosynthetic Analysis

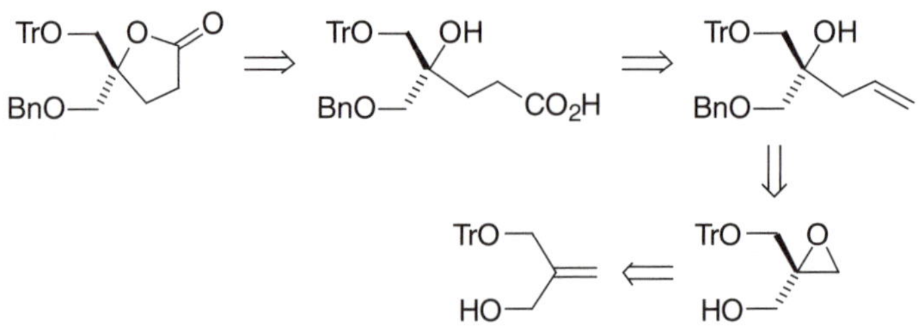

Chapter 6

1b. House, H. O., Lee, J. H. C., VanDerveer, D., Wissinger, J. E. *J. Org. Chem.* **1983**, *48*, 5285.

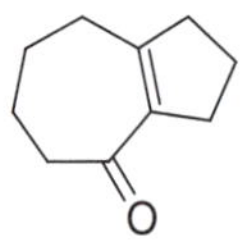

1d. Nantz, M. H., Radisson, X., Fuchs, P. L. *Synth. Commun.* **1987**, *17*, 55.

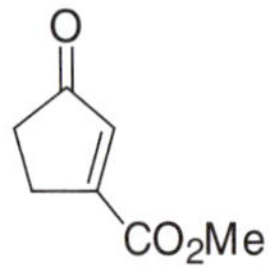

1f. Falck, J. R., Manna, S., Chandrasekhar, S., Alcaraz, L., Mioskowski, C. *Tetrahedron Lett.* **1994**, *35*, 2013.

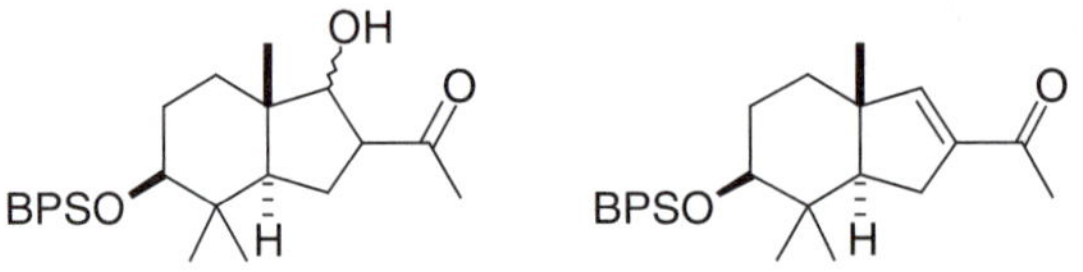

F1   **F2**

1h. Durham, T. B., Miller, M. J. *J. Org. Chem.* **2003**, *68*, 35.

1j.

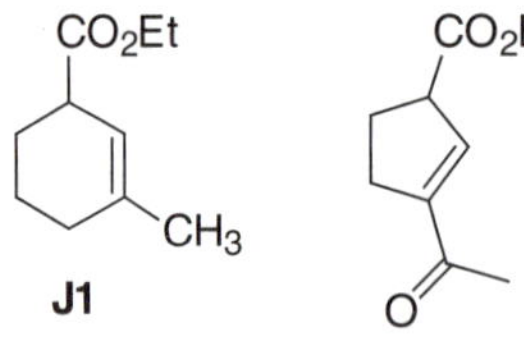

J1   **J2**

1k. Gaul, C., Njardarson, J. T., Danishefsky, S. J. *J. Am. Chem. Soc.* **2003**, *125*, 6042.

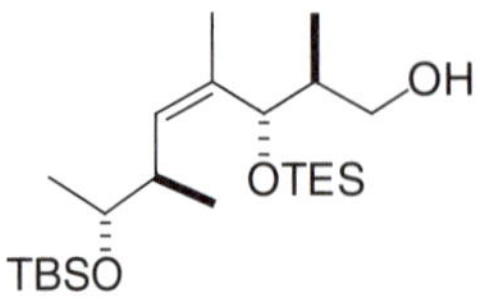

2a. Mehta, G. *Pure & Appl. Chem.* **1990**, *62*, 1263.

2c. Trost, B. M., Shuey, C. D., DiNinno, F., Jr., McElvain, S. S. *J. Am. Chem. Soc.* **1979**, *101*, 1284.

2e. Hartung, R., Paquette, L. A. *J. Org. Chem.* **2005**, *70*, 1597.

3a. Paterson, I., Wallace, D. J., Cowden, C. J. *Synthesis* **1998**, 639.

3c. Shi, B., Hawryluk, N. A., Snider, B. B. *J. Org. Chem.* **2003**, *68*, 1030.

3e. Evans, D. A., Bender, S. L., Morris, J. *J. Am. Chem. Soc.* **1988**, *110*, 2506.

3f. King, S. A. *J. Org. Chem.* **1994**, *59*, 2253.

Base catalyzed reaction : saponification followed by alkylation with dimethylsulfate does not invert the chiral center.

Acid catalyzed reaction : Dioxocarbenium ion activation of the lactone carbonyl group followed by attack of methanol (S$_N$2 inversion) at the methyl-substituted γ-carbon.

4a. Padwa, A., Lee, H. I., Rashatasakhon, P., Rose, M. *J. Org. Chem.* **2004**, *69*, 8209.

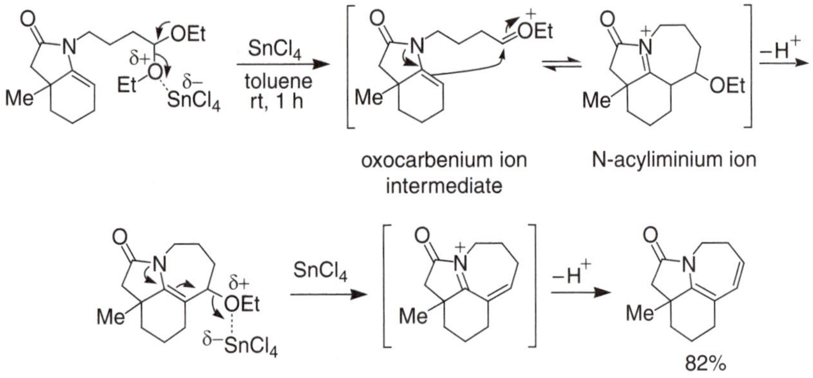

oxocarbenium ion intermediate

N-acyliminium ion

82%

4c. Boeckman, R. K., Jr., Bruza, K. J. *Tetrahedron Lett.* **1977**, 4187.

4e. Morihira, K., Seto, M., Furukawa, T., Horiguchi, Y., Kuwajima, I. *Tetrahedron Lett.* **1993**, *34*, 345.

4f. Amato, J. S., Chung, J. Y. L., Cvetovich, R. J., Gong, X., McLaughlin, M., Reamer, R. A. *J. Org. Chem.* **2005**, *70*, 1930.

5b. McDermott, T. S., Mortlock, A. A., Heathcock, C. H. *J. Org. Chem.* **1996**, *61*, 700.

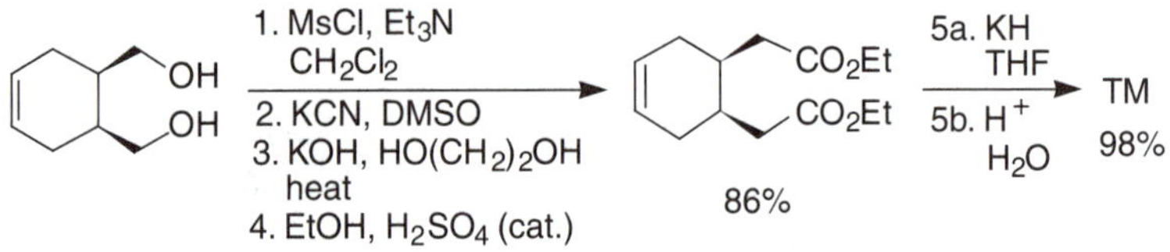

5d. Clark, R. D., Heathcock, C. H. *J. Org. Chem.* **1976**, *41*, 1396.

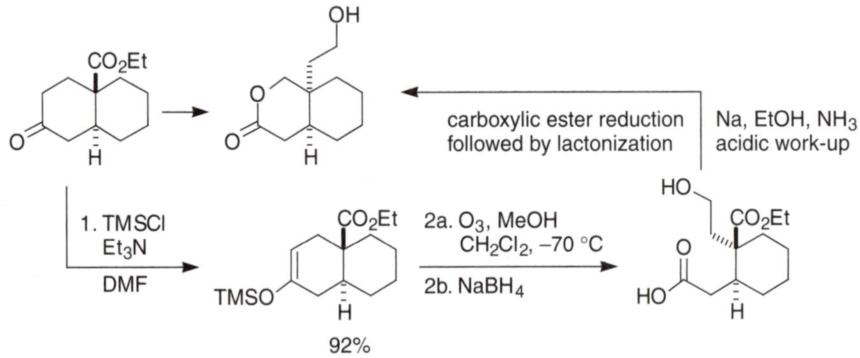

5f.

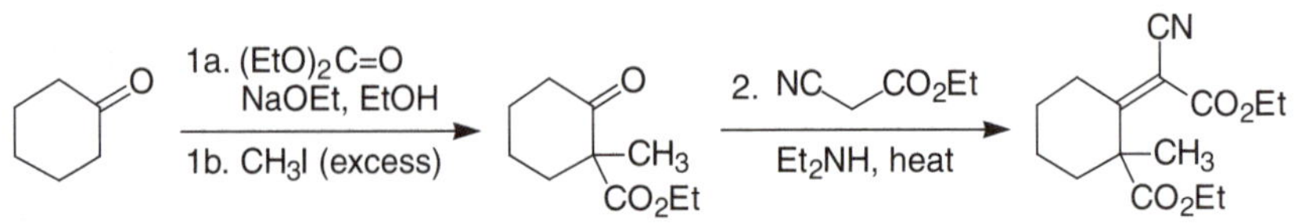

5h.

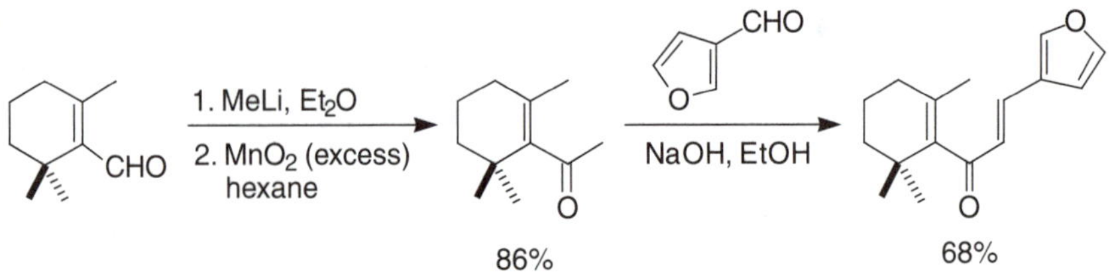

5i. Helal, C. J., Kang, Z., Lucas, J. C., Bohall, B. R. *Org. Lett.* **2004**, *6*, 1853.

5j. Fleck, T. J., McWhorter, W. W., Jr., DeKam, R. N., Pearlman, B. A. *J. Org. Chem.* **2003**, *68*, 9612.

6a. Clive, D. L. J., Wang, J. *J. Org. Chem.* **2004**, *69*, 2773.

Retrosynthetic Analysis

Chapter 7

1a. Negishi, E., Bagheri, V., Chatterjee, S., Luo, F.-T., Miller, J. R., Stoll, A. T. *Tetrahedron Lett.* **1983**, *24*, 5181.

1c. Piscopio, A. D., Minowa, N., Chakraborty, T. K., Koide, K., Bertinato, P., Nicolaou, K. C. *J. Chem. Soc., Chem. Commun.* **1993**, 617.

1e. Su, Z., Paquette, L. A. *J. Org. Chem.* **1995**, *60*, 764.

1f. Rodeschini, V., Boiteau, J.-G., Van de Weghe, P., Tarnus, C., Eustache, J. *J. Org. Chem.* **2004**, *69*, 357.

1h.

1i. Zhang, Y., Wu, G., Agnel, G., Negishi, E. *J. Am. Chem. Soc.* **1990**, *112*, 8590.

1k. Alibés, R., Ballbé, M., Busqué, F., de March, P., Elias, L., Figueredo, M., Font, J. *Org. Lett.* **2004**, *6*, 1813.

1l. Taillier, C., Gille, B., Bellosta, V., Cossy, J. *J. Org. Chem.* **2005**, *70*, 2097.
1n. Tsuji, J. *Tetrahedron* **1986**, *42*, 4361.

1p. Burke, S. D., Piscopio, A. D., Kort, M. E., Matulenko, M. A., Parker, M. H., Armistead, D. M., Shankaran, K. *J. Org. Chem.* **1994**, *59*, 332.

2b. Hoye, T. R., Humpal, P. E., Jiménez, J. I., Mayer, M. J., Tan, L., Ye, Z. *Tetrahedron Lett.* **1994**, *35*, 7517.

2c. Pelter, A., Smith, K., Hutchings, M. G., Rowe, K. *J. Chem. Soc., Perkin Trans. I* **1975**, 129.

3a. Nagamitsu, T., Takano, D., Fukuda, T., Otoguro, K., Kuwajima, I., Harigaya, Y., Omura, S. *Org. Lett.* **2004**, *6*, 1865.

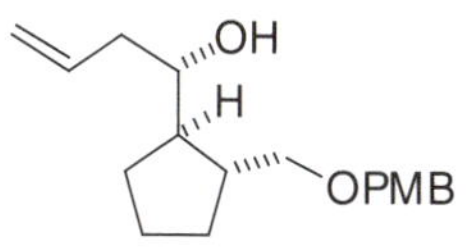

A

3c. Vyryan, J. R., Peterson, E. A., Stephan, M. L. *Tetrahedron Lett.* **1999**, *40*, 4947.

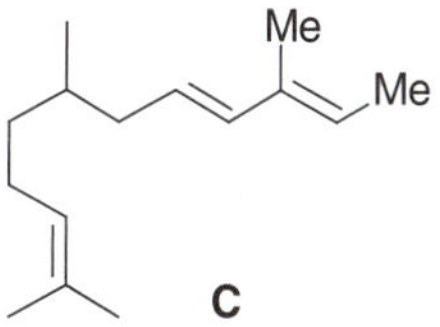

C

3e. McCombie, S. W., Ortiz, C., Cox, B., Ganguly, A. K. *Synlett* **1993**, 541.

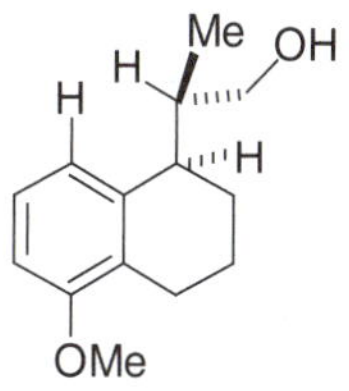

E

3g. Kouklovsky, C., Ley, S. V., Marsden, S. P. *Tetrahedron Lett.* **1994**, *35*, 2091.

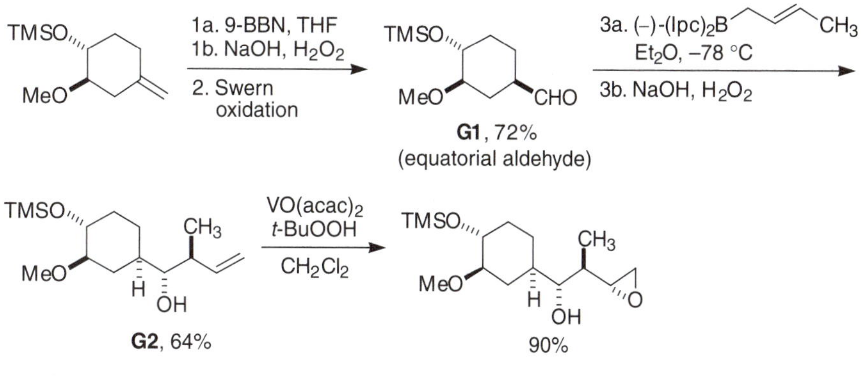

4b. Zhang, Y., Negishi, E. *J. Am. Chem. Soc.* **1989**, *111*, 3454.

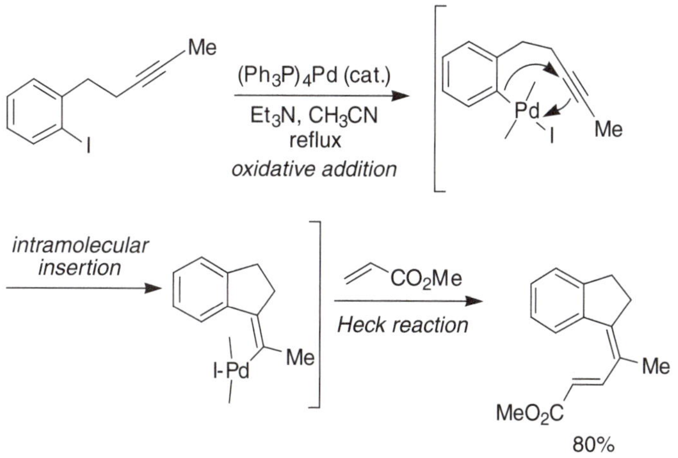

4d. Majetich, G., Song, J.-S., Leigh, A. J., Condon, S. M. *J. Org. Chem.* **1993**, *58*, 1030.

5a.

5c. Takano, S., Sugihara, T., Ogasawara, K. *Synlett* **1990**, 453, see also Takano, S., Murakami, T., Samizu, K., Ogasawara, K. *Heterocycles* **1994**, *39*, 67.

5e. Imamoto, T., Sugiura, Y., Takiyama, N. *Tetrahedron Lett.* **1984**, *25*, 4233.

5g. Anderson, J. C., Ley, S. V., Marsden, S. P. *Tetrahedron Lett.* **1994**, *35*, 2087.

5i. Kitagawa, Y., Itoh, A., Hashimoto, S., Yamamoto, H., Nozaki, H. *J. Am. Chem. Soc.* **1977**, *99*, 3864.

6a. Soderquist, J. A., Rane, A. M. *Tetrahedron Lett.* **1993**, *34*, 5031.

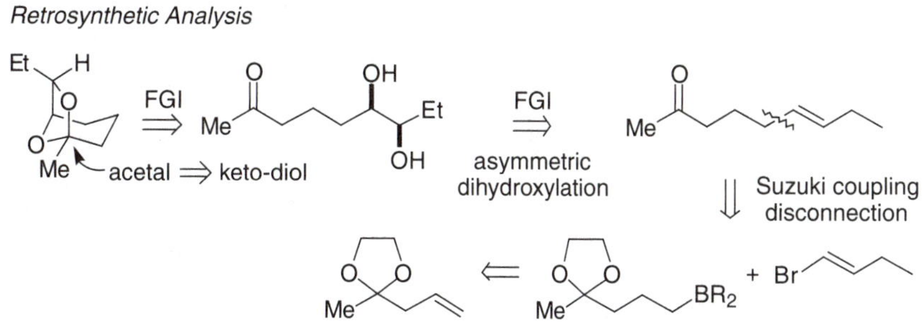

Retrosynthetic Analysis

6b. Miyaura, N., Ishikawa, M., Suzuki, A. *Tetrahedron Lett.* **1992**, *33*, 2571.

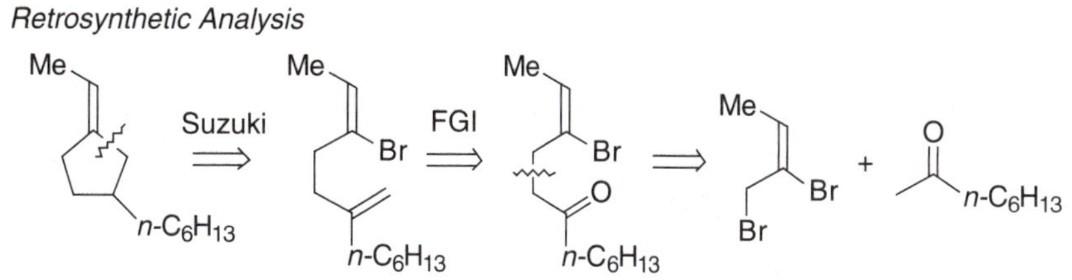

Retrosynthetic Analysis

Chapter 8

1a. Moman, E., Nicoletti, D., Mourino, A. *J. Org. Chem.* **2004**, *69*, 4615.

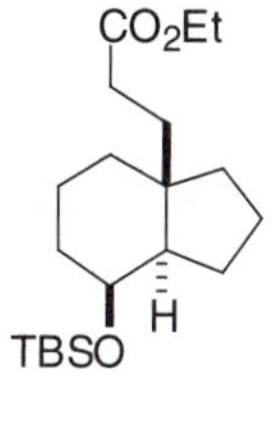

1c.

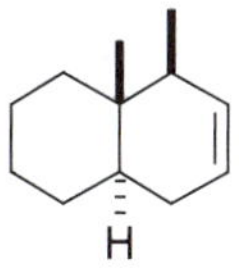

1e. Quéron, E., Lett, R. *Tetrahedron Lett.* **2004**, *45*, 4527.

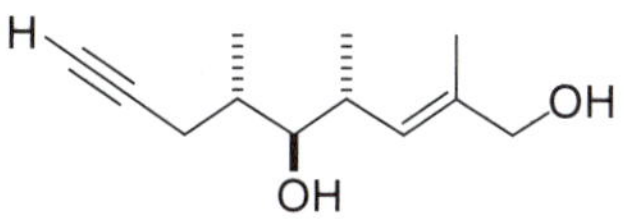

1g. Piers, E., Abeysekera, B., Scheffer, J. R. *Tetrahedron Lett.* **1979**, 3279.

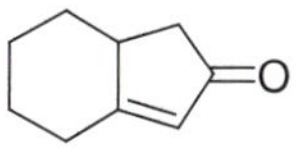

1i. Barrow, R. A., Hemscheidt, T., Liang, J., Paik, S., Moore, R. E., Tius, M. A. *J. Am. Chem. Soc.* **1995**, *117*, 2479.

I1   **I2**

Step 3. HWE-modification of the Wittig reaction, tetramethyl-guanidine functions as a base, much like DBU.

1k. Yoshino, T., Nagata, Y., Itoh, E., Hashimoto, M., Katoh, T., Terashima, S. *Tetrahedron Lett.* **1996**, *37*, 3475.

1l. Boyd, V. A., Drake, B. E., Sulikowski, G. A. *J. Org. Chem.* **1993**, *58*, 3191.

2a. Marron, B. E., Nicolaou, K. C. *Synthesis* **1989**, 537.

2b. Trost, B. M., Verhoeven, T. R. *J. Am. Chem. Soc.* **1977**, *99*, 3867.

2c. Powell, N. A., Roush, W. R. *Org. Lett.* **2001**, *3*, 453.

3a. Byrne, B., Lawter, L. M. L., Wengenroth, K. J. *J. Org. Chem.* **1986**, *51*, 2607.

3c. Clive, D. L. J., Magnuson, S. R., Manning, H. W., Mayhew, D. L. *J. Org. Chem.* **1996**, *61*, 2095.

4a. Büchi, G., Pawlak, M. *J. Org. Chem.* **1975**, *40*, 100.

4c. Zutterman, F., Krief, A. *J. Org. Chem.* **1983**, *48*, 1135.

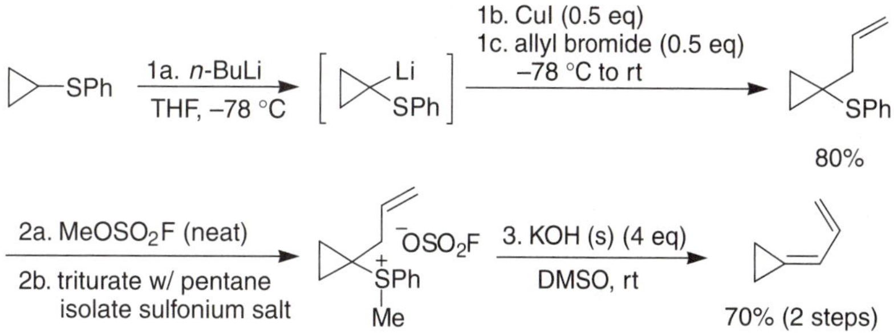

80%

70% (2 steps)

Step 1a. The enhanced acidity of cyclopropyl over normal secondary hydrogens permits the facile deprotonation using *n*-BuLi; see *J. Am. Chem. Soc.* **1973**. *95*, 3068.

Step 2a. Methylation using magic methyl (caution, carcinogen!).

Step 3. Regioselective E2-elimination (no cyclopropene is formed due to the ring strain).

5c. Mehta, G., Krishnamurthy, N., Karra, S. R. *J. Am. Chem. Soc.* **1991**, *113*, 5765.

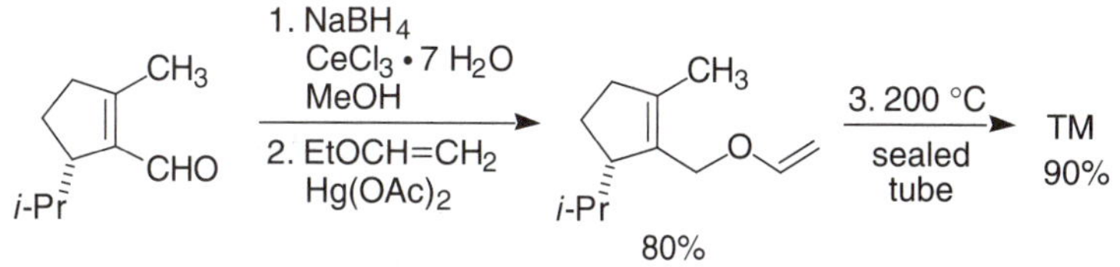

80%

5e. Normant, J. F., Cahiez, G., Chuit, C., Villieras, J. *J. Organomet. Chem.* **1974**, *77*, 281.

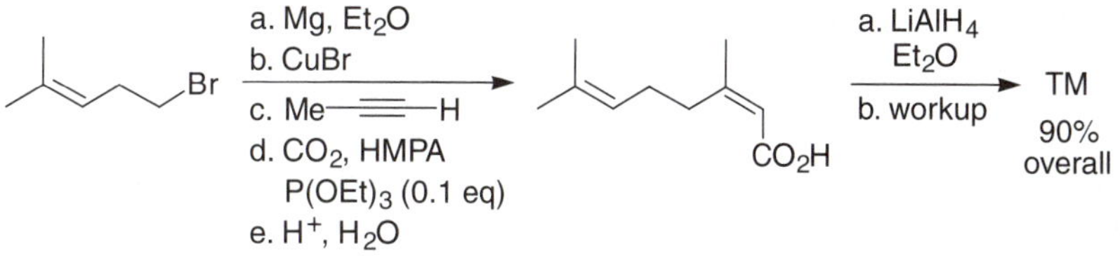

5g. Reyes, E. D., Carballeira, N. M. *Synthesis* **1997**, 1195.

5i. White, J. D., Jeffrey, S. C. *J. Org. Chem.* **1996**, *61*, 2600.

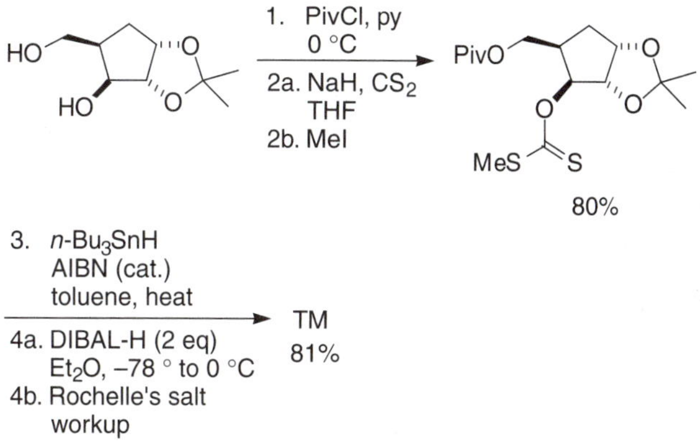

80%

Chapter 9

1a. Taillier, C., Gille, B., Bellosta, V., Cossy, J. *J. Org. Chem.* **2005**, *70*, 2097.

BnO
O
O

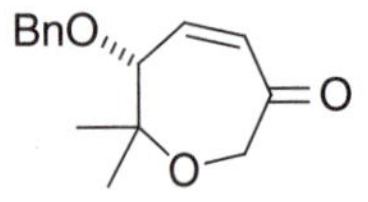

1c. Yamamoto, H., Sham, H. L. *J. Am. Chem. Soc.* **1979**, *101*, 1609.

Me
O

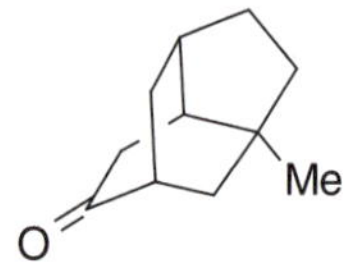

1e. Jones, T. K., Denmark, S. E. *Helv. Chim. Acta* **1983**, *66*, 2377.

HO
SiMe3

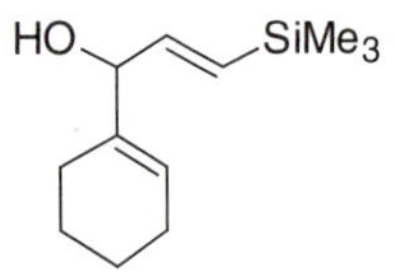

1g. Oppolzer, W., Robyr, C. *Tetrahedron* **1994**, *50*, 415.

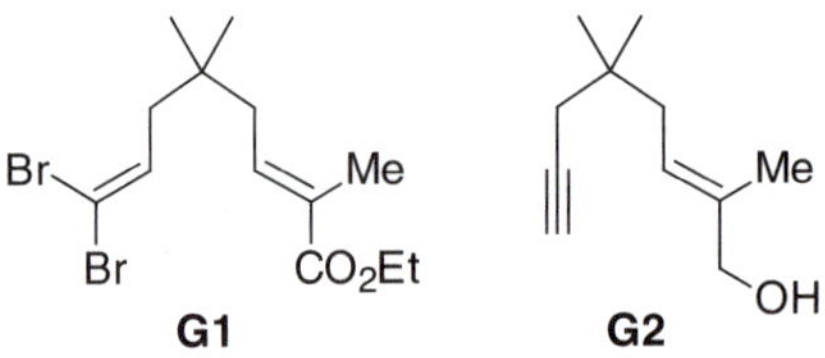

Br
Br
Me
CO2Et
G1
Me
OH
G2

1i. Overman, L. E., Jessup, P. J. *J. Am. Chem. Soc.* **1978**, *100*, 5179.

2a. Woodward, R. B., Sondheimer, F., Taub, D., Heusler, K., McLamore, W. M. *J. Am. Chem. Soc.* **1952**, *74*, 4223.

O
Me
MeO
H
O

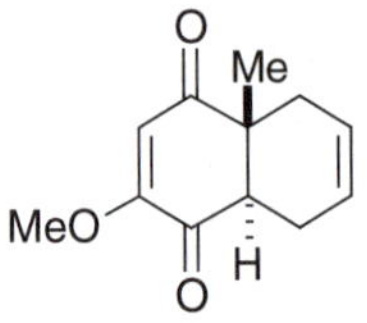

trans-isomer

2c. Mikami, K., Takahashi, K., Nakai, T. *J. Am. Chem. Soc.* **1990**, *112*, 4035.

2e. Kim, D., Lee, J., Shim, P. J., Lim, J. I., Jo, H., Kim, S. *J. Org. Chem.* **2002**, *67*, 764.

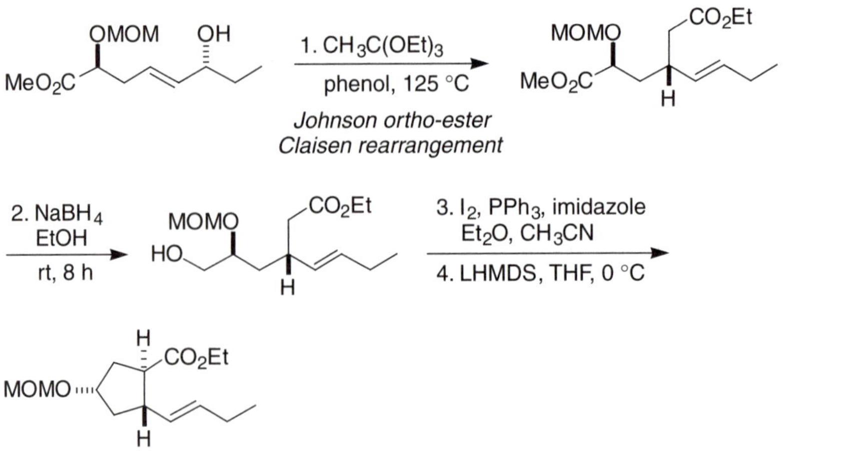

OMOM
OH
MeO2C
1. CH3C(OEt)3
phenol, 125 °C
Johnson ortho-ester
Claisen rearrangement
MOMO
CO2Et
MeO2C
H
2. NaBH4
EtOH
rt, 8 h
MOMO
CO2Et
HO
H
3. I2, PPh3, imidazole
Et2O, CH3CN
4. LHMDS, THF, 0 °C
H
CO2Et
MOMO
H

Step 2. Chemoselective reduction of the α-alkoxy ester
Step 4. Ester enolate alkylation (5-exo-tet cyclization)

3a. Kan, T., Hosokawa, S., Nara, S., Oikawa, M., Ito, S., Matsuda, F., Shirahama, H. *J. Org. Chem.* **1994**, *59*, 5532.

3c. Burke, S. D., Pacofsky, G. J. *Tetrahedron Lett.* **1986**, *27*, 445.
3e. Li, W.-D. Z., Yang, J.-H. *Org. Lett.* **2004**, *6*, 1849.

4b. Volkmann, R. A., Andrews, G. C., Johnson, W. S. *J. Am. Chem. Soc.* **1975**, *97*, 4777.

tertiary
carbenium ion

4d. Chen, C. Y., Hart, D. J. *J. Org. Chem.* **1993**, *58*, 3840.
4e. Willmore, N. D., Goodman, R., Lee, H. H., Kennedy, R. M. *J. Org. Chem.* **1992**, *57*, 1216.

5a. Baumstark, A. L., McCloskey, C. J., Witt, K. E. *J. Org. Chem.* **1978**, *43*, 3609.

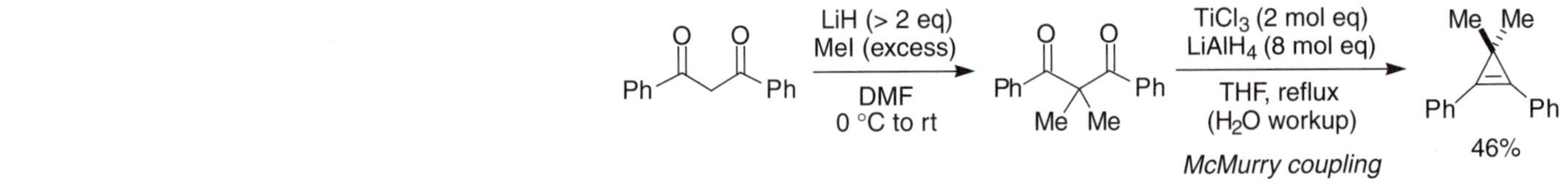

McMurry coupling

5c. Bassetti, M., D'Annibale, A., Fanfoni, A., Minissi, F. *Org. Lett.* **2005**, *7*, 1805.

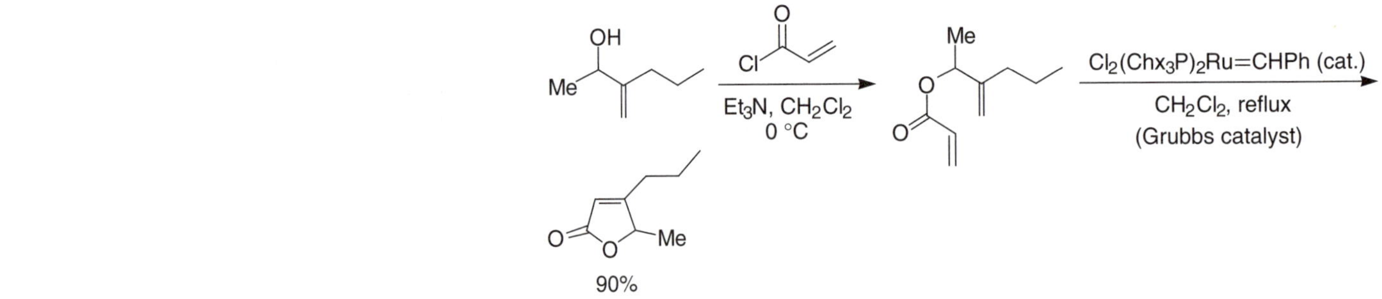

90%

The ring-closing metathesis reaction (step 2) required dilute conditions (0.01M substrate) and employed a syringe pump to introduce the substrate at a controlled rate.

5e. Corey, E. J., Danheiser, R. L., Chandrasekaran, S. *J. Org. Chem.* **1976**, *41*, 260.

5f. Lee, W.-D., Kim, K., Sulikowski, G. A. *Org. Lett.* **2005**, *7*, 1687.

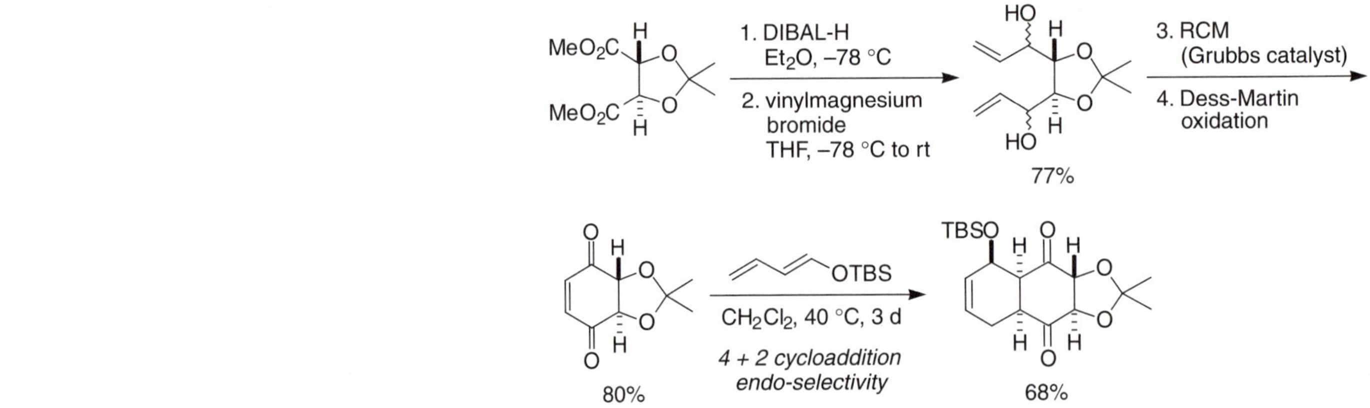

77%

80%   *4 + 2 cycloaddition endo-selectivity*   68%

Steps 1 and 2. See *Tetrahedron* **2000**, *56*, 2195.
Step 4. The Swern oxidation resulted in aromatization of the 1,4-quinone.

6a. Arisawa, M., Takezawa, E., Nishida, A., Mori, M., Nakagawa, M. *Synlett* **1997**, 1179.

Retrosynthetic Analysis

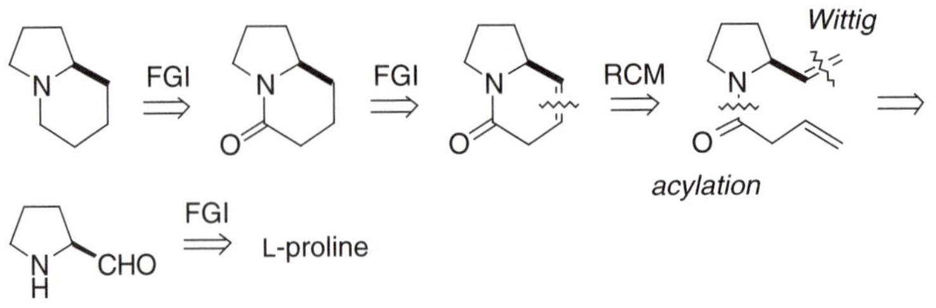

INDEX

Note: Page numbers followed by t indicate tables.

A

$A^{1,2}$-strain, 45, 239, 260
$A^{1,3}$-strain, 45–46, 53, 247
A values, in substituted
    cyclohexanes, 36, 38t
Acetals
    in alcohol protection,
        66–69
    in carbonyl group
        protection, 71–78
    in diol protection, 69–71
Acetoacetic ester synthesis,
    220
Acetonide, 70
Acetylene zipper reaction,
    402–403
Acid-catalyzed benzylation,
    62
Acidity, pK_a values, 213t
Acyclic systems. *See also*
    specific systems
    allylic strain, 45–46
    conformations, 31–33
    diastereofacial
        selectivity, 118–124
    chelation-controlled
        addition
        reactions, 122–
        123
    Cram's rule, 120
    double asymmetric
        induction, 119–
        120
    Felkin-Anh model,
        120–121, 121t
    hydroxyl-directed
        reduction, 123–
        124
Acyl anions, 8
    in Umpolung
        cyanohydrin-derived,
            12–14
        dithiane-derived, 9–11
        enol ether–derived, 14,
            282

lithium acetylide–
    derived, 14
nitroalkane-derived,
    11–12
Acylation, of
    alcohols, 68
    alkynylides, 400–401
    amines, 59–60
    enolates, 229, 262
    organocopper reagents,
        291
Acylium ion, 3t, 314
Acyloin condensation,
    412–414
Acyloin oxidation, 96–97
N-Acyloxazolidones,
    253–254
Acylsilanes, 320–322
Adam's catalyst, 139
AD-mix, 184
AIBN, 78, 187, 363
Alcohols
    dehydration, 359–362
    functional group
        interconversions,
        20t
    oxidation, 88–97
        acyloins, 96–97
        to aldehydes and
            ketones, 88–93
        barium manganate, 94
        ceric ammonium
            nitrate, 95
        chemoselective agents,
            93–96
        Collins-Ratcliff
            reagent, 89t, 90
        Dess-Martin
            periodinane, 89t,
            92
        Fetizon's reagent,
            94–95
        Jones reagent, 89, 89t
        manganese dioxide,
            93–94

pyridinium
    chlorochromate,
    89t, 90
pyridinium
    dichromate, 89t,
    90–91
silver carbonate,
    94–95
sodium hypochlorite,
    96
Swern, 89t, 91–92
TPAP (tetrapropyl-
    ammonium
    perruthenate),
    89t, 93
TEMPO, 95
tertiary allylic
    alcohols, 97
triphenylcarbenium
    tetrafluoroborate,
    95–96
propargylic, reduction of,
    199–200
protection, 60–69
stereochemical inversion,
    117–118
synthesis, from
    alkenes, 151–159,
        181–185
    allylic oxidation of
        alkenes, 99–101
    carbonylation, 307
    carbonyl compounds,
        104t, 106–109,
        111–117, 115t,
        117t
    epoxide ring-opening,
        168–172, 179–180
Aldehydes
    deoxygenation, 77–78
    diastereo- and
        enantiotopicity,
        118–120
    as dienophiles in Diels-
        Alder reactions, 425

functional group
    interconversions, 21t
nucleophilic additions,
    118–123, 291
oxidation, 98
protection, 71–78
reduction, 112–114, 113t
    enantioselective,
        124–129, 125t
    in the presence of
        ketones, 113–114
synthesis, from
    acid chlorides, 105
    alcohols, 88–95, 97
    alkenes (via
        ozonolysis),
        188–190
    alkynes, 200
    amides, 105, 254
    carbonylation,
        305–306
    Claisen rearrangement,
        390–392, 395
    epoxides, 180
    esters, 105, 110, 123
    glycol cleavage,
        191–192
    nitriles, 105, 110
    α,β-unsaturated, via
        aldol reaction, 245
    alkenes by SeO_2
        oxidation, 100
    alkenyl halides, 276
    Shapiro reaction, 387–
        389
    terminal alkynes, 401
Alder, Kurt, 421
Aldol reactions, 240–255
    diastereoselective,
        245–251
    (E)- vs. (Z)-
        stereochemistry,
        246–248
    syn-anti selectivity,
        248–252

Aldol reactions (*continued*)
enantioselective, 252–255
intermolecular, 240–241
intramolecular, 242
mixed (crossed), 243–245
Claisen-Schmidt, 243
imine, 244–245
Mukaiyama, 243
α–silyloxyketones in, 252–253
Alkenes
allylic oxidation, 99–101
cleavage, 188–193
dihydroxylation, 181–185
asymmetric, 184–185
osmium tetroxide, 181–183
potassium permanganate, 184
dissolving metal reduction, 143–145
epoxidation, 160–166, 173–178
diastereoselective, 173–178
DMDO (dimethyldioxirane), 164–165
enantioselective, 176–178
halohydrin approach, 165–167
hydrogen peroxide, 163–164
hydroxyl-directed, 173–176
Jacobsen, 181
peroxy acids, 160–163
Sharpless asymmetric epoxidation, 176–181
functional group interconversions, 22t
halolactonization, 185–188
hydration, 151–159
hydroboration-oxidation, 151–158
oxymercuration-demercuration, 158–159

hydroboration, 151–158
asymmetric, 156–158
chemoselective, 154, 155t
diastereoselective, 157–158
organoborane oxidation, 155–156
regiochemistry, 154, 154t
stereochemistry, 154–155
hydrogenation, 139–143
oxidation, 160–166, 173–192
lead tetraacetate, 191–192
Lemieux-Johnson, 190–191
osmium tetroxide, 190–191
potassium permanganate, 191
ruthenium tetroxide, 192
oxymercuration-demercuration, 158–159
ozonolysis, 188–190
stereochemical inversion, 170
synthesis, 359–396
amine oxide pyrolysis, 363
β-elimination reactions, 50–51, 359–362
Chugaev reaction, 362–363
Claisen rearrangement, 390–396
Cope elimination, 363
from alkynes, 193–200, 365–372
from enol phosphates, 295
from palladium-catalyzed coupling reactions, 322–345
Horner-Wadsworth-Emmons reaction, 301, 378–381

Julia reaction, 385–387
metathesis, 433–435
McMurray reaction, 415–417
Peterson reaction, 381–384
pyrolytic *syn* elimination reactions, 362–365
selenoxide elimination, 364–365
Shapiro reaction, 387–389
sulfoxide pyrolysis, 363–364
Wittig reaction, 372–378
xanthate pyrolysis, 362–363
transposition, 390–396
Wacker reaction, 201
Alkenyl halide synthesis, 325, 325t
Alkenylboranes
cleavage, 197–198
synthesis, 195–196
Alkenyllithium reagents, 275–276
Alkenylsilanes, 315–316
Alkyl bromides, 19t
Alkyl chlorides, 19t
Alkyl halides
in organolithium reagent preparation, 273–275
reduction to hydrocarbon, 107, 186–187
Alkyl iodides, 19t
Alkyllithium reagents, 278–281
Alkynes
alkylation, 399–402
carboalumination, 331, 369–370
carbocupration, 370–372
dissolving metal reduction, 198–200
functional group interconversions, 22t
hydration, 200–202

hydride reduction, 199–200
hydroalumination, 330, 367–368
hydroboration, 200–201, 366–368
hydrozirconation, 331–332
isomerization, 402–403
methylalumination, 369
oxidation, 99
oxymercuration, 201–202
reduction via alkenylborane protonolysis, 195–198
semi-hydrogenation, 193–195
synthesis, 396–404
alkynylide alkylation, 399–401
allene isomerization, 403
Corey-Fuchs bromomethylenation, 397–398
dehydrohalogenation, 396–404
enol phosphate elimination, 398
from aldehydes, 397–398, 403–404
Fuchs alkynylation, 401–402
Gilbert's reagent, 403–404
Alkynylide anions, in alkyne synthesis, 398–399
Alkynylsilanes, 313–315
Allenes, 403
π-Allyl palladium complexes, 343–345
Allyl vinyl ethers, 395–396
Allylic alcohols
epoxidation, 173–181
diastereoselective, 173–178
enantioselective, 176–177
hydroxyl-directed, 173–176
Jacobsen, 181
Sharpless asymmetric epoxidation, 176–180

oxidation of *tert*-allylic alcohols, 97
synthesis
  oxidation of alkenes, 99–101
  reduction of propargylic alcohols, 199–200
resolution, 180
Allylic boranes, 309–312
Allylic bromides, functional group interconversions, 19t
Allylic metalation, 279
Allylic strain, 45–46
Allylic substitution, palladium-catalyzed, 343–345
Allylsilanes, 317–319
Alpine-Borane, 125–126, 125t
Alternating polarity disconnections, 4–8
Aluminum hydrides, 103–106, 104t
Amides
  alkylation of amide enolates, 225
  in NH group protection, 59
  pK_a, 213t
  reduction, 104t, 104–105, 254
  Weinreb amide, 254
Amine oxides, pyrolysis of, 363
Amines
  functional group interconversions, 20t
  oxides, 155, 362–363
  synthesis, 104t, 109–110, 112t
Amino acids, 58, 106–107, 252–253, 257
Ammonium nitrate, 95
Ando-Wittig reaction, 381
Anomeric effect, 40–41
Anti conformation, 32–33
Arbuzov reaction, 378–379
Aromatic compounds
  reduction, 145–151. *See also* Birch reduction
  synthesis, 151

Aryl esters, in carboxyl group protection, 80–81
Aryllithium reagents, 276–277
Asymmetric induction, 103, 119–120
Asymmetric reactions
  aldol, 252–255
  alkylation, 237–238
  allylation, 311–312
  allylic alcohol resolution, 180
  allylic oxidation, 101
  cycloaddition, 431–432
  dihydroxylation, 184–185
  epoxidation, 176–179, 181
  hydroboration, 156–157
  hydrogenation, 142–143
  ketone reduction, 124–126, 125t
  Robinson reactions, 262–263

B

Babler oxidation, 97
Baeyer, Adolf von, 162
Baeyer-Villiger reaction, 162–163
Baldwin's rules, 231–234
Barium manganate, 94
Barton, Derek H.R., 31
Barton-McCombie deoxygenation, 345, 363
9-BBN (9-borabicyclo[3.3.1]-nonane), 125, 153, 154t, 155t, 156, 196, 336–337
Benzamides, 59
Benzyl esters, 80
Benzyl ethers, 61–63
N-Benzylamine, in NH group protection, 58–59
Benzylation, acid-catalyzed, 62
Benzylic metalation, 279
Benzyloxycarbonyl protective group, 60
Bestmann reagent, 377
β-effect, 313
β-elimination reactions, 359–362

Bicyclo[4.4.0]decane, 41–43
Bicyclo[2.2.1]heptane, 43–44
Bicyclo[4.3.0]nonane, 41
BINAL-H, 125t, 128
BINAP, 143
Birch reduction, 145–151
  chiral quaternary carbon formation, 150–151
  1,3-cyclohexadiene formation, 149
  cyclohexenone formation, 149–150
  reaction conditions, 146–147
  regiochemistry, 147–149
Bismuth sesquioxide, 96
Boat conformation, 35
Bond dissociation energies, 313t
9-Borabicyclo[3.3.1]nonane, 125, 153, 154t, 155t, 156, 196, 336–337
Borane-dimethyl sulfide, 151
Borane-tetrahydrofuran, 151
Borohydrides, 106–109
Boron enolates, 248–255
Boronic ester homologation, 308–309
Brassard's diene, 422
Bredt's rule, 44
Brook rearrangement, 321–322
Brown, Herbert C., 151
Brown² gasimeter, 194
Brown's asymmetric crotylboration, 309–312
Bürgi-Dunitz trajectory, 116
Butane, 32
t-Butoxycarbonyl protective group (Boc), 60
t-Butyl esters, 80
t-Butyl ethers, 61
t-Butyl peroxybenzoate, 101
t-Butyldimethylsilyl ethers, 64–65
t-Butyldiphenylsilyl ethers, 65–66

C

Cadiot-Chodkiewicz reaction, 342
Camphorsulfonic acid (CSA), 71
CAN (ceric ammonium nitrate), 62, 95
Carbamates, in NH group protection, 59
Carbenium ion, 3t
Carbenoids, 167, 301, 303–304
Carboalumination, 331, 368–369
Carbocupration, 370–372
Carbonyl groups
  acetalization
    with *O,O*-acetals, 71–76
    with *S,S*-acetals, 76–77
  polarity reversal (Umpolung), 8–14
  reduction, 101–129
    diisobutylaluminum hydride, 109–111, 112t
    lithium borohydride, 106–107
    lithium trialkylboro-hydride, 107–108
    Rosemund procedure, 105
    sodium borohydride, 106
    sodium cyanoborohydride, 107–108
    terminology, 101–102
    zinc borohydride, 107
Carbonylation, 305–307
Carboxyl groups, protection of, 78–82
Carboxylic acids,
  enolate alkylation, 225
  functional group interconversion, 21t
  reduction, 104t, 107
    using borane, 111–112, 114
  synthesis, from
    alcohols, 88–91
    aldehydes, 98
    alkynes, 99, 401
    glycol oxidative cleavage, 191–193

Carboxylic acids (*continued*)
Grignard reagents, 274, 275t, 284
silyl enol ether ozonolysis, 190
Carroll-Claisen rearrangement, 392
Cascade reactions, 26
Castro-Stevens reaction, 339–341
Catecholborane, 334
Cation-π cyclizations, 314, 417–420, 418t
CBS reagents, 125t, 127–128, 431–432, 432t
$CeCl_3$, 114, 287–288, 382
Ceric ammonium nitrate, 62, 95
Chair conformation, 34–35
Chelation control, 122–124, 244
Chemoselective reactions, 48, 102
epoxidation, 161
hydroboration, 154–155
hydrogenation, 194
oxidation, 93–96
reduction, 112–115
B-Chlorodiisopinocampheylborane, 125t, 126
Chloroiodomethane, 167
Chlorotris(triphenylphosphine)rhodium, 141–142
Chromic acid oxidation, 89
Chromic anhydride, 100
Chugaev reaction, 362–363
Claisen condensation, 217
Claisen rearrangement, 390–396
Carroll-Claisen variant, 392
Eschenmoser variant, 394
Ireland variant, 394
Johnson ortho-ester variant, 392–394
Yamamoto asymmetric variant, 395
Claisen-Schmidt reaction, 243
Clemmensen reduction, 78
Collins-Ratcliff reagent, 89t, 90
Computer-assisted molecular modeling, 46–47

Conformational analysis
acyclic systems, 31–33
anomeric effect, 40–41
boat, 35
bridged bicyclic systems, 43–44
chair, 34
cyclohexyl systems with sp^2-hybridized atoms, 44–46
destabilization energies, 38–39
eclipsed, 31–32
envelope, 34
gauche, 32–33
LHASA program, 38–39
nonbonded interactions, 35–39
of ring systems, 33–44
of six-member heterocyclic systems, 40–41, 46
skew, 31
staggered, 31–32
tricyclic systems, 43–44
twist boat, 35
Conjugate addition (1,4-addition)
of dithianes, 10
of hydrides, 112–113
Michael-type reactions, 215–216, 258–262
of organocopper reagents, 229, 292t, 292–294, 371–372
of sodium benzensulfinate, 386
stereochemistry of, 296–297
Consonant patterns, 5–7
Convergent synthesis, 23–24
Cope elimination, 363
Corey, E.J., 1
Corey-Fuchs bromomethylenation, 397–398
Corey lactone, 187
Crabtree's catalyst, 142
Cram's rule, 120
Crotylboration, 309–312
Crotyllithium reagents, 279–281
Crown ethers, 278
Cupric acetate, 96
Curtin-Hammett principle, 47

Cyanidation, 307–308
Cyanocuprate reagents, 289
Cyanohydrins, 12–14, 98
Cyclization reactions, 231–234, 412–435
in aldol condensations, 233
Baldwin's rules, 231–234
cation-π, 417–420, 418t
intramolecular free radical, 412–417
acyloin condensation, 412
McMurray reaction, 414, 415–417
Pinacol coupling, 414–415
pericyclic
Diels-Alder cycloaddition, 421–432
1,3-dipolar cycloaddition, 432–433
ring-closing olefin metathesis, 433–434
Cyclobutane, 33–34
Cyclohexanes
conformations, 34–40
destabilization energies, 38–39, 38t
Cyclohexanones, conformations, 44
Cyclohexenes
allylic strain, 45–46
conformations, 44–46
Cyclopentadienones, 424
Cyclopentane, 34
2-Cyclopentenone, 113t
Cyclopropanation, 303–305
Cyclopropane, 33

D

Danishefsky's diene, 422
DBU, 257, 361, 381
DDQ, 62–63
DEAD (diethyl azodicarboxylate), 79–80, 118
Decalin, 41–43
Decarboxylation, 215, 221–222
Demercuration, 158–159
Deoxygenation, via reduction of

propargylic carbonates, 345
thioacetals, 77–78
tosylates, 69, 107–108
tosylhydrazones, 109
xanthates, 363
Dess-Martin periodinane (DMP), 89t, 92
Destabilization energies, 38–39, 38t
DHQ, DHQD, 184
Dianions
β-keto ester, 221
carboxylic acid, 255
propargylic, 401
Diastereomeric excess (de), 103
Diastereotopicity, 119–120
Diazomethane, 79
DIBAL-H, 109–111, 112t, 367–368
1,3-Dicarbonyl compounds
acidity, 213–214, 213t
synthesis and reactions, 213–223
Dichloromethyl methyl ether reaction (DCME), 305, 308
Dicyclohexylborane, 99, 153, 196–198
Dicyclohexylcarbodiimide (DCC), 79
Dieckmann condensation, 217–219
Diels, Otto, 421
Diels-Alder reaction, 421–432
asymmetric, 431–432
electron-demand, 422
frontier molecular orbitals, 421
intramolecular, 429–430
Lewis acid–catalyzed, 424–425, 431
oxazaborolidine-catalyzed, 431–432, 432t
regiochemistry, 425–426
reversibility, 421
stereochemistry, 426–429
Dihydroxylation, 181–185
asymmetric, 184–185
osmium tetroxide, 181–183
Woodward procedure, 183

Diisobutylaluminum hydride, 109–111, 112t, 367–368
Diisopinocampheylborane, 153, 156
Diketene, 219, 392
1,3-Diketones, 222–223
4-(N,N-Dimethylamino)-pyridine (DMAP), 63
Dimethylborolane, 125t, 126–127, 157
Dimethyldioxirane (DMDO), 164–165
Dimethyloxosulfonium methylide, 166–167
Dimethylsulfonium methylide, 166
Dimsyl anion, 375
1,2-Diols
  acetalization, 69–71
  oxidation, 91–92, 94
  oxidative cleavage, 190–193
  synthesis, 181–185, 414–415
DIPAMP, 143
DIP-Cl (B-Chlorodiisopinocam-pheylborane), 125t, 126
1,3-Dipolar cycloaddition reactions, 432–433
Disconnections, table, 15t
Disiamylborane, 153, 196–198
Dissolving metal reductions, 143–151
Dissonant patterns, 5, 7–8
1,3-Dithianes, 9–11, 320
Diynes, 194–195, 341–343
DMPU, 224, 278
Doebner condensation, 216
Domino reactions, 26
Double asymmetric induction, 255

E
E1 eliminations, 359–360
E2 eliminations, 360–362
  reactivity and product differentiation in, 50–51
Eclipsed conformation, 31–32
18 electron rule, 323
Electronegativity values, 273t

Electrophilic reducing agents, 109–111, 112t
Elimination reactions
  alkene synthesis, 359–365
  alkyne synthesis, 396–398
Enamines, 238–240
Enantiomeric excess (ee), 103
Enantioselective reactions. *See* Asymmetric reactions
Enantiotopicity, 119–120
Enediynes, 341
Enol borinates, 248–255
Enol ethers
  acyl anion equivalent, 14
  dihydropyran, 66, 281
  α-metalation, 281–283
  2-methoxypropene, 68, 70
  ozonolysis of, 190
  silyl, 226–227, 243–244
  synthesis via Birch reduction, 149–150
Enol phosphates, 295, 398
Enol triflates, 328, 337–339
Enolate trapping, 229, 294–296, 338–339, 365
Enolates, 213–263
  alkylation, 223–231
  amide-derived, 225
  boron, 248–255
  C- vs. O-alkylation, 222–223
  carboxylic acid–derived, 225
  condensation reactions, 256–260
  (E)-(O)- vs. (Z)-(O)-stereochemistry, 246–248
  kinetic vs. thermodynamic, 226–229, 339
  intramolecular alkylation, 231–234
  nitrile-derived, 225
Enones
  conjugate reduction, 112t, 112–113, 140–141, 143–145
  protection, 73–76
  synthesis, 257, 297–298, 316, 336
  transposition, 97

Enynes, 194–196
Enzymatic reductions, 129
Epothilone D, synthesis of, 443–444
Epoxidation
  alkenes, 160–173
  allylic alcohols, 173–181
Epoxides
  reactions
    acid-catalyzed ring opening, 171–172
    base-induced elimination, 170–171
    intramolecular ring-opening, 169–170
    Lewis acid-mediated rearrangement, 172–173, 180
    nucleophilic ring-opening, 168, 178–179
    organocopper cleavage, 292, 371
  synthesis, from
    alkenes, 160–166, 173–181
    halohydrins, 165–166
    ketones, 166–167
Eschenmoser-Claisen rearrangement, 394
Eschenmoser salt, 257
Ester enolates, alkylation of, 223
Esters
  in alcohol protection, 68–69
  in carboxyl group protection, 79–82
  pyrolysis of, 362
  reduction, 104t, 104, 109–111, 114–115
  synthesis, via
    aldehyde oxidation, 98–99
    allylic alcohol oxidation, 98
    carboxylic acid esterification, 79–81
    Claisen rearrangement, 390–394
    enolate acylation, 219, 225, 230
    enol ether ozonolysis, 189

Esterification, rate of, 48
Ethane, 31–32
Ethers, in alcohol protection, 60–66
Evans aldol reaction, 253–254

F
Felkin-Anh model, 120–121
Fetizon's Reagent, 94
Formyl anions, 8
  Umpolung, dithiane-derived, 9–11
Frank-Condon effect, 50
Frontier molecular orbital theory, 426
Fuchs alkynylation, 401–402
Functional group interconversions, 19t–22t
Furakawa's reagent, 301, 304

G
G values, for substituted cyclohexanes, 38t
Gauche conformation, 32
Gilbert reagent, 403–404
Gilman reagents, 289
Glaser reaction, 341–343
Glycols, formation and cleavage, 190–193
Grignard, Victor, 283
Grignard-copper reagents, 289–290
Grignard reagents, 283–286
Grubbs, R., 433
Grubbs catalyst, 433–434

H
Halohydrins, 165–166
Halolactonization, 185–188
Hammond postulate, 47
Hassel, Odd, 31
Heck reaction, 326–329
Henry-Nef reaction, 11
Heterocuprate reagents, 289
Heterolytic disconnections
  alternating-polarity, 4–8
  in retrosynthetic analysis, 2–8
Hexamethylphosphoramide (HMPA), 224, 227, 278

Hoffman eliminations, 359
Homochirality, 103
Homocuprate reagents, 289
Homolytic disconnections, 2
Horner-Wadsworth-Emmons
    reaction, 301, 378–381
Huang-Minlon reduction, 78
Hudrlik-Peterson diene
    synthesis, 319
Hydration, of
    alkenes, 151–158
    alkynes, 200–202
Hydrazone anions, 237–238
Hydrazones, 78, 237–238,
    387–389
Hydrindane, 41
Hydroalumination, 330,
    367–368
Hydroboration, of
    alkenes, 151–158
    alkynes, 195–196,
        200–201, 366–368
    enynes, 195–196
Hydrogen peroxide, 163–164
Hydrogenation
    of alkenes, 139–143
    of alkynes, 193–195
    directed, 142
    asymmetric, 142
Hydrogenolysis, of
    allylic groups, 141
    benzylic groups, 59–62,
        141
    cyclopropanes, 33
    halides, 141, 187
    thioacetals, 77
Hydrosilylation, 315
β-Hydroxy ketones,
    reduction of, 123–124
Hydrozirconation, 331–332

I
IBX, 92, 297
Imine aldol reactions,
    244–245
Imine anions, 236–237
Intramolecular reactions
    aldol reaction, 242,
        260–263
    alkylation, 215, 223,
        231–234
    cation-π cyclization, 314,
        417–420, 418t
    diester condensation
        acyloin, 219, 412–414
    Dieckmann, 217–218

Diels-Alder cycloaddition,
    429–430
    epoxide cleavage,
        169–170, 172,
        178–180, 418t
    free radical cyclizations,
        412–417
    halonium ion capture,
        185–188
    Michael addition,
        259–260, 303
    oxymercuration, 159
    Prins reaction, 418t
    Wittig reaction, 376
Iodoform, 299
Iodolactonization, 185–188
Iodotrimethylsilane, 59
Ireland-Claisen
    rearrangement, 394
Isomerization, in alkyne
    synthesis, 402–403
Isopropylidene acetals,
    70–71
Isoxazolines, 433

J
Jacobsen epoxidation, 181
Johnson ortho-ester
    rearrangement,
    392–394
Jones reagent, 89, 89t
Julia olefination, 385–387
    Kocienski modification
        of, 386
Juvenile hormone,
    443–444

K
KAPA (potassium 3-
    aminopropylamide),
    402–403
Katsuki, T., 176
β-Keto esters
    alkylation, 220–221
    decarboxylation,
        221–222
    preparation
        of acetoacetic esters,
            219
        by C-acylation,
            219–220
        by Claisen
            condensation, 217
        by Dieckmann
            condensation,
            217–219

Ketone enamines, 238–240
Ketone enolates
    alkylation, 230–236
    formation, 225–230
Ketones
    1,2-addition, 118–123,
        291
    deoxygenation, 77–78
    diastereo- and
        enantiotopicity,
        118–120
    epoxide preparation
        from, 166–167
    functional group
        interconversions, 21t
    methyl ketone syntheses,
        201–202
    protection, 71–78
    reduction
        diastereoselective,
            115–116, 117t
        dissolving metal,
            143–145
        enantioselective,
            124–129, 125t
        enzymatic, 129
        hydroxyl-directed,
            123–124
        in presence of
            aldehydes or
            esters, 114
    synthesis, via
        alcohol oxidation,
            88–97
        alkene cleavage,
            188–192
        alkyne hydration,
            200–201
        allylic oxidation,
            100
        carbonylation,
            306–307
        cyanidation, 307–308
        enol ether hydrolysis,
            149–150
        thioesters, 251
Kinetic enolate formation,
    226–229
Kinetic resolution, 180
Knoevenagel condensation,
    216
Kocienski-modified Julia
    olefination, 386
K-Selectride (potassium tri-
    sec-butylborohydride),
    107–108, 113, 116

L
Lactones, from
    Baeyer-Villiger
        oxidation, 162–163
    lactol oxidation, 94–95
Larock annulation,
    328–329
Latent polarity, 4–5
Lead tetraacetate,
    191–192
Lemieux-Johnson
    oxidation, 190–191
Lewis acid–catalyzed
    reactions
    allylic alcohol
        epoxidation, 180
    allylsilane addition,
        318–319
    crossed Aldol, 243–244
    Diels-Alder, 424–425,
        431
    epoxide rearrangement,
        172–173, 179–180
Lindlar catalyst, 193–194
Linear synthesis, 23–24
Lipschutz reagents, 289
Lithium acetylide, 14,
    398–399
Lithium alkynylides,
    preparation, 283,
    397–401
Lithium aluminum hydride,
    103–105, 104t
Lithium borohydride,
    106–107
Lithium diisopropylamide
    (LDA), 278
Lithium
    hexamethyldisilazide
    (LHMDS), 278
Lithium
    2,2,6,6–tetramethylpip
    eridide (LTMP), 278
Lithium tri-t-
    butoxyaluminum
    hydride, 104t, 105
Lithium
    triethoxyaluminum
    hydride, 104t, 105
Lithium-halogen exchange,
    275–277
L-Selectride (lithium tri-
    sec-butylborohydride),
    107–108, 115t, 116,
    117t, 121t
Luche reagent, 112–114

M

Malonic acid esters, 214–216
Manganese dioxide, 93–94
Mannich reaction, 256–257
Martin sulfurane, 361
Matteson's boronic ester homologation, 308–309
McMurray reaction, 414–417
mCPBA (meta-chloroperoxybenzoic acid), 160–163
Meerwein-Ponndorf-Verley reduction, 115, 115t
Mesylates
  based-induced elimination, 359–361
  preparation, 360–361
  reduction, 69
Metalation, 277–283
  of α-heteroatom substituted alkenes, 281–283
  of 1-alkynes, 283, 397–401
  allylic or benzylic, 279–281
  reagents for, 278–281
p-Methoxybenzyl (PMB) ethers, 62–63
Methoxymethyl (MOM) ethers, 66–67
Methyl ketones, preparation of, 201–202, 251
3-Methylcyclohexanone, 115, 115t
Michael addition, 258–260
  disconnections in, 15t
  intramolecular, 259–260
  in Robinson annulation, 26
  stereochemistry of, 260
  stereoelectronic factors in, 50
Michael-type reactions, 215–216, 258–262
Mitsunobu reaction, 79–80, 117–118, 387
Monoisopinocampheyl-borane (IpcBH₂), 157
Mukaiyama reaction, 243
MVK (methyl vinyl ketone), 260–263

N

Nef reaction, 11
Negishi reaction, 329–332, 340–341
Ni-boride, in
  hydrogenation, 139, 140
  thioacetal reduction, 77
Nitriles
  enolate alkylation, 225
  functional group interconversions, 20t
  as ketone precursors, 275t
  pKₐ values, 12, 213t
  reduction (DIBAL-H), 110
Nitroalkanes, 11–12
NMO (N-methylmorpholine-N-oxide), 182–183
2,5-Norbornadiene-Rh(I) catalyst, 142
Norbornane, 43–44
Nozaki-Hiyama reaction, 298–299
Nozaki-Takai-Hiyama-Kishi coupling, 299–300
Nucleophilic reducing agents, 103–109

O

Octalactin B, synthesis of, 443–444
Olefin metathesis, 433–435
Oppenauer oxidation, 115
Organoalanes, 330, 367–368
Organoboranes, oxidation of, 155–156
Organoboron reagents, 305–312
  in asymmetric crotylboration, 309–312
  in cyanidation, 307–308
  in dichloromethyl methyl ether reaction, 308
  in Matteson's boronic ester homologation, 308–309
  in Suzuki reaction, 332–337
Organocerium reagents, 287–288, 288t

Organochromium reagents, 298–300
  in alcohol oxidation, 88
Organocopper reagents, 288–297
  Grignard-copper, 289–290
  heterocuprates (Gilman), 289
  higher-order cyanocuprates (Lipshutz), 289
  homocuprates, 289
  in ketone preparation, 297–298
  preparation of, 289–290
  reactions of, 288–297
Organolithium reagents, 273–285
  alkenyllithium, 275–276
  alkyllithium, 278–281
  aryllithium, 276–281
  conjugate addition reactions of, 81, 283
  crotyllithium, 279–281
Organomagnesium reagents, 283–286
Organomercury compounds, 159
Organopalladium species, 322–345
Organosilicon reagents, 312–322
  acylsilanes, 320–322
  alkenylsilanes, 315–316
  alkynylsilanes and, 313–315
  allylsilanes, 317–319
  α- and β-stabilization, 313
  electronegativity, 312–313, 313t
Organotin reagents, in Stille reaction, 337–339
Organotitanium reagents, 286–287, 287t
Organotransition metal complexes
  ligand coordination number, 323
  oxidation state, 323–324
Organozinc reagents
  in palladium-catalyzed cross-coupling reactions, 340–341
  preparation, 300–301
  reactions, 301–305

Organozirconium reagents, 331–332
Ortho ester rearrangement, 392–394
Ortho esters, in carboxyl protection, 81–82
Ortho-lithiation, 282–283
Osmium tetroxide
  in alkene dihydroxylation, 181–183
  in alkene oxidation, 190–191
Oxazaborolidines
  in asymmetric Diels-Adler reactions, 431–432, 432t
  in reduction of aldehydes or ketones, 125t, 127–128
Oxazolidinones, 253–254
Oxazolines, 81
Oxidation, of
  acyloins, 96–97
  alcohols, 88–97
  aldehydes, 98
  alkanes, 165
  alkenes (allylic), 99–101
  alkynes, 99
  enones, 163–164
  ketones, 162–163
  organoboranes, 155–156
Oxone, 92, 98–99, 164
Oxymercuration
  of alkenes, 158–159
  of alkynes, 201–202
Ozonolysis, 188–190

P

Palladium-catalyzed coupling reactions, 322–345
  allylic substitution, 343–345
  β-hydride elimination, 326
  Cadiot-Chodkiewicz reaction, 342
  Castro-Stevens reaction, 339–341
  conjugated enediyne synthesis, 341
  Glaser reaction, 341–343
  Heck reaction, 326–329
  Larock annulation, 328–329

Palladium-catalyzed coupling reactions (*continued*)
Negishi reaction, 329–339, 340–341
oxidative addition, 324–326
Pd(0)-catalyzed olefin insertion, 326–329
reductive elimination, 326
Sonogashira reaction, 339–340
with sp-carbons, 339–343
Suzuki reaction, 332–337
transmetalation, 324, 326
Trost-Tsuji reaction, 343–345
Payne rearrangement, 178–179
PCC, 90, 97
PDC, 90–91, 97–98
Pericyclic reactions, 421–433
Periodates, 190–193
Peroxy acids, 160–163, 173–176
Peterson olefination, 381–384
Phase transfer catalysis, 99, 223
Phosphine ligands, 324, 324t
Phosphonate reagents, 378–381
Pinacol coupling, 414–416
p*K*a values, 213t, 277
Polycyclic ring systems, 41–44
Potassium caroate, 164
Potassium permanganate, 184
PPTS, 66
Prochiral centers, 102
Progesterone, 443–444
Proline, 257
Propane, 32
Propargylic alcohols, reduction of, 199–200
Propargylic alkylation, 401
Protective groups, of alcohols, 20t, 60–69
carbonyl groups in aldehydes and ketones, 71–78
carboxyl groups, 78–82

diols, 69–71
double bonds, 82
NH groups, 58–60
triple bonds, 82
Pyridinium chlorochromate (PCC), 89t, 90
Pyridinium dichromate (PDC), 89t, 90–91
Pyrolysis, 362–365

Q
Quinidines, 184
Quinones, 423–424

R
Ra-Ni (Raney nickel), 77, 139–140
Rawal's diene, 422
Re face, 118
Red-Al, 105–106, 180, 200
Reduction, of
alkenes, 139–151
alkynes, 193–200
aromatic compounds (dissolving metal), 145–151
carbonyl groups, 101–129
chemoselective, 114–115
enantioselective, 124–128
cyclic ketones, 115–117
cyclopropanes, 33
enones, 112–113, 140–141, 143–145
enzymatic, 129
nitriles, 110
sulfonate esters, 69
vinyl esters, 145
Reductive amination, 109
Reformatsky reaction, 301
Retrosynthetic analysis, 1–26
carbon skeleton construction in, 15–18
convergent vs. linear synthesis, 23–25
synthetic equivalents, 2, 15t
synthons, 2–4, 3t
Umpolung in, 8–14
Rieke magnesium, 284

Rieke zinc, 300–303
Ring-closing olefin metathesis, 433–435
Robinson, Robert, 260
Robinson annulation, 26, 260–263
Rochelle's salt, 104
Rosenmund reduction, 105
Ruthenium tetroxide, 192–193

S
Saegusa-Ito oxidation, 297–298
Schlenk equilibrium, 285
Schlosser-modified Wittig olefination, 375–376
Schlosser's base, 280
Schrock, R., 433
Schwartz reagent, 331
Secondary orbital overlap, 427–428
Selectivity, terminology of, 48
Selenium dioxide, 99–100
Selenoxides, *syn*-elimination of, 364–365
Shapiro reaction, 387–389
Sharpless, K.B., 176
Sharpless asymmetric dihydroxylation, 184–185
Sharpless asymmetric epoxidation, 176–181
resolution of racemic allylic alcohols, 180
Si face, 118
[3,3]-Sigmatropic rearrangement, 390–395
Silver carbonate, 94–95
Silyl enol ethers, 226–227, 243–244, 294–295, 322, 414
Silyl esters, in carboxyl protection, 81
Silyl ethers, 63–64
α–Silyloxyketones, in aldol reactions, 252–253
Simmons-Smith cyclopropanation, 303–305
Skew conformation, 31
S$_N$2 reactions, 15t, 49–50

Sodium bis(2-methoxyethoxy)aluminum hydride (Red-Al), 105–106
Sodium borohydride, 106
Sodium chlorite, 98
Sodium cyanoborohydride, 108–109
Sodium hypochlorite, 96
Solvomercuration-demercuration, 159
Sonogashira reaction, 339–340
Staggered conformation, 31–33
Stereoelectronic effects, 47–54
Stereogenic carbon, 102
Stereoisomers, 48, 102
Stereoselective reactions, 48, 102
Stereospecific reactions, 48, 102
Steric strain, 32
Stetter reaction, 13
Stille reaction, 338–339
Still-Gennari olefination, 380
Sulfones, 213t, 260, 385–387
Sulfoxides, pyrolysis of, 363–364
Sulfur ylides, 166–167
Super Base, 280
Suzuki reaction, 332–337
Swern oxidation, 89t, 91–92
Synthetic equivalents, 2
Synthons, 2–4, 3t, 15t

T
Tandem reactions, 26, 341
TBAF, 63–64
TEMPO, 95
Tetrahydropyrans, conformations of, 40
Tetrahydropyranyl ethers, 66
Tetramethylammonium triacetoxyborohydride, 123–124
Thermodynamic enolate formation, 226–229
Thexylborane, 152–153
Thexylchloroborane, 153
Thexyldimethylsilyl ethers, 65
Thioacetals, reduction of, 77–78

Thioesters, 251
TMEDA
    (tetramethylethylene-
    diamine), 278
Torsional strain, 32
Tosyl hydrazone, 109,
    387–388
Tosylates, preparation of,
    360
TPAP
    (tetrapropylammonium
    perruthenate)
    oxidation, 89t, 93
Transforms, 2. *See also*
    Disconnections;
    Functional group
    interconversions
Transition states, early vs.
    late, 47
Transmetalation, 277
Trapp mixture, 276
Trichloroacetimidates, 62
Triflating agents, 338
Trifluoromethanesulfon-
    imide (Tf$_2$NH), 431
Triisopropylsilyl ethers, 65

3,3,5-Trimethyl-
    cyclohexanone, 116,
    117t
Trimethylsilyl ethers, 64
Tri-*n*-butyltin hydride
    in halide reduction,
        186–187
    in thioacetal reduction,
        78
    in xanthate reduction,
        363
Triphenylcarbenium
    tetrafluoroborate,
    95–96
Trisylhydrazones,
    388–389
Trityl ethers, 63
Trost-Tsuji reaction,
    343–345
Twist boat conformation,
    35

U
U values, for substituted
    cyclohexanes, 38t
Umpolung, 2, 8–14

V
Vilsmeier-Haack
    formylation, 151
Vinyl halides, 291, 316,
    325t, 367
Vinyl lithium, 275–276,
    387–389
Vinyl triflates, 328,
    337–339
Vinylic boranes, protonolysis
    of, 197–198
Vinylsilanes, 315–316
VO(acac)$_2$, 175

W
Wacker oxidation, 201
Weinreb amide, 254
Wilkinson's catalyst,
    141–142
Wittig, Georg, 372
Wittig reaction, 372–378
    Schlosser modification,
        375–376
    ylides
        nonstabilized, 373–376
        stabilized, 376–378

Wolff-Kishner reduction,
    78
Woodward-Hoffmann rules,
    421

X
Xanthates, pyrolysis of,
    362–363

Y
Yamamoto asymmetric
    Claisen rearrangement,
    395
Ylides
    phosphonium, 373–378
    sulfonium, 166–167

Z
Zimmerman-Traxler
    transition state model,
    249, 253–254, 311
Zinc borohydride, 107
Zinc carbenoids, 301
Zirconium reagents, in
    hydrozirconation,
    331–332